Stochastic Processes and Random Matrices

Lecture Notes of the Les Houches Summer School:

Volume 104, 6th – 31st July 2015

Stochastic Processes and Random Matrices

Edited by

Grégory Schehr, Alexander Altland, Yan V. Fyodorov,
Neil O'Connell, Leticia F. Cugliandolo

OXFORD
UNIVERSITY PRESS

OXFORD
UNIVERSITY PRESS

Great Clarendon Street, Oxford, OX2 6DP,
United Kingdom

Oxford University Press is a department of the University of Oxford.
It furthers the University's objective of excellence in research, scholarship,
and education by publishing worldwide. Oxford is a registered trade mark of
Oxford University Press in the UK and in certain other countries

First Edition published in 2017

Impression: 2

Published in the United States of America by Oxford University Press
198 Madison Avenue, New York, NY 10016, United States of America

British Library Cataloguing in Publication Data
Data available

Library of Congress Control Number: 2017943724

ISBN 978-0-19-879731-9

Printed and bound by
CPI Group (UK) Ltd, Croydon, CR0 4YY

École de Physique des Houches

Service inter-universitaire commun
à l'Université Joseph Fourier de Grenoble
et à l'Institut National Polytechnique de Grenoble

Subventionné par l'Université Joseph Fourier de Grenoble,
le Centre National de la Recherche Scientifique,
le Commissariat à l'Énergie Atomique

Directeur:
Leticia F. Cugliandolo, Sorbonne Universités, Université Pierre et Marie Curie
Laboratoire de Physique Théorique et Hautes Energies, Paris, France

Directeurs scientifiques de la session:
Grégory Schehr, Université Paris-Sud, Laboratoire de Physique Théorique et
Modèles Statistique, Orsay, France
Alexander Altland, Universität zu Köln, Institute for Theoretical Physics, Köln,
Germany
Yan V. Fyodorov, King's College London, Department of Mathematics, London,
United Kingdom
Neil O'Connell, School of Mathematics, University of Bristol, Bristol, United
Kingdom
Leticia F. Cugliandolo, Sorbonne Universités, Université Pierre et Marie Curie
Laboratoire de Physique Théorique et Hautes Energies, Paris, France

Previous sessions

Publishers

- Session VIII: Dunod, Wiley, Methuen
- Sessions IX and X: Herman, Wiley
- Session XI: Gordon and Breach, Presses Universitaires
- Sessions XII–XXV: Gordon and Breach
- Sessions XXVI–LXVIII: North Holland
- Session LXIX–LXXVIII: EDP Sciences, Springer
- Session LXXIX–LXXXVIII: Elsevier
- Session LXXXIX– : Oxford University Press

Preface

The field of stochastic processes and RMT has been a rapidly evolving subject during the last fifteen years where the continuous development and discovery of new tools, connections and ideas have led to an avalanche of new results. These breakthroughs have been made possible thanks, to a large extent, to the recent development of various new techniques in Random Matrix Theory (RMT). Matrix models have been playing an important role in theoretical physics for a long time and they are currently also a very active domain of research in mathematics. An emblematic example of these recent advances concerns the theory of growth phenomena in the Kardar-Parisi-Zhang (KPZ) universality class where the joint efforts of physicists and mathematicians during the last twenty years have unveiled the beautiful connections between this fundamental problem of statistical mechanics and the theory of random matrices, namely the fluctuations of the largest eigenvalue of certain ensemble of random matrices. These lecture notes not only cover this topic in detail but also present more recent developments that have emerged from these discoveries, for instance in the context of low dimensional heat transport (on the physics side) or in context of integrable probability (on the mathematical side). More widely, our goal, in organizing this school in Les Houches, was to present the latest developments on these topics at the interface between theoretical physics and mathematics, with a special emphasis on the large spectrum of techniques and applications of RMT. By following the lectures in this volume the reader will surely be able to appreciate the breadth and beauty of the subject.

The school was held in July 2015. It started with a historical introductory lectures on the applications of RMT given by H. Weidenmüller and consisted of more than fifty 90-minutes lectures, covering a wide range of topics. The school comprised, on the one hand, five long courses (five lectures each) on rather general subjects, and, on the other hand, ten shorter courses (two to three lectures each) on more specialized topics. This volume presents the lecture notes prepared by the speakers.

This volume starts with the lecture notes of three of the five long lectures: A. Borodin on Integrable Probability, A. Guionnet on Free Probability and H. Spohn on the Kardar-Parisi-Zhang equation (we regret that we were not able to include the lectures notes by P. Le Doussal and B. Virag in this volume). They are followed up by the lecture notes on more specialized topics: G. Akemann on RMT and quantum chromodynamics, J.-Ph. Bouchaud on RMT and (Big) data analysis, B. Eynard on Random matrices and loop equations, J. P. Keating on Random matrices and number theory, A. L. Moustakas on the applications of RMT to modern telecommunications, H. Schomerus on random matrix approaches to open quantum systems, Y. Tourigny (and A. Comtet) on impurity models and products of random matrices, V. Vargas (and R. Rhodes) Gaussian multiplicative chaos and Liouville Quantum Gravity and A. Zabrodin on Quantum spin chains and classical integrable systems. We also had

the pleasure to listen to a seminar by J.-P. Eckmann on M. Hairer's work and the Kardar-Parisi-Zhang equation.

The students, half of them physicists and the other half mathematicians, were full of enthusiasm both inside and outside the classroom. Many of them had the opportunity to present their work during two poster sessions. This school in Les Houches has certainly been a very good opportunity for them to interact scientifically with other students as well as with the lecturers, who were all of them extremely open to discussions with the students during their stay in Les Houches.

We are deeply grateful to the lecturers work in preparing the lecture notes, which will be useful in the future to the whole community (both physicists and mathematicians) working on stochastic processes and RMT. We also wish to thank Piotr Warchol for his pictures of the lecturers and Sunčana Dulić for her beautiful drawings of the mountains around Les Houches. Finally, we want to warmly thank the staff of the School of Physics in Les Houches, who made a fantastic work during this school, which was really a great moment for all of us.

Drawings by Sunčana Dulić

Contents

List of participants

ORGANIZERS

SCHEHR GRÉGORY
Université Paris-Sud, France

ALTLAND ALEXANDER
University of Köln, Germany

FYODOROV YAN V.
King's College London, Department of Mathematics, London WC2R 2LS,
United Kingdom

O'CONNELL
University of Bristol, School of Mathematics, Howard House, Queen's Ave, Bristol
BS8 1SD, United Kingdom

CUGLIANDOLO LETICIA E.
Sorbonnes Universités, France

LECTURERS

AKEMANN GERNOT
Bielefeld University, Germany

BORODIN ALEXEI
MIT, Cambridge, MA, USA

BOUCHAUD JEAN-PHILIPPE
CFM/Ecole polytechnique, Paris, France

ECKMANN JEAN-PIERRE
University of Geneva, Switzerland

EYNARD BERTRAND
IPHT CEA Saclay, Gif sur Yvette, France

GUIONNET ALICE
MIT, Cambridge, MA, USA
Université de Lyon, Ecole Normale Supérieure, Lyon, France

KEATING JON
University of Bristol, UK

LE DOUSSAL PIERRE
ENS, Paris, France

MOUSTAKAS ARIS
Universtiy of Athens, Greece

SCHOMERUS HENNING
Lancaster University, UK

SPOHN HERBERT
Technical University Munich, Germany

TOURIGNY YVES
University of Bristol, UK

VARGAS VINCENT
ENS, Paris, France

VIRAG BALINT
University of Toronto, Canada

WEIDENMÜLLER HANS
Max Planck Institut, Heidelberg, Germany

ZABRODIN ANTON
ITEP, Moscow, Russia

PARTICIPANTS

ALLEGRA NICOLAS
Université de Lorraine, Nancy, France

ANDRAUS SERGIO
University of Tokyo, Japan

ANOKHINA ALEXANDRA
ITEP, Moscow, Russia

ASSIOTIS THEODOROS
University of Warwick, Coventry, UK

AUGERI FANNY
Université Paul Sabatier, Toulouse, France

BALDWIN CHRISTOPHER
University of Washington, Seattle, WA, USA

BENITO-MATIAS ENRIQUE
CFMAC, Madrid, Spain

BENOIST TRISTAN
McGill University, Montreal, Canada

BUTEZ RAPHAËL
Université Paris-Dauphine, Paris, France

CHECINSKI TOMASZ
University of Bielefeld, Germany

CUNDEN FABIO DEELAN
University of Bristol, UK

DE NARDIS JACOPO
University of Amsterdam, NL

DOERAENE Antoine
Université Catholique de Louvain, Belgium

DUCLUT Charlie
LPTMC, Paris, France

EMRAH Elnur
University of Wisconsin, Madison, WI, USA

FAHS Benjamin
University of Louvain, Louvain-la-Neuve, Belgium

FUKAI Yosuke
Sano Laboratory, University of Tokyo, Japan

GROUX Benjamin
Laboratoire de Mathématiques de Versailles, France

HOLCOMB Diane
University of Arizona, Tucson, AZ, USA

HUANG Jiaoyang
Harvard University, Cambridge, MA, USA

IPSEN Jesper
Bielefeld University, Germany

JANJIGIAN Christopher
University of Wisconsin, Madison, WI, USA

JOYNER Christopher
Queen Mary University, London, UK

JUNNILA Janne
University of Helsinki, Finland

KIEFF Max
Columbia University, New York, USA

LAMBERT Gaultier
KTH Stockholm Sweden

LI Yiting
Brandeis University, Waltham, MA, USA

LIU Dang-Zheng
University of China, Hefei, China

MARINO Ricardo
LPTMS, Université Paris-Sud, Paris, France

MUDUTE-NDUMBE Steve
Imperial College, London, UK

NEMISH Yuriy
Université Paul Sabatier, Toulouse, France

NGUYEN Vu-Lan
LPMA, Paris, France

NOCK André
Queen Mary University, London, UK

OCHAB Jeremi
Jagiellonian University, Krakow, Poland

PAQUETTE Elliot
Weizmann Institute of Science, Rehovot, Israel

PEREZ CASTILLO Isaac
UNAM, Mexico

POPLAVSKYI Mihail
University of Warwick, Coventry, UK

PRAT Tony
Laboratoire Kastler-Brossel, Paris, France

REYNOLDS Alexi
University of Bristol, UK

RODRIGUEZ-LOPEZ Pablo
LPTMS—Université Paris Sud, Orsay, France

SIMM Nicholas
University of London, UK

SLOWMAN Alexander
SUPA, University of Edinburgh, UK

SWIECH Artur
University of Cologne Germany

TARNOWSKI Wojciech
Jagiellonian University, Krakow, Poland

TARPIN Malo
LPMMC, Grenoble, France

THIERY Thimothée
LPT-ENS, Paris, France

TUISKU Petri
University of Helsinki, Finland

TURUNEN Joonas
University of Helsinki, Finland

VADAKKE VEETTIL Prasad
Raman Research Institute, Bangalore, India

WARCHOL Piotr
Jagiellonina University, Krakow, Poland

XU Yuanyuan
University of California, Davis, CA, USA

1
History—an overview

Oriol BOHIGAS[1] and Hans A. WEIDENMÜLLER[2]

[1]LPTMS, CNRS, Univ. Paris-Sud, Université Paris-Saclay, 91405 Orsay, France; deceased
[2]Max-Planck-Institut für Kemphysik, Heidelberg, P.O. Box 103980, 69029 Heidelberg, Germany

Bohigas, O. and Weidenmüller, H.A., 'History – an overview' first published in *The Oxford Handbook of Random Matrix Theory*. Edited: Gernot Akemann, Jinho Baik, Philippe Di Francesco, Oxford University Press (2011). © Oxford University Press 2011. Reproduced as 'History' in Stochastic *Processes and Random Matrices*. Edited by: Grégory Schehr et al, Oxford University Press (2017). © Oxford University Press 2017. DOI 10.1093/oso/9780198797319.003.0001

Chapter Contents

Preface

When asked to write an overview over the history of random matrix theory (RMT), we faced the difficulty that in the past 20 years RMT has experienced very rapid development. As witnessed by the table of contents of this book, RMT has expanded into a number of areas of physics and mathematics. It has turned out to be impossible to account for this development in the space of the 20 or so pages allocated to us. We decided to focus attention on the first four decades of the history of RMT, and to follow only some lines of development until recent times. Our choice was determined by personal preference and subject knowledge. We have omitted, for instance, the connections between RMT and low-energy field theory, those between RMT and integrability, works on extreme-value statistics, and applications of RMT to wireless communication and to stock-market data. For our reconstruction of the historical line of development, we have only used published literature on the subject. In addition to the original references given in this chapter, there are several reviews and/or reprint collections [Por65, Meh67, Bro81, Boh84b, Gia91, Guh98, Wei09] that are helpful in studying the history of the field.

1.1   Bohr's concept of the compound nucleus

The discovery of narrow resonances in the scattering of slow neutrons by Fermi et al. [Fer34, Fer35] and others in the 1930s came as a big surprise to nuclear physicists. It contradicted earlier ideas of independent-particle motion in nuclei and led Bohr to formulate the idea of the 'compound nucleus' [Boh37]. Bohr wrote

In the atom and in the nucleus we have indeed to do with two extreme cases of mechanical many-body problems for which a procedure of approximation resting on a combination of one-body problems, so effective in the former case, loses any validity in the latter where we, from the very beginning, have to do with essential collective aspects of the interplay between the constituent particles.

And:

The phenomena of neutron capture thus force us to assume that a collision between a ... neutron and a heavy nucleus will in the first place result in the formation of a compound system of remarkable stability. The possible later breaking up of this intermediate system by the ejection of a material particle, or its passing with the emission of radiation to a final stable state, must in fact be considered as separate competing processes which have no immediate connection with the first stage of the encounter.

Bohr's view was generally adopted. Attempts at developing a theory of compound-nuclear reactions could, thus, not be based upon a single-particle approach but had instead to assume that in the compound nucleus, the constituents interact strongly. That fact and the almost complete lack of knowledge of the nucleon–nucleon interaction led to the development of formal theories of nuclear resonance reactions. The most influential of these was the R–matrix theory due to Wigner and Eisenbud [Wig47, Lan58]. The cross-section was parameterized in terms of an R–matrix or, in the single-channel case, an R–function. That function contains the eigenvalues

E_μ of the nuclear Hamiltonian as unknown parameters and is singular whenever the collision energy E equals one of the E_μ, giving rise to a resonance in the cross-section. At the time (and even now) it seemed hopeless to determine the E_μ from a dynamical calculation. Instead of such a calculation, in a series of papers published around 1951 (see the reprint collection [Por65]) Wigner sought to determine characteristic features of the R–function such as its average and its fluctuation about the mean in terms of the distribution of the E_μ. That search and, as he says (see p. 225 of Ref. [Por65]), the 'accidental' discovery of the Wishart ensemble of random matrices in an early version of the book by Wilks [Wil62] motivated Wigner to use random matrices [Wig55, Wig57a]. Thus, Bohr's idea of the compound nucleus is at the root of the use of random matrices in physics.

1.2   Spectral properties

The years following Wigner's introduction of random matrices saw a rapid development of the theory of spectral fluctuations. The Wishart ensemble [Wis28] consists of matrices H that can be written as $H = AA^{\mathrm{T}}$, where T denotes the transpose, and A is real and Gaussian distributed. That ensemble has only positive eigenvalues. In addition to the Wishart ensemble, Wigner considered also an ensemble of real and symmetric matrices H with elements that have a Gaussian zero-centred distribution. Gaussian ensembles with a probability density proportional to $\exp[-(N/\lambda^2)\mathrm{Tr}H^2]$ have since played a dominant role in physics applications of random matrices. Here $N \gg 1$ is the matrix dimension and λ a parameter that scales the average level density $\rho(E)$. As a function of energy E and for $N \to \infty$, $\rho(E)$ has the form [Wig55, Wig58]

$$\rho(E) = \frac{\pi\lambda}{N} \sqrt{1 - (E/2\lambda)^2}. \tag{1.1}$$

This is Wigner's 'semicircle law'. That law can be derived in a number of ways. The method used by Pastur [Pas72] is particularly illuminating and can be generalized. Calculating the distribution $\mathcal{P}(s)$ of spacings of neighbouring levels (the 'nearest-neighbour spacing (NNS) distribution') turned out to be much more difficult. Using the result for random matrices of dimension $N = 2$, Wigner [Wig57b] guessed that $\mathcal{P}(s)$ has the form

$$\mathcal{P}(s) = (\pi/2)s \exp[-(\pi/4)s^2]. \tag{1.2}$$

Here s is the actual spacing in units of the mean spacing $d = 1/\rho$. The linear rise of $\mathcal{P}(s)$ for small s is due to quantum-mechanical level repulsion first considered in 1929 by von Neumann and Wigner [Von29] and, in the present context, by Landau and Smorodinsky, see [Por65]. The 'Wigner surmise' (1.2) was in agreement with results of computer simulations [Ros58] but a definitive experimental test was not possible at the time. There were not enough data. In heavy atoms there was evidence [Ros60] in favour of the distribution (1.2).

For a full theoretical analysis of the eigenvalue distribution, one rewrites the integration measure $\prod_{\mu \leq \nu} \mathrm{d}H_{\mu\nu}$ of the Gaussian real ensemble in terms of the

N eigenvalues λ_μ and of the elements of the diagonalizing orthogonal matrix. Wigner [Wig55, Wig57a] obtained that

$$\prod_{\mu \le \nu} \mathrm{d}H_{\mu\nu} \propto \mathrm{d}\mathcal{O} \prod_{\mu < \nu} |\lambda_\mu - \lambda_\nu| \prod_\sigma \mathrm{d}\lambda_\sigma. \qquad (1.3)$$

Here $\mathrm{d}\mathcal{O}$ is the Haar measure of the orthogonal group in N dimensions. This transition to 'polar coordinates' was first considered by Hua [Hua53] but his book [Hua63] only became available in English in 1963. Equation (1.3) shows that the eigenvalues and eigenfunctions of the Gaussian ensemble are statistically uncorrelated. The Haar measure implies [Por60] that for $N \gg 1$ the projections of the eigenvectors onto an arbitrary direction in Hilbert space have a Gaussian distribution centred at 0, and that the partial widths of the neutron resonances have a χ-squared distribution with one degree of freedom (the 'Porter–Thomas distribution'). This fact had been inferred earlier both from the analysis of neutron resonance data [Por56] and from numerical simulations [Blu58].

Calculation of the NNS distribution in the limit $N \to \infty$ was made possible by introduction of the method of orthogonal polynomials by Mehta and Gaudin [Meh60a, Meh60b, Meh60c]. The key was the recognition that the factor $\prod_{\mu < \nu} |\lambda_\mu - \lambda_\nu|$ in expression (1.3) can be expressed in terms of the Vandermonde determinant of the eigenvalues, and that that determinant can be rewritten in terms of Hermite polynomials $H_m(x)$. The latter are mutually orthogonal, $\int \mathrm{d}x H_m(x) H_n(x) \exp[-x^2] \propto \delta_{mn}$, with respect to the Gaussian weight. That weight factor arises when the probability density $\exp[-(N/\lambda^2)\mathrm{Tr}H^2]$ for the Gaussian ensemble is expressed in terms of the eigenvalues. The determinantal structure so essential for the method is most simply displayed for the ensemble of Hermitian (nonreal) random matrices H of dimension N, defined by a Gaussian weight factor and the measure $\prod_{\mu \le \nu} \mathrm{d}\mathrm{Re}H_{\mu\nu} \prod_{\mu < \nu} \mathrm{d}\mathrm{Im}H_{\mu\nu}$. After transforming to polar coordinates, that measure is $\mathrm{d}U \prod_{\mu < \nu}(\lambda_\mu - \lambda_\nu)^2 \prod_\sigma \mathrm{d}\lambda_\sigma$, where $\mathrm{d}U$ is the Haar measure of the unitary group in N dimensions. The product of the Gaussian weight factor and the factor $\prod_{\mu < \nu}(\lambda_\mu - \lambda_\nu)^2$ can be written in the form $\det K_N(\lambda_\mu, \lambda_\nu)_{\mu,\nu=1,\ldots,N}$, and the function $K_N(x,y) = \sum_{k=0}^{N-1} \phi_k(x)\phi_k(y)$ is given in terms of the harmonic oscillator functions $\phi_k(x)$ (products of Hermite polynomials and a Gaussian). Results for $N \gg 1$ are obtained by using the asymptotic form of ϕ_k for large k. In this way, Gaudin [Gau61] obtained the NNS distribution for the Gaussian ensemble of real symmetric matrices in the form of an infinite product. He found that the Wigner surmise (1.2) is a very good approximation to the exact answer. The method of orthogonal polynomials [Meh67] is not restricted to a Gaussian probability density and can be used for other cases provided the orthogonal polynomials for those cases are known.

In a series of papers, Dyson [Dys62a, Dys62b, Dys62c, Dys62d, Dys62e] contributed to the rapid development of random-matrix theory. Dyson introduced 'circular ensembles' of unitary matrices of dimension N. The eigenvalues are located on the unit circle in the complex plane and in the limit of large matrix dimension have the same spectral fluctuation properties as the eigenvalues of the corresponding Gaussian ensemble located on the real axis. Using group theory he showed that there can be only three

types of such ensembles (the 'threefold way'). The orthogonal ensemble ($\beta = 1$) applies when time-reversal invariance holds, the unitary ensemble ($\beta = 2$) when time-reversal invariance is violated, and the symplectic ensemble ($\beta = 4$) when the system is time-reversal invariant, has half-odd integral spin, and is not rotationally invariant. These statements are not restricted to the circular ensembles but likewise apply to Gaussian ensembles (the GOE with $\beta = 1$, the GUE with $\beta = 2$, and the GSE with $\beta = 4$) and to ensembles with non-Gaussian probability densities. Level repulsion in the three ensembles is governed by the factor $\prod_{\mu<\nu} |\lambda_\mu - \lambda_\nu|^\beta$ which arises from the polar-coordinate representation of the invariant measure. Using orthogonal polynomials, Dyson found the n-level correlation functions for the eigenvalues of the three circular ensembles. He paid special attention to the two-level correlation function. That function relates to the rigidity of the spectrum. In the form of the Δ_3-statistic introduced by Mehta and Dyson [Meh63], that function has, in addition to the NNS distribution, become an important measure for the statistical analysis of empirical or numerically generated eigenvalue distributions. Dyson rederived Gaudin's result for the NNS distribution and calculated the distribution of spacings of next-nearest neighbours. Dyson connected the distribution of the eigenvalues in the Gaussian ensembles with the properties of a classical 'Coulomb gas'. The product $\exp[-(N/\lambda^2)\sum_\mu \lambda_\mu^2]\prod_{\mu<\nu} |\lambda_\mu - \lambda_\nu|^\beta$ is written as $\exp[-\beta W]$ with $W = -\sum_{\mu<\nu} \ln|\lambda_\mu - \lambda_\nu| + \sum_\mu (N/\beta\lambda^2)\lambda_\mu^2$. That distribution is identical to the thermodynamic equilibrium distribution at temperature $kT = 1/\beta$ of the positions of N point charges moving in one dimension under the influence of mutual two-dimensional Coulomb repulsion and an attractive harmonic oscillator potential. Thus, the term 'level repulsion' gains direct physical meaning. Dyson also generalized this static analogy to a dynamic one: the eigenvalues undergo Brownian motion.

For a long time Dyson's classification leading to the three canonical ensembles GOE, GUE, GSE was considered complete. Since the beginning of the 1990s, however, new ensembles (the 'chiral' ensembles) of random matrices have been studied both in disordered systems [Gad91, Gad93, Sle93] and in elementary-particle physics [Shu93, Ver94]. The Hamiltonians have the form

$$H = \begin{pmatrix} 0 & h \\ h^\dagger & 0 \end{pmatrix}. \tag{1.4}$$

There are three symmetry classes as the ensemble may be orthogonally, unitarily, or symplectically invariant. Depending on the topology of the problem the matrix h may be rectangular. Another four classes (the 'Bogoliubov–de Gennes' ensembles) were discovered [Opp90, Alt97] in the context of superconducting systems. The Hamiltonians have the form

$$H = \begin{pmatrix} h & \Delta \\ -\Delta^* & -h^{\mathrm{T}} \end{pmatrix}. \tag{1.5}$$

Here $h = h^\dagger$ and $\Delta = -\Delta^{\mathrm{T}}$. There are four such ensembles because spin rotations have an impact even when time-reversal invariance is violated. Dyson's canonical ensembles are the only ones that occur if the ensembles are postulated to be stationary in energy.

The new ensembles arise in systems with a special and distinct energy value like the origin or the Fermi energy. They play a role when the energy of the system is near that energy. These altogether ten matrix ensembles are complete. That was shown [Zir96, Hei05] with the help of Cartan's classification of symmetric spaces first used in that context by Dyson [Dys70]. (When h in Eq. (1.4) is rectangular rather than square, that classification scheme will not suffice.)

For a meaningful comparison with experimental or numerically generated data, the spectral fluctuation measures of random-matrix theory (RMT) must be stationary, ergodic, and universal. Both stationarity and ergodicity for RMT were first addressed in [Fre78, Pan79]; see also [Bro81]. Stationarity holds if the fluctuation measures (typically local entities defined over an energy interval small in comparison with the total range of the spectrum) do not depend on the centroid energy for which they are calculated. Without this property one would face the (arbitrary) choice of the centroid energy. Stationarity holds [Pan79] for both the Gaussian and the circular ensembles because all n-level correlation functions depend on s, the ratio of the actual level spacing and the mean spacing, and have the same analytical form throughout the spectrum. This statement likewise applies to the S-matrix correlation functions considered in Section 1.7. For a given observable $\mathcal{O}(E)$, ergodicity assures the equality of the ensemble average $\langle\mathcal{O}(E)\rangle$ (a theoretically accessible entity) and the running average $\overline{\mathcal{O}(E)}$ of $\mathcal{O}$ over the spectrum of a single realization of the ensemble. Only the latter is accessible experimentally as one always deals with a specific system. Within RMT, the equality $\overline{\mathcal{O}(E)} = \langle\mathcal{O}(E)\rangle$ cannot be proved as there is no way to evaluate the left-hand side theoretically. But the weaker condition [Fre78, Pan79]

$$\overline{\left(\overline{\mathcal{O}(E)} - \langle\mathcal{O}(E)\rangle\right)^2} = 0 \tag{1.6}$$

involves ensemble averages throughout, can be worked out theoretically, and implies that ergodicity holds for almost all members of the ensemble (i.e., for all except for a set of measure zero, the measure being the integration measure defining the ensemble). Equation (1.6) is met if the autocorrelation function of $\mathcal{O}(E)$ falls off sufficiently rapidly with increasing energy difference. That condition applies [Pan79] for all n-level correlation functions, and the spectral fluctuations of RMT are, therefore, ergodic. For the S-matrix, ergodicity was first proved [Ric77] under the assumption that the underlying random-matrix ensemble has that property. From Eq. (1.6), ergodicity for the S-matrix has been demonstrated only in the Ericson regime of strongly overlapping resonances [Fre78, Bro81]. As regards universality, the weight factor defining the Gaussian ensembles can be justified by a maximum-entropy argument [Bal68] but implies the physically implausible form (1.1) of the average spectrum. More realistic forms of the spectrum can be obtained [Bal68] by replacing the Gaussian weight factor by $\exp[-V(H)]$ where $V(H)$ is some invariant and positive-definite function of H. It is, therefore, important to ascertain that the local spectral fluctuation measures derived for the Gaussian ensembles hold universally, i.e., also for the non-Gaussian ensembles, except for the replacement of the local average level density (1.1) by its counterpart

defined by the weight function $\exp[-V(H)]$. In [Fox64] that fact was demonstrated for those unitary ensembles for which orthogonal polynomials were available. It was conjectured repeatedly (see [Dys72, Boh92, Meh67]) that universality holds quite generally. In [Bre93, Bee93], universality was proved for the local two-level correlation function and for $\beta = 1, 2, 4$ provided that function $V(H)$ confines the spectrum to a finite interval of the real energy axis. For a very general class of observables, a proof valid for all universality classes and for weight functions $V(H)$ obeying that same constraint was given in [Hac95]. That proof clearly displays the root of universality: for $N \to \infty$, global and local spectral properties become independent. The technique of proof used in that paper contained in embryonic form what is now referred to as superbosonization [Bun07, Lit08].

The breaking of a symmetry or an invariance has been an important issue for RMT from the early years. In the simplest case the generic Hamiltonian matrix model for symmetry violation [Ros60] has two block-diagonal entries. Both belong to the GOE and are uncorrelated. They model states with two different quantum numbers. The entries in the nondiagonal blocks have smaller variances and model symmetry breaking. The model has been widely employed in nuclear physics to describe iso-spin violation by the Coulomb interaction. The violation of time-reversal invariance is simulated by writing the Hamiltonian as $H_{\mu\nu} + iA_{\mu\nu}$. Here $H_{\mu\nu}$ is a member of the GOE and A is a real and antisymmetric Gaussian random matrix. Significant violation of symmetry or invariance (perceptible in the spectral fluctuation properties) occurs when the root-mean-square (rms) values of the perturbing matrix elements are on the order of the mean level spacing, i.e., of relative order $1/\sqrt{N}$ (with respect to the unperturbed matrix elements). Following [Dys62d] this extreme sensitivity of the spectral fluctuation measures was first investigated in [Pan81]. It was used [Fre85] to deduce from the NNS distribution of energy levels an upper bound on the strength of time-reversal-invariance breaking in nuclei.

Ginibre [Gin65] extended random matrix theory to the case of non-Hermitian matrices. If the elements are Gaussian-distributed zero-centred uncorrelated random variables, the eigenvalues uniformly fill a circle centred at the origin of the complex plane.

1.3   Data

Precise data on nuclear resonances started to become available in the 1960s from Rainwater's group at Columbia working at the Nevis synchrocylotron. The authors wrote [Rai60]

Although the Nevis synchrocylotron is primarily a tool for high-energy physics, ..., it has also been organized from its inception as an exceptionally strong source of pulsed neutrons for slow neutron spectroscopy.

With the help of time-of-flight spectroscopy, the authors measured the total neutron cross-section versus energy for a number of medium-weight and heavy nuclei and identified up to 200 resonances per nucleus. For a spin 0 target nucleus, all resonances are expected to have identical quantum numbers

because a slow neutron carrying zero angular momentum can excite only resonances with spin $\frac{1}{2}$. For target nuclei with spins different from 0, the resonances can have two different total spins, and in some cases these can be separated. The number 200 was too small for a statistically meaningful comparison with RMT predictions. In addition, spurious levels (strong p-wave resonances mistaken for weak s-wave resonances) and missing levels (s-wave resonances with very small widths) could cause errors in the analysis. Thus, in 1963 Dyson and Mehta [Dys63] wrote, 'Unfortunately, our model is as yet neither proved nor disproved.' In the 1970s more data [Lio72a, Lio72b], and also some from proton scattering on light nuclei below the Coulomb barrier [Wil75, Wat81], became available. In 1982 Bohigas, Haq, and Pandey [Haq82, Boh83] combined all data then available into what they called the 'nuclear data ensemble' comprising 1726 levels. The analysis of that data set showed very good agreement with the RMT predictions for the nearest-neighbour spacing distribution, for spectral rigidity, and for other statistical measures. This work established the view that complex nuclear spectra obey Wigner–Dyson statistics. Since 1982, more sophisticated experimental techniques have become available. This opens up the possibility for renewed critical analysis of a new data set.

1.4   Many-body theory

In the early 1970s, a link between the many-body theory in the form of the nuclear shell model and random matrix theory was established. The shell model is based upon a mean-field description (a set of single-particle states and single-particle energies) supplemented by a 'residual interaction'. The latter is usually assumed to comprise two-body interactions only. French and collaborators had successfully promoted statistical nuclear spectroscopy as a tool for calculating average spectral properties of nuclei (an early summary of this work is given in [Fre66]). These are essentially determined by the normalized traces of powers of the shell-model Hamiltonian. It was found that the spectral density of the shell model is nearly Gaussian in shape (while it is semicircular for RMT). This fact motivated French and Wong [Fre70] and Bohigas and Flores [Boh71] to introduce and investigate the two-body random ensemble (TBRE), a random matrix version of the nuclear shell model: the matrix elements of the residual interaction are not determined from the nuclear dynamics but taken to be uncorrelated Gaussian random variables. Implementation of that model into existing computer codes for shell-model calculations was straightforward. The calculations were done in the sd–shell. Here the maximum number of particles (holes) is six. The spectral density of the TBRE was found to be Gaussian. Increasing the rank k of the k-body interaction resulted in a slow transition to semicircular spectral shape. In all cases the spectral fluctuation properties agreed with those of RMT. This suggested agreement of spectral fluctuation properties with RMT predictions both in the dilute and in the dense limits (number of particles or holes small compared to or on the same order as the number of shell-model single-particle states, respectively).

In [Mon73a] a simplified version of the TBRE that avoids the complexities due to angular momentum of this ensemble was proposed. In the embedded k-body random

ensemble EGE(k) one considers m spinless fermions in Ω degenerate single-particle states that carry no further quantum numbers. The particles interact via a random k-body interaction that is orthogonally, unitarily, or symplectically invariant, as the case may be. Averaging traces of powers of EGE(k) one finds that with increasing k the form of the average spectrum changes from Gaussian to semicircle. This is because both cases are dominated by different Wick contraction patterns. For $k = 2$, results from the orthogonal EGE(2) can be used to reliably predict average results of shell-model calculations [Bro81]. It is not known whether EGE(k) is stationary, ergodic, or universal. Numerical simulations suggest that the spectral fluctuation properties coincide with those of the canonical ensembles but an analytical proof of that assertion does not exist.

1.5   Chaos

In 1890, Poincaré became aware of the existence of chaotic motion in the astronomical three-body problem. Around the middle of the twentieth century, Kolmogorov and his school and other mathematicians and astronomers investigated classical chaos in considerable depth. But it was not until the computer became a universal research tool that physicists at large became familiar with chaotic motion in classical systems. It came as a surprise that chaos occurs in systems with few degrees of freedom, so that complexity is not synonymous with many degrees of freedom. The insight that chaos is a generic feature of classical systems spread in the 1960s and 1970s and naturally led to the question whether there is a difference between quantum systems that are integrable and those that are fully chaotic in the classical limit. Einstein [Ein17] had anticipated that question when he realized that semiclassical quantization can be applied only to integrable systems. The question was mainly studied on systems with few degrees of freedom. While Chirikov and collaborators [Cas87] studied the time evolution of wave packets and discovered the quantum suppression of chaos, others focused attention on spectral properties of quantum systems. Using the correspondence principle and semiclassical arguments, Percival [Per73] put forward the notion that each discrete level of a quantum system can be associated with regular or chaotic classical motion. Berry and Tabor [Ber77] showed that the eigenvalues of a regular quantum system are uncorrelated and, thus, generically possess a Poissonian spectrum. Chaotic quantum systems also received much attention [McD79, Cas80, Ber81]. Using numerical results on about 1000 eigenvalues of the Sinai billiard (classically, a fully chaotic system) and the refined statistical analysis first developed for neutron resonances, Bohigas et al. [Boh84a] demonstrated agreement of some fluctuation measures with RMT predictions and formulated the following conjecture: the spectral fluctuation measures of a generic classically chaotic system coincide with those of the canonical random matrix ensemble that has the same symmetry (unitary, orthogonal, or symplectic). That conjecture was soon supported by numerical studies of other chaotic systems.

Insight into the validity of the conjecture and the difference between integrable and chaotic systems came mainly from Gutzwiller's trace formula [Gut70, Gut90]; see also [Bal70]. The study of manifolds with negative curvature by Schmit (see [Gia91]) and

by Balazs and Voros [Bal86] played a particular role. Spectral fluctuation measures can be written in terms of the level density and the latter, in turn, as a Feynman path integral. Using the stationary-phase approximation in the path integral, one expresses the level density in terms of a sum over periodic orbits. In its essentials, Gutzwiller's trace formula has the form

$$\sum_i \delta(E - E_i) \approx \sum_{\text{p. o.}} A \exp[iS]. \tag{1.7}$$

The left-hand side is the level density (a quantum object), written as the sum over the eigenvalues E_i of the system, with E the energy. Except for the appearance of Planck's constant $\hbar$, the right-hand side contains only classical information: the sum is over all periodic orbits of the system, S is the action of the trajectory (in units of $\hbar$), and A is an amplitude that contains information on the period and the stability of the orbit. Regular systems possess families of periodic orbits while in fully chaotic systems the periodic orbits are isolated. As shown in [Han84] with the help of a sum rule, in chaotic systems only the short periodic orbits are system-specific while the long periodic orbits have universal properties. It is these features that led to an understanding [Ber85] and, eventually, to a demonstration [Sie01, Heu07] of the Bohigas–Giannoni–Schmit conjecture. On the scale of the mean level spacing, the long periodic orbits give rise to universal spectral fluctuation properties. By the uncertainty relation, the shortest periodic orbit determines the scale of the energy interval within which these fluctuations coincide with those of RMT.

1.6 Number theory

In the 1970s a connection of RMT with number theory was discovered. The Riemann zeta function, an analytic function in the entire complex s-plane (except for the point $s = 1$) and defined in terms of prime numbers, plays a fundamental role in number theory. With trivial exceptions, the zeros of this function are known to lie in the strip limited by the two lines $0 + i\alpha$ and $1 + i\alpha$ with α real and $-\infty < \alpha < +\infty$. The Riemann hypothesis says that in fact all these zeros are located on the line $(\frac{1}{2}) + iE$ with $-\infty < E < \infty$. Using that hypothesis, Montgomery [Mon73b] calculated the asymptotic form of the two-point correlation function Y_2 of these zeros. Dyson realized that the expression for Y_2 is the same as for the two–level correlation function of the unitary ensemble in the limit of large matrix dimension. Starting in the 1980s, Odlyzko [Odl87] determined numerically large numbers of zeros of the Riemann zeta function with large imaginary parts. All of these obey the Riemann conjecture. As a function of E, the fluctuating part of the density of the zeros on the line $(\frac{1}{2}) + iE$ is given by

$$-\frac{1}{\pi} \sum_p \sum_{r=1}^{\infty} \frac{\ln p}{p^{r/2}} \cos(Er \ln p). \tag{1.8}$$

Here p denotes the prime numbers and r counts the repetitions. Odlyzko found that the local fluctuation properties of the zeros agree with those of the unitary ensemble of RMT (Montgomery–Odlyzko conjecture) except for modifications due to small prime numbers. Using the analogy to Gutzwiller's trace formula (1.7), Berry and Keating [Ber99] proposed a dynamical interpretation of expression (1.8). The sum over p is like the sum over all periodic orbits, $\ln p$ stands for the period of the periodic orbit, $p^{r/2}$ for the stability, and $E \ln p$ for the action. The deviations found by Odlyzko are then interpreted as system-specific departures from universal behaviour. The connection between RMT and the Riemann zeta function is strengthened by the fact that Bogomolny and Keating [Bog95] have shown that the Hardy–Littlewood conjecture (existence of weak correlations between prime numbers) combined with an expansion of the type (1.8) implies the Montgomery–Odlyzko conjecture. More recently, Keating and Snaith [Kea00] have shown that RMT is able to suggest exact formulas for moments of the Riemann zeta function that were not known and that still have to be proved.

1.7   Scattering theory

Parallel to the developments in statistical spectroscopy triggered by Bohr's paper, there was also important work in nuclear reaction theory. Bohr had postulated the independence of formation and decay of the compound nucleus (CN); see Section 1.1. Hauser and Feshbach [Hau52] used that postulate to write the compound-nucleus cross-section in factorized form,

$$\sigma_{ab} \propto T_a \frac{T_b}{\sum_c T_c}. \tag{1.9}$$

One factor (the 'transmission coefficient' T_a) gives the probability of formation of the CN from the entrance channel a; the other factor $T_b / \sum_c T_c$ gives the normalized probability of decay of the CN into one of the available exit channels b. Both factors are linked by microscopic reversibility, $\sigma_{ab} = \sigma_{ba}$. Originally, the CN was considered a black box and the formation probability of the CN was put equal to unity, $T_a = 1$ for all channels. That view changed with the advent of the nuclear shell model. The concept of a black box was replaced by that of a partially transparent nucleus, formulated in terms of the optical model of elastic scattering [Fes54]. The imaginary part of the optical-model potential (an extension of the concept of the shell-model potential to scattering processes) describes absorption and, thus, formation of the CN. The transmission coefficients can be calculated from the optical model and may be smaller than unity. The average total width Γ of CN resonances was estimated in terms of these formation probabilities as $\Gamma = (d/(2\pi)) \sum_c T_c$ ('Weisskopf estimate' [Bla52]). Here d is the average spacing of the CN resonances

Isolated CN resonances as studied in the time-of-flight experiments by Rainwater and collaborators occur in the scattering of very slow neutrons: only elastic scattering is possible, and Γ is small compared to d. As the bombarding energy is increased, ever more channels open up (inelastic neutron scattering leaving the residual nucleus

in an excited state becomes possible, other breakup channels open up). As a result, Γ grows strongly with energy. At the same time, the average spacing d of resonances shrinks because (like in any many-body fermionic system) the average density $\rho(E) = 1/d$ of states in the CN grows with energy E like $\exp\{\sqrt{aE}\}$, where a is a mass-dependent constant. Thus, with increasing energy the resonances begin to overlap, and for neutrons with bombarding energies of several MeV the CN is in the regime of strongly overlapping resonances ($\Gamma \gg d$). It had been held for a long time [Bla52] that in this regime, the numerous resonances contributing randomly at each energy would yield a scattering amplitude that is smooth in energy. Ericson [Eri60, Eri63] and Brink and Stephen [Bri63] realized that this is not the case. They argued that for $\Gamma \gg d$ the CN cross-section should display strong fluctuations with energy (fluctuations that are as big as the average cross-section, with a correlation width given by Γ), that the Bohr assumption and the Hauser–Feshbach formula Eq. (1.9) hold only for the average cross-section (as opposed to the cross-section at fixed bombarding energy), and that the elements of the scattering matrix are complex random processes with a Gaussian probability distribution. With the advent of electrostatic accelerators of sufficient energy resolution, such fluctuations of the CN cross-section could actually be measured, and the theoretical predictions were verified [Eri66]. Ericson fluctuations (as the phenomenon has come to be called) have since surfaced in many areas of physics. These fluctuations are now understood to describe the generic features of wave scattering by chaotic systems in the regime $\Gamma \gg d$.

These theoretical developments, although inspired by Bohr's idea and informed by statistical arguments, were not linked directly to RMT. The obvious question was: is it possible to derive the Hauser–Feshbach formula (1.9), the Weisskopf estimate for Γ, and Ericson fluctuations from a random-matrix model that is consistent with known or anticipated properties of isolated resonances, i.e., from the GOE? The answer could obviously be given only on the basis of a theory of nuclear resonance reactions. Because of its dependence on a large number of arbitrary parameters (channel radii and boundary condition parameters), the formal R-matrix theory of Wigner and Eisenbud [Wig47] did not qualify for that purpose. The development of nuclear-structure theory in terms of the shell model led to a parameter-free dynamical theory of nuclear resonance reactions [Mah69]. In that theory, the scattering matrix $S_{ab}(E)$ (a unitary symmetric matrix in the space of Λ open channels) is given explicitly in terms of the nuclear Hamiltonian $H_{\mu\nu}$, a matrix in the N-dimensional space of quasibound states. In the simplest case S and H are related by

$$S_{ab}(E) = \delta_{ab} - 2i\pi \sum_{\mu\nu} W_{a\mu}(D^{-1})_{\mu\nu}W_{\nu b}, \qquad (1.10)$$

where

$$D_{\mu\nu}(E) = \delta_{\mu\nu}E - H_{\mu\nu} + i\pi \sum_{c} W_{\mu c}W_{c\nu}. \qquad (1.11)$$

Here $W_{\mu a} = W_{a\mu}$ are real matrix elements that couple the quasibound states $\mu = 1, \ldots, N$ to the open channels $a = 1, \ldots, \Lambda$. The dynamical S-matrix of Eqs. (1.10)

and (1.11) becomes a stochastic matrix when we replace the real and symmetric Hamiltonian matrix $H_{\mu\nu}$ in Eq. (1.11) with the N-dimensional GOE matrix $H_{\mu\nu}^{\mathrm{GOE}}$. The cross-section σ_{ab} is proportional to $|S_{ab}(E)|^2$, and it is the aim of the theory to calculate average values, fluctuations, and correlation functions of cross-sections. This is done by averaging over the ensemble defined in Eqs. (1.10) and (1.11) by the replacement $H_{\mu\nu} \to H_{\mu\nu}^{\mathrm{GOE}}$. The limit $N \to \infty$ is always taken. Ensemble averages are related to experimentally accessible energy averages by ergodicity.

The S-matrix of Eq. (1.11) is written as $S_{ab} = \langle S_{ab}\rangle + S_{ab}^{\mathrm{fl}}$. The average part $\langle S_{ab}\rangle$ and the fluctuating part S_{ab}^{fl} relate to very different time scales of the reaction. By ergodicity $\langle S_{ab}\rangle$ corresponds to an average over energy encompassing very many resonances and, thus, to the fast part of the reaction. Therefore, simple models involving only a few degrees of freedom like the optical model of elastic scattering can be used to calculate $\langle S_{ab}\rangle$ which serves as input for the statistical model of Eqs. (1.10) and (1.11). The fluctuating part S_{ab}^{fl} describes the slow processes (formation and decay of the N resonances) and is the object of interest of the statistical model. The characteristic time scale is the average lifetime of the resonances.

Equations (1.10) and (1.11) contain the $W_{\mu a}$ as parameters. Because of the orthogonal invariance of the GOE in the space of quasibound states, the distribution of S-matrix elements depends only on the orthogonal invariants $(1/N) \sum_\mu W_{a\mu} W_{\mu b}$. For simplicity we choose a basis in channel space for which $\langle S_{ab}\rangle$ is diagonal. Then $(1/N) \sum_\mu W_{a\mu} W_{\mu b} = \delta_{ab} v_a^2$, where v_a^2 has dimension energy and measures the strength of the average coupling of the quasibound states to channel a. The elements of the S-matrix are dimensionless, and the dimensionless invariants have the form $x_a = \pi N v_a^2/\lambda = v_a^2/d$. Here 2λ is the radius of the GOE semicircle and d is the average level spacing (taken in the centre of that semicircle). The average S-matrix can be worked out and has the form $\langle S_{ab}\rangle = \delta_{ab}(1 - x_a)/(1 + x_a)$, and the transmission coefficients are given by $T_a = 1 - |\langle S_{aa}\rangle|^2 = 4x_a/(1 + x_a)^2$. As mentioned earlier, the transmission coefficients are determined phenomenologically in terms of the optical model and serve as input parameters. The distribution of S-matrix elements defined by Eqs. (1.10) and (1.11) is then completely determined. The task of theory consists in working out that distribution explicitly.

Finding the complete distribution of S-matrix elements involves an integration over the $N(N+1)/2$ random variables of the GOE Hamiltonian H^{GOE}. That is a difficult task which has not been fully accomplished. The method of orthogonal polynomials that is so successful for the calculation of spectral fluctuation properties, fails for the scattering problem. A reduced aim is to determine low moments and correlation functions of the S-matrix as these can be compared directly to experimental data. Analyticity and causality imply that for an arbitrary set of real energies E_i and pairs of channels $\{a_i b_i\}$ with $i = 1, \ldots, n$ we have $\langle \prod_i S_{a_i b_i}(E_i)\rangle = \prod_i \langle S_{a_i b_i}(E_i)\rangle$. Therefore, only correlation functions involving at least one factor S and one factor S^* need be calculated. For isolated ($\Gamma \ll d$) and strongly overlapping ($\Gamma \gg d$) resonances first results were obtained by rewriting Eqs. (1.10) and (1.11) in the diagonal representation of H^{GOE}. Let $\mathcal{O}$ be the orthogonal transformation that diagonalizes H^{GOE}, let E_μ with $\mu = 1, \ldots, N$ be the eigenvalues, and let $\tilde{W}_{a\mu} = \sum_\rho \mathcal{O}_{\mu\rho} W_{a\rho}$ be the transformed coupling matrix elements. The latter are Gaussian-distributed random variables. The

S-matrix is expanded in a Born series with respect to the nondiagonal elements of the transformed width matrix $2\pi \sum_a \tilde{W}_{\mu a} \tilde{W}_{a\nu}$; see Eq. (1.10). For $\Gamma \ll d$ these are small in comparison with d, and correlations among the E_μ are irrelevant. Thus, the average cross-section is calculated easily [Lan57, Mol64]. For $\Gamma \gg d$, it is assumed [Aga75] that the eigenvalues E_μ have constant spacings d (this 'picket fence model' neglects GOE correlations among the E_μ). The average over the Gaussian-distributed $\tilde{W}_{\mu a}$ is calculated with the help of Wick contraction. The contraction patterns can be written diagrammatically and are ordered according to the power in d/Γ to which they contribute. The terms of low order can be resummed and given explicitly. (That same scheme was later used by Brézin and Zee [Bre94] for the calculation of level correlations in disordered systems.) As a result, one obtains [Aga75] the Hauser–Feshbach formula (1.9) as the leading term in the asymptotic expansion of the average cross-section. To leading order in d/Γ, the S-matrix elements are found to be Gaussian-distributed random processes with a Lorentzian correlation function. The correlation width Γ is given by the Weisskopf estimate. Thus, for $\Gamma \gg d$ the distribution of S-matrix elements is fully known and consistent with the predictions of [Eri60, Eri63, Bri63]. The neglect of GOE level correlations in these calculations is physically plausible and was later justified when the replica trick was used [Wei84] to calculate the S-matrix correlation function $\langle S_{ab}(E) S_{cd}^*(E + \varepsilon) \rangle$ as a function of the energy difference ε. Without further approximation, that calculation generates an asymptotic expansion in powers of d/Γ, the first few terms of which agree with the results of [Aga75]. Exact calculation of the full correlation function $\langle S_{ab}(E) S_{cd}^*(E + \varepsilon) \rangle$ for all values of Γ/d was possible only with the help of the supersymmetry technique [Ver85a]. The result is an integral representation of the correlation function in terms of a threefold integral over real integration variables. For $\Gamma \ll d$ and $\Gamma \gg d$ the result agrees [Ver86] with the perturbative [Lan57, Mol64] and asymptotic results [Aga75, Wei84] obtained earlier. The result [Ver85a] has been widely used in the analysis of chaotic scattering, i.e., for compound-nucleus reactions, for electron transport through disordered mesoscopic samples [Alh00, Bee97], and for the passage of electromagnetic waves through microwave cavities [Fyo05]. Higher moments and correlation functions of S are not known in general. The supersymmetry method becomes too complex for such a calculation.

Dyson's circular ensembles of unitary S-matrices introduced in Section 1.2 correspond to the case of a very large number of channels all with $T_c = 1$. They have not been used in nuclear reaction theory.

1.8   Replica trick and supersymmetry

The method of orthogonal polynomials, so successful for the calculation of spectral fluctuation measures, faced severe difficulties in the case of scattering problems, and for a long time the calculation of the correlation function of the S-matrix given by Eqs. (1.10) and (1.11) remained an open problem. The situation changed when it was recognized [Ver84] that random-matrix theory corresponds to an Anderson model of dimensionality zero. This established a link between random-matrix theory and the theory of disordered solids that has become ever more important since. In particular,

the link suggested that methods developed for calculating ensemble averages in the theory of disordered solids could be used to advantage also in random-matrix problems. The two important methods taken from the theory of disordered solids were the replica trick due to Edwards and Anderson [Edw75] and the supersymmetry method due to Efetov [Efe83b]. Both methods show that in the limit of large matrix dimension, local fluctuation measures (taken on the scale of the mean level spacing) and global (mean) properties of the spectrum separate. This is the basic reason why random-matrix results for the spectral fluctuation measures are universal while that is not generally true for the global properties.

In both approaches it is assumed that the observable $\mathcal{O}$ that is to be averaged can be written as a suitable derivative of the logarithm of a determinant,

$$\mathcal{O} = \frac{\partial}{\partial j} \ln \det(E\delta_{\mu\nu} - H_{\mu\nu} + M_{\mu\nu}(j)) \Big|_{j=0}, \tag{1.12}$$

or as the derivative of the logarithm of a product of such determinants. The observable usually involves the resolvent of the Hamiltonian, and the logarithm is needed to remove the determinant from the final expression. We have used the same notation as in Eq. (1.11). The matrix $M(j)$ specifies the observable. In addition, $M(j)$ may describe the coupling to open channels, etc. The inverse determinant can be written as a Gaussian integral over N complex integration variables ψ_μ. Averaging $\mathcal{O}$ over the ensemble is then tantamount to averaging the logarithm of the generating functional

$$Z(j) = \int_{-\infty}^{+\infty} \prod_{\rho=1}^{N} d\psi_\rho \, \exp\{(i/2) \sum_{\mu\nu} \psi_\mu^*[E\delta_{\mu\nu} - H_{\mu\nu} + M_{\mu\nu}(j)]\psi_\nu\}, \tag{1.13}$$

or of a product of such functionals. Formally speaking, Eq. (1.13) establishes the connection to a field theory with commuting (bosonic) fields. Instead of the commuting integration variables ψ_μ appearing in Eq. (1.13), one may use anticommuting (fermionic) integration variables χ_μ, and the average of $\mathcal{O}$ is obtained by averaging a generating functional of the form (1.13) but with anticommuting integration variables. That yields a relation to a field theory with fermionic fields. The use of anticommuting integration variables goes back to Berezin [Ber61].

Averaging the logarithm of $Z(j)$ is very complicated for both commuting and anticommuting integration variables. The difficulty is circumvented with the help of the replica trick, originally introduced in [Edw75] for spin-glass problems. Here one determines $\langle \ln Z \rangle$ from the identity

$$\langle \ln Z \rangle = \lim_{n \to 0} \left(\frac{\langle Z^n \rangle - 1}{n} \right). \tag{1.14}$$

Equation (1.14) holds if $\langle Z^n \rangle$ is known analytically as a function of complex n in the vicinity of $n = 0$. Actually, one calculates $\langle Z^n \rangle$ for positive or negative integer n (using fermionic or bosonic integration variables, respectively) and then uses Eq. (1.14). Thus, the approach is not exact and is called the replica trick (rather

than the replica method). The replica trick with commuting variables was first used in random-matrix theory to calculate the shape of the average spectrum of the GOE (the 'Wigner semicircle law') [Edw76]. In two seminal papers, Wegner [Weg79] and Schäfer and Wegner [Sch80] analysed the structure of the underlying theory for bosonic fields. Evaluating the integrals with the help of the saddle-point approximation, they showed that for the two-point function (an average involving both the resolvent and its complex conjugate), the symmetry of the theory is broken, leading to a Goldstone Boson. The single saddle-point encountered for the one-point function becomes a saddle-point manifold with hyperbolic symmetry. After integration over the massive modes (the modes orthogonal to the saddle-point manifold), the theory is equivalent to a nonlinear sigma model. When applied to the two-point function of the stochastic scattering matrix, again with commuting integration variables, the replica trick yields an asymptotic expansion in powers of d/Γ [Wei84] but, unfortunately, not the full answer.

A new and different approach to ensemble averaging in the theory of disordered metals was initiated by Efetov [Efe82], first applied to disordered systems by Efetov and Larkin [Efe83a]) and then extended by Efetov [Efe83b]. In that approach the use of the replica trick is avoided by writing (similarly to Eq. (1.12)) the observable as the derivative of the ratio of two determinants or, equivalently, of the product of a generating functional $Z^{(1)}$ with fermionic integration variables and of a generating functional $Z^{(-1)}$ with bosonic integration variables. (It seems that fermionic integration variables were first used for condensed-matter problems in [Efe80].) At $j = 0$ that product equals unity, the observable is simply the ordinary derivative of the product, and averaging the latter is simple. The price one must pay consists in working with both commuting and anticommuting integration variables. A similar combination of bosonic and fermionic fields occurs in the theory of elementary particles [Wes74]. Here, a special relativistic fermion-boson symmetry is denoted as 'supersymmetry'. That same term is used for Efetov's method although the method lacks that special relativistic symmetry. The method can be formulated for the cases of unitary, orthogonal, and symplectic invariance.

Efetov's approach leads generally to a supersymmetric nonlinear sigma model. With the help of Efetov's method, the two-point function of the scattering matrix defined in Eqs. (1.10) and (1.11) could be worked out exactly [Ver85a]. In [Efe83b, Ver85a], the results of [Sch80] on the structure of the saddle-point manifold were extended to the supersymmetry approach. The saddle–point manifold was shown to involve both compact and noncompact integration manifolds, stemming from the fermionic and bosonic integration variables, respectively. The supersymmetry method has since seen an ever-growing number of applications to problems in random-matrix theory [Efe97]. It applies to all 10 random–matrix ensembles [Zir96, Eve08].

The supersymmetry method gave rise to a critique of the replica trick. In [Ver85b] it was argued that the asymptotic result found in [Wei84] is generic in the sense that the replica trick always yields an asymptotic or perturbative (and not the exact) result. The claim was substantiated by analysis of the replica-trick calculation of the spectral two-point function for the GUE. It was shown that the use of only bosonic or only fermionic degrees of freedom fails to account for the full complexity of the saddle-point

manifold which is obtained in the framework of the supersymmetry method and is needed to obtain the exact result. The paper was followed by an animated debate in the community on the validity and limitations of the replica trick. In [Kam99, Yur99] a cure for its failure was seen in the breaking of the replica symmetry and hitherto neglected saddle-point contributions; see [Zir99]. Reference [Kan02] cast a new light on the problem. In the framework of matrix ensembles with unitary symmetry it was argued that the deficiencies of the replica trick were caused by an approximate evaluation of $\langle Z^{(n)} \rangle$ for integer n. The problem and the explicit calculation of these functions were circumvented by showing that in the fermionic case (n positive integer) the $\langle Z^{(n)} \rangle$ are closely related to functions $\tau_n(E)$ known from the theory of integrable systems. These functions obey a set of coupled nonlinear differential equations in the parameter E (the energy), the Toda lattice equations. The same functions also appear in the Hamiltonian formulation of the six Painlevé equations. There, however, the replica index n plays the role of a parameter. That connection shows that it is possible and meaningful to continue the Toda lattice equations analytically in n. Extrapolating the result to $n = 0$ yields the exact result for the one-point and two-point functions. In addition to showing that the replica trick can be made exact (at least for the case of unitary symmetry), the work of [Kan02] established a connection between random-matrix theory and integrable nonlinear systems. The arguments in [Kan02] were much simplified in [Spl04]. The authors showed that $\langle Z^{(n)} \rangle$ obeys the Toda lattice equations separately for positive and for negative integer values of n and assumed that these equations also hold at $n = 0$. Then $\langle Z^{(0)} \rangle$ (the quantity of interest) is given by the product $\langle Z^{(1)} \rangle \langle Z^{(-1)} \rangle$.

This shows that '*the factorization of the two-point function into a bosonic and a fermionic partition function is not an accident but rather the consequence of the relation between random-matrix theories and integrable hierarchies*' [Spl04].

The assumption made in [Spl04]—validity of the Toda lattice equations across the point $n = 0$—was later justified with the help of orthogonal polynomials [Ake05]. Thus, it is now established (at least for the unitary case) that when handled properly the replica trick yields exact results, too.

1.9   Disordered solids

In order to describe the response of small disordered metallic particles to external fields, Gor'kov and Eliashberg [Gor65] needed the two-level eigenvalue correlation function. They argued that the Wigner–Dyson statistic should apply in this case as well as in atomic nuclei. That was the first application of RMT in condensed-matter physics. Almost 20 years later Efetov [Efe83b] derived their statistical hypothesis from a generic model for a quasi one-dimensional disordered solid in the limit of small system length using the supersymmetry technique. Independently, in [Ver84] it was recognized that random-matrix theory as used in nuclear physics corresponds to an Anderson model of dimensionality zero. The resulting link between random-matrix theory and the theory of disordered solids has become ever more important since. As described in the

following paragraphs, that link has played a role in designing random-matrix models for Anderson localization and for electron transport through disordered mesoscopic samples. It has likewise been important in formulating a theoretical approach to Andreev scattering in disordered conductors where it gave rise to the discovery of the four Bogoliubov–de Gennes ensembles. But the first and obvious question was: how is the RMT behaviour valid for small disordered samples modified when we consider larger samples? Al'tshuler and Shklovskii [Alt86] considered disordered systems in the diffusive regime and showed that the range of validity of RMT in energy is limited by the Thouless energy $E_c = \hbar \mathcal{D}/L^2$. Here $\mathcal{D}$ is the diffusion constant and L is the length of the sample. They also calculated corrections to RMT predictions.

In 1959, Anderson showed that the eigenfunctions of noninteracting electrons in disordered solids may be localized (i.e., do not extend uniformly over the entire system). An important link to chaotic quantum systems was established when Fishman et al. [Fis82], see also [Chi81], recognized that such systems (like, for instance, the kicked rotor) display localization in momentum space in a manner closely related to the localization that occurs in ordinary space for disordered solids. In [Sel86] it was proposed to use random band matrices (for which the variances of the nondiagonal elements that have a distance $\geq b$ from the diagonal are suppressed) as vehicles to study localization phenomena in quasi one-dimensional geometries. In his original papers on RMT, Wigner had named such matrices 'bordered matrices'. As models for random linear chains (which play a role in a variety of physical situations) they were studied early on by Dyson [Dys53], Wigner [Wig55, Wig57a], and Engleman [Eng58]. After being used numerically both for the study of the kicked rotor and in the form of transfer matrices for the Anderson localization problem, random band matrices were first investigated analytically in [Fyo91]. With the help of supersymmetry it was shown that random band matrices are fully equivalent to random quasi one-dimensional systems. The parameter relevant for localization is the square of the band width b divided by the matrix dimension N, as had been expected on numerical grounds. Since then power-law random band matrices (where the variances of the off-diagonal matrix elements $\langle H_{ij}^2 \rangle$ decrease with an inverse power of $|i-j|$) have been used to model the metal-insulator transition in disordered solids [Mir96]; see also [Eve08].

Interest in the energy eigenvalue statistics of disordered mesoscopic samples (metallic or semiconducting devices of sufficiently small size so that at low temperature the electron's inelastic mean free path is larger than the sample size) was stimulated by the experimental discovery of universal conductance fluctuations (UCF). Following the seminal work of Al'tshuler and Shklovskii [Alt86] and Imry's application [Imr86] of RMT to UCF, a random-matrix description of quantum transport was developed in [Mut87]. The energy levels are linked to the transmission eigenvalues. Alternatively, electron transport was described in terms of the scattering approach of Eqs. (1.10) and (1.11). The linear extension of the sample was accounted for by replacing the GOE Hamiltonian with a random band matrix [Iid90]. In either of these forms, RMT has been widely applied to electron transport through mesoscopic systems [Bee97].

1.10   Interacting fermions and field theory

A dynamical extension of Dyson's Coulomb gas (see Section 1.2) was considered by Calogero [Cal69a, Cal69b]. The Hamiltonian

$$H_C = -\frac{\hbar^2}{2m} \left[\sum_{i=1}^{N} \frac{\partial^2}{\partial \lambda_i^2} - \frac{1}{2} \sum_{i<j} \frac{\beta(\beta - 2)}{(\lambda_i - \lambda_j)^2} \right] + \sum_i V(\lambda_i) \qquad (1.15)$$

describes the dynamics of N particles in one dimension with position variables λ_i interacting via Coulomb forces and under the influence of a common potential V. For a harmonic-oscillator potential V, that Hamiltonian is integrable. The ground-state wave function defines a probability distribution of the particle positions which for $\beta = 1, 2, 4$ coincides with the joint probability densities for the eigenvalues (in units of the mean level spacing) of Dyson's three canonical Gaussian ensembles. An analogous construction for the circular ensembles is due to Sutherland [Sut71]. A link between the parametric level correlation functions of RMT and the time-dependent particle density correlation functions of the Sutherland Hamiltonian was conjectured and later proved in [Sim93a, Sim93b, Sim93c, Sim94]. In recent years, the connection between interacting fermions in one dimension and RMT has given rise to substantial research activity; see [Kor93].

Random-matrix theory has been useful in model studies of power-series expansions of field theories with internal SU(N) symmetry. The terms arising in these expansions can be expressed as Feynman diagrams, and counting the leading terms in an expansion in inverse powers of N (the 'planar diagrams') can be accomplished with the help of RMT [Bre78]. An analogous statement applies to quantum gravity in two dimensions.

Starting from the low-energy effective Lagrangean of quantum chromodynamis (QCD), Leutwyler and Smilga [Leu92] derived sum rules for the inverse eigenvalues of the Dirac operator. These same sum rules can also be derived from chiral random-matrix theories [Shu93, Ver93, Hal95]. This fact gave rise to the insight that in the low-energy limit, QCD is equivalent to a chiral random-matrix theory that has the same symmetries. Similar equivalence relations exist for other field theories.

Acknowledgements

The authors are grateful to J. J. M. Verbaarschot for a critical reading of parts of the manuscript and for helpful suggestions.

References

[Aga75] D. Agassi, H. A. Weidenmüller, and G. Mantzouranis, *Phys. Rep.* **22** (1975), 145.

[Ake05] G. Akemann, J. C. Osborn, K. Splittorf, J. J. M. Verbaarschot, *Nucl. Phys.* B **712** (2005), 287.

[Alh00] Y. Alhassid, *Rev. Mod. Phys.* **72** (2000), 895.

[Alt97] A. Altland and M. R. Zirnbauer, *Phys. Rev.* **B 55** (1997), 1142.

[Alt86] B. L. Altshuler and B. I. Shklovskii, *Zh. Eksp. Teor. Fiz.* **91** (1986), 220 [*Sov. Phys. JETP* **64** (1986), 127].

[Bal86] N. L. Balazs and A. Voros, *Phys. Rep.* **143** (1986), 109.

[Bal68] R. Balian, *Nuov. Cim.* **LVII B** (1968), 183.

[Bal70] R. Balian and C. Bloch, *Ann. Phys. (N. Y.)* **60** (1970), 401.

[Bee93] C. W. Beenakker, *Phys. Rev. Lett.* **70** (1993), 1155.

[Bee97] C. W. Beenakker, *Rev. Mod. Phys.* **69** (1997), 731.

[Ber61] F. A. Berezin, *Dokl. Akad. Nauk SSR* **137** (1961), 31.

[Ber77] M. V. Berry and M. Tabor, *Proc. R. Soc. London, Ser. A* **356** (1977), 375.

[Ber81] M. V. Berry, *Ann. Phys. (N.Y.)* **131** (1981), 163.

[Ber85] M. V. Berry, *Proc. R. Soc. Ser. A* **400** (1985), 229.

[Ber99] M. V. Berry and J. P. Keating, *SIAM Rev.* **41** (1999), 236.

[Bla52] J. M. Blatt and V. F. Weisskopf, *Theoretical Nuclear Physics*, Wiley, New York, 1952.

[Blu58] S. Blumberg and C. E. Porter, *Phys. Rev.* **110** (1958), 786.

[Bog95] E. B. Bogomolny and J. P. Keating, *Nonlinearity* **8** (1995), 1115; ibid. **9** (1996) 911.

[Boh71] O. Bohigas and J. Flores, *Phys. Lett.* **34 B** (1971), 261.

[Boh83] O. Bohigas, R. U. Haq, and A. Pandey, in *Nuclear Data for Science and Technology*, ed. K. H. Böckhoff, p. 809. Reidel, Dordrecht, 1983.

[Boh84a] O. Bohigas, M.-J. Giannoni, and C. Schmit, *Phys. Rev. Lett.* **52** (1984), 1.

[Boh84b] O. Bohigas and M.-J. Giannoni, in *Lecture Notes in Physics*, vol. 209, p. 1, Springer–Verlag, Berlin, 1984.

[Boh92] O. Bohigas, in *Chaos and Quantum Physics*, ed. .M.-J. Giannoni, A. Voros, and J. Zinn-Justin. Elsevier, Amsterdam, 1992.

[Boh37] N. Bohr, *Nature* **137** (1936), 344.

[Bre78] E. Brézin, C. Itzykson, G. Parisi, J. Zuber, *Comm. Math. Phys.* **59** (1978), 35.

[Bre93] E. Brézin and A. Zee, *Nucl. Phys.* **B 402** (1993), 613.

[Bre94] E. Brézin and A. Zee, *Phys. Rev.* **E 49** (1994), 2588.

[Bri63] D. M. Brink and R. O. Stephen, *Phys. Lett.* **5** (1963), 77.

[Bro81] T. A. Brody, J. Flores, J. B. French, P. A. Mello, A. Pandey, and S. S. M. Wong, *Rev. Mod. Phys.* **53** (1981), 385.

[Bun07] J. E. Bunder, K. B. Efetov, V. E. Kravtsov, O. M. Yevtushenko, and M. R. Zirnbauer, *J. Stat. Phys.* **129** (2007), 809.

[Cas87] G. Casati, B. V. Chirikov, I. Guarneri, and D. L. Shepelyansky, *Phys. Rep.* **154** (1987), 77.

[Cas80] G. Casati, F. Valz-Gris, and I. Guarneri, *Lett. Nuovo Cimento Soc. Ital. Fis.* **28** (1980), 279.

[Chi81] B. V. Chirikov, F. M. Izrailev, and D. L. Shepelyansky, *Sov. Sci. Rev.* **C2** (1981), 209.

[Cal69a] F. Calogero, *J. Math. Phys.* **10** (1969), 2191.

[Cal69b] F. Calogero, *J. Math. Phys.* **10** (1969), 2197.

[Dys53] F. J. Dyson, *Phys. Rev.* **92** (1953), 1331.

[Dys62a] F. J. Dyson, *J. Math. Phys.* **3** (1962), 140.

[Dys62b] F. J. Dyson, *J. Math. Phys.* **3** (1962), 157.

[Dys62c] F. J. Dyson, *J. Math. Phys.* **3** (1962), 166.

[Dys62d] F. J. Dyson, *J. Math. Phys.* **3** (1962), 1191.

[Dys62e] F. J. Dyson, *J. Math. Phys.* **3** (1962), 1199.

[Dys63] F. J. Dyson and M. L. Mehta, J. Math. Phys. **4** (1963), 701.

[Dys70] F. J. Dyson, *Commun. Math. Phys.* **19** (1970), 235.

[Dys72] F. J. Dyson, *J. Math. Phys.* **13** (1972), 90.

[Edw75] S. F. Edwards and P. W. Anderson, *J. Phys. F: Metal Physics* **5** (1975), 965.

[Edw76] S. F. Edwards and R. C. Jones, *J. Phys. A: Math. Gen.* **9** (1976), 1595.

[Efe80] K. B. Efetov, A. I. Larkin, and D. E. Khmelnitzky, *Zh. Exp. Teor. Fiz.* **79** (1980), 1120 (*Sov. Phys. JETP* **52** (1980), 568).

[Efe82] K. B. Efetov, *Zh. Exp. Teor. Fiz.* **82** (1982), 872 (*Sov. Phys. JETP* **55** (1982), 514).

[Efe83a] K. B. Efetov and A. I. Larkin, *Zh. Exp. Teor. Fiz.* **85** (1983) 764 (*Sov. Phys. JETP* **58** (1980), 444).

[Efe83b] K. B. Efetov, *Adv. Phys.* **32** (1983), 53.

[Efe97] K. B. Efetov, *Supersymmetry in Disorder and Chaos*. Cambridge University Press, Cambridge, 1997.

[Ein17] A. Einstein, *Verh. Deutsch. Phys. Ges.* **19** (1917), 82.

[Eng58] R. Engleman, *Nuovo Cimento* **10** (1958), 615.

[Eri60] T. Ericson, *Phys. Rev. Lett.* **5** (1960), 430.

[Eri63] T. Ericson, *Ann. Phys. (N.Y.)* **23** (1963), 390.

[Eri66] T. Ericson and T. Mayer–Kuckuk, *Ann. Rev. Nucl. Sci.* **16** (1966), 183.

[Eve08] F. Evers and A. D. Mirlin, *Rev. Mod. Phys.* **80** (2008), 1355.

[Fer34] E. Fermi *et al.*, *Proc. Roy. Soc.* **A 148** (1934), 483.

[Fer35] E. Fermi *et al.*, *Proc. Roy. Soc.* **A 149** (1935), 522.

[Fes54] H. Feshbach, C. E. Porter, and V. F. Weisskopf, *Phys. Rev.* **96** (1954), 448.

[Fis82] S. Fishman, D. R. Grempel, and R. E. Prange, *Phys. Rev. Lett.* **49** (1982), 509.

[Fox64] D. Fox and P. B. Kahn, *Phys. Rev.* **134** (1964), B1151.

[Fre66] J. B. French, in *Many–body Description of Nuclei and Reactions*, International School of Physics Enrico Fermi, Course **36**, ed. C. Bloch, p. 278. Academic Press, New York, 1966.

[Fre70] J. B. French and S. S. M. Wong, *Phys. Lett.* **B 33** (1970), 449.

[Fre78] J. B. French, P. A. Mello, and A. Pandey, *Phys. Lett.* **B 80** (1978), 17.

[Fre85] J. B. French, V. K. B. Kota, A. Pandey, and S. Tomsovic, *Phys. Rev. Lett.* **54** (1985), 2313.

[Fyo91] Y. V. Fyodorov and A. D. Mirlin, *Phys. Rev. Lett.* **67** (1991), 2405.

[Fyo05] Y. V. Fyodorov, D. V. Savin, and H. J. Sommers, *J. Phys. A: Math. Gen.* **38** (2005), 10731.

[Gad91] R. Gade and F. Wegner, *Nucl. Phys.* **B 360** (1991), 213.

[Gad93] R. Gade, *Nucl. Phys.* **B 398** (1993), 499.

[Gau61] M. Gaudin, *Nucl. Phys.* **25** (1961), 447.

[Gia91] M.-J. Giannoni, A. Voros, and J. Zinn–Justin (eds). *Chaos and Quantum Physics*. North Holland, Amsterdam, 1991.

[Gin65] J. Ginibre, *J. Math. Phys.* **6** (1965), 440.

[Gor65] L. P. Gor'kov and G. M. Eliashberg, *Zh. Exp. Teor. Fiz.* **48** (1965), 1407.

[Guh98] T. Guhr, A. Müller–Groeling, and H. A. Weidenmüller, *Phys. Rep.* **299** (1998), 189.

[Gut70] M. Gutzwiller, *J. Math. Phys.* **11** (1970), 1791.

[Gut90] M. Gutzwiller, *Chaos in Classical and Quantum Mechanics*. Springer–Verlag, Berlin, 1990.

[Hac95] G. Hackenbroich and H. A. Weidenmüller, *Phys. Rev. Lett.* **74** (1995), 4118.

[Hal95] M. A. Halasz and J. J. M. Verbaarschot, *Phys. Rev.* **D 3** (1995), 2563.

[Han84] J. H. Hannay and A. M. Ozorio de Almeida, *J. Phys. A* **17** (1984), 3429.

[Haq82] R. U. Haq, A. Pandey, and O. Bohigas, *Phys. Rev. Lett.* **48** (1982), 1086.

[Hau52] W. Hauser and H. Feshbach, *Phys. Rev.* **87** (1952) 366.

[Hei05] P. Heinzner, A. Huckleberry, and M. R. Zirnbauer, *Comm. Math. Phys.* **257** (2005), 725.

[Heu07] S. Heusler, S. Müller, A. Altland, P. Braun, and F. Haake, *Phys. Rev. Lett.* **98** (2007), 044103.

[Hua53] L. K. Hua, *J. Chinese Math. Soc.* **2** (1953), 288.

[Hua63] L. K. Hua, *Harmonic Analysis*. American Mathematical Society, Providence, R I., 1963. [Originally published in Chinese in 1958, translated from the Russian translation that appeared in 1959.]

[Iid90] S. Iida, H. A. Weidenmüller, and J. A. Zuk, *Ann. Phys. (N.Y.)* **200** (1990), 219.

[Imr86] Y. Imry, *Europhys. Lett.* **1** (1986), 249.

[Kam99] A. Kamenev and M. Mezard, *Phys. Rev.* **B 60** (1999), 3944.

[Kan02] E. Kanzieper, *Phys. Rev. Lett.* **89** (2002), 250201.

[Kea00] J. P. Keating and N. C. Snaith, *Comm. Math. Phys.* **214** (2000), 57.

[Kor93] V. E. Korepin, N. M. Bogoljubov, and A. G. Izergin. *Quantum Inverse Scattering Method and Correlation Functions*. Cambridge University Press, Cambridge, 1993.

[Lan57] A. M. Lane and J. E. Lynn, *Proc. Roy. Soc.* **A 70** (1957), 557.

[Lan58] A. M. Lane and R. G. Thomas, *Rev. Mod. Phys.* **30** (1958), 257.

[Leu92] H. Leutwyler and A. Smilga, *Phys. Rev.* **D 46** (1992), 5607.

[Lio72a] H. L. Liou, H. S. Camarda, S. Wynchank, M. Slagowitz, G. Hacken, F. Rahn, and J. Rainwater, *Phys. Rev.* **C 5** (1972), 974.

[Lio72b] H. L. Liou, H. S. Camarda, and F. Rahn, *Phys. Rev.* **C 5** (1972), 1002.

[Lit08] M. Littelmann, H.–J. Sommers, and M. R. Zirnbauer, *Comm. Math. Phys.* **283** (2008), 343.

[Mah69] C. Mahaux and H. A. Weidenmüller. *Shell-Model Approach to Nuclear Reactions*. North–Holland, Amsterdam, 1969.

[McD79] S. W. McDonald and A. N. Kaufman, *Phys. Rev. Lett.* **42** (1979), 1189.

[Meh60a] M. L. Mehta, *Nucl. Phys.* **18** (1960), 395.

[Meh60b] M. L. Mehta and M. Gaudin, *Nucl. Phys.* **18** (1960), 420.

[Meh60c] M. L. Mehta and S. P. H. Rapport (Saclay) No. 658 (1960) (reprinted in Ref. [Por65]).

[Meh63] M. L. Mehta and F. J. Dyson, *J. Math. Phys.* **4** (1963), 713.

[Meh67] M. L. Mehta. *Random Matrices*. Academic Press, New York, 1967; 2nd edn, 1991; 3rd edn, Elsevier, Amsterdam, 2004.

[Mir96] A. D. Mirlin, Y. V. Fyodorov, F.-M. Dittes, J. Quesada, and T. H. Seligman, *Phys. Rev.* **E 54** (1996), 3221.

[Mol64] P. A. Moldauer, *Phys. Rev.* **135 B** (1964), 642.

[Mon73a] K. F. Mon and J. B. French, *Ann. Phys. (N.Y.)* **95** (1975), 90.

[Mon73b] H. Montgomery, *Proc. Sympos. Pure Math.* **24** (1973), 181.

[Mut87] K. A. Muttalib, J.-L. Pichard, and A. D. Stone, *Phys. Rev. Lett.* **59** (1987), 2475.

[Odl87] A. M. Odlyzko, *Math. Comp.* **48** (1987), 273.

[Opp90] R. Oppermann, *Physica* **A 167** (1990), 301.

[Pan79] A. Pandey, *Ann. Phys. (N.Y.)* **119** (1979), 170.

[Pan81] A. Pandey, *Ann. Phys. (N.Y.)* **134** (1981), 110.

[Pas72] L. A. Pastur, *Theor. Math. Phys.* **10** (1972), 67.

[Per73] I. C. Percival, *J. Phys.* **B6** (1973), L229.

[Por56] C. E. Porter and R. G. Thomas, *Phys. Rev.* **104** (1956), 483.

[Por60] C. E. Porter and N. Rosenzweig, *Suomalaisen Tiedeakatemian Toimituksia (Ann. Akad. Sci. Fennicae)* **AVI** (1960), no. 44, reprinted in Ref. [Por65].

[Por65] C. E. Porter, *Statistical Theories of Spectra: Fluctuations*. Academic Press, New York, 1965.

[Rai60] J. Rainwater, W. W. Havens, Jr., D. S. Desjardins, and J. L. Rosen, *Rev. Sci. Instr.* **31** (1960), 481.

[Ric77] J. Richert and H. A. Weidenmüller, *Phys. Rev.* **C 16** (1977), 1309.

[Ros58] N. Rosenzweig, *Phys. Rev. Lett.* **1** (1958), 24.

[Ros60] N. Rosenzweig and C. E. Porter, *Phys. Rev.* **120** (1960), 1698.

[Sch80] L. Schäfer and F. Wegner, *Z. Phys.* **B 38** (1980), 113.

[Sel86] T. H. Seligman and J. J. M. Verbaarschot, in *Proceedings of the Fourth International Conference on Quantum Chaos and the 2nd Colloquium on Statistical Nuclear Physics*, ed. T. H. Seligman and H. Nishioka, p. 131. Springer, Berlin, 1986.

[Shu93] E. V. Shuryak and J. J. M. Verbaarschot, *Nucl. Phys.* **A 560** (1993), 306.

[Sie01] M. Sieber and K. Richter, *Physica Scripta* **T 90** (2001), 128.

[Sim93a] B. D. Simons, P. A. Lee, and B. L. Altshuler, *Phys. Rev.* **B 48** (1993), 11450.

[Sim93b] B. D. Simons, P. A. Lee, and B. L. Altshuler, *Phys. Rev. Lett.* **70** (1993), 4122.

[Sim93c] B. D. Simons, P. A. Lee, and B. L. Altshuler, *Nucl. Phys.* **B 409** (1993), 487.

[Sim94] B. D. Simons, P. A. Lee, and B. L. Altshuler, *Phys. Rev. Lett.* **72** (1993), 64.

[Sle93] K. Slevin and T. Nagao, *Phys. Rev. Lett.* **70** (1993), 635.

[Spl04] K. Splittorf and J. J. M. Verbaarschot, *Nucl. Phys.* **B 683** (2004), 467.

[Sut71] B. Sutherland, *J. Math. Phys.* **12** (1971), 246.

[Ver86] J. J. M. Verbaarschot, *Ann. Phys, (N.Y.)* **168** (1986), 368.

[Ver84] J. J. M. Verbaarschot, H. A. Weidenmüller, and M. R. Zirnbauer, *Phys. Rev. Lett.* **52** (1984), 1597.

[Ver85a] J. J. M. Verbaarschot, H. A. Weidenmüller, and M. R. Zirnbauer, *Phys. Rep.* **129** (1985), 367.

[Ver85b] J. J. M. Verbaarschot and M. R. Zirnbauer, *J. Phys. A: Math. Gen.* **17** (1985), 1093.

[Ver93] J. J. M. Verbaarschot and I. Zahed, *Phys. Rev. Lett.* **70** (1993), 3852.

[Ver94] J. J. M. Verbaarschot, *Phys. Rev. Lett.* **72** (1994), 2531.

[Von29] J. von Neumann and E. P. Wigner, *Phys. Zschr.* **30** (1929), 465.

[Wat81] W. A. Watson III, E. G. Bilpuch, and G. E. Mitchell, *Z. Phys.* **A 300** (1981), 89.

[Weg79] F. Wegner, *Phys. Rev.* **B 19** (1979), 783.

[Wei84] H. A. Weidenmüller, *Ann. Phys. (N.Y.)* **158** (1984), 120.

[Wei09] H. A. Weidenmüller and G. E. Mitchell, *Rev. Mod. Phys.* **81** (2009), 539.

[Wes74] G. Wess and B. Zumino, *Nucl. Phys.* **B 70** (1974), 39.

[Wig55] E. P. Wigner, *Ann. Math.* **62** (1955), 548.

[Wig57a] E. P. Wigner, *Ann. Math.* **65** (1957), 203.

[Wig57b] E. P. Wigner, in *Conference on Neutron Physics by Time–of–Flight*, Oak Ridge National Laboratory Report 2309 (1957), p. 59.

[Wig58] E. P. Wigner, *Ann. Math.* **67** (1958), 325.

[Wig47] E. P. Wigner and L. Eisenbud, *Phys. Rev.* **72** (1947), 29.

[Wil62] S. S. Wilks, *Mathematical Statistics.* Wiley, New York, 1962.

[Wil75] W. M. Wilson, E. G. Bilpuch, and G. E. Mitchell, *Nucl. Phys.* **A 245** (1975), 285.

[Wis28] J. Wishart, *Biometrika* **20** A (1928), 32.

[Yur99] I. V. Yurkevich and I. V. Lerner, *Phys. Rev.* **B 60** (1999), 3955.

[Zir96] M. R. Zirnbauer, *J. Math. Phys.* **37** (1996), 4986.

[Zir99] M. R. Zirnbauer, preprint cond-mat/9903338 (1999).

2

Integrable probability: stochastic vertex models and symmetric functions

Alexei BORODIN[1] and Leonid PETROV[2]

[1]Department of Mathematics, Massachusetts Institute of Technology,
77 Massachusetts Ave., Cambridge, MA 02139, USA; and
Institute for Information Transmission Problems, Bolshoy Karetny
per. 19, Moscow, 127994, Russia
[2]Department of Mathematics, University of Virginia, 141 Cabell Drive, Kerchof Hall,
P.O. Box 400137, Charlottesville, VA 22904, USA; and
Institute for Information Transmission Problems, Bolshoy Karetny
per. 19, Moscow, 127994, Russia

Borodin, A. and Petrov, L., 'Integrable Probability: Stochastic Vertex Models and Symmetric
Functions' in *Stochastic Processes and Random Matrices*. Edited by: Grégory Schehr et al, Oxford
University Press (2017). © Oxford University Press 2017. DOI 10.1093/oso/9780198797319.003.0002

Chapter Contents

We consider a homogeneous stochastic higher spin six-vertex model in a quadrant. For this model we derive concise integral representations for multipoint q-moments of the height function and for the q-correlation functions. At least in the case of the step initial condition, our formulas degenerate in appropriate limits to many known formulas of such type for integrable probabilistic systems in the (1+1)d KPZ (Kardar–Parisi–Zhang) universality class, including the stochastic six-vertex model, asymmetric simple exclusion process (ASEP), various q-deformed totally asymmetric simple exclusion processes (q-TASEPs), and associated zero range processes. Our arguments are largely based on properties of a family of symmetric rational functions (introduced in Borodin (2014)) that can be defined as partition functions of the higher spin six-vertex model for suitable domains; they generalize classical Hall–Littlewood and Schur polynomials. A key role is played by Cauchy-like summation identities for these functions, which are obtained as a direct corollary of the Yang–Baxter equation for the higher spin six-vertex model. This chapter is derived from lecture notes for a course given by A.B. at the Ecole de Physique des Houches in July 2015. All the results and proofs presented here generalize to the setting of the fully inhomogeneous higher spin six-vertex model; see Borodin and Petrov (2016a) for a detailed exposition of the inhomogeneous case.

2.1   Introduction

2.1.1   Preface

The past two decades have seen remarkable progress in understanding the so-called KPZ universality class in (1+1) dimensions. This is a rather broad and somewhat vaguely defined class of probabilistic systems describing random interface growth, named after a seminal physics paper by Kardar et al. (1986). A key conjectural property of the systems in this class is that the large time fluctuations of the interfaces should be the same for all of them. See Corwin (2012) for an extensive survey.

While proving such a universality principle remains largely out of reach, by now many concrete systems have been found, for which the needed asymptotics have been actually computed (the universality principle appears to hold so far).

The first wave of these solved systems started in the late 1990s with the papers Johansson (2001) and Baik et al. (1999), and the key to their solvability, or *integrability*, was in (highly non-obvious) reductions to what physicists would call *free-fermion models*—probabilistic systems, many of whose observables are expressed in terms of determinants and Pfaffians. Another domain where free-fermion models are extremely important is random matrix theory. Perhaps not surprisingly, the large time fluctuations of the (1+1)d KPZ models are very similar to those arising in (largest eigenvalues of) random matrices with real spectra.

The second wave of integrable (1+1)d KPZ systems started in late 2000s. The reasons for their solvability are harder to see, but one way or another they can be traced to quantum integrable systems. For example, looking at the earlier papers of the second wave we see that: (a) the pioneering work of Tracy and Widom (2008; 2009a,b) on the ASEP was based on the famous idea of Bethe (1931) of looking for eigenfunctions of a quantum many-body system in the form of superposition of those for noninteracting bodies (coordinate Bethe ansatz); (b) the work of O'Connell (2012)

and Borodin and Corwin (2014) on semi-discrete Brownian polymers utilized properties of eigenfunctions of the Macdonald–Ruijsenaars quantum integrable system—the celebrated Macdonald polynomials and their degenerations; (c) the physics papers Dotsenko (2010) and Calabrese et al. (2010) and a later work of Borodin et al. (2014) used a duality trick to show that certain observables of infinite-dimensional models solve finite-dimensional quantum many-body systems that are, in their turn, solvable by the coordinate Bethe ansatz; etc.

In fact, most currently known integrable (1+1)d KPZ models of the second wave come from *one and the same* quantum integrable system. Corwin and Petrov (2016) recently showed that they can be realized as suitable limits of what they called a *stochastic higher spin vertex model*; see the introduction to their paper for a description of degenerations.[1] They used duality and coordinate Bethe ansatz to show the integrability of their model; the Bethe ansatz part relied on previous works of Borodin et al. (2015a, b) and Borodin (2014).

The main subject of the present chapter is exactly that higher spin six-vertex model. Our main result is an integral representation for certain multipoint q-moments of this model. Such formulas are well known to be a source of meaningful asymptotic results, but we leave asymptotic questions outside of the scope of these lectures.

The results and the proofs presented in these notes generalize to the setting of the fully inhomogeneous higher spin six-vertex model; those are presented in our recent work Borodin and Petrov (2016a).

The core of our proofs consists of the so-called (skew) Cauchy identities for certain rational symmetric functions which were introduced in Borodin (2014). For special parameter values, these functions turn into (skew) Hall–Littlewood and Schur symmetric functions, and the name 'Cauchy identities' is borrowed from the theory of those, cf. Macdonald (1995).

Our symmetric rational functions can be defined as partition functions of the higher spin six-vertex model for domains with special boundary conditions. Following Borodin (2014), we use the *Yang–Baxter equation*, or rather its infinite-volume limit, to derive the Cauchy identities. A similar approach to Cauchy-like identities for the Hall–Littlewood polynomials was also realized in Wheeler and Zinn-Justin (2016).

Remarkably, the Cauchy identities themselves are essentially sufficient to define our probabilistic models, show their connection to KPZ interfaces, prove orthogonality and completeness of our symmetric rational functions, and evaluate averages of a large family of observables with respect to our measures. The last bit can be derived from comparing Cauchy identities with different sets of parameters.

While the Cauchy identities also played an important role in the theory of Schur and Macdonald processes, they were never the main heroes there. Here they really take centre stage. Given their direct relation to the Yang–Baxter equation, one could thus say that the *integrability of the (1+1)d KPZ models takes its origin in the Yang–Baxter integrability of the six-vertex model.*

---

[1] The integrable (1+1)d KPZ models that have not yet been shown to arise as limits of the stochastic vertex models are various versions of PushTASEP, cf. Corwin and Petrov (2015) and Matveev and Petrov (2015). This appears to be simply an oversight as those models are diagonalized by the same wave functions, which means that needed reductions should also exist.

Our present approach circumvents the duality trick that has been so powerful in treating the integrable (1+1)d KPZ models (including that of Corwin and Petrov (2016)). We do explain how the duality can be discovered from our results, but we do not prove or rely on it. Unfortunately, for the moment the use and success of the duality approach remains somewhat mysterious and ad hoc; the form of the duality functional needs to be guessed from previously known examples (some of which have better explanations, cf. Schütz (1997), Borodin and Corwin (2015)). We hope that the path that we present here is more straightforward, and that it can be used to shed further light on the existence of nontrivial dualities.

Let us now describe our results in more detail.

2.1.2   Our model in a quadrant

Consider an ensemble $\mathcal{P}$ of infinite oriented up-right paths drawn in the first quadrant $\mathbb{Z}^2_{\geq 1}$ of the square lattice, with all the paths starting from a left-to-right arrow entering at each of the points $\{(1, m) : m \in \mathbb{Z}_{\geq 1}\}$ on the left boundary (no path enters through the bottom boundary). Assume that no two paths share any horizontal piece (but common vertices and vertical pieces are allowed). See Fig. 2.1.

Define a probability measure on the set of such path ensembles in the following Markovian way. For any $n \geq 2$, assume that we already have a probability distribution on the intersections $\mathcal{P}_n$ of $\mathcal{P}$ with the triangle $T_n = \{(x, y) \in \mathbb{Z}^2_{\geq 1} : x + y \leq n\}$. We are going to increase n by 1. For each point (x, y) on the upper boundary of T_n, i.e., for $x + y = n$, every $\mathcal{P}_n$ supplies us with two inputs: (1) the number of paths that enter (x, y) from the bottom—denote it by $i_1 \in \mathbb{Z}_{\geq 0}$; and (2) the number of paths $j_1 \in \{0, 1\}$ that enter (x, y) from the left. Now choose, independently for all (x, y) on the upper boundary of T_n, the number of paths i_2 that leave (x, y) in the upward direction, and the number of paths j_2 that leave (x, y) in the rightward direction, using the probability distribution with weights of the transitions $(i_1, j_1) \to (i_2, j_2)$ given by (throughout the text $\mathbf{1}_A$ stands for the indicator function of the event A)

$$\mathsf{Prob}((i_1, 0) \to (i_2, 0)) = \frac{1 - q^{i_1} s u_y}{1 - s u_y} \mathbf{1}_{i_1 = i_2},$$

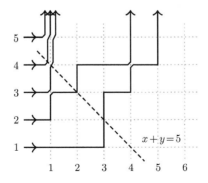

Fig. 2.1   A path collection $\mathcal{P}$.

$$\mathsf{Prob}((i_1, 0) \to (i_2, 1)) = \frac{(q^{i_1} - 1)su_y}{1 - su_y} \mathbf{1}_{i_1 = i_2 + 1},$$

$$\mathsf{Prob}((i_1, 1) \to (i_2, 1)) = \frac{s^2 q^{i_1} - su_y}{1 - su_y} \mathbf{1}_{i_1 = i_2}, \qquad (2.1)$$

$$\mathsf{Prob}((i_1, 1) \to (i_2, 0)) = \frac{1 - s^2 q^{i_1}}{1 - su_y} \mathbf{1}_{i_1 = i_2 - 1}.$$

Assuming that all these expressions are nonnegative, which happens, e.g., if $q \in (0, 1)$, $u_y > 0$, $s \in (-1, 0)$, this procedure defines a probability measure on the set of all $\mathcal{P}$'s because we always have $\sum_{i_2, j_2} \mathsf{Prob}((i_1, j_1) \to (i_2, j_2)) = 1$, and $\mathsf{Prob}((i_1, j_1) \to (i_2, j_2))$ vanishes unless $i_1 + j_1 = i_2 + j_2$.

The right-hand sides in (2.1) are closely related to matrix elements of the R-matrix for $U_q(\widehat{\mathfrak{sl}_2})$, with one representation being an arbitrary Verma module and the other one being tautological.

Parameter s is related to the value of the *spin*. If $s^2 = q^{-I}$ for a positive integer I, then we have $\mathsf{Prob}((I, 1) \to (I + 1, 0)) = 0$, which means that no more than I paths can share the same vertical piece. This corresponds to replacing the arbitrary Verma module in the R-matrix with its $(I + 1)$-dimensional irreducible quotient. The spin-$\frac{1}{2}$ situation $s = q^{-\frac{1}{2}}$ gives rise to the stochastic six vertex introduced over 20 years ago by Gwa–Spohn (1992) (see Borodin et al. (2016a) for its detailed treatment).

The spin parameter s is related to columns, and there is no similar row parameter: recall that no two paths can share the same horizontal piece. This restriction can be repaired using the procedure of *fusion* that goes back to Kulish et al. (1981). In plain words, fusion means grouping the spectral parameters $\{u_y\}$ into subsequences of the form $\{u, qu, \dots, q^{J-1}u\}$ and collapsing the corresponding J rows onto a single one. Here the positive integer J plays the same role as I in the previous paragraph. The reason we did not use the second spin parameter in (2.1) is that the transition probabilities then become rather cumbersome, and one needs to specialize other parameters to achieve simpler expressions. A detailed exposition of the fusion is contained in §2.5.

Let us also note that there are several substantially different possibilities of making the weights (2.1) nonnegative; some of those we consider in detail. Since our techniques are algebraic, our results actually apply to any generic parameter values, with typically only minor modifications needed in case of some denominators vanishing.

2.1.3   The main result

Encode each path ensemble $\mathcal{P}$ by a *height function* $\mathfrak{h} : \mathbb{Z}_{\geq 1}^2 \to \mathbb{Z}_{\geq 0}$, which assigns to each vertex (x, y) the number $\mathfrak{h}(x, y)$ of paths in $\mathcal{P}$ that pass through or to the right of this vertex.

Theorem 2.1.1 (Theorem 2.9.8 in the text).   Assume that $q \in (0, 1)$, $s \in (-1, 0)$, $u_y > 0$ for all $y \in \mathbb{Z}_{\geq 1}$, and $u_i \neq qu_j$ for any $i, j \geq 1$. Then for any integers $x_1 \geq \dots \geq x_\ell \geq 1$ and $y \geq 1$,

$$\mathbb{E} \prod_{i=1}^{\ell} q^{\mathfrak{h}(x_i, y)} = q^{\frac{\ell(\ell-1)}{2}} \oint_{\gamma[\bar{u}|1]} \frac{dw_1}{2\pi \mathbf{i}} \cdots \oint_{\gamma[\bar{u}|\ell]} \frac{dw_\ell}{2\pi \mathbf{i}} \prod_{1 \le \alpha < \beta \le \ell} \frac{w_\alpha - w_\beta}{w_\alpha - q w_\beta}$$

$$\times \prod_{i=1}^{\ell} \left(w_i^{-1} \left(\frac{1 - s w_i}{1 - s^{-1} w_i} \right)^{x_i - 1} \prod_{j=1}^{y} \frac{1 - q u_j w_i}{1 - u_j w_i} \right), \qquad (2.2)$$

where the expectation is taken with respect to the probability measure defined in Section 2.1.2, and the integration contours are described in Definitions 2.8.12 and 2.9.4 and pictured in Fig. 2.26 later in the chapter.

Let us emphasize that the inequalities on the parameters here are exceedingly restrictive; the statement can be analytically continued with suitable modifications of the contours and the integrand. Examples of such analytic continuation can be found in Section 2.10, where they are used to degenerate the result in (2.2) to various q-versions of the T(otally)ASEP.

We also prove integral formulas similar to (2.2) for another set of observables of our model that we call *q-correlation functions*. The two are related, but in a rather nontrivial way, and one set of formulas does not immediately imply the other.

While at the moment averages (2.2) seem more useful for asymptotic analysis (and that is the reason we list them as our main result), it is entirely possible that the q-correlation functions will become useful for other asymptotic regimes. The definition and the expressions for the q-correlation functions can be found in Section 2.8.

2.1.4 Symmetric rational functions

One consequence of the Yang–Baxter integrability of our model is that one can explicitly compute the distribution of intersection points of the paths in $\mathcal{P}$ with any horizontal line. More exactly, let $X_1 \ge \ldots \ge X_n \ge 1$ be the x-coordinates of the points where our paths intersect the line $y = \text{const}$ with $n < \text{const} < n + 1$; there are exactly n of those, counting the multiplicities. Then

$$\text{Prob}\{X_1 = \nu_1 + 1, \ldots, X_n = \nu_n + 1\} = \prod_{k \ge 0} \frac{(-s)^{n_k} (s; q)_{n_k}}{(q; q)_{n_k}} \cdot \mathsf{F}_\nu(u_1, \ldots, u_n), \qquad (2.3)$$

where $(a; q)_m = (1 - a)(1 - aq) \ldots (1 - aq^{m-1})$ are the q-Pochhammer symbols, n_k is the multiplicity of k in the sequence $\nu = (\nu_1 \ge \ldots \ge \nu_n) = 0^{n_0} 1^{n_1} \cdots$, and for any $\mu = (\mu_1 \ge \cdots \ge \mu_M \ge 0)$ we define

$$\mathsf{F}_\mu(u_1, \ldots, u_M) = \frac{(1 - q)^M}{\prod_{i=1}^{M}(1 - s u_i)} \sum_{\sigma \in \mathcal{S}_M} \sigma \left(\prod_{1 \le \alpha < \beta \le M} \frac{u_\alpha - q u_\beta}{u_\alpha - u_\beta} \prod_{i=1}^{M} \left(\frac{u_i - s}{1 - s u_i} \right)^{\mu_i} \right).$$

$$(2.4)$$

Here $\mathcal{S}_M$ is the symmetric group of degree M, and its elements σ permute the variables $\{u_i\}_{i=1}^{M}$ in the right-hand side of (2.4).

The symmetric rational functions $\{\mathsf{F}_\mu\}$ play a central role in our work. The right-hand side of (2.4) can be viewed as a coordinate Bethe ansatz expression for the

eigenfunctions of the transfer matrix of the higher spin six-vertex model. Note that one would need to additionally impose Bethe equations on the u's for periodic in the x-direction boundary conditions.

The probabilistic interpretation of F_ν given above is equivalent to saying that F_ν is the partition function for ensembles of n up-right lattice paths that enter the semi-infinite strip $\mathbb{Z}_{\geq 0} \times \{1, \ldots, n\}$ at the left boundary at $(0, 1), \ldots, (0, n)$ and exit at the top of the strip at locations $(\nu_1, n), \ldots, (\nu_n, n)$. The weight of such an ensemble is equal to the product of weights over all vertices of the strip. The vertex weights for F_ν itself are slightly modified right-hand sides of (2.1) given by (2.6). Allowing some paths to enter at the bottom boundary gives a definition of the skew functions $\mathsf{F}_{\mu/\lambda}$; removing the paths entering from the left gives a definition of the skew functions $\mathsf{G}_{\mu/\lambda}$ and non-skew $\mathsf{G}_\mu = \mathsf{G}_{\mu/0^M}$, cf. Fig. 2.9. A symmetrization formula for G_μ which is similar to (2.4) is given in Theorem 2.4.12.

The functions F and G were introduced in Borodin (2014). As explained there, further degenerations turn them into skew and non-skew Hall–Littlewood and Schur symmetric polynomials.

2.1.5   Cauchy identities

A basic fact about functions F and G that we heavily use is the following skew Cauchy identity. Let $u, v \in \mathbb{C}$ satisfy

$$\left| \frac{u-s}{1-su} \cdot \frac{v-s}{1-sv} \right| < 1.$$

Then for any nonincreasing integer sequences λ and ν as above, we have

$$\sum_\kappa \frac{\mathsf{c}(\kappa)}{\mathsf{c}(\lambda)} \mathsf{G}_{\kappa/\lambda}(v) \mathsf{F}_{\kappa/\nu}(u) = \frac{1 - quv}{1 - uv} \sum_\mu \mathsf{F}_{\lambda/\mu}(u) \frac{\mathsf{c}(\nu)}{\mathsf{c}(\mu)} \mathsf{G}_{\nu/\mu}(v), \qquad (2.5)$$

where $\mathsf{c}(\alpha) = \prod_{i \geq 0} \dfrac{(s^2; q)_{a_i}}{(q; q)_{a_i}}$ for $\alpha = 0^{a_0} 1^{a_1} \cdots$. This identity is a direct consequence of the Yang–Baxter equation for the R-matrix of the higher spin six-vertex model. It involves only two spectral parameters u and v and corresponds to permuting two single-row transfer matrices. Identity (2.5) can be immediately iterated to include any finite number of u's and v's, and also to involve non-skew functions (by setting ν to $\varnothing$ and/or λ to 0^L).

We put different versions of Cauchy identities to multiple uses:

1. The fact that probabilities (2.3) add up to 1 is a limiting instance of a Cauchy identity. Thus, we can think of the weights of the probability measures we are interested in as of (normalized) terms in a Cauchy identity. Such an interpretation (for other Cauchy identities) lies at the basis of the theory of Schur and Macdonald measures and processes (Okounkov 2001; Okounkov and Reshetikhin 2003; Borodin and Corwin 2014).

2. Markov chains that connect measures of the form (2.3) with different values of n are instances of skew Cauchy identities. Such an interpretation was also

previously used in the Schur/Macdonald setting, cf. Borodin (2011); Borodin and Corwin (2014); Borodin and Ferrari (2014).

3. Comparing two Cauchy identities which differ by adding a few extra variables leads to the average of an observable with respect to the measure whose weights are given by the terms of the other identity. This fact by itself is a triviality, but we show that it can be used to extract nontrivial consequences. To our knowledge, such use of Cauchy identities is new.

4. In extracting those consequences, a key role is played by a Plancherel theory for the functions $\{F_\mu\}$, and Cauchy identities can be employed to establish certain orthogonality properties of the F_μ's. These orthogonality relations were first proved in Borodin et al. (2015a), and the link to Cauchy identities goes back to Borodin (2014).

2.1.6   Organization of the chapter

In Section 2.2 we define the higher spin six-vertex model in the language which is used throughout the chapter. The Yang–Baxter integrability of our model is discussed in Section 2.3. In Section 2.4 we take the infinite volume limit of the Yang–Baxter equation, introduce functions F and G, and derive Cauchy identities and symmetrization-type formulas for them. Fusion—a procedure of collapsing several horizontal rows with suitable spectral parameters onto a single one—is discussed in Section 2.5. In Section 2.6 we use skew Cauchy identities to define various Markov dynamics for our model, and also show how known integrable (1+1)d KPZ models can be obtained from those. In Section 2.7 we prove the Plancherel isomorphisms (equivalently, two types of orthogonality relations for the F_μ's). In Section 2.8 we derive integral representations for the q-correlation functions. In Section 2.9 we prove our main result—the integral formula (2.2) for the q-moments of the height function. The final Section 2.10 demonstrates how our main result degenerates to various similar known results for the models which are hierarchically lower: stochastic six-vertex model, ASEP, various q-TASEPs, and associated zero range processes.

2.2   Vertex weights

2.2.1   Higher spin six-vertex model

The higher spin six-vertex model can be viewed as a way of assigning weights to collections of up-right paths in a finite region of $\mathbb{Z}^2$, subject to certain boundary conditions. An example of such a collection of paths is given in Fig. 2.2. The *weight* of a path collection is equal to the product of weights of all the vertices that belong to the paths. We will always assume that the weight of the empty vertex ⋅⋮⋅ is equal to 1. Thus, the weight of a path collection can be equivalently defined as the product of weights of all vertices in $\mathbb{Z}^2$.

Note that the weight of a collection of paths is in general not equal to the product of weights of individual paths (defined in an obvious way). But if the paths in a collection

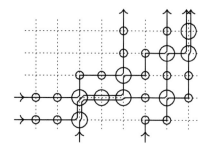

Fig. 2.2 An example of a collection of up-right paths in a region in $\mathbb{Z}^2$. Note that several paths are allowed to pass along the same edge. At each vertex the total number of incoming arrows (= coming from the left or from below) must be equal to the total number of outgoing ones (= pointing to the right or upwards), cf. Fig. 2.3. The circles indicate nonempty vertices which contribute to the weight of the path collection.

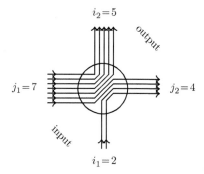

Fig. 2.3 Incoming and outgoing vertical and horizontal arrows at a vertex which we will denote by $(i_1, j_1; i_2, j_2) = (2, 7; 5, 4)$.

have no vertices in common, then the weight of this collection will in fact be equal to the product of weights of individual paths.

2.2.2   Vertex weights

We choose the weights of vertices in a special way. First, we postulate that the number of incoming arrows $i_1 + j_1$ into any vertex must be the same as the number of outgoing arrows $i_2 + j_2$; see Fig. 2.3.

Remark 2.2.1.   This *arrow preservation* condition obviously fails at the boundaries, so one should either fix boundary conditions in some way, or specify weights on the boundary independently.

The vertex weights will depend on two (generally speaking, complex) parameters that we denote by q and s, and on an additional *spectral parameter* $u \in \mathbb{C}$. All these

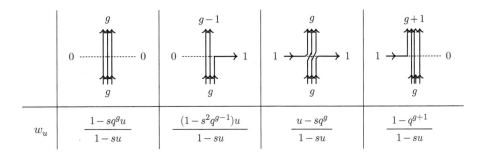

Fig. 2.4   Vertex weights (2.6). Here $g \in \mathbb{Z}_{\geq 0}$ and by agreement, $w_u(0, 0; -1, 1) = 0$.

parameters are assumed to be *generic*.[2] The vertex weights are explicitly given by (see also Fig. 2.4)

$$
w_u(g, 0; g, 0) := \frac{1 - sq^g u}{1 - su}, \qquad w_u(g + 1, 0; g, 1) := \frac{(1 - s^2 q^g)u}{1 - su},
$$

$$
\tag{2.6}
$$

$$
w_u(g, 1; g, 1) := \frac{u - sq^g}{1 - su}, \qquad w_u(g, 1; g + 1, 0) := \frac{1 - q^{g+1}}{1 - su},
$$

where g is any nonnegative integer. All other weights are assumed to be 0. Note that the weight of the empty vertex ┄┼┄ (that is, $(0, 0; 0, 0)$) is indeed equal to 1. Throughout the text, the parameters q and s are assumed fixed, and u will be regarded as an indeterminate, which is reflected in the notation.

Observe that the weights w_u (2.6) are nonzero only for $j_1, j_2 \in \{0, 1\}$; that is, the multiplicities of horizontal edges are bounded by 1. This restriction will be removed later (in Section 2.5).

2.2.3   Motivation

Weights defined in (2.6) are closely related to matrix elements of the higher spin R-matrix associated with $U_q(\widehat{\mathfrak{sl}_2})$ (e.g., see Mangazeev (2014) and also Baxter (2007) and Reshetikhin (2010) for a general introduction). Because of this, they satisfy a version of the Yang–Baxter equation which we discuss in Section 2.3. The exact connection of weights (2.6) with R-matrices is written down in Borodin (2014, Section 2), and here we follow the notation of that paper.

For the weights (2.6), the R-matrix in question corresponds to one of the highest weight representations (the 'vertical' one) being a generic Verma module (associated with the parameter s), while the other representation ('horizontal') is two-dimensional. This choice of the 'horizontal' representation dictates the restriction on the horizontal multiplicities $j_1, j_2 \in \{0, 1\}$. Vertex weights associated with other 'horizontal' representations (finite-dimensional of dimension $J + 1$, or generic Verma modules) are discussed in Section 2.5.

---

[2] That is, vanishing of certain algebraic expressions in the parameters may make some of our statements meaningless. We will not focus on these special cases.

If we set $s^2 = q^{-I}$ with I being a positive integer, then matrix elements of the generic Verma module turn into those of the $(I+1)$-dimensional highest weight representation (of weight I), and thus the multiplicities of vertical edges will be bounded by I. In particular, setting $I = 1$ leads to the well-known six-vertex model (we discuss it in Section 2.6.5). Throughout most of the text we will assume, however, that the parameter s is generic, and so there is no restriction on the vertical multiplicity.

2.2.4 Conjugated weights and stochastic weights

Here we write down two related versions of the vertex weights which will be useful later for probabilistic applications.

Throughout the text we will employ the q-Pochhammer symbols

$$(z;q)_n := \begin{cases} \prod_{k=0}^{n-1}(1 - zq^k), & n > 0, \\ 1, & n = 0, \\ \prod_{k=0}^{-n-1}(1 - zq^{n+k})^{-1}, & n < 0. \end{cases}$$

If $|q| < 1$ and $n = +\infty$, then the q-Pochhammer symbol $(z;q)_\infty$ also makes sense. We will also use the q-binomial coefficients

$$\binom{n}{k}_q := \frac{(q;q)_n}{(q;q)_k(q;q)_{n-k}}.$$

Define the *conjugated vertex weights*

$$w_u^{\mathsf{c}}(i_1, j_1; i_2, j_2) := \frac{(s^2;q)_{i_2}}{(q;q)_{i_2}} \frac{(q;q)_{i_1}}{(s^2;q)_{i_1}} w_u(i_1, j_1; i_2, j_2). \tag{2.7}$$

Also define

$$\mathsf{L}_u(i_1, j_1; i_2, j_2) := (-s)^{j_2} w_u^{\mathsf{c}}(i_1, j_1; i_2, j_2). \tag{2.8}$$

These quantities are given in Fig. 2.5. Note that for any $i_1 \in \mathbb{Z}_{\geq 0}$, $j_1 \in \{0,1\}$ we have

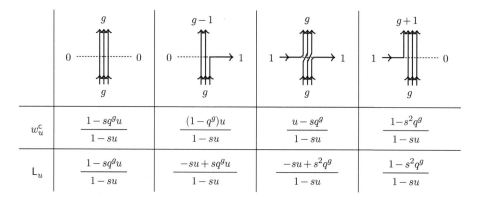

	g	$g-1$	g	$g+1$
w_u^{c}	$\dfrac{1 - sq^g u}{1 - su}$	$\dfrac{(1 - q^g)u}{1 - su}$	$\dfrac{u - sq^g}{1 - su}$	$\dfrac{1 - s^2 q^g}{1 - su}$
L_u	$\dfrac{1 - sq^g u}{1 - su}$	$\dfrac{-su + sq^g u}{1 - su}$	$\dfrac{-su + s^2 q^g}{1 - su}$	$\dfrac{1 - s^2 q^g}{1 - su}$

Fig. 2.5 Vertex weights (2.7) and (2.8). Note that they automatically vanish at the forbidden configuration $(0, 0; -1, 1)$.

$$\sum_{i_2,j_2\in\mathbb{Z}_{\geq0}:\ i_2+j_2=i_1+j_1} \mathsf{L}_u(i_1,j_1;i_2,j_2)=1. \qquad (2.9)$$

Therefore, if the L_u's are nonnegative, they can be interpreted as defining a *probability distribution* on all possible output configurations $\{(i_2,j_2)\colon i_2+j_2=i_1+j_1\}$ given the input configuration (i_1,j_1), cf. Fig. 2.3. We will discuss values of parameters leading to nonnegative L_u's in Section 2.5.1.

A motivation for introducing the conjugated weights w_u^c can be found in Section 2.4.3.

2.3 The Yang–Baxter equation

2.3.1 The Yang–Baxter equation in coordinate language

The Yang–Baxter equation deals with vertex weights at two vertices connected by a vertical edge, with spectral parameters u_1,u_2. Define the two-vertex weights by

$$w_{u_1,u_2}^{(m,n)}(k_1,k_2;k_1',k_2'):=\sum_{l\geq0}w_{u_1}(m,k_1;l,k_1')w_{u_2}(l,k_2;n,k_2'),\qquad k_1,k_2,k_1',k_2'\in\{0,1\}. \qquad (2.10)$$

The expression (2.10) is the weight of the two-vertex configuration as in Fig. 2.6, left, with numbers of incoming and outgoing arrows $m,n,k_{1,2},k_{1,2}'$ fixed. The number of arrows $l\geq0$ along the inside edge is arbitrary, but due to the arrow preservation, no more than one value of l contributes to the sum.

Also define

$$\widetilde{w}_{u_1,u_2}^{(m,n)}(k_1,k_2;k_1',k_2'):=w_{u_1,u_2}^{(m,n)}(k_2,k_1;k_2',k_1'); \qquad (2.11)$$

this is the weight of the configuration as in Fig. 2.6, right.

Let us organize the weights (2.10) into 4×4 matrices

$$w_{u_1,u_2}^{(m,n)}=\begin{bmatrix} w_{u_1,u_2}^{(m,n)}(0,0;0,0) & w_{u_1,u_2}^{(m,n)}(0,0;0,1) & w_{u_1,u_2}^{(m,n)}(0,0;1,0) & w_{u_1,u_2}^{(m,n)}(0,0;1,1) \\ w_{u_1,u_2}^{(m,n)}(0,1;0,0) & w_{u_1,u_2}^{(m,n)}(0,1;0,1) & w_{u_1,u_2}^{(m,n)}(0,1;1,0) & w_{u_1,u_2}^{(m,n)}(0,1;1,1) \\ w_{u_1,u_2}^{(m,n)}(1,0;0,0) & w_{u_1,u_2}^{(m,n)}(1,0;0,1) & w_{u_1,u_2}^{(m,n)}(1,0;1,0) & w_{u_1,u_2}^{(m,n)}(1,0;1,1) \\ w_{u_1,u_2}^{(m,n)}(1,1;0,0) & w_{u_1,u_2}^{(m,n)}(1,1;0,1) & w_{u_1,u_2}^{(m,n)}(1,1;1,0) & w_{u_1,u_2}^{(m,n)}(1,1;1,1) \end{bmatrix},$$

and similarly for $\widetilde{w}_{u_1,u_2}^{(m,n)}$.

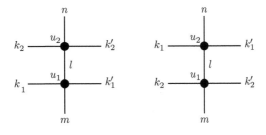

Fig. 2.6 Two-vertex configurations corresponding to (2.10) and (2.11), respectively.

Proposition 2.3.1 (The Yang–Baxter equation).   We have
$$\widetilde{w}^{(m,n)}_{u_2,u_1} = X w^{(m,n)}_{u_1,u_2} X^{-1}, \qquad (2.12)$$

where the matrix X depending on u_1 and u_2 is given by

$$X = \begin{bmatrix} u_1 - qu_2 & 0 & 0 & 0 \\ 0 & q(u_1 - u_2) & (1-q)u_1 & 0 \\ 0 & (1-q)u_2 & u_1 - u_2 & 0 \\ 0 & 0 & 0 & u_1 - qu_2 \end{bmatrix}. \qquad (2.13)$$

Note that X is independent of m and n, and it is this fact that makes the weights w_u (2.6) very special. Note also that X matters only up to an overall factor (which can depend on u_1 and u_2).

Proof This equation can be checked directly. Alternatively, as shown in Borodin (2014, Prop. 2.5), it can be derived from the Yang–Baxter equation for the R-matrices.

The conjugated and the stochastic weights ((2.7) and (2.8), respectively) also satisfy certain versions of the Yang–Baxter equation: see, e.g. Corwin and Petrov (2016, Appendix C).

Remark 2.3.2.   The matrix X (2.13) itself can be viewed as a version of the R-matrix corresponding to both representations being two-dimensional (details may be found in the proof of Proposition 2.5 in Borodin (2014)).

2.3.2   The Yang–Baxter equation in operator language

Before drawing corollaries from the Yang–Baxter equation, let us restate it in a different language, which is sometimes more convenient.

Consider a vector space $V = \operatorname{span}\{e_i : i = 0, 1, 2, \ldots\}$, and linear operators $\mathsf{A}(u)$, $\mathsf{B}(u)$, $\mathsf{C}(u)$, $\mathsf{D}(u)$ on this space which depend on a spectral parameter $u \in \mathbb{C}$ and act in this basis as follows (cf. §2.2.2):

$$\mathsf{A}(u)\, e_g := w_u\!\left(0 \cdots\!\!\!\!\!\! \begin{array}{c} g \\[-2pt] \| \\[-2pt] g \end{array}\!\!\!\!\!\! \cdots 0 \right) e_g = \frac{1 - sq^g u}{1 - su}\, e_g,$$

$$\mathsf{B}(u)\, e_g := w_u\!\left(1 \rightarrow\!\!\!\!\!\! \begin{array}{c} g+1 \\[-2pt] \| \\[-2pt] g \end{array}\!\!\!\!\!\! \cdots 0 \right) e_{g+1} = \frac{1 - q^{g+1}}{1 - su}\, e_{g+1},$$

$$\mathsf{C}(u)\, e_g := w_u\!\left(0 \cdots\!\!\!\!\!\! \begin{array}{c} g-1 \\[-2pt] \| \\[-2pt] g \end{array}\!\!\!\!\!\! \rightarrow 1 \right) e_{g-1} = \frac{(1 - s^2 q^{g-1})u}{1 - su}\, e_{g-1},$$

$$\mathsf{D}(u)\, e_g := w_u\!\left(1 \rightarrow\!\!\!\!\!\! \begin{array}{c} g \\[-2pt] \| \\[-2pt] g \end{array}\!\!\!\!\!\! \rightarrow 1 \right) e_g = \frac{u - sq^g}{1 - su}\, e_g,$$

$$(2.14)$$

where $g \in \mathbb{Z}_{\geq 0}$, and, by agreement, $w_u(0, 0; -1, 1) = 0$. Note that in every vertex $(i_1, j_1; i_2, j_2)$, i_1 corresponds to the index of the vector that the operator is applied to, and i_2 corresponds to the index of the image vector. The four possibilities for $j_1, j_2 \in \{0, 1\}$ correspond to the four operators.

These four operators are conveniently united into a 2×2 matrix with operator entries

$$\mathsf{T}(u) := \begin{bmatrix} \mathsf{A}(u) & \mathsf{B}(u) \\ \mathsf{C}(u) & \mathsf{D}(u) \end{bmatrix},$$

known as the *monodromy matrix*. It can be viewed as an operator $\mathsf{T}(u) : \mathbb{C}^2 \otimes V \to \mathbb{C}^2 \otimes V$. The space $\mathbb{C}^2$ is often called the *auxiliary space*, and V is referred to as the *physical*, or *quantum space*.

In terms of the monodromy matrices, the Yang–Baxter equation (Proposition 2.3.1) takes the form

$$\big(\mathsf{T}(u_1) \otimes \mathsf{T}(u_2)\big) = Y \big(\mathsf{T}(u_2) \otimes \mathsf{T}(u_1)\big) Y^{-1}, \tag{2.15}$$

where

$$Y := (X^{-1})^{\text{transpose}}$$

$$= \frac{1}{(u_1 - qu_2)(u_2 - qu_1)} \begin{bmatrix} u_2 - qu_1 & 0 & 0 & 0 \\ 0 & u_2 - u_1 & (1-q)u_2 & 0 \\ 0 & (1-q)u_1 & q(u_2 - u_1) & 0 \\ 0 & 0 & 0 & u_2 - qu_1 \end{bmatrix},$$

with X given by (2.13).

The tensor product in both sides of (2.15) is taken with respect to the two different auxiliary spaces corresponding to (k_1, k_2) in Fig. 2.6. Namely, we have

$$\mathsf{T}(u_1) \otimes \mathsf{T}(u_2) = \begin{bmatrix} \mathsf{A}(u_1)\mathsf{A}(u_2) & \mathsf{A}(u_1)\mathsf{B}(u_2) & \mathsf{B}(u_1)\mathsf{A}(u_2) & \mathsf{B}(u_1)\mathsf{B}(u_2) \\ \mathsf{A}(u_1)\mathsf{C}(u_2) & \mathsf{A}(u_1)\mathsf{D}(u_2) & \mathsf{B}(u_1)\mathsf{C}(u_2) & \mathsf{B}(u_1)\mathsf{D}(u_2) \\ \mathsf{C}(u_1)\mathsf{A}(u_2) & \mathsf{C}(u_1)\mathsf{B}(u_2) & \mathsf{D}(u_1)\mathsf{A}(u_2) & \mathsf{D}(u_1)\mathsf{B}(u_2) \\ \mathsf{C}(u_1)\mathsf{C}(u_2) & \mathsf{C}(u_1)\mathsf{D}(u_2) & \mathsf{D}(u_1)\mathsf{C}(u_2) & \mathsf{D}(u_1)\mathsf{D}(u_2) \end{bmatrix}, \tag{2.16}$$

and similarly,

$$\mathsf{T}(u_2) \otimes \mathsf{T}(u_1) = \begin{bmatrix} \mathsf{A}(u_2)\mathsf{A}(u_1) & \mathsf{B}(u_2)\mathsf{A}(u_1) & \mathsf{A}(u_2)\mathsf{B}(u_1) & \mathsf{B}(u_2)\mathsf{B}(u_1) \\ \mathsf{C}(u_2)\mathsf{A}(u_1) & \mathsf{D}(u_2)\mathsf{A}(u_1) & \mathsf{C}(u_2)\mathsf{B}(u_1) & \mathsf{D}(u_2)\mathsf{B}(u_1) \\ \mathsf{A}(u_2)\mathsf{C}(u_1) & \mathsf{B}(u_2)\mathsf{C}(u_1) & \mathsf{A}(u_2)\mathsf{D}(u_1) & \mathsf{B}(u_2)\mathsf{D}(u_1) \\ \mathsf{C}(u_2)\mathsf{C}(u_1) & \mathsf{D}(u_2)\mathsf{C}(u_1) & \mathsf{C}(u_2)\mathsf{D}(u_1) & \mathsf{D}(u_2)\mathsf{D}(u_1) \end{bmatrix}. \tag{2.17}$$

See also Fig. 2.7 for an example.

The Yang–Baxter equation (2.15) in the matrix form makes it possible to extract individual commutation relations among the operators A, B, C, and D. Let us write down relations which will be useful in what follows. Comparing matrix elements $(1, 1)$ on both sides of (2.15) implies that

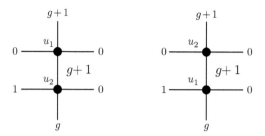

Fig. 2.7 The operator $A(u_1)B(u_2)$ applied to the basis vector e_g corresponds to the configuration on the left, and $A(u_1)B(u_2)\, e_g = \widetilde{w}^{(g,g+1)}_{u_2,u_1}(0,1;0,0)\, e_{g+1}$. Similarly, we have $A(u_2)B(u_1)\, e_g = w^{(g,g+1)}_{u_1,u_2}(1,0;0,0)\, e_{g+1}$, which corresponds to the configuration on the right.

$$A(u_1)A(u_2) = A(u_2)A(u_1). \tag{2.18}$$

Similarly, looking at matrix elements $(1,4)$ and $(4,4)$ gives rise to

$$B(u_1)B(u_2) = B(u_2)B(u_1), \tag{2.19}$$
$$D(u_1)D(u_2) = D(u_2)D(u_1), \tag{2.20}$$

respectively. Looking at matrix elements $(2,4)$ leads to

$$B(u_1)D(u_2) = \frac{u_1 - u_2}{qu_1 - u_2}D(u_2)B(u_1) + \frac{(1-q)u_2}{u_2 - qu_1}B(u_2)D(u_1). \tag{2.21}$$

Finally, considering matrix elements $(2,1)$ and $(1,3)$ implies, respectively, that

$$A(u_1)C(u_2) = \frac{u_1 - u_2}{qu_1 - u_2}C(u_2)A(u_1) + \frac{(1-q)u_2}{u_2 - qu_1}A(u_2)C(u_1), \tag{2.22}$$

$$B(u_1)A(u_2) = \frac{u_1 - u_2}{u_1 - qu_2}A(u_2)B(u_1) + \frac{(1-q)u_2}{u_1 - qu_2}B(u_2)A(u_1). \tag{2.23}$$

2.3.3 Attaching vertical columns

A very important property of the Yang–Baxter equation is that it survives when one attaches several vertical columns on the side, with the requirement that the conjugating matrix is the same across the vertical columns. Let us consider the situation of two vertical columns as in Fig. 2.8. Attaching these columns involves summing over all possible intermediate numbers of arrows k'_1 and k'_2; i.e., this corresponds to taking the product of two 4×4 matrices $w^{(m_1,n_1)}_{u_1,u_2}$ and $w^{(m_2,n_2)}_{u_1,u_2}$. Clearly, for this product the Yang–Baxter equation (2.12) is not going to change. One can similarly attach an arbitrary finite number of vertical columns, and the Yang–Baxter equation will continue to hold.

In the operator language attaching two vertical columns is equivalent to taking a tensor product $V = V_1 \otimes V_2$ of two different physical spaces V_1 and V_2 with the same conjugating matrix X. The monodromy matrix in the space V has the form

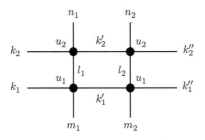

Fig. 2.8 Attaching two vertical columns.

$$\mathsf{T} = \begin{bmatrix} \mathsf{A} & \mathsf{B} \\ \mathsf{C} & \mathsf{D} \end{bmatrix} = \begin{bmatrix} \mathsf{A}_2 & \mathsf{B}_2 \\ \mathsf{C}_2 & \mathsf{D}_2 \end{bmatrix} \begin{bmatrix} \mathsf{A}_1 & \mathsf{B}_1 \\ \mathsf{C}_1 & \mathsf{D}_1 \end{bmatrix} = \begin{bmatrix} \mathsf{A}_2\mathsf{A}_1 + \mathsf{B}_2\mathsf{C}_1 & \mathsf{A}_2\mathsf{B}_1 + \mathsf{B}_2\mathsf{D}_1 \\ \mathsf{C}_2\mathsf{A}_1 + \mathsf{D}_2\mathsf{C}_1 & \mathsf{C}_2\mathsf{B}_1 + \mathsf{D}_2\mathsf{D}_1 \end{bmatrix}, \qquad (2.24)$$

where the lower index 1 or 2 in the operators corresponds to the vertical (= physical) space in which they act; i.e., $\mathsf{A}_2 = \mathsf{A}_2(u)$ acts in the second vertical column, and $\mathsf{A}_1 = \mathsf{A}_1(u)$ acts in the same way in the first vertical column, and similarly for $\mathsf{B}_{1,2}$, $\mathsf{C}_{1,2}$, and $\mathsf{D}_{1,2}$. (All operators in (2.24) depend on the same spectral parameter u, and we have omitted it in the notation). Note that any two operators with different lower indices commute.

The monodromy matrix $\mathsf{T} = \mathsf{T}(u)$ in (2.24) corresponds to one horizontal row of vertices. That is, the four matrix elements of $\mathsf{T}(u)$ correspond to the four configurations

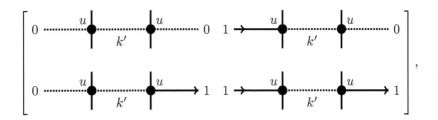

and the two summands in each matrix element in (2.24) correspond to k' being 0 or 1.

Furthermore, tensor products of two monodromy matrices like (2.16) and (2.17) correspond to configurations as in Fig. 2.8. As follows from this discussion, these tensor products satisfy the same Yang–Baxter equation (2.15).

2.4   Symmetric rational functions

We will now discuss how the setup of Section 2.3 can be applied to the physical space corresponding to the semi-infinite horizontal strip. This will lead to certain symmetric rational functions which are one of the main objects of these notes.

2.4.1 Signatures

Let us first introduce some necessary notation. By a *signature* of length N we mean a sequence $\lambda = (\lambda_1 \geq \lambda_2 \geq \ldots \geq \lambda_N)$, $\lambda_i \in \mathbb{Z}$. The set of all signatures of length N will be denoted by Sign_N, and Sign_N^+ will stand for the set of signatures with $\lambda_N \geq 0$. By agreement, by $\mathsf{Sign}_0 = \mathsf{Sign}_0^+$ we will denote the set consisting of the single empty signature $\varnothing$ of length 0. Also let $\mathsf{Sign}^+ := \bigsqcup_{N \geq 0} \mathsf{Sign}_N^+$ denote the set of all possible nonnegative signatures (including the empty one). We will also use the multiplicative notation $\mu = 0^{m_0} 1^{m_1} 2^{m_2} \ldots \in \mathsf{Sign}^+$ for signatures, which means that $m_j := |\{i \colon \mu_i = j\}|$ is the number of parts in μ that are equal to j (m_j is called the *multiplicity* of j).

2.4.2 Semi-infinite operators A and B and definition of symmetric rational functions

Let us consider the physical space $V = V_0 \otimes V_1 \otimes V_2 \otimes \ldots$, i.e., a tensor product of countably many 'elementary' physical spaces (each of the latter has basis $\{e_j\}_{j \geq 0}$ marked by $\mathbb{Z}_{\geq 0}$). We will think that V corresponds to the semi-infinite (to the right) row of vertices attached to one another on the side. We will make sense of the infinite tensor product V by requiring that we only consider *finitary* vectors $V^{\mathrm{fin}} \subset V$; i.e., those in which almost all tensor factors are equal to e_0. Therefore, a natural basis in the space V^{fin} is indexed by nonnegative signatures:

$$e_\mu = e_{m_0} \otimes e_{m_1} \otimes e_{m_2} \otimes \ldots, \qquad \mu = 0^{m_0} 1^{m_1} 2^{m_2} \ldots \in \mathsf{Sign}^+$$

($m_0 + m_1 + \ldots$ is the length of the signature μ which is finite). We will work in the space $\bar{V}^{\mathrm{fin}}$ of all possible linear combinations of e_μ with complex coefficients.

Defining the operators A and B acting in $\bar{V}^{\mathrm{fin}}$ causes no problems. Indeed, we have for any $N \in \mathbb{Z}_{\geq 0}$ and $\lambda \in \mathsf{Sign}_N^+$:

$$\mathsf{A}(u)\, e_\lambda = \sum_{\mu \in \mathsf{Sign}_N^+} \mathrm{weight}_u \left(\ldots \right) e_\mu, \tag{2.25}$$

and

$$\mathsf{B}(u)\, e_\lambda = \sum_{\mu \in \mathsf{Sign}_{N+1}^+} \mathrm{weight}_u \left(\ldots \right) e_\mu. \tag{2.26}$$

That is, in (2.25) and (2.26) we sum over all possible signatures μ, and for each fixed μ the coefficient is equal to the weight of the unique path collection connecting the

arrow configuration λ to the configuration μ, as shown pictorially (the coefficient is 0 if no admissible path collection exists).[3] The difference is that in (2.25) the path collection contains N paths connecting λ_j to μ_j, $j = 1, \ldots, N$, and in (2.26) there is one additional path starting horizontally at the left boundary, and ending at μ_{N+1}.

Let us denote the coefficients in the sums in (2.25) and (2.26) by $\mathsf{G}_{\mu/\lambda}(u)$ and $\mathsf{F}_{\mu/\lambda}(u)$, respectively (here u is the spectral parameter we are using).

Remark 2.4.1.   Each coefficient $\mathsf{G}_{\mu/\lambda}(u)$ and $\mathsf{F}_{\mu/\lambda}(u)$ in the semi-infinite setting is the same as if we took it in a finite tensor product, with the number of factors $\geq \mu_1 + 1$. It follows that the semi-infinite operators (2.25) and (2.26) satisfy the same commutation relations (2.18) and (2.19). Indeed, to check the commutation relations, apply them to e_λ and read off the coefficient by each e_μ. One readily sees that each such coefficient by e_μ involves only finite summation.

Similarly, we define the coefficients $\mathsf{G}_{\mu/\lambda}(u_1, \ldots, u_n)$ and $\mathsf{F}_{\mu/\lambda}(u_1, \ldots, u_n)$ arising from products of our operators in the following way:

$$\mathsf{A}(u_1) \ldots \mathsf{A}(u_n)\,\mathsf{e}_\lambda = \sum_{\mu \in \mathrm{Sign}_N^+} \mathsf{G}_{\mu/\lambda}(u_1, \ldots, u_n)\,\mathsf{e}_\mu, \tag{2.27}$$

$$\mathsf{B}(u_1) \ldots \mathsf{B}(u_n)\,\mathsf{e}_\lambda = \sum_{\mu \in \mathrm{Sign}_{N+n}^+} \mathsf{F}_{\mu/\lambda}(u_1, \ldots, u_n)\,\mathsf{e}_\mu, \tag{2.28}$$

where $N \in \mathbb{Z}_{\geq 0}$ and $\lambda \in \mathrm{Sign}_N^+$ are arbitrary.

Equivalently, the quantities $\mathsf{G}_{\mu/\lambda}(u_1, \ldots, u_n)$ and $\mathsf{F}_{\mu/\lambda}(u_1, \ldots, u_n)$ can be defined as certain partition functions in the higher spin six-vertex model:

Definition 2.4.2.   *Let* $N, n \in \mathbb{Z}_{\geq 0}$, $\lambda, \mu \in \mathrm{Sign}_N^+$. *Assign to each vertex* $(x, y) \in \mathbb{Z} \times \{1, 2, \ldots, n\}$ *the spectral parameter* u_y. *Define* $\mathsf{G}_{\mu/\lambda}(u_1, \ldots, u_n)$ *to be the sum of the weights of all possible collections of* N *up-right paths such that they*

- *start with* N *vertical edges* $(\lambda_i, 0) \to (\lambda_i, 1)$, $i = 1, \ldots, N$, *and*
- *end with* N *vertical edges* $(\mu_i, n) \to (\mu_i, n + 1)$, $i = 1, \ldots, N$.

See Fig. 2.9, left. We will also use the abbreviation $\mathsf{G}_\mu := \mathsf{G}_{\mu/(0,0,\ldots,0)}$, *which corresponds to the decomposition of* $\mathsf{A}(u_1) \ldots \mathsf{A}(u_n)(\mathsf{e}_N \otimes \mathsf{e}_0 \otimes \mathsf{e}_0 \otimes \ldots)$.

Definition 2.4.3.   *Let* $N, n \in \mathbb{Z}_{\geq 0}$, $\lambda \in \mathrm{Sign}_N^+$, $\mu \in \mathrm{Sign}_{N+n}^+$. *As before, assign to each vertex* $(x, y) \in \mathbb{Z} \times \{1, 2, \ldots, n\}$ *the spectral parameter* u_y. *Define* $\mathsf{F}_{\mu/\lambda}(u_1, \ldots, u_n)$ *to be the sum of the weights of all possible collections of* $N + n$ *up-right paths such that they*

---

[3] Recall that the weight of a path collection is defined as the product of weights of all (nonempty) vertices in the corresponding region of $\mathbb{Z}^2$, and that the weight of the empty vertex ⋯ is 1.

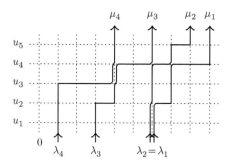 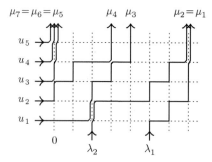

Fig. 2.9 Path collections used in the definitions of $\mathsf{G}_{\mu/\lambda}$ (left) and $\mathsf{F}_{\mu/\lambda}$ (right). The weight of a path collection is the product of weights of all nonempty vertices (cf. Fig. 2.2).

- *start with N vertical edges $(\lambda_i, 0) \to (\lambda_i, 1)$, $i = 1, \ldots, N$, and with n horizontal edges $(-1, y) \to (0, y)$, $y = 1, \ldots, n$, and*
- *end with $N + n$ vertical edges $(\mu_i, n) \to (\mu_i, n + 1)$, $i = 1, \ldots, N + n$.*

See Fig. 2.9, right. We will also use the abbreviation $\mathsf{F}_\mu := \mathsf{F}_{\mu/\varnothing}$, which corresponds to the decomposition of $\mathsf{B}(u_1) \ldots \mathsf{B}(u_n)(\mathsf{e}_0 \otimes \mathsf{e}_0 \otimes \ldots)$.

In both Definitions 2.4.2 and 2.4.3, if a collection of paths has no interior vertices, we define its weight to be 1. Also, the weight of an empty collection of paths is 0.

Clearly, both quantities $\mathsf{G}_{\mu/\lambda}(u_1, \ldots, u_n)$ and $\mathsf{F}_{\mu/\lambda}(u_1, \ldots, u_n)$ depend on the spectral parameters $u_1, \ldots, u_n$ in a *rational* way. These definitions first appeared in Borodin (2014, §3).

Proposition 2.4.4.   The rational functions $\mathsf{F}_{\mu/\lambda}(u_1, \ldots, u_n)$ and $\mathsf{G}_{\mu/\lambda}(u_1, \ldots, u_n)$ defined here are symmetric with respect to permutations of the u_j's.

Proof This immediately follows from the commutation relations (2.18) and (2.19) (cf. Remark 2.4.1).

The functions $\mathsf{F}_{\mu/\lambda}$ and $\mathsf{G}_{\mu/\lambda}$ satisfy the following branching rules.

Proposition 2.4.5.

1. For any $N, n_1, n_2 \in \mathbb{Z}_{\geq 0}$, $\lambda \in \mathrm{Sign}_N^+$, and $\mu \in \mathrm{Sign}_{N+n_1+n_2}^+$, one has

$$\mathsf{F}_{\mu/\lambda}(u_1, \ldots, u_{n_1+n_2}) = \sum_{\kappa \in \mathrm{Sign}_{N+n_1}^+} \mathsf{F}_{\mu/\kappa}(u_{n_1+1}, \ldots, u_{n_1+n_2}) \mathsf{F}_{\kappa/\lambda}(u_1, \ldots, u_{n_1}).$$

$$(2.29)$$

2. For any $N, n_1, n_2 \in \mathbb{Z}_{\geq 0}$, and $\lambda, \mu \in \mathsf{Sign}_N^+$, one has

$$\mathsf{G}_{\mu/\lambda}(u_1, \dots, u_{n_1+n_2}) = \sum_{\kappa \in \mathsf{Sign}_N^+} \mathsf{G}_{\mu/\kappa}(u_{n_1+1}, \dots, u_{n_1+n_2})\mathsf{G}_{\kappa/\lambda}(u_1, \dots, u_{n_1}).$$

$$(2.30)$$

Proof This proof follows from the definitions (2.27) and (2.28) in a straightforward way. In other words, identities (2.29) and (2.30) simply mean the splitting of summation over path collections in $\mathsf{F}_{\mu/\lambda}$ and $\mathsf{G}_{\mu/\lambda}$, such that the signature κ keeps track of the cross-section of the path collection at height n_1.

2.4.3   Semi-infinite operator D

It is slightly more difficult to define the action of the other two operators, C and D, in the semi-infinite context. We will not need the operator C, so let us focus on $\mathsf{D} = \mathsf{D}(u)$. The action of D (in a finite tensor product) corresponds to the following configuration (cf. (2.25) and (2.26)):

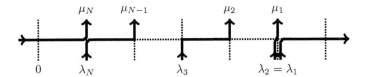

For the semi-infinite horizontal strip, the weight of this configuration would involve an infinite product of the form $w_u(0, 1; 0, 1)^\infty = \left(\frac{u-s}{1-su}\right)^\infty$. This means that one cannot define the operator $\mathsf{D}(u)$ in the semi-infinite setting directly.

However, the definition of D can be easily corrected, by considering strips of finite length $L + 1$ and the operators D in $V_0 \otimes \dots \otimes V_L$. For a fixed L denote such an operator by $\mathsf{D}_L = \mathsf{D}_L(u)$. Dividing D_L by $w_u(0, 1; 0, 1)^{L+1}$, and sending $L \to +\infty$, we would arrive at a meaningful object. Indeed, under these transformations the weights of individual vertices will turn into

$$\frac{1}{w_u(0, 1; 0, 1)} w_u \left(0 \begin{array}{c} \vphantom{g} \\ g \\ \vphantom{g} \end{array} 0 \right) = \frac{1 - sq^g u}{u - s}$$

$$= w_{u-1}\left(1 \to \begin{array}{c} g \\ \\ g \end{array} \to 1 \right) = w_{u-1}^c\left(1 \to \begin{array}{c} g \\ \\ g \end{array} \to 1 \right),$$

$$\frac{1}{w_u(0,1;0,1)}\,w_u\!\left(\,0\ \longrightarrow\ 1\,\right)=\frac{(1-s^2q^g)u}{u-s}$$

$$=w_{u^{-1}}\!\left(\,1\ \longrightarrow\ 0\,\right)\frac{1-s^2q^g}{1-q^{g+1}}=w^{\mathsf c}_{u^{-1}}\!\left(\,1\ \longrightarrow\ 0\,\right),$$

$$\frac{1}{w_u(0,1;0,1)}\,w_u\!\left(\,1\ \longrightarrow\ 1\,\right)=\frac{u-sq^g}{u-s}$$

$$=w_{u^{-1}}\!\left(\,0\ \longrightarrow\ 0\,\right)=w^{\mathsf c}_{u^{-1}}\!\left(\,0\ \longrightarrow\ 0\,\right),$$

$$\frac{1}{w_u(0,1;0,1)}\,w_u\!\left(\,1\ \longrightarrow\ 0\,\right)=\frac{1-q^{g+1}}{u-s}$$

$$=w_{u^{-1}}\!\left(\,0\ \longrightarrow\ 1\,\right)\frac{1-q^{g+1}}{1-s^2q^g}=w^{\mathsf c}_{u^{-1}}\!\left(\,0\ \longrightarrow\ 1\,\right),$$

$$\tag{2.31}$$

where we have used the conjugated weights (2.7). Note that the numbers of vertical incoming and outgoing arrows at a vertex were *swapped* under these transformations. Therefore, for any $L\ge\lambda_1+1$ we have

$$\frac{[\text{coefficient of } e_\mu \text{ in } D_L(u)\,e_\lambda]}{w_u(0,1;0,1)^{L+1}}=[\text{coefficient of } e_\lambda \text{ in } A(u^{-1})\,e_\mu]\cdot\frac{\mathsf c(\lambda)}{\mathsf c(\mu)},$$

where for any signature $\nu\in\mathsf{Sign}^+$ we have denoted

$$\mathsf c(\nu):=\prod_k\frac{(s^2;q)_{n_k}}{(q;q)_{n_k}},\qquad \nu=0^{n_0}1^{n_1}2^{n_2}\cdots\tag{2.32}$$

(this product has finitely many factors not equal to 1). The operator $A(u^{-1})$ can be regarded as acting either in a finite tensor product or in the semi-infinite space $\bar V^{\mathrm{fin}}$, since matrix elements corresponding to (e_μ,e_λ) of these two versions of $A(u^{-1})$ coincide for fixed μ,λ, and large enough L (cf. Remark 2.4.1).

We see that it is natural to define the normalized operator

$$\bar D(u):=\lim_{L\to+\infty}\frac{D_L(u)}{w_u(0,1;0,1)^{L+1}},\tag{2.33}$$

where the limit is taken in the sense of matrix elements corresponding to the basis vectors $\{e_\lambda\}_{\lambda \in \text{Sign}^+}$. The matrix elements of $\overline{D}(u)$ are (cf. (2.25))

$$\overline{D}(u) e_\lambda = \sum_{\mu \in \text{Sign}_N^+}^{\bullet} \frac{c(\lambda)}{c(\mu)} G_{\lambda/\mu}(u^{-1}) e_\mu.$$

Observe that this sum over μ is *finite*, in contrast with the operators (2.25) and (2.26). From (2.33) and (2.20) it follows that the operators $\overline{D}(u)$ commute for different u.

In what follows we will use the notation

$$F_{\lambda/\mu}^c := \frac{c(\lambda)}{c(\mu)} F_{\lambda/\mu}, \qquad G_{\lambda/\mu}^c := \frac{c(\lambda)}{c(\mu)} G_{\lambda/\mu}.$$

2.4.4 Cauchy-type identities from the Yang–Baxter commutation relations

This part closely follows Borodin (2014, Section 4).

Let us consider the semi-infinite limit as $L \to +\infty$ (similar to what was done in Section 2.4.3) of the Yang–Baxter commutation relation (2.21). Looking at (2.21), we immediately face the question of what we need to normalize the two sides by: $w_{u_1}(0,1;0,1)^{L+1}$ or $w_{u_2}(0,1;0,1)^{L+1}$? Since out of the three terms in (2.21) two require the normalization involving u_2, let us use that one. To be able to take the limit as $L \to +\infty$, we will also require that

$$\left| \frac{w_{u_1}(0,1;0,1)}{w_{u_2}(0,1;0,1)} \right| = \left| \frac{u_1 - s}{1 - su_1} \cdot \frac{1 - su_2}{u_2 - s} \right| < 1. \tag{2.34}$$

Under (2.34), we can take the normalized (by $w_{u_2}(0,1;0,1)^{L+1}$) limit of the relation (2.21) and, using (2.33), conclude that

$$B(u_1)\overline{D}(u_2) = \frac{u_1 - u_2}{qu_1 - u_2} \overline{D}(u_2)B(u_1). \tag{2.35}$$

Indeed, before the limit the normalized second term of (2.21) contains

$$\frac{D_L(u_1)}{w_{u_2}(0,1;0,1)^{L+1}} = \frac{D_L(u_1)}{w_{u_1}(0,1;0,1)^{L+1}} \left(\frac{w_{u_1}(0,1;0,1)}{w_{u_2}(0,1;0,1)} \right)^{L+1},$$

which converges to 0 by (2.34).

Using the notation $F_{\mu/\lambda}$ and $G_{\mu/\lambda}$ introduced in Section 2.4.2, relation (2.35) becomes

$$\sum_{\mu \in \text{Sign}^+} F_{\lambda/\mu}(u_1) G_{\nu/\mu}^c(u_2^{-1}) = \frac{u_1 - u_2}{qu_1 - u_2} \sum_{\kappa \in \text{Sign}^+} G_{\kappa/\lambda}^c(u_2^{-1}) F_{\kappa/\nu}(u_1). \tag{2.36}$$

Therefore, we have established the following fact.

Proposition 2.4.6. Let $u, v \in \mathbb{C}$ satisfy

$$\left| \frac{u - s}{1 - su} \cdot \frac{v - s}{1 - sv} \right| < 1. \tag{2.37}$$

Then for any $\lambda, \nu \in \mathsf{Sign}^+$ we have

$$\sum_{\kappa \in \mathsf{Sign}^+} \mathsf{G}^c_{\kappa/\lambda}(v) \mathsf{F}_{\kappa/\nu}(u) = \frac{1 - quv}{1 - uv} \sum_{\mu \in \mathsf{Sign}^+} \mathsf{F}_{\lambda/\mu}(u) \mathsf{G}^c_{\nu/\mu}(v). \tag{2.38}$$

Proof Indeed, this is just (2.36) under the replacement of (u_1, u_2) by (u, v^{-1}). ∎

Identity (2.38) is nontrivial only if $\nu \in \mathsf{Sign}^+_N$ and $\lambda \in \mathsf{Sign}^+_{N+1}$. In this case the sum in the right-hand side of (2.38) is over $\mu \in \mathsf{Sign}^+_N$ and is finite, while in the left-hand side it is over $\kappa \in \mathsf{Sign}^+_{N+1}$ and is infinite (but converges due to (2.37)).

We will call (2.38) the *skew Cauchy identity* for the symmetric functions $\mathsf{F}_{\mu/\lambda}$ and $\mathsf{G}_{\mu/\lambda}$ because of its similarity with the skew Cauchy identities for the Schur, Hall–Littlewood, and Macdonald symmetric functions (Macdonald 1995, Ch. I.5, Ex. 26, and Ch. VI.7, Ex. 6). In fact, if $s = 0$, our identity (2.38) becomes the skew Cauchy identity for the Hall–Littlewood symmetric functions. Further letting $q \to 0$, we recover the Schur case.

Definition 2.4.7. *Let us say that two complex numbers $u, v \in \mathbb{C}$ are admissible, denoted $(u, v) \in \mathsf{Adm}$, if (2.37) holds. Note that this relation is symmetric in u and v.*

The skew Cauchy identity can obviously be iterated with the following result.

Corollary 2.4.8. Let $u_1, \dots, u_M$ and $v_1, \dots, v_N$ be complex numbers such that $(u_i, v_j) \in \mathsf{Adm}$ for all $i = 1, \dots, M$ and $j = 1, \dots, N$. Then for any $\lambda, \nu \in \mathsf{Sign}^+$ one has

$$\sum_{\kappa \in \mathsf{Sign}^+} \mathsf{G}^c_{\kappa/\lambda}(v_1, \dots, v_N) \mathsf{F}_{\kappa/\nu}(u_1, \dots, u_M)$$

$$= \prod_{i=1}^{M} \prod_{j=1}^{N} \frac{1 - qu_i v_j}{1 - u_i v_j} \sum_{\mu \in \mathsf{Sign}^+} \mathsf{F}_{\lambda/\mu}(u_1, \dots, u_M) \mathsf{G}^c_{\nu/\mu}(v_1, \dots, v_N). \tag{2.39}$$

Furthermore, the skew Cauchy identity (2.39) can be simplified by specializing some of the indices. Recall the abbreviations G_μ and F_μ from Definitions 2.4.2 and 2.4.3. The identity of Corollary 2.4.8 readily implies the following facts:

Corollary 2.4.9.

1. For any $N \in \mathbb{Z}_{\geq 0}$, $\lambda \in \mathsf{Sign}^+_N$, and any complex $u_1, \dots, u_N$ and v such that $(u_i, v) \in \mathsf{Adm}$ for all i, we have

$$\sum_{\kappa \in \mathsf{Sign}_N^+} \mathsf{G}_{\kappa/\lambda}^{\mathsf{c}}(v) \mathsf{F}_\kappa(u_1, \dots, u_N) = \prod_{i=1}^{N} \frac{1 - q u_i v}{1 - u_i v} \mathsf{F}_\lambda(u_1, \dots, u_N). \qquad (2.40)$$

2. For any $N, n \in \mathbb{Z}_{\geq 0}$ any $\nu \in \mathsf{Sign}_N^+$, and any complex u and $v_1, \dots, v_n$ such that $(u, v_j) \in \mathsf{Adm}$ for all j, we have

$$\sum_{\kappa \in \mathsf{Sign}_{N+1}^+} \mathsf{G}_\kappa^{\mathsf{c}}(v_1, \dots, v_n) \mathsf{F}_{\kappa/\nu}(u) = \frac{1 - q^{N+1}}{1 - su} \prod_{j=1}^{n} \frac{1 - q u v_j}{1 - u v_j} \mathsf{G}_\nu^{\mathsf{c}}(v_1, \dots, v_n). \qquad (2.41)$$

Proof   Identity (2.40) follows from (2.39) by taking $\nu = \varnothing$ and a single v-variable. Then the sum over μ in the right-hand side of (2.39) reduces to just $\mu = \varnothing$.

Identity (2.41) follows by taking $\lambda = 0^{N+1}$ and a single u-variable in (2.39), and observing that $\mathsf{F}_{0^{N+1}/\mu}(u) = \frac{1 - q^{N+1}}{1 - su} \mathbf{1}_{\mu = 0^N}$ by the very definition of F.

Identities (2.40) and (2.41) are analogous to the *Pieri rules* for Schur, Hall–Littlewood, and Macdonald symmetric functions (Macdonald 1995, Ch. I.5, formula (5.16), and Ch. VI.6).

Remark 2.4.10.   Identity (2.40) shows that the functions $\{\mathsf{F}_\lambda(u_1, \dots, u_N)\}_{\lambda \in \mathsf{Sign}_N^+}$ for each set of the u's form an eigenvector of the transfer matrix $\{\mathsf{G}_{\nu/\lambda}^{\mathsf{c}}(v)\}_{\lambda, \nu \in \mathsf{Sign}_N^+}$ viewed as acting in the spatial variables corresponding to signatures (i.e., with rows indexed by λ and columns indexed by ν). Equivalently, $\{\mathsf{F}_\lambda^{\mathsf{c}}(u_1, \dots, u_N)\}_{\lambda \in \mathsf{Sign}_N^+}$ is an eigenvector of the transfer matrix $\{\mathsf{G}_{\nu/\lambda}(v)\}_{\lambda, \nu \in \mathsf{Sign}_N^+}$ (i.e., the conjugation "c" can be moved). This statement is parallel (and simpler) to the fact that on a finite lattice, the vector $\mathsf{B}(u_1) \dots \mathsf{B}(u_n)(\mathsf{e}_0 \otimes \dots \otimes \mathsf{e}_0)$ is an eigenvector of the operator $\mathsf{A}(v) + \mathsf{D}(v)$ given certain nonlinear *Bethe equations* on $u_1, \dots, u_N$. In our case the Bethe equations disappeared, and only one of the terms in $\mathsf{A}(v) + \mathsf{D}(v)$ has survived.

One can also obtain analogous statements when the number of v-variables in (2.40) is greater than 1—this would correspond to applying a sequence of transfer matrices with varying spectral parameters.

Taking $\nu = \varnothing$ and $\lambda = 0^M$ in (2.39) and noting that

$$\mathsf{F}_{0^M}(u_1, \dots, u_M) = \frac{(q; q)_M}{\prod_{i=1}^{M}(1 - su_i)}, \qquad (2.42)$$

we arrive at the following analogue of the *usual (non-skew) Cauchy identity* (see Macdonald (1995, Ch. I.4, formula (4.3), and Ch. VI.4, formula (4.13)) for the corresponding Schur and Macdonald Cauchy identities):

Corollary 2.4.11.   For $M, N \geq 0$ and complex numbers $u_1, \ldots, u_M$ and $v_1, \ldots, v_N$ such that $(u_i, v_j) \in \mathsf{Adm}$ for all i and j, one has

$$\sum_{\mu \in \mathsf{Sign}_M^+} \mathsf{F}_\mu(u_1, \ldots, u_M) \mathsf{G}_\mu^c(v_1, \ldots, v_N) = \frac{(q;q)_M}{\prod_{i=1}^M (1 - su_i)} \prod_{i=1}^M \prod_{j=1}^N \frac{1 - qu_i v_j}{1 - u_i v_j}. \qquad (2.43)$$

2.4.5 Symmetrization formulas

So far, our definition of the symmetric functions $\mathsf{F}_{\mu/\lambda}$ and $\mathsf{G}_{\nu/\lambda}$ was not too explicit—they were defined as large sums over all possible path collections with certain boundary conditions (see Definitions 2.4.2 and 2.4.3). However, it turns out that the non-skew symmetric functions F_μ and G_ν can be evaluated more explicitly.

Theorem 2.4.12.

1. For any $M \geq 0$, any $\mu \in \mathsf{Sign}_M^+$, and any $u_1, \ldots, u_M \in \mathbb{C}$ we have[4]

$$\mathsf{F}_\mu(u_1, \ldots, u_M) = \frac{(1-q)^M}{\prod_{i=1}^M (1 - su_i)} \sum_{\sigma \in \mathcal{S}_M} \sigma \left(\prod_{1 \leq \alpha < \beta \leq M} \frac{u_\alpha - qu_\beta}{u_\alpha - u_\beta} \prod_{i=1}^M \left(\frac{u_i - s}{1 - su_i} \right)^{\mu_i} \right).$$

$$(2.44)$$

2. For any $n \geq 0$, $\nu \in \mathsf{Sign}_n^+$, let k be the number of zero coordinates in ν, i.e., $\nu_{n-k+1} = \ldots = \nu_n = 0$. Then for any $N \geq n - k$ and any $v_1, \ldots, v_N \in \mathbb{C}$ we have

$$\mathsf{G}_\nu(v_1, \ldots, v_N) = \frac{(s^2; q)_n}{(q;q)_{N-n+k}(s^2;q)_k} \frac{(1-q)^N}{\prod_{j=1}^N (1 - sv_j)}$$

$$\times \sum_{\sigma \in \mathcal{S}_N} \sigma \left(\prod_{1 \leq \alpha < \beta \leq N} \frac{v_\alpha - qv_\beta}{v_\alpha - v_\beta} \prod_{j=1}^{n-k} \left(\frac{v_j - s}{1 - sv_j} \right)^{\nu_j} \prod_{i=1}^{n-k} \frac{v_i}{v_i - s} \prod_{j=n-k+1}^N (1 - sq^k v_j) \right).$$

$$(2.45)$$

If $N < n - k$, the function $\mathsf{G}_\nu(v_1, \ldots, v_N)$ vanishes for trivial reasons.

This theorem was established in Borodin (2014). Here we present a different proof which involves the operators $\mathsf{A}, \mathsf{B}, \mathsf{C}, \mathsf{D}$ from Section 2.3.2, and closely follows the algebraic Bethe ansatz framework (Korepin et al. 1993; Felder and Varchenko 1996). Let us first discuss certain straightforward corollaries of Theorem 2.4.12. For $\mu \in \mathsf{Sign}_M^+$ and $r \in \mathbb{Z}_{\geq 0}$, let $\mu + r^M$ denote the shifted signature $(\mu_1 + r, \mu_2 + r, \ldots, \mu_M + r)$.

---

[4] In both formulas (2.44) and (2.45) the permutation σ (belonging, respectively, to $\mathcal{S}_M$ or $\mathcal{S}_N$) acts by permuting the indeterminates u_i or v_j, respectively. The same convention is used throughout the text.

Corollary 2.4.13.

1. For any $\mu \in \mathsf{Sign}_M^+$ and any $r \in \mathbb{Z}_{\geq 0}$ one has

$$\mathsf{F}_{\mu+r^M}(u_1,\ldots,u_M) = \prod_{i=1}^{M}\left(\frac{u_i - s}{1 - su_i}\right)^r \cdot \mathsf{F}_\mu(u_1,\ldots,u_M). \qquad (2.46)$$

2. For any $\nu \in \mathsf{Sign}_N^+$ with $\nu_N \geq 1$ one has

$$\mathsf{G}_\nu(v_1,\ldots,v_N) = (s^2;q)_N\left(\prod_{i=1}^{N}\frac{v_i}{v_i - s}\right)\mathsf{F}_\nu(v_1,\ldots,v_N).$$

That is, when $k = 0$ and $N = n$ in (2.45), the function $\mathsf{G}_\nu(v_1,\ldots,v_N)$ almost coincides with F_ν.

Proof A straightforward verification using (2.44) and (2.45). Alternatively, this immediately follows from the definitions of the functions F and G as partition functions of path collections (Definitions 2.4.3 and 2.4.2).

The next corollary utilizes the explicit formulas (2.44) and (2.45) in an essential way:

Corollary 2.4.14.

1. For any $M \geq 0$, $\mu \in \mathsf{Sign}_M^+$, and $u \in \mathbb{C}$ we have

$$\mathsf{F}_\mu(u, qu, \ldots, q^{M-1}u) = \frac{(q;q)_M}{(su;q)_M}\left(\prod_{j=0}^{\mu_i-1}\frac{q^{i-1}u - s}{1 - sq^{i-1}u}\right)^{\mu_i}. \qquad (2.47)$$

2. For any $n \geq 0$ and $\nu \in \mathsf{Sign}_n^+$ with k zero coordinates, any $N \geq n - k$, and any $v \in \mathbb{C}$ we have

$$\mathsf{G}_\nu(v, qv, \ldots, q^{N-1}v) = \frac{(q;q)_N}{(q;q)_{N-n+k}}\frac{(sv;q)_{N+k}}{(sv;q)_n(sv;q)_N}$$

$$\times \frac{(s^2;q)_n}{(s^2;q)_k}\frac{1}{(s/v;q^{-1})_{n-k}}\prod_{j=1}^{N}\left(\frac{q^{j-1}v - s}{1 - sq^{j-1}v}\right)^{\nu_j}. \qquad (2.48)$$

Substituting a geometric sequence with ratio q into a function F or G will be referred to as the *principal specialization* of these symmetric functions.

Proof The substitutions of geometric sequences into F or G make all terms except the one with $\sigma = \mathrm{id}$ vanish due to the presence of the cross term $\sigma\left(\prod_{1\leq\alpha<\beta\leq M}\frac{u_\alpha - qu_\beta}{u_\alpha - u_\beta}\right)$. For $\sigma = \mathrm{id}$ this cross term is equal to $(q;q)_M/(1-q)^M$. The rest is obtained in a straightforward way by evaluating the remaining parts of the formulas.

The proof of Theorem 2.4.9 occupies the rest of this part.

Proof of (2.44)   **Step 1.** To obtain an explicit formula for $F_\mu(u_1, \ldots, u_M)$, we need to understand how the operator $B(u_1) \ldots B(u_M)$ acts on the vector $(e_0 \otimes e_0 \otimes \ldots)$. Let us first consider what happens in the physical space containing just two tensor factors, which puts us into the setting described in Section 2.3.3. We have from (2.24):

$$B(u) = B_1(u)A_2(u) + D_1(u)B_2(u), \tag{2.49}$$

where the lower indices in the operators in the right-hand side stand for the spaces in which they act. The operators in the right-hand side act as in (2.14). Recall that any two operators with different lower indices commute.

When we multiply together a number of operators $B(u)$ (with different spectral u-parameters) and open the parentheses, we collect several factors B_1 and D_1, and several other factors A_2 and B_2. Using the Yang–Baxter commutation relations (2.21) and (2.23), we can swap these operators at the expense of picking certain prefactors, and also this swapping of operators could lead to an exchange of their spectral parameters. Therefore, we can write $B(u_1) \ldots B(u_M)(e_0 \otimes e_0)$ as a linear combination of vectors of the form

$$B_1(u_{k_1}) \ldots B_1(u_{k_{M-r}})D_1(u_{\ell_1}) \ldots D_1(u_{\ell_r}) e_0$$
$$\otimes B_2(u_{i_1}) \ldots B_2(u_{i_r})A_2(u_{j_1}) \ldots A_2(u_{j_{M-r}}) e_0, \tag{2.50}$$

with

$$\mathcal{I} = \{i_1 < \ldots < i_r\}, \qquad \mathcal{J} = \{j_1 < \ldots < j_{M-r}\}, \qquad \mathcal{I} \sqcup \mathcal{J} = \{1, \ldots, M\},$$
$$\mathcal{K} = \{k_1 < \ldots < k_{M-r}\}, \quad \mathcal{L} = \{\ell_1 < \ldots < \ell_r\}, \qquad \mathcal{K} \sqcup \mathcal{L} = \{1, \ldots, M\}.$$

Step 2. The coefficients of the vectors (2.50) are computed using only the commutation relations (2.21) and (2.23), and we argue that these coefficients *do not depend* on how exactly we apply the commutation relations to reach the result. This property is based on the fact that for generic spectral parameters, there exists a representation of $\begin{bmatrix} A(u) & B(u) \\ C(u) & D(u) \end{bmatrix}$ subject to the same commutation relations, and a *highest weight vector* v_0 in that representation,[5] such that vectors $\left(\prod_{j \in \mathcal{J}} B(u_j) \right) v_0$, with $\mathcal{J}$ ranging over all subsets of $\{1, 2, \ldots, M\}$, are linearly independent. This is shown in Felder and Varchenko (1996, Lemma 14), and we will not repeat the argument here.

Knowing this fact, if we have two ways of applying the commutation relations which yield different coefficients of the vectors (2.50), then we can apply these commutation relations in the highest weight representation, and this would lead to a contradiction with the linear independence property.

Step 3. Our next goal is to show that the coefficient of each vector of the form (2.50) vanishes unless $\mathcal{I} \cap \mathcal{K} = \varnothing$. We argue by induction on M. For $M = 1$, the

---

[5] Meaning that v_0 is annihilated by $C(u)$ and is an eigenvector for $A(u)$ and $D(u)$.

application of the operator (2.49) (with $u = u_1$) to $e_0 \otimes e_0$ obviously has this property. When we apply the next operator $B(u_2)$, we see that the sets $\mathcal{I}$ and $\mathcal{K}$ could grow by the element 2, and that they can also lose the element 1 in the process of commuting the D's and the A's to the right. However, the sets $\mathcal{I}$ and $\mathcal{K}$ cannot gain the element 1. This means that $1 \notin \mathcal{I} \cap \mathcal{K}$. However, we could have applied $B(u_1)B(u_2) = B(u_2)B(u_1)$ in the opposite order, which implies (by the uniqueness of the coefficients) that $2 \notin \mathcal{I} \cap \mathcal{K}$. Therefore, $\mathcal{I} \cap \mathcal{K} = \varnothing$ for $M = 2$. Clearly, we can continue this argument with more factors in the same way, and conclude that $\mathcal{I} \cap \mathcal{K} = \varnothing$ for any M.

Step 4. Since $\mathcal{I} \sqcup \mathcal{J} = \mathcal{K} \sqcup \mathcal{L} = \{1, \ldots, M\}$, we see that $\mathcal{I} = \mathcal{L}$ and $\mathcal{K} = \mathcal{J}$. This implies that the desired action of a product of the B operators takes the form

$$
B(u_1) \ldots B(u_M)(e_0 \otimes e_0)
$$
$$
= \sum_{\mathcal{K} \subseteq \{1,\ldots,M\}} C_\mathcal{K} \left(\prod_{k \in \mathcal{K}} B_1(u_k) \prod_{\ell \notin \mathcal{K}} D_1(u_\ell) \right) e_0 \otimes \left(\prod_{\ell \notin \mathcal{K}} B_2(u_\ell) \prod_{k \in \mathcal{K}} A_2(u_k) \right) e_0,
$$
(2.51)

with some uniquely defined coefficients $C_\mathcal{K}(u_1, \ldots, u_M)$, where $\mathcal{K} \subseteq \{1, \ldots, M\}$.

Now, since we obviously can permute the spectral parameters u_j without changing the desired action (2.51), by uniqueness of the coefficients we must have

$$
C_\mathcal{K}(u_{\sigma(1)}, \ldots, u_{\sigma(M)}) = C_{\sigma(\mathcal{K})}(u_1, \ldots, u_M) \qquad \text{for all } \sigma \in \mathcal{S}_M.
$$

Thus, it suffices to compute these coefficients for $\mathcal{K} = \{1, 2, \ldots, r\}$ for each $r = 1, 2, \ldots, M$. This can be done by simply opening the parentheses in

$$
\big(B_1(u_1)A_2(u_1) + D_1(u_1)B_2(u_1)\big) \cdots \big(B_1(u_M)A_2(u_M) + D_1(u_M)B_2(u_M)\big), \quad (2.52)
$$

because the only way to end up with the vector

$$
B_1(u_1) \cdots B_1(u_r)D_1(u_{r+1}) \cdots D_1(u_M) e_0 \otimes B_2(u_{r+1}) \cdots B_2(u_M)A_2(u_1) \cdots A_2(u_r) e_0
$$

is to use the first summand in (2.52) for $j = 1, \ldots, r$, the second summand for $j = r+1, \ldots, M$, and commute all the A_2's through the B_2's without swapping the spectral parameters. From (2.23) we readily have

$$
A(w_1)B(w_2) = \frac{w_2 - qw_1}{w_2 - w_1} B(w_2)A(w_1) - \frac{(1-q)w_1}{w_2 - w_1} B(w_1)A(w_2), \quad (2.53)
$$

and we are only interested in the first summand in (2.53). Our commutations thus give the coefficient

$$
C_{\{1,2,\ldots,r\}}(u_1, \ldots, u_M) = \prod_{\alpha=1}^{r} \prod_{\beta=r+1}^{M} \frac{u_\beta - qu_\alpha}{u_\beta - u_\alpha},
$$

and so we have

$$\mathsf{B}(u_1)\ldots\mathsf{B}(u_M)(\mathsf{e}_0\otimes\mathsf{e}_0)$$

$$= \sum_{\mathcal{K}\subseteq\{1,\ldots,M\}}\prod_{\substack{\alpha\in\mathcal{K}\\\beta\notin\mathcal{K}}}\frac{u_\beta-qu_\alpha}{u_\beta-u_\alpha}\left(\prod_{k\in\mathcal{K}}\mathsf{B}_1(u_k)\prod_{\ell\notin\mathcal{K}}\mathsf{D}_1(u_\ell)\right)\mathsf{e}_0$$

$$\otimes\left(\prod_{\ell\notin\mathcal{K}}\mathsf{B}_2(u_\ell)\prod_{k\in\mathcal{K}}\mathsf{A}_2(u_k)\right)\mathsf{e}_0. \tag{2.54}$$

Recall that e_0 is an eigenvector for D_1 and A_2, and introduce the notation $\mathsf{a}_{1,2}$ and $\mathsf{d}_{1,2}$ by

$$\mathsf{D}_j(u)\,\mathsf{e}_0 = \mathsf{d}_j(u)\,\mathsf{e}_0, \qquad \mathsf{A}_j(u)\,\mathsf{e}_0 = \mathsf{a}_j(u)\,\mathsf{e}_0, \qquad j=1,2. \tag{2.55}$$

Thus, $\mathsf{a}_{1,2}$ and $\mathsf{d}_{1,2}$ are eigenvalues (scalars).[6] Hence our final result (2.54) for two tensor factors can be rewritten in the form

$$\mathsf{B}(u_1)\ldots\mathsf{B}(u_M)(\mathsf{e}_0\otimes\mathsf{e}_0)$$

$$= \sum_{\mathcal{K}\subseteq\{1,\ldots,M\}}\mathsf{d}_1(\mathcal{K}^c)\mathsf{a}_2(\mathcal{K})\prod_{\substack{\alpha\in\mathcal{K}\\\beta\notin\mathcal{K}}}\frac{u_\beta-qu_\alpha}{u_\beta-u_\alpha}\,(\mathsf{B}_1(\mathcal{K})\,\mathsf{e}_0)\otimes(\mathsf{B}_2(\mathcal{K}^c)\,\mathsf{e}_0), \tag{2.56}$$

where we have abbreviated

$$\mathcal{K}^c := \{1,\ldots,M\}\setminus\mathcal{K}, \qquad f(\mathcal{K}) := \prod_{k\in\mathcal{K}}f(u_k). \tag{2.57}$$

Step 5. In this form the formula (2.56) for two tensor factors can be immediately extended to arbitrarily many tensor factors. Indeed, let us think of the second vector e_0 as $\tilde{\mathsf{e}}_0\otimes\tilde{\mathsf{e}}_0$. Then we can use (2.56) to evaluate $\mathsf{B}_2(\mathcal{K}^c)\,\mathsf{e}_0 = \mathsf{B}_2(\mathcal{K}^c)(\tilde{\mathsf{e}}_0\otimes\tilde{\mathsf{e}}_0)$, split the second $\tilde{\mathsf{e}}_0$ again, and so on.

Therefore, we obtain the final formula for the action of $\mathsf{B}(u_1)\ldots\mathsf{B}(u_M)$ on the vector $(\mathsf{e}_0\otimes\mathsf{e}_0\otimes\ldots)$:

$$\mathsf{B}(u_1)\ldots\mathsf{B}(u_M)(\mathsf{e}_0\otimes\mathsf{e}_0\otimes\ldots)$$

$$= \sum_{\substack{\mathcal{K}_0,\mathcal{K}_1,\ldots\subseteq\{1,\ldots,M\}\\\mathcal{K}_0\sqcup\mathcal{K}_1\sqcup\ldots=\{1,2,\ldots,M\}}}\prod_{0\leq i<j}\mathsf{d}_i(\mathcal{K}_j)\mathsf{a}_j(\mathcal{K}_i)$$

$$\times\prod_{\substack{\alpha\in\mathcal{K}_i\\\beta\in\mathcal{K}_j}}\frac{u_\beta-qu_\alpha}{u_\beta-u_\alpha}\Big(\mathsf{B}_0(\mathcal{K}_0)\,\mathsf{e}_0\otimes\mathsf{B}_1(\mathcal{K}_1)\,\mathsf{e}_0\otimes\ldots\Big). \tag{2.58}$$

To finish the derivation of (2.44), we need to recall the action (2.14) of the operators A,B, and D in the 'elementary' physical space $\mathrm{span}\{\mathsf{e}_i : i=0,1,2,\ldots\}$. We have

---

[6] When e_0 is the highest weight vector in the representation V of Section 2.3.2, these eigenvalues can be read off (2.14). However, in Step 5 we will use notation (2.55) for highest weight vectors of representations obtained by tensoring several such V's.

$$a_j(\mathcal{K}) = 1;$$

$$
\begin{aligned}
\mathsf{B}_j(\mathcal{K})\, e_0 &= \frac{(q;q)_{|\mathcal{K}|}}{\prod_{k \in \mathcal{K}}(1 - su_k)}\, e_{|\mathcal{K}|} \\
&= \frac{(1-q)^{|\mathcal{K}|}}{\prod_{k \in \mathcal{K}}(1 - su_k)} \left(\sum_{\sigma \in \mathcal{S}(\mathcal{K})} \sigma \left(\prod_{\substack{\alpha < \beta \\ \alpha,\beta \in \mathcal{K}}} \frac{u_\alpha - qu_\beta}{u_\alpha - u_\beta} \right) \right) e_{|\mathcal{K}|};
\end{aligned}
\tag{2.59}
$$

$$d_j(\mathcal{K}) = \prod_{k \in \mathcal{K}} \frac{u_k - s}{1 - su_k},$$

where for $\mathsf{B}_j(\mathcal{K})$ we have used the symmetrization formula (Macdonald, 1995, Ch. III.1, formula (1.4))[7] to insert an additional sum over permutations of $\mathcal{K}$ (here $\mathcal{S}(\mathcal{K})$ denotes the group of permutations of $\mathcal{K}$, and σ acts by permuting the corresponding variables).

To read off the coefficient of $e_\mu = e_{m_0} \otimes e_{m_1} \otimes e_{m_2} \otimes \dots$, $\mu \in \mathsf{Sign}_M^+$, in (2.58), we must have $|\mathcal{K}_i| = m_i$ for all $i \geq 0$. Let us fix one such partition $\mathcal{K}_0 \sqcup \mathcal{K}_1 \sqcup \dots = \{1, \dots, M\}$. For each $\alpha \in \{1, \dots, M\}$, let $\mathsf{k}(\alpha) \in \mathbb{Z}_{\geq 0}$ denote the number j such that $\alpha \in \mathcal{K}_j$. Then we can write

$$\prod_{0 \leq i < j} d_i(\mathcal{K}_j) = \prod_{r=1}^{M} \left(\frac{u_r - s}{1 - su_r} \right)^{\mathsf{k}(r)}.$$

Note that this product does not change if we permute the u_i's within the sets $\mathcal{K}_j$ as in (2.59). Furthermore, we can also write

$$\prod_{0 \leq i < j} \prod_{\substack{\alpha \in \mathcal{K}_i \\ \beta \in \mathcal{K}_j}} \frac{u_\beta - qu_\alpha}{u_\beta - u_\alpha} = \prod_{\substack{1 \leq \alpha,\beta \leq M \\ \mathsf{k}(\alpha) < \mathsf{k}(\beta)}} \frac{u_\beta - qu_\alpha}{u_\beta - u_\alpha}.
\tag{2.60}$$

We then combine this with the coefficients coming from $\mathsf{B}_j(\mathcal{K}_j)\, e_0$ which involve summations over permutations within the sets $\mathcal{K}_j$, and compare the result with the desired formula (2.44).

Clearly, fixing a partition into the $\mathcal{K}_j$'s corresponds to considering only permutations $\sigma \in \mathcal{S}_M$ in (2.44) which place each $i \in \{1, \dots, M\}$ into $\mathcal{K}_{\mathsf{k}(i)}$. This is the mechanism which gives rise to the summations over permutations within the sets $\mathcal{K}_j$. One can readily check that the summands agree, and thus (2.44) is established.

---

[7] That is, for any $r \in \mathbb{Z}_{\geq 1}$, we have

$$\sum_{w \in \mathcal{S}_r} \prod_{1 \leq \alpha < \beta \leq r} \frac{u_{w(\alpha)} - qu_{w(\beta)}}{u_{w(\alpha)} - u_{w(\beta)}} = \frac{(q;q)_r}{(1-q)^r}.$$

Example. To illustrate the last step of the proof in which we match our computation to (2.44), let us consider a concrete example $\mu = (2, 2, 1)$. By (2.44), we have

$$F_\mu(u_1, u_2, u_3) = \frac{(1 - q)^3}{(1 - su_1)(1 - su_2)(1 - su_3)}$$

$$\times \sum_{\sigma \in S_3} \sigma \left(\frac{(u_1 - qu_2)(u_1 - qu_3)(u_2 - qu_3)}{(u_1 - u_2)(u_1 - u_3)(u_2 - u_3)} \left(\frac{u_1 - s}{1 - su_1} \right)^2 \left(\frac{u_2 - s}{1 - su_2} \right)^2 \left(\frac{u_3 - s}{1 - su_3} \right) \right).$$

Fix the partition $\{1, 2, 3\} = \mathcal{K}_1 \sqcup \mathcal{K}_2$, where $\mathcal{K}_1 = \{2\}$ and $\mathcal{K}_2 = \{1, 3\}$, which corresponds to permutations $\sigma = (132)$ and $\sigma = (312)$. These permutations yield the following cross terms, respectively:

$$\frac{(u_1 - qu_3)(u_1 - qu_2)(u_3 - qu_2)}{(u_1 - u_3)(u_1 - u_2)(u_3 - u_2)}, \qquad \frac{(u_3 - qu_1)(u_3 - qu_2)(u_1 - qu_2)}{(u_3 - u_1)(u_3 - u_2)(u_1 - u_2)}.$$

The factor $\frac{(u_1 - qu_2)(u_3 - qu_2)}{(u_1 - u_2)(u_3 - u_2)}$ can be matched to (2.60), and the remaining factors $\frac{u_1 - qu_3}{u_1 - u_3}$ and $\frac{u_3 - qu_1}{u_3 - u_1}$ correspond to the summation within $\mathcal{K}_2$ coming from (2.59).   □

Remark 2.4.15.   Formula (2.44) that we just established links the algebraic and the coordinate Bethe ansatz. Its proof given earlier closely follows the proof of Theorem 5 in Section 8 of Felder and Varchenko (1996). The key relation (2.58) without proof can be found in Korepin et al. (1993, Appendix VII.2).

Proof of (2.45)   **Step 1.** We will use the same approach as in the proof of (2.44) to get an explicit formula for $G_\nu(v_1, \ldots, v_N)$. That is, we need to compute

$$A(v_1) \ldots A(v_N)(e_n \otimes e_0 \otimes e_0 \otimes \ldots).$$

We start with just two tensor factors, and consider the application of this operator to $e_n \otimes e_0$. After that we will use (2.44) to turn the second e_0 into $e_0 \otimes e_0 \otimes \ldots$.
    For two tensor factors we have from (2.24)

$$A(v) = A_1(v)A_2(v) + C_1(v)B_2(v). \tag{2.61}$$

Taking the product $A(v_1) \ldots A(v_N)$ and opening the parentheses, we can use the commutation relations (2.22) and (2.23) to express the result as a linear combination of vectors of the form

$$A_1(v_{k_1}) \ldots A_1(v_{k_{N-r}}) C_1(v_{\ell_1}) \ldots C_1(v_{\ell_r}) e_n \otimes B_2(v_{i_1}) \ldots B_2(v_{i_r}) A_2(v_{j_1}) \ldots A_2(v_{j_{N-r}}) e_0, \tag{2.62}$$

with

$$\mathcal{I} = \{i_1 < \ldots < i_r\}, \qquad \mathcal{J} = \{j_1 < \ldots < j_{N-r}\}, \qquad \mathcal{I} \sqcup \mathcal{J} = \{1, \ldots, N\},$$
$$\mathcal{K} = \{k_1 < \ldots < k_{N-r}\}, \qquad \mathcal{L} = \{\ell_1 < \ldots < \ell_r\}, \qquad \mathcal{K} \sqcup \mathcal{L} = \{1, \ldots, N\}.$$

Step 2. Again, the key point is that the coefficients by vectors of the form (2.62) are uniquely determined by the commutation relations, and do not depend on the order of commuting. The uniqueness argument here is very similar to that in Step 2 of the proof of (2.44), and we will not repeat it.

Step 3. We now observe that we must have $\mathcal{I} = \mathcal{L}$ and $\mathcal{J} = \mathcal{K}$. Indeed, let us show that $\mathcal{I} \cap \mathcal{K} = \varnothing$, which would imply the claim. We argue by induction. The case of $N = 1$ is obvious. When we then apply the next operator $\mathsf{A}(v_2)$ to (2.61) (with $v = v_1$) and use the commutation relations to write all vectors in the required form (2.62), neither $\mathcal{I}$ nor $\mathcal{K}$ can gain index 1, exactly in the same way as in Step 3 of the proof of (2.44). The fact that $1 \notin \mathcal{I} \cap \mathcal{K}$ does not change after we apply all other operators $\mathsf{A}(v_j)$, $j = 3, \ldots, N$. Since the order of factors in $\mathsf{A}(v_1) \ldots \mathsf{A}(v_N)$ does not matter, we conclude that $\mathcal{I} \cap \mathcal{K} = \varnothing$ for any N.

Step 4. We thus conclude that

$$\mathsf{A}(v_1) \ldots \mathsf{A}(v_N)(e_n \otimes e_0) = \sum_{\mathcal{K} \subseteq \{1,2,\ldots,N\}} C_{\mathcal{K}} \big(\mathsf{A}_1(\mathcal{K}) \mathsf{C}_1(\mathcal{K}^c)\big) e_n \otimes \big(\mathsf{B}_2(\mathcal{K}^c) \mathsf{A}_2(\mathcal{K})\big) e_0,$$

(2.63)

where we are using the abbreviation (2.57). Here the coefficients $C_{\mathcal{K}}$ are uniquely determined, and satisfy

$$C_{\mathcal{K}}(v_{\sigma(1)}, \ldots, v_{\sigma(N)}) = C_{\sigma(\mathcal{K})}(v_1, \ldots, v_N) \qquad \text{for all } \sigma \in \mathcal{S}_N.$$

Thus, we need to compute only the coefficients $C_{\mathcal{K}}$ for $\mathcal{K} = \{r+1, \ldots, N\}$, where $r = 1, 2, \ldots, N$. Since $\nu \in \mathsf{Sign}_n^+$ has exactly k zero coordinates (see (2.45)), we must have $|\mathcal{K}^c| = r = n - k$. Indeed, this is because $\mathsf{C}_1(\mathcal{K}^c)$ is responsible for moving some of n arrows to the right from the location 0.

The coefficients $C_{\{r+1,\ldots,N\}}$ can thus be computed by simply opening the parentheses in

$$\big(\mathsf{A}_1(v_N)\mathsf{A}_2(v_N) + \mathsf{C}_1(v_N)\mathsf{B}_2(v_N)\big) \ldots \big(\mathsf{A}_1(v_1)\mathsf{A}_2(v_1) + \mathsf{C}_1(v_1)\mathsf{B}_2(v_1)\big),$$

and noting that there is a unique way of reaching $\mathcal{K} = \{r+1, \ldots, N\}$: pick the first summands in the first $N - r = N - n + k$ factors and the second summands in the last $r = n - k$ factors, and after that move $\mathsf{A}_2(\mathcal{K})$ to the right of $\mathsf{B}_2(\mathcal{K}^c)$ without swapping the spectral parameters in the process of commuting. Using (2.53) (where we are interested only in the first term in the right-hand side), we thus get the product

$$C_{\{n-k+1,\ldots,N\}}(v_1, \ldots, v_N) = \prod_{\alpha=1}^{n-k} \prod_{\beta=n-k+1}^{N} \frac{v_\alpha - q v_\beta}{v_\alpha - v_\beta}.$$

Next, we note that $\mathsf{A}_2(\mathcal{K}) e_0 = e_0$ and that (from (2.14))

$$\mathsf{A}_1(\mathcal{K})\mathsf{C}_1(\mathcal{K}^c) e_n = \frac{(1 - s^2 q^{n-1}) \ldots (1 - s^2 q^k) v_1 \ldots v_{n-k}}{(1 - s v_1) \ldots (1 - s v_{n-k})} \prod_{j=n-k+1}^{N} \frac{1 - s q^k v_j}{1 - s v_j} \cdot e_k.$$

Step 5. What remains unaccounted for in (2.63) is $B_2(\mathcal{K})\, e_0 = B_2(v_1) \ldots B_2(v_{n-k})\, e_0$. But this was computed earlier in the proof of (2.44), and we can also immediately take the second vector to be $e_0 \otimes e_0 \otimes \ldots$ instead of just e_0. Thus, by (2.28), we have

$$B(v_1) \ldots B(v_{n-k})\big(e_0 \otimes e_0 \otimes \ldots\big) = \sum_{\kappa \in \mathrm{Sign}^+_{n-k}} F_\kappa(v_1, \ldots, v_{n-k})\, e_\kappa, \qquad (2.64)$$

and so the coefficient of e_ν (with $\nu \in \mathrm{Sign}^+_n$ having exactly k zero coordinates) in $A(v_1) \ldots A(v_N)(e_n \otimes e_0 \otimes e_0 \otimes \ldots)$ is equal to

$$\frac{(s^2;q)_n}{(s^2;q)_k} \sum_{\substack{\mathcal{K}^c \subseteq \{1,\ldots,N\} \\ |\mathcal{K}^c| = n-k}} \prod_{i \in \mathcal{K}^c} \frac{v_i}{1 - s v_i} \prod_{j \in \mathcal{K}} \frac{1 - s q^k v_j}{1 - s v_j}$$

$$\times \prod_{\substack{\alpha \in \mathcal{K}^c \\ \beta \in \mathcal{K}}} \frac{v_\alpha - q v_\beta}{v_\alpha - v_\beta} F_{(\nu_1 - 1, \ldots, \nu_{n-k} - 1)}\big(\{v_i\}_{i \in \mathcal{K}^c}\big). \qquad (2.65)$$

Indeed, the signature κ in (2.64) corresponds to nonzero parts in ν, and coordinates in κ are counted starting from location 1 (hence the shifts $\nu_i - 1$).

To match (2.65) to (2.45), we use formula (2.44) to write $F_{(\nu_1 - 1, \ldots, \nu_{n-k} - 1)}$ as a sum over permutations of $\mathcal{K}^c$, and insert an additional symmetrization over $\mathcal{K}$ (see footnote 7):

$$1 = \frac{(1-q)^{N-n+k}}{(q;q)_{N-n+k}} \sum_{\substack{\sigma: \mathcal{K} \to \mathcal{K} \\ \sigma \text{ is a bijection}}} \sigma \left(\prod_{\substack{\alpha,\beta \in \mathcal{K} \\ \alpha < \beta}} \frac{v_\alpha - q v_\beta}{v_\alpha - v_\beta} \right).$$

After that one readily checks that (2.65) coincides with the desired expression. This concludes the proof of Theorem 2.4.12. □

2.5 Stochastic weights and fusion

One key object of the present notes is the set of probability measures afforded by the Cauchy identities of Section 2.4.4. We describe and study them in Section 2.6. Here we discuss the *fusion procedure* on which some of the constructions of Section 2.6 are based.

2.5.1 Stochastic weights L_u

If we assume that

$$0 < q < 1 \text{ and } -1 < s < 0, \qquad (2.66)$$

and, moreover, that

$$\text{all spectral parameters } u_i \text{ are nonnegative}, \qquad (2.67)$$

then all the vertex weights w_u, w_u^c, and L_u (see Figs. 2.4 and 2.5) are nonnegative. Under these assumptions, (2.9) implies that the stochastic weights $L_u(i_1, j_1; i_2, j_2)$, where $i_1, i_2 \in \mathbb{Z}_{\geq 0}$ and $j_1, j_2 \in \{0, 1\}$, define a *probability distribution* on all possible output arrow configurations $\{(i_2, j_2) \in \mathbb{Z}_{\geq 0} \times \{0, 1\}: i_2 + j_2 = i_1 + j_1\}$ given the input arrow configuration (i_1, j_1). We will use conditions (2.66)–(2.67) to define Markov dynamics in §2.6.

The conditions (2.66)–(2.67) are sufficient but not necessary for the nonnegativity of the L_u's; for other conditions see Corwin and Petrov (2016, Prop. 2.3) and also §2.6.5 and Section 2.6.6.

Remark 2.5.1.   We will always assume that the parameters q and s are nonzero. In fact, without this assumption the weights L_u may still define probability distributions. If q or s vanish, then some of our statements remain valid and simplify, but we will not focus on the necessary modifications.

Remark 2.5.2.   Since the stochastic weights L_u depend on s and u only through su and s^2, they are invariant under the simultaneous change of sign of both s and u. We have chosen s to be negative, and u will be nonnegative.

2.5.2   Fusion of stochastic weights

For each $J \in \mathbb{Z}_{\geq 1}$, we will now define certain more general stochastic vertex weights $L_u^{(J)}(i_1, j_1; i_2, j_2)$, where $(i_1, j_1), (i_2, j_2) \in \mathbb{Z}_{\geq 0} \times \{0, 1, \ldots, J\}$. That is, we want to relax the restriction that the horizontal arrow multiplicities are bounded by 1, and consider multiplicities bounded by any fixed $J \geq 1$. When $J = 1$, the vertex weights $L_u^{(J)}$ will coincide with L_u. Of course, we want the new weights $L_u^{(J)}$ to share some of the nice properties of the L_u's; most importantly, the $L_u^{(J)}$'s should satisfy a version of the Yang–Baxter equation. The construction of the weights $L_u^{(J)}$ follows the so-called *fusion procedure*, which was invented in a representation-theoretic context Kulish et al. (1981) (see also Kirillov and Reshetikhin (1987)) to produce higher-dimensional solutions of the Yang–Baxter equation from lower-dimensional ones. Following Corwin and Petrov (2016), here we describe the fusion procedure in purely combinatorial/probabilistic terms.

We will need the following definition.

Definition 2.5.3.   *A probability distribution P on $\{0, 1\}^J$ is called q-exchangeable if the probability weights $P(\vec{h})$, $\vec{h} = (h^{(1)}, \ldots, h^{(J)}) \in \{0, 1\}^J$, depend on $\vec{h}$ in the following way:*

$$P(\vec{h}) = \tilde{P}(j) \cdot \frac{q^{\sum_{r=1}^{J}(r-1)h^{(r)}}}{Z_j(J)}, \qquad j := \sum_{r=1}^{J} h^{(r)}, \qquad (2.68)$$

where $\tilde{P}$ is a probability distribution on $\{0, 1, \ldots, J\}$. In words, for a fixed sum of coordinates j, the weights of the conditional distribution of $\vec{h}$ are proportional to the

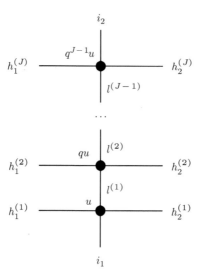

Fig. 2.10 Attaching J vertices with spectral parameters $u, qu, \ldots, q^{J-1}u$ vertically.

product of the factors q^{r-1} for each coordinate '1' at location $r \in \{1, \ldots, J\}$. The normalization constant $Z_j(J)$ is given by the following expression involving the q-binomial coefficient:[8]

$$Z_j(J) = q^{\frac{j(j-1)}{2}} \binom{J}{j}_q = q^{\frac{j(j-1)}{2}} \frac{(q;q)_J}{(q;q)_j (q;q)_{J-j}}. \tag{2.69}$$

The name 'q-exchangeable' refers to the fact that any exchange in $\vec{h}$ of the form $10 \to 01$ multiplies the weight of $\vec{h}$ by q. See Gnedin and Olshanski 2009; 2010 for a detailed treatment of q-exchangeable distributions.

Returning to vertex weights, a key probabilistic feature observed in Corwin and Petrov (2016) which triggers the fusion procedure is the following. Attach vertically J vertices with spectral parameters $u, qu, \ldots, q^{J-1}u$ (see Fig. 2.10), and assign to them the corresponding weights $\mathsf{L}_{q^i u}$ given by (2.8). Fixing the numbers i_1 and i_2 of arrows at the bottom and at the top, we see that this vertex configuration maps probability distributions P_1 on incoming arrows $h_1^{(1)}, \ldots, h_1^{(J)}$ to probability distributions P_2 on outgoing arrows $h_2^{(1)}, \ldots, h_2^{(J)}$.

Proposition 2.5.4.   The mapping $P_1 \mapsto P_2$ preserves the class of q-exchangeable distributions.

---

[8] Indeed, $Z_j(J)$ is the sum of $q^{\sum_{r=1}^{J}(r-1)h^{(r)}}$ over all $\vec{h} \in \{0,1\}^J$ with $\sum_{r=1}^{J} h^{(r)} = j$. Considering two cases $h^{(J)} = 1$ or $h^{(J)} = 0$, we see that it satisfies the recursion $Z_j(J) = q^{J-1} Z_{j-1}(J-1) + Z_j(J-1)$, with $Z_0(J) = 1$. This recursion is solved by (2.69).

Proof Let us fix the numbers $i_1, i_2 \in \mathbb{Z}_{\geq 0}$ of bottom and top arrows, as well as the total number $j_1 = \sum_{\ell=1}^{J} h_1^{(\ell)} \in \{0, 1, \ldots, J\}$ of incoming arrows from the left. Under these conditions, the incoming q-exchangeable distribution P_1 is unique (its partition function is $Z_{j_1}(J)$). It suffices to show that for any $\vec{h}_2 \in \{0, 1\}^J$ with $h_2^{(r)} = 0$, $h_2^{(r+1)} = 1$ for some r, we have

$$P_2(h_2^{(1)}, \ldots, h_2^{(r)}, h_2^{(r+1)}, \ldots h_2^{(J)}) = q \cdot P_2(h_2^{(1)}, \ldots, h_2^{(r+1)}, h_2^{(r)}, \ldots h_2^{(J)}).$$

Since this property involves only two neighboring vertices, it suffices to consider the case $J = 2$. The desired statement now follows from the relations (here g is arbitrary)

In each of the relations the right-hand side differs by moving the outgoing arrow down, and 'weight' means the product of the weights L_u at the bottom vertex and L_{qu} at the top vertex. These relations are readily verified from the definition of L_u (2.8) (see also Fig. 2.5).

This proposition implies that for any fixed $i_1, i_2 \in \mathbb{Z}_{\geq 0}$, the Markov operator mapping P_1 to P_2 (where P_1, P_2 are probability distributions on $\{0, 1\}^J$) can be projected to another Markov operator which maps $\tilde{P}_1$ to $\tilde{P}_2$ (cf. (2.68)), i.e., acts on probability distributions on the smaller space $\{0, 1, \ldots, J\}$. We will denote the matrix elements of this 'collapsed' Markov operator by $L_u^{(J)}(i_1, j_1; i_2, j_2)$, where $(i_1, j_1), (i_2, j_2) \in \mathbb{Z}_{\geq 0} \times \{0, 1, \ldots, J\}$.

The definition of $L_u^{(J)}$ implies that these matrix elements satisfy a certain recursion relation in J. This relation is obtained by considering two cases, whether there is a

left-to-right arrow at the very bottom, or not (i.e., $h_1^{(1)} = 0$ or $h_1^{(1)} = 1$). Therefore, we obtain the recursion

$$\mathsf{L}_u^{(J)}(i_1, j_1; i_2, j_2) = \sum_{a,b \in \{0,1\}} \sum_{l \geq 0} P(h_1^{(1)} = a)\, \mathsf{L}_u(i_1, a; l, b) \mathsf{L}_{qu}^{(J-1)}(l, j_1 - a; i_2, j_2 - b).$$

$$(2.70)$$

Here the probability $P(h_1^{(1)} = a)$ corresponds to our division into two cases. It can be readily computed using Definition 2.5.3:

$$P(h_1^{(1)} = 0) = \frac{q^{j_1} Z_{j_1}(J-1)}{Z_{j_1}(J)} = \frac{q^{j_1} - q^J}{1 - q^J},$$

$$P(h_1^{(1)} = 1) = \frac{q^{j_1-1} Z_{j_1-1}(J-1)}{Z_{j_1}(J)} = \frac{1 - q^{j_1}}{1 - q^J}.$$

The recursion relation (2.70) has a solution expressible in terms of terminating q-hypergeometric functions (here we follow the notation of Mangazeev (2014) and Borodin (2014)):

$$_{r+1}\bar{\varphi}_r \left(\begin{matrix} q^{-n}; a_1, \ldots, a_r \\ b_1, \ldots, b_r \end{matrix} \middle| q, z \right) := \sum_{k=0}^{n} z^k \frac{(q^{-n}; q)_k}{(q; q)_k} \prod_{i=1}^{r} (a_i; q)_k (b_i q^k; q)_{n-k}$$

$$= \prod_{i=1}^{r} (b_i; q)_n \cdot {}_{r+1}\varphi_r \left(\begin{matrix} q^{-n}, a_1, \ldots, a_r \\ b_1, \ldots, b_r \end{matrix} \middle| q, z \right),$$

where here $n \in \mathbb{Z}_{\geq 0}$. The solution $\mathsf{L}_u^{(J)}$ looks as follows:

$$\mathsf{L}_u^{(J)}(i_1, j_1; i_2, j_2) = \mathbf{1}_{i_1+j_1=i_2+j_2} \frac{(-1)^{i_1} q^{\frac{1}{2}i_1(i_1+2j_1-1)} u^{i_1} s^{j_1+j_2-i_2} (us^{-1}; q)_{j_2-i_1}}{(q; q)_{i_2} (su; q)_{i_2+j_2} (q^{J+1-j_1}; q)_{j_1-j_2}}$$

$$\times {}_4\bar{\varphi}_3 \left(\begin{matrix} q^{-i_2}; q^{-i_1}, suq^J, qs/u \\ s^2, q^{1+j_2-i_1}, q^{J+1-i_2-j_2} \end{matrix} \middle| q, q \right). \quad (2.71)$$

Formula (2.71) for fused vertex weights is essentially due to Mangazeev (2014). In the present form (2.71) it was obtained in Corwin and Petrov (2016, Thm. 3.15) by matching the recursion (2.70) to the recursion relation for the classical q-Racah orthogonal polynomials.[9] About the latter see Koekoek and Swarttouw (1996, Ch. 3.2).

2.5.3 Principal specializations of skew functions

The fused stochastic vertex weights discussed in Section 2.5.2 can be used to describe principal specializations of the skew functions $\mathsf{F}_{\mu/\lambda}$ and $\mathsf{G}_{\mu/\lambda}$, in analogy to the nonskew principal specializations of Corollary 2.4.14.

---

[9] The parameters in (2.71) match those in Corwin and Petrov (2016, Thm. 3.15) as $\beta = \alpha q^J$, $\alpha = -su$, and $\nu = s^2$.

Mimicking (2.7)–(2.8), we will use the general J stochastic weights $\mathsf{L}_u^{(J)}$ (2.71) to define the weights which are general J versions of the w_u's (2.6):

$$w_u^{(J)}(i_1, j_1; i_2, j_2) := \frac{1}{(-s)^{j_2}} \frac{(q;q)_{i_2}}{(s^2;q)_{i_2}} \frac{(s^2;q)_{i_1}}{(q;q)_{i_1}} \mathsf{L}_u^{(J)}(i_1, j_1; i_2, j_2), \qquad (2.72)$$

where $i_1, i_2 \in \mathbb{Z}_{\geq 0}$ and $j_1, j_2 \in \{0, 1, \dots, J\}$ are such that $i_1 + j_1 = i_2 + j_2$ (otherwise the weight is set to 0). These weights are expressed via the q-hypergeometric function as follows:

$$w_u^{(J)}(i_1, j_1; i_2, j_2) = \frac{(-1)^{i_1+j_2} q^{\frac{1}{2}i_1(i_1+2j_1-1)} s^{j_2-i_1} u^{i_1} (q;q)_{j_1} (us^{-1};q)_{j_1-i_2}}{(q;q)_{i_1} (q;q)_{j_2} (us;q)_{i_1+j_1}}$$

$$\times {}_4\bar{\phi}_3 \left(\begin{matrix} q^{-i_1}; q^{-i_2}, q^J su, qsu^{-1} \\ s^2, q^{1+j_1-i_2}, q^{1+J-i_1-j_1} \end{matrix} \middle| q, q \right). \qquad (2.73)$$

We see that the weights $w_u^{(J)}$ depend on the spectral parameter u in a rational manner. One can check that for $J = 1$, the weights $w_u^{(J)}$ turn into (2.6).

Proposition 2.5.5.

1. For any $J \in \mathbb{Z}_{\geq 1}$, $N \in \mathbb{Z}_{\geq 0}$, $\lambda \in \mathsf{Sign}_N^+$, $\mu \in \mathsf{Sign}_{N+J}^+$, and $u \in \mathbb{C}$, the principal specialization of the skew function

$$\mathsf{F}_{\mu/\lambda}(u, qu, \dots, q^{J-1}u)$$

   is equal to the weight of the unique collection of $N+J$ up-right paths in the semi-infinite horizontal strip of height 1, with vertex weights equal to $w_u^{(J)}$ (so that at most J horizontal arrows per edge are allowed). The paths in the collection start with N vertical edges $(\lambda_i, 0) \to (\lambda_i, 1)$ and with J horizontal edges $(-1, 1) \to (0, 1)$, and end with $N + J$ vertical edges $(\mu_j, 1) \to (\mu_j, 2)$, see Fig. 2.11, top.
2. For any $J \in \mathbb{Z}_{\geq 1}$, any $\lambda, \mu \in \mathsf{Sign}_N$, and any $v \in \mathbb{C}$, the principal specialization of the skew function

$$\mathsf{G}_{\mu/\lambda}(v, qv, \dots, q^{J-1}v)$$

   is equal to the weight of the unique collection of N up-right paths in the semi-infinite horizontal strip of height 1, with vertex weights $w_v^{(J)}$ (so that at most J horizontal arrows per edge are allowed). The paths in the collection start with N vertical edges $(\lambda_i, 0) \to (\lambda_i, 1)$ and end with N vertical edges $(\mu_i, 1) \to (\mu_i, 2)$, see Fig. 2.11, bottom.

Proof Let us focus on the second claim. Relation (2.7)–(2.8) between the $J = 1$ weights w_u and L_u readily implies that for any $z \in \mathbb{C}$, any $L \in \mathbb{Z}_{\geq 0}$ and $\nu, \kappa \in \mathsf{Sign}_L^+$, the quantity

$$(-s)^{|\kappa|-|\nu|} \frac{\mathsf{c}(\kappa)}{\mathsf{c}(\nu)} \mathsf{G}_{\kappa/\nu}(z) = (-s)^{|\kappa|-|\nu|} \mathsf{G}_{\kappa/\nu}^{\mathsf{c}}(z)$$

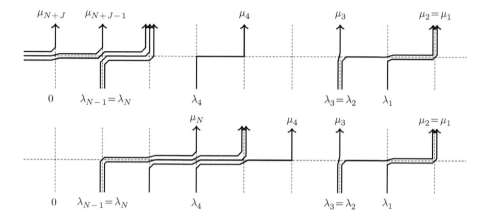

Fig. 2.11 A unique path collection with horizontal multiplicities bounded by $J = 3$ corresponding to the function $\mathsf{F}_{\mu/\lambda}(u, qu, \ldots, q^{J-1}u)$ (top) or $\mathsf{G}_{\mu/\lambda}(v, qv, \ldots, q^{J-1}v)$ (bottom).

is equal to the weight of the unique collection of L paths in the semi-infinite horizontal strip of height 1 connecting ν to κ, but with horizontal arrow multiplicities bounded by 1. The vertex weights in this path collection are the stochastic weights L_z.

Therefore, by Proposition 2.4.5, the quantity

$$(-s)^{|\mu|-|\lambda|} \mathsf{G}^{\mathsf{c}}_{\mu/\lambda}(v, qv, \ldots, q^{J-1}v) \tag{2.74}$$

is equal to the sum of weights of collections of N paths in $\{1, 2, \ldots, J\} \times \mathbb{Z}_{\geq 0}$ connecting λ to μ (as in Definition 2.4.2), with vertex weights $\mathsf{L}_{q^{j-1}v}$ at the jth horizontal line (hence the horizontal arrow multiplicities are bounded by 1). In (2.74), the configuration of input horizontal arrows at location 0 is empty, and hence its distribution is q-exchangeable. Thus, we may use the fusion of stochastic weights from Section 2.5.2 to collapse the J horizontals into one, with horizontal arrow multiplicities bounded by J, and with fused vertex weights $\mathsf{L}_v^{(J)}$. In this way all path collections in $\{1, 2, \ldots, J\} \times \mathbb{Z}_{\geq 0}$ connecting λ to μ map to the unique collection in $\{1\} \times \mathbb{Z}_{\geq 0}$ with horizontal edge multiplicities bounded by J. Using (2.72), we conclude that the second claim holds.

The first claim about $\mathsf{F}_{\mu/\lambda}(u, qu, \ldots, q^{J-1}u)$ is analogous, because the corresponding configuration of input arrows in $\{1, 2, \ldots, J\} \times \mathbb{Z}_{\geq 0}$ is the fully packed one, whose distribution is also q-exchangeable. This completes the proof.

Remark 2.5.6. Since the general J vertex weights $w_u^{(J)}(i_1, j_1; i_2, j_2)$ (2.73) depend on q^J in a rational manner, they make sense for q^J being an arbitrary (generic) complex parameter. Thus, we can consider the principal specializations $\mathsf{G}_{\mu/\lambda}(v, qv, \ldots, q^{J-1}v)$ for a generic $q^J \in \mathbb{C}$. In other words, this quantity admits an analytic continuation in q^J. The second part of Proposition 2.5.5 thus states that when $J \in \mathbb{Z}_{\geq 1}$, the function $\mathsf{G}_{\mu/\lambda}(v, qv, \ldots, q^{J-1}v)$ can be expressed as a substitution of the values $(v, qv, \ldots, q^{J-1}v)$ into the symmetric function $\mathsf{G}_{\mu/\lambda}(v_1, v_2, \ldots, v_J)$.

In contrast with the functions $\mathsf{G}_{\mu/\lambda}$ in which the number of indeterminates does not depend on λ and μ, the number of arguments in the functions $\mathsf{F}_{\mu/\lambda}$ is completely determined by the signatures λ and μ. Therefore, the parameter J in $\mathsf{F}_{\mu/\lambda}(u, qu, \ldots, q^{J-1}u)$ must remain a positive integer.

Remark 2.5.7. The weights (2.72) are related to the weights $\widetilde{w}_v^{(J)}$ of Borodin (2014, (6.8)) via

$$w_v^{(J)}(i_1, j_1; i_2, j_2) = q^{\frac{1}{4}(j_2^2 - j_1^2)}(-s)^{j_2 - j_1}\frac{(q;q)_{j_1}}{(q;q)_{j_2}}\widetilde{w}_v^{(J)}(i_1, j_1; i_2, j_2).$$

The structure of the path collection for $\mathsf{G}_{\mu/\lambda}$ implies that for the purposes of computing $\mathsf{G}_{\mu/\lambda}$, any prefactors in the vertex weights of the form $f(j_1)/f(j_2)$ are irrelevant. Therefore, the principal specializations $\mathsf{G}_{\mu/\lambda}(v, qv, \ldots, q^{J-1}v)$ coincide with those in Borodin (2014, Section 6). Note that, however, factors of the form $f(j_1)/f(j_2)$ in the vertex weights do make a difference for the functions $\mathsf{F}_{\mu/\lambda}(u, qu, \ldots, q^{J-1}u)$.

2.6 Markov kernels and stochastic dynamics

Here we describe probability distributions on signatures arising from the Cauchy identity (Corollary 2.4.11), as well as discrete time stochastic systems (i.e., discrete time Markov chains) which act nicely on these measures.

2.6.1 Probability measures associated with the Cauchy identity

The idea that summation identities for symmetric functions lead to interesting probability measures dates back at least to Fulman (1997) and Okounkov (2001), and it was further developed in Okounkov and Reshetikhin (2003); Vuletic (2007); Borodin (2011); Borodin and Corwin (2014); Borodin et al. (2016b). Similar ideas in our context lead to the definition of the following probability measures which are analogous to the Schur and Macdonald measures:

Definition 2.6.1. Let $M, N \in \mathbb{Z}_{\geq 0}$,[10] and the parameters q, s, and $\boldsymbol{u} = (u_1, \ldots, u_M)$, $\boldsymbol{v} = (v_1, \ldots, v_N)$ satisfy (2.66)–(2.67).[11] Moreover, assume that $(u_i, v_j) \in \mathsf{Adm}$ for all i, j (for the admissibility it is enough to require that all u_i and v_j are sufficiently small, cf. Definition 2.4.7). Define the probability measure on Sign_M^+ via

$$\mathscr{M}_{\boldsymbol{u};\boldsymbol{v}}(\nu) := \frac{1}{Z(\boldsymbol{u};\boldsymbol{v})}\mathsf{F}_\nu(u_1, \ldots, u_M)\mathsf{G}_\nu^c(v_1, \ldots, v_N), \qquad \nu \in \mathsf{Sign}_M^+, \tag{2.75}$$

where the normalization constant is given by

---

[10] Here and in the following if $M = 0$, then Sign_M^+ consists of the single empty signature $\varnothing$, and thus all probability measures and Markov operators on this space are trivial.

[11] These conditions are assumed throughout Section 2.6 except Sections 2.6.5 and 2.6.6.

$$Z(\boldsymbol{u};\boldsymbol{v}) := (q;q)_M \prod_{i=1}^{M} \left(\frac{1}{1-su_i} \prod_{j=1}^{N} \frac{1-qu_iv_j}{1-u_iv_j} \right). \tag{2.76}$$

The fact that the unnormalized weights $Z(\mathbf{u};\mathbf{v})\,\mathcal{M}_{\mathbf{u};\mathbf{v}}(\nu)$ are nonnegative follows from (2.66)–(2.67). Indeed, these conditions imply that the vertex weights w_u and w_u^c are nonnegative, and hence so are the functions F_ν and G_ν^c. The form (2.76) of the normalization constant follows from the Cauchy identity (2.43).[12] Note that the length of the tuple $\mathbf{u}$ determines the length of the signatures on which the measure $\mathcal{M}_{\mathbf{u};\mathbf{v}}$ lives. In contrast, the length of the tuple $\mathbf{v}$ may be arbitrary.

In two degenerate cases, $\mathcal{M}_{\varnothing;\mathbf{v}}$ is the delta measure at the empty configuration (for any $\mathbf{v}$), and $\mathcal{M}_{(u_1,\dots,u_M);\varnothing}$ is the delta measure at the configuration 0^M (that is, all M particles are at 0).

The measures $\mathcal{M}_{\mathbf{u};\mathbf{v}}$ can be represented pictorially, see Fig. 2.12. Let us look at the bottom part of the path collection as in Fig. 2.12, and let us denote the positions of the vertical edges at the kth horizontal by $\nu_i^{(k)}$, $1 \le i \le k \le M$ see Fig. 2.13, bottom. In the top part of the path collection, let us denote the coordinates of the vertical

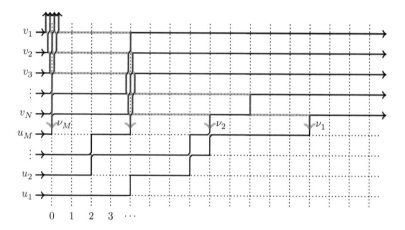

Fig. 2.12 Probability weights $\mathcal{M}_{\mathbf{u};\mathbf{v}}(\nu)$ as partition functions. The bottom part of height M corresponds to $\mathsf{B}(u_1)\dots\mathsf{B}(u_M)$, and the top half of height N — to $\mathsf{D}(v_1^{-1})\dots\mathsf{D}(v_N^{-1})$. The initial configuration at the bottom is empty, and the final configuration of the solid (black) paths at the top is $\mathbf{e}_{(0^M)} = \mathbf{e}_M \otimes \mathbf{e}_0 \otimes \mathbf{e}_0 \dots$. The locations where the solid paths cross the horizontal division line correspond to the signature $\nu = (\nu_1,\dots,\nu_M)$. The opaque (grey) paths complement the solid (black) paths; in the top part; this corresponds to the renormalization (2.31) employed in the passage from the operators $\mathsf{D}(v_j^{-1})$ to $\overline{\mathsf{D}}(v_j^{-1})$ (the latter involves coefficients $\mathsf{G}_{\lambda/\mu}^c(v_j)$). After the renormalization, we let the width of the grid go to infinity.

---

[12] The sum of the unnormalized weights converges due to the admissibility conditions, and hence the normalization constant $Z(\mathbf{u};\mathbf{v})$ is nonnegative. This constant is positive whenever the measure $\mathcal{M}_{\mathbf{u};\mathbf{v}}$ is nontrivial.

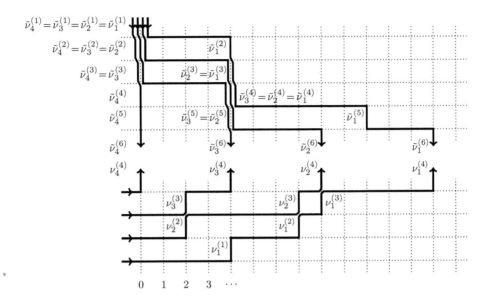

Fig. 2.13   A pair of interlacing arrays from path collections. Horizontal parts of the paths are for illustration.

edges by $\tilde{\nu}_j^{(\ell)}$, $1 \le \ell \le N$, $1 \le j \le M$ (see Fig. 2.13, top). We have $\nu_i^{(M)} = \tilde{\nu}_i^{(N)} = \nu_i$, $i = 1, \dots, M$. By construction, these coordinates satisfy *interlacing constraints*

$$\nu_i^{(k)} \le \nu_{i-1}^{(k-1)} \le \nu_{i-1}^{(k)}, \qquad \tilde{\nu}_j^{(\ell)} \le \tilde{\nu}_{j-1}^{(\ell-1)} \le \tilde{\nu}_{j-1}^{(\ell)} \tag{2.77}$$

for all meaningful values of k, i and ℓ, j. Arrays of the form $\{\nu_i^{(k)}\}_{1 \le i \le k \le M}$ satisfying the interlacing properties are also sometimes called *Gelfand–Tsetlin schemes/patterns*. By the very definition of the skew F and G functions, the distribution of the sequence of signatures $(\nu^{(1)}, \dots, \nu^{(M)} = \tilde{\nu}^{(N)}, \dots, \tilde{\nu}^{(1)})$ has the form

$$\mathcal{M}_{\mathbf{u};\mathbf{v}}(\nu^{(1)}, \dots, \nu^{(M)} = \tilde{\nu}^{(N)}, \dots, \tilde{\nu}^{(1)})$$

$$= \frac{1}{Z(\mathbf{u};\mathbf{v})} \, \mathsf{F}_{\nu^{(1)}}(u_1) \mathsf{F}_{\nu^{(2)}/\nu^{(1)}}(u_2) \dots \mathsf{F}_{\nu^{(M)}/\nu^{(M-1)}}(u_M)$$

$$\times \, \mathsf{G}^c_{\tilde{\nu}^{(1)}}(v_1) \mathsf{G}^c_{\tilde{\nu}^{(2)}/\tilde{\nu}^{(1)}}(v_2) \dots \mathsf{G}^c_{\tilde{\nu}^{(N)}/\tilde{\nu}^{(N-1)}}(v_N). \tag{2.78}$$

The probability distribution (2.78) on interlacing arrays is an analogue of Schur and Macdonald processes of Okounkov and Reshetikhin (2003) and Borodin and Corwin (2014). It readily follows from the Pieri rules (Corollary 2.4.9) that under (2.78), the marginal distribution of $\nu^{(k)}$ for any $k = 1, \dots, M$ is $\mathcal{M}_{(u_1, \dots, u_k);(v_1, \dots, v_N)}$, and similarly the marginal distribution of $\tilde{\nu}^{(\ell)}$, $\ell = 1, \dots, N$, is $\mathcal{M}_{(u_1, \dots, u_M);(v_1, \dots, v_\ell)}$.

2.6.2 Four Markov kernels

Let us now define four Markov kernels which map the measure $\mathscr{M}_{\mathbf{u};\mathbf{v}}$ to a measure of the same form, but with modified parameters $\mathbf{u}$ or $\mathbf{v}$.

The first two Markov kernels, Λ and Λ°, correspond to taking conditional distributions given $\nu^{(M)} = \tilde{\nu}^{(N)}$ of $\nu^{(k)}$ or $\tilde{\nu}^{(\ell)}$, respectively, in the ensemble (2.78). Namely, let us define for any m,

$$\Lambda_{u\,|\,\mathbf{u}}(\nu \to \mu) := \frac{F_\mu(u_1, \ldots, u_m)}{F_\nu(u_1, \ldots, u_m, u)}\, F_{\nu/\mu}(u), \qquad (2.79)$$

where $\mathbf{u} = (u_1, \ldots, u_m)$, and $\nu \in \mathsf{Sign}^+_{m+1}$, $\mu \in \mathsf{Sign}^+_m$. Also, let us define for any n,

$$\Lambda^\circ_{v\,|\,\mathbf{v}}(\lambda \to \mu) := \frac{\mathsf{G}^{\mathsf{c}}_\mu(v_1, \ldots, v_n)}{\mathsf{G}^{\mathsf{c}}_\lambda(v_1, \ldots, v_n, v)}\, \mathsf{G}^{\mathsf{c}}_{\lambda/\mu}(v), \qquad (2.80)$$

where $\mathbf{v} = (v_1, \ldots, v_n)$, and $\lambda, \mu \in \mathsf{Sign}^+_m$ for some m. The facts that the quantities (2.79) and (2.80) sum to 1 (over all $\mu \in \mathsf{Sign}^+_m$; note that these sums are finite) follow from the branching rules (Proposition 2.4.5). Hence, $\Lambda_{u\,|\,\mathbf{u}} \colon \mathsf{Sign}^+_{m+1} \dashrightarrow \mathsf{Sign}^+_m$ and $\Lambda^\circ_{v\,|\,\mathbf{v}} \colon \mathsf{Sign}^+_m \dashrightarrow \mathsf{Sign}^+_m$ define Markov kernels.[13] Note that in (2.79) and (2.80) one can replace all functions by F^{c} or G, respectively, and get the same kernels.

The kernels Λ and Λ° act on the measures (2.75) as

$$\mathscr{M}_{\mathbf{u}\cup u;\mathbf{v}}\Lambda_{u\,|\,\mathbf{u}} = \mathscr{M}_{\mathbf{u};\mathbf{v}}, \qquad \mathscr{M}_{\mathbf{u};\mathbf{v}\cup v}\Lambda^\circ_{v\,|\,\mathbf{v}} = \mathscr{M}_{\mathbf{u};\mathbf{v}}, \qquad (2.81)$$

this follows from the Pieri rules (Corollary 2.4.9). The matrix products in Eq. (2.81) are understood in a natural way, for example, $(\mathscr{M}_{\mathbf{u}\cup u;\mathbf{v}}\Lambda_{u\,|\,\mathbf{u}})(\mu) = \sum_\nu \mathscr{M}_{\mathbf{u}\cup u;\mathbf{v}}(\nu)\Lambda_{u\,|\,\mathbf{u}}(\nu \to \mu)$.

Remark 2.6.2 (Gibbs measures). Conditioned on any $\nu^{(k)}$ (where $k = 1, \ldots, M$), the distribution of the lower levels $\nu^{(1)}, \ldots, \nu^{(k-1)}$ under (2.78) is independent of $\mathbf{v}$ and is given by

$$\Lambda_{u_k\,|\,(u_1, \ldots, u_{k-1})}(\nu^{(k)} \to \nu^{(k-1)}) \ldots \Lambda_{u_3\,|\,(u_1, u_2)}(\nu^{(3)} \to \nu^{(2)})\Lambda_{u_2\,|\,(u_1)}(\nu^{(2)} \to \nu^{(1)}), \qquad (2.82)$$

and a similar expression can be written for conditioning on $\tilde{\nu}^{(\ell)}$, yielding a distribution which is independent of $\mathbf{u}$ and involves the kernels Λ°.

It is natural to call a probability measure on a sequence of interlacing signatures $(\nu^{(1)}, \ldots, \nu^{(M)})$ whose conditional distributions are given by (2.82), a *Gibbs measure* (with respect to the $\mathbf{u}$ parameters). In fact, when $q = s = 0$ and $u_i \equiv 1$, this Gibbs property turns into the following: conditioned on any $\nu^{(k)}$, the distribution of the

---

[13] We use the notation '$\dashrightarrow$' to indicate that $\Lambda_{u\,|\,\mathbf{u}}$ and $\Lambda^\circ_{v\,|\,\mathbf{v}}$ are Markov kernels; i.e., they are functions in the first variable (belonging to the space on the left of '$\dashrightarrow$') and probability distributions in the second variable (belonging to the space on the right of '$\dashrightarrow$').

lower levels $\nu^{(1)}, \ldots, \nu^{(k-1)}$ is *uniform* among all sequences of signatures satisfying the interlacing constraints (2.77).

This Gibbs property (as well as commutation relations discussed later) can be used to construct 'multivariate' Markov kernels on arrays of interlacing signatures which act nicely on distributions of the form (2.78), but we will not address this construction here (about similar constructions see references given in Remark 2.6.7). For details of such constructions in the case of Macdonald processes see Borodin and Petrov (2016b) and Matveev and Petrov (2015).

The other two Markov kernels, $\mathscr{Q}^+$ and $\mathscr{Q}^\circ$, increase the number of parameters in the measures $\mathscr{M}_{\mathbf{u};\mathbf{v}}$, as opposed to (2.81), where the number of parameters is decreased. These kernels are defined as follows. For any $n, m \in \mathbb{Z}_{\geq 0}$, define

$$\mathscr{Q}^+_{u;\mathbf{v}}(\lambda \to \nu) := \frac{1 - su}{1 - q^{m+1}} \left(\prod_{j=1}^{n} \frac{1 - uv_j}{1 - quv_j} \right) \frac{\mathsf{G}^{\mathsf{c}}_\nu(v_1, \ldots, v_n)}{\mathsf{G}^{\mathsf{c}}_\lambda(v_1, \ldots, v_n)} \mathsf{F}_{\nu/\lambda}(u), \tag{2.83}$$

where $\mathbf{v} = (v_1, \ldots, v_n)$ such that $(u, v_j) \in \mathsf{Adm}$ for all j, with $\lambda \in \mathsf{Sign}^+_m$ and $\nu \in \mathsf{Sign}^+_{m+1}$. Also, for any $m \in \mathbb{Z}_{\geq 0}$, define

$$\mathscr{Q}^\circ_{\mathbf{u};v}(\mu \to \nu) := \left(\prod_{i=1}^{m} \frac{1 - u_i v}{1 - q u_i v} \right) \frac{\mathsf{F}_\nu(u_1, \ldots, u_m)}{\mathsf{F}_\mu(u_1, \ldots, u_m)} \mathsf{G}^{\mathsf{c}}_{\nu/\mu}(v), \tag{2.84}$$

where $\mathbf{u} = (u_1, \ldots, u_M)$ such that $(u_i, v) \in \mathsf{Adm}$ for all i, with $\mu, \nu \in \mathsf{Sign}^+_m$. By the Pieri rules of Corollary 2.4.9, $\mathscr{Q}^+_{u;\mathbf{v}} \colon \mathsf{Sign}^+_m \dashrightarrow \mathsf{Sign}^+_{m+1}$ and $\mathscr{Q}^\circ_{\mathbf{u};v} \colon \mathsf{Sign}^+_m \dashrightarrow \mathsf{Sign}^+_m$ define Markov kernels (i.e., they sum to one in the second argument).

Remark 2.6.3. In $\mathscr{Q}^\circ_{\mathbf{u};v}$ (2.84), moving the conjugation from the function G to both functions F does not change the kernel. However, doing so in $\mathscr{Q}^+_{u;\mathbf{v}}$ (2.83) requires modifying the prefactor:

$$\mathscr{Q}^+_{u;\mathbf{v}}(\lambda \to \nu) = \frac{1 - su}{1 - s^2 q^m} \left(\prod_{j=1}^{n} \frac{1 - uv_j}{1 - quv_j} \right) \frac{\mathsf{G}_\nu(v_1, \ldots, v_n)}{\mathsf{G}_\lambda(v_1, \ldots, v_n)} \mathsf{F}^{\mathsf{c}}_{\nu/\lambda}(u).$$

From the branching rules (Proposition 2.4.5) it readily follows that the kernels $\mathscr{Q}^+$ and $\mathscr{Q}^\circ$ act on the measures (2.75) as

$$\mathscr{M}_{\mathbf{u};\mathbf{v}} \mathscr{Q}^+_{u;\mathbf{v}} = \mathscr{M}_{\mathbf{u} \cup u;\mathbf{v}}, \qquad \mathscr{M}_{\mathbf{u};\mathbf{v}} \mathscr{Q}^\circ_{\mathbf{u};v} = \mathscr{M}_{\mathbf{u};\mathbf{v} \cup v}. \tag{2.85}$$

The Markov kernels defined in (2.83) and (2.84) enter the following commutation relations.

Proposition 2.6.4.

1. For any $\mathbf{u} = (u_1, \ldots, u_m)$ and $u, v \in \mathbb{C}$ such that $(u, v) \in \mathsf{Adm}$, we have $\mathscr{Q}^\circ_{\mathbf{u} \cup u;v} \Lambda_{u \mid \mathbf{u}} = \Lambda_{u \mid \mathbf{u}} \mathscr{Q}^\circ_{\mathbf{u};v}$ (as Markov kernels $\mathsf{Sign}^+_{m+1} \dashrightarrow \mathsf{Sign}^+_m$), or, in more detail,

$$\sum_{\lambda \in \mathsf{Sign}^+_{m+1}} \mathscr{Q}^\circ_{\mathbf{u} \cup u;v}(\nu \to \lambda) \Lambda_{u \,|\, \mathbf{u}}(\lambda \to \mu) = \sum_{\kappa \in \mathsf{Sign}^+_m} \Lambda_{u \,|\, \mathbf{u}}(\nu \to \kappa) \mathscr{Q}^\circ_{\mathbf{u};v}(\kappa \to \mu),$$

where $\nu \in \mathsf{Sign}^+_{m+1}$ and $\mu \in \mathsf{Sign}^+_m$.

2. For any $\mathbf{v} = (v_1, \ldots, v_n)$, $u, v \in \mathbb{C}$ such that $(u, v) \in \mathsf{Adm}$, we have $\mathscr{Q}^+_{u;\mathbf{v} \cup v} \Lambda^\circ_{v \,|\, \mathbf{v}} = \Lambda^\circ_{v \,|\, \mathbf{v}} \mathscr{Q}^+_{u;\mathbf{v}}$ (as Markov kernels $\mathsf{Sign}^+_m \dashrightarrow \mathsf{Sign}^+_{m+1}$), which is unabbreviated in the same way as the first relation.

Proof A straightforward corollary of the skew Cauchy identity (Proposition 2.4.6).

Remark 2.6.5. In the context of Schur functions, the Markov kernels $\mathscr{Q}^+$ and Λ are often referred to as *transition* and *cotransition* probabilities. In Borodin (2011, §9) and Borodin and Corwin (2014, Section 2.3.3) similar kernels are denoted by $p^\uparrow$ and $p^\downarrow$, respectively. The kernels $\mathscr{Q}^+$ and Λ involve the skew functions F in the $\mathbf{u}$ parameters, and similarly $\mathscr{Q}^\circ$ and Λ° correspond to the G's in the $\mathbf{v}$ parameters. The latter operators differ from the former ones because (unlike in the Schur or Macdonald setting) the functions F and G are not proportional to each other.

We will treat the Markov kernels $\mathscr{Q}^\circ_{\mathbf{u};v}$ and $\mathscr{Q}^+_{u;\mathbf{v}}$ as one-step transition operators of certain discrete time Markov chains.

Remark 2.6.6. One can readily write down eigenfunctions of $\mathscr{Q}^\circ_{\mathbf{u};v}$ viewed as an operator on functions on Sign^+_m. Here we mean algebraic (or formal) eigenfunctions, i.e., we do not address the question of how they decay at infinity. We have for any $\mu \in \mathsf{Sign}^+_m$

$$\left(\mathscr{Q}^\circ_{\mathbf{u};v} \Psi^{\mathbf{u}}_\bullet(z_1, \ldots, z_m) \right)(\mu) = \sum_{\nu \in \mathsf{Sign}^+_m} \mathscr{Q}^\circ_{\mathbf{u};v}(\mu \to \nu) \Psi^{\mathbf{u}}_\nu(z_1, \ldots, z_m)$$

$$= \left(\prod_{i=1}^m \frac{1 - q z_i v}{1 - z_i v} \frac{1 - u_i v}{1 - q u_i v} \right) \Psi^{\mathbf{u}}_\mu(z_1, \ldots, z_m), \qquad (2.86)$$

where the eigenfunction $\Psi^{\mathbf{u}}_\lambda(\mathbf{z})$ depends on the spectral variables $\mathbf{z} = (z_1, \ldots, z_m)$ satisfying the admissibility conditions $(z_i, v) \in \mathsf{Adm}$ for all i, and is defined as

$$\Psi^{\mathbf{u}}_\lambda(z_1, \ldots, z_m) := \frac{1}{\mathsf{F}_\lambda(u_1, \ldots, u_m)} \mathsf{F}_\lambda(z_1, \ldots, z_m). \qquad (2.87)$$

Relation (2.86) readily follows from the Pieri rules (Corollary 2.4.9). This eigenrelation can be employed to write down a spectral decomposition of the operator $\mathscr{Q}^\circ$, see Remark 2.7.13.

2.6.3 Specializations

Let us now discuss special choices of parameters $\mathbf{u}$ and $\mathbf{v}$ which greatly simplify the Markov kernels $\mathscr{Q}^\circ_{\mathbf{u};v}$ and $\mathscr{Q}^+_{u;\mathbf{v}}$, respectively. First, observe that for any $m \in \mathbb{Z}_{\geq 0}$ and any $\mu \in \mathsf{Sign}^+_m$ we have

$$F_\mu(0^m) = F_\mu(\underbrace{0,0,\dots,0}_{m \text{ times}}) = (-s)^{|\mu|}(q;q)_m,$$

where the last equality is due to (2.47) because we can take $u = 0$ in that formula (note that by (2.44), the function $F_\mu(u_1,\dots,u_m)$ is continuous at $\mathbf{u} = 0^m$).

A similar limit for the functions G_μ is given in the next proposition.

Proposition 2.6.7. For any $n \in \mathbb{Z}_{\geq 0}$ and $\nu \in \mathsf{Sign}_n^+$, we have

$$G_\nu(\varrho) := \lim_{\epsilon \to 0} \left(G_\nu(\epsilon, q\epsilon, \dots, q^{J-1}\epsilon) \Big|_{q^J = 1/(s\epsilon)} \right) = \begin{cases} (-s)^{|\nu|}(s^2;q)_n s^{-2n}, & \text{if } \nu_n > 0; \\ 0, & \text{if } \nu_n = 0. \end{cases}$$

$$(2.88)$$

Proof Let k be the number of zero coordinates in ν. From (2.48) we have for $J \geq n-k$

$$G_\nu(\epsilon, q\epsilon, \dots, q^{J-1}\epsilon) = \frac{(q;q)_J}{(q;q)_{J-n+k}} \frac{(s\epsilon;q)_{J+k}}{(s\epsilon;q)_n} \frac{(s^2;q)_n}{(s^2;q)_k} \frac{1}{(s/\epsilon;q^{-1})_{n-k}}$$

$$\times \frac{1}{(sq^{n-k}\epsilon;q)_{J-n+k}} \frac{1}{(s\epsilon;q)_{n-k}} \prod_{j=1}^{n-k} \left(\frac{q^{j-1}\epsilon - s}{1 - sq^{j-1}\epsilon} \right)^{\nu_j}.$$

The $\epsilon \to 0$ limit of the product over j (which is independent of q^J) gives $(-s)^{|\nu|}$. We can rewrite the remaining factors as

$$\frac{(q;q)_J(s^2;q)_n(s\epsilon;q)_{J+k}}{(q;q)_{J-n+k}(s^2;q)_k(s\epsilon;q)_n(s/\epsilon;q^{-1})_{n-k}(sq^{n-k};q)_{J-n+k}} \frac{1}{(s\epsilon;q)_{n-k}}$$

$$= \frac{(s^2q^k;q)_{n-k}}{(s/\epsilon;q^{-1})_{n-k}(s\epsilon;q)_n}(q^{J+1+k-n};q)_{n-k}(s\epsilon q^J;q)_k.$$

One readily sees that this quantity depends on q^J in a rational manner.[14] This makes it possible to analytically continue in q^J, and set $q^J = 1/(s\epsilon)$. Observe that the result involves $(1;q)_k$, which vanishes unless $k = 0$. For $k = 0$ we obtain

$$\frac{(s^2;q)_n}{(s/\epsilon;q^{-1})_n(s\epsilon;q)_n}((s\epsilon)^{-1}q^{1-n};q)_n(s\epsilon;q)_n = \frac{(s^2;q)_n}{(s/\epsilon;q^{-1})_n}((s\epsilon)^{-1}q^{1-n};q)_n,$$

and in the $\epsilon \to 0$ limit this turns into $(s^2;q)_n s^{-2n}$, which completes the proof.

Remark 2.6.8. An alternative proof of Proposition 2.6.7 (and in fact a computation of a more general specialization) using the integral formula of Corollary 2.7.16 is discussed in Section 2.8.2.

---

[14] This can also be thought of as a consequence of the fusion procedure (§2.5.3), but the statement of the proposition does not require fusion.

Let us substitute the specializations 0^m and ϱ into the Markov kernels. The kernel $\mathcal{Q}^\circ_{0^m;v}$ looks as follows:

$$\mathcal{Q}^\circ_{0^m;v}(\mu \to \nu) = (-s)^{|\nu|-|\mu|}\mathsf{G}^c_{\nu/\mu}(v), \qquad \mu, \nu \in \mathsf{Sign}^+_m. \tag{2.89}$$

Similarly, the kernel $\mathcal{Q}^+_{u;\varrho}$ has the form

$$\mathcal{Q}^+_{u;\varrho}(\lambda \to \nu) = \frac{1-su}{s(s-u)}(-s)^{|\nu|-|\lambda|}\mathsf{F}^c_{\nu/\lambda}(u),$$

where $\lambda \in \mathsf{Sign}^+_m$, $\nu \in \mathsf{Sign}^+_{m+1}$ are such that $\lambda_m, \nu_{m+1} > 0$. Because of this latter condition, we can subtract 1 from all parts of λ and ν, and rewrite $\mathcal{Q}^+_{u;\varrho}$ as

$$\mathcal{Q}^+_{u;\varrho}(\lambda \to \nu) = (-s)^{|\nu|-|\lambda|}\frac{1}{\mathsf{L}_u(0,1;0,1)}\,\mathsf{F}^c_{\nu/\lambda}(u) = (-s)^{|\tilde{\nu}|-|\tilde{\lambda}|}\mathsf{F}^c_{\tilde{\nu}/\tilde{\lambda}}(u), \tag{2.90}$$

where $\tilde{\nu} = \nu - 1^{m+1}$ and $\tilde{\lambda} = \lambda - 1^m$.

2.6.4 Interacting particle systems

Fix $M \in \mathbb{Z}_{\geq 0}$. Let us interpret Sign^+_M as the space of M-particle configurations on $\mathbb{Z}_{\geq 0}$, in which putting an arbitrary number of particles per site is allowed (particles are assumed to be identical). That is, each $\lambda = 0^{\ell_0}1^{\ell_1}2^{\ell_2}\ldots \in \mathsf{Sign}^+_M$ corresponds to having ℓ_0 particles at site 0, ℓ_1 particles at site 1, and so on.

We can interpret the Markov kernels $\mathcal{Q}^\circ_{u;v}$ and $\mathcal{Q}^+_{u;v}$ (for any $\mathbf{u}$ or $\mathbf{v}$) as one-step transition operators of two discrete time Markov chains. Denote these Markov chains by $\mathcal{X}^\circ_{\mathbf{u};\{v_t\}}$ and $\mathcal{X}^+_{\{u_t\};\mathbf{v}}$, respectively. Here $\{v_t\}_{t\in\mathbb{Z}_{\geq 0}}$ and $\{u_t\}_{t\in\mathbb{Z}_{\geq 0}}$ are time-dependent parameters which are added during one step of $\mathcal{X}^\circ$ or $\mathcal{X}^+$, respectively (we tacitly assume that all parameters u_i and v_j satisfy the necessary admissibility conditions as in Section 2.6.2).

For generic $\mathbf{u}$ and $\mathbf{v}$ parameters, the Markov chains $\mathcal{X}^\circ_{\mathbf{u};\{v_t\}}$ and $\mathcal{X}^+_{\{u_t\};\mathbf{v}}$, respectively, are *nonlocal*; i.e., transitions at a given location depend on the whole particle configuration. However, taking $\mathbf{u} = 0^m$ or $\mathbf{v} = \varrho$ in the corresponding chain makes them *local* (in fact, we will get certain *sequential update* rules).

Remark 2.6.9. The origin of nonlocality in these Markov chains is the conjugation of the skew functions that is necessary for the transition probabilities to add up to 1 (cf. (2.83) and (2.84)). This conjugation may be viewed as an instance of the classical Doob's h-transform (we refer to, e.g., König et al. (2002) and König (2005) for details).

Another way of introducing locality to Markov chains $\mathcal{X}^\circ_{\mathbf{u};\{v_t\}}$ and $\mathcal{X}^+_{\{u_t\};\mathbf{v}}$ that works for generic $\mathbf{u}$ and $\mathbf{v}$, respectively, could be to consider 'multivariate' chains on whole interlacing arrays (similarly to, e.g., O'Connell 2003a, b), O'Connell (2003b); Borodin and Ferrari (2014); Borodin and Petrov (2016b); Matveev and Petrov (2015), with Borodin and Bufetov (2015) providing an application to the six-vertex model on the torus), but we will not discuss this here.

Let us discuss update rules of the dynamics $\mathscr{X}^{\circ}_{0M;\{v_t\}}$ and $\mathscr{X}^{+}_{\{u_t\};\varrho}$ in detail. They follow from (2.89) and (2.90) combined with the interpretation of functions F and G as partition functions of path collections with stochastic vertex weights (2.8).

Dynamics $\mathscr{X}^{\circ}_{0M;\{v_t\}}$

Fix $M \in \mathbb{Z}_{\geq 0}$. During each time step $t \to t+1$ of the chain $\mathscr{X}^{\circ}_{0M;\{v_t\}}$, the current configuration $\mu = 0^{m_0} 1^{m_1} 2^{m_2} \ldots \in \mathsf{Sign}^+_M$ is randomly changed to $\nu = 0^{n_0} 1^{n_1} 2^{n_2} \ldots \in \mathsf{Sign}^+_M$ according to the following sequential (left to right) update. First, choose $n_0 \in \{0, 1, \ldots, m_0\}$ from the probability distribution

$$\mathsf{L}_{v_{t+1}}(m_0, 0; n_0, m_0 - n_0),$$

and set $h_1 := m_0 - n_0 \in \{0, 1\}$, Then, having h_1 and m_1, choose $n_1 \in \{0, 1, \ldots, m_1 + h_1\}$ from the probability distribution

$$\mathsf{L}_{v_{t+1}}(m_1, h_1; n_1, m_1 + h_1 - n_1),$$

and set $h_2 := m_1 + h_1 - n_1 \in \{0, 1\}$. Continue in the same manner for $x = 2, 3, \ldots$ by choosing $n_x \in \{0, 1, \ldots, m_x + h_x\}$ from the distribution

$$\mathsf{L}_{v_{t+1}}(m_x, h_x; n_x, m_x + h_x - n_x),$$

and setting $h_{x+1} := m_x + h_x - n_x \in \{0, 1\}$. Since at each step the probability that $h_{x+1} = 1$ is strictly less than 1, eventually for some $x > \mu_1$ we will have $h_{x+1} = 0$, which means that the update will terminate (all these choices are independent). See Fig. 2.14, top, for an example.

Dynamics $\mathscr{X}^{+}_{\{u_t\};\varrho}$

During each time step $t \to t+1$ of the chain $\mathscr{X}^{+}_{\{u_t\};\varrho}$, the current configuration $\mu = 1^{m_1} 2^{m_2} \ldots \in \mathsf{Sign}^+_M$ is randomly changed to $\nu = 1^{n_1} 2^{n_2} \ldots \in \mathsf{Sign}^+_{M+1}$ according to the following sequential (left to right) update (note that here M is increased with time, and also that there cannot be any particles at location 0).

First, choose $n_1 \in \{0, 1, \ldots, m_1 + 1\}$ from the probability distribution

$$\mathsf{L}_{u_{t+1}}(m_1, 1; n_1, m_1 + 1 - n_1),$$

and set $h_2 := m_1 + 1 - n_1 \in \{0, 1\}$. The fact that $j_1 = 1$ in this stochastic vertex weight accounts for the incoming arrow from 0. For $x = 2, 3, \ldots$ continue in the same way, for each x choosing $n_x \in \{0, 1, \ldots, m_x + h_x\}$ from the probability distribution

$$\mathsf{L}_{u_{t+1}}(m_x, h_x; n_x, m_x + h_x - n_x),$$

and setting $h_{x+1} = m_x + h_x - n_x \in \{0, 1\}$. The update will eventually terminate when $h_{x+1} = 0$ for some $x > \mu_1$. See Fig. 2.14, bottom, for an example.

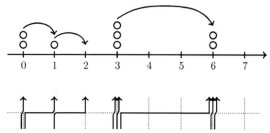

$$\mathsf{L}_v(2,0;1,1)\ \mathsf{L}_v(1,1;1,1)\ \mathsf{L}_v(0,1;1,0)\ \mathsf{L}_v(3,0;2,1)\ \mathsf{L}_v(0,1;0,1)\ \mathsf{L}_v(0,1;0,1)\ \mathsf{L}_v(2,1;3,0)$$

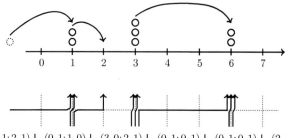

$$\mathsf{L}_u(2,1;2,1)\ \mathsf{L}_u(0,1;1,0)\ \mathsf{L}_u(3,0;2,1)\ \mathsf{L}_u(0,1;0,1)\ \mathsf{L}_u(0,1;0,1)\ \mathsf{L}_u(2,1;3,0)$$

Fig. 2.14 Top: a possible move under the chain $\mathscr{X}^{\circ}_{0^M;\{v_t\}}$ with $M = 8$ (depicted in terms of particle and path configurations). Bottom: a possible move under the chain $\mathscr{X}^{+}_{\{u_t\};\varrho}$ with $M = 7$ (so that the resulting configuration has 8 particles). The probabilities of these moves are also given, where $v = v_{t+1}$ and $u = u_{t+1}$.

Properties of dynamics

We will now list a number of immediate properties of the Markov chains $\mathscr{X}^{\circ}_{0^M;\{v_t\}}$ and $\mathscr{X}^{+}_{\{u_t\};\varrho}$ described earlier.

- Under both dynamics, particles move only to the right. Moreover, at most one particle can leave any given stack of particles and it can move only as far as the next nonempty stack of particles.
- The property that at most one particle can leave any given stack of particles is a $J = 1$ feature. One can readily define *fused dynamics* involving stochastic vertex weights $\mathsf{L}^{(J)}_z$ for any $J \in \mathbb{Z}_{\geq 1}$ (see Section 2.5.2). In these general J dynamics, at most J particles can leave any given stack. One step of a general J dynamics (say, an analogue of $\mathscr{X}^+$) can be thought of as simply combining J steps of the $J = 1$ dynamics with parameters $u_t, qu_t, \ldots, q^{J-1}u_t$. Results of Section 2.5.2 show that one can then forget about the intermediate configurations during these J steps, and still obtain a Markov chain. We will utilize these general J Markov chains in Section 2.6.6.
- If the dynamics $\mathscr{X}^{\circ}_{\mathbf{u};\{v_t\}}$ is started from the initial configuration 0^M (that is, all M particles are at 0), then at any time t the distribution of the particle configuration is given by $\mathscr{M}_{\mathbf{u};(v_1,\ldots,v_t)}$. Similarly, if $\mathscr{X}^{+}_{\{u_t\};\mathbf{v}}$ starts from the empty initial

configuration, then at any time t the distribution of the particle configuration is given by $\mathcal{M}_{(u_1,\ldots,u_t);\mathbf{v}}$. This follows from (2.85).

- Let us return to local dynamics. As follows from (2.86)–(2.87), the eigenfunctions of the transition operator $\mathcal{Q}^\circ_{0^M;v}$ corresponding to the dynamics $\mathscr{X}^\circ$ (on M-particle configurations) are

$$\Psi_\lambda(z_1,\ldots,z_M) = \frac{1}{(q;q)_M(-s)^{|\lambda|}} \, F_\lambda(z_1,\ldots,z_M),$$

with eigenrelations

$$\mathcal{Q}^\circ_{0^M;v}\Psi_\lambda(\mathbf{z}) = \left(\prod_{j=1}^M \frac{1-qz_jv}{1-z_jv} \right) \Psi_\lambda(\mathbf{z})$$

(here and in the following for $\mathbf{u} = 0^M$ we write Ψ_λ instead of $\Psi^{\mathbf{u}}_\lambda$).

- The dynamics $\mathscr{X}^\circ_{0^M;\{v_t\}}$ on M-particle configurations appeared in Corwin and Petrov (2016) (under the name $J = 1$ *higher spin zero range process*), and certain duality relations for it were established in that paper.[15] Some of the results in Corwin and Petrov (2016) also deal with infinite-particle process like $\mathscr{X}^\circ_{0^M;\{v_t\}}$, which starts from the initial configuration $0^\infty 1^0 2^0 \ldots$ (interpreting the zero range process as an exclusion process, this would correspond to the most well-studied *step initial data*). In this case, during each time step $t \to t+1$, one particle can escape the location 0 with probability $\mathsf{L}_{v_{t+1}}(\infty,0;\infty,1) = (-sv_{t+1})/(1-sv_{t+1})$ (note that under (2.66)–(2.67) this number is between 0 and 1). In Section 2.6.6 we will discuss how this initial condition can be obtained by a straightforward limit transition from the dynamics $\mathscr{X}^+_{\{u_t\};\varrho}$. Thus, considering the latter dynamics without this limit transition adds a new boundary condition, under which during each time step a new particle is *always* added at the leftmost location.

2.6.5   Degeneration to the six-vertex model and the ASEP

Here we do not assume that our parameters satisfy (2.66)–(2.67). However, all algebraic statements discussed previously in this part (e.g., Proposition 2.6.4) continue to hold without this assumption—they just become statements about linear operators. Moreover, one can say that these are statements about *formal* Markov operators, i.e., in which the matrix elements sum up to one along each row, but are not necessarily nonnegative.

Observe that taking $s^2 = q^{-I}$ for $I \in \mathbb{Z}_{\geq 1}$ makes the weight

$$\mathsf{L}_u(I,1;I+1,0) = \frac{1-s^2q^I}{1-su}$$

---

[15] Similar duality results also appeared earlier in Borodin et al. (2014); Borodin and Corwin (2015) and Corwin (2014) for q-TASEP and q-Hahn degenerations of the general higher spin six-vertex model.

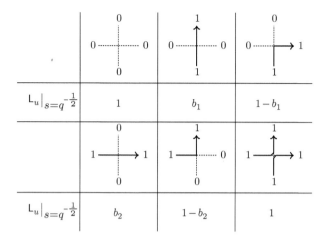

Fig. 2.15 All six stochastic vertex weights corresponding to $s = q^{-\frac{1}{2}}$. The weights $b_{1,2}$ are expressed through u and q as $b_1 = \dfrac{1 - uq^{\frac{1}{2}}}{1 - uq^{-\frac{1}{2}}}$ and $b_2 = \dfrac{-uq^{-\frac{1}{2}} + q^{-1}}{1 - uq^{-\frac{1}{2}}}$.

vanish, regardless of u. If, moreover, all other weights $L_u(i_1, j_1; i_2, j_2)$ with $i_{1,2} \in \{0, 1, \ldots, I\}$ and $j_{1,2} \in \{0, 1\}$ are nonnegative, then we can restrict our attention to path ensembles in which the multiplicities of all vertical edges are bounded by I, and still talk about interacting particle systems as in Section 2.6.4.

Let us consider the simplest case and take $I = 1$, so $s = q^{-\frac{1}{2}}$. For this choice of s, there are six possible arrow configurations at a vertex, and their weights are given in Fig. 2.15. These weights are nonnegative if either $0 < q < 1$ and $u \geq q^{-\frac{1}{2}}$, or $q > 1$ and $0 \leq u \leq q^{-\frac{1}{2}}$ (these are the new nonnegativity conditions replacing (2.66)–(2.67) for $s = q^{-\frac{1}{2}}$). Observe the following symmetry of these degenerate vertex weights:

$$L_u(i_1, j_1; i_2, j_2) = L_{u^{-1}}(1 - i_1, 1 - j_1; 1 - i_2, 1 - j_2)\Big|_{q \to q^{-1}}, \qquad i_1, j_1, i_2, j_2 \in \{0, 1\}. \tag{2.91}$$

These vertex weights define the *stochastic six-vertex model* introduced in Gwa and Spohn (1992) and studied recently in Borodin et al. (2016a). A simulation of a random configuration of the stochastic six-vertex model with boundary conditions considered in Borodin et al. (2016a) is given in Fig. 2.16. Note that the latter paper deals with the stochastic six-vertex model in which the vertical arrows are entering from below, and no arrows enter from the left. Moreover, to get a nontrivial limit shape as in Fig. 2.16 one should take $q > 1$. However, with the help of the symmetry (2.91) (leading to the swapping of arrows with empty edges), these boundary conditions are equivalent to considering the process $\mathscr{X}^+_{\{u_t\}; \varrho}$ with $0 < q < 1$, which is our usual assumption throughout the text. Simulations of the latter dynamics can be obtained from the pictures in Fig. 2.16 by reflecting them with respect to the diagonal of the first quadrant.

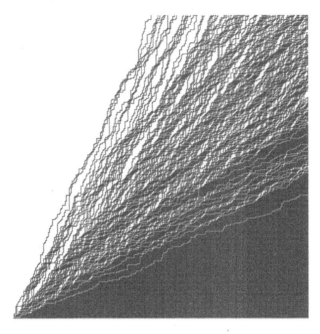

Fig. 2.16 A simulation of the stochastic six-vertex model of size 300 with boundary conditions as in Borodin et al. (2016a) and parameters $\mathsf{L}(0,1;0,1) = 0.3$, $\mathsf{L}(1,0;1,0) = 0.7$.

Let us briefly discuss two continuous time limits of the stochastic six-vertex model. Here we restrict our attention to systems of the type $\mathscr{X}^{\circ}_{0M;\{v_t\}}$, i.e., with a fixed finite number of particles (about other boundary and initial conditions see also Section 2.10.1). The first of the limits is the well-known ASEP introduced in Spitzer (1970) (see Fig. 2.17), which is obtained as follows. Observe that for $u = q^{-\frac{1}{2}} + (1-q)q^{-\frac{1}{2}}\epsilon$, we have as $\epsilon \searrow 0$:

$$\mathsf{L}_u(0,1;0,1) = \epsilon + O(\epsilon^2), \qquad \mathsf{L}_u(1,0;1,0) = q\epsilon + O(\epsilon^2).$$

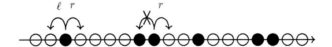

Fig. 2.17 The ASEP is a continuous time Markov chain on particle configurations on $\mathbb{Z}$ (in which there is at most one particle per site). Each particle has two exponential clocks of rates r and ℓ, respectively (all exponential clocks in the process are assumed independent). When the 'r' clock of a particle rings, it immediately tries to jump to the right by one, and similarly for the 'ℓ' clock and left jumps. If the destination of a jump is already occupied, then the jump is blocked. (This describes the ASEP with finitely many particles, but one can also construct the infinite-particle ASEP following, e.g., the graphical method of Harris (1978).)

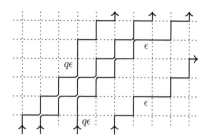

Fig. 2.18 Limit of the six-vertex model to the ASEP.

Therefore, for ϵ small, the particles in the stochastic six-vertex model will mostly travel to the right by 1 at every step. If we subtract this deterministic shift and look at times of order ϵ^{-1}, then the rescaled discrete time process will converge to the continuous time ASEP with $r = 1$ and $\ell = q$; see Fig. 2.18 (note that multiplying both r and ℓ by a constant is the same as a deterministic rescaling of the continuous time in the ASEP, and thus is a harmless operation).

Another continuous time limit is obtained by setting

$$q = \frac{1 - \epsilon}{\alpha}, \qquad u = \frac{\epsilon \alpha^{\frac{1}{2}}}{1 - \alpha},$$

where $0 < \alpha < 1$, so that as $\epsilon \to 0$ we have

$$L_u(0, 1; 0, 1) = \alpha + O(\epsilon^2), \qquad L_u(1, 0; 1, 0) = 1 - \epsilon + O(\epsilon^2).$$

At times of order ϵ^{-1}, the system behaves as follows. Each particle at a location j has an exponential clock with rate 1. When the clock rings, the particle wakes up and performs a jump to the right having the geometric distribution with parameter α. However, if in the process of the jump this particle runs into another particle (i.e., its first neighbour on the right), then the moving particle stops at this neighbour's location, and the neighbour wakes up (and subsequently performs a geometrically distributed jump). See Fig. 2.19.

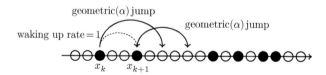

Fig. 2.19 A possible jump in the second limit of the stochastic six-vertex model. The particle at x_k wakes up at rate 1 and decides to jump by 5 with probability $(1 - \alpha)\alpha^4$ (waking up means that the particle will jump by at least 1). However, x_{k+1} is closer than the intended jump of x_k, and so x_k stops at the location of x_{k+1}, and the latter particle wakes up. Then x_{k+1} decides to jump by 4 with probability $(1 - \alpha)\alpha^3$.

2.6.6 Degeneration to q-Hahn and q-Boson systems

We will now consider another family of degenerations of the higher spin six-vertex model which puts no restrictions on the vertical multiplicities. For these degenerations we will need to employ the general J stochastic vertex weights $\mathsf{L}_u^{(J)}(i_1, j_1; i_2, j_2)$ described in Section 2.5.2.

Proposition 2.6.10. When $u = s$, formula (2.71) for the weights $\mathsf{L}_u^{(J)}$ simplifies to the product form

$$\mathsf{L}_s^{(J)}(i_1, j_1; i_2, j_2)$$

$$= \mathbf{1}_{i_1+j_1=i_2+j_2} \cdot \mathbf{1}_{j_2 \leq i_1} \cdot (s^2 q^J)^{j_2} \frac{(q^{-J}; q)_{j_2}(s^2 q^J; q)_{i_1-j_2}}{(s^2; q)_{i_1}} \frac{(q; q)_{i_1}}{(q; q)_{j_2}(q; q)_{i_1-j_2}}. \quad (2.92)$$

Proof To show this, one can directly check that (2.92) satisfies the corresponding recursion relation for $u = s$ (2.70). Alternatively, one can transform the $_4\bar{\varphi}_3$ q-hypergeometric function to the desired form. We refer to Borodin (2014, Prop. 6.7) for the complete proof following the second approach. $\qquad \blacksquare$

We see that this degeneration turns the higher spin interacting particle systems described in Section 2.6.4 with sequential update into simpler systems with parallel update.

Distribution $\phi_{q,\mu,\nu}$

Before discussing interacting particle systems arising from the vertex weights (2.92), let us focus on the q-*deformed beta-binomial distribution* appearing in the right-hand side of that formula:

$$\phi_{q,\mu,\nu}(j \,|\, m) := \mu^j \frac{(\nu/\mu; q)_j (\mu; q)_{m-j}}{(\nu; q)_m} \frac{(q; q)_m}{(q; q)_j (q; q)_{m-j}}, \qquad j \in \{0, 1, \dots, m\}. \quad (2.93)$$

Here $m \in \mathbb{Z}_{\geq 0} \cup \{+\infty\}$, and the case $m = +\infty$ corresponds to a straightforward limit of (2.93); see (2.97). If the parameters belong to one of the following families,

1. $0 < q < 1$, $0 \leq \mu \leq 1$, and $\nu \leq \mu$;
2. $0 < q < 1$, $\mu = q^J \nu$ for some $J \in \mathbb{Z}_{\geq 0}$, and $\nu \leq 0$;
3. m is finite, $q > 1$, $\mu = q^{-J}\nu$ for some $J \in \mathbb{Z}_{\geq 0}$, and $\nu \leq 0$;
4. m is finite, $q > 0$, $\mu = q^{\tilde{\mu}}$, and $\nu = q^{\tilde{\nu}}$ with $\tilde{\mu}, \tilde{\nu} \in \mathbb{Z}$, such that
   * either $\tilde{\mu}, \tilde{\nu} \geq 0$, and $\tilde{\nu} \geq \tilde{\mu}$,
   * or $\tilde{\mu}, \tilde{\nu} \leq 0$, $\tilde{\nu} \leq -m$, and $\tilde{\nu} \leq \tilde{\mu}$,

$\quad (2.94)$

then the weights (2.93) are nonnegative.[16] The conditions (2.94) replace the nonnegativity conditions (2.66)–(2.67) for this degeneration.

---

[16] These are sufficient conditions for nonnegativity, and in fact some of these families intersect nontrivially. We do not attempt to list all the necessary conditions (as, for example, for $q < 0$ there also exist values of μ and ν leading to nonnegative weights).

We will now discuss several interpretations of the distribution (2.93) which, in particular, will justify its name. The significance of the probability distribution $\phi_{q,\mu,\nu}$ for interacting particle systems was first realized in Povolotsky (2013), who showed that it corresponds to the most general 'chipping model' (i.e., a particle system as in Fig. 2.14 with possibly multiple particles leaving a given stack at a time) having parallel update, product-form steady state, and such that the system is solvable by the coordinate Bethe ansatz. He also provided an algebraic interpretation of this distribution.

Proposition 2.6.11 ((Povolotsky 2013, Thm. 1)). Let A and B be two letters satisfying the quadratic commutation relation

$$BA = \alpha A^2 + \beta AB + \gamma B^2, \qquad \alpha + \beta + \gamma = 1.$$

Then

$$\left(pA + (1-p)B\right)^m = \sum_{j=0}^{m} \phi_{q,\mu,\nu}(j \mid m) A^j B^{m-j}, \qquad (2.95)$$

where

$$\alpha = \frac{\nu(1-q)}{1-q\nu}, \qquad \beta = \frac{q-\nu}{1-q\nu}, \qquad \gamma = \frac{1-q}{1-q\nu}, \qquad \mu = p + \nu(1-p).$$

In particular, taking $A = B = 1$ in (2.95) implies that the weights (2.93) sum to 1 over $j = 0, 1, \ldots, m$. The proof of Eq. (2.95) is nontrivial, and we will not reproduce it here.

Another interpretation of the q-deformed beta-binomial distribution can be given via a q-version of the Pólya's urn process due to Gnedin and Olshanski (2009). Consider the Markov chain on the Pascal triangle

$$\bigsqcup_{m=0}^{\infty} \{(k, \ell) \in \mathbb{Z}_{\geq 0}^2 : k + \ell = m\}$$

with the following transition probabilities (here $m = k + \ell$ is the time in this chain)

$$(k, \ell) \quad \xrightarrow{\frac{1-q^{b+\ell}}{1-q^{a+b+m}}} \quad (k, \ell+1)$$

$$(k, \ell) \quad \xrightarrow{q^{\ell+b}\frac{1-q^{a+k}}{1-q^{a+b+m}}} \quad (k+1, \ell)$$

Then the distribution of this Markov chain (started from the initial vertex $(0,0)$) at time m is

$$\mathsf{Prob}\left((k, \ell)\right) = \phi_{q,q^b,q^{a+b}}(k \mid m), \qquad k = 0, 1, \ldots, m.$$

More general Markov chains (on the space of interlacing arrays) based on the distributions $\phi_{q,q^b,q^{a+b}}$ with negative a and b which have a combinatorial significance (they are q-deformations of the classical Robinson–Schensted–Knuth insertion algorithm) were constructed recently in Matveev and Petrov (2015).

Another feature of the distribution $\phi_{q,\mu,\nu}$ is that it is the weight function for the so-called *q-Hahn orthogonal polynomials*. See Koekoek and Swarttouw (1996, §3.6) about the polynomials, and Borodin et al. (2015a, Section 5.2) for the exact matching between $\phi_{q,\mu,\nu}$ and the notation related to the q-Hahn polynomials.

q-Hahn particle system

We will now discuss what the dynamics $\mathscr{X}^{\circ}_{0M;\{v_t\}}$ and $\mathscr{X}^{+}_{\{u_t\};\varrho}$ look like under the degeneration described in Proposition 2.6.10. We will first consider the dynamics $\mathscr{X}^{\circ}_{0M;\{v_t\}}$ which lives on particle configurations with a fixed number of particles (say, $M \in \mathbb{Z}_{\geq 0}$), and then will deal with $\mathscr{X}^{+}_{\{u_t\};\varrho}$. The resulting dynamics will be commonly referred to as the *q-Hahn particle system* with different initial or boundary conditions.

In order to perform the desired degeneration of $\mathscr{X}^{\circ}$, fix $J \in \mathbb{Z}_{\geq 1}$ and take the time-dependent parameters $\{v_t\}$ to be

$$(v_1, v_2, \ldots) = (s, qs, \ldots, q^{J-1}s, s, qs, \ldots, q^{J-1}s, \ldots).$$

We will consider the *fused dynamics* in which one time step corresponds to J steps of the original dynamics. The fused dynamics is Markovian due to the results of Section 2.5.2. As follows from Section 2.6.4, each time step $\mu = 0^{m_0} 1^{m_1} 2^{m_2} \ldots \to \nu = 0^{n_0} 1^{n_1} 2^{n_2} \ldots$ ($\mu, \nu \in \mathsf{Sign}^{+}_M$) of the fused dynamics looks as follows. For each location $x \in \mathbb{Z}_{\geq 0}$, sample $j_x \in \{0, 1, \ldots, m_x\}$ independently of other locations according to the probability distribution $\phi_{q,q^J s^2, s^2}(j_x \mid m_x)$ (clearly, $j_x = m_x = 0$ for all large enough x). Then, in parallel, move j_x particles from location x to location $x + 1$ for each $x \in \mathbb{Z}_{\geq 0}$; that is, set $n_x = m_x - j_x + j_{x+1}$. Denote this dynamics by $\mathscr{X}^{\circ}_{q\text{-Hahn}}$ (see Fig. 2.20, top).

Note that the weights $\phi_{q,q^J s^2, s^2}$ are nonnegative for $J \in \mathbb{Z}_{\geq 1}$ if $s^2 \leq 0$ (case 2 in (2.94)), and we are assuming this in our construction.

Remark 2.6.12. For $J \in \mathbb{Z}_{\geq 1}$, at most J particles can leave any given location during one time step. However, since the weights of the distribution $\phi_{q,q^J s^2, s^2}$ depend on q^J in a rational way, we may analytically continue $\mathscr{X}^{\circ}_{q\text{-Hahn}}$ from the case $q^J \in q^{\mathbb{Z}_{\geq 1}}$ and $s^2 \leq 0$, and let the parameters $\mu = q^J \nu$ and $\nu = s^2$ belong to one of the other families in (2.94). If $\mu/\nu \notin q^{\mathbb{Z}_{\geq 1}}$, then an arbitrary number of particles can leave any given location during one time step.

Let us now discuss the degeneration of the dynamics $\mathscr{X}^{+}_{\{u_t\};\varrho}$. Fix $J \in \mathbb{Z}_{\geq 1}$ and take the time-dependent parameters $\{u_t\}$ to be

$$(u_1, u_2, \ldots) = (s, qs, \ldots, q^{J-1}s, s, qs, \ldots, q^{J-1}s, \ldots). \tag{2.96}$$

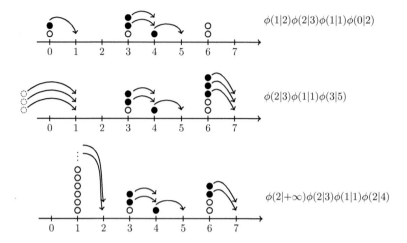

$\phi(1|2)\phi(2|3)\phi(1|1)\phi(0|2)$

$\phi(2|3)\phi(1|1)\phi(3|5)$

$\phi(2|+\infty)\phi(2|3)\phi(1|1)\phi(2|4)$

Fig. 2.20 Possible transitions of the q-Hahn particle system, with probabilities given on the right (here $\phi \equiv \phi_{q,q^J s^2,s^2}$). Top: dynamics $\mathscr{X}^{\circ}_{q\text{-Hahn}}$ living on configurations with a fixed number of particles. Middle: dynamics $\mathscr{X}^{+}_{q\text{-Hahn}}$, in which at each time step J new particles are added at location 1 ($J = 3$ on the picture). Bottom: dynamics $\mathscr{X}^{\infty}_{q\text{-Hahn}}$ with the initial condition $1^{\infty}2^{0}3^{0}\ldots$.

The corresponding fused dynamics is very similar to $\mathscr{X}^{\circ}_{q\text{-Hahn}}$, and the only difference is in the behaviour at locations 0 and 1. Namely, location 0 cannot be occupied, and at each time step, exactly J new particles are added at location 1. Denote this degeneration of $\mathscr{X}^{+}_{\{u_t\};\varrho}$ by $\mathscr{X}^{+}_{q\text{-Hahn}}$ (see Fig. 2.20, middle).

Because $J \in \mathbb{Z}_{\geq 1}$ particles are added to the configuration at each time step, dynamics $\mathscr{X}^{+}_{q\text{-Hahn}}$ cannot be analytically continued in J similarly to Remark 2.6.12. However, we can simplify this dynamics, by generalizing (2.96) to

$$(u_1, u_2, \ldots) = (s, qs, \ldots, q^{K-1}s, s, qs, \ldots, q^{J-1}s, s, qs, \ldots, q^{J-1}s, \ldots),$$

where $K \in \mathbb{Z}_{\geq 1}$ is a new parameter. If we start the corresponding fused dynamics from the empty initial configuration $1^{0}2^{0}\ldots$, then after the first step of this dynamics the configuration will be simply $1^{K}2^{0}3^{0}\ldots$. Moreover, during the evolution, the number of particles at location 1 will always be $\geq K$. Assume that $|q| < 1$, and take $K \to +\infty$. Under the limiting dynamics, at all subsequent times the number of particles leaving location 1 has the distribution

$$\lim_{i_1 \to +\infty} \mathsf{L}_s^{(J)}(i_1, j_1; i_2, j_2) = \mathbf{1}_{i_2 = +\infty} \cdot \phi_{q,q^J s^2, s^2}(j_2 \,|\, +\infty),$$

$$\phi_{q,\mu,\nu}(j \,|\, +\infty) = \mu^j \frac{(\nu/\mu; q)_j}{(q; q)_j} \frac{(\mu; q)_\infty}{(\nu; q)_\infty}. \tag{2.97}$$

The limit as $K \to +\infty$ clearly does not affect probabilities of particle jumps at all other locations. We will denote the limiting dynamics by $\mathscr{X}^{\infty}_{q\text{-Hahn}}$ (see Fig. 2.93, bottom).

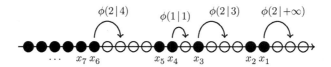

Fig. 2.21 The transition of the q-Hahn system $\mathscr{X}^{\infty}_{q\text{-Hahn}}$ from Fig. 2.20, bottom, interpreted in terms of the q-Hahn TASEP (here $\phi \equiv \phi_{q,q^J s^2, s^2}$).

This particle system was introduced in Povolotsky (2013). The system $\mathscr{X}^{\infty}_{q\text{-Hahn}}$ readily admits an analytic continuation from $q^J \in q^{\mathbb{Z}_{\geq 1}}$ and $s^2 \leq 0$ as in Remark 2.6.12.

Thanks to infinitely many particles at location 1, the system $\mathscr{X}^{\infty}_{q\text{-Hahn}}$ admits another nice particle interpretation. Namely, consider right-finite particle configurations $\{x_1 > x_2 > x_3 > \ldots\}$ in $\mathbb{Z}$, in which there can be at most one particle at a given location. For a configuration $\lambda = 1^{\infty}2^{\ell_2}3^{\ell_3}\ldots$ of $\mathscr{X}^{\infty}_{q\text{-Hahn}}$, let $\ell_j = x_{j-1} - x_j - 1$ (with $x_0 = +\infty$) be the number of empty spaces between consecutive particles. Let the process start from the *step initial configuration* $x_i(0) = -i$ for all i (corresponding to $\lambda = 1^{\infty}2^0 3^0 \ldots$). Then during each time step, each particle x_i jumps to the right according to the distribution $\phi_{q,q^J s^2, s^2}(\cdot \mid \mathrm{gap}_i)$, where $\mathrm{gap}_i = \ell_i$ is the distance to the nearest right neighbor of x_i (the first particle employs the distribution with $\mathrm{gap}_1 = +\infty$). This system is called the *q-Hahn TASEP*, see Fig. 2.21. It was introduced in Povolotsky (2013), and duality relations for this process were obtained in Corwin (2014).

Remark 2.6.13. In all the above q-Hahn systems, one can clearly let the parameter J to depend on time.

q-TASEP and q-Boson

Let us now perform a further degeneration of the q-Hahn TASEP corresponding to the parameters q, μ, and ν, by setting $\mu = q\nu$ (that is, we take $J = 1$, and thus must consider $\nu \leq 0$). Then (2.93) implies that $\phi_{q,\mu,\nu}(j \mid m)$ vanishes unless $j \leq 1$, and

$$\phi_{q,\mu,\nu}(0 \mid m) = \frac{1 - q^m \nu}{1 - \nu}, \qquad\qquad \phi_{q,\mu,\nu}(1 \mid m) = \frac{-\nu(1 - q^m)}{1 - \nu},$$

$$\phi_{q,\mu,\nu}(0 \mid +\infty) = \frac{1}{1 - \nu}, \qquad\qquad \phi_{q,\mu,\nu}(1 \mid +\infty) = \frac{-\nu}{1 - \nu}.$$

Taking $\nu = -\epsilon$ and speeding up the time by ϵ^{-1}, we arrive at the *q-TASEP*—a continuous time particle system on configurations $\{x_1 > x_2 > x_3 > \ldots\}$ on $\mathbb{Z}$ (with no more than one particle per location) in which each particle x_i jumps to the right by 1 at rate $1 - q^{\mathrm{gap}_i}$, where, as before, $\mathrm{gap}_i = x_{i-1} - x_i - 1$ is the distance to the right neighbour of x_i.

The q-TASEP was introduced in Borodin and Corwin (2014) (see also Borodin et al. (2014)), and an 'arrow' interpretation of the q-TASEP (on configurations $\lambda = 1^{\infty}2^{\mathrm{gap}_2}3^{\mathrm{gap}_3}\ldots$) had been considered much earlier (Bogoliubov et al. 1994, 1998; Sasamoto and Wadati 1998) under the name of the *(stochastic) q-Boson system*.

It is worth noting that the q-Hahn system has a variety of other degenerations; see Povolotsky 2013 and Barraquand and Corwin (2016) for examples.

2.7 Orthogonality relations

In this part we recall two types of (bi)orthogonality relations for the symmetric rational functions F_λ (from Section 2.4) which first appeared in Borodin et al. (2015a). These relations imply certain Plancherel isomorphism theorems. We also apply biorthogonality to get an integral representation for the functions G_μ. These results provide us with tools which will eventually make it possible to explicitly evaluate averages of certain observables of the interacting particle systems described in Section 2.6.

2.7.1 Spatial biorthogonality

First, we will need the following general statement.

Lemma 2.7.1. *Let $\{f_m(u)\}$, $\{g_\ell(u)\}$ be two families of rational functions in $u \in \mathbb{C}$ such that there exist two disjoint finite sets $P_1, P_2 \subset \mathbb{C} \cup \{\infty\}$ and positively oriented pairwise nonintersecting closed contours $c_1, \ldots, c_k$ with the following properties:*

- *All singularities of all the functions $f_m(u), g_\ell(u)$ lie inside $P_1 \cup P_2$.*
- *The product $f_m(u)g_\ell(u)$ does not have singularities in P_1 if $m < \ell$, and the same product does not have singularities in P_2 if $m > \ell$.*
- *For any $i > j$ the contour c_i can be shrunk to P_1 without intersecting the contour[17] $q^{-1}c_j$ (equivalently, for any $j < i$ the contour c_j can be shrunk to P_2 without intersecting $q \cdot c_i$). Shrinking takes place on the Riemann sphere $\mathbb{C} \cup \{\infty\}$.*

Fix $k \in \mathbb{Z}_{\geq 1}$ and two signatures $\mu, \lambda \in \mathsf{Sign}_k^+$. If $\mu \neq \lambda$, then for any permutation $\sigma \in \mathcal{S}_k$ we have

$$\oint_{c_1} du_1 \ldots \oint_{c_k} du_k \prod_{1 \leq \alpha < \beta \leq k} \left(\frac{u_\alpha - u_\beta}{u_\alpha - qu_\beta} \cdot \frac{u_{\sigma(\alpha)} - qu_{\sigma(\beta)}}{u_{\sigma(\alpha)} - u_{\sigma(\beta)}} \right) \prod_{j=1}^k f_{\mu_j}(u_j)g_{\lambda_{\sigma^{-1}(j)}}(u_j) = 0.$$

$$(2.98)$$

Proof This is a straightforward generalization of Lemma 3.5 in Borodin et al. (2015a). Let us outline the steps of the proof.

We will assume that the integral (2.98) is nonzero, and will show that then it must be that $\lambda = \mu$. First, we observe that, by our assumed structure of the poles,

- if it is possible to shrink the contour c_i to P_1, then for the integral to be nonzero we must have $\mu_i \geq \lambda_{\sigma^{-1}(i)}$.
- if it is possible to shrink the contour c_i to P_2, then for the integral to be nonzero we must have $\mu_i \leq \lambda_{\sigma^{-1}(i)}$.

---

[17] Here and in the following $R\gamma$ denotes the image of the contour γ under the multiplication by R.

Next, using

$$\prod_{1\le\alpha<\beta\le k}\frac{u_\alpha-u_\beta}{u_\alpha-qu_\beta}\cdot\frac{u_{\sigma(\alpha)}-qu_{\sigma(\beta)}}{u_{\sigma(\alpha)}-u_{\sigma(\beta)}}=\mathrm{sgn}(\sigma)\prod_{\alpha<\beta\,:\,\sigma(\alpha)>\sigma(\beta)}\frac{u_{\sigma(\alpha)}-qu_{\sigma(\beta)}}{u_\alpha-qu_\beta}\qquad(2.99)$$

and the properties of the integration contours, we see that

- if for some $i\in\{1,\ldots,k\}$ one has $\sigma(i)>\max(\sigma(1),\ldots,\sigma(i-1))$, then the numerator in the left-hand side of (2.99) contains all terms of the form $(u_{\sigma(i)}-qu_{\sigma(i)+1}),(u_{\sigma(i)}-qu_{\sigma(i)+2}),\ldots,(u_{\sigma(i)}-qu_k)$, and thus the expression in the right-hand side of (2.99) does not have poles at $u_{\sigma(i)}=qu_j$ for all $j>\sigma(i)$. This means that we can shrink the contour $c_{\sigma(i)}$ to P_1 without picking any residues. This implies that for the integral (2.98) to be nonzero, we must have $\mu_{\sigma(i)}\ge\lambda_i$.
- similarly, if for some $i\in\{1,\ldots,k\}$ one has $\sigma(i)<\min(\sigma(i+1),\ldots,\sigma(k))$, then the contour $c_{\sigma(i)}$ can be shrunk to P_2, and so the integral (2.98) can be nonzero only if $\mu_{\sigma(i)}\le\lambda_i$.

This completes the argument analogous to Step I of the proof of Borodin et al. (2015a, Lemma 3.5). Further steps of the proof have a purely combinatorial nature and can be repeated without change. We will not reproduce the full combinatorial argument here, but will illustrate it on a concrete example.

Take $\sigma=(3,2,4,5,1,8,6,7)$, and consider an arrow diagram as in Fig. 2.22. That is, think of the bottom row as $\lambda_1\ge\lambda_2\ge\ldots\ge\lambda_8$ and of the top row as $\mu_1\ge\mu_2\ge\ldots\ge\mu_8$, and draw the corresponding horizontal arrows from larger to smaller integers. Labels in the nodes correspond to the permutation σ itself.

As we read this permutation from left to right, we see running maxima $\sigma(1)$, $\sigma(3)>\max(\sigma(1),\sigma(2))$, $\sigma(4)>\max(\sigma(1),\sigma(2),\sigma(3))$, and $\sigma(6)>\max(\sigma(1),\sigma(2),\sigma(3),\sigma(4),\sigma(5))$, and correspondingly we add dashed arrows $\mu_{\sigma(i)}\to\lambda_i$. Similarly, reading the permutation from right to left, we see running minima $\sigma(8)$, $\sigma(7)<\sigma(8)$, and $\sigma(5)<\min(\sigma(6),\sigma(7),\sigma(8))$, and we add solid arrows $\lambda_i\to\mu_{\sigma(i)}$. As one examines the arrow diagram, it becomes obvious that for the integral to be nonzero, we must have

$$\lambda_1=\ldots=\lambda_5=\mu_1=\ldots=\mu_5,\qquad\lambda_6=\lambda_7=\lambda_8=\mu_6=\mu_7=\mu_8.$$

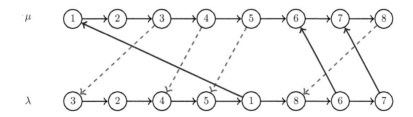

Fig. 2.22 Arrow diagram for $\sigma=(3,2,4,5,1,8,6,7)$.

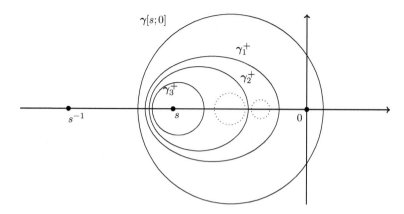

Fig. 2.23 A possible choice of nested integration contours $\gamma_i^+ = \gamma_i^+[s]$, $i = 1, 2, 3$ (Definition 2.7.2). Contours $q\gamma_3^+[s]$ and $q^2\gamma_3^+[s]$ are shown dotted. The large contour $\gamma[s; 0]$ is also shown.

We also see that any permutation σ yielding a nonzero integral (2.98) splits into two blocks permuting $\{1, 2, 3, 4, 5\}$ and $\{5, 6, 7\}$, which are the clusters of equal parts in λ and μ. A similar clustering occurs in the general situation, too.

This implies that for the integral (2.98) to be nonzero, we must have $\lambda = \mu$, as desired.

For the orthogonality statements in the following we assume that (2.66) holds.

Definition 2.7.2.   *For any $k \geq 1$, let $\gamma_1^+[s], \dots, \gamma_k^+[s]$ be positively oriented closed contours such that (see Fig. 2.23)*

- *each contour $\gamma_\alpha^+[s]$ encircles all the points of the set $P_1 := \{s\}$, while leaving outside all the points of $P_2 := \{s^{-1}\}$.*
- *for any $\beta > \alpha$, the interior of $\gamma_\alpha^+[s]$ contains the contour $q\gamma_\beta^+[s]$.*
- *the contour $\gamma_k^+[s]$ is sufficiently small so that it does not intersect with $q\gamma_k^+[s]$.*

Also, let $\gamma[s; 0]$ be a positively oriented closed contour encircling $s, qs, \dots, q^{k-1}s$, which also contains $q\gamma[s; 0]$ and leaves s^{-1} outside. Note that 0 must be inside this contour.

Remark 2.7.3.   The superscript '+' in the contours of Definition 2.7.2 refers to the property that they are q-nested, as opposed to q^{-1}-nested contours $\gamma_j^-[\bullet]$ which we will consider later in Section 2.8. Throughout the text points encircled by a contour will be explicitly indicated in the square brackets.

Theorem 2.7.4. For any $k \in \mathbb{Z}_{\geq 1}$ and $\lambda, \mu \in \mathrm{Sign}_k^+$, we have

$$
\oint_{\gamma_1^+[s]} \frac{du_1}{2\pi\mathrm{i}} \cdots \oint_{\gamma_k^+[s]} \frac{du_k}{2\pi\mathrm{i}} \prod_{1 \leq \alpha < \beta \leq k} \frac{u_\alpha - u_\beta}{u_\alpha - qu_\beta}
$$

$$
\times \, \mathsf{F}_\lambda^\mathsf{c}(u_1, \ldots, u_k) \prod_{i=1}^{k} \frac{1}{u_i - s} \left(\frac{1 - su_i}{u_i - s} \right)^{\mu_i} = \mathbf{1}_{\lambda = \mu}, \tag{2.100}
$$

where F_λ is as in Section 2.4.5.

An immediate corollary of Theorem 2.7.4 is the following 'spatial' biorthogonality property of the functions F_μ.

Corollary 2.7.5. For any $k \in \mathbb{Z}_{\geq 1}$ and $\lambda, \mu \in \mathrm{Sign}_k^+$, we have

$$
\frac{\mathsf{c}(\lambda)}{(1-q)^k k!} \oint_{\gamma[s;0]} \frac{du_1}{2\pi\mathrm{i}u_1} \cdots \oint_{\gamma[s;0]} \frac{du_k}{2\pi\mathrm{i}u_k} \prod_{1 \leq \alpha \neq \beta \leq k} \frac{u_\alpha - u_\beta}{u_\alpha - qu_\beta}
$$

$$
\times \, \mathsf{F}_\lambda(u_1, \ldots, u_k) \mathsf{F}_\mu(u_1^{-1}, \ldots, u_k^{-1}) = \mathbf{1}_{\lambda = \mu}, \tag{2.101}
$$

with $\mathsf{c}(\cdot)$ as in (2.32).

We call this property *spatial biorthogonality* because for, say, fixed λ and varying μ, the right-hand side is the delta function in the *spatial* variables μ. This should be compared to the *spectral biorthogonality* of Theorem 2.7.11 with delta functions in spectral variables in the right-hand side.

Theorem 2.7.4 and Corollary 2.7.5 were conjectured in Povolotsky (2013) and proved in Borodin et al. (2015a, Section 3) (see also Borodin (2014, Thm. 7.2)). Here we present an outline of the proof in which the abstract combinatorial part (Lemma 2.7.1) is separated from a concrete form of integration contours. This makes the proof applicable in more general situations which will be discussed in a subsequent publication.

Proof of Corollary 2.7.5 Assuming Theorem 2.7.4, deform the integration contours $\gamma_1^+[s], \ldots, \gamma_k^+[s]$ (in this order) to $\gamma[s;0]$. One readily sees that this does not lead to any additional residues. Next, observe that the integral over all $u_j \in \gamma[s;0]$ is invariant under permutations of the u_j's, and thus one can perform the symmetrization and divide by $k!$. This leads to

$$
\mathbf{1}_{\lambda = \mu} = \frac{(1-q)^{-k}}{k!} \oint_{\gamma[s;0]} \frac{du_1}{2\pi\mathrm{i}u_1} \cdots \oint_{\gamma[s;0]} \frac{du_k}{2\pi\mathrm{i}u_k} \prod_{1 \leq \alpha \neq \beta \leq k} \frac{u_\alpha - u_\beta}{u_\alpha - qu_\beta} \mathsf{F}_\lambda^\mathsf{c}(u_1, \ldots, u_k)
$$

$$
\times \underbrace{\sum_{\sigma \in S_k} \prod_{1 \leq \beta < \alpha \leq k} \frac{u_{\sigma(\alpha)} - qu_{\sigma(\beta)}}{u_{\sigma(\alpha)} - u_{\sigma(\beta)}} \prod_{i=1}^{k} \frac{1-q}{1 - su_{\sigma(i)}^{-1}} \left(\frac{1 - su_{\sigma(i)}}{u_{\sigma(i)}^{-1} - s} \right)^{\mu_i}}_{\mathsf{F}_\mu(u_1^{-1}, \ldots, u_k^{-1}) \text{ by } (2.44)},
$$

as desired. $\qquad\square$

Proof of Theorem 2.7.4   Fix k and signatures λ and μ. In order to apply Lemma 2.98, set

$$f_m(u) := \frac{1-q}{u-s}\left(\frac{1-su}{u-s}\right)^m, \qquad g_\ell(u) := \frac{1-q}{1-su}\left(\frac{u-s}{1-su}\right)^\ell,$$

and use $P_1 = \{s\}$ and $P_2 = \{s^{-1}\}$. If $m > \ell$, then all singularities of $f_m(u)g_\ell(u)$ are in P_1, and if $m < \ell$, then all singularities of this product are in P_2. By virtue of the symmetrization formula for F_λ (2.44), we see that the integrand in (2.100) is (up to a multiplicative constant) the same as that in Lemma 2.7.1 (with the above specialization of f_m and g_ℓ). The structure of our integration contours $\gamma_j^+[s]$ and the fact that $\infty \notin P_1 \cup P_2$ (so the contours can be dragged through infinity without picking the residues) implies that the third condition of Lemma 2.7.1 is also satisfied. Thus, we conclude that the integral in (2.100) vanishes unless $\mu = \lambda$.

Example 2.7.6.   To better illustrate the application of Lemma 2.7.1 here, consider contours $\gamma_1^+[s], \gamma_2^+[s]$, and $\gamma_3^+[s]$ as in Fig. 2.23. Depending on σ, the denominator in the integrand in (2.98) contains some of the factors $u_1 - qu_2$, $u_1 - qu_3$, and $u_2 - qu_3$. The contour $\gamma_3^+[s]$ can always be shrunk to P_1 without picking residues at $u_3 = q^{-1}u_1$ and $u_3 = q^{-1}u_2$ (this is the assumption on the contours in Lemma 2.7.1). Moreover, if, for example, the permutation σ provides a cancellation of the factor $u_2 - qu_3$ in the denominator, then the contour $\gamma_2^+[s]$ can also be shrunk to P_1 without picking the residue at $u_2 = qu_3$. Similarly, the contour $\gamma_1^+[s]$ can always be expanded ('shrunk' on the Riemann sphere) to P_2 (recall that infinity does not supply any residues), and if the factor $u_1 - qu_2$ in the denominator is cancelled for a certain σ, then $\gamma_2^+[s]$ can also be expanded to P_2 without picking the residue at $u_2 = q^{-1}u_1$.

Now we must consider the case when $\mu = \lambda$; that is, evaluate the 'squared norm' of F_μ. Arguing similarly to the example in the proof of Lemma 2.98 (see Borodin et al. (2015a, Lemma 3.5) for more detail), we see that the integral

$$\sum_{\sigma \in S_k} \oint_{\gamma_1^+[s]} \frac{du_1}{2\pi\mathrm{i}} \cdots \oint_{\gamma_k^+[s]} \frac{du_k}{2\pi\mathrm{i}} \prod_{1 \le \alpha < \beta \le k} \left(\frac{u_\alpha - u_\beta}{u_\alpha - qu_\beta} \frac{u_{\sigma(\alpha)} - qu_{\sigma(\beta)}}{u_{\sigma(\alpha)} - u_{\sigma(\beta)}}\right)$$

$$\times \prod_{i=1}^k \frac{1}{(u_i - s)(1 - su_i)}\left(\frac{u_i - s}{1 - su_i}\right)^{\lambda_{\sigma^{-1}(i)} - \lambda_i} \tag{2.102}$$

(this is the same as the left-hand side of (2.100) with $\mu = \lambda$, up to the constant $(1-q)^k c(\lambda)$) vanishes unless $\sigma \in S_k$ permutes within clusters of the signature λ; that is, σ must preserve each maximal set of indices $\{a, a+1, \ldots, b\} \subseteq \{1, \ldots, k\}$ for which $\lambda_a = \lambda_{a+1} = \ldots = \lambda_b$. Let c_λ be the number of such clusters in λ. Denote the set of all permutations permuting within clusters of λ by $S_k^{(\lambda)}$. Any permutation $\sigma \in S_k^{(\lambda)}$ can be represented as a product of c_λ permutations $\sigma_1, \ldots, \sigma_{c_\lambda}$, with each σ_i fixing all elements of $\{1, \ldots, k\}$ except those belonging to the ith cluster of λ. We will denote the set of indices within the ith cluster by $C_i(\lambda)$, and write $\sigma_i \in S_k^{(\lambda,i)}$. For example,

if $k = 5$ and $\lambda_1 = \lambda_2 = \lambda_3 > \lambda_4 = \lambda_5$, then $c_\lambda = 2$, and the permutation $\sigma \in \mathcal{S}_k^{(\lambda)}$ must stabilize the sets $C_1(\lambda) = \{1, 2, 3\}$ and $C_2(\lambda) = \{4, 5\}$; $\mathcal{S}_k^{(\lambda,1)}$ is the set of all permutations of $\{1, 2, 3\}$, and $\mathcal{S}_k^{(\lambda,2)}$ is the set of all permutations of $\{4, 5\}$.

Therefore, the sum in (2.102) is only over $\sigma \in \mathcal{S}_k^{(\lambda)}$. Let us now compute it. We have

$$\sum_{\sigma \in \mathcal{S}_k^{(\lambda)}} \prod_{1 \le \alpha < \beta \le k} \frac{u_{\sigma(\alpha)} - q u_{\sigma(\beta)}}{u_{\sigma(\alpha)} - u_{\sigma(\beta)}} \prod_{i=1}^k \frac{1}{(u_i - s)(1 - su_i)} \left(\frac{u_i - s}{1 - su_i} \right)^{\lambda_{\sigma^{-1}(i)} - \lambda_i}$$

$$= \prod_{r=1}^k \frac{1}{(u_r - s)(1 - su_r)} \cdot \prod_{1 \le i < j \le c_\lambda} \prod_{\substack{\alpha \in C_i(\lambda) \\ \beta \in C_j(\lambda)}} \frac{u_\alpha - q u_\beta}{u_\alpha - u_\beta}$$

$$\times \prod_{i=1}^{c_\lambda} \sum_{\sigma_i \in \mathcal{S}_k^{(\lambda,i)}} \prod_{\substack{1 \le \alpha < \beta \le k \\ \alpha, \beta \in C_i(\lambda)}} \frac{u_{\sigma_i(\alpha)} - q u_{\sigma_i(\beta)}}{u_{\sigma_i(\alpha)} - u_{\sigma_i(\beta)}}.$$

For each sum over $\sigma_i \in \mathcal{S}_k^{(\lambda,i)}$ we can use the symmetrization identity (footnote 7). Thus, (2.102) becomes

$$\sum_{\sigma \in \mathcal{S}_k} \oint_{\gamma_1^+[s]} \frac{du_1}{2\pi i} \cdots \oint_{\gamma_k^+[s]} \frac{du_k}{2\pi i} \prod_{\alpha < \beta} \left(\frac{u_\alpha - u_\beta}{u_\alpha - q u_\beta} \frac{u_{\sigma(\alpha)} - q u_{\sigma(\beta)}}{u_{\sigma(\alpha)} - u_{\sigma(\beta)}} \right)$$

$$\times \prod_{i=1}^k \frac{1}{(u_i - s)(1 - su_i)} \left(\frac{u_i - s}{1 - su_i} \right)^{\lambda_{\sigma^{-1}(i)} - \lambda_i}$$

$$= (1-q)^{-k} \prod_{i \ge 0} (q; q)_{\ell_i} \oint_{\gamma_1^+[s]} \frac{du_1}{2\pi i} \cdots \oint_{\gamma_k^+[s]} \frac{du_k}{2\pi i} \prod_{i=1}^{c_\lambda} \prod_{\substack{1 \le \alpha < \beta \le k \\ \alpha, \beta \in C_i(\lambda)}} \frac{u_\alpha - u_\beta}{u_\alpha - q u_\beta}$$

$$\times \prod_{r=1}^k \frac{1}{(u_r - s)(1 - su_r)},$$

where we have used the usual multiplicative notation $\lambda = 0^{\ell_0} 1^{\ell_1} 2^{\ell_2} \dots$. The integration variables above corresponding to each cluster are now independent, and thus the integral reduces to a product of c_λ smaller nested contour integrals of similar form. Each of these smaller integrals can be computed as[18]

$$\oint_{\gamma_1^+[s]} \frac{du_1}{2\pi i} \oint_{\gamma_2^+[s]} \frac{du_2}{2\pi i} \cdots \oint_{\gamma_\ell^+[s]} \frac{du_\ell}{2\pi i} \prod_{1 \le \alpha < \beta \le \ell} \frac{u_\alpha - u_\beta}{u_\alpha - q u_\beta} \prod_{i=1}^\ell \frac{1}{(u_i - s)(1 - su_i)}$$

---

[18] In fact, a more general integral of this sort can also be computed; see Borodin et al. (2015a), Prop. 3.7).

$$= \frac{1}{1-s^2} \oint_{\gamma_2^+[s]} \frac{du_2}{2\pi\mathrm{i}} \cdots \oint_{\gamma_\ell^+[s]} \frac{du_\ell}{2\pi\mathrm{i}} \prod_{j=2}^{\ell} \frac{1-su_j}{1-qsu_j}$$

$$\times \prod_{2\le\alpha<\beta\le\ell} \frac{u_\alpha - u_\beta}{u_\alpha - qu_\beta} \prod_{i=2}^{\ell} \frac{1}{(u_i - s)(1 - su_i)}$$

$$= \frac{1}{1-s^2} \oint_{\gamma_2^+[s]} \frac{du_2}{2\pi\mathrm{i}} \cdots \oint_{\gamma_\ell^+[s]} \frac{du_\ell}{2\pi\mathrm{i}} \prod_{2\le\alpha<\beta\le\ell} \frac{u_\alpha - u_\beta}{u_\alpha - qu_\beta} \prod_{i=2}^{\ell} \frac{1}{(u_i - s)(1 - qsu_i)}$$

$$= \text{etc.}$$

$$= \frac{1}{(s^2; q)_\ell}.$$

Indeed, there is only one u_1-pole outside the contour $\gamma_1^+[s]$, namely, $u_1 = s^{-1}$. Evaluating the integral over u_1 by taking the minus residue at $u_1 = s^{-1}$ leads to a smaller similar integral with the outside pole s^{-1} replaced by $(qs)^{-1}$. Continuing in the same way with integration over $u_2, \dots, u_\ell$, we obtain $1/(s^2; q)_\ell$. Putting together all of these components, we see that we have established the desired claim. □

2.7.2 Plancherel isomorphisms and completeness

Here we discuss Plancherel isomorphism results related to the functions F_λ. This material is essentially a citation from Borodin et al. (2015a) and Borodin (2014).

Let us fix the number of variables $n \in \mathbb{Z}_{\ge 1}$ and assume (2.66), as usual. Extend the definition of the functions $F_\lambda(u_1, \dots, u_n)$ to all $\lambda \in \mathsf{Sign}_n$ using the same symmetrization formula (2.44). These functions also satisfy the shifting property (2.46).

Definition 2.7.7 (Function spaces). *Denote the space of functions $f(\lambda)$ on Sign_n with finite support by $\mathcal{W}^n$. Also, denote by $\mathcal{C}^n$ the space of symmetric Laurent polynomials $R(u_1, \dots, u_n)$ in the variables $\frac{u_i - s}{1 - su_i}$, $i = 1, \dots, n$, with the additional requirement that $R(u_1, \dots, u_n)$ converges to zero as $|u_i| \to \infty$ for any i.*

Note that as functions in λ, the $F_\lambda(u_1, \dots, u_n)$'s clearly do not belong to $\mathcal{W}^n$. However, as functions in the u_i's, they belong to $\mathcal{C}^n$, because we can use

$$\frac{1}{1-su} = \frac{1}{1-s^2} + \frac{s}{1-s^2}\frac{u-s}{1-su}$$

to rewrite the prefactor in (2.44) containing $\prod_{i=1}^n (1 - su_i)^{-1}$ as a Laurent polynomial in the desired variables. (A Laurent polynomial in $\frac{u_i - s}{1 - su_i}$ is clearly bounded at infinity, and this simple computation shows that it can decay, too.) The sum over σ readily produces a symmetric Laurent polynomial in $\frac{u_i - s}{1 - su_i}$ because all the factors in the denominator of the form $u_\alpha - u_\beta$ cancel out. Finally, $F_\lambda(u_1, \dots, u_n)$ clearly goes to 0 as $|u_i| \to \infty$.

Definition 2.7.8 (Plancherel transforms). *The direct transform $\mathscr{F}$ maps a function f from $\mathcal{W}^n$ to $\mathscr{F}f \in \mathcal{C}^n$ and acts as*

$$(\mathscr{F}f)(u_1,\ldots,u_n) := \sum_{\lambda \in \mathsf{Sign}_n} f(\lambda)\mathsf{F}_\lambda(u_1,\ldots,u_n).$$

The inverse transform $\mathscr{J}$ takes $R \in \mathcal{C}^n$ to $\mathscr{J}R \in \mathcal{W}^n$ and acts as

$$(\mathscr{J}R)(\lambda) := \mathsf{c}(\lambda) \oint_{\gamma_1^+[s]} \frac{du_1}{2\pi\mathrm{i}} \cdots \oint_{\gamma_n^+[s]} \frac{du_n}{2\pi\mathrm{i}} \prod_{1 \le \alpha < \beta \le n} \frac{u_\alpha - u_\beta}{u_\alpha - q u_\beta}$$

$$\times R(u_1,\ldots,u_n) \prod_{i=1}^{n} \frac{1}{u_i - s} \left(\frac{1 - su_i}{u_i - s} \right)^{\lambda_i},$$

where $\mathsf{c}(\lambda)$ is defined by (2.32). Let us explain why $\mathscr{J}R$ has finite support in λ. If $\lambda_1 \ge M$ for sufficiently large $M > 0$, then the integrand has no pole at s^{-1} outside $\gamma_1^+[s]$, and thus vanishes. (It is crucial that R vanishes at $u_1 = \infty$, so that the integrand has no residue at $u_1 = \infty$.) Similarly, if $\lambda_n \le -M$, then there is no u_n-pole at s inside $\gamma_n^+[s]$, and so the integral also vanishes. Clearly, the bound M depends on the function R.

Remark 2.7.9.   Similarly to Borodin et al. (2015a, Proposition 3.2), the nested contours in the transform $\mathscr{J}$ can be replaced by two different families of identical contours, which makes it possible to symmetrize under the integral and interpret $\mathscr{J}R$ as a bilinear pairing between R and $\mathsf{F}_\lambda^c(u_1^{-1},\ldots,u_n^{-1})$. One of the choices of these identical contours is $\gamma[s;0]$, cf. Corollary 2.7.5. Another one is the small contour $\gamma_1^+[s]$ around s, but the formula for $\mathscr{J}$ would then involve string specializations of the u_i's. We refer to Borodin et al. (2015a, b) for details.

The following result Borodin et al. (2015a, Theorems 3.4 and 3.9) shows that the transforms $\mathscr{F}$ and $\mathscr{J}$ are mutual inverses:

Theorem 2.7.10 (Plancherel isomorphisms).   The operator $f \mapsto \mathscr{J}(\mathscr{F}f)$ acts as the identity on $\mathcal{W}^n$. The operator $R \mapsto \mathscr{F}(\mathscr{J}R)$ acts as the identity on $\mathcal{C}^n$.

The first statement is clearly equivalent to Theorem 2.7.4 (note that by (2.46), identities (2.100) and (2.101) are invariant under simultaneous shifts in λ and μ, and thus also hold for all $\lambda, \mu \in \mathsf{Sign}_n$). The second statement follows from the *spectral biorthogonality* which can be established independently.

Theorem 2.7.11.   The functions F_λ satisfy the spectral biorthogonality relation

$$\sum_{\lambda \in \text{Sign}_n} \oint \cdots \oint \frac{d\mathbf{u}}{(2\pi\mathbf{i})^n} \oint \cdots \oint \frac{d\mathbf{v}}{(2\pi\mathbf{i})^n} \varphi(\mathbf{u})\psi(\mathbf{v})$$

$$\times \, \mathsf{F}_\lambda(\mathbf{u})\mathsf{F}^{\mathsf{c}}_\lambda(\bar{\mathbf{v}}) \prod_{1 \le \alpha < \beta \le n} (u_\alpha - u_\beta)(v_\alpha - v_\beta)$$

$$= (-1)^{\frac{n(n-1)}{2}} \oint \cdots \oint \frac{d\mathbf{u}}{(2\pi\mathbf{i})^n} \prod_{1 \le \alpha, \beta \le n} (u_\alpha - qu_\beta) \cdot \varphi(\mathbf{u}) \sum_{\sigma \in \mathcal{S}_n} \text{sgn}(\sigma)\psi(\sigma\mathbf{u}),$$

where φ and ψ are test functions, and each integration is performed over one and the same sufficiently small contour encircling s while leaving s^{-1} outside.[19]

Informally, the spectral biorthogonality can be written as

$$\prod_{1 \le \alpha < \beta \le n} (u_\alpha - u_\beta)(v_\alpha - v_\beta) \sum_{\lambda \in \text{Sign}_n} \mathsf{F}_\lambda(\mathbf{u})\mathsf{F}^{\mathsf{c}}_\lambda(\bar{\mathbf{v}})$$

$$= (-1)^{\frac{n(n-1)}{2}} \prod_{1 \le \alpha, \beta \le n} (u_\alpha - qu_\beta) \cdot \det[\delta(v_i - u_j)]^n_{i,j=1}.$$

Two different ways of making Theorem 2.7.11 precise (by choosing suitable classes of test functions φ and ψ) are presented in Borodin et al. (2015a, Section 4) and Borodin (2014, Section 7), along with two corresponding proofs. These proofs differ significantly, and the proof in the latter paper directly utilizes the Cauchy identity (Proposition 2.4.11).

Example 2.7.12. Let us illustrate the connections between spatial and spectral biorthogonality statements and the Cauchy identity in the simplest one-variable case. For that, let us consider the following variant of the Cauchy identity:

$$\sum_{n=0}^{\infty} \frac{z^n}{w^{n+1}} = \frac{1}{w - z}, \qquad \left|\frac{z}{w}\right| < 1.$$

By shifting the summation index towards $-\infty$, we can write

$$\sum_{n=-M}^{\infty} \frac{z^n}{w^{n+1}} = \frac{w^M}{z^M} \frac{1}{w - z}, \qquad \left|\frac{z}{w}\right| < 1.$$

Now take contour integrals in z and w (over positively oriented circles with $|z| < |w|$) of both sides of this relation multiplied by $P(z)Q(w)$, where P and Q are Laurent polynomials. Then in the left-hand side we obtain the same sum for any $M \gg 1$, and

---

[19] Throughout the text we will use the abbreviated notation $\bar{\mathbf{v}} = (v_1^{-1}, \ldots, v_n^{-1})$, and $d\mathbf{v}$ stands for $dv_1 \ldots dv_n$. Similarly for $\bar{\mathbf{u}}$ and $d\mathbf{u}$.

in the right-hand side the w contour can be shrunk to 0, thus picking the residue at $w = z$. Therefore, we have

$$\sum_{n=-\infty}^{\infty} \oint \oint \frac{z^n}{w^{n+1}} P(z)Q(w) \frac{dz}{2\pi i} \frac{dw}{2\pi i} = \oint P(z)Q(z) \frac{dz}{2\pi i}.$$

The convergence condition $|z| < |w|$ is irrelevant for the left-hand side because the sum over n now contains only finitely many terms. The resulting spectral biorthogonality can be informally written as

$$\sum_{n=-\infty}^{\infty} \frac{z^n}{w^{n+1}} = \delta(w - z).$$

To get the other biorthogonality relation, integrate both sides the earlier identity against $w^m \frac{dw}{2\pi i}$, $m \in \mathbb{Z}$. Since $\{z^n\}_{n \in \mathbb{Z}}$ are linearly independent, we obtain

$$\oint_{|w|=\text{const}} w^m \frac{1}{w^{n+1}} \frac{dw}{2\pi i} = \mathbf{1}_{m=n}.$$

This is the spatial biorthogonality relation. This identity also readily follows from the Cauchy integral formula. Similar considerations work for Cauchy identities in several variables. The second type of biorthogonality relations can often be verified independently in a simpler fashion (as in the proof of Theorem 2.7.4).

Plancherel isomorphism results (Theorem 2.7.10) imply that the (coordinate) Bethe ansatz yielding the eigenfunctions F_λ of the transfer matrices is *complete*. That is, any function $f \in \mathcal{W}^n$ can be mapped into the spectral space, and then reconstructed back from its image. One of the ways to write down this completeness statement (using the orthogonality relation (2.101)) is

$$f(\lambda) = \frac{1}{(1-q)^n n!} \oint_{\gamma[s;0]} \frac{du_1}{2\pi i u_1} \cdots \oint_{\gamma[s;0]} \frac{du_n}{2\pi i u_n} \prod_{1 \le \alpha \ne \beta \le n} \frac{u_\alpha - u_\beta}{u_\alpha - q u_\beta}$$

$$\times (\mathscr{F}f)(u_1, \ldots, u_n) \mathsf{F}_\lambda^{\mathsf{c}}(u_1^{-1}, \ldots, u_n^{-1}).$$

Remark 2.7.13 (Spectral decomposition of $\mathscr{Q}_{0^m;v}^\circ$). Relation (2.101) also implies a spectral decomposition of the operator $\mathscr{Q}_{0^m;v}^\circ$ acting on functions on Sign_m^+, cf. Remark 2.6.6:

$$\mathscr{Q}_{0^m;v}^\circ(\mu \to \nu) = \frac{(-s)^{|\nu|-|\mu|}}{(1-q)^m m!} \oint_{\gamma[s;0]} \frac{dz_1}{2\pi i z_1} \cdots \oint_{\gamma[s;0]} \frac{dz_m}{2\pi i z_m} \prod_{1 \le \alpha \ne \beta \le m} \frac{z_\alpha - z_\beta}{z_\alpha - q z_\beta}$$

$$\times \left(\prod_{i=1}^m \frac{1 - q z_i v}{1 - z_i v} \right) \mathsf{F}_\mu(z_1, \ldots, z_m) \mathsf{F}_\nu^{\mathsf{c}}(z_1^{-1}, \ldots, z_m^{-1}). \tag{2.103}$$

Indeed, by (2.86) this operator has eigenfunctions

$$\Psi_\lambda(z_1, \ldots, z_m) = (-s)^{-|\lambda|} F_\lambda(z_1, \ldots, z_m)$$

with eigenvalues $\prod_{i=1}^{m} \dfrac{1 - q z_i v}{1 - z_i v}$ (the constant $(q; q)_m$ can be ignored). Thus, (2.103) follows by multiplying the eigenrelation (2.86) by $(-s)^{|\nu|} F_\nu^c(z_1^{-1}, \ldots, z_m^{-1})$ and integrating as in (2.101). Since the identity (2.86) requires the admissibility $(z_i, v) \in \mathsf{Adm}$ (Definition 2.4.7) before the integration, in (2.103) the point v^{-1} should be outside the integration contour $\gamma[s; 0]$ (the argument for this is similar to the proof of Proposition 2.7.15 in Section 2.7.3).

Remark 2.7.14. Function spaces $\mathcal{W}^n$ and $\mathcal{C}^n$ are far from being optimal. This is because we only address algebraic aspects of Plancherel isomorphisms. An extension of the first Plancherel isomorphism to larger spaces is described in Corwin and Petrov (2016, Appendix A).

2.7.3 An integral representation for G_μ

Using the orthogonality result of Theorem 2.7.4 and the Cauchy identity, we can obtain relatively simple nested contour integral formulas for the skew functions $\mathsf{G}_{\mu/\kappa}$ (and, in particular, for G_μ), which will be useful later in Sections 2.8 and 2.9.

Proposition 2.7.15. For any $k, N \in \mathbb{Z}_{\geq 1}$, $\mu, \kappa \in \mathsf{Sign}_k^+$, and $v_1, \ldots, v_N$ such that the v_i^{-1}'s are outside all of the integration contours $\gamma_j^+[s]$ of Definition 2.7.2, we have

$$\mathsf{G}_{\mu/\kappa}(v_1, \ldots, v_N) = \oint_{\gamma_1^+[s]} \frac{du_1}{2\pi \mathbf{i}} \cdots \oint_{\gamma_k^+[s]} \frac{du_k}{2\pi \mathbf{i}} \prod_{1 \leq \alpha < \beta \leq k} \frac{u_\alpha - u_\beta}{u_\alpha - q u_\beta}$$

$$\times \, F_\kappa^c(u_1, \ldots, u_k) \prod_{i=1}^{k} \frac{1}{u_i - s} \left(\frac{1 - s u_i}{u_i - s} \right)^{\mu_i} \prod_{\substack{1 \leq i \leq k \\ 1 \leq j \leq N}} \frac{1 - q u_i v_j}{1 - u_i v_j}. \qquad (2.104)$$

When $\kappa = (0^k)$, with the help of (2.42), formula (2.38) reduces to the following.

Corollary 2.7.16 (Borodin, 2014, Prop. 7.3). Under the assumptions of Proposition 2.7.15,

$$\mathsf{G}_\mu(v_1, \ldots, v_N) = (s^2; q)_k \oint_{\gamma_1^+[s]} \frac{du_1}{2\pi \mathbf{i}} \cdots \oint_{\gamma_k^+[s]} \frac{du_k}{2\pi \mathbf{i}} \prod_{1 \leq \alpha < \beta \leq k} \frac{u_\alpha - u_\beta}{u_\alpha - q u_\beta}$$

$$\times \prod_{i=1}^{k} \frac{1}{(u_i - s)(1 - s u_i)} \left(\frac{1 - s u_i}{u_i - s} \right)^{\mu_i} \prod_{\substack{1 \leq i \leq k \\ 1 \leq j \leq N}} \frac{1 - q u_i v_j}{1 - u_i v_j}. \qquad (2.105)$$

Proof of Proposition 2.7.15 Fix $\mu, \kappa \in \mathsf{Sign}_k^+$, multiply both sides of (2.100) by the skew function $\mathsf{G}_{\lambda/\kappa}(v_1, \ldots, v_N)$, and sum over $\lambda \in \mathsf{Sign}_k^+$. The right-hand side obviously equals $\mathsf{G}_{\mu/\kappa}(v_1, \ldots, v_N)$, while in the left-hand side we have

$$
\sum_{\lambda \in \mathsf{Sign}_k^+} \oint_{\gamma_1^+[s]} \frac{du_1}{2\pi\mathbf{i}} \cdots \oint_{\gamma_k^+[s]} \frac{du_k}{2\pi\mathbf{i}} \prod_{1 \le \alpha < \beta \le k} \frac{u_\alpha - u_\beta}{u_\alpha - qu_\beta}
$$
$$
\times \prod_{i=1}^k \frac{1}{u_i - s} \left(\frac{1 - su_i}{u_i - s} \right)^{\mu_i} \mathsf{F}_\lambda^{\mathsf{c}}(u_1, \ldots, u_k) \mathsf{G}_{\lambda/\kappa}(v_1, \ldots, v_N).
$$

If one can perform the (infinite) summation over λ inside the integral, then by the (iterated) Corollary 2.4.9.1 (which follows from the Cauchy identity), one readily gets the desired formula for the symmetric function $\mathsf{G}_{\mu/\kappa}(v_1, \ldots, v_N)$. It remains to justify that we indeed can interchange summation and integration.

The (absolutely convergent) summation can be performed inside the integral if $(u_i, v_j) \in \mathsf{Adm}$ for u_i on the contours, i.e.,

$$
\left| \frac{u_i - s}{u_i - s^{-1}} \right| \cdot \left| \frac{v_j^{-1} - s^{-1}}{v_j^{-1} - s} \right| < 1.
$$

Therefore, deforming the contours so that all u_i are closer to s than to s^{-1}, and taking $|v_j|$ sufficiently small ensures this inequality. Performing the summation over λ, we obtain the desired identity (2.104) for these deformed contours and restricted v_j. However, we may deform contours and drop these restrictions as long as the right-hand side of (2.104) represents the same rational function, because the left-hand side is a priori rational. This implies that the desired formula holds as long as the v_j^{-1} are outside the integration contours, which establishes the proposition. $\square$

As shown in Borodin (2014, Prop. 7.3), the nested contour integral formula for G_μ (2.105) may be used as an alternative way for deriving the symmetrization formula for G_μ of Theorem 2.4.12. Note that (2.105) in turn follows from the Cauchy identity plus the spatial orthogonality of the F_λ's, and the latter is implied by the symmetrization formula for F_λ. We will not reproduce the argument here.

Another use of formula (2.105) is a straightforward alternative proof of Proposition 2.6.7 (computation of the specialization $\mathsf{G}_\mu(\varrho)$), which can also be generalized to other specializations of G_μ. This will be a starting point for averaging observables in Section 2.8.2.

2.8 *q*-correlation functions

Here we compute q-correlation functions of the stochastic dynamics $\mathscr{X}_{\{u_t\};\varrho}^+$ of Section 2.6.4 assuming it starts from the empty initial configuration.

2.8.1 Computing observables via the Cauchy identity

Let us first briefly explain main ideas behind our computations. We are interested only in single-time observables (i.e., those which depend on the state of $\mathscr{X}^+_{\{u_t\};\varrho}$ at a single time moment, say, $t = n$), and getting them is equivalent to computing expectations $\mathbb{E}_{\mathbf{u};\varrho} f(\nu)$ of certain functions $f(\nu)$ of the configuration $\nu \in \mathsf{Sign}_n^+$ with respect to the probability measure

$$\mathscr{M}_{\mathbf{u};\varrho}(\nu) = \frac{1}{Z(\mathbf{u};\varrho)} \mathsf{F}_\nu(u_1, \ldots, u_n) \mathsf{G}_\nu^{\mathsf{c}}(\varrho) = \mathbf{1}_{\nu_n \geq 1} \cdot (-s)^{|\nu| - n} \cdot \mathsf{F}_{\nu - 1^n}^{\mathsf{c}}(u_1, \ldots, u_n),$$

(2.106)

where $\mathbf{u} = (u_1, \ldots, u_n)$. The measure $\mathscr{M}_{\mathbf{u};\mathbf{v}}$ (2.75) takes this form for $\mathbf{v} = \varrho$ due to (2.90).

The weights (2.106) are nonnegative if the parameters satisfy (2.66)–(2.67). To ensure that (2.106) defines a probability distribution on the infinite set Sign_n^+, we need to impose admissibility conditions (cf. Definition 2.6.1). The latter are implied by

$$\left| s \frac{u_j - s}{1 - s u_j} \right| < 1.$$

(2.107)

Indeed, these conditions ensure (2.37) for very small v (limit $v \to 0$ is a part of the specialization ϱ). Alternatively, interpret the probability weight

$$(-s)^{|\nu| - n} \cdot \mathsf{F}_{\nu - 1^n}^{\mathsf{c}}(u_1, \ldots, u_n)$$

in (2.106) as a partition function of path collections, and fix ν with large ν_1 (other parts can be large, too). The only vertex weight which enters the weight of a particular path collection a large number of times is $\mathsf{L}_{u_j}(0, 1; 0, 1) = (-s u_j + s^2)/(1 - s u_j)$, which is bounded in absolute value by (2.107). One readily sees that conditions (2.66)–(2.67) (which, in particular, require $u_i \geq 0$) automatically imply (2.107).

Cauchy identity (2.43) suggests a large family of observables of the measure $\mathscr{M}_{\mathbf{u};\varrho}$ whose averages can be computed right away. Namely, let us fix variables $w_1, \ldots, w_k$, and set

$$f(\nu) := \frac{\mathsf{G}_\nu(\varrho, w_1, \ldots, w_k)}{\mathsf{G}_\nu(\varrho)},$$

(2.108)

where $(\varrho, w_1, \ldots, w_k)$ means that we add the variables $w_1, \ldots, w_k$ to the specialization $(\epsilon, q\epsilon, \ldots, q^{J-1}\epsilon)$, then set $q^J = 1/(s\epsilon)$, and finally send $\epsilon \to 0$, cf. (2.88). Note that one can replace both G_ν in (2.108) by $\mathsf{G}_\nu^{\mathsf{c}}$ without changing $f(\nu)$. The $\mathbb{E}_{\mathbf{u};\varrho}$ expectation of the function (2.108) takes the form

$$\mathbb{E}_{\mathbf{u};\varrho} f(\nu) = \sum_{\nu \in \mathsf{Sign}_n^+} \frac{1}{Z(\mathbf{u};\varrho)} \mathsf{F}_\nu(u_1, \ldots, u_n) \mathsf{G}_\nu^{\mathsf{c}}(\varrho) \frac{\mathsf{G}_\nu^{\mathsf{c}}(\varrho, w_1, \ldots, w_k)}{\mathsf{G}_\nu^{\mathsf{c}}(\varrho)}$$

$$= \frac{Z(\mathbf{u}; \varrho, w_1, \ldots, w_k)}{Z(\mathbf{u}; \varrho)} = \prod_{\substack{1 \leq i \leq n \\ 1 \leq j \leq k}} \frac{1 - q u_i w_j}{1 - u_i w_j},$$

(2.109)

where the ratio of the partition functions $Z(\cdots)$ is computed via the corresponding ϱ limit of (2.76). We will discuss admissibility conditions (necessary for the convergence of the sum in (2.109) in Section 2.8.4.

One now needs to understand the dependence of (2.108) on ν. Using the integral formula (2.105) for G_ν, we can compute for $k = 1$:

$$\frac{G_\nu(\varrho, w)}{G_\nu(\varrho)} = q^n + \sum_{i=1}^{n} \frac{q^{i-1}}{(-s)^{\nu_i}} \frac{1-q}{1-s^{-1}w} \left(\frac{w-s}{1-sw}\right)^{\nu_i}, \qquad \nu_n \geq 1. \qquad (2.110)$$

A general result of this sort is given in Proposition 2.8.2.

Next, by a suitable contour integration in w one can extract the term in (2.110) with $\nu_i = m$ for any fixed $m \geq 1$. Therefore, the same integration of the right-hand side of (2.109) will yield a contour integral formula for

$$\mathbb{E}_{\mathbf{u};\varrho} \sum_{i=1}^{n} q^i \mathbf{1}_{\nu_i=m},$$

which can be viewed as a *q-analogue of the density function* of the random configuration ν. Higher *q-correlation functions* (defined in Section 2.8.4) can be computed in a similar way by working with (2.108) with general k. Therefore, for general k the right-hand side of (2.109) should be regarded as a generating function (in $w_1, \ldots, w_k$) for the *q*-correlation functions, and the latter can be extracted by integrating over the w_j's.

2.8.2   Computation of $G_\nu(\varrho, w_1, \ldots, w_k)$

Let us fix $n \geq k \geq 0$ and a signature $\nu \in \mathrm{Sign}_n^+$, and compute the specialization $G_\nu(\varrho, w_1, \ldots, w_k)$. The result of this computation is a general k version of (2.110), and it is given in Proposition 2.8.2.

For the computation we will assume that w_p, $p = 1, \ldots, k$, are pairwise distinct and are such that the points w_p^{-1} are outside the integration contours $\gamma_j^+[s]$ of Definition 2.7.2. We can readily take the ϱ limit (2.88) in (2.105), and write

$$G_\nu(\varrho, w_1, \ldots, w_k) = (s^2; q)_n \oint_{\gamma_1^+[s]} \frac{du_1}{2\pi\mathbf{i}} \cdots \oint_{\gamma_n^+[s]} \frac{du_n}{2\pi\mathbf{i}} \prod_{1 \leq \alpha < \beta \leq n} \frac{u_\alpha - u_\beta}{u_\alpha - qu_\beta}$$

$$\times \prod_{i=1}^{n} \left(\frac{1}{-s(1-su_i)}\right) \left(\frac{1-su_i}{u_i-s}\right)^{\nu_i} \prod_{j=1}^{k} \frac{1-qu_iw_j}{1-u_iw_j}. \qquad (2.111)$$

If $\nu_n = 0$, then the integral (2.111) vanishes because there are no u_n-poles inside $\gamma_n^+[s]$. We will thus assume that $\nu_n \geq 1$ (so all $\nu_i \geq 1$), and explicitly compute this integral. Denote it by $\mathcal{I}_{1,\ldots,n}^{w_1,\ldots,w_k}$.

We aim to peel off the contours $\gamma_1^+[s], \ldots, \gamma_n^+[s]$ (in this order), and take minus residues at poles outside these contours. Observe that the integrand in u_1 has only two

types of simple poles outside $\gamma_1^+[s]$, namely, $u_1 = \infty$ and $u_1 = w_p^{-1}$ for $p = 1, \ldots, k$ (there is no singularity at s^{-1}). We have

$$-\mathrm{Res}_{u_1 = \infty}\left(\frac{1}{-s(u_1 - s)}\left(\frac{1 - su_1}{u_1 - s}\right)^{\nu_1 - 1}\right) = (-s)^{\nu_1 - 2}.$$

Thus, the whole minus residue of (2.111) at $u_1 = \infty$ is equal to $(1 - s^2 q^{n-1})(-s)^{\nu_1 - 2} q^k \cdot \mathcal{I}_{2,\ldots,n}^{w_1,\ldots,w_k}$, where q^k comes from the product involving the w_p's.

Next, for any $p = 1, \ldots, k$, the pole w_p^{-1} yields

$$- \mathrm{Res}_{u_1 = w_p^{-1}} = (1 - s^2 q^{n-1})(-s)^{-2}\frac{1 - q}{1 - s^{-1}w_p}\left(\frac{w_p - s}{1 - sw_p}\right)^{\nu_1}\prod_{\substack{1 \le j \le k \\ j \ne p}}\frac{w_p - qw_j}{w_p - w_j}$$

$$\times (s^2; q)_{n-1} \oint_{\gamma_2^+[s]}\frac{du_2}{2\pi \mathbf{i}} \cdots \oint_{\gamma_n^+[s]}\frac{du_n}{2\pi \mathbf{i}}\prod_{2 \le \alpha < \beta \le n}\frac{u_\alpha - u_\beta}{u_\alpha - qu_\beta}$$

$$\times \prod_{i=2}^{n}\left(\frac{1}{-s(u_i - s)}\left(\frac{1 - su_i}{u_i - s}\right)^{\nu_i - 1}\prod_{\substack{1 \le j \le k \\ j \ne p}}\frac{1 - qu_iw_j}{1 - u_iw_j}\right).$$

Therefore, taking the minus residue at $u_1 = w_p^{-1}$ leads to

$$(1 - s^2 q^{n-1})(-s)^{-2}\frac{1 - q}{1 - s^{-1}w_p}\left(\frac{w_p - s}{1 - sw_p}\right)^{\nu_1}\prod_{\substack{1 \le j \le k \\ j \ne p}}\frac{w_p - qw_j}{w_p - w_j}\cdot \mathcal{I}_{2,\ldots,n}^{w_1,\ldots,w_{p-1},w_{p+1},\ldots,w_k}.$$

One can continue with similar computations for $\gamma_2^+[s], \ldots, \gamma_n^+[s]$. Let us write down the final integral with the only remaining contour $\gamma_n^+[s]$:

$$\mathcal{I}_n^{w_1,\ldots,w_k} = (1 - s^2)\oint_{\gamma_n^+[s]}\frac{du_n}{2\pi \mathbf{i}}\frac{1}{-s(u_n - s)}\left(\frac{1 - su_n}{u_n - s}\right)^{\nu_n - 1}\prod_{j=1}^{k}\frac{1 - qu_nw_j}{1 - u_nw_j}$$

$$= (1 - s^2)(-s)^{\nu_n - 2}q^k + (1 - s^2)(-s)^{-2}\sum_{p=1}^{k}\frac{1 - q}{1 - s^{-1}w_p}\left(\frac{w_p - s}{1 - sw_p}\right)^{\nu_n}\prod_{\substack{1 \le j \le k \\ j \ne p}}\frac{w_p - qw_j}{w_p - w_j}.$$

In general, the integral in (2.111) is equal to a summation of the following sort. For every $\ell = 0, \ldots, k$, choose two collections of indices $\mathcal{I} = \{i_1 < \ldots < i_\ell\} \subseteq \{1, \ldots, n\}$ and $\mathcal{J} = (j_1, \ldots, j_\ell) \subseteq \{1, \ldots, k\}$ (note that the order of the j_p's in $\mathcal{J}$ matters, while the i_p's are assumed already ordered). We will take residues at $u_{i_p} = w_{j_p}^{-1}$, $1 \le p \le \ell$, and the remaining residues at $u_i = \infty$ for $i \notin \mathcal{I}$. Denote the summand corresponding to these residues by $\mathrm{Res}_{\mathcal{I}, \mathcal{J}}$. All these summands have a common

prefactor $(-s)^{-2n}(s^2; q)_n$. The contribution to $\mathrm{Res}_{\mathcal{I},\mathcal{J}}$ from residues at infinity is equal to

$$q^{k(i_1-1)+(k-1)(i_2-i_1-1)+\ldots+(k-\ell+1)(i_\ell-i_{\ell-1}-1)+(k-\ell)(n-i_\ell)} \prod_{i \notin \mathcal{I}} (-s)^{\nu_i}$$

$$= q^{-\frac{1}{2}\ell(2k+1-\ell)+n(k-\ell)} q^{i_1+\ldots+i_\ell} \prod_{i \notin \mathcal{I}} (-s)^{\nu_i}. \qquad (2.112)$$

The residues at $u_{i_p} = w_{j_p}^{-1}$ contribute

$$\prod_{p=1}^{\ell} \left(\frac{1-q}{1-s^{-1}w_{j_p}} \left(\frac{w_{j_p}-s}{1-sw_{j_p}} \right)^{\nu_{i_p}} \right) \prod_{j \in \{1,\ldots,k\}\setminus\mathcal{J}} \frac{w_{j_p}-qw_j}{w_{j_p}-w_j} \prod_{1\le\alpha<\beta\le\ell} \frac{w_{j_\alpha}-qw_{j_\beta}}{w_{j_\alpha}-w_{j_\beta}}.$$

$$(2.113)$$

We see that (2.112) depends only on the choice of $\mathcal{I}$, while (2.113) depends on both $\mathcal{I}$ and $\mathcal{J}$. Thus, for a fixed $\mathcal{I}$, one can first sum $\mathrm{Res}_{\mathcal{I},\mathcal{J}}$ over all subsets $\mathcal{J} = \{j_1 < \ldots < j_\ell\} \subseteq \{1,\ldots,k\}$, and, for a fixed such $\mathcal{J}$, over all its permutations. This summation over $\mathcal{J}$ is performed using the following lemma.

Lemma 2.8.1. *Let $f_i(\zeta)$ be arbitrary functions in $\zeta \in \mathbb{C}$. For any $m \ge 1$ and $\ell \le m$, we have*

$$\sum_{\sigma \in \mathcal{S}_m} \sigma \left(\prod_{i=1}^{\ell} f_i(\zeta_i) \prod_{1\le\alpha<\beta\le m} \frac{\zeta_\alpha - q\zeta_\beta}{\zeta_\alpha - \zeta_\beta} \right)$$

$$= \frac{(q;q)_{m-\ell}}{(1-q)^{m-\ell}} \sum_{\mathcal{J}=\{j_1<\ldots<j_\ell\}\subseteq\{1,\ldots,m\}} \left(\prod_{\substack{j \in \mathcal{J} \\ r \notin \mathcal{J}}} \frac{\zeta_j - q\zeta_r}{\zeta_j - \zeta_r} \sum_{\sigma' \in \mathcal{S}_\ell} \prod_{i=1}^{\ell} f_i(\zeta_{j_{\sigma'(i)}}) \right.$$

$$\left. \times \prod_{1\le\alpha<\beta\le\ell} \frac{\zeta_{j_{\sigma'(\alpha)}} - q\zeta_{j_{\sigma'(\beta)}}}{\zeta_{j_{\sigma'(\alpha)}} - \zeta_{j_{\sigma'(\beta)}}} \right),$$

where the permutation σ in the left-hand side acts on the variables ζ_j.

Proof For each $\sigma \in \mathcal{S}_m$, let $\mathcal{J}$ be the ordered list of elements of $\{\sigma(1),\ldots,\sigma(\ell)\}$. The left-hand side of the desired claim contains

$$\prod_{i=1}^{\ell} f_i(\zeta_{\sigma(i)}) \prod_{1\le\alpha<\beta\le m} \frac{\zeta_{\sigma(\alpha)} - q\zeta_{\sigma(\beta)}}{\zeta_{\sigma(\alpha)} - \zeta_{\sigma(\beta)}}$$

$$= \prod_{i=1}^{\ell} f_i(\zeta_{j_i}) \prod_{1\le\alpha<\beta\le\ell} \frac{\zeta_{j_\alpha} - q\zeta_{j_\beta}}{\zeta_{j_\alpha} - \zeta_{j_\beta}} \underbrace{\prod_{\substack{j \in \mathcal{J} \\ r \notin \mathcal{J}}} \frac{\zeta_j - q\zeta_r}{\zeta_j - \zeta_r} \prod_{\ell+1\le\alpha<\beta\le m} \frac{\zeta_{\sigma(\alpha)} - q\zeta_{\sigma(\beta)}}{\zeta_{\sigma(\alpha)} - \zeta_{\sigma(\beta)}}}_{\text{symmetric in } \mathcal{J} \text{ and } \mathcal{J}^c}.$$

Symmetrizing over $\sigma \in S_m$ can be done in two steps: first, choose a subset $\mathcal{J}$ of $\{1, \ldots, m\}$ of size ℓ, and then symmetrize over indices inside and outside $\mathcal{J}$. For the symmetrization outside $\mathcal{J}$ we can use the symmetrization identity of footnote 7. This yields the result.

Applying this lemma to (2.113) with $m = k$, we arrive at the following formula for our specialization, which is the first step towards q-correlation functions.

Proposition 2.8.2. For $n \geq k \geq 0$ and $\nu \in \mathsf{Sign}_n^+$, we have

$$\mathsf{G}_\nu(\varrho, w_1, \ldots, w_k) = \underbrace{1_{\nu_n \geq 1}(-s)^{|\nu|} \frac{(s^2; q)_n}{s^{2n}}}_{\mathsf{G}_\nu(\varrho)}$$

$$\times \sum_{\ell=0}^{k} \frac{q^{-\frac{1}{2}\ell(2k+1-\ell)+n(k-\ell)}(1-q)^{k-\ell}}{(q;q)_{k-\ell}} \sum_{\mathcal{I}=\{i_1 < \ldots < i_\ell\} \subseteq \{1, \ldots, n\}} q^{i_1 + \ldots + i_\ell} \prod_{i \in \mathcal{I}} (-s)^{-\nu_i}$$

$$\times \sum_{\sigma \in S_k} \sigma \left(\prod_{1 \leq \alpha < \beta \leq k} \frac{w_\alpha - q w_\beta}{w_\alpha - w_\beta} \prod_{p=1}^{\ell} \frac{1-q}{1 - s^{-1}w_p} \left(\frac{w_p - s}{1 - s w_p} \right)^{\nu_{i_p}} \right), \tag{2.114}$$

where the permutation σ acts on $w_1, \ldots, w_k$.

Proof Identity (2.114) was established under certain restrictions on the w_p's. Observe that both sides of the identity (2.114) are a priori rational functions. This is clear for the right-hand side, and the left-hand side of (2.114) is also rational because using the branching rule of Proposition 2.4.5 one can separate the specialization ϱ and the variables w_p (skew G-functions in the w_j's are rational by the very definition), and then evaluate the specialization ϱ by Proposition 2.6.7. Thus, we can drop any restrictions on the w_p's.

Note that in the particular case $k = 0$ the above proposition reduces to Proposition 2.6.7.

2.8.3 Extracting terms by integrating over w_i

Observe now that when $k = \ell$, the summation over σ in (2.114) produces

$$(-s)^k \mathsf{F}_{(\nu_{i_1}-1, \ldots, \nu_{i_\ell}-1)}(w_1, \ldots, w_k).$$

Indeed, see (2.44) and recall that each $\nu_{i_p} \geq 1$. This observation motivates our next step in computation of the q-correlation functions: we will utilize orthogonality of the functions F_μ (similar to Theorem 2.7.4), and integrate (2.114) over the w_i's to extract certain terms. We will need the following nested integration contours.

Definition 2.8.3. For any $k \geq 1$, let $\gamma_1^-[s^{-1}], \ldots, \gamma_k^-[s^{-1}]$ be positively oriented closed contours such that

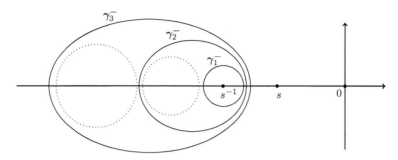

Fig. 2.24 A possible choice of nested integration contours $\gamma_i^- = \gamma_i^-[s^{-1}]$, $i = 1, 2, 3$ (Definition 2.8.3). Contours $q^{-1}\gamma_1^-[s^{-1}]$ and $q^{-2}\gamma_1^-[s^{-1}]$ are shown dotted.

- *each contour $\gamma_\alpha^-[s^{-1}]$ encircles s^{-1}, while leaving s outside,*
- *for any $\beta > \alpha$, the interior of $\gamma_\beta^-[s^{-1}]$ contains the contour $q^{-1}\gamma_\alpha^-[s^{-1}]$, and*
- *the contour $\gamma_1^-[s^{-1}]$ is sufficiently small so that it does not intersect with the contour $q^{-1}\gamma_1^-[s^{-1}]$.*

See Fig. 2.24. The superscript '−' refers to the property that the contours are q^{-1}-nested.

Remark 2.8.4.   The integration contours $\gamma_j^-[s^{-1}]$ can be obtained from the contours $\gamma_j^+[s]$ of Definition 2.7.2 by dragging $\gamma_1^+[s], \dots, \gamma_k^+[s]$ (in this order) through infinity, if this operation is allowed for a particular integrand (i.e., it must have no residues at infinity).

We will use the following integral transform.

Definition 2.8.5.   *For $k \geq 1$, let $R(w_1, \dots, w_k)$ be a symmetric rational function with singularities occurring only when some of the w_j's belong to $\{s^{\pm 1}\}$. Let $\vartheta \in \mathsf{Sign}_k^+$. Define*

$$\left(\mathscr{T}^{(k)}R\right)(\vartheta) := (-1)^k \mathsf{c}(\vartheta) \oint_{\gamma_1^-[s^{-1}]} \frac{dw_1}{2\pi i} \cdots \oint_{\gamma_k^-[s^{-1}]} \frac{dw_k}{2\pi i} \prod_{1 \leq \alpha < \beta \leq k} \frac{w_\alpha - w_\beta}{w_\alpha - q w_\beta}$$

$$\times R(w_1, \dots, w_k) \prod_{i=1}^{k} \frac{1}{w_i - s} \left(\frac{1 - s w_i}{w_i - s}\right)^{\vartheta_i},$$

where the integration contours are described in Definition 2.8.3.

Let us denote for any $\lambda \in \mathsf{Sign}_\ell^+$

$$R_\lambda^{(\ell)}(w_1, \ldots, w_k)$$

$$:= \mathbf{1}_{\lambda_\ell \geq 1} \sum_{\sigma \in S_k} \sigma \left(\prod_{1 \leq \alpha < \beta \leq k} \frac{w_\alpha - qw_\beta}{w_\alpha - w_\beta} \prod_{p=1}^{\ell} \frac{1-q}{1-s^{-1}w_p} \left(\frac{w_p - s}{1 - sw_p} \right)^{\lambda_p} \right);$$

these are the w_j-dependent summands in (2.114). As mentioned earlier, for $\ell = k$,

$$R_\lambda^{(k)}(w_1, \ldots, w_k) = (-s)^k \mathsf{F}_{\lambda - 1^k}(w_1, \ldots, w_k).$$

Moreover, the action of the transform $\mathscr{T}^{(k)}$ on the $R_\lambda^{(\ell)}$'s for any ℓ and λ turns out to be very simple.

Lemma 2.8.6. *For any $\ell = 0, \ldots, k$ and any $\lambda \in \mathsf{Sign}_\ell^+$ we have*

$$\left(\mathscr{T}^{(k)} R_\lambda^{(\ell)} \right)(\vartheta) = (-s)^k \mathbf{1}_{\ell=k} \mathbf{1}_{\vartheta=\lambda-1^k}, \qquad \vartheta \in \mathsf{Sign}_k^+.$$

Proof We may assume that $\lambda_\ell \geq 1$. Let us complement λ by zeroes so that it has length k; i.e., write $\lambda = (\lambda_1, \ldots, \lambda_\ell, 0, \ldots, 0)$. We have

$$\left(\mathscr{T}^{(k)} R_\lambda^{(\ell)} \right)(\vartheta)$$

$$= (-1)^k c(\vartheta) \sum_{\sigma \in S_k} \oint_{\gamma_1^-[s^{-1}]} \frac{dw_1}{2\pi \mathbf{i}} \cdots \oint_{\gamma_k^-[s^{-1}]} \frac{dw_k}{2\pi \mathbf{i}} \prod_{1 \leq \alpha < \beta \leq k} \frac{w_\alpha - w_\beta}{w_\alpha - qw_\beta} \frac{w_{\sigma(\alpha)} - qw_{\sigma(\beta)}}{w_{\sigma(\alpha)} - w_{\sigma(\beta)}}$$

$$\times (-s)^\ell \prod_{p=1}^{\ell} \underbrace{\frac{1-q}{1-sw_p} \left(\frac{w_p - s}{1-sw_p} \right)^{\lambda_p - 1}}_{g_{\lambda_p}(w_{\sigma(p)})} \cdot \prod_{i=1}^{k} \underbrace{\frac{1}{w_i - s} \left(\frac{1-sw_i}{w_i - s} \right)^{\vartheta_i}}_{f_{\vartheta_i + 1}(w_i)}. \tag{2.115}$$

We now wish to apply Lemma 2.7.1 with

$$f_m(w) = \begin{cases} \dfrac{1}{w-s} \left(\dfrac{1-sw}{w-s} \right)^{m-1}, & m \geq 1; \\ 1, & m = 0, \end{cases}$$

$$g_l(w) = \begin{cases} \dfrac{1-q}{1-sw} \left(\dfrac{w-s}{1-sw} \right)^{l-1}, & l \geq 1; \\ 1, & l = 0, \end{cases}$$

and $P_1' = \{s, \infty\}$, $P_2' = \{s^{-1}\}$ (we use notation $P_{1,2}'$ to distinguish from $P_{1,2}$ in Definition 2.7.2). Indeed, since in our integral we always have $m \geq 1$, all singularities of $f_m(w)g_l(w)$ are in $P_1' \cup P_2'$. Moreover, for $m < l$, all poles of this product are in P_2', and for $m > l$ (which may include $l = 0$) all poles are in P_1'. Finally, observe that we can deform the integration contours as in the hypothesis of Lemma 2.7.1.

Therefore, since $\vartheta_i + 1 \geq 1$ for all i in our integral, we can apply Lemma 2.7.1, and conclude that it must be that $\lambda_i = \vartheta_i + 1$ for all $i = 1, \ldots, k$, in order for the integral to be nonzero. In particular, the integral can be nonzero only for $\ell = k$.

When $\ell = k$, the desired claim for $\vartheta = \lambda - 1^k$ follows by analogy with the last computation in the proof of Theorem 2.7.4. Namely, we first sum over σ (the integral vanishes unless σ permutes within clusters of ϑ), and then compute the resulting smaller integrals by taking residues at $w_1 = s^{-1}$, $w_2 = q^{-1}s^{-1}$, etc. This leads to the desired result.

Remark 2.8.7. Note that for $\ell = k$ in the proof for Lemma 2.8.6 we could simply drag the integration contours $\gamma_j^-[s^{-1}]$ through infinity to the negatively oriented $\gamma_j^+[s]$ (cf. Remark 2.8.4). Indeed, this is because for $\ell = k$ the integrand in (2.115) is regular at $w_i = \infty$ for all i. The passage to the contours $\gamma_j^+[s]$ eliminates the sign $(-1)^k$, and the desired claim for $\vartheta = \lambda - 1^k$ directly follows from Theorem 2.7.4. In other words, the transform $\mathscr{T}^{(k)}$ acts essentially as the inverse Plancherel transform (Definition 2.7.8). It is, however, crucial that the former is defined using the contours $\gamma_j^-[s^{-1}]$ and not $\gamma_j^+[s]$ because of nontrivial residues at infinity in (2.115) for $\ell < k$.

Therefore, applying the transform $\mathscr{T}^{(k)}$ to the rational function $\mathsf{G}_\nu(\varrho, w_1, \ldots, w_k)$ and using Proposition 2.8.2, we arrive at the following statement summarizing the second step of the computation of the q-correlation functions.

Proposition 2.8.8. For any $n \geq k \geq 0$, $\nu \in \mathrm{Sign}_n^+$, and $\vartheta \in \mathrm{Sign}_k^+$, we have

$$\left(\mathscr{T}^{(k)} \mathsf{G}_\nu(\varrho, \bullet) \right)(\vartheta) = \mathsf{G}_\nu(\varrho) \frac{q^{-\frac{1}{2}k(k+1)}}{(-s)^{|\vartheta|}} \sum_{\substack{\mathcal{I} = \{i_1 < \ldots < i_k\} \subseteq \{1,\ldots,n\} \\ \nu_{i_1} = \vartheta_1 + 1, \ldots, \nu_{i_k} = \vartheta_k + 1}} q^{i_1 + \ldots + i_k}, \qquad (2.116)$$

where '$\bullet$' stands for the variables $w_1, \ldots, w_k$ in which the transform $\mathscr{T}^{(k)}$ is applied. Note that for (2.116) to be nontrivial one must have $\nu_n \geq 1$.

2.8.4 q-correlation functions

The structure of formula (2.116) suggests the following definition. For any $n \geq k \geq 0$, any $\vartheta = (\vartheta_1 \geq \vartheta_2 \geq \ldots \geq \vartheta_k \geq 0) \in \mathrm{Sign}_k^+$ and any $\nu \in \mathrm{Sign}_n^+$, set

$$\mathcal{Q}_\vartheta(\nu) := \sum_{\substack{\mathcal{I} = \{i_1 < \ldots < i_k\} \subseteq \{1,\ldots,n\} \\ \nu_{i_1} = \vartheta_1, \ldots, \nu_{i_k} = \vartheta_k}} q^{i_1 + \ldots + i_k}. \qquad (2.117)$$

If $n < k$, then the expression is zero, by agreement. We will now employ Proposition 2.8.8 to compute the expectations

$$\mathbb{E}_{\mathbf{u};\varrho}(\mathcal{Q}_\vartheta) = \sum_{\nu \in \mathrm{Sign}_n^+} \mathscr{M}_{\mathbf{u};\varrho}(\nu) \mathcal{Q}_\vartheta(\nu). \qquad (2.118)$$

Note that by (2.88), the summation ranges only over $\nu \in \mathrm{Sign}_n^+$ with $\nu_n \geq 1$. The sum converges if the parameters u_i satisfy (2.107).

Remark 2.8.9. Expectations (2.118) can be viewed as *q-analogues of the correlation functions* of $\mathcal{M}_{u;\varrho}$. Indeed, when $q = 1$ and all ϑ_i's are distinct, (2.117) turns into

$$\mathcal{Q}_\vartheta(\nu)\Big|_{q=1} = \prod_{i=1}^{k} \#\{j \colon 1 \le j \le n \text{ and } \nu_j = \vartheta_i\}.$$

When all the ν_j's are also pairwise distinct, let us interpret them as coordinates of distinct particles on $\mathbb{Z}_{\ge 0}$. In this case the above expression further simplifies to

$$\mathcal{Q}_\vartheta(\nu)\Big|_{q=1} = \mathbf{1}_{\{\text{there is a particle of the configuration } \nu \text{ at each of the locations } \vartheta_1, \dots, \vartheta_k\}}.$$

(2.119)

A probability distribution on configurations ν of n distinct particles on $\mathbb{Z}_{\ge 0}$ is often referred to as the (*n*-)*point process* on $\mathbb{Z}_{\ge 0}$. An expectation of (2.119) with respect to a point process on $\mathbb{Z}_{\ge 0}$ is known as the (*k*th) *correlation function* of this point process.

For the purpose of analytic continuation in the parameter space, it is useful to establish that our *q*-correlation functions are a priori rational.

Lemma 2.8.10. *Fix $n \in \mathbb{Z}_{\ge 0}$, and let (2.107) hold. Then for any fixed $k = 0, \dots, n$ and $\vartheta \in \mathsf{Sign}_k^+$, the expectation $\mathbb{E}_{u;\varrho}(\mathcal{Q}_\vartheta)$ is a rational function in $u_1, \dots, u_n$ and the parameters q and s.*

Proof Write

$$\mathbb{E}_{u;\varrho}(\mathcal{Q}_\vartheta) = \sum_{\mathcal{I} = \{i_1 < \dots < i_k\} \subseteq \{1, \dots, n\}} q^{i_1 + \dots + i_k} \sum_{\substack{\nu \in \mathsf{Sign}_n^+ \\ \nu_{i_1} = \vartheta_1, \dots, \nu_{i_k} = \vartheta_k}} \mathcal{M}_{u;\varrho}(\nu),$$

and observe that only the second sum is infinite. Therefore, we may fix $\mathcal{I}$ and consider only the summation over ν. By (2.106) and (2.90), the sum over ν is the same as the sum of products of stochastic vertex weights L_{u_j} over certain collections of n paths in $\{0, 1, 2, \dots\} \times \{1, 2, \dots, n\}$, as in Definition 2.4.3. Namely, these paths start with n horizontal edges $(-1, t) \to (0, t)$, $t = 1, \dots, n$ and end with n vertical edges $(\nu_i, n) \to (\nu_i, n+1)$ (note that $\nu_n \ge 1$), and the end edges are partially fixed by the condition $\nu_{i_1} = \vartheta_1, \dots, \nu_{i_k} = \vartheta_k$. Therefore, only the coordinates $\nu_1, \dots, \nu_{i_1 - 1}$ belong to the infinite range $\{\vartheta_1 + 1, \vartheta_1 + 2, \dots\}$.

Assume that $r \le i_1 - 1$ out of our n paths go strictly to the right of ϑ_1 (i.e., we have $\nu_r > \vartheta_1$ and $\nu_{r+1} = \dots = \nu_{i_1} = \vartheta_1$). Let these paths contain edges $(\vartheta_1, j_i) \to (\vartheta_1 + 1, j_i)$ for some $\mathcal{J} = \{j_1 < \dots < j_r\} \subseteq \{1, \dots, n\}$. Fixing $r \le i_1 - 1$ and such $\mathcal{J}$ (there are only finitely many ways to choose these data), we may now split the summation over our n paths to paths in $\{0, 1, \dots, \vartheta_1\} \times \{1, \dots, n\}$ and in $\{\vartheta_1 + 1, \vartheta_1 + 2, \dots\} \times \{1, \dots, n\}$. The first sum over paths is also finite. See Fig. 2.25 for an illustration of this splitting of paths.

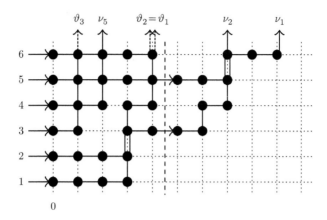

Fig. 2.25 Splitting of summation over path collections in the proof of Lemma 2.8.10 for $n = 6$, $k = 3$, and $r = 2$. The dotted arrows on the top correspond to fixed vertical edges prescribed by ϑ, and here $\nu_3 = \vartheta_1$, $\nu_4 = \vartheta_2$, $\nu_6 = \vartheta_3$, and $\mathcal{J} = \{3, 5\}$.

Since finite sums clearly produce rational functions, it now suffices to fix r and $\mathcal{J}$ as we've just shown and consider the corresponding sum over collections of r paths in $\{\vartheta_1 + 1, \vartheta_1 + 2, \dots\} \times \{1, \dots, n\}$ starting with horizontal edges $(\vartheta_1 + 1, j_i) \to (\vartheta_1 + 1, j_i)$ and ending with vertical edges $(\nu_i, n) \to (\nu_i, n + 1)$, $i = 1, \dots, r$. Because this final infinite sum involves stochastic vertex weights and is over all unrestricted path collections, it is simply equal to 1 (recall that the u_i's satisfy (2.107), so this sum converges). This completes the proof of the lemma.

We are now in a position to compute the q-correlation functions (2.118). First, we will obtain a nested contour integration formula when the points u_i^{-1}, $i = 1, \dots, n$, are inside the integration contour $\gamma_1^-[s^{-1}]$ of Definition 2.8.3. Note that this requires $\Re(u_i) < 0$ for all i, which is incompatible with our usual assumption $u_i \geq 0$. However, in the former case the correlation functions (2.118) are clearly well-defined as sums of possibly negative terms, and we have the following formula for them.

Proposition 2.8.11.   Fix $n \geq k \geq 0$, and let $u_1, \dots, u_n$ satisfy (2.107) and be such that the points u_i^{-1} are inside the integration contour $\gamma_1^-[s^{-1}]$. Then for any $\vartheta = (\vartheta_1 \geq \vartheta_2 \geq \dots \geq \vartheta_k \geq 0) \in \mathsf{Sign}_k^+$ we have

$$\mathbb{E}_{\mathbf{u};\varrho}\left(\mathcal{Q}_{\vartheta+1^k}\right) = (-1)^k q^{\frac{k(k+1)}{2}} (-s)^{|\vartheta|} \mathsf{c}(\vartheta)$$

$$\times \oint_{\gamma_1^-[s^{-1}]} \frac{dw_1}{2\pi i} \cdots \oint_{\gamma_k^-[s^{-1}]} \frac{dw_k}{2\pi i} \prod_{1 \leq \alpha < \beta \leq k} \frac{w_\alpha - w_\beta}{w_\alpha - qw_\beta}$$

$$\times \prod_{i=1}^{k} \left(\frac{1}{w_i - s}\left(\frac{1 - sw_i}{w_i - s}\right)^{\vartheta_i} \prod_{j=1}^{n} \frac{1 - qu_j w_i}{1 - u_j w_i}\right), \tag{2.120}$$

where $\mathsf{c}(\vartheta)$ is defined by (2.32).

After proving this proposition, we will relax the conditions on the u_i's (to include $u_i \geq 0$) by suitably deforming the integration contours.

Proof The desired identity (2.120) formally follows from Proposition 2.8.8 combined with the Cauchy identity summation (2.109). However, one needs to justify that this summation can be performed inside the integral. This is possible if $(u_i, w_j) \in \mathsf{Adm}$ for all i, j, which can be written as

$$\left| \frac{u_i^{-1} - s^{-1}}{u_i^{-1} - s} \cdot \frac{w_j - s}{w_j - s^{-1}} \right| < 1.$$

In other words, this inequality suggests that the u_i^{-1}'s should be closer to s^{-1} than to s, while the w_j's should be closer to s. Moreover, (2.107) is equivalent to the first factor being less than 1 in the absolute value. Therefore, if

$$\left| \frac{u_i^{-1} - s^{-1}}{u_i^{-1} - s} \right| < r, \qquad \left| \frac{w_j - s}{w_j - s^{-1}} \right| < \frac{1}{r}, \qquad \text{for some } 0 < r < 1,$$

then we can sum under the integral, and (2.107) also holds. Thus, the points u_i^{-1} must be inside a certain disc around s^{-1}, while the contours $\gamma_j^-[s^{-1}]$ must encircle this disc. However, for the contours to exist, this disc must not intersect with itself multiplied by q^{-1}. By restricting the values of u_i so that the u_i^{-1}'s are sufficiently close to s^{-1}, one can ensure that the latter condition holds. Therefore, for the restricted values of the u_i's, we can perform the summation inside the integral and obtain the desired formula (2.120).

Since the left-hand side of (2.120) is a rational function in the u_i's, we can continue identity (2.120) of rational functions by dropping the restrictions on the u_i's, as long as (2.107) holds and the integral in the right-hand side of (2.120) represents the same rational function. The latter requirement leads to the assumption that the u_i^{-1}'s must be inside the integration contours. $\qquad\blacksquare$

To state our final result for the q-correlation functions, we need the following integration contour:

Definition 2.8.12. *Let* $u_1, \ldots, u_n > 0$, *and assume that* $u_i \neq q u_j$ *for any* i, j. *Define the contour* $\gamma[\bar{u}]$ *to be a union of sufficiently small positively oriented circles around all the points* $\{u_i^{-1}\}$, *such that the interior of* $\gamma[\bar{u}]$ *does not intersect with* $q^{\pm 1} \gamma[\bar{u}]$, *and* s *is outside the contour* $\gamma[\bar{u}]$.

For $u_i > 0$ and $q \cdot \max_i u_i < \min_i u_i$, let the q^{-1}-nested contours $\gamma_j^-[\bar{u}]$, $j = 1, \ldots, k$, be defined analogously to $\gamma_j^-[s^{-1}]$ of Definition 2.8.3 (but the $\gamma_j^-[\bar{u}]$'s encircle the points u_i^{-1}). In this case the contour $\gamma_1^-[\bar{u}]$ can also play the role of $\gamma[\bar{u}]$ of Definition 2.8.12.

With these contours we can now formulate the final result of the computation in Sections 2.8.2–2.8.4:

Theorem 2.8.13.   Assume (2.66). The nested contour integral formula (2.120) for the q-correlation functions of the dynamics $\mathscr{X}^+_{\{u_t\};\varrho}$ at time $= n$ holds in each of the following three cases:

1. Let the points u_i^{-1} be inside the integration contour $\gamma_1^-[s^{-1}]$, and the u_i's satisfy (2.107). Then (2.120) holds with the integration contours $w_j \in \gamma_j^-[s^{-1}]$ of Definition 2.8.3.
2. Let $u_i > 0$ for all i and $u_i \neq qu_j$ for any i, j. Then (2.120) holds when all the integration contours are the same, $w_j \in \gamma[\bar{u}]$ (described in Definition 2.8.12). In this case we can symmetrize the integrand similarly to the proof of Corollary 2.7.5, and the formula takes the form

$$\mathbb{E}_{\mathbf{u};\varrho}(\mathcal{Q}_{\vartheta+1^k}) = \frac{(-1)^k q^{\frac{k(k+1)}{2}}}{(1-q)^k k!} \oint_{\gamma[\bar{u}]} \frac{dw_1}{2\pi\mathbf{i}} \cdots \oint_{\gamma[\bar{u}]} \frac{dw_k}{2\pi\mathbf{i}} \prod_{1\leq\alpha\neq\beta\leq k} \frac{w_\alpha - w_\beta}{w_\alpha - qw_\beta}$$

$$\times (-s)^{|\vartheta|} \frac{\mathsf{F}^c_\vartheta(w_1^{-1}, \ldots, w_k^{-1})}{w_1 \ldots w_k} \prod_{i=1}^{k}\prod_{j=1}^{n} \frac{1 - qu_j w_i}{1 - u_j w_i}. \tag{2.121}$$

3. Let $\mathbf{u}$ have the form

$$\mathbf{u} = (u_1, qu_1, \ldots, q^{J-1}u_1, u_2, qu_2, \ldots, q^{J-1}u_2, \ldots, u_{n'}, qu_{n'}, \ldots, q^{J-1}u_{n'}),$$

where $n = Jn'$ with some $J \in \mathbb{Z}_{\geq 1}$, $u_i > 0$, and $q \cdot \max_i u_i < \min_i u_i$. $\tag{2.122}$

Then (2.120) holds with the integration contours $w_j \in \gamma_j^-[\bar{u}]$. In this case the double product in the integrand takes the form

$$\prod_{i=1}^{k}\prod_{j=1}^{n'} \frac{1 - q^J u_j w_i}{1 - u_j w_i}.^{20}$$

Proof

1. This is Proposition 2.8.11.
2. To prove the second case, start with (2.120) with contours $w_j \in \gamma_j^-[s^{-1}]$ and q, s, and $\{u_i\}$ fixed, and observe the following effect. The integrand

$$\prod_{1\leq\alpha<\beta\leq k} \frac{w_\alpha - w_\beta}{w_\alpha - qw_\beta} \prod_{i=1}^{k} \left(\frac{1}{w_i - s}\left(\frac{1 - sw_i}{w_i - s}\right)^{\vartheta_i} \prod_{j=1}^{n} \frac{1 - qu_j w_i}{1 - u_j w_i}\right) \tag{2.123}$$

has only the poles $w_1 = u_j^{-1}$, $j = 1, \ldots, n$, inside the contour $\gamma_1^-[s^{-1}]$, because the other poles ∞ and s are outside $\gamma_1^-[s^{-1}]$. Deform the integration contour $\gamma_2^-[s^{-1}]$

---

[20] Since the definition of the contours $\gamma_j^-[\bar{u}]$ does not depend on $J \in \mathbb{Z}_{\geq 1}$, we can analytically continue the nested contour integral formula in q^J (similarly to the discussion in Section 2.6.6). We will employ this continuation in Section 2.10.2.

so that it becomes the same as $\gamma_1^-[s^{-1}]$, thus picking the residue at $w_2 = q^{-1}w_1$. We see that

$$\text{Res}_{w_2=q^{-1}w_1} \frac{w_1-w_2}{w_1-qw_2} \left(\prod_{j=1}^n \frac{1-qu_jw_1}{1-u_jw_1} \frac{1-qu_jw_2}{1-u_jw_2} \right) = (1-q)q^{-2}w_1 \prod_{j=1}^n \frac{q(1-qu_jw_1)}{q-u_jw_1}.$$

Hence, the residue at $w_2 = q^{-1}w_1$ is regular in w_1 on the contour $\gamma_1^-[s^{-1}]$, and thus vanishes after the w_1 integration. Continuing this argument in a similar way, we may deform all integration contours to be $\gamma_1^-[s^{-1}]$. In other words, we see that the integral (2.120) can be computed by taking only the residues at points $w_i = u_{j_i}^{-1}$, where $i = 1, \ldots, k$ and $\{j_1, \ldots, j_k\} \subseteq \{1, \ldots, n\}$.

Next, observe that s^{-1} is not a pole of the integrand (2.123), and so the requirement that the contour $\gamma_1^-[s^{-1}]$ encircles this point can be dropped. Thus, we may take $u_i > 0$, and the integration contours to be $\gamma[\bar{u}]$ instead of $\gamma_1^-[s^{-1}]$. Symmetrizing the integration variables finishes the second case.

3. This case can be obtained as a limit of the second case. Namely, under assumptions of case 2 and also assuming $q \cdot \max_i u_i < \min_i u_i$, let us first pass to the q^{-1}-nested contours $\gamma_j^-[\bar{u}]$, which start from $\gamma[\bar{u}] = \gamma_1^-[\bar{u}]$. This can be done following the argument in case 2, because the integration in both cases $\gamma[\bar{u}]$ and $\gamma_j^-[\bar{u}]$ involves the same residues.

Next, if $u_2 \neq qu_1$, then the integrand in (2.120) has nonzero residues at both $(w_1, w_2) = (u_1^{-1}, u_2^{-1})$ and $(w_1, w_2) = (u_2^{-1}, u_1^{-1})$, and so both $u_{1,2}^{-1}$ must be inside $\gamma_1^-[\bar{u}]$ to produce the correct rational function. However, if $u_2 = qu_1$, then the second residue vanishes due to the presence of the factor $u_2 - qu_1$. One can readily check that the same effect occurs when we first move u_2^{-1} outside the contour $\gamma_1^-[\bar{u}]$ (but still inside $\gamma_2^-[\bar{u}]$), and then set $u_2 = qu_1$. This agrees with the presence of the factors $\frac{1-q^2 u_1 w_i}{1-u_1 w_i}$ in the integrand after setting $u_2 = qu_1$, which do not have poles at $u_2^{-1} = (qu_1)^{-1}$. One also sees that after taking residue at $w_1 = u_1^{-1}$, the pole at $w_2 = u_1^{-1}$ disappears, but there is a new pole at $w_2 = (qu_1)^{-1}$.

Continuing on, we see that for the contours $\gamma_j^-[\bar{u}]$ we can specialize $\mathbf{u}$ to (2.122), and the contour integration will still yield the correct rational function (i.e., the corresponding specialization of the left-hand side of (2.120)). This establishes the third case.

2.8.5   Remark. From observables to duality, and back

Formula (2.121) for the q-correlation functions readily suggests a certain *self-duality* relation associated with the stochastic higher spin six-vertex model. Denote $H(\nu; \vartheta) := \mathcal{Q}_{\vartheta+1^k}(\nu)$. Then (2.121) implies that

$$\sum_{\eta \in \text{Sign}_k^+} T(\vartheta \to \eta) \mathbb{E}_{\mathbf{u} \cup u; \varrho} H(\bullet; \eta) = \mathbb{E}_{\mathbf{u}; \varrho} H(\bullet; \vartheta), \tag{2.124}$$

where $T(\vartheta \to \eta) := q^{-k}(-s)^{|\vartheta|-|\eta|} \mathsf{G}_{\eta/\vartheta}((qu)^{-1})$. In (2.124) by '$\bullet$' we mean the variables in which the expectation is applied. To see (2.124), apply the operator T inside the integral, and note that (2.40) is equivalent to

$$\sum_{\eta \in \mathsf{Sign}_k^+} T(\vartheta \to \eta)(-s)^{|\eta|}\mathsf{F}_\eta^{\mathsf{c}}(w_1^{-1}, \dots, w_k^{-1}) = \prod_{i=1}^{k} \frac{1 - uw_i}{1 - quw_i}(-s)^{|\vartheta|}\mathsf{F}_\vartheta^{\mathsf{c}}(w_1^{-1}, \dots, w_k^{-1}).$$

$$(2.125)$$

On the other hand, adding the new parameter u to the specialization $\mathbf{u}$ in (2.124) corresponds to time evolution, i.e., to the application of the operator $\mathscr{D}_{u;\varrho}^+$ (2.90). That is, the left-hand side of (2.124) can be written as

$$\sum_{\lambda \in \mathsf{Sign}_{n+1}^+} \sum_{\eta \in \mathsf{Sign}_k^+} T(\vartheta \to \eta)\mathscr{M}_{\mathbf{u} \cup u;\varrho}(\lambda)H(\lambda; \eta)$$

$$= \sum_{\mu \in \mathsf{Sign}_n^+} \mathscr{M}_{\mathbf{u};\varrho}(\mu) \sum_{\eta \in \mathsf{Sign}_k^+} T(\vartheta \to \eta) \sum_{\lambda \in \mathsf{Sign}_{n+1}^+} \mathscr{D}_{u;\varrho}^+(\mu \to \lambda)H(\lambda; \eta).$$

Since the right-hand side of (2.124) involves the expectation with respect to the same measure $\mathscr{M}_{\mathbf{u};\varrho}$ and since identity (2.124) holds for arbitrary u_i's, this suggests the *duality relation*

$$\mathscr{D}_{u;\varrho}^+ H T^{\text{transpose}} = H, \qquad (2.126)$$

where the operators $\mathscr{D}_{u;\varrho}^+$ and $T^{\text{transpose}}$ are applied in the first and the second variable in H, respectively. Similar duality relations can be written down by considering q-moments which are computed in Theorem 2.9.8.

It is worth noting that (self-)dualities like (2.126) can sometimes be independently proven from the very definition of the dynamics, and then utilized to produce nested contour integral formulas for the observables of these dynamics. This can be thought of as an alternative way to proving results like Theorem 2.8.13. Let us outline this argument. Applying $(\mathscr{D}_{u;\varrho}^+)^n$ to (2.126) gives

$$(\mathscr{D}_{u;\varrho}^+)^{n+1}H T^{\text{transpose}} = (\mathscr{D}_{u;\varrho}^+)^n H.$$

Taking the expectation in both sides above, we arrive back at our starting point (2.124):

$$(\mathbb{E}_{(u^{n+1});\varrho}H)T^{\text{transpose}} = \mathbb{E}_{(u^n);\varrho}H, \qquad (u^m) := \underbrace{(u, \dots, u)}_{m \text{ times}}, \qquad (2.127)$$

where, as before, the expectation of $H = H(\nu; \vartheta)$ is taken with respect to the probability distribution in ν, and the operator $T^{\text{transpose}}$ acts on ϑ. Thus, knowing (2.126) and passing to (2.127), one gets a closed system of linear equations for the observables $\mathbb{E}_{u^n;\varrho}H(\bullet; \vartheta)$, where n runs over $\mathbb{Z}_{\geq 0}$, and ϑ — over Sign_k^+. This system can sometimes be reduced to a simpler system of free evolution equations subject to certain two-body boundary conditions, and the latter can be solved explicitly in terms of nested contour integrals.

This alternative route towards explicit formulas for averaging of observables was taken (for various degenerations of the higher spin six vertex model) in Borodin et al.

(2014); Borodin and Corwin (2015) and Corwin (2014). Duality for the higher spin six-vertex model started from infinitely many particles at the leftmost location was considered in Corwin and Petrov (2016).

Remark 2.8.14.   An advantage of this alternative route starting from duality (2.126) is that it implies Eqs (2.127) for arbitrary (sufficiently nice) initial conditions, because one can take an arbitrary expectation in the last step leading to (2.127).[21] This argument could lead to nested contour integral formulas for arbitrary initial conditions, similarly to what is done in Borodin et al. (2015a, b). We will not discuss duality relations or formulas with arbitrary initial conditions here.

2.9   *q*-moments of the height function

Here we compute another type of observables of the stochastic dynamics $\mathscr{X}^{+}_{\{u_t\};\varrho}$ started from the empty initial configuration—the *q*-moments of its height function. These moment formulas could be viewed as the main result of the present notes.

2.9.1   Height function and its *q*-moments

Let $\nu \in \mathsf{Sign}^{+}_{n}$. Define the *height function* corresponding to ν as

$$\mathfrak{h}_{\nu}(x) := \#\{j \colon \nu_j \geq x\}, \qquad x \in \mathbb{Z}.$$

Clearly, $\mathfrak{h}_{\nu}(x)$ is a nonincreasing function of x, $\mathfrak{h}_{\nu}(0) = n$, and $\mathfrak{h}_{\nu}(+\infty) = 0$. Our goal now is to compute the (multipoint) *q*-moments

$$\mathbb{E}_{\mathbf{u};\varrho} \prod_{i=1}^{\ell} q^{\mathfrak{h}_{\nu}(x_i)} = \sum_{\nu \in \mathsf{Sign}^{+}_{n}} \mathscr{M}_{\mathbf{u};\varrho}(\nu) \prod_{i=1}^{\ell} q^{\mathfrak{h}_{\nu}(x_i)}$$

of the height function, where $x_1 \geq \ldots \geq x_{\ell} \geq 1$ are arbitrary. Note that this summation ranges only over signatures with $\nu_n \geq 1$.

Lemma 2.9.1.   *Fix* $n \in \mathbb{Z}_{\geq 0}$ *and* $u_1, \ldots, u_n$ *satisfying* (2.107). *Then for any* ℓ *and* $x_1 \geq \ldots \geq x_{\ell} \geq 1$, *the q-moments* $\mathbb{E}_{\mathbf{u};\varrho} \prod_{i=1}^{\ell} q^{\mathfrak{h}_{\nu}(x_i)}$ *are rational functions in the* u_i *'s and the parameters q and s.*

Proof This is established similarly to Lemma 2.8.10, because if $\mathfrak{h}_{\nu}(x_1) \in \{0, 1, \ldots, n\}$ is fixed, then there is a fixed number of the coordinates of ν belonging to an infinite range, and the summation over them produces a rational function.

We will first use the *q*-correlation functions discussed in Section 2.8 to compute one-point *q*-moments $\mathbb{E}_{\mathbf{u};\varrho} q^{\ell \, \mathfrak{h}_{\nu}(x)}$. The formula for these one-point *q*-moments allows to

---

[21] This is in contrast with (2.124) which is implied by (2.121), and thus holds only for the dynamics $\mathscr{X}^{+}$ started from the empty initial configuration.

formulate an analogous multi-point statement, and we will then present its verification proof. Thus, the one-point formula will be proven in two different ways.

One-point q-moments from q-correlations

Let us first establish an algebraic identity connecting one-point q-moments with q-correlation functions. In fact, the identity holds even before taking the expectation:

Lemma 2.9.2.   *For any $x \geq 1$, $\ell \geq 0$, and a signature ν, we have*

$$q^{\ell \, \mathfrak{h}_\nu(x)} = \sum_{k=0}^{\ell} (-q)^{-k} \binom{\ell}{k}_q (q; q)_k \sum_{\vartheta_1 \geq \ldots \geq \vartheta_k \geq x} \mathcal{Q}_{(\vartheta_1, \ldots, \vartheta_k)}(\nu). \tag{2.128}$$

Proof Denote $\Delta \mathfrak{h}_\nu(x) := \mathfrak{h}_\nu(x) - \mathfrak{h}_\nu(x+1)$; this is the number of parts of ν that are equal to x. First, let us express the quantities $\mathcal{Q}_\vartheta(\nu)$ through the height function. We start with the case $\vartheta = (x^\ell) = (x, \ldots, x)$. We have

$$\mathcal{Q}_{(x^\ell)}(\nu) = \sum_{\substack{\mathcal{I} = \{i_1 < \ldots < i_\ell\} \subseteq \{1, \ldots, n\} \\ \nu_{i_1} = \ldots = \nu_{i_\ell} = x}} q^{i_1 + \ldots + i_\ell}$$

$$= q^{\frac{\ell(\ell+1)}{2}} q^{\ell \, \mathfrak{h}_\nu(x+1)} \binom{\Delta \mathfrak{h}_\nu(x)}{\ell}_q$$

$$= q^{\frac{\ell(\ell+1)}{2}} \frac{(q^{\Delta \mathfrak{h}_\nu(x)}; q^{-1})_\ell}{(q; q)_\ell} q^{\ell \, \mathfrak{h}_\nu(x+1)},$$

where the second equality follows similarly to the computation of the partition function (2.69).

For general

$$\vartheta = (x_1^{\ell_1}, \ldots, x_m^{\ell_m}) := (\underbrace{x_1, \ldots, x_1}_{\ell_1 \text{ times}}, \ldots, \underbrace{x_m, \ldots, x_m}_{\ell_m \text{ times}}),$$

where $x_1 > \ldots > x_m \geq 0$ and $\ell = (\ell_1, \ldots, \ell_m) \in \mathbb{Z}_{\geq 0}^m$, the summation over $\mathcal{I}$ in (2.117) is clearly equal to the product of individual summations corresponding to each x_j, $j = 1, \ldots, m$. Therefore,

$$\mathcal{Q}_{(x_1^{\ell_1}, \ldots, x_m^{\ell_m})}(\nu) = \prod_{j=1}^{m} q^{\frac{\ell_j(\ell_j+1)}{2}} \frac{(q^{\Delta \mathfrak{h}_\nu(x_j)}; q^{-1})_{\ell_j}}{(q; q)_{\ell_j}} q^{\ell_j \, \mathfrak{h}_\nu(x_j+1)}. \tag{2.129}$$

Our next goal is to invert relation (2.129). Let us write down certain abstract inversion formulas which will lead us to the desired statement. In these formulas, we will assume that $A, B, A_0, A_1, A_2, \ldots$ are indeterminates. Let us also denote that

$$T_i(A) := q^{\frac{i(i+1)}{2}} \frac{(A; q^{-1})_i}{(q; q)_i}, \qquad R_i := \frac{(-q)^i}{(q; q)_i}.$$

Note that by the very definition, $T_i(A) = 0$ for $i < 0$, $T_0(A) = 1$, and $T_i(1) = 1_{i=0}$. Moreover, $R_i = 0$ for $i < 0$, and $R_0 = 1$.

The first inversion formula is

$$A^n R_n = \sum_{k=0}^{n} T_k(A) R_{n-k}. \tag{2.130}$$

Indeed, multiply this identity by B^n, and sum over $n \geq 0$. The left-hand side gives, by the q-binomial theorem,

$$\sum_{n=0}^{\infty} (AB)^n \frac{(-q)^n}{(q;q)_n} = \frac{1}{(-ABq;q)_\infty},$$

and in the right-hand side we first sum over $n \geq k$ and then over $k \geq 0$, which yields

$$\sum_{k=0}^{\infty} q^{\frac{k(k+1)}{2}} \frac{(A;q^{-1})_k}{(q;q)_k} B^k \sum_{n=k}^{\infty} \frac{(-q)^{n-k}}{(q;q)_{n-k}} B^{n-k} = \frac{1}{(-Bq;q)_\infty} \sum_{k=0}^{\infty} q^{\frac{k(k+1)}{2}} \frac{(A;q^{-1})_k}{(q;q)_k} B^k$$

$$= \frac{1}{(-Bq;q)_\infty} \sum_{k=0}^{\infty} \frac{(A^{-1},q)_k}{(q;q)_k} (-q)^k A^k B^k$$

$$= \frac{1}{(-Bq;q)_\infty} \frac{(-Bq;q)_\infty}{(-ABq;q)_\infty} = \frac{1}{(-ABq;q)_\infty},$$

where we have used the q-binomial theorem twice. This establishes (2.130), because the generating series of both its sides coincide.[22]

Replace A by A_1 in (2.130), multiply it by $A_2 = A_2^k A_2^{n-k}$, and apply (2.130) to $A_2^{n-k} R_{n-k}$ in the right-hand side. Continuing this process with $A_3, \ldots, A_N$, we obtain for any $\ell \geq 0$ and $N \geq 1$:

$$(A_1 \ldots A_N)^\ell = \sum_{k \in \mathbb{Z}_{\geq 0}^N} \frac{R_{\ell - |k|}}{R_\ell} \prod_{j=1}^{N} T_{k_j}(A_j) (A_{j+1} A_{j+2} \ldots A_N)^{k_j}, \tag{2.131}$$

where the sum is over all (unordered) nonnegative integer vectors $k = (k_1, \ldots, k_N)$ of length N. Here and in the following $|k|$ stands for $k_1 + \ldots + k_N$. Clearly, the sum ranges only over k with $|k| \leq \ell$. Note that if only finitely many of the indeterminates A_j differ from 1, then one can send $N \to +\infty$ in (2.131) and sum over integer vectors k of arbitrary length.

If we set $A_j := q^{\Delta \mathfrak{h}_\nu (x+j-1)}$ and send $N \to +\infty$, the left-hand side of (2.131) becomes $q^{\ell \mathfrak{h}_\nu (x)}$, and in the right-hand side we obtain

$$T_{k_j}(A_j) (A_{j+1} A_{j+2} \ldots)^{k_j} = q^{\frac{k_j(k_j+1)}{2}} \frac{(q^{\Delta \mathfrak{h}_\nu (x+j-1)}; q^{-1})_{k_j}}{(q;q)_{k_j}} q^{k_j \mathfrak{h}_\nu (x+j)}.$$

---

[22] In these manipulations with infinite series we assume that $0 \leq q < 1$ and that A and B are sufficiently small. Alternatively, it is enough to think that we are working with formal power series.

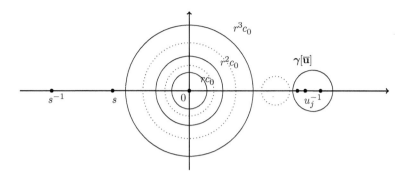

Fig. 2.26 A possible choice of integration contours $\gamma[\bar{u}|1] = \gamma[\bar{u}] \cup rc_0$, $\gamma[\bar{u}|2] = \gamma[\bar{u}] \cup r^2c_0$, and $\gamma[\bar{u}|3] = \gamma[\bar{u}] \cup r^3c_0$ for $\ell = 3$ in Definition 2.9.4. Contours $q\gamma[\bar{u}]$ and $q\gamma[\bar{u}|2]$, $q\gamma[\bar{u}|3]$ are shown dotted.

Therefore, the product of these quantities in (2.131) matches formula (2.129) for $\mathcal{Q}_\vartheta(\nu)$, where the point $x + j - 1$ enters the signature ϑ with multiplicity $k_j \geq 0$. This yields the desired formula.

Remark 2.9.3.   Using a similar approach as in lemma 2.9.2, one can write down more complicated formulas expressing $\prod_{i=1}^{\ell} q^{\mathfrak{h}_\nu(x_i)}$ for any $x_1 \geq \ldots \geq x_\ell \geq 1$ through the quantities $\mathcal{Q}_\vartheta(\nu)$. However, except for the one-point case, these expressions do not seem to be convenient for computing the q-moments. Therefore, in Section 2.9.1 we present a verification-style proof for the multipoint q-moments.

Definition 2.9.4.   Fix $\ell \in \mathbb{Z}_{\geq 1}$. Let $u_i > 0$ and $u_i \neq qu_j$ for any i, j. Then the integration contour $\gamma[\bar{u}]$ encircling all u_i^{-1} is well-defined (see Definition 2.8.12). Let also c_0 be a positively oriented circle around 0 which is sufficiently small. Let $r > q^{-1}$ be such that $q\gamma[\bar{u}]$ does not intersect $r^\ell c_0$, and $r^\ell c_0$ does not encircle s. Denote $\gamma[\bar{u}|j] := \gamma[\bar{u}] \cup r^j c_0$, where $j = 1, \ldots, \ell$. See Fig. 2.26.

We are now in a position to compute the one-point q-moments.

Proposition 2.9.5.   Let $u_i > 0$ and $u_i \neq qu_j$ for any i, j. Then for any $\ell \in \mathbb{Z}_{\geq 0}$ and $x \in \mathbb{Z}_{\geq 1}$ we have

$$
\mathbb{E}_{\mathbf{u};\varrho} q^{\ell \, \mathfrak{h}_\nu(x)} = q^{\frac{\ell(\ell-1)}{2}} \oint_{\gamma[\bar{u}|1]} \frac{dw_1}{2\pi \mathbf{i}} \cdots \oint_{\gamma[\bar{u}|\ell]} \frac{dw_\ell}{2\pi \mathbf{i}} \prod_{1 \leq \alpha < \beta \leq \ell} \frac{w_\alpha - w_\beta}{w_\alpha - qw_\beta}
$$

$$
\times \prod_{i=1}^{\ell} \left(w_i^{-1} \left(\frac{1 - sw_i}{1 - s^{-1}w_i} \right)^{x-1} \prod_{j=1}^{n} \frac{1 - qu_j w_i}{1 - u_j w_i} \right). \tag{2.132}
$$

Proof Taking the expectation with respect to $\mathscr{M}_{\mathbf{u};\varrho}$ in both sides of (2.128) and using (2.121) in the right-hand side, we obtain

$$\mathbb{E}_{\mathbf{u};\varrho} q^{\ell \, \mathfrak{h}_\nu(x)} = \sum_{k=0}^{\ell} \binom{\ell}{k}_q \frac{q^{\frac{k(k-1)}{2}} (q;q)_k}{(1-q)^k k!}$$

$$\times \sum_{\vartheta_1 \geq \dots \geq \vartheta_k \geq x-1} \oint_{\gamma[\bar{u}]} \frac{dw_1}{2\pi i} \cdots \oint_{\gamma[\bar{u}]} \frac{dw_k}{2\pi i} \prod_{1 \leq \alpha \neq \beta \leq k} \frac{w_\alpha - w_\beta}{w_\alpha - q w_\beta}$$

$$\times (-s)^{|\vartheta|} \frac{\mathsf{F}_\vartheta^{\mathsf{c}}(w_1^{-1}, \dots, w_k^{-1})}{w_1 \dots w_k} \prod_{i=1}^{k} \prod_{j=1}^{n} \frac{1 - q u_j w_i}{1 - u_j w_i}.$$

Because $\vartheta_k \geq x-1$, we can subtract $(x-1)$ from all parts of ϑ. We readily have

$$(-s)^{|\vartheta|} \mathsf{F}_\vartheta^{\mathsf{c}}(w_1^{-1}, \dots, w_k^{-1})$$

$$= (-s)^{k(x-1)} \prod_{i=1}^{k} \left(\frac{w_i^{-1} - s}{1 - s w_i^{-1}} \right)^{x-1} \underbrace{(-s)^{|\vartheta|-k(x-1)} \mathsf{F}_{\vartheta-(x-1)k}^{\mathsf{c}}(w_1^{-1}, \dots, w_k^{-1})}_{\mathscr{M}_{(w_1^{-1}, \dots, w_k^{-1});\varrho}(\vartheta - (x-2)^k) \text{ by } (2.106)},$$

where we have also used the fact that $\vartheta_k - (x-2) \geq 1$. The probability weight $\mathscr{M}_{(w_1^{-1}, \dots, w_k^{-1});\varrho}$ is the only thing which now depends on ϑ, and the summation over all ϑ of these weights gives 1. This summation can be performed under the integral because on the contour $\gamma[\bar{u}]$ we have $\Re(w_j) > 0$, and so conditions (2.107) with u_j replaced by w_j^{-1} hold. We see that this summation over ϑ yields

$$\mathbb{E}_{\mathbf{u};\varrho} q^{\ell \, \mathfrak{h}_\nu(x)} = \sum_{k=0}^{\ell} \binom{\ell}{k}_q q^{\frac{k(k-1)}{2}} \oint_{\gamma[\bar{u}]} \frac{dw_1}{2\pi i} \cdots \oint_{\gamma[\bar{u}]} \frac{dw_k}{2\pi i} \prod_{1 \leq \alpha < \beta \leq k} \frac{w_\alpha - w_\beta}{w_\alpha - q w_\beta}$$

$$\times \frac{1}{w_1 \dots w_k} \prod_{i=1}^{k} \left(\frac{1 - s w_i}{1 - s^{-1} w_i} \right)^{x-1} \prod_{i=1}^{k} \prod_{j=1}^{n} \frac{1 - q u_j w_i}{1 - u_j w_i}.$$

Here we have applied the symmetrization formula (footnote 7) to rewrite $(q;q)_k/(1-q)^k$, which canceled the factor $1/k!$ and half of the product over $\alpha \neq \beta$.

Finally, the summation over k in the above formula can be eliminated by changing the integration contours with the help of Borodin et al. (2014, Lemma 4.21) (which we recall as Lemma 2.9.6 for convenience). This completes the proof of the desired identity (2.132).

Lemma 2.9.6 (Borodin et al. 2014, Lemma 4.21). *Let $\ell \geq 1$ and $f(w)$ with $f(0) = 1$ be a meromorphic function in $\mathbb{C}$ having no poles in a disc around 0. Then we have*

$$\oint_{\gamma[\bullet|1]} \frac{dw_1}{2\pi i} \cdots \oint_{\gamma[\bullet|\ell]} \frac{dw_\ell}{2\pi i} \prod_{1 \leq \alpha < \beta \leq \ell} \frac{w_\alpha - w_\beta}{w_\alpha - q w_\beta} \prod_{i=1}^{\ell} \frac{f(w_i)}{w_i}$$

$$= \sum_{k=0}^{\ell} \binom{\ell}{k}_q q^{\frac{1}{2}k(k-1) - \frac{1}{2}\ell(\ell-1)} \oint_{\gamma[\bullet]} \frac{dw_1}{2\pi i} \cdots \oint_{\gamma[\bullet]} \frac{dw_k}{2\pi i} \prod_{1 \leq \alpha < \beta \leq k} \frac{w_\alpha - w_\beta}{w_\alpha - q w_\beta} \prod_{i=1}^{k} \frac{f(w_i)}{w_i},$$

where as $\gamma[\bullet]$ *we can take an arbitrary closed contour not encircling* 0, *and all other contours and conditions on them are analogous to Definition 2.9.4.*

Remark 2.9.7 (Fredholm determinants). Using a general approach outlined in the paper Borodin et al. (2014), the one-point q-moment formula of Proposition 2.9.5 (as well as its degenerations discussed in Section 2.10) can be employed to obtain Fredholm determinantal expressions for the q-Laplace transform

$$\mathbb{E}_{\mathbf{u};\varrho}\left(\frac{1}{(\zeta q^{\mathfrak{h}_\nu(x)};q)_\infty}\right)$$

of the height function, which may be suitable for asymptotic analysis. We will not pursue this direction here.

Multipoint q-moment formula

By analogy with existing multipoint q-moment formulas for related systems,[23] we can formulate a generalization of Proposition 2.9.5:

Theorem 2.9.8. Let $u_i > 0$ and $u_i \neq qu_j$ for any $i,j = 1,\ldots,n$. Then for any integers $x_1 \geq \ldots \geq x_\ell \geq 1$ the corresponding q-moment of the dynamics $\mathscr{X}^+_{\{u_t\};\varrho}$ at time $= n$ is given by

$$\mathbb{E}_{\mathbf{u};\varrho}\prod_{i=1}^{\ell}q^{\mathfrak{h}_\nu(x_i)} = q^{\frac{\ell(\ell-1)}{2}}\oint_{\gamma[\bar{u}|1]}\frac{dw_1}{2\pi\mathrm{i}}\cdots\oint_{\gamma[\bar{u}|\ell]}\frac{dw_\ell}{2\pi\mathrm{i}}\prod_{1\leq\alpha<\beta\leq\ell}\frac{w_\alpha-w_\beta}{w_\alpha-qw_\beta}$$

$$\times\prod_{i=1}^{\ell}\left(w_i^{-1}\left(\frac{1-sw_i}{1-s^{-1}w_i}\right)^{x_i-1}\prod_{j=1}^{n}\frac{1-qu_jw_i}{1-u_jw_i}\right),\tag{2.133}$$

where the integration contours are described in Definition 2.9.4.

Corollary 2.9.9. Let the parameters $\mathbf{u}$ have the form (2.122). Then the q-moments of the height function $\mathbb{E}_{\mathbf{u};\varrho}\prod_{i=1}^{\ell}q^{\mathfrak{h}_\nu(x_i)}$ are given by the same formula as (2.133), but with integration contours $w_j \in \gamma_j^-[\bar{u}|j] := \gamma_j^-[\bar{u}] \cup r^j c_0$.

Proof of Corollary 2.9.9 Assume that Theorem 2.9.8 holds. We argue as in the proof of case 3 in Theorem 2.8.13, by first taking $\mathbf{u}$ with $u_i > 0$ and $q \cdot \max_i u_i < \min_i u_i$, which makes it possible to immediately pass to the nested contours $\gamma_j^-[\bar{u}|j]$ in (2.133). Then we can move u_2^{-1} outside $\gamma_1^-[\bar{u}]$ but still inside $\gamma_2^-[\bar{u}]$, set $u_2 = qu_1$,

---

[23] Namely, q-TASEPs Borodin et al. (2014); Borodin and Corwin (2015), q-Hahn TASEP (Corwin 2014), and stochastic higher spin six-vertex model Corwin and Petrov (2016). Note that, however, all these systems start with infinitely many particles at the leftmost location, and in our system a new particle is always added at location 1, so that the corresponding degenerations of Theorem 2.9.8 do not follow from those works.

and continue specializing the rest of **u** to (2.122) in a similar way. This specialization inside the integral will coincide with the same specialization of the left-hand side of (2.133), and thus the corollary is established.   □

The rest of this part it devoted to the proof of Theorem 2.9.8. The proof is of verification type: we start with the nested contour integral in the right-hand side of (2.133), and rewrite it as an expectation with respect to $\mathcal{M}_{\mathbf{u};\varrho}$.

Lemma 2.9.10.   *Under the assumptions of Theorem 2.9.8, the collection of identities* (2.133) *(for all $\ell \geq 1$ and all $x_1 \geq \ldots \geq x_\ell \geq 1$) follows from a collection of identities of the form*

$$\mathbb{E}_{u;\varrho} \prod_{i=1}^{\ell} (q^{i-1} - q^{\mathfrak{h}_\nu(x_i)}) = (-1)^\ell q^{\frac{\ell(\ell-1)}{2}} \oint_{\gamma[\bar{u}]} \frac{dw_1}{2\pi i} \cdots \oint_{\gamma[\bar{u}]} \frac{dw_\ell}{2\pi i} \prod_{1 \leq \alpha < \beta \leq \ell} \frac{w_\alpha - w_\beta}{w_\alpha - qw_\beta}$$

$$\times \prod_{i=1}^{\ell} \left(w_i^{-1} \left(\frac{1 - sw_i}{1 - s^{-1}w_i} \right)^{x_i - 1} \prod_{j=1}^{n} \frac{1 - qu_j w_i}{1 - u_j w_i} \right). \tag{2.134}$$

That is, removing the parts of the contours around 0 leads to a modification of the left-hand side, as shown earlier.

This lemma should also hold in the opposite direction (that identities (2.133) imply (2.134)), but we do not need this statement.

Proof The right-hand side of (2.133) can be written as

$$q^{\frac{\ell(\ell-1)}{2}} \oint_{\gamma[\bar{u}|1]} \frac{dw_1}{2\pi i} \cdots \oint_{\gamma[\bar{u}|\ell]} \frac{dw_\ell}{2\pi i} \prod_{1 \leq \alpha < \beta \leq \ell} \frac{w_\alpha - w_\beta}{w_\alpha - qw_\beta} \prod_{i=1}^{\ell} \frac{f_{x_i}(w_i)}{w_i}, \tag{2.135}$$

where each $f_x(w)$, $x \in \mathbb{Z}_{\geq 1}$, is a meromorphic (in fact, rational) function without poles in a disc around 0, and $f_x(0) = 1$.

Split the integral in (2.135) into 2^n integrals indexed by subsets $\mathcal{I} \subseteq \{1, \ldots, \ell\}$ determining that w_i for $i \notin \mathcal{I}$ are integrated around 0, while other w_i's are integrated over $\gamma[\bar{u}]$. Let $|\mathcal{I}| = k$, $k = 0, \ldots, \ell$, and also denote $\|\mathcal{I}\| := \sum_{i \in \mathcal{I}} i$. Let $\mathcal{I} = \{i_1 < \ldots < i_k\}$ and $\{1, 2, \ldots, \ell\} \setminus \mathcal{I} = \{p_1 < \ldots < p_{\ell-k}\}$. The contours around 0 (corresponding to w_{p_j}) can be shrunk to 0 in the order $p_1, \ldots, p_{\ell-k}$ without crossing any other poles, and each such integration produces the factor $q^{-(\ell-p_j)}$ coming from the cross-product over $\alpha < \beta$. Thus, (2.135) becomes (after renaming $w_{i_j} = z_j$)

$$\sum_{k=0}^{\ell} \sum_{\mathcal{I}=\{i_1<\ldots<i_k\}\subseteq\{1,\ldots,\ell\}} q^{k\ell-\|\mathcal{I}\|} \oint_{\gamma[\bar{u}]} \frac{dz_1}{2\pi i} \cdots \oint_{\gamma[\bar{u}]} \frac{dz_k}{2\pi i} \prod_{1 \leq \alpha < \beta \leq k} \frac{z_\alpha - z_\beta}{z_\alpha - qz_\beta} \prod_{j=1}^{k} \frac{f_{x_{i_j}}(z_j)}{z_j}.$$

$$\tag{2.136}$$

The summation (2.136) now involves integrals as in the right-hand side of (2.134). If the latter identity holds, then we can rewrite each such integral as a certain expectation as in the left-hand side of (2.134). Relation (2.133) now follows from a formal identity in indeterminates $\hat{q}, X_1, \ldots, X_\ell$,

$$\sum_{k=0}^{\ell} \sum_{\mathcal{I}=\{i_1<\ldots<i_k\}\subseteq\{1,\ldots,\ell\}} \hat{q}^{\frac{(\ell-k)(\ell-k+1)}{2}} (X_{i_1} - \hat{q}^{i_1})$$

$$\times (X_{i_2} - \hat{q}^{i_2-1}) \ldots (X_{i_k} - \hat{q}^{i_k-k+1}) = X_1 \ldots X_\ell, \qquad (2.137)$$

where we have matched $\hat{q}$ to q^{-1} in (2.135) and (2.136). To establish (2.137), observe that both its sides are linear in X_1, and so it suffices to show that the identity holds at two points, say, $X_1 = \hat{q}$ and $X_1 = \infty$. Substituting each of these values into (2.137) leads to an equivalent identity with ℓ replaced by $\ell - 1$. Namely, for $X_1 = \hat{q}$ we obtain

$$\sum_{k=0}^{\ell-1} \sum_{\mathcal{I}=\{1<i_1<\ldots<i_k\}\subseteq\{1,\ldots,\ell\}} \hat{q}^{\frac{(\ell-1-k)(\ell-1-k+1)}{2}}$$

$$\times (\hat{q}^{-1}X_{i_1} - \hat{q}^{i_1-1}) \ldots (\hat{q}^{-1}X_{i_k} - \hat{q}^{i_k-k}) = \hat{q}^{1-\ell}X_2 \ldots X_\ell,$$

which becomes (2.137) with $\ell - 1$ after setting $Y_i = \hat{q}^{-1}X_{i+1}$. Dividing (2.137) by X_1 and letting $X_1 \to \infty$, we obtain

$$\sum_{k=1}^{\ell} \sum_{\mathcal{I}=\{1=i_1<\ldots<i_k\}\subseteq\{1,\ldots,\ell\}} \hat{q}^{\frac{((\ell-1)-(k-1))((\ell-1)-(k-1)+1)}{2}}$$

$$\times (X_{i_2} - \hat{q}^{i_2-1}) \ldots (X_{i_k} - \hat{q}^{i_k-k+1}) = X_2 \ldots X_\ell,$$

which becomes (2.137) with $\ell - 1$ after setting $Y_i = X_{i+1}$. Thus, (2.137) follows by induction, which implies the lemma.

Below in our proof of Theorem 2.9.8 we will assume that the u_j's are pairwise distinct. If (2.133) and (2.134) hold for distinct u_j's, then when some of the u_j's coincide the same formulas can be obtained by a simple substitution. Indeed, this is because both sides of each of the identities are a priori rational functions in the u_j's belonging to a certain domain (cf. the proof of Proposition 2.7.15).

Denote the right-hand side of (2.134) by $R(u_1, \ldots, u_n)$. First, let us show that there exists a decomposition of $R(u_1, \ldots, u_n)$ into the functions F^c_λ.

Lemma 2.9.11. *If $q \in (0,1)$ is sufficiently close to 1 and all the u_i's are sufficiently close to s, then the integral in the right-hand side of (2.134) can be written as*

$$R(u_1, \ldots, u_n) = \sum_{\lambda \in \mathsf{Sign}_n^+} r_\lambda \mathsf{F}^c_\lambda(u_1, \ldots, u_n), \qquad (2.138)$$

where the sum over λ converges uniformly in $u_j \in \gamma_j^+[s]$, $j = 1, \ldots, n$.

Proof Write the product in (2.134) as a sum over $\mu \in \mathsf{Sign}_n^+$ using the Cauchy identity (Corollary 2.4.11):

$$\prod_{i=1}^{\ell}\prod_{j=1}^{n}\frac{1-qu_jw_i}{1-u_jw_i}=\frac{1}{(s^2;q)_n}\prod_{i=1}^{n}\frac{1-su_i}{1-s^{-1}u_i}$$

$$\times \sum_{\mu\in\mathsf{Sign}_n^+:\,\mu_n\geq 1} \mathsf{F}_\mu^c(u_1,\ldots,u_n)\mathsf{G}_\mu(\varrho,w_1,\ldots,w_\ell), \qquad (2.139)$$

where we also used the specialization ϱ, cf. (2.88) and Proposition 2.8.2. This is possible if $(u_i,w_j)\in\mathsf{Adm}$ for all i,j, and the u_i's satisfy (2.107). These conditions can be achieved by a deformation of the contour $\gamma[\bar{u}]$ because the u_i's can be taken close to s (the argument is similar to the proof of Proposition 2.8.11).

This also implies that the sum over μ in (2.139) converges uniformly in w_i belonging to the deformed contours. Thus, the integration in the w_j's can be performed for each μ separately. These integrals involving $\mathsf{G}_\mu(\varrho,w_1,\ldots,w_\ell)$ obviously do not introduce any new dependence on the u_i's.

Therefore, the right-hand side of (2.139) depends on the u_i's only through the functions $\mathsf{F}_{\mu-1^n}^c(u_1,\ldots,u_n)$ (cf. (2.46)), which yields expansion (2.138). Moreover, we see that

$$r_{\mu-1^n}=\frac{(-1)^\ell(-s)^n q^{\frac{\ell(\ell-1)}{2}}}{(s^2;q)_n}\oint_{\gamma[s^{-1}]}\frac{dw_1}{2\pi\mathbf{i}}\cdots\oint_{\gamma[s^{-1}]}\frac{dw_\ell}{2\pi\mathbf{i}}\prod_{1\leq\alpha<\beta\leq\ell}\frac{w_\alpha-w_\beta}{w_\alpha-qw_\beta}$$

$$\times\mathsf{G}_\mu(\varrho,w_1,\ldots,w_\ell)\prod_{i=1}^{\ell}w_i^{-1}\left(\frac{1-sw_i}{1-s^{-1}w_i}\right)^{x_i-1}.$$

One can readily check that these coefficients grow in μ not faster than of order $\exp\{c|\mu|\}$ for some constant $c>0$. Thus, by taking the u_i's closer to s if necessary, one can ensure that the sum in (2.138) converges uniformly in the u_i's. If q is sufficiently close to 1, these u_i's can be chosen on the contours $\gamma_i^+[s]$.

The integral formula for the coefficients r_λ in the proof of Lemma 2.9.11 does not seem to be convenient for their direct computation. We will instead rewrite $R(u_1,\ldots,u_n)$ by integrating over the w_i's in (2.134), and then employ orthogonality of the functions F_λ to extract the r_λ's. This will imply that $R(u_1,\ldots,u_n)$ is equal to the left-hand side of (2.134), yielding Theorem 2.9.8.

The integral in (2.134) can be computed by taking residues at $w_i=u_{\sigma(i)}^{-1}$ for all $i=1,\ldots,\ell$, where σ runs over all maps $\{1,\ldots,\ell\}\to\{1,\ldots,n\}$ (we will see later that other residues do not participate). Denote the residue corresponding to σ by Res_σ, and also denote $\mathcal{J}:=\sigma(\{1,\ldots,\ell\})$. Because of the factors $w_\alpha-w_\beta$, the same $u_{\sigma(i)}^{-1}$ cannot participate twice, so σ must be injective. Thus, in contrast with (2.133), the integral in the right-hand side of (2.134) vanishes if $\ell>n$. Note, however, that since $\mathfrak{h}_\nu(x)\leq n$ for all $x\in\mathbb{Z}_{\geq 1}$, the product in the left-hand side

also vanishes for $\ell > n$, as it should be. Therefore, it suffices to consider only the case $\ell \leq n$.

The integral in (2.134) can be written in the form

$$(-1)^\ell q^{\frac{\ell(\ell-1)}{2}} \oint_{\gamma[\bar{u}]} \frac{dw_1}{2\pi i} \cdots \oint_{\gamma[\bar{u}]} \frac{dw_\ell}{2\pi i} \prod_{1 \leq \alpha < \beta \leq \ell} \frac{w_\alpha - w_\beta}{w_\alpha - q w_\beta}$$

$$\times \prod_{i=1}^{\ell} \left(f_{x_i}(w_i; \sigma) \prod_{j=1}^{\ell} \frac{1 - q u_{\sigma(j)} w_i}{1 - u_{\sigma(j)} w_i} \right),$$

with $f_x(w; \sigma) = w^{-1} \left(\frac{1 - sw}{1 - s^{-1}w} \right)^{x-1} \prod_{j \notin \mathcal{J}} \frac{1 - q u_j w}{1 - u_j w}$. Taking residues at the points $w_1 = u_{\sigma(1)}^{-1}, \ldots, w_\ell = u_{\sigma(\ell)}^{-1}$ (in this order), we see that

$$\text{Res}_\sigma = q^{\frac{\ell(\ell-1)}{2}} f_{x_1}(u_{\sigma(1)}^{-1}; \sigma) \frac{1-q}{u_{\sigma(1)}} (-1)^{\ell-1} \prod_{j=2}^{\ell} \frac{u_{\sigma(1)} - q u_{\sigma(j)}}{u_{\sigma(1)} - u_{\sigma(j)}}$$

$$\times \text{Res}_{w_\ell = u_{\sigma(\ell)}^{-1}} \cdots \text{Res}_{w_2 = u_{\sigma(2)}^{-1}} \prod_{2 \leq \alpha < \beta \leq \ell} \frac{w_\alpha - w_\beta}{w_\alpha - q w_\beta} \prod_{i=2}^{\ell} \left(f_{x_i}(w_i; \sigma) \prod_{j=2}^{\ell} \frac{1 - q u_{\sigma(j)} w_i}{1 - u_{\sigma(j)} w_i} \right)$$

$$= q^{\frac{\ell(\ell-1)}{2}} f_{x_1}(u_{\sigma(1)}^{-1}; \sigma) f_{x_2}(u_{\sigma(2)}^{-1}; \sigma) \frac{(1-q)^2}{u_{\sigma(1)} u_{\sigma(2)}}$$

$$\times (-1)^{\ell-2} \prod_{j=2}^{\ell} \frac{u_{\sigma(1)} - q u_{\sigma(j)}}{u_{\sigma(1)} - u_{\sigma(j)}} \prod_{j=3}^{\ell} \frac{u_{\sigma(2)} - q u_{\sigma(j)}}{u_{\sigma(2)} - u_{\sigma(j)}}$$

$$\times \text{Res}_{w_\ell = u_{\sigma(\ell)}^{-1}} \cdots \text{Res}_{w_3 = u_{\sigma(3)}^{-1}} \prod_{3 \leq \alpha < \beta \leq \ell} \frac{w_\alpha - w_\beta}{w_\alpha - q w_\beta} \prod_{i=3}^{\ell} \left(f_{x_i}(w_i; \sigma) \prod_{j=3}^{\ell} \frac{1 - q u_{\sigma(j)} w_i}{1 - u_{\sigma(j)} w_i} \right)$$

$$= \text{etc.}$$

$$= (1-q)^\ell q^{\frac{\ell(\ell-1)}{2}} \prod_{i=1}^{\ell} \frac{f_{x_i}(u_{\sigma(i)}^{-1}; \sigma)}{u_{\sigma(i)}} \prod_{1 \leq \alpha < \beta \leq \ell} \frac{u_{\sigma(\alpha)} - q u_{\sigma(\beta)}}{u_{\sigma(\alpha)} - u_{\sigma(\beta)}}$$

$$= (1-q)^\ell q^{\frac{\ell(\ell-1)}{2}} \prod_{i=1}^{\ell} \left(\frac{u_{\sigma(i)} - s}{u_{\sigma(i)} - s^{-1}} \right)^{x_i - 1}$$

$$\times \prod_{\alpha \in \mathcal{J}, \beta \notin \mathcal{J}} \frac{u_\alpha - q u_\beta}{u_\alpha - u_\beta} \prod_{1 \leq \alpha < \beta \leq \ell} \frac{u_{\sigma(\alpha)} - q u_{\sigma(\beta)}}{u_{\sigma(\alpha)} - u_{\sigma(\beta)}}.$$

In particular, we see that each step does not introduce any new poles inside the integration contours besides u_j^{-1}. Therefore,

$$R(u_1, \ldots, u_n) = \sum_{\sigma : \{1, \ldots, \ell\} \to \{1, \ldots, n\}} \text{Res}_\sigma(u_1, \ldots, u_n), \qquad (2.140)$$

with Res_σ as defined in the previous equation. This identity clearly holds for generic complex $u_1, \ldots, u_n$ (and not only for $u_i > 0$) because both sides are rational functions in the u_i's.

Let us now apply the inverse Plancherel transform $\mathscr{J}$ (Definition 2.7.8) without the factor $\mathsf{c}(\lambda)$ to $R(u_1, \ldots, u_n)$ to recover the coefficients r_λ in (2.138). This is possible for q sufficiently close to 1 because the series in the right-hand side of (2.138) converges uniformly in the u_i's on the contours involved in $\mathscr{J}$. The application of this slightly modified transform to $R(u_1, \ldots, u_n)$ written as (2.140) can be performed separately for each σ, and the result is the following.

Lemma 2.9.12. *For any $\sigma \colon \{1, \ldots, \ell\} \to \{1, \ldots, n\}$ and $\lambda \in \mathrm{Sign}_n^+$, we have*

$$
\oint_{\gamma_1^+[s]} \frac{du_1}{2\pi i} \cdots \oint_{\gamma_n^+[s]} \frac{du_n}{2\pi i} \prod_{1 \le \alpha < \beta \le n} \frac{u_\alpha - u_\beta}{u_\alpha - q u_\beta} \, Res_\sigma(u_1, \ldots, u_n) \prod_{i=1}^n \frac{1}{u_i - s} \left(\frac{1 - s u_i}{u_i - s} \right)^{\lambda_i}
$$

$$
= \mathbf{1}_{\{\lambda_{\sigma(i)} \ge x_i - 1 \text{ for all } i\}} \cdot (-s)^{|\lambda|} q^{inv(\sigma)} q^{\sigma(1) + \cdots + \sigma(\ell)} q^{-\ell} (1 - q)^\ell. \tag{2.141}
$$

Here the integration contours are described in Definition 2.7.2, and $inv(\sigma)$ is the number of inversions in σ, i.e., the number of pairs (i, j) with $i < j$ and $\sigma(i) > \sigma(j)$.

Proof We need to compute

$$
\oint_{\gamma_1^+[s]} \frac{du_1}{2\pi i} \cdots \oint_{\gamma_n^+[s]} \frac{du_n}{2\pi i} \prod_{1 \le \alpha < \beta \le n} \frac{u_\alpha - u_\beta}{u_\alpha - q u_\beta} \prod_{\alpha \in \mathcal{J}, \, \beta \notin \mathcal{J}} \frac{u_\alpha - q u_\beta}{u_\alpha - u_\beta}
$$

$$
\times \prod_{1 \le \alpha < \beta \le \ell} \frac{u_{\sigma(\alpha)} - q u_{\sigma(\beta)}}{u_{\sigma(\alpha)} - u_{\sigma(\beta)}} \prod_{i=1}^\ell \left(\left(\frac{u_{\sigma(i)} - s}{u_{\sigma(i)} - s^{-1}} \right)^{x_i - 1} \frac{1}{u_{\sigma(i)} - s} \left(\frac{1 - s u_{\sigma(i)}}{u_{\sigma(i)} - s} \right)^{\lambda_{\sigma(i)}} \right)
$$

$$
\times \prod_{\beta \notin \mathcal{J}} \frac{1}{u_\beta - s} \left(\frac{1 - s u_\beta}{u_\beta - s} \right)^{\lambda_\beta}.
$$

Observe the following:

- If $x_i \ge \lambda_{\sigma(i)} + 2$, then the integrand does not have a pole at s inside the contour $\gamma_{\sigma(i)}^+[s]$. In this case, if we can shrink this contour without picking residues at $u_{\sigma(i)} = q u_\beta$ for any $\beta > \sigma(i)$, then the whole integral vanishes.
- If $x_i \le \lambda_{\sigma(i)} + 1$, then the integrand does not have a pole at s^{-1} outside the contour $\gamma_{\sigma(i)}^+[s]$. Note also that for $\beta \notin \mathcal{J}$, the integrand also does not have a pole at s^{-1} outside $\gamma_\beta^+[s]$. The integrand, however, has simple poles at each $u_i = \infty$.

If $\sigma(i) > \max\left(\sigma(1), \ldots, \sigma(i-1)\right)$ is a running maximum, then the contour $\gamma_{\sigma(i)}^+[s]$ can be shrunk without picking residues at $u_{\sigma(i)} = q u_\beta$. Indeed, the factors $u_{\sigma(i)} - q u_\beta$ in the denominator with $\beta \notin \mathcal{J}$ are cancelled out by the product over $\alpha \in \mathcal{J}$ and $\beta \notin \mathcal{J}$, and all the factors $u_{\sigma(i)} - q u_{\sigma(j)}$ with $\sigma(i) < \sigma(j)$ are present in the other product

over $1 \le \alpha < \beta \le \ell$. Therefore, the whole integral vanishes unless $x_i \le \lambda_{\sigma(i)} + 1$ for each such running maximum $\sigma(i)$.

Next, if the latter condition holds, then we also have $x_j \le \lambda_{\sigma(j)} + 1$ for all $j = 1, \dots, \ell$. Indeed, if $\sigma(j)$ is not a running maximum, then there exists $i < j$ with $\sigma(i) > \sigma(j)$ (as $\sigma(i)$ we can take the previous running maximum), and it remains to recall that both the λ_p's and the x_p's are ordered:

$$x_j \le x_i \le \lambda_{\sigma(i)} + 1 \le \lambda_{\sigma(j)} + 1.$$

Assuming now that $x_j \le \lambda_{\sigma(j)} + 1$ for all j, we can expand the integration contours $\gamma_1^+[s], \dots, \gamma_n^+[s]$ (in this order) to infinity, and evaluate the integral by taking minus residues at that point. The single products over $i = 1, \dots, \ell$ and $\beta \notin \mathcal{J}$ produce the factor $(-s)^{|\lambda|}$. Let the ordered sequence of elements of $\mathcal{J}$ be $\mathcal{J} = \{j_1 < \dots < j_\ell\}$. One can readily see that the three remaining cross-products lead to the factor

$$q^{\mathrm{inv}(\sigma)} q^{\ell(j_1 - 1) + (\ell - 1)(j_2 - j_1 - 1) + \dots + (j_\ell - j_{\ell-1} - 1)} = q^{\mathrm{inv}(\sigma)} q^{j_1 + \dots + j_\ell} q^{-\frac{\ell(\ell+1)}{2}}.$$

This completes the proof. $\quad\blacksquare$

The coefficient r_λ in (2.138) is thus equal to the sum of the right-hand sides of (2.141) over all $\sigma: \{1, \dots, \ell\} \to \{1, \dots, n\}$. This sum can be computed using the following lemma.

Lemma 2.9.13.   *Let $X_1, X_2, \dots$ be indeterminates, $k \in \mathbb{Z}_{\ge 1}$, and $n_1, \dots, n_k \in \mathbb{Z}_{\ge 1}$ be arbitrary. We have the identity*

$$\sum_{\substack{1 \le i_1 \le n_1, \dots, 1 \le i_k \le n_k \\ (i_1, \dots, i_k) \text{ pairwise distinct}}} X_{i_1} X_{i_2 + \mathrm{inv}_{\le 2}} \cdots X_{i_k + \mathrm{inv}_{\le k}}$$

$$= (X_1 + \dots + X_{n_1})(X_2 + \dots + X_{n_2}) \cdots (X_k + \dots + X_{n_k}),$$

where $\mathrm{inv}_{\le p} := \#\{j < p: i_j > i_p\}$. By agreement, the right-hand side is zero if one of the sums is empty.

Proof It suffices to show that the map

$$(i_1, \dots, i_k) \mapsto (i_1, i_2 + \mathrm{inv}_{\le 2}, \dots, i_k + \mathrm{inv}_{\le k})$$

is a bijection between the sets

$$\{1 \le i_1 \le n_1, \dots, 1 \le i_k \le n_k, \ (i_1, \dots, i_k) \text{ pairwise distinct}\}$$
$$\text{and}\quad \{1 \le j_1 \le n_1, 2 \le j_2 \le n_2, \dots, k \le j_k \le n_k\}.$$

By induction, this statement will follow if we show that for any pairwise distinct $1 \le i_1, \dots, i_{k-1} \le n_{k-1}$, the map $i_k \mapsto i_k + \mathrm{inv}_{\le k}$ is a bijection between

$$\{1 \le j \le n_k: j \notin \{i_1, \dots, i_{k-1}\}\} \quad \text{and} \quad \{k, k+1, \dots, n_k\}.$$

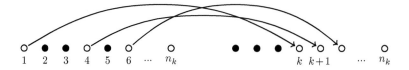

Fig. 2.27 A bijection used in the proof of Lemma 2.9.13.

But the latter fact is evident from Fig. 2.27, as the map $i_k \mapsto i_k + \text{inv}_{\leq k}$ simply corresponds to stacking together the elements of $\{1 \leq j \leq n_k : j \notin \{i_1, \ldots, i_{k-1}\}\}$.

By Lemma 2.9.12, the right-hand side of (2.141) takes the form

$$r_\lambda = q^{-\ell}(1-q)^\ell(-s)^{|\lambda|} \sum_{\substack{\sigma(1),\ldots,\sigma(\ell)\in\{1,\ldots,n\} \\ \text{pairwise distinct}}} q^{\sigma(1)+\ldots+\sigma(\ell)+\text{inv}(\sigma)} \mathbf{1}_{\lambda_{\sigma(i)} \geq x_i - 1 \text{ for all } i}$$

$$= q^{-\ell}(1-q)^\ell(-s)^{|\lambda|} \sum_{\substack{1\leq\sigma(1)\leq\mathfrak{h}_\lambda(x_1-1),\ldots,1\leq\sigma(\ell)\leq\mathfrak{h}_\lambda(x_\ell-1) \\ \text{pairwise distinct}}} q^{\sigma(1)+\ldots+\sigma(\ell)+\text{inv}(\sigma)}.$$

where we have recalled that the height function is defined as $\mathfrak{h}_\lambda(x-1) = \max\{j: \lambda_j \geq x-1\}$. We can now apply Lemma 2.9.13 with $X_i = q^i$, and conclude that this sum factorizes as

$$r_\lambda = q^{-\ell}(1-q)^\ell(-s)^{|\lambda|} \prod_{i=1}^{\ell}(q^i + q^2 + \ldots + q^{\mathfrak{h}_\lambda(x_i-1)}) = (-s)^{|\lambda|} \prod_{i=1}^{\ell}(q^{i-1} - q^{\mathfrak{h}_\lambda(x_i-1)}).$$

$$(2.142)$$

Therefore, we have finally computed the right-hand side of (2.134), and it is equal to

$$\sum_{\lambda\in\text{Sign}_n^+} (-s)^{|\lambda|} \mathsf{F}_\lambda^c(u_1,\ldots,u_n) \prod_{i=1}^{\ell}(q^{i-1} - q^{\mathfrak{h}_\lambda(x_i-1)})$$

$$= \sum_{\lambda\in\text{Sign}_n^+} \prod_{i=1}^{\ell}(q^{i-1} - q^{\mathfrak{h}_\lambda(x_i)}) \mathscr{M}_{\mathbf{u};\varrho}(\lambda) \qquad (2.143)$$

(because $\mathscr{M}_{\mathbf{u};\varrho}$ is given by (2.106)), which is the same as the left-hand side of (2.134) by the very definition. Identity (2.134) is thus established for q close to 1 and the u_i's close to s. However, as both sides of this identity are rational functions in the u_i's and q (cf. Lemma 2.9.1 for the left-hand side and formula (2.140) for the right-hand side), we conclude that these restrictions can be dropped as long as the sum over λ in (2.143) converges. This implies Theorem 2.9.8.

2.10 Degenerations of moment formulas

Here we apply q-moment formulas from Section 2.9 to rederive q-moment formulas for the stochastic six-vertex model, the ASEP, q-Hahn, and q-Boson systems obtained earlier in the literature (see references).

2.10.1   Moment formulas for the stochastic six-vertex model and the ASEP

Recall the stochastic six-vertex model described in §2.6.5. That is, we consider the dynamics $\mathscr{X}^+_{\{u_t\};\varrho}$ in which at each discrete time step, a new particle is born at location 1 (Section 2.6.4). For this dynamics to be an honest Markov process (i.e., with nonnegative transition probabilities), we require that

$$0 < q < 1, \quad \text{and} \quad u_i > q^{-\frac{1}{2}}. \tag{2.144}$$

(Another range with $q > 1$ will lead to a trivial limit shape for the stochastic six-vertex model, see the discussion in Section 2.6.5.)

Corollary 2.10.1.   Assume that (2.144) holds. Moreover, let $u_i \neq qu_j$ for any $i, j = 1, \ldots, n$. Then the q-moments of the height function of the stochastic six-vertex model $\mathscr{X}^+_{\{u_t\};\varrho}$ at time $= n$ are given by

$$\mathbb{E}^{\text{six vertex}}_{\mathbf{u};\varrho} \prod_{i=1}^{\ell} q^{b_\nu(x_i)} = q^{\frac{\ell(\ell-1)}{2}} \oint_{\gamma[\bar{u}|1]} \frac{dw_1}{2\pi\mathbf{i}} \cdots \oint_{\gamma[\bar{u}|\ell]} \frac{dw_\ell}{2\pi\mathbf{i}} \prod_{1 \leq \alpha < \beta \leq \ell} \frac{w_\alpha - w_\beta}{w_\alpha - qw_\beta}$$

$$\times \prod_{i=1}^{\ell} \left(w_i^{-1} \left(\frac{1 - q^{-\frac{1}{2}} w_i}{1 - q^{\frac{1}{2}} w_i} \right)^{x_i - 1} \prod_{j=1}^{n} \frac{1 - qu_j w_i}{1 - u_j w_i} \right) \tag{2.145}$$

for any $\ell \in \mathbb{Z}_{\geq 1}$ and $x_1 \geq \ldots \geq x_\ell \geq 1$. The integration contours are as in Definition 2.9.4.

This formula is essentially equivalent to Borodin et al. (2016a, Thm. 4.12), which was proven by a different method.

Proof   The claim follows from Theorem 2.9.8 because both sides of the identity (2.133) are rational functions in all parameters, and, moreover, the integrations in the right-hand sides of (2.133) and (2.145) are sums over the same sets of residues. Indeed, our conditions (2.144) imply that $u_i^{-1} < s^{-1} < s$ (where $s = q^{-\frac{1}{2}}$), and so the contours $\gamma[\bar{u}|j]$ exist and yield the same residues. Note also that even though now $s = q^{-1/2} > 1$ instead of belonging to $(-1, 0)$, conditions (2.107) (ensuring the existence of the measure $\mathscr{M}_{\mathbf{u};\varrho}$) readily follow from (2.144).

Let us now consider the continuous time limit of the stochastic six-vertex model to the ASEP. In Section 2.6.5 we have described this limit in the case of a fixed number of particles, but the dynamics $\mathscr{X}^+_{\{u_t\};\varrho}$ (in which at each time step a new particle is born at location 1) in this limit also produces a meaningful initial condition for the ASEP. Indeed, setting $u_i = q^{-\frac{1}{2}} + (1-q)q^{-\frac{1}{2}}\epsilon$, we see that for $\epsilon = 0$ the configuration $\lambda \in \mathsf{Sign}^+_n$ of the stochastic six-vertex model at time $= n$ is simply $\lambda = (n, n-1, \ldots, 1)$. For small $\epsilon > 0$, at times $n = \lfloor t\epsilon^{-1} \rfloor$, the configuration of the stochastic six-vertex model will be a finite perturbation of $(n, n-1, \ldots, 1)$ near the diagonal. Thus, shifting

the lattice coordinate as $\lambda_i = n+1+y_i$ with $y_i \in \mathbb{Z}$, we see that in the limit $\epsilon \searrow 0$ the initial condition for the ASEP becomes $y_1(0) = -1$, $y_2(0) = -2, \ldots$, which is known as the *step initial configuration*.

Corollary 2.10.2. For $0 < q < 1$, any $\ell \in \mathbb{Z}_{\geq 1}$, and $(x_1 \geq \ldots \geq x_\ell) \in \mathbb{Z}^\ell$, the q-moments of the height function of the ASEP $\mathfrak{h}_{\mathrm{ASEP}}(x) := \#\{i : y_i \geq x\}$ started from the step initial configuration $y_i(0) = -i$ are given by

$$
\mathbb{E}^{\mathrm{ASEP, \, step}} \prod_{i=1}^{\ell} q^{\mathfrak{h}_{\mathrm{ASEP}}(x_i)} = q^{\frac{\ell(\ell-1)}{2}} \oint_{\gamma[\sqrt{q}|1]} \frac{dw_1}{2\pi \mathbf{i}} \cdots \oint_{\gamma[\sqrt{q}|\ell]} \frac{dw_\ell}{2\pi \mathbf{i}} \prod_{1 \leq \alpha < \beta \leq \ell} \frac{w_\alpha - w_\beta}{w_\alpha - q w_\beta}
$$

$$
\times \prod_{i=1}^{\ell} \left(w_i^{-1} \left(\frac{1 - q^{-\frac{1}{2}} w_i}{1 - q^{\frac{1}{2}} w_i} \right)^{x_i} \exp \left\{ - \frac{(1-q)^2}{(1 - q^{\frac{1}{2}} w_i)(1 - q^{\frac{1}{2}}/w_i)} t \right\} \right), \tag{2.146}
$$

for any time $t \geq 0$. Each integration contour $\gamma[\sqrt{q}|j]$ consists of two positively oriented circles—one is a small circle around $\sqrt{q}$, and another one is a circle $r^j c_0$ around zero as in Definition 2.9.4.

Proof Start with the moment formula (2.145) for $u_i = q^{-\frac{1}{2}} + (1-q)q^{-\frac{1}{2}}\epsilon$, $x_i = n + 1 + x_i'$, and $n = \lfloor t\epsilon^{-1} \rfloor$. Here $x_i' \in \mathbb{Z}$ are the shifted labels of moments. As $\epsilon \searrow 0$, the contours $\gamma[\bar{u}|j]$ will still contain the same parts around 0, and the part $\gamma[\bar{u}]$ encircling the u_i^{-1}'s will turn into a small circle $\gamma[\sqrt{q}]$ around $\sqrt{q}$ (because $\sqrt{q}$ is the limit of u^{-1} as $\epsilon \searrow 0$). Let us now look at the integrand. We have

$$
\left(\frac{1 - q^{-\frac{1}{2}} w}{1 - q^{\frac{1}{2}} w} \right)^{x_i - 1} \left(\frac{1 - quw}{1 - uw} \right)^n
$$

$$
= \left(\frac{1 - q^{-\frac{1}{2}} w}{1 - q^{\frac{1}{2}} w} \right)^{n + x_i'} \left(\frac{1 - q^{\frac{1}{2}} w}{1 - q^{-\frac{1}{2}} w} - \epsilon \frac{(1-q)^2}{(1 - q^{-\frac{1}{2}} w)(1 - q^{\frac{1}{2}}/w)} + O(\epsilon^2) \right)^n
$$

$$
= \left(\frac{1 - q^{-\frac{1}{2}} w}{1 - q^{\frac{1}{2}} w} \right)^{x_i'} \left(1 - \epsilon \frac{(1-q)^2}{(1 - q^{\frac{1}{2}} w)(1 - q^{\frac{1}{2}}/w)} + O(\epsilon^2) \right)^{\lfloor t\epsilon^{-1} \rfloor}.
$$

In the limit as $\epsilon \searrow 0$, the second factor turns into the exponential. Thus, renaming x_i' back to x_i, we arrive at the desired claim. $\qquad \blacksquare$

When $x_1 = \ldots = x_\ell$, formula (2.146) essentially coincides with that obtained in Borodin et al. (2014, Thm. 4.20) using duality. The multipoint generalization (2.146) of that formula seems to be new.

The paper Borodin et al. (2014) also deals with other multipoint observables. Namely, denote

$$
\tilde{Q}_x := q^{\mathfrak{h}_{\mathrm{ASEP}}(x+1)} \mathbf{1}_{\text{there is a particle at location } x}.
$$

The expectations $\mathbb{E}^{\text{ASEP}}(\tilde{Q}_{x_1} \ldots \tilde{Q}_{x_k})$, $x_1 > \ldots > x_k$ (for the step, and in fact also for the step-Bernoulli initial conditions), were computed in Borodin et al. (2014, Cor. 4.14). The duality statement pertaining to these observables dates back to Schütz (1997). Then the expectation of $q^{\ell\, \mathfrak{h}_{\text{ASEP}}(x)}$ was recovered from these multipoint observables Borodin et al. (2014, Thm. 4.20).

Note that for the ASEP, expectations of $\tilde{Q}_{x_1} \ldots \tilde{Q}_{x_k}$ are essentially the same as the q-correlation functions (Section 2.8.4). In fact, our proof of Proposition 2.9.5 (recovering one-point q-moments from the q-correlation functions) somewhat mimics the ASEP approach mentioned earlier, but dealing with a higher spin system introduces the need for the more complicated observables (2.117).

2.10.2 Moment formulas for q-Hahn and q-Boson systems

We will now consider the q-Hahn particle system $\mathscr{X}^\infty_{q\text{-Hahn}}$ depending on $J \in \mathbb{Z}_{\geq 1}$ which starts from infinitely many particles at 1. As explained in Section 2.6.6, this q-Hahn system is obtained from the process $\mathscr{X}^+_{\{u_t\};\varrho}$ by taking the $\mathbf{u}$ parameters

$$(u_1, u_2, \ldots) = (s, qs, \ldots, q^{K-1}s, s, qs, \ldots, q^{J-1}s, s, qs, \ldots, q^{J-1}s, \ldots),$$

and sending $K \to +\infty$. When J is an integer, the q-Hahn system has nonnegative transition probabilities for $s^2 \leq 0$. This system can also be analytically continued to generic values of the parameter q^J.

Corollary 2.10.3. Let $J \in \mathbb{Z}_{\geq 1}$ and $s^2 < 0$ for all j. Then for any $\ell \in \mathbb{Z}_{\geq 1}$ and $x_1 \geq \ldots \geq x_\ell \geq 1$, the moments of $\mathscr{X}^\infty_{q\text{-Hahn}}$ at time $= n$ have the form

$$\mathbb{E}^{q\text{-Hahn}} \prod_{i=1}^{\ell} q^{\mathfrak{h}_\nu(x_i)} = (-1)^\ell q^{\frac{\ell(\ell-1)}{2}} \oint_{\gamma_1^+[s]} \frac{dw_1}{2\pi\mathbf{i}} \cdots \oint_{\gamma_\ell^+[s]} \frac{dw_\ell}{2\pi\mathbf{i}} \prod_{1\leq\alpha<\beta\leq\ell} \frac{w_\alpha - w_\beta}{w_\alpha - qw_\beta}$$

$$\times \prod_{i=1}^{\ell} \left(\frac{1}{w_i(1 - sw_i)} \left(\frac{1 - sw_i}{1 - s^{-1}w_i} \right)^{x_i - 1} \left(\frac{1 - q^J sw_i}{1 - sw_i} \right)^n \right), \tag{2.147}$$

where the integration contours $\gamma_j^+[s]$ are q-nested around s and leave 0 and s^{-1} outside.

This formula is equivalent to that obtained in Corwin (2014) using duality.

Proof First, note that if $x_\ell = 1$, then the integrand has no w_ℓ-poles inside the corresponding integration contour and thus vanishes, as it should be for the left-hand side of (2.147) because $\mathfrak{h}_\nu(1) = +\infty$.

Since before the $K \to +\infty$ limit the parameters $\mathbf{u}$ have the form (2.122), we must use Corollary 2.9.9 instead of Theorem 2.9.8, and take the integration contours to be $\gamma_j^-[\bar{\mathbf{u}}|j]$, $j = 1, \ldots, \ell$. Let us first take

$$(u_1, u_2, \ldots) = (s', qs', \ldots, q^{K-1}s', s', qs', \ldots, q^{J-1}s', s', qs', \ldots, q^{J-1}s', \ldots),$$

for some other parameter $s' \neq s$, and take the limit as $K \to +\infty$. Then the integration contours $\boldsymbol{\gamma}_j^-[\bar{\mathbf{u}}|j]$ will be q^{-1}-nested around $1/s'$, leave s and $1/s$ outside, contain parts $r^j c_0$ around 0 (cf. Definition 2.9.4), and will not change as $K \to +\infty$. The limiting integrand will contain the product

$$\prod_{i=1}^{\ell} \frac{1}{1 - s'w_i},$$

which means that after taking the limit $K \to +\infty$ the integrand will be regular at infinity.

Thus, one can drag the integration contours $\boldsymbol{\gamma}_\ell^-[\bar{\mathbf{u}}|\ell], \ldots, \boldsymbol{\gamma}_1^-[\bar{\mathbf{u}}|1]$ (in this order) through infinity, and they turn into q-nested and negatively oriented contours $\boldsymbol{\gamma}_j^+[s]$ around s, which leave $1/s$, $1/s'$, and 0 outside. These contours make it possible to set $s' = s$, which brings us to the desired formula.

Let us now turn to the stochastic q-Boson system which is obtained from the q-Hahn process by setting $J = 1$ and $s^2 = -\epsilon$ and speeding up the time by a factor of ϵ^{-1} (see Section 2.6.6).

Corollary 2.10.4. The q-moments of the height function of the q-Boson process (started with infinitely many particles at 1) have the form

$$\mathbb{E}^{q\text{-Boson}} \prod_{i=1}^{\ell} q^{\mathfrak{h}_\nu(x_i)}$$

$$= (-1)^\ell q^{\frac{\ell(\ell-1)}{2}} \oint_{\boldsymbol{\gamma}_1^+[-1]} \frac{dw_1}{2\pi \mathbf{i}} \cdots \oint_{\boldsymbol{\gamma}_\ell^+[-1]} \frac{dw_\ell}{2\pi \mathbf{i}} \prod_{1 \leq \alpha < \beta \leq \ell} \frac{w_\alpha - w_\beta}{w_\alpha - q w_\beta} \prod_{i=1}^{\ell} \frac{e^{(1-q)tw_i}}{w_i(1 + w_i)^{x_i-1}},$$

$$(2.148)$$

where $t \geq 0$ is the time and $\ell \in \mathbb{Z}_{\geq 1}$ and $x_1 \geq x_2 \geq \ldots \geq x_\ell \geq 1$ are arbitrary. The integration contours $\boldsymbol{\gamma}_j^+[-1]$ are q-nested around -1 and do not contain 0.

The moment formula (2.148) first appeared in the papers Borodin and Corwin (2014) and Borodin et al. (2014).

Proof Set $J = 1$, $s^2 = -\epsilon$, and $n = \lfloor t\epsilon^{-1} \rfloor$ in (2.147), and change the variables as $w_i = \epsilon s^{-1} w_i'$. Then the integral in (2.147) becomes (without the prefactor)

$$\oint_{\boldsymbol{\gamma}_1^+[-1]} \frac{dw_1'}{2\pi \mathbf{i}} \cdots \oint_{\boldsymbol{\gamma}_\ell^+[-1]} \frac{dw_\ell'}{2\pi \mathbf{i}} \prod_{1 \leq \alpha < \beta \leq \ell} \frac{w_\alpha' - w_\beta'}{w_\alpha' - q w_\beta'}$$

$$\times \prod_{i=1}^{\ell} \left(\frac{1}{w_i'(1 - \epsilon w_i')} \left(\frac{1 - \epsilon w_i'}{1 + w_i'} \right)^{x_i-1} \left(\frac{1 - q\epsilon w_i'}{1 - \epsilon w_i'} \right)^{\lfloor t\epsilon^{-1} \rfloor} \right).$$

Sending $\epsilon \searrow 0$ and renaming w_i' back to w_i, we arrive at the desired formula.

Acknowledgements

We are grateful to Ivan Corwin and Vadim Gorin for valuable discussions. A.B. was partially supported by the NSF Grant DMS-1056390.

References

Baik, J., Deift, P., and Johansson, K. (1999). On the distribution of the length of the longest increasing subsequence of random permutations. *J. Amer. Math. Soc.* **12**(4), 1119–78. arXiv:math/9810105 [math.CO].

Barraquand, G., and Corwin, I. (2016). Random-walk in Beta-distributed random environment. *Probab. Theory Relat. Fields* arXiv:1503.04117 [math.PR].

Baxter, R. (2007). *Exactly solved models in statistical mechanics.* Courier Dover, London.

Bethe, H. (1931). Zur Theorie der Metalle. I. Eigenwerte und Eigenfunktionen der linearen Atomkette. (On the theory of metals. I. Eigenvalues and eigenfunctions of the linear atom chain). *Z. Phys.* **71**, 205–226.

Bogoliubov, N., Bullough, R., and Timonen, J. (1994). Critical behavior for correlated strongly coupled boson systems in 1+1 dimensions. *Phys. Rev. Lett.* **72**(25), 3933–6.

Bogoliubov, N., Izergin, A., and Kitanine, N. (1998). Correlation functions for a strongly correlated boson system. *Nucl. Phys. B* **516**(3), 501–28. arXiv:solv-int/9710002.

Borodin, A. (2011). Schur dynamics of the Schur processes. *Adv. Math.* **228**(4), 2268–91. arXiv:1001.3442 [math.CO].

Borodin, A. (2014). On a family of symmetric rational functions. arXiv.preprint arXiv:1410.0976 [math.CO].

Borodin, A., and Bufetov, Al. (2015). An irreversible local Markov chain that preserves the six vertex model on a torus. arXiv:1509.05070 [math-ph]. To appear in *Ann. Inst. Henri Poineare*

Borodin, A., and Corwin, I. (2014). Macdonald processes. *Prob. Theory Rel. Fields* **158**, 225–400. arXiv:1111.4408 [math.PR].

Borodin, A., and Corwin, I. (2015). Discrete time q-TASEPs. *Intern. Math. Res. Not.* **2015** (2), 499–537. arXiv:1305.2972 [math.PR].

Borodin, A., Corwin, I., and Gorin, V. (2016a). Stochastic six-vertex model. *J. Math.* **165**(3), 563–624. arXiv:1407.6729 [math.PR].

Borodin, A., Corwin, I., Gorin, V., and Shakirov, S. (2016b). Observables of Macdonald processes. *Trans. Amer. Math. Soc.* **368**(3), 1517–58. arXiv:1306.0659 [math.PR].

Borodin, A., Corwin, I., Petrov, L., and Sasamoto, T. (2015a). Spectral theory for interacting particle systems solvable by coordinate Bethe ansatz. *Comm. Math. Phys.* **339**(3), 1167–1245. arXiv:1407.8534 [math-ph].

Borodin, A., Corwin, I., Petrov, L., and Sasamoto, T. (2015b). Spectral theory for the q-Boson particle system. *Comp. Math.* **151**(1), 1–67. arXiv:1308.3475 [math-ph].

Borodin, A., Corwin, I., and Sasamoto, T. (2014). From duality to determinants for q-TASEP and ASEP. *Ann. Probab.* **42**(6), 2314–2382. arXiv:1207.5035 [math.PR].

Borodin, A., and Ferrari, P. (2014). Anisotropic growth of random surfaces in 2+1 dimensions. *Comm. Math. Phys.* **325**, 603–684. arXiv:0804.3035 [math-ph].

Borodin, A., and Petrov, L. (2016a). Higher spin six vertex model and symmetric rational functions. arXiv preprint. arXiv:1601.05770 [math.PR], doi:10.1007/s00029-016-0301-7. To appear in *Selecta Mathematica*, New Series.

Borodin, A., and Petrov, L. (2016b). Nearest neighbor Markov dynamics on Macdonald processes. *Adv. Math.* **300**, 71–155. arXiv:1305.5501 [math.PR].

Calabrese, P., Le Doussal, P., and Rosso, A. (2010). Free-energy distribution of the directed polymer at high temperature. *Euro. Phys. Lett.* **90**(2), 20002.

Corwin, I. (2012). The Kardar-Parisi-Zhang equation and universality class. *Random Matrices Theory Appl.* **1**. arXiv:1106.1596 [math.PR].

Corwin, I. (2014). The *q*-Hahn Boson process and *q*-Hahn TASEP. *Intern. Math. Res. Not.*, rnu094. arXiv:1401.3321 [math.PR].

Corwin, I. and Petrov, L. (2015). The q-PushASEP: A new integrable model for traffic in 1+1 Dimension. *J. Stat. Phys.* **160**(4), 1005–26. arXiv:1308.3124 [math.PR].

Corwin, I., and Petrov, L. (2016). Stochastic higher spin vertex models on the line. *Comm. Math. Phys.* **343**(2), 651–700. arXiv:1502.07374 [math.PR].

Dotsenko, V. (2010). Replica Bethe ansatz derivation of the Tracy-Widom distribution of the free energy fluctuations in one-dimensional directed polymers. *J. Stat. Mech. Theory Exp.* (07), P07010. arXiv:1004.4455 [cond-mat.dis-nn].

Felder, G., and Varchenko, A. (1996). Algebraic Bethe ansatz for the elliptic quantum group $E_{\tau,\eta}(\mathrm{sl}_2)$. *Nucl. Phys. B* **480**(1–2), 485–503. arXiv:q-alg/9605024.

Fulman, J. (1997). Probabilistic measures and algorithms arising from the Macdonald symmetric functions. arXiv:preprint. arXiv:math/9712237 [math.CO].

Gnedin, A., and Olshanski, G. (2009, May). A q-analogue of de Finetti's theorem. *Electr. J. Combin.* **16**, R16. arXiv:0905.0367 [math.PR].

Gnedin, A., and Olshanski, G. (2010). q-exchangeability via quasi-invariance. *Ann. Probab.* **38**(6), 2103–35. arXiv:0907.3275 [math.PR].

Gwa, L.-H., and Spohn, H. (1992). Six-vertex model, roughened surfaces, and an asymmetric spin Hamiltonian. *Phys. Rev. Lett.* **68**(6), 725–8.

Harris, T. E. (1978). Additive set-valued Markov processes and graphical methods. *Ann. Probab.* **6**(3), 355–78.

Johansson, K. (2001). Discrete orthogonal polynomial ensembles and the Plancherel measure. *Ann. Math.* **153**(1), 259–96. arXiv:math/9906120 [math.CO].

Kardar, M., Parisi, G., and Zhang, Y. (1986). Dynamic scaling of growing interfaces. *Phy. Rev. Lett.* **56**(9), 889.

Kirillov, A.N., and Reshetikhin, N. (1987). Exact solution of the integrable XXZ Heisenberg model with arbitrary spin. I. The ground state and the excitation spectrum. *J. Phys. A* **20**(6), 1565–85.

Koekoek, R., and Swarttouw, R.F. (1996). The Askey-scheme of hypergeometric orthogonal polynomials and its q-analogue. Technical report, Delft University of Technology and Free University of Amsterdam.

König, W. (2005). Orthogonal polynomial ensembles in probability theory. *Probab. Surv.* **2**, 385–447. arXiv:math/0403090 [math.PR].

König, W., O'Connell, N., and Roch, S. (2002). Non-colliding random walks, tandem queues, and discrete orthogonal polynomial ensembles. *Electron. J. Probab.* **7**(5), 1–24.

Korepin, V., Bogoliubov, N., and Izergin, A. (1993). *Quantum Inverse Scattering Method and Correlation Functions.* Cambridge University Press, Cambridge, UK.

Kulish, P., Reshetikhin, N., and Sklyanin, E. (1981). Yang-Baxter equation and representation theory: I. *Lett. Math. Phys.* **5**(5), 393–403.

Macdonald, I. G. (1995). *Symmetric functions and Hall polynomials*, 2nd edn. Oxford University Press, Oxford.

Mangazeev, V. (2014). On the Yang–Baxter equation for the six-vertex model. *Nucl. Phys. B* **882**, 70–96. arXiv:1401.6494 [math-ph].

Matveev, K., and Petrov, L. (2015). q-randomized Robinson–Schensted–Knuth correspondences and random polymers. arXiv preprint. arXiv:1504.00666 [math.PR]. To appear in *Ann. Inst. Henri Poincare*.

O'Connell, N. (2003a). A path-transformation for random walks and the Robinson-Schensted correspondence. *Trans. Amer. Math. Soc.* **355**(9), 3669–97.

O'Connell, N. (2003b). Conditioned random walks and the RSK correspondence. *J. Phys. A* **36**(12), 3049–66.

O'Connell, N. (2012). Directed polymers and the quantum Toda lattice. *Ann. Probab.* **40**(2), 437–58. arXiv:0910.0069 [math.PR].

Okounkov, A. (2001). Infinite wedge and random partitions. *Selecta Math. New Series* **7**(1), 57–81. arXiv:math/9907127 [math.RT].

Okounkov, A., and Reshetikhin, N. (2003). Correlation function of Schur process with application to local geometry of a random 3-dimensional Young diagram. *J. Amer. Math. Soc.* **16**(3), 581–603. arXiv:math/0107056 [math.CO].

Povolotsky, A. (2013). On integrability of zero-range chipping models with factorized steady state. *J. Phys. A* **46**, 465205.

Reshetikhin, N. (2010). Lectures on the integrability of the 6-vertex model. In *Exact Methods in Low-dimensional Statistical Physics and Quantum Computing*, pp. 197–266. Oxford University Press, Oxford. arXiv:1010.5031 [math-ph].

Sasamoto, T., and Wadati, M. (1998). Exact results for one-dimensional totally asymmetric diffusion models. *J. Phys. A* **31**, 6057–71.

Schütz, G. (1997). Duality relations for asymmetric exclusion processes. *J. Stat. Phys.* **86**(5-6), 1265–87.

Spitzer, F. (1970). Interaction of Markov processes. *Adv. Math.* **5**(2), 246–90.

Tracy, C., and Widom, H. (2008). Integral formulas for the asymmetric simple exclusion process. *Commun. Math. Phys.* **279**, 815–44. arXiv:0704.2633 [math.PR]. Erratum: *Commun. Math. Phys.* **304**, 875–8, 2011.

Tracy, C., and Widom, H. (2009a). Asymptotics in ASEP with step initial condition. *Comm. Math. Phys.* **290**, 129–54. arXiv:0807.1713 [math.PR].

Tracy, C., and Widom, H. (2009b). On ASEP with step Bernoulli initial condition. *J. Stat. Phys.* **137**, 825–38. arXiv:0907.5192 [math.PR].

Vuletic, M. (2007). Shifted Schur process and asymptotics of large random strict plane partitions. *Intern. Math. Res. Not.*, **2007**(rnm043). arXiv:math-ph/0702068.

Wheeler, M., and Zinn-Justin, P. (2016). Refined Cauchy/Littlewood identities and six-vertex model partition functions: III. Deformed bosons. *Adv. Math.* **299**, 543–600. arXiv:1508.02236 [math-ph].

3
Free probability

Alice GUIONNET

ENS Lyon, Université de Lyon, UMPA, UMR 5669 CNRS, 46 allée d'Italie, 69364 Lyon Cedex 07, France

Guionnet, A., 'Free Probability' in *Stochastic Processes and Random Matrices*. Edited by: Grégory Schehr et al, Oxford University Press (2017). © Oxford University Press 2017. DOI 10.1093/oso/9780198797319.003.0003

Chapter Contents

3.1 Introduction

Free probability was introduced by D. Voiculescu as a theory of noncommutative random variables (similar to integration theory) equipped with a notion of freeness very similar to independence. In fact, it is possible in this framework to define natural 'free' counterpart of central limit theorem, Gaussian distribution, Brownian motion, stochastic differential calculus, entropy, etc. Free probability also appears as the natural setup for studying large random matrices as their sizes go to infinity and hence is central in the study of random matrices as their sizes go to infinity. It is also, and first of all, designed to solve problems related to free group factors, because the notion of freeness is intimately related to the notion of freeness in group theory. Indeed, one of the central goals of the theory concerns the classification of von Neumann algebras, and settling whether free group factors are isomorphic. At this point, one could wonder why physicists should care and why the organizers of the summer school in Les Houches in July 2015 asked me to lecture on this subject. I must admit I was myself surprised by the request. Not because it is not well founded, but rather because the formalism may discourage the reader.

Why is it well founded? First, because random matrices are of interest in physics for many reasons, starting from Wigner (1958). Free probability is the natural framework in which several random matrices live as their dimensions go to infinity. Understanding it is, therefore, crucial to understanding many questions where several matrices are involved. We shall, for instance, discuss the asymptotics of the sum or the multiplication of two generic random matrices $X + Y$ or XY in section 3.4. But there are many other questions where one deals with 'several' random matrices at a time. For instance, Gaussian Wishart matrices with nontrivial covariance can be written as $G\Sigma G$, where G is a matrix with independent standard Gaussian entries and Σ a covariance matrix. This single matrix can be decomposed as the multiplication of two simpler matrices and free probability offers tools for studying its spectral measure (Shlyakhtenko 1996). Of course, such results can be derived directly; however, free probability provides general results for solving them. Free probability can also be used to more complicated questions. For instance, we shall mention the case of nonnormal matrices of the form UDV, where D is a diagonal matrix and U and V are unitary Haar distributed matrices. Such a matrix arises, for instance, when one considers a square non-Hermitian matrix submitted to a potential $V(XX^*)$: the singular values D of such a matrix are easy to analyze, and X decomposes as UDV and U, V Haar distributed unitary matrices. The latter model was considered by Feinberg and Zee (1997), who conjectured that their spectrum is distributed in a ring. Such a result can be rigorously derived thanks to the computation of Haagerup and Larsen (2000), who used free probability techniques; see section 3.5.2. More generally, one can wonder what is the spectral measure of polynomials and words in several independent random matrices as their sizes go to infinity. It can be shown, for instance, that if two matrices follow the Gaussian unitary ensemble (GUE), any self-adjoint polynomial in these two matrices will have a connected support: this uses nontrivial results from K-theory.

A second reason why free probability should attract more attention from the physics community is that it shares quite a few interests with physics. For instance, matrix

integrals were popularized in physics by 't Hooft and Brézin–Parisi–Itzykson and Zuber when they related them with the enumeration of maps via the so-called topological expansion. Random matrices became a tool for attacking random maps and random topology. More recently, topological recursions were shown to be related with Virasoro or W-algebra constraints in conformal field theory (CFT), in the work of Eynard and Orantin (2009). The interest in the asymptotics of matrix integrals in free probability is that it provides the construction of interesting noncommutative laws. In fact, it is conjectured that any noncommutative law can be constructed as the limit of matrix integrals. We give an example of this two-sided motivation based on the $O(n)$ model in the last part of this chapter. It is well known that such models can be represented in terms of matrix integrals, at least for integer fugacities, which eventually can be used to compute partition functions or other quantities. On the other hand, such models are related to planar algebras and with Jones (2000) and Shlyakhtenko (1996) we generalized such a construction to build noncommutative laws on planar algebras which are factors (that is, have a trivial centre).

In this chapter we shall introduce the free probability framework, show how it naturally shows up in the random matrices asymptotics via the so-called 'asymptotic freeness' and discuss the connection with combinatorics and the enumeration of planar maps, including loop models. I hope that this chapter will be sufficiently reader-friendly. The first part sets the framework and is therefore rather formal. Applications to random matrices and the enumeration of maps are described in the sections that follow.

3.2   Free probability setting

Free probability is a probability theory for noncommutative variables, equipped with a notion of freeness which resembles the notion of independence in classical probability theory. It follows the viewpoint that noncommutative variables being the central object, we shall only be interested in expectation of functions in these random variables. Hence, probability measures can be thought of as linear functionals on a set of test functions. This point of view encompasses classical probability theory, but also spectral theory (variables being given by operators or matrices) and quantum mechanics (physical observables being given by operators). Another point of view on the theory is to see it as the natural asymptotic framework for studying random matrices. The main object at hand when dealing with several matrices $\mathbf{X^N} = (X_1^N, \ldots, X_k^N)$ is their 'joint moments' given by

$$\frac{1}{N}\mathrm{Tr}(X_{i_1}^N \cdots X_{i_\ell}^N)$$

or more generally their joint $*$-moments

$$\frac{1}{N}\mathrm{Tr}((X_{i_1}^N)^{\epsilon_1} \cdots (X_{i_\ell}^N)^{\epsilon_\ell}),$$

with $\epsilon_i = 1$ or $*$. Imagine that a sequence of matrices $\mathbf{X^N}$ is chosen such that these moments converge towards some limit $\tau(X_{i_1} \cdots X_{i_\ell})$ for all choices of ϵ_i and $i_p \in$

$\{1, \ldots, k\}$. What can we say about the limit τ? In this chapter we detail the framework in which this object lives.

Let us assume that the matrices X_i^N are self-adjoint and let us consider

$$\mathbb{C}\langle X_1, \ldots, X_k \rangle = \{ \lambda_0 + \sum_p \lambda_p X_{i_1^p} \cdots X_{i_{\ell_p}^p}, \lambda_i \in \mathbb{C}, i_j^p \in \{1, \ldots, k\} \}$$

to be the set of polynomials in k noncommutative letters endowed with the involution

$$(z X_{i_1} \cdots X_{i_\ell})^* = \bar{z} X_{i_k} \cdots X_{i_1}.$$

Then, τ can be seen as a linear form on $\mathbb{C}\langle X_1, \ldots, X_k \rangle$ such that

- $\tau(1) = 1$,
- $\tau(PQ) = \tau(QP)$ for all $P, Q \in \mathbb{C}\langle X_1, \ldots, X_k \rangle$,
- $\tau(PP^*) \geq 0$ for all $P \in \mathbb{C}\langle X_1, \ldots, X_k \rangle$,
- $\tau(P^*) = \overline{\tau(P)}$.

We shall see in this section that these properties allow us to define an appropriate setup for a noncommutative probability theory. We shall first abstract the required properties in a more general setup where $\mathbb{C}\langle X_1, \ldots, X_k \rangle$ is replaced by a general algebra A and then introduce the notion of freeness, analogous to the notion of independence in the classical framework. We will see in section 3.3 that this property naturally shows up at the large N limit of random matrices.

3.2.1   Noncommutative probability space

Noncommutative probability spaces

The following is an algebraic noncommutative analogue of a classical probability space.

Definition 3.2.1.   *A noncommutative probability space is a pair (A, ϕ), where A is a unital algebra (over $\mathbb{C}$), and $\phi : A \to \mathbb{C}$ is a linear functional, $\phi(1) = 1$. A noncommutative random variable on a noncommutative probability space (A, ϕ) is an element $x \in A$.*

It is also convenient to consider noncommutative $*$-probability spaces (A, ϕ).

Definition 3.2.2.   *A noncommutative $*$-probability space (A, ϕ) is a noncommutative probability space (A, ϕ) for which A is endowed with an antilinear involution $*$ so that $\phi(x^*) = \overline{\phi(x)}$.*

The involution $*$ is the analogue of complex conjugation for complex-valued classical random variables. They are two important assumptions on ϕ that we shall make.

Definition 3.2.3.   *Let (A, ϕ) be a noncommutative probability space. Then*

1. ϕ is a trace iff for all $a, b \in A$, $\phi(ab) = \phi(ba)$.
2. ϕ is faithful iff for all $a \in A$, $\phi(aa^*) = 0$ implies that $a = 0$.

Example. We shall mention some natural examples of noncommutative probability spaces. As it happens, all of our examples have a natural involution $*$.

Example 3.2.4.

1. If $(X, \mathcal{B}, \mu)$ is a classical probability space, we can take, e.g., $A = L^\infty(X, \mu)$, $\phi(f) = \int f d\mu$, or even $A = \bigcap_{p<\infty} L^p(X, \mu)$, $\phi(f) = \int f d\mu$. In the case $X = \mathbb{R}$, $d\mu(t) = \frac{1}{\sqrt{2\pi}} \exp(-t^2/2) dt$, the function $f(t) = t$ is called the Gaussian random variable, and $\mu_f = \mu$.

2. Let H be a Hilbert space and $\xi \in H$ be a vector of norm 1. Let $B(H)$ be the set of bounded linear operators on H. For $T \in B(H)$, set $\phi(T) = \langle T\xi, \xi \rangle$. Then $(B(H), \phi)$ is a noncommutative probability space, and any $T \in B(H)$ is a random variable. In fact, (A, ϕ) is a noncommutative probability space for any unital subalgebra $A \subset B(H)$. The involution $*$ is the usual operation of taking adjoints on $B(H)$.

3. Let Γ be a discrete group. We consider the group algebra $\mathbb{C}\Gamma$ given by

$$\mathbb{C}\Gamma = \left\{ \sum_{g \in \Gamma} \alpha_g g \,\middle|\, \alpha_g \in \Gamma, \text{ only finitely } \alpha_g \neq 0 \right\}.$$

We equip it with the operations of multiplication and involution

$$\left(\sum_{g \in \Gamma} \alpha_g g \right) \cdot \left(\sum_{g \in \Gamma} \beta_g g \right) = \sum_{g, h \in \Gamma} \alpha_g \beta_h gh$$

$$\left(\sum_{g \in \Gamma} \alpha_g g \right)^* = \sum_{g \in \Gamma} \bar{\alpha}_g g^{-1}.$$

The functional τ_Γ from $\mathbb{C}\Gamma$ into $\mathbb{C}$ given by

$$\tau_\Gamma \left(\sum \alpha_g g \right) = \alpha_e$$

is called the canonical trace or von-Neumann trace. You can check that it is indeed a trace.

4. $A = M_{N \times N}(\mathbb{C})$ the algebra of $N \times N$ matrices with complex entries and $\Phi(x) = \frac{1}{N} \mathrm{Tr}(x)$ with $\mathrm{Tr}(x) = \sum_{i=1}^{N} x_{ii}$. We denote this noncommutative probability space by $(M_{N \times N}(\mathbb{C}), \frac{1}{N} \mathrm{Tr})$. This notion can be extended to random matrices. We let $(\Omega, \mathcal{B}, P)$ be a classical probability space; consider $A = M_{N \times N}(L^\infty(P))$ as the space on $N \times N$ matrices with random entries uniformly bounded under P. Then we set

$$\phi(x) = \int \frac{1}{N} \mathrm{Tr}(x(\omega)) dP(\omega).$$

In the following, we shall always state convergence for random matrices in the later space; that is, consider convergence of traces in expectation. However, most of our results also hold for almost sure convergence as can be checked by concentration of measure phenomenon (see, e.g., Lemma 3.3.3), controls on covariances (see, e.g., (3.41)), and the Borel-Cantelli Lemma.

Distribution of noncommutative random variables

Definition 3.2.5.   *The joint distribution of an n-tuple of self-adjoint noncommutative random variables $x_1, \dots, x_n$ on a noncommutative probability space (A, ϕ) is the linear functional*

$$\mu_{x_1,\dots,x_n} : \mathbb{C}\langle X_1, \dots, X_n\rangle \to \mathbb{C}$$

given by $\mu_{x_1,\dots,x_n}(p) = \phi(p(x_1, \dots, x_n))$.
   If (A, ϕ) is a noncommutative $*$ probability space, we can also define the $*$- joint distribution of an n-tuple $(x_1, \dots, x_n)$ as the joint distribution of $(x_1, \dots, x_n, x_1^*, x_2^*, \dots, x_n^*)$, that is,

$$\mu_{x_1,\dots,x_n} : \mathbb{C}\langle X_1, \dots, X_n, X_{n+1}, \dots, X_{2n}\rangle \to \mathbb{C}$$

given by $\mu_{x_1,\dots,x_n}(p) = \phi(p(x_1, \dots, x_n, x_1^*, \dots, x_n^*))$.

Note that the $i_1, \dots, i_p$th moment of $(x_1, \dots, x_n)$ is just $\phi(x_{i_1} \cdots x_{i_p})$. In particular, for a single variable x, the kth moment $\phi(x^k)$ is the same as its classical definition. Observe that since any element of A can be decomposed as the sum of two self-adjoint variables, we can always assume that the variables are self-adjoint. Yet, it is sometimes clearer to keep non-self-adjoint elements, for instance when we will deal with unitary matrices. The involution over the set of polynomial functions is defined according to this choice.

Positivity

At this point, the notion of distribution misses two important properties of probability measures: the positivity and the continuity with respect to a norm. Indeed, these two properties are fundamental to many computations in classical analysis such as extending the domain of definition of a law to bounded continuous functions or getting bounds such as Cauchy–Schwartz or Hölder inequalities. In classical probability theory, the Riesz representation theorem asserts that if a linear functional on the set of bounded continuous functions on $[-R, R]$ is positive, in the sense that for any non-negative function f, $\mu(f)$ is nonnegative, then there exists a positive Borel measure on $[-R, R]$ finite such that for all $f \in C_b([-R, R])$ we have

$$\phi(f) = \int f(x)\mathrm{d}\mu(x) \,.$$

Hence, we see that the existence of a ∗-distribution is ensured under two conditions: τ extends to bounded continuous functions and τ is nonnegative. For the first condition, it is enough to assume that $|\tau(f)| \leq \sup_{x \in [-R,R]} |f(x)|$ by density of the polynomial functions in the set of bounded continuous functions on $[-R, R]$. Hence we see that a necessary and sufficient condition for the existence of a ∗-distribution is that

1. Continuity : $|\tau(f)| \leq \sup_{x \in [-R,R]} |f(x)|$
2. Positivity: $\tau(f) \geq 0$ if $f \geq 0$.

The analogue of this condition is the following.

Definition 3.2.6. *Let (A, ϕ) be a noncommutative ∗-probability space. Then ϕ is positive iff $\phi(aa^*) \geq 0$ for all $a \in A$. We also say that ϕ is a state.*

An important consequence of positivity is the Cauchy–Schwarz inequality

$$|\phi(ab^*)| \leq |\phi(aa^*)|^{1/2}|\phi(bb^*)|^{1/2} . \tag{3.1}$$

Another important consequence is given by the Gelfand–Neimark–Seigal construct which ensures that we can represent random variables as bounded linear operators on a Hilbert space, which is somehow a noncommutative analogue to the Riesz theorem. To do so, we need a priori to assume that τ extends to a family analogous to bounded continuous functions, that is given by C^*-algebras.

Definition 3.2.7. *A C^*-algebra is a unital algebra equipped with a norm $\| \cdot \|$ inducing a complete distance, and an involution $*$ so that*

$$\|xy\| \leq \|x\|\|y\|, \quad \|a^*a\| = \|a\|^2 .$$

The Gelfand–Neimark–Segal construction then makes it possible to represent non-commutative variables as bounded linear operators on a Hilbert space. Namely, we have the following.

Theorem 3.2.8. *(Gelfand–Segal–Naimark construction) Let $(A, \phi, \|.\|)$ be a non-commutative ∗-probability space with A a C^* algebra. Assume that ϕ is a state. Then there exists a Hilbert space H equipped with a scalar product $\langle \cdot, \cdot \rangle$, a unit vector $\xi \in H$, and a norm decreasing ∗-homomorphism $\pi : A \to B(H)$, so that*

1. $\phi(x) = \langle \xi, \pi(x)\xi \rangle$ for all $x \in A$,
2. $\pi(A)\xi$ is dense in H,
3. if ϕ is faithful, then π is injective.

Note that, in fact, if we consider k self-adjoint random variables $x_1, \ldots, x_k$ and ϕ a state on $\mathbb{C}\langle X_1, \ldots, X_k \rangle$ so that $x_1, \ldots, x_k$ are uniformly bounded in the sense that there exists $R < \infty$ such that for all $i_1, \ldots, i_\ell \in \{1, \ldots, k\}$

$$|\phi(x_{i_1} \cdots x_{i_\ell})| \leq R^\ell$$

then we can always construct a C^* algebra (depending on R) such that $x_1, \ldots, x_k$ lives in A (see Guionnet et al. (2011, Section 5)). Hence, if we are given uniformly bounded random variables we can apply the Gelfand–Neimark–Segal construction. This makes it possible to use inequalities such as Hölder's inequality.

3.2.2   Weak *-topology

It is convenient to have a notion of convergence of laws, similar to weak convergence. Hereafter I, J denote countable sets.

Definition 3.2.9. *Let* $(\mathcal{A}_N, \phi_N)$, $N \in \mathbb{N} \cup \{\infty\}$, *be noncommutative probability spaces, and let* $\{a_i^N\}_{i \in J}$ *be a sequence of elements of* $\mathcal{A}_N$. *Then* $\{a_i^N\}_{i \in J}$ *converges in law to* $\{a_i^\infty\}_{i \in J}$ *if and only if for all* $P \in \mathbb{C}\langle X_i | i \in J \rangle$,

$$\lim_{N \to \infty} \mu_{\{a_i^N\}_{i \in J}}(P) = \mu_{\{a_i^\infty\}_{i \in J}}(P).$$

We also say in such a situation that $\{a_i^N\}_{i \in J}$ *converges in moments to* $\{a_i^\infty\}_{i \in J}$.

We shall later on use repeatedly the fact that if we have a uniformly bounded sequence of noncommutative variables $\{a_i^N\}_{i \in J}$ of $(\mathcal{A}_N, \phi_N)$, that is that there exists $R < \infty$ such that for any N, for any $j_1, \ldots, j_k \in J$,

$$|\phi_N(a_{j_1}^N \cdots a_{j_k}^N)| \le R^k,$$

then the laws of $\{a_i^N\}_{i \in J}$ are tight, in the sense that we can take subsequences which converge in moments. Showing that there is at most one possible limit point then ensures convergence.

3.2.3   Freeness

Definition 3.2.10. *Let* τ *be a linear function on the set* $\mathbb{C}\langle X_1, \ldots, X_m \rangle$ *of polynomials in* m *(self-adjoint) variables. Assume that* τ *is tracial and unital* $\tau(1) = 1$. *Then, we say that for* $p \le m$, $X_1, \ldots, X_p$ *are free from* $X_{p+1}, \ldots, X_m$ *iff for any* $P_1, \ldots, P_k \in \mathbb{C}\langle X_1, \ldots, X_p \rangle$ *and* $Q_1, \ldots, Q_k \in \mathbb{C}\langle X_{p+1}, \ldots, X_m \rangle$

$$\tau(P_1(X_1, \ldots, X_p)Q_1(X_{p+1}, \ldots, X_m)P_2 \cdots Q_k(X_{p+1}, \ldots, X_m)) = 0$$

as soon as $\tau(P_i(X_1, \ldots, X_p)) = 0$ *for* $2 \le i \le k$ *and* $\tau(Q_i(X_{p+1}, \ldots, X_m)) = 0$ *for* $1 \le i \le k$.

Property 3.2.11. If $\mathbf{X} = (X_1, \ldots, X_p)$ and $\mathbf{Y} = (Y_1, \ldots, Y_k)$ are free, their joint moments are determined uniquely by the joint moments of $\mathbf{X}$ and $\mathbf{Y}$, respectively.

The proof easily follows by induction. Indeed, $\tau(P(\mathbf{X}, \mathbf{Y}))$ is known for all words P of degree equal to 0 or 1. Let us assume it is known for words of total degree smaller than r. Then it is determined uniquely if this word only contains letters in $\mathbf{X}$ or $\mathbf{Y}$.

Otherwise, consider a reduced word in **X** and **Y** and recenter the words in **X** and **Y**, respectively. By assumption, the term where all words are recentered vanishes. The rest contains traces of words with smaller degree, and therefore, by induction, is known.

In fact, it is more natural somehow to speak about freeness of the algebras generated by the variables $(x_1, \ldots, x_n)$ and $(y_1, \ldots, y_m)$. The definition then becomes the following.

Definition 3.2.12. *A family of subalgebras $A_i \subset A$ is called freely independent if*

$$\phi(a_1 \cdots a_n) = 0$$

whenever $a_i \in A_{i(j)}$, $i(1) \neq i(2), \ldots, i(n-1) \neq i(n)$ and $\phi(a_i) = 0$, $i = 1, \ldots, n$.

A sequence of families $F_i = (x_1^{(i)}, \ldots, x_{n(i)}^{(i)})$ of noncommutative random variables on A is called freely independent if the algebras A_i generated by $(x_1^{(i)}, \ldots, x_{n(i)}^{(i)})$ are freely independent.

Two sequences $\{a_i^N\}_{i \in I}$ and $\{b_j^N\}_{j \in J}$ are said to be asymptotically free iff their joint noncommutative distribution converges towards the joint distribution of $\{a_i\}_{i \in I}$ and $\{b_j\}_{j \in J}$ and that these variables are free.

Being given noncommutative probability spaces $(A_i, \varphi_i)_{1 \leq i \leq n}$, one can always construct a bigger noncommutative probability space (A, φ) so that $(A_i, \varphi_i)_{1 \leq i \leq n}$ are free inside (A, φ), namely $A_i \subset A$, $\varphi|_{A_i} = \varphi_i$ and $(A_i)_{1 \leq i \leq n}$ are free in A. In the next section we shall see that free variables arise as well at the large N limit of matrices with eigenvectors 'as independent' as possible.

3.3 Asymptotic freeness of random matrices

In this section, we show that random matrices are asymptotically free if their eigenvectors are 'as independent as possible'. We first consider Gaussian matrices and then matrices whose eigenvectors are uniformly distributed.

3.3.1 Independent GUE matrices are asymptotically free

The aim of this section is to prove that if $\{\mathbf{X}^{N,\ell}, 1 \leq \ell \leq m\}$ are m independent Wigner matrices such that

$$\mathbb{E}[\mathbf{X}_{ij}^{N,\ell}] = 0, \forall 1 \leq i, j \leq N, 1 \leq \ell \leq m, \quad \lim_{N \to \infty} \max_{1 \leq i, j \leq N} |N\mathbb{E}[|\mathbf{X}_{ij}^{N,\ell}|^2] - 1| = 0$$

and such that the moments of $\sqrt{N}\mathbf{X}_{ij}^{N,\ell}$ are uniformly bounded, then they are asymptotically free. We first study the convergence of their joint moments and afterwards identify the limit as the law of m free semi-circular variables.

Theorem 3.3.1. Assume that for all $k \in \mathbb{N}$,

$$B_k := \sup_{1 \leq \ell \leq m} \sup_{N \in \mathbb{N}} \sup_{ij \in \{1, \cdots, N\}^2} \mathbb{E}[|\sqrt{N}\mathbf{X}_{ij}^{N,\ell}|^k] < \infty. \tag{3.2}$$

Then, for any $\ell_j \in \{1, \cdots, m\}, 1 \leq j \leq k$,

$$\lim_{N \to \infty} \mathbb{E}[\frac{1}{N} \mathrm{Tr}\left(\mathbf{X}^{N, \ell_1} \mathbf{X}^{N, \ell_2} \cdots \mathbf{X}^{N, \ell_k}\right)] = \sigma^m(X_{\ell_1} \cdots X_{\ell_k}).$$

$\sigma^m(X_{\ell_1} \cdots X_{\ell_k})$ is the number $|NP(X_{\ell_1} \cdots X_{\ell_k})|$ of noncrossing pair partitions of k labelled points $S(X_{\ell_1} \cdots X_{\ell_k})$ given by $(1, \ell_1), (2, \ell_2), \ldots, (k, \ell_k)$ so that every block contains points with the same label.

In Theorem 3.3.1, a noncrossing pair partitions of $S(X_{\ell_1} \cdots X_{\ell_k})$ is a collection of $k/2$ blocks $((i_s, \ell_{i_s}), (i_t, \ell_{i_t}))$ so that $\ell_{i_s} = \ell_{i_t}$ and for any two blocks $((i_s, \ell_{i_s}), (i_t, \ell_{i_t}))$ and $((i_r, \ell_{i_r}), (i_w, \ell_{i_w}))$ so that $i_s < i_r$, either $i_s < i_t < i_r < i_w$ or $i_s < i_r < i_w < i_t$. The convergence also holds almost surely: as this will be the consequence of a concentration of measure phenomenon that we shall soon describe, we do not detail it here.

Proof Setting $\mathbf{Y}^N(\ell) = \sqrt{N} \mathbf{X}^{N, \ell}$, we have

$$\mathbb{E}\left[\frac{1}{N} \mathrm{Tr}\left(\mathbf{X}^{N, \ell_1} \cdots \mathbf{X}^{N, \ell_k}\right)\right] = \sum_{i_1, \cdots, i_\ell = 1}^{N} N^{-\frac{k}{2} - 1} \mathbb{E}[Y_{i_1 i_2}(\ell_1) Y_{i_2 i_3}(\ell_2) \cdots Y_{i_k i_1}(\ell_k)], \quad (3.3)$$

where $Y_{ij}(\ell), 1 \leq i, j \leq N$, denote the entries of $\mathbf{Y}^N(\ell)$ (which may eventually depend on N). We denote $\mathbf{i} = (i_1, \cdots, i_k)$ and set

$$P^N(\mathbf{i}) := \mathbb{E}[Y_{i_1 i_2}(\ell_1) Y_{i_2 i_3}(\ell_2) \cdots Y_{i_k i_1}(\ell_k)]. \quad (3.4)$$

Note that P^N depends on N through $\mathbf{Y}^N$. By (3.2) and Hölder's inequality, $P(\mathbf{i})$ is bounded uniformly by B_k, independently of $\mathbf{i}$ and N. Since the random variables $(Y_{ij}(\ell), 1 \leq i \leq j \leq N)$ are independent and centered, $P^N(\mathbf{i})$ equals 0 unless for any pair (i_p, i_{p+1}), $p \in \{1, \cdots, k\}$, there exists $l \neq p$ such that $(i_p, i_{p+1}) = (i_l, i_{l+1})$ or (i_{l+1}, i_l). Here, we used the convention $i_{k+1} = i_1$. To find more precisely which set of indices contributes to the first order in the right-hand side of (3.3), we next provide some combinatorial insight into the sum over the indices.

Connected graphs and trees

$V(\mathbf{i}) = \{i_1, \cdots, i_k\}$ will be called the vertices. We identify i_l and i_p iff they are equal. An edge is a pair (i, j) with $i, j \in \{1, \cdots, N\}^2$. At this point, edges are directed in the sense that we distinguish (i, j) from (j, i) when $j \neq i$ and we shall point out later when we consider undirected edges. We denote by $E(\mathbf{i})$ the collection of the k edges $(e_p)_{p=1}^{k} = (i_p, i_{p+1})_{p=1}^{k}$.

We consider the graph $G(\mathbf{i}) = (V(\mathbf{i}), E(\mathbf{i}))$, see, e.g., Fig. 3.1. $G(\mathbf{i})$ is connected by construction. Note that $G(\mathbf{i})$ may contain loops (i.e., cycles, for instance edges of type (i, i)) and multiple undirected edges.

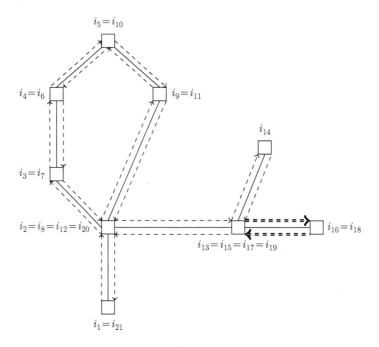

Fig. 3.1 $G(\mathbf{i})$ (in dash) versus $\tilde{G}(\mathbf{i})$ (in bold), $|\tilde{E}(i)| = 9, |\tilde{V}(\mathbf{i})| = 9$.

The skeleton $\tilde{G}(\mathbf{i})$ of $G(\mathbf{i})$ is the graph $\tilde{G}(\mathbf{i}) = \left(\tilde{V}(\mathbf{j}), \tilde{E}(\mathbf{i})\right)$ where vertices in $\tilde{V}(\mathbf{i})$ appear only once and edges in $\tilde{E}(\mathbf{i})$ are undirected and appear only once. We, however, keep a root in $\tilde{G}(\mathbf{i})$ given by an oriented edge: we choose the oriented edge corresponding to (i_1, i_2).

In other words, $\tilde{G}(\mathbf{i})$ is the graph $G(\mathbf{i})$ where multiplicities and orientation have been erased. It is connected, as is $G(\mathbf{i})$.

We now state and prove a well-known inequality concerning undirected connected graphs $G = (V, E)$. If we let, for a discrete finite set A, $|A|$ be the number of its distinct elements, we have the inequality

$$|V| \le |E| + 1. \tag{3.5}$$

Let us prove this inequality and that equality holds only if G is a tree at the same time. This relation is straightforward when $|V| = 1$ and can be proven by induction as follows. Assume $|V| = n$ and consider one vertex v of V. This vertex is contained in l edges of E which we denote $(e_1, \cdots, e_l)$ and with $l \ge 1$ by connectedness. The graph G then decomposes into $(\{v\}, \{e_1, \cdots, e_l\})$ and $r \le l$ undirected connected graphs $(G_1, \cdots, G_r)$. We denote $G_j = (V_j, E_j)$, for $j \in \{1, \cdots, r\}$. We have

$$|V| - 1 = \sum_{j=1}^{r} |V_j|, \quad |E| - l = \sum_{j=1}^{r} |E_j|. \tag{3.6}$$

Applying the induction hypothesis to the graphs $(G_j)_{1 \le j \le r}$ gives

$$|V| - 1 \le \sum_{i=1}^{r}(|E_j| + 1)$$
$$= |E| + r - l \le |E|, \tag{3.7}$$

which proves (3.5). In the case where $|V| = |E| + 1$, we claim that G is a tree, namely G does not have a loop. In fact, for equality to hold, we need to have equalities when performing the previous decomposition of the graph, a decomposition which can be reproduced until all vertices have been considered. If the graph contains a loop, the first time that we erase a vertex of this loop when performing this decomposition we will create one connected component less than the number of edges we erased and so a strict inequality occurs in the right-hand side of (3.7) (i.e., $r < l$).

Convergence in expectation

Since we noted that $P(\mathbf{i})$ equals 0 unless each edge in $E(\mathbf{i})$ is repeated at least twice, we have that

$$|\tilde{E}(\mathbf{i})| \le 2^{-1}|E(\mathbf{i})| = \frac{k}{2}, \tag{3.8}$$

and so by (3.5) applied to the skeleton $\tilde{G}(\mathbf{i})$ we find that

$$|\tilde{V}(\mathbf{i})| \le \lfloor \frac{k}{2} \rfloor + 1, \tag{3.9}$$

where $\lfloor x \rfloor$ is the integer part of x. Thus, since the indices are chosen in $\{1, \cdots, N\}$, there are at most $N^{\lfloor \frac{k}{2} \rfloor + 1}$ indices which contribute to the sum (3.3) and so we have

$$\left| \mathbb{E}\left[\frac{1}{N} \mathrm{Tr} \left(\mathbf{X}^{N,\ell_1} \cdots \mathbf{X}^{N,\ell_k} \right) \right] \right| \le B_k N^{\lfloor \frac{k}{2} \rfloor - \frac{k}{2}}, \tag{3.10}$$

where we used (3.2) and Hölder's inequality. In particular, if k is odd,

$$\lim_{N \to \infty} \mathbb{E}\left[\frac{1}{N} \mathrm{Tr} \left(\mathbf{X}^{N,\ell_1} \cdots \mathbf{X}^{N,\ell_k} \right) \right] = 0. \tag{3.11}$$

If k is even, the only indices which will contribute to the first-order asymptotics in the sum are those such that

$$|\tilde{V}(\mathbf{i})| = \frac{k}{2} + 1, \tag{3.12}$$

since the other indices will be such that $|\tilde{V}(\mathbf{i})| \le \frac{k}{2}$ and so will contribute at most by a term $N^{\frac{k}{2}} B_k N^{-\frac{k}{2}-1} = O(N^{-1})$. Let us consider indices such that $|V(\mathbf{i})| = \frac{k}{2} + 1$. Observe that there are $N(N-1) \cdots (N - |V|) \simeq N^{|V|}$ choices of indices associated

with the same graph, up to a relabeling of the vertices. This number simplifies with the normalization and the expectation hence turns out as a sum over the graphs which will appear for a given label. We then investigate which graphs will contribute. By the previous considerations, when $|\tilde{V}(\mathbf{i})| = \frac{k}{2} + 1$, we have that:

1. $\tilde{G}(\mathbf{i})$ is a tree,
2. $|\tilde{E}(\mathbf{i})| = 2^{-1}|E(\mathbf{i})| = \frac{k}{2}$ and so each edge in $E(\mathbf{i})$ appears exactly twice.

We can explore $G(\mathbf{i})$ by following the path P of edges $i_1 \rightarrow i_2 \rightarrow i_3 \cdots \rightarrow i_k \rightarrow i_1$. We then see that each pair of directed edges corresponding to the same undirected edge in $\tilde{E}(\mathbf{i})$ consists of taking both directions (since otherwise the path of edges has to form a loop to return to i_0). We label the steps of the path P by labeling the first step $i_1 \rightarrow i_2$ by ℓ_1, the second $i_2 \rightarrow i_3$ by ℓ_2, etc., till the last by ℓ_k. For $e \in \tilde{G}(\mathbf{i})$, choosing a direction $e = (s,t)$, we let $l_1(e)$ (resp. $l_2(e)$) be the label of the step when P goes through (s,t) (resp. (t,s)). Therefore, for these indices, we have

$$P^N(\mathbf{i}) = \prod_{e \in \tilde{G}(\mathbf{i})} \mathbb{E}[Y_e(l_1(e))\bar{Y}_e(l_2(e))] = \prod_{e \in \tilde{G}(\mathbf{i})} 1_{l_1(e)=l_2(e)}.$$

Hence, the indices $\mathbf{i}$ contribute only if the path P goes twice through each edge, at two steps with the same label. To summarize,

$$\lim_{N \rightarrow \infty} \mathbb{E}\left[\frac{1}{N}\mathrm{Tr}\left(\mathbf{X}^{N,\ell_1} \cdots \mathbf{X}^{N,\ell_k}\right)\right] = \sum_{\tilde{G} \in \mathcal{S}_k} \sum_{P \in \mathcal{P}(\tilde{G})} \prod_{e \in \tilde{G}(\mathbf{i})} 1_{l_1(e)=l_2(e)},$$

where $\mathcal{S}_k$ is the set of rooted trees with $k/2$ edges, $\mathcal{P}(\tilde{G})$ is the set of paths with the k steps, with first step given by the root, and passing twice through each edge.

Finally, observe that $\tilde{G}$ gives a pair partition of the edges of the path P (since each undirected edge has to appear exactly twice) and that this partition is noncrossing (as can be seen by unfolding the path and keeping track of the pairing between edges by drawing an arc between paired edges) and colored. Moreover, this pair partition is the same for $\mathbf{j}$ and $\mathbf{i}$ iff the graphs $G(\mathbf{j})$ and $G(\mathbf{i})$ are isomorphic (that is, they correspond to a different labeling of the vertices of the same rooted graph).

We finally show that the limit is the law of free variables:

Property. σ^m is the law of m free variables with semi-circular distribution

$$\sigma(dx) = \frac{1}{2\pi}\sqrt{4 - x^2}dx .$$

Proof We leave the reader to check that the kth moments of the semi-circular law is given by the number of noncrossing pair partitions of k ordered points. It is then enough to show that

$$\sigma^m(P_1(X_{i_1}) \cdots P_k(X_{i_k})) = 0$$

for all choices of $i_j \in \{1, \ldots, m\}$ so that $i_j \neq i_{j+1}$ and all centred monomials P_i such that $\sigma(P_i(x)) = 0$. Note that we can relax the assumption by saying that at most one of the P_i's is not centered. We can proceed by induction over k. It is clearly true for $k = 1$ and we can assume it is true up to $k = \ell$. Let us consider the case $k = \ell + 1$ and take $P_i(x) = x^{n_i} - \sigma(x^{n_i})$. We may assume without loss of generality that $i_1 \neq i_{\ell+1}$ by traciality. We use the definition of $\sigma^m(P_1(X_{i_1}) \cdots P_k(X_{i_k}))$ in terms of colored noncrossing pair partitions. If all the labelled points $L_1 = \{(1, i_1), \ldots (n_1, i_1)\}$ are paired together, then the contribution disappears because of the centring. Otherwise, one point (m, i_1) of L_1 must be paired with a point (n, i_p) in $L_p = \{(n_1 + \cdots + n_{p-1} + 1, i_p), \ldots, (n_1 + \cdots + n_p, i_p)\}$ for some $p \in \{2, \ldots, k\}$. But then we must have $i_1 = i_p$ and all points between (m, i_1) and (n, i_p) must be paired together since the partition is noncrossing. We are thus reduced to count the contribution of $\sigma^m(X_{i_1}^{n_1 - m} P_2(X_{i_2}) \cdots P_{p-1}(X_{i_{p-1}}) X_{i_1}^{n - n_1 + \cdots + n_{p-1}})$. By traciality we can put the last monomial with the first and we can use our induction hypothesis to see that this contribution vanishes.

3.3.2   Matrices conjugated by Haar unitary matrices are asymptotically free

Observe that if X_N is a GUE matrix, then $U X_N U^*$ is also a GUE matrix for any deterministic unitary matrix U. It turns out that one can extend the convergence of joint words in independent GUE matrices into the convergence of words in unitary Haar distributed random matrices and deterministic matrices, and show that the limit is free.

Theorem 3.3.2.   Let $\mathbf{D}^N = \{D_i^N\}_{1 \leq i \leq p}$ be a sequence of Hermitian (possibly random) $N \times N$ matrices. Assume that their empirical distribution converges to a noncommutative law μ. Assume also that there exists a deterministic $D < \infty$ such that, for all $k \in \mathbb{N}$ and all $N \in \mathbb{N}$,

$$\frac{1}{N} \mathrm{Tr}((D_i^N)^{2k}) \leq D^{2k}, \ a.s.$$

Let $\mathbf{U}^N = \{U_i^N\}_{1 \leq i \leq p}$ be independent unitary matrices with Haar law $\rho_{U(N)}$, independent from $\{D_i^N\}_{1 \leq i \leq p}$. Then the subalgebras $\mathcal{U}_i^N$ generated by the matrices $\{U_i^N, (U_i^N)^*\}_{1 \leq i \leq p}$, and the subalgebra $\mathcal{D}^N$ generated by the matrices $\{D_i^N\}_{1 \leq i \leq p}$ are asymptotically free. Moreover, the algebras $\mathcal{U}_i^N$ are asymptotically free. Finally, for all $i \in \{1, \ldots, p\}$, the limit law of $\{U_i^N, (U_i^N)^*\}$ is given by

$$\tau((UU^* - 1)^2) = 0, \quad \tau(U^n) = \tau((U^*)^n) = \mathbf{1}_{n=0}.$$

Again, convergence holds both in expectation and almost surely. Corollary 3.4.1 provides a comparison between independence (respectively, standard convolution) and freeness (respectively, free convolution) in terms of random matrices. If D_1^N and D_2^N are two diagonal matrices whose eigenvalues are independent and equidistributed, the spectral measure of $D_1^N + D_2^N$ converges to a standard convolution. At the other

extreme, if the eigenvectors of a matrix A_1^N are 'very independent' from those of a matrix A_2^N in the sense that the joint distribution of the matrices can be written as the distribution of $(A_1^N, U^N A_2^N (U^N)^*)$, then free convolution will describe the limit law of $A_1^N + A_2^N$.

Proof of Theorem 3.3.2 We denote by $\hat{\mu}_N := \mu_{\{D_i^N, U_i^N, (U_i^N)^*\}_{1 \leq i \leq p}}$ the joint empirical distribution of $\{D_i^N, U_i^N, (U_i^N)^*\}_{1 \leq i \leq p}$. $\mathbb{E}[\hat{\mu}_N]$ is tight as for any monomial q, $\mathbb{E}[\hat{\mu}_N(q)]$ is bounded by $D^{\deg(q)}$. Hence, we can take converging subsequences and consider their limit points. The strategy of the proof will be to show that these limit points satisfy a Schwinger–Dyson equation. This is a system of equations defined by an appropriate noncommutative derivative, and will be derived from the invariance by multiplication of the Haar measure. We will then show that this limit is exactly the law of free variables. Finally, concentration inequalities will allow us to extend the result to the almost sure convergence of $\{\hat{\mu}_N\}_{N \in \mathbb{N}}$.

- *Schwinger–Dyson equation.* We consider a limit point τ of $\{E[\hat{\mu}_N]\}_{N \in \mathbb{N}}$. Because we have $\hat{\mu}_N((U_i(U_i)^* - 1)^2) = 0$ and $\hat{\mu}_N(PQ) = \hat{\mu}_N(QP)$ for any $P, Q \in \mathbb{C}\langle D_i, U_i, U_i^* | 1 \leq i \leq p\rangle$, almost surely, we know by taking the large N limit that

$$\tau(PQ) = \tau(QP), \quad \tau((U_i U_i^* - 1)^2) = 0, \; 1 \leq i \leq p. \tag{3.13}$$

Since τ is a tracial state, the second equality in (3.13) implies that $U_i U_i^* = 1$ (as bounded linear operators as constructed in the Gelfand–Seigal–Neimark construction).

By definition, the Haar measure $\rho_{U(N)}$ is invariant under multiplication by a unitary matrix. In particular, if $P \in \mathbb{C}\langle D_i, U_i, U_i^* | 1 \leq i \leq p\rangle$, we have for all $k, l \in \{1, \ldots, N\}$,

$$\partial_t \int \left(P(D_j, U_j e^{tB_j}, e^{-tB_j} U_j^*) \right)(k,l) d\rho_{U(N)}(U_1) \cdots d\rho_{U(N)}(U_p) = 0$$

for any anti-Hermitian matrices B_j ($B_j^* = -B_j$), $1 \leq j \leq p$, since then $e^{tB_i} \in U(N)$. Taking $B_j(rs) = 1_{j=i_0} i(1_{rs=rq} + 1_{rs=qr})$ and $B_j(rs) = 1_{j=i_0}(1_{rs=qr} - 1_{rs=rq})$ we deduce after summation of the two resulting equalities that

$$\int (\partial_{i_0} P)(\{D_i, U_i, U_i^*\}_{1 \leq i \leq p})(k,r,q,l) d\rho_{U(N)}(U_1) \cdots d\rho_{U(N)}(U_p) = 0$$

with ∂_i being the derivative which obeys the Leibniz rules

$$\partial_i(PQ) = \partial_i P \times 1 \otimes Q + P \otimes 1 \times \partial_i Q,$$
$$\partial_i U_j = 1_{j=i} U_j \otimes 1, \partial_i U_j^* = -1_{j=i} 1 \otimes U_j^*,$$

where we used the notation $(A \otimes B)(k,r,q,l) := A(k,r)B(q,l)$. Taking $k = r$ and $q = l$ and summing over r, q gives

$$E\left[\hat{\mu}_N \otimes \hat{\mu}_N(\partial_i P)\right] = 0. \tag{3.14}$$

We now use concentration inequalities. It turns out that $SU(N)$ can be viewed as a compact submanifold of the set of $N \times N$ matrices with positive Ricci curvature from which it can be inferred that the Haar measure on U(N) satisfies the following bounds; see Anderson et al. (2010). $\square$

Lemma 3.3.3.   *Let $X_i \in Mat_{N \times N}(\mathbb{C})$ for $i = 1, \ldots, k$ be a collection of nonrandom matrices and let D be a constant bounding all singular values of these matrices. Let $p = p(t_1, \ldots, t_{k+2})$ be a polynomial in $k + 2$ noncommuting variables with complex coefficients, and for $X \in U(N)$, define $f(X) = Trp(X, X^*, X_1, \ldots, X_k)$. Then there exist positive constants $N_0 = N_0(p)$ and $c = c(p, D)$ such that, for any $\delta > 0$ and $N > N_0(p)$,*

$$\rho_{U(N)} \left(|f - \rho_{U(N)} f| > \delta N \right) \leq 2e^{-cN^2\delta^2} . \tag{3.15}$$

Using Lemma 3.3.3 inductively (on the number p of independent unitary matrices), we find that, for any polynomial $P \in \mathbb{C}\langle D_i, U_i, U_i^* | 1 \leq i \leq p\rangle$, there exists a positive constant $c(P)$ such that

$$\rho_{U(N)}^{\otimes p} \left(|\mathrm{Tr}P(\{D_i^N, U_i^N, (U_i^N)^*\}_{1\leq i\leq p}) - E\mathrm{Tr}P| > \delta \right) \leq 2e^{-c(P)\delta^2} ,$$

and therefore

$$E[|\mathrm{Tr}P - E\mathrm{Tr}P|^2] \leq \frac{2}{c(P)} .$$

Writing $\partial_i P = \sum_{j=1}^{M} P_j \otimes Q_j$ for appropriate integer M and polynomials $P_j, Q_j \in \mathbb{C}\langle D_i, U_i, U_i^* | 1 \leq i \leq p\rangle$, we deduce by the Cauchy–Schwarz inequality that

$$|E\left[(\hat{\mu}_N - E[\hat{\mu}_N]) \otimes (\hat{\mu}_N - E[\hat{\mu}_N])(\partial_i P)\right]|$$

$$\leq \sum_{j=1}^{M} E\left[|(\hat{\mu}_N - E[\hat{\mu}_N])(P_j)|^2\right]^{1/2} E[|(\hat{\mu}_N - E[\hat{\mu}_N])(Q_j)|^2]^{1/2}$$

$$\leq \frac{2M}{N^2} \max_{1\leq j\leq p} \max\{\frac{1}{c(P_j)}, \frac{1}{c(Q_j)}\} \to_{N\to\infty} 0 .$$

We thus deduce from (3.14) that

$$\lim_{N\to\infty} E\left[\hat{\mu}_N\right] \otimes E\left[\hat{\mu}_N\right](\partial_i P) = 0 .$$

Therefore, the limit point τ satisfies the Schwinger–Dyson equation

$$\tau \otimes \tau(\partial_i P) = 0 , \tag{3.16}$$

for all $i \in \{1, \ldots, p\}$ and $P \in \mathbb{C}\langle D_i, U_i, U_i^* | 1 \leq i \leq p\rangle$.

- *Uniqueness of the solution to* (3.16). Let τ be a solution to (3.13) and (3.16), and let P be a monomial in $\mathbb{C}\langle D_i, U_i, U_i^* | 1 \leq i \leq p\rangle$. We show by induction over the total degree n of P in the variables U_i and U_i^* that $\tau(P)$ is uniquely determined

by (3.13) and (3.16). Note that if $P \in \mathbb{C}\langle D_i | 1 \le i \le p \rangle$, $\tau(P) = \mu(P)$ is uniquely determined. If $P \in \mathbb{C}\langle D_i, U_i, U_i^* | 1 \le i \le p \rangle \backslash \mathbb{C}\langle D_i | 1 \le i \le p \rangle$ is a monomial, we can always write $\tau(P) = \tau(QU_i)$ or $\tau(P) = \tau(U_i^*Q)$ for some monomial Q and some $i \in \{1, \dots, p\}$, by the tracial property (3.13). We study the first case, the second being equivalent by taking the adjoint. If $\tau(P) = \tau(QU_i)$,

$$\partial_i(QU_i) = \partial_i Q \times 1 \otimes U_i + (QU_i) \otimes 1 \,,$$

and so (3.16) gives

$$\tau(QU_i) = -\tau \otimes \tau(\partial_i Q \times 1 \otimes U_i)$$
$$= -\sum_{Q=Q_1 U_i Q_2} \tau(Q_1 U_i)\tau(Q_2 U_i) + \sum_{Q=Q_1 U_i^* Q_2} \tau(Q_1)\tau(Q_2) \,,$$

where we used the fact that $\tau(U_i^* Q_2 U_i) = \tau(Q_2)$ by (3.13). Each term in the right side is the trace under τ of a polynomial of degree strictly smaller in U_i and U_i^* than QU_i. Hence, this relation defines τ uniquely by induction. In particular, taking $P = U_i^n$ we get, for all $n \ge 1$,

$$\sum_{k=1}^{n} \tau(U_i^k)\tau(U_i^{n-k}) = 0 \,,$$

from which we deduce by induction that $\tau(U_i^n) = 0$ for all $n \ge 1$ since $\tau(U_i^0) = \tau(1) = 1$. Moreover, as τ is a state, $\tau((U_i^*)^n) = \tau(((U_i)^n)^*) = \overline{\tau(U_i^n)} = 0$ for $n \ge 1$.

- *The solution is the law of free variables.* It is enough to show by the previous point that the joint law μ of the two free p-tuples $\{U_i, U_i^*\}_{1 \le i \le p}$ and $\{D_i\}_{1 \le i \le p}$ satisfies (3.16). So take $P = U_{i_1}^{n_1} B_1 \cdots U_{i_p}^{n_p} B_p$ with some B_k's in the algebra generated by $\{D_i\}_{1 \le i \le p}$ and $n_i \in \mathbb{Z} \backslash \{0\}$ (where we observed that $U_i^* = U_i^{-1}$). We wish to show that, for all $i \in \{1, \dots, p\}$,

$$\mu \otimes \mu(\partial_i P) = 0. \tag{3.17}$$

Note that, by linearity, it is enough to prove this equality when $\mu(B_j) = 0$ for all j. Now, by definition, we have

$$\partial_i P = \sum_{k:i_k=i, n_k>0} \sum_{l=1}^{n_k} U_{i_1}^{n_1} B_1 \cdots B_{k-1} U_i^l \otimes U_i^{n_k-l} B_k \cdots U_{i_p}^{n_p} B_p$$

$$- \sum_{k:i_k=i, n_k<0} \sum_{l=0}^{n_k-1} U_{i_1}^{n_1} B_1 \cdots B_{k-1} U_i^{-l} \otimes U_i^{n_k+l} B_k \cdots U_{i_p}^{n_p} B_p \,.$$

Taking the expectation on both sides, since $\mu(U_j^i) = 0$ and $\mu(B_j) = 0$ for all $i \ne 0$ and j, we see that freeness implies that the trace of the right-hand side vanishes (recall here that, in the definition of freeness, two consecutive elements must be in free algebras but the first and last elements can be in the same algebra). Thus, $\mu \otimes \mu(\partial_i P) = 0$, which proves the claim.

3.4   Free convolution

We have the following corollary of Theorem 3.3.2.

Corollary 3.4.1.   Let $\{D_i^N\}_{1\leq i\leq p}$ be a sequence of uniformly bounded real diagonal matrices with empirical measure of diagonal elements converging to μ_i, $i = 1,\ldots,p$, respectively. Let $\{U_i^N\}_{1\leq i\leq p}$ be independent unitary matrices following the Haar measure, independent from $\{D_i^N\}_{1\leq i\leq p}$.

1. The noncommutative variables $\{U_i^N D_i^N (U_i^N)^*\}_{1\leq i\leq p}$ are asymptotically free with marginals given by the μ_i.
2. The empirical measure of the eigenvalues of $D_1^N + U_N D_2^N U_N^*$ converges weakly almost surely as N goes to infinity to the law $\mu_1 \boxplus \mu_2$ of $A + B$, A, B being free and with distribution μ_1 and μ_2, respectively. $\mu_1 \boxplus \mu_2$ is caleed the free additive convolution of μ_1 and μ_2.
3. Assume that D_1^N is nonnegative for all $N \geq 0$. Then, the empirical measure of eigenvalues of

$$(D_1^N)^{\frac{1}{2}} U_N D_2^N U_N^* (D_1^N)^{\frac{1}{2}}$$

converges weakly almost surely to the distribution $\mu_1 \boxtimes \mu_2$ of $A^{1/2} B A^{1/2}$, A, B being free and with distribution μ_1 and μ_2, respectively. $\mu_1 \boxtimes \mu_2$ is caleed the free multiplicative convolution of μ_1 and μ_2.

Proof Proof of Corollary 3.4.1 The only point to prove is the first. By Theorem 3.3.2, we know that the normalized trace of any polynomial P in $\{U_i^N D_i^N (U_i^N)^*\}_{1\leq i\leq p}$ converges to $\tau(P(\{U_i D_i U_i\}_{1\leq i\leq p}))$ with the subalgebras generated by $\{D_i\}_{1\leq i\leq p}$ and $\{U_i, U_i^*\}_{1\leq i\leq p}$ being free. Thus, if

$$P(\{X_i\}_{1\leq i\leq p}) = Q_1(X_{i_1})\cdots Q_k(X_{i_k}), \quad \text{with } i_{\ell+1} \neq i_\ell, 1 \leq \ell \leq k-1$$

and $\tau(Q_\ell(X_{i_\ell})) = \tau(Q_\ell(D_{i_\ell})) = 0$, then

$$\tau(P(\{U_i D_i U_i\}_{1\leq i\leq p})) = \tau(U_{i_1} Q_1(D_{i_1}) U_{i_1}^* \cdots U_{i_k} Q_k(D_{i_k}) U_{i_k}^*) = 0\,,$$

since $\tau(Q_\ell(D_{i_\ell})) = 0$ and $\tau(U_i) = \tau(U_i^*) = 0$.

3.4.1   The R-transform

We show next how free additive convolution can be analyzed/computed thanks to the analogue of the Fourier transform, the so-called R-transform which is defined as follows. Remember that $G_\mu(z) = \int (z - x)^{-1}\mathrm{d}\mu(x)$ can be inverted in a neighborhood of infinity as it behaves like $1/z$ (by the implicit function theorem). We let K_μ be this inverse, at least defined in a neighborhood of the origin. We then set

$$R_\mu(z) = K_\mu(z) - \frac{1}{z}\,,$$

for z small.

Proposition 3.4.2. Let μ_1, μ_2 be two compactly supported probability measures on the real line. Then, for z in a neighborhood of the origin, we have

$$R_{\mu_1 \boxplus \mu_2}(z) = R_{\mu_1}(z) + R_{\mu_2}(z).$$

Proof We follow here the proof proposed by Tao (2012). Take μ compactly supported on K. Take $z \neq 0 \in \mathbb{C}$ small enough. Then the definition of $K_\mu(z)$ is equivalent to the fact that

$$zE_\mu^z(x) = \frac{1}{K_\mu(z) - x} - z$$

is measurable and centered under $d\mu(x)$. Note here that $E_\mu^z(x)$ is bounded uniformly. In fact, as $K_\mu(z) \simeq 1/z$ is large and x uniformly bounded, there exists a finite constant C and $\epsilon > 0$ so that for $|z| \leq \epsilon$, $|E_\mu^z(x)| \leq C|z|$. Moreover, it is a converging power series of x. In other words,

$$K_\mu(z) - x = \frac{1}{z(1 + E_\mu^z(x))}.$$

Let μ_1, μ_2 be compactly supported probability measures, and X, Y be free with distribution μ_1 and μ_2 under the state ϕ. As E_μ^z is a converging power series which is smaller than $\frac{1}{2}$ for z small enough, we easily define $E_{\mu_1}^z(X)$ and $E_{\mu_2}^z(Y)$ and the random variable A so that

$$\frac{1}{z(1 + A)} = \frac{1}{z(1 + E_{\mu_1}^z(X))} + \frac{1}{z(1 + E_{\mu_2}^z(Y))} - \frac{1}{z}.$$

Then, we deduce that

$$X + Y - R_{\mu_1}(z) - R_{\mu_2}(z) + \frac{1}{z} = X + Y - K_{\mu_1}(z) - K_{\mu_2}(z) - \frac{1}{z} = \frac{1}{z(1 + A)}.$$

Hence, if $\phi(A) = 0$, $R_{\mu_1}(z) + R_{\mu_2}(z) - \frac{1}{z}$ is the inverse of the Cauchy transform of $X + Y$, that is the expected result. We find that

$$A = (1 + E_{\mu_2}^z(Y))(1 - E_{\mu_1}^z(X)E_{\mu_2}^z(Y))^{-1}E_{\mu_1}^z(X).$$

But, for z small enough $|E_{\mu_1}^z(X)E_{\mu_2}^z(Y)| \leq 1/4$ so that we can expand the above inverse. We then note that freeness implies that for all $n \geq 0$,

$$\phi((E_{\mu_1}^z(X)E_{\mu_2}^z(Y))^{n+1}) = \phi(E_{\mu_2}^z(Y)(E_{\mu_1}^z(X)E_{\mu_2}^z(Y))^n) \tag{3.18}$$
$$= \phi((E_{\mu_1}^z(X)E_{\mu_2}^z(Y))^n E_{\mu_1}^z(X)) = 0 \tag{3.19}$$

so that $\phi(A) = 0$.

A similar result holds for multiplicative convolution. In this case one considers the S-transform. To define it, consider a compactly supported probability measure μ and assume that $\mu(x) \neq 0$. We then set

$$m_\mu(z) = \int \frac{zx}{1-zx} d\mu(x).$$

m_μ is invertible in a neighborhood of the origin by the implicit function theorem since $\mu(x) \neq 0$. We then set

$$S_\mu(z) = \frac{1+z}{z} m_\mu^{-1}(z).$$

Note that the S-transform determines uniquely the moments of μ and hence μ. We then have the following.

Theorem 3.4.3.   Let μ_1, μ_2 be compactly supported probability measures on $\mathbb{R}$ with nonvanishing first moment. Then, the law $\mu_1 \boxtimes \mu_2$ of $x_1 x_2$, x_1, x_2 being free and a distributed according to μ_1 and μ_2, respectively, is given by

$$S_{\mu_1 \boxtimes \mu_2}(z) = S_{\mu_1}(z) \times S_{\mu_2}(z)$$

for z in a neighborhood of the origin.

Note that if μ_1 is supported on $\mathbb{R}^+$, $x_1^{1/2} x_2 x_1^{1/2}$ is a self-adjoint operator whose spectral measure has the same moments as $\mu_1 \boxtimes \mu_2$.

Proof   We simply sketch the proof which is similar to the previous one. Denote s_X the inverse of m_μ, with X with law μ. Then

$$\int \frac{x s_X(z)}{1 - x s_X(z)} d\mu(x) = z,$$

which is equivalent to the existence of E_X centered so that

$$\frac{X s_\mu(z)}{1 - X s_\mu(z)} = z + E_X \Leftrightarrow S_X(z) X = \frac{1 + \frac{1}{z} E_X}{1 + \frac{1}{z+1} E_X}.$$

We deduce that

$$S_Y(z) S_X(z) XY = \frac{1 + \frac{1}{z} E_{XY}}{1 + \frac{1}{z+1} E_{XY}} \quad \text{with } E_{XY} = \frac{E_X + E_Y + \frac{2z+1}{z(z+1)} E_X E_Y}{z(z+1) - E_X E_Y}.$$

E_{XY} is centered by freeness. Hence, $S_{XY} = S_X S_Y$. Finally, the moments of $X^{1/2} Y X^{1/2}$ are the same as those of XY, so that $S_{X^{1/2} Y X^{1/2}} = S_{XY}$.

3.4.2 The free central limit theorem

The semi-circle distribution plays the role of the Gaussian law in the sense that it governs the fluctuations of the sum of free variables.

Theorem 3.4.4. Let $X_j \in (M, \phi)$ be a family of freely independent self-adjoint random variables, so that (i) $\phi(X_j) = 0$; (2) $\phi(X_j^2) = 1$; (3) $\sup_j |\phi(|X_j|^k)| < \infty$ for all k, then the random variables

$$Z_N = \frac{1}{\sqrt{N}} \sum_{j=1}^{N} X_j$$

converge in moments to the semi-circular law σ.

Proof In view of the properties of the R-transform, we get that

$$R_{\mu_{Z_N}}(z) = \sum_{j=1}^{N} N^{-\frac{1}{2}} R_{\mu_{X_j}}(N^{-\frac{1}{2}} z).$$

As we assumed that the variables have a finite third moment, for ϵ small, uniformly in z in $\Gamma_{\alpha,\beta} = \{|\Re(z)| \leq \alpha \Im z, \Im z \geq \beta\}$ with β large enough,

$$G_{\mu_{\epsilon X_j}}(z) = \frac{1}{z}(1 + \frac{\epsilon^2}{z^2}) + O(\epsilon^3).$$

This implies that

$$K_{\mu_{\epsilon X_j}}(\frac{1}{z}) = z(1 + \frac{\epsilon^2}{z^2}) + O(\epsilon^3),$$

that is, for z in a a neighbourhood of the origin

$$R_{\mu_{\epsilon X_j}}(z) = \epsilon^2 z + O(\epsilon^3).$$

We deduce that for such z

$$R_{\mu_{Z_N}}(z) = z + O(\frac{1}{\sqrt{N}}) \Leftrightarrow K_{\mu_{Z_N}}(z) = z + \frac{1}{z} + O(\frac{1}{\sqrt{N}}).$$

As $G_{\mu_{Z_N}}$ goes to 0 uniformly as z goes to infinity, we deduce that for $z \in \Gamma_{\alpha,\beta}$, β large enough, it solves

$$z = G_{\mu_{Z_N}}(z) + G_{\mu_{Z_N}}(z)^{-1} + O(\frac{1}{\sqrt{N}})$$

uniformly. Solving this quadratic equation shows that for $z \in \Gamma_{\alpha,\beta}$, β large enough,

$$\lim_{N \to \infty} G_{\mu_{Z_N}}(z) = \frac{z - \sqrt{z^2 - 4}}{2},$$

This convergence is enough to conclude that the moments of Z_N converge, as announced. Indeed, this implies that the above convergence holds on the upper-half plane as the family $G_{\mu_{Z_N}}$ is tight by the Arzela–Ascoli theorem and any limit point is analytic by Montel's theorem. This also implies the convergence of the derivatives of $G_{\mu_{Z_N}}$ in the upper-half plane. To show that this entails the convergence in moments, we first observe that our assumption implies that all moments of Z_N are uniformly bounded. Indeed, for all integer number k,

$$\phi(Z_N^{2k}) = \frac{1}{N^k} \sum_{i_1,\ldots,i_{2k}=1}^{N} \phi(X_{i_1} \cdots X_{i_{2k}}).$$

But $\phi(X_{i_1} \cdots X_{i_{2k}})$ vanishes as soon as there is a ℓ so that $i_\ell \cap \{i_p, p \neq \ell\} = \emptyset$. Hence each index must be repeated at least twice. Moreover, each of these moment is bounded by $\sup_j \phi(X_j^{2k})$. Therefore, $\sup_N \phi(Z_N^{2k})$ is finite. As a consequence,

$$\left| \partial_z^p \frac{1}{z} G_{\mu_{Z_N}}(1/z) - \phi(Z_N^p) \right| = \left| \phi\left(\frac{Z_N^p}{1-zZ_N} - Z_N^p \right) \right| \leq |z| \phi(|Z_N|^{p+1})$$

goes to 0 as z goes to the origin. The convergence of $\partial_z^p \frac{1}{z} G_{\mu_{Z_N}}(1/z)$ in the upper-half plane therefore entails the convergence of $\phi(Z_N^p)$ towards the limit of $\partial_z^p \frac{1}{z} G_\sigma(1/z)$ when z goes to the origin.

Note here that the result could have been proved directly as well by combinatorial arguments based on the computation of moments. However, this proof, as far as the convergence of $G_{\mu_{Z_N}}$ is concerned, only required bounds on moments of order $2 + \varepsilon$, $\varepsilon > 0$: it already implies the convergence of the spectral measure of Z_N for the vague topology.

3.5 Nonnormal operators

Recall that nonnormal matrices are matrices which do not commute with their adjoint, which is equivalent to the fact that they cannot be diagonalized in an orthonormal basis. The spectrum of such matrices is well known to be very unstable: the addition of very tiny matrices may change it drastically. For instance, if one considers the matrix

$$\Xi^{(N)} = \begin{pmatrix} 0 & 0 & \cdots & 0 & 0 \\ 1 & 0 & \cdots & 0 & 0 \\ 0 & 1 & \ddots & \vdots & \vdots \\ \vdots & \ddots & \ddots & 0 & 0 \\ 0 & \cdots & 0 & 1 & 0 \end{pmatrix},$$

the spectrum of $\Xi^{(N)}$ is concentrated at the origin. But if you add a very small entry for instance N^{-10^6} at the right up position, the spectrum will be asymptotically uniformly spread on the unit disc.

Indeed, direct inspection shows that the eigenvalues of such matrix are the N solutions of $\lambda^N N^{-10^6} = 1$.

In contrast, the spectrum of normal matrices is very stable, as can, for instance, be seen thanks to the following inequality.

Theorem 3.5.1 (Hoffman–Wielandt). Let A, B be two normal $N \times N$ matrices with complex-valued entries with eigenvalues $(\lambda_i(A))_{1 \leq i \leq N}$ and $(\lambda_i(B))_{1 \leq i \leq N}$, respectively. Then

$$\min_{\pi \text{ permutation}} \sum_{i=1}^{N} |\lambda_i(A) - \lambda_{\pi(i)}(B)|^2 \leq \operatorname{Tr}(A - B)(A - B)^* .$$

Hence, if the entries of $A - B$ are small, the distribution of the spectrum of A and B will be close. Moreover, the spectral measure is described by moments as if $A = UDU^*$, $UU^* = I$, and D is the diagonal matrix with entries $(\lambda_i)_{1 \leq i \leq N}$,

$$\frac{1}{N} \operatorname{Tr}(A^k (A^*)^p) = \frac{1}{N} \sum_{i=1}^{N} \lambda_i^k \bar{\lambda}_i^p .$$

However, one can hope that the lack of stability of the spectrum is a 'rare' event and that, in fact, random matrices will behave more nicely. In fact, using Green's formula we find that if X_N has eigenvalues λ_i^N,

$$\sum_{i=1}^{N} \psi(\lambda_i^N) = \frac{1}{2\pi} \int_{\mathbb{C}} \Delta\psi(z) \log |\prod_{i=1}^{N}(z - \lambda_i^N)| dz$$

$$= \frac{N}{4\pi} \int_{\mathbb{C}} \Delta\psi(z) \frac{1}{N} \operatorname{Tr}(\log(z - X_N)(z - X_N)^*) dz,$$

depending on the empirical measures of the Hermitian matrices $(z - X_N)(z - X_N)^*$. The only cause of instability in the right-hand side of the previous equation comes from the singularity of the logarithm. In particular, if X_N converges in $*$-moments to X, for any smooth function ϕ_ε vanishing on $[0, \varepsilon]$,

$$\lim_{N \to \infty} \int_{\mathbb{C}} \Delta\psi(z) \frac{1}{N} \operatorname{Tr}(\phi_\varepsilon((z - X_N)(z - X_N)^*) \log((z - X_N)(z - X_N)^*)) dz$$

$$= \int_{\mathbb{C}} \Delta\psi(z) \tau(\phi_\varepsilon((z - X)(z - X)^*) \log((z - X)(z - X)^*)) dz . \qquad (3.20)$$

A natural guess for the limiting spectral measure is, therefore, when X_N converges in $*$-moments to X, given by the Brown measure μ_X defined, for any C^2 function ψ, by

$$\int \psi(z) d\mu_X(z) = \frac{1}{4\pi} \int_{\mathbb{C}} \Delta\psi(z) \tau(\log[(z - X)(z - X)^*]) dz$$

Particularly, natural nonnormal operators are the so-called R-diagonal operators which can be decomposed as $X = UDV$ with U, V free unitaries, free from D Hermitian. Note that its spectral distribution is the same as that for $X = UD$, U unitary, free from D. Indeed, any nonnormal operator can be decomposed as UDV with D the diagonal operator with eigenvalues given by the singular values of X, and U, V unitaries. Assuming that they are Haar distributed is then quite natural. It turns out that the Brown measure of such operators, called R-diagonal operators, was computed by Haagerup and Larsen (2000).

Theorem 3.5.2 (Haagerup–Larsen (2000)). Take $X_D := UDV$ or UD, with U, V, D free, D self-adjoint with law μ_D, and U, V Haar distributed unitaries. The Brown measure of X_D is rotation invariant and radially supported on an annulus

$$\mu_{X_D}(B(0, f(t))) = t,$$

where $f(t) = 1/\sqrt{S_{D^2}(t-1)}$ if S_{D^2} is the S-transform of D^2. Moreover the support of μ_{X_D} is a ring given by

$$\text{supp}(\mu_{X_D}) = \{re^{i\theta}, r \in [(\mu_D(x^{-2}))^{-\frac{1}{2}}, (\mu_D(x^2))^{\frac{1}{2}}], \theta \in [0, 2\pi[\}.$$

The strategy to prove the convergence of the spectral measure is simply to show that one can remove the function ϕ_ε in (3.20). To do that, the arguments are as follows.

1. In general the law of $(z - X)(z - X)^*$ will be absolutely continuous with respect to the Lebesgue measure, and hence the singularity of the logarithm will not matter.
2. Prove that the smallest eigenvalue of $(z - X_N)(z - X_N)^*$ is greater than some $N^{-\xi}$ with overwhelming probability for all $z \in \mathbb{C}$.
3. Prove, for any complex number z, the 'local law' for the matrix $W_N^z := (z - X_N)(z - X_N)^*$. Namely, there exists $\delta > 0$ so that for any $\varepsilon > 0$, for any a, b so that $b - a > N^{-\delta}$, for N large enough we have with overwhelming probability

$$\left|\frac{1}{N}\text{Tr}(1_{W_N^z \in [a,b]}) - \tau(1_{W_z \in [a,b]})\right| \le \varepsilon|b - a|,$$

where $W_z = (z - X)(z - X)^*$. This sort of result can be derived by estimating the Stieltjes transform of the matrix W_z^N for z with imaginary part of order $N^{-\delta}$, thanks to the loop equations.

We leave as an exercise (observe that we do not need any relation between δ and ξ) to show that the two last points imply that with overwhelming probability

$$\lim_{\varepsilon \to 0} \lim_{N \to \infty} \frac{1}{N}\text{Tr}(1_{(z-X_N)(z-X_N)^* \le \varepsilon} \log((z - X_N)(z - X_N)^*)) = 0.$$

This approach could be successfully used in the following models.

3.5.1 The circular law

Let X_N be a $N \times N$ matrix with independent centered complex entries, $\sqrt{N} X_N(ij), 1 \leq i, j \leq N$ i.i.d with law μ so that $\int x^2 d\mu(x) = 1$.

Theorem 3.5.3 (Girko (10), Bai (2), Gotze-Tikhomirov (11), Tao-Vu (26)). $L_{X_N} = \frac{1}{N} \sum \delta_{\lambda_i^N}$ converges almost surely towards the uniform measure on the unit disc, which is the Brown measure of $\frac{1}{\sqrt{2}}(S + i\tilde{S})$, with $S, \tilde{S}$ free semicircles.

3.5.2 The single ring theorem

Theorem 3.5.4 (Guionnet–Krishnapur–Zeitouni (2011), Rudelson–Vershynin (2014)). Take $X_N = U_N D_N$ with U_N being unitary, Haar distributed, independent of D_N. Then, if $L_{D_N} \to \mu_D$, $\mu_D((z - x)^{-1})$ being bounded and D_N uniformly bounded, we have

- The spectral measure L_{X_N} converges weakly in probability towards the Brown measure of UD, (U, D) being free, as described in Theorem 3.5.2.
- The support of the limiting measure is a ring as described in Theorem 3.5.2.
- There is no eigenvalue away from this support with probability going to 1.

The last statement is slightly more refined than the previous one and was obtained in Guionnet and Zeitouni (2012). Generalizations of this theorem to the orthogonal case were obtained in 3.1. An application is the actual proof of a result of Feinberg and Zee (1997):

Corollary 3.5.5. Take V to be a polynomial going to infinity faster than $\log |x|$ and consider the matrix X_N with law

$$P_N^V(dX_N) = \exp\{-NV(XX^*)\}dX_N \,,$$

with dX_N the Lebesgue measure over the complex entries. Then the spectral measure of X_N converges weakly almost surely towards the Brown measure μ_{X_D} of D with law μ which is the unique minimizer of

$$I(\mu) = \int \left(V(x^2) + V(y^2) - 2 \log |x^2 - y^2| \right) d\mu(x) d\mu(y)$$

over the set of probability measures on $\mathbb{R}^+$.

3.6 Loop equations and topological expansions

To prove asymptotic freeness for unitary matrices, we developed a Stein method by showing that their noncommutative law was tight, and then showing that any limit point satisfies an equation, basically as a consequence of the underlying invariance of the Haar measure. We shall pursue the same route in this section to study the law

of Hermitian matrices in interaction via a small potential. Again, the limit law will be described by an equation reflecting the invariance by translation of the Lebesgue measure, in the form of integration by parts. In fact, such equations are of central interest in free probability. Indeed, there is no notion of density in free probability, but there is a notion of integration by parts and conjugate variable, taking the place of the gradient of the log density.

3.6.1   The free difference quotient and the conjugate variable

We will consider hereafter noncommutative laws in k self-adjoint variables.

We define the free difference quotient $\partial_i : \mathbb{C}\langle X_1,\ldots,X_k\rangle \to \mathbb{C}\langle X_1,\ldots,X_k\rangle \otimes \mathbb{C}\langle X_1,\ldots,X_k\rangle$ by

$$\partial_i(PQ) = \partial_i P \times 1 \otimes Q + P \otimes 1 \times \partial_i P, \qquad \partial_i X_j = 1_{i=j}1 \otimes 1,$$

where $a \otimes b \times c \otimes d = ac \otimes bd$. We say that $(\xi_i)_{1\le i\le k}$ is the conjugate variable of a noncommutative measure τ iff for any polynomial $P \in \mathbb{C}\langle X_1,\ldots,X_k\rangle$, any $i \in \{1,\ldots,k\}$ we have

$$\tau \otimes \tau\,(\partial_i P) = \tau(\xi_i P).$$

We next define the cyclic derivative D_i by $D_i = m \circ \partial_i$, where $m(a \otimes b) = ba$. These two definitions amount to set for any $i_1,\ldots,i_\ell \in \{1,\ldots,k\}$,

$$\partial_i X_{i_1}\cdots X_{i_\ell} = \sum_{p:i_p=i} X_{i_1}\cdots X_{i_{p-1}} \otimes X_{i_{p+1}}\cdots X_{i_\ell},$$

$$D_i X_{i_1}\cdots X_{i_\ell} = \sum_{p:i_p=i} X_{i_{p+1}}\cdots X_{i_\ell} X_{i_1}\cdots X_{i_{p-1}}.$$

We say that τ satisfies the Schwinger–Dyson equation with potential W iff $(D_i W)_{1\le i\le k}$ is the conjuguate variable of τ: that is, that for all polynomial P, all $i \in \{1,\ldots,k\}$,

$$\tau \otimes \tau(\partial_i P) = \tau(D_i W P). \tag{3.21}$$

Voiculescu proved that if $\xi_i, 1 \le i \le k$, are polynomial, then they must be of the form $(D_i V)_{1\le i\le k}$ for some polynomial V. We shall see that matrix integrals asymptotically satisfy such an equation.

3.6.2   Matrix model and integration by parts

We shall consider the distribution on k $N \times N$ Hermitian matrices given by

$$d\mathbb{P}_N^V(X_1^N,\ldots,X_k^N) = \frac{1}{Z_N^V}\exp\{N\mathrm{Tr}(V(X_1^N,\ldots,X_k^N))\}d\mathbb{P}_N(X_1^N)\cdots d\mathbb{P}_N(X_k^N),$$

$$\tag{3.22}$$

where $\mathbb{P}_N$ is the distribution of the GUE matrices given by

$$d\mathbb{P}_N(X^N) = \frac{1}{Z_N} \exp\{-\frac{N}{2} \text{Tr}(X_N^2)\} dX_N .$$

As we already pointed out, by a simple integration by part, we can prove that

$$\int \frac{1}{N} \text{Tr}(P(X_i - D_i V)) d\mathbb{P}_V^N(X_1^N, \ldots, X_d^N) = \int \frac{1}{N} \text{Tr} \otimes \frac{1}{N} \text{Tr}(\partial_i P) d\mathbb{P}_V^N(X_1^N, \ldots, X_d^N).$$

$$(3.23)$$

This equation is called a Schwinger–Dyson or loop equation. In fact it is intimately related to the previous equation since if we assume that traces of polynomials self-average, we see that limit points of the trace of polynomial will satisfy (3.21) with $W = \frac{1}{2} \sum X_i^2 - V$.

3.6.3 Topological expansions

Let us for the time being consider the case of a single matrix ($d = 1$) with no potential ($V = 0$), that is, a matrix from the GUE. Topological expansions built upon the formula for moments of Gaussian (GUE) matrices read

$$\mathbb{E}[\frac{1}{N} \text{Tr}((X^N)^p)] = \sum_{\substack{\text{graph 1 vertex} \\ \text{degree p}}} N^{-2g} = \sum_{g \geq 0} N^{-2g} M_g(p),$$

$$(3.24)$$

where $M(g, p)$ is the number of maps with genus g built over a vertex of degree p, that is, the number of graphs built over one vertex of degree p which can be properly embedded onto a surface of genus g (but not in a smaller genus surface). This formula will be proved in (3.26) based on Wick calculus. 't Hooft and Brézin–Itzykson–Parisi–Zuber (1978) had the idea in the seventies to use further this remarkable relation between matrix moments and the enumeration of graphs to enumerate maps with several vertices. Topological expansions have since beats then used in diverse contexts in physics or mathematics; after the enumeration of triangulations following Brézin et al. (1978), it was used to study the enumeration of meanders (Di Francesco et al. 2000), the enumeration of loop configurations and the O(n) model (Kostov and Staudacher 1992, Eynard and Kristjansen 1995, Guionnet et al. 2012), and its application to knot theory (Zinn-Justin and Zuber 2011). The full topological expansions have been used in mathematics since the work of Harer and Zagier (1986) in their article on the Euler characteristics of the moduli space of curves, and the famous work of Kontsevich. It has also been as a tool for constructing invariants based on its relation with algebraic geometry and topological string theory (the famous Dijkgraaf–Vafa conjecture states that Gromov–Witten invariant generating functions should be matrix integrals).

It turns out that such topological expansions are closely related with the loop (or Schwinger–Dyson) equations which are satisfied by matrix models but also can be seen as topological recursion relations. At the first order, these equations are

just given by some type of noncommutative integration by parts formula. The next order equations appear as derivatives of the first loop equation taken at finite dimension, and makes it possible to describe the full topological expansion. Hence, as put forward by Eynard, the loop equations can be used as the key to construct topological expansions and therefore interesting geometric quantities and invariants.

In this section we describe more precisely the relation among matrix integrals, topological expansions, and loop equations. Based on this relation, we show that topological expansions can be derived in much greater generality than those related to matrices with Gaussian entries and Feynmann diagrams, namely in models for which loop equations given by a noncommutative derivative are valid.

3.6.4 Formal topological expansions

Topological expansions refer to (formal) expansion of matrix integrals, or other large dimension integrals, in terms of the dimension where the coefficient corresponding to the term N^{-g} counts graphs which are embedded in a surface with genus g. We shall first describe how this relation can be put forward in the case of GUE matrices.

Maps

Let us first describe the graphs that will be enumerated. In this section we restrict ourselves to one-matrix integrals, and therefore to noncolored graphs. A map is a connected, oriented diagram which is embedded into a surface. Its genus g is by definition the genus of a surface in which it can be embedded in such a way that edges do not cross and the faces of the graph (which are defined by following the boundary of the graph) are homeomorphic to a disc. The vertices of the maps we shall consider will have the structure of a star; a star of type q, for some monomial $q = X^k$, or equivalently a star with valence k, is a vertex with valence k and a rooted edge. We can also think about the edges as half-edges that will be matched to create a map. These half-edges are drawn on the surface, or equivalently we will count maps with labeled half-edges: two maps are the same iff they were obtained by gluing half-edges with the same labels, regardless of symmetries. $M_g((q_i, k_i)_{1 \leq i \leq k})$ is then the number of maps with k_i stars of type q_i, $1 \leq i \leq n$. Observe that a star of type $q = x^k$ is in bijection with the set $S(q)$ of k ordered points.

Wick calculus

Let X^N be a matrix following the Gaussian unitary ensemble, that is, a $N \times N$ Hermitian matrix with i.i.d. centered Gaussian entries with covariance N^{-1};

$$X^N(k\ell) = \bar{X}^N(\ell k) = \frac{1}{\sqrt{2}}(x_{k\ell} + iy_{k\ell}), \quad 1 \leq k < \ell \leq N,$$

$$X^N(kk) = x_{kk}, \quad 1 \leq k \leq N,$$

with

$$dP_N(X^N) = \frac{1}{Z^N} \exp\{-\frac{N}{2}\sum_{k,\ell}(x_{k\ell}^2 + y_{k\ell}^2)\} \prod dx_{k\ell}dy_{k\ell}$$

$$= \frac{1}{Z^N} \exp\{-\frac{N}{2}\mathrm{Tr}((X^N)^2)\}dX_N.$$

The main result of this section is the following.

Theorem 3.6.1. For any integer number $p \geq 0$

$$\mathbb{E}[\frac{1}{N}\mathrm{Tr}((X^N)^p)] = \sum_{g\geq 0}\frac{1}{N^{2g}}M_g(p), \qquad (3.25)$$

where $M_g(p) = M_g(x^p)$ is the number of maps with genus g built on one star of degree p.

One of the corollaries of this theorem, which is valid under much more general assumptions on the entries of X^N, see Theorem 3.3.1, is Wigner theorem (Wigner 1958).

Corollary 3.6.2 (Wigner). For any $p \geq 0$,

$$\lim_{N\to\infty}\mathbb{E}[\frac{1}{N}\mathrm{Tr}((X^N)^p)] = C_{p/2} = \int x^p d\sigma(x),$$

with $d\sigma(x) = \frac{1}{2\pi}\sqrt{4-x^2}1_{|x|\leq 2}dx$ being the semi-circle distribution and $C_{p/2} = 0$ if p is odd and otherwise equals the Catalan number, that is, the number of noncrossing pair partitions of p points.

The proof of this result relies on Wick calculus, which, in fact, provides not only the asymptotic of traces of words in random matrices but also the whole N expansion. Indeed, the Wick formula shows that if $(G_1, \ldots, G_{2p})$ is a centered Gaussian vector, then

$$\mathbb{E}[G_1 \cdots G_{2p}] = \sum_{\pi}\prod_{(s,r) \text{ block of } \pi}\mathbb{E}[G_sG_r],$$

where we sum over all pair partitions π. To prove the theorem, one simply expands the trace in terms of the matrix entries

$$\mathbb{E}[\frac{1}{N}\mathrm{Tr}((X^N)^p)] = \frac{1}{N}\sum_{i(1),\ldots,i(p)=1}^{N}\mathbb{E}[X_{i(1)i(2)}^N X_{i(2)i(3)}^N \cdots X_{i(p)i(1)}^N].$$

Using the Wick formula with $\mathbb{E}[X_{ij}^N X_{k\ell}^N] = N^{-1}1_{ij=\ell k}$, one gets

$$\mathbb{E}[X_{i(1)i(2)}^N \cdots X_{i(p)i(1)}^N] = \frac{1}{N^{p/2}} \sum_\pi \prod_{\substack{(k,\ell) \text{ block of } \pi}} 1_{i(k)i(k+1)=i(\ell+1)i(\ell)},$$

where the sum over π runs over pair partitions of p ordered points. The later matchings can be conveniently represented by seeing the Gaussian entries as the end points of half-edges of a vertex with valence p with one marked vertex (that is, a star of type x^p)

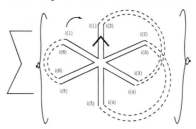

A face is obtained by cutting the graph along the edges. As $\mathbb{E}[X_{ij}^N X_{k\ell}^N] = N^{-1}1_{ij=\ell k}$, only matchings such that indices are constant along the boundary of the faces contribute to the sum. Since indices are constant along the boundaries of the faces and take any value between 1 and N, we conclude that

$$\mathbb{E}[\frac{1}{N}\text{Tr}((X^N)^p)] = \sum_{\substack{\text{graph 1 vertex} \\ \text{degree } p}} N^{\#\text{faces}-p/2-1}. \qquad (3.26)$$

But, by Euler's formula, any connected graph satisfies that its genus is given by

$$2 - 2g = \#\{\text{vertices}\} + \#\{\text{faces}\} - \#\{\text{edges}\} = 1 + \#\{\text{faces}\} - p/2,$$

such that

$$\#\{\text{faces}\} - p/2 - 1 = -2g \le 0$$

with equality only if the graph is planar. This proves the theorem and then taking the limit N going to infinity yields

$$\lim_{N\to\infty} \mathbb{E}[\frac{1}{N}\text{Tr}((X^N)^p)] = \#\{\text{planar graph with 1 vertex with degree } p\}.$$

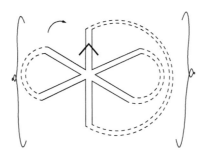

Topological expansions for several vertices

The first natural idea to count maps with several vertices is to consider the expectation of a product of traces of words as follows. Let $q_1, \cdots, q_n$ be monomials in one variable, $q_i(x) = x^{j(i)}$. Then, applying Gaussian calculus (that is, the Wick formula), we find that

$$\int \prod_{i=1}^{n} (N\mathrm{Tr}(q_i(X))) \mathrm{d}\mathbb{P}_N(X) = \sum_{g \in \mathbb{N}} \sum_{c \geq 1} \frac{1}{N^{2g-2c}} \#\{G_{g,c}((j(i),1), 1 \leq i \leq n)\}, \quad (3.27)$$

where $\#\{G_{g,c}((j(i),1), 1 \leq i \leq n\}$ is the number of graphs (up to homeomorphism) that can be built on vertices of degree $j(i)$, $1 \leq i \leq n$ (which is rooted and with labeled edges) with exactly c connected components such that the sum of their genera is equal to g.

Hence, such expectations are related to the enumeration of graphs with several vertices but unfortunately do not sort the connected graphs. We next show how this can be done.

Matrix models and topological expansions

To enumerate connected graphs, and more precisely maps, the idea of Brézin et al. (1978) is to consider partition functions instead of moments, that is, the logarithm of Laplace transforms of traces of monomials.

Consider $q_1, \cdots, q_n$ monomials. Then Brézin et al. (1978) shows that

$$\log Z_N^{\sum t_i x^{j(i)}} := \log \left(\int e^{\sum_{i=1}^{n} t_i N\mathrm{Tr}(X^{j(i)})} \mathrm{d}\mathbb{P}_N(X) \right)$$

$$= \sum_{g \geq 0} \frac{1}{N^{2g-2}} \sum_{k_1, \cdots, k_n \in \mathbb{N}} \prod_{i=1}^{n} \frac{(t_i)^{k_i}}{k_i!} M_g((x^{j(i)}, k_i), 1 \leq i \leq n), \quad (3.28)$$

where the equality is formal, which means that derivatives of all orders at $t_i = 0, 1 \leq i \leq n$, match. The proof of this formula is simply done by developing the exponential, using (3.27) and recalling that the logarithmic function will yield cumulants, and therefore connected graphs. Adding in the potential a term tq, taking a formal derivative in t at the origin shows that if $V = \sum t_i q_i(X)$ then for any monomial q

$$\int \frac{1}{N} \mathrm{Tr}((X^N)^d) \mathrm{d}\mathbb{P}_N^V(X^N)$$

$$= \sum_{g \geq 0} \frac{1}{N^{2g-2}} \sum_{k_1, \cdots, k_n \in \mathbb{N}} \prod_{i=1}^{n} \frac{(t_i)^{k_i}}{k_i!} M_g((x^d, 1); (x^{j(i)}, k_i), 1 \leq i \leq n), \quad (3.29)$$

where $M_g((x^d, 1); (x^{j(i)}, k_i), 1 \leq i \leq n)$ is the number of maps of genus g built on one vertex of degree d and k_i is the vertices of degree $j(i)$, $1 \leq i \leq n$ (with labeled 'half-edges'), and $\mathrm{d}\mathbb{P}_N^V(X^N)$ is given by

$$d\mathbb{P}_N^V(X^N) = \frac{1}{Z_N^V}\exp\{N\mathrm{Tr}(V(X^N))\}d\mathbb{P}_N(X^N).$$

Colored maps

To consider the enumeration of colored maps, we introduce the notion of stars which is a 'vertex with colored half-edges'. Moreover, a star is in bijection with words in noncommutative letters. Namely, associate with

$$q(X_1,\ldots,X_d) = X_{i_1}X_{i_2}\cdots X_{i_p},$$

a 'star of type q' given by the vertex with p colored half-edges drawn on the sphere so that the first branch has color i_1, the second of color i_2, etc., until the last which has color i_p. For instance, if $q(X_1, X_2) = X_1^2 X_2^2 X_1^4 X_2^2$ and 1 is associated with red, whereas 2 is associated with blue, then the star of type q is a vertex with, first, two half-edges which are red, then two blue, four red, and, finally, two blue.

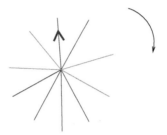

We will denote for $\mathbf{k} = (k_1, \cdots, k_n)$ and monomials $q_1, \ldots, q_n$,

$$M_g((q_i, k_i), 1 \le i \le n) = \#\{\text{ maps with genus } g$$

$$\text{and } k_i \text{ stars of type } q_i, 1 \le i \le n\},$$

the number of maps with genus g built on k_i stars of type q_i, $1 \le i \le n$, by matching the half-edges of the stars which have the same color. The enumeration is done up to homeomorphisms. By convention, we will denote $M_0(1) = 1$.

Note that stars can also be seen by duality as polygons with colored sides and one mark side, where each end point of the half-edge is replaced by a perpendicular segment of the same color. Maps are then 'polygonizations' of a surface with a given genus by polygons of a prescribed nature. For instance, for the matrix model with $q(X) = X^4$, the stars are vertices with valence four, which in the dual picture are just square. We are thus counting quadrangulations of a surface with a given genus and a given number of squares. The counting is done with labeled sides.

Random matrices and the enumeration of colored maps

Consider $q_1, \cdots, q_n$ monomials. Then Brézin et al. (1978) shows that

$$\log\left(\int e^{\sum_{i=1}^n t_i N \mathrm{Tr}(q_i(X_1, \cdots, X_m))} d\mathbb{P}_N(X_1)\cdots d\mathbb{P}_N(X_m)\right)$$

$$= \sum_{g \geq 0} \frac{1}{N^{2g-2}} \sum_{k_1,\cdots,k_n \in \mathbb{N}} \prod_{i=1}^{n} \frac{(t_i)^{k_i}}{k_i!} M_g((q_i, k_i), 1 \leq i \leq n), \qquad (3.30)$$

where the equality is formal. The proof of this formula is similar to the one-color case. Adding in the potential a term tq, taking a formal derivative in t at the origin shows that if $V = \sum t_i q_i(X_1, \ldots, X_m)$ and $\mathbb{P}_N^V$ is given by (3.22), then for any monomial q

$$\int \frac{1}{N} \mathrm{Tr}(q(X_1^N, \ldots, X_d^N)) \mathrm{d}\mathbb{P}_N^V(X_1^N, \ldots, X_d^N)$$

$$= \sum_{g \geq 0} \frac{1}{N^{2g-2}} \sum_{k_1,\cdots,k_n \in \mathbb{N}} \prod_{i=1}^{n} \frac{(t_i)^{k_i}}{k_i!} M_g((q, 1); (q_i, k_i), 1 \leq i \leq n), \qquad (3.31)$$

where $M_g((q, 1); (q_i, k_i), 1 \leq i \leq n)$ is the number of maps of genus g built on one star of type q and k_i stars of type q_i, $1 \leq i \leq n$.

At this point the equality is formal but it can, in fact, be made asymptotic as soon as reasonable assumptions are made to ensure that the integral converges and that the t_i are small enough to guarantee the convergence of the series. Equality (3.29) given asymptotically up to any order of correction N^{-k} is called an asymptotic topological expansion. We discuss this issue next, and how the loop equations can play a key role in deriving the asymptotic topological expansions.

3.6.5 Loop equations and asymptotic expansions

It is possible to prove topological expansions by using functional calculus (namely integration by parts) rather than the Wick formula and Gaussian calculus. This approach turns out to allow the proof of asymptotic topological expansions but also to generalize to different settings which are not related to any Gaussian variables, such as the integration over the unitary group or under the so-called β ensemble. We first describe the strategy for the law $\mathbb{P}_V^N$. As we already pointed out, by a simple integration by parts, we can prove that

$$\int \frac{1}{N} \mathrm{Tr}(P(X_i - D_i V)) \mathrm{d}\mathbb{P}_V^N(X_1^N, \ldots, X_d^N) \qquad (3.32)$$

$$= \int \frac{1}{N} \mathrm{Tr} \otimes \frac{1}{N} \mathrm{Tr}(\partial_i P) \mathrm{d}\mathbb{P}_V^N(X_1^N, \ldots, X_d^N).$$

In the case where V is uniformly strictly concave (in the sense that $\mathrm{Hess}(\mathrm{Tr}V) \leq -cI$ for some $c > 0$), we can argue by the Brascamp–Lieb inequality (see Guionnet (2009)) that there exists a finite constant C (which only depends on c) so that for any monomial q of degree less than $\sqrt{N}$

$$\int |\frac{1}{N} \mathrm{Tr}(q(X_1^N, \ldots, X_d^N))| \mathrm{d}\mathbb{P}_N^V(X_1^N, \ldots, X_d^N) \leq C^{\deg q}. \qquad (3.33)$$

Moreover, by standard concentration of measure result (due to Bakry–Emery, see Guionnet et al. (2011)) we find that there exists ϵ_N going to 0 with N, $c, D > 0$, C finite, such that for any $\delta \geq 0$

$$\mathbb{P}_N^V \left(\max_{\substack{1 \leq i_1,\dots,i_\ell \leq k \\ \ell \leq \sqrt{N}}} D^{-\ell} \left| \mathrm{Tr}(X_{i_1} \cdots X_{i_\ell}) - \mathbb{P}_N^V(\mathrm{Tr}(X_{i_1} \cdots X_{i_\ell})) \right| \geq \delta + \epsilon_N \right)$$
$$\leq C e^{-c\delta^2} + e^{-N}.$$

As a consequence, the family $\{\int \frac{1}{N} \mathrm{Tr}(q(X_1^N,\dots,X_d^N)) \mathrm{d}\mathbb{P}_N^V(X_1^N,\dots,X_d^N), q\}$ is tight. Any limit point $\{\tau_V(q), q\}$ satisfies the Schwinger–Dyson equation

$$\tau_V(P(X_i - D_i V)) = \tau_V \otimes \tau_V(\partial_i P) \tag{3.34}$$

with $\tau_V(I) = 1$. Here τ_V is extended linearly to polynomials. Moreover, for any monomial q, we deduce from (3.33) that

$$|\tau_V(q)| \leq C^{\deg(q)}. \tag{3.35}$$

As a consequence, when $V = \sum t_i q_i$ with the t_i small enough, there exists a unique solution to (3.34) so that $\tau_V(1) = 1$. Indeed, when $t_i = 0$, the moments are just defined inductively by (3.34). When the t_i are small enough, the equation still has a unique solution. Indeed, taking two solutions $\tau, \tilde{\tau}$ and denoting

$$\Delta_k := \sup_{q:\deg(q) \leq k} |\tau(q) - \tilde{\tau}(q)|,$$

where the supremum is taken on monomials of degree smaller or equal to k, we have by using (3.34), (3.37) and $\Delta_0 = 0$, if $D + 1 = \max \deg(q_i)$,

$$\Delta_{k+1} = \max_i \sup_{q:\deg q \leq k} |\tau(X_i q) - \tilde{\tau}(X_i q)|$$

$$|\tau(X_i q) - \tilde{\tau}(X_i q)| \leq |\tau \otimes \tau(\partial_i q) - \tilde{\tau} \otimes \tilde{\tau}(\partial_i q)| + D \sum t_j \Delta_{k+D}$$

$$\leq \sum_{l=1}^k \Delta_l C^{k-l} + D \sum t_j \Delta_{k+D}.$$

Hence,

$$\Delta_{k+1} \leq \sum_{l=1}^k \Delta_l C^{k-l} + D \sum t_j \Delta_{k+D}, \Delta_k \leq 2C^k$$

so that for $\gamma < 1/C$,

$$\Delta_\gamma := \sum_{k \geq 1} \gamma^k \Delta_k \leq \frac{\gamma}{1 - C\gamma} \Delta_\gamma + \frac{D \sum |t_j|}{\gamma^D} \Delta_\gamma \tag{3.36}$$

which entails that $\Delta_\gamma = 0$ for $\gamma < 1/C\Lambda$ so that

$$\frac{\gamma}{1 - C\gamma} + \frac{D\sum |t_j|}{\gamma^D} < 1.$$

As soon as there exists such a γ, that is, when the t_i's are small enough, we conclude that $\tau = \tilde{\tau}$.

As a result, we see the following.

Theorem 3.6.3. Set for q a word in $X_1, \ldots, X_m$ and t_i complex numbers

$$M_{\mathbf{t}}(q) = \sum_{\mathbf{k} \in \mathbb{N}^n} \prod_{i=1}^n \frac{(-t_i)^{k_i}}{k_i!} M_0((q,1); (q_i, k_i), 1 \le i \le n).$$

Then, if the t_i are small enough, $M_{\mathbf{t}}$ is solution of Eq. (3.34). Therefore, if $V = \sum t_i q_i$ is self-adjoint, for all monomial q,

$$\tau_V(q) = M_{\mathbf{t}}(q).$$

Let us remark that by definition of τ_V, for all polynomials P, Q,

$$\tau_V(PP^*) \ge 0 \quad \tau_V(PQ) = \tau_V(QP).$$

As a consequence, $M_{\mathbf{t}}$ also satisfy these equations: for all P, Q

$$M_{\mathbf{t}}(PP^*) \ge 0, \quad M_{\mathbf{t}}(PQ) = M_{\mathbf{t}}(QP), \quad M_{\mathbf{t}}(1) = 1.$$

This means that $M_{\mathbf{t}}$ is a tracial state. The traciality property can easily be derived by symmetry properties of the maps. However, the positivity property $M_{\mathbf{t}}(PP^*) \ge 0$ is not easy to prove by combinatorial arguments, and hence matrix models are a nice way to derive it. This property may be seen to be useful for actually solving the combinatorial problem (i.e., find an explicit formula for $M_{\mathbf{t}}$).

Proof Let us denote, in short, for $\mathbf{k} = (k_1, \ldots, k_n)$ and a monomial q by $M_{\mathbf{k}}(q) = M_0((q,1); (q_i, k_i), 1 \le i \le n)$ the number of planar maps with k_i stars of type q_i and one of type q. We generalize this definition to polynomials P by linearity. We let

$$M_{\mathbf{t}} = \sum_{\mathbf{k} \in \mathbb{N}^n} \prod_{i=1}^n \frac{(-t_i)^{k_i}}{k_i!} M_{\mathbf{k}}(q).$$

This series is a priori formal but we shall see in the following that, in fact, there exists a finite constant C so that for any monomials q_i

$$M_{\mathbf{k}}(q) \le \prod k_i! C^{\sum k_i \deg(q_i)}. \tag{3.37}$$

Hence, $M_{\mathbf{t}}$ converges for $|t_i| < 1/C$. $M_{\mathbf{t}}$ satisfies (3.34) if and only if for every $\mathbf{k}$ and P

$$M_{\mathbf{k}}(X_i P) = \sum_{\substack{0 \le p_j \le k_j \\ 1 \le j \le n}} \prod_{j=1}^n C_{k_j}^{p_j} M_{\mathbf{p}} \otimes M_{\mathbf{k}-\mathbf{p}}(\partial_i P) + \sum_{1 \le j \le n} k_j M_{\mathbf{k}-1_j}([D_i q_j] P), \qquad (3.38)$$

where $1_j(i) = 1_{i=j}$ and $M_{\mathbf{k}}(1) = 1_{\mathbf{k}=\mathbf{0}}$.

- We first check (3.38) for $\mathbf{k} = \mathbf{0} = (0, \cdots, 0)$. By convention, $M_{\mathbf{0}}(1) = 1$. We now check that

$$M_{\mathbf{0}}(X_i P) = M_{\mathbf{0}} \otimes M_{\mathbf{0}}(\partial_i P) = \sum_{P = p_1 X_i p_2} M_{\mathbf{0}}(p_1) M_{\mathbf{0}}(p_2).$$

But in any planar map with only one star of type $X_i P$, the half-edge corresponding to X_i must be glued with another half-edge of P. If X_i is glued with the half-edge X_i coming from the decomposition $P = p_1 X_i p_2$, the map is then split into two (independent) planar maps with stars p_1 and p_2, respectively. (Note here that p_1 and p_2 inherit the structure of stars since they inherit the orientation from P as well as a marked half-edge corresponding to the first neighbour of the glued X_i.) Hence the relation is satisfied.
- We now proceed by induction over $\mathbf{k}$ and the degree of P; we assume that (3.38) is true for $\sum k_i \le M$ and all monomials, and for $\sum k_i = M+1$ when $\deg(P) \le L$. Note that $M_{\mathbf{k}}(1) = 0$ for $|\mathbf{k}| \ge 1$ since we cannot glue a vertex with no half-edges with any star. Hence, this induction can be started with $L = 0$. Now, consider $R = X_i P$ with P of degree less than L and the set of planar maps with a star of type $X_i P$ and k_j stars of type q_j, $1 \le j \le n$, with $|\mathbf{k}| = \sum k_i = M + 1$. Then:
  ◇ either the half-edge corresponding to X_i is glued with an half-edge of P, say to the half-edge corresponding to the decomposition $P = p_1 X_i p_2$; we then can use the argument as above; the map M is cut into two disjoint planar maps M_1 (containing the star p_1) and M_2 (resp. p_2), the stars of type q_i being distributed in either one or the other of these two planar maps; there will be $r_i \le k_i$ stars of type q_i in M_1, the rest in M_2. Since all stars are labeled, there will be $\prod C_{k_i}^{r_i}$ ways to assign these stars in M_1 and M_2.

    Hence, the total number of planar maps with a star of type $X_i P$ and k_i stars of type q_i, such that the marked half-edge of $X_i P$ is glued with an half-edge of P, is

$$\sum_{P = p_1 X_i p_2} \sum_{\substack{0 \le r_i \le k_i \\ 1 \le i \le n}} \prod_{i=1}^n C_{k_i}^{r_i} M_{\mathbf{r}}(p_1) M_{\mathbf{k}-\mathbf{r}}(p_2). \qquad (3.39)$$

  ◇ or the half-edge corresponding to X_i is glued with an half-edge of another star, say q_j; let's say with the edge coming from the decomposition of q_j into $q_j = q_j^1 X_i q_j^2$. Then, once we are giving this gluing of the two edges, we can replace the two stars $X_i P$ and $q_j^1 X_i q_j^2$ glued by their X_i by the star $q_j^2 q_j^1 P$.

    We have k_j ways to choose the star of type q_j and the total number of such maps is

$$\sum_{q_j = q_j^1 X_i q_j^2} k_j M_{\mathbf{k}-1_j}(q_j^2 q_j^1 P).$$

Summing over j, we obtain by linearity of $M_{\mathbf{k}}$

$$\sum_{j=1}^{n} k_j M_{\mathbf{k}-1_j}([D_i q_j] P). \tag{3.40}$$

Equations (3.39) and (3.40) give (3.38). Moreover, it is clear that (3.38) defines uniquely $M_{\mathbf{k}}(P)$ by induction. In addition, we see that the solution to (3.38) satisfies (3.37). Indeed this is true for $\mathbf{k} = 0$ as free semi-circle variables are bounded by 2 and then follows for large $\mathbf{k}$ by induction over $\sum k_i$.

It turns out that this strategy can be followed up for each genera to consider a family of loop equations which are obtained by differentiating the first one with respect to small additional potentials. The first point is to derive the second-order Schwinger–Dyson equation by varying V into $V + \varepsilon W$ and differentiating at $\varepsilon = 0$ the first-order loop equation (3.32), hence getting equations for the cumulants. We refer the interested reader to Guionnet and Maurel-Segala (2007) and Maurel-Segala (2006) for full details, but outline the approach in the following. The first point is to prove an a priori rough estimate by showing that there exists a finite constant $C > 0$ so that for all t_i's small enough, all monomials q of degree less than $N^{1/2-\varepsilon}$ for $\varepsilon > 0$, we have

$$\left| \mathbb{E}[\frac{1}{N}\mathrm{Tr}[q]] - \tau_V(q) \right| \le \frac{C^{\deg(q)}}{N^2}. \tag{3.41}$$

The proof elaborates on the ideas developed around (3.36) to prove uniqueness of the solution to the Schwinger–Dyson equation and the concentration inequalities (3.41), which give a fine control on the error term in the loop equation satisfied by $\mathbb{E}[\tau_{X_N}]$ with respect to the Schwinger–Dyson equation. Once we have this a priori estimate, we write the second loop equation by making a small change in the potential $V \to V + \epsilon N^{-1} W$ and identifying the linear term in ε in the first-order loop equation. We denote by

$$W_2^V(P, Q) = \mathbb{E}[(\mathrm{Tr}P - \mathbb{E}\mathrm{Tr}P)(\mathrm{Tr}Q - \mathbb{E}[\mathrm{Tr}Q])]$$
$$= \partial_\varepsilon \mathbb{P}^{V-\epsilon N^{-1}Q}(\mathrm{Tr}P)|_{\varepsilon=0}$$
$$W_3^V(P, Q, R) = \partial_\varepsilon W_2^{V-\epsilon N^{-1}R}(P, Q)|_{\varepsilon=0}.$$

We denote by $\bar{\delta}^N(P) = \mathbb{E}[\mathrm{Tr}(P)] - N\tau_V(P)$. Note that Eq. (3.32) can be written as

$$\mathbb{E}[\mathrm{Tr}(\Xi_i P)] = \frac{1}{N} W_2(\partial_i P) + \frac{1}{N} \bar{\delta}^N \otimes \bar{\delta}^N(\partial_i P), \tag{3.42}$$

where

$$\Xi_i P = (X_i \otimes 1 - \partial_i V) \# P - (\tau_V \otimes I + I \otimes \tau_V)\partial_i P.$$

By our a priori estimate on $\bar{\delta}_N$ the last term is at most of order N^{-3}. Hence, to estimate the first-order correction, we would like to estimate the asymptotics of W_2 as well as 'invert' Ξ_i. Note that Ξ_i is defined up to cyclic symmetries $X_{i_1} \cdots X_{i_\ell} \mapsto X_{i_\ell} X_{i_1} \cdots X_{i_{\ell-1}}$ since we always consider it under the trace. It turns out that even though Ξ_i is hardly invertible, a combination Ξ of the $\Xi_i \circ D_i$ (up to cyclic symmetries) is invertible on polynomials with vanishing constant term, namely

$$\Xi P = \sum_i \partial_i P \sharp X_i - \sum_i \partial_i P \sharp D_i V - (\tau_V \otimes I + I \otimes \tau_V) \partial_i \circ D_i P.$$

Indeed, if P is a monomial, $\sum_i \partial_i P \sharp X_i = \deg(P) P$ is invertible, whereas the last term reduces the degree, so that their sum is invertible. Finally, for t_i small, the second term can be seen to be a perturbation, so that the sum of all these operators, that is Ξ, is invertible.

To estimate W_2^V, we obtain the second loop equation by changing $V \to V - \epsilon N^{-1} Q$ in (3.32) and identifying the linear terms in ϵ; we find after taiing $P \mapsto D_i P$ and summing over i,

$$W_2^V(\Xi P, Q) = \sum_i \mathbb{E}[\frac{1}{N} \mathrm{Tr}(D_i P D_i Q)]$$

$$+ N^{-1} \sum_i W_3^V(\partial_i D_i P, Q) + (W_2^V \otimes \bar{\delta}^N + \bar{\delta}^N \otimes W_2^V)(\partial_i D_i P, Q).$$

It turns out that the term in W_3^V is bounded by concentration inequalities, whereas $\bar{\delta}^N$ is of order N^{-1} by our previous rough estimate. Hence we see that

$$\lim_{N \to \infty} W_2^V(\Xi_i P, Q) = \sum_i \tau(D_i P D_i Q)$$

for all P. We conclude that

$$\lim_{N \to \infty} W_2^V(P, Q) = \tau_V(\sum_i D_i \Xi^{-1} P \times D_i Q) =: w_2(P, Q)$$

and therefore plugging this back into (3.42) (applied with $P \mapsto D_i P$ and summing over i) we deduce the first-order correction

$$\mathbb{E}[\frac{1}{N} \mathrm{Tr}(P)] = \tau_V(P) + \frac{1}{N^2} w_2[\sum_i \partial_i D_i \Xi^{-1} P] + o(N^{-2}).$$

It can be checked that this first-order correction is the generating function for the enumeration of maps with genus one, in the same spirit that what we just did for the genus zero. The next orders of the asymptotic expansion can be found similarly.

3.6.6 Extension to Haar distributed matrices

We have already seen in the proof of Theorem 3.3.2 that moments in unitary Haar distributed matrices could be characterized with some appropriate limiting Schwinger–Dyson equation. In fact, this approach could be as well extended to the case where these matrices are in (small) interaction, that is to study the asymptotics of the moments of unitary matrices U_i and deterministic matrices $(D_i^N)_{1 \leq i \leq p}$ under

$$\mathrm{d}\mu_V^N(U_1,\ldots,U_p) = \frac{1}{Z_V^N} \exp\{\epsilon N \mathrm{Tr}(P(U_i, U_i^*, D_i, 1 \leq i \leq k))\} \prod \mathrm{d}U_i,$$

where P is a self-adjoint polynomial, ϵ a small enough real constant, and dU the Haar measure on the unitary group. This is done in Collins et al. (2009) and Guionnet and Novak (2015); see also Chatterjee (2015).

3.7 Applications to loop models

In this last section, we show another example of matrix integrals which come from physics, as they are related to the so-called $O(n)$-model, but are also interesting in operator algebra theory since they allow the construction of tracial states on the so-called planar algebras, which have nice properties such as defining factors and towers of factors. A nice application of this two-sided interest is to generalize constructions coming from physics to define mateix models for the $O(n)$ models with all possible fugacities.

3.7.1 Application to planar algebras and loop models

We have already seen in the previous section that random matrices could be used to enumerate planar graphs. In this section we show how this point can be specified to enumerate loop models.

In the following we shall consider loop models with vertices given by Temperley–Lieb elements.

The Temperley–Lieb elements are boxes with boundary points connected by nonintersecting strings, equipped with a shading and a marked boundary point. For instance, we will consider the following Temperley–Lieb element B:

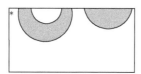

The easier loop models are those with only one vertex and the task one may undertake is, being given a Temperley–Lieb element, to count the number of planar matching of the end points of the Temperley–Lieb element so that there are exactly n loops. The picture below shows the case of 2 loops:

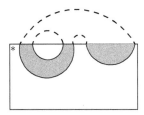

This question was related with random matrices for a long time in the physics literature, see, e.g., Kostov and Staudacher (1992), Eynard and Kristjansen (1995), and Bertola (2011). For a Temperley–Lieb element B, we denote $p \overset{B}{\sim} \ell$ if a string joins the pth boundary point with the ℓth boundary point in B; then we associate with B with k strings the polynomial

$$q_B(X) = \sum_{\substack{i_j = i_p \text{ if } j \overset{B}{\sim} p \\ 1 \le i_j \le n}} X_{i_1} \cdots X_{i_{2k}}.$$

For instance, if B is given by two embedded cups followed by a cup as in the above drawing, we have

$$q_B(X) = \sum_{i,j,k=1}^{n} X_i X_j X_j X_i X_k X_k.$$

Theorem 3.7.1.   If ν^M denotes the law of n independent GUE matrices,

$$\lim_{M \to \infty} \int \frac{1}{M} \mathrm{Tr}\,(q_B(X)) \, \nu^M(\mathrm{d}X) = \sum n^{\sharp \text{ loops}},$$

where we sum over all planar maps that can be built on B.

Proof By Voiculescu's theorem, if B is as just described,

$$\lim_{M \to \infty} \int \frac{1}{M} \mathrm{Tr}\,(q_B(X)) \, \nu^M(\mathrm{d}X)$$

$$= \sum_{i,j,k=1}^{n} \lim_{M \to \infty} \int . \frac{1}{M} \mathrm{Tr}\,(X_i X_j X_j X_i X_k X_k) \, \nu^M(\mathrm{d}X)$$

$$= \sum n^{\sharp \text{ loops}}$$

because the indices must be constant along loops. The proof extends to all Temperley–Lieb elements.

The problem with the previous theorem is that moments of random matrices can only be used so far as generating function for the enumeration of loop configurations taken at integer values of the fugacity. This is enough to characterize polynomials but not the series we shall consider later.

In Jones (2000), the author proposed a construction of a planar algebra associated with a bipartite graph. It was used in Guionnet et al. (2010) to overcome this point.

The idea is to take random matrices which are indexed by the edges of a bipartite graph instead of the integer number and to modify the polynomial q_B in such a way that the fugacity is the Perron–Frobenius eigenvalue of the adjacency matrix of the graph.

To be more precise, let $\Gamma = (V = V_+ \cup V_-, E)$ be a bipartite graph with oriented edges so that if $e \in E$, its opposite e^o is also in E. Assume that the adjacency matrix of Γ has a Perron–Frobenius eigenvalue. Note that this restricts the possible values of δ to $\{2\cos(\frac{\pi}{n}), n \geq 3\} \cup [2, +\infty[$ which is, however, a set which contains limit points.

Now, let us define for a Temperley-Lieb element B the polynomial

$$q_B^v(X) = \sum_{e_j = e_p^o \text{ if } j \overset{B}{\sim} p} \sigma_B(w) X_{e_1} \cdots X_{e_{2k}},$$

where we recall that $p \overset{B}{\sim} j$ if a string joins the pth boundary point with the jth boundary point in the TL element B. The sum runs over loops $w = e_1 \cdots e_{2k}$ in Γ which starts at $v \in V$. $v \in V_+$ iff $*$ is in a white region. σ_B is defined as follows. Denote $(\mu_v)_{v \in V}$ with $\mu_v \geq 0$ the eigenvector of Γ for the Perron–Frobenius eigenvalue δ and set, if $\sigma(e) := \sqrt{\frac{\mu_{t(e)}}{\mu_{s(e)}}}$, $e = (s(e), t(e))$,

$$\sigma_B(e_1 \cdots e_{2p}) = \prod_{\substack{i \overset{B}{\sim} j \\ i < j}} \sigma(e_i)$$

to be the sum over products of $\sigma(e)$ so that each string of B brings $\sigma(e)$ with e the edge which labels the start of the string.

For $e \in E$, $e = (s(e), t(e))$, let X_e^M be independent (except $X_{e^o} = X_e^*$) $[M\mu_{s(e)}] \times [M\mu_{t(e)}]$ matrices with i.i.d-centered Gaussian entries with variance $1/(M\sqrt{\mu_{s(e)}\mu_{t(e)}})$.

Theorem 3.7.2 (Guionnet et al. 2010).   Let Γ be a bipartite graph whose adjacency matrix has δ as a Perron–Frobenius eigenvalue. Let B be a Temperley–Lieb element so that $*$ is in an unshaded region. Then, for all $v \in V^+$,

$$\tau_\delta(B) := \lim_{M \to \infty} E[\frac{1}{M\mu_v} \text{Tr}(q_B^v(X^M))] = \sum \delta^{\sharp \text{ loops}},$$

where the sum runs above all planar maps built on B.

Maybe the best proof is by trying examples.
If B is as described earlier, for all $v \in V^+$

$$E[\frac{1}{M\mu_v} \text{Tr}(\sum_{e:s(e)=v} \sigma(e) X_e X_{e^o}))] = \frac{1}{M\mu_v} \sum_{e:s(e)=v} \sqrt{\frac{\mu_{t(e)}}{\mu_v}} \frac{M\mu_v M\mu_{t(e)}}{M\sqrt{\mu_{t(e)}\mu_{s(e)}}}$$

$$= \frac{1}{\mu_v} \sum_{e:s(e)=v} \mu_{t(e)} = \delta$$

If B is as described earlier, for all $v \in V^+$

$$\lim_{M \to \infty} \mathbb{E}[\frac{1}{M\mu_v} \mathrm{Tr}(\sum_{\substack{e:s(e)=v \\ s(f)=v}} \sigma(e)\sigma(f)X_e X_{e^0} X_f X_{f^0})]$$

$$= \delta^2 + \frac{1}{\mu_v} \sum_{e=f} \frac{\mu_{t(e)}}{\mu_v} \frac{\mu_v^2 \mu_{t(e)}}{\mu_{t(e)}\mu_v} = \delta^2 + \delta.$$

More generally, the edges are constant along the loops and brings the contribution $\mu_{t(e)}/\mu_v$, hence leading after summation to δ.

As in the previous section we can make these enumeration questions more interesting by adding a potential, and in turn enumerating loop models with several Temperley–Lieb vertices. Let B_i be Temperley–Lieb elements with $*$ with color $\sigma_i \in \{+, -\}$, $1 \le i \le p$. Let Γ be a bipartite graph whose adjacency matrix has eigenvalue δ as before. Let ν^M be the law of the previous independent rectangular Gaussian matrices and set

$$d\nu_{(B_i)_i}^M(X_e) = \frac{1_{\|X_e\|_\infty \le L}}{Z_B^M} e^{M\mathrm{Tr}(\sum_{i=1}^p \beta_i \sum_{v \in V_{\sigma_i}} \mu_v q_{B_i}^v(X))} d\nu^M(X_e).$$

Theorem 3.7.3 (Guionnet et al. 2012). For any $L > 2$, for β_i small enough real numbers, for any Temperley–Lieb element B with color σ, any $v \in V_\sigma$,

$$\tau_{\delta,\beta}(B) := \lim_{M \to \infty} \int \frac{1}{M\mu_v} \mathrm{Tr}(q_B^v(X)) d\nu_{(B_i)_i}^M(X) = \sum_{n_i \ge 0} \sum \delta^{\sharp \text{ loops}} \prod_{i=1}^p \frac{\beta_i^{n_i}}{n_i!},$$

where we sum over the planar maps built on n_i Temperley–Lieb elements B_i and one B.

The proof is based, as in the previous section, on Schwinger–Dyson's equation and concentration of measure.

3.7.2 Loop models and subfactors

Another point of view on the previous section is the subfactor theory. In fact, Temperley–Lieb algebra can be viewed as a special case of planar algebra and $\tau_{\delta,\beta}$ are tracial states on this planar algebra if they are equipped with the multiplication

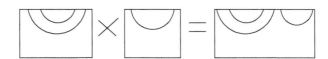

and the involution which is given by taking the symmetric picture of the element.

Theorem 3.7.4 (Guionnet et al. 2010).   Take $\delta \in \{2\cos(\pi/n), n \geq 3\} \cup [2, +\infty[$. Then

- $\tau_{\delta,0}$ is a tracial state on the Temperley–Lieb algebra.
- The von Neumann algebra associated with the Gelfand–Naimark–Segal construction is a factor; namely, its center is trivial. A tower of subfactors with index δ^2 can be built.

The tower is built by changing the multiplication so that the nearest boundary points of both Temperley–Lieb elements are capped. The construction presented here can be generalized to any planar algebra. Hence, it shows that there is a canonical way to construct a tower of subfactors from any subfactor planar algebra. It is still unknown whether the von Neumann algebra associated with $\tau_{\delta,\beta}$ are factors for $\beta \neq 0$.

References

Anderson, G., Guionnet, A., and Zeitouni, O. (2010). *An Introduction to Random Matrices*. Cambridge University Press, Cambridge.

Bai, Z. D. (2013). Circular law. *Ann. Probab.* **25**, 1, 494–529.

Basak, A., and Dembo, A. (1997). Limiting spectral distribution of sums of unitary and orthogonal matrices. *Electron. Commun. Probab.* **18**, no. 69, 19.

Bertola, M. (2011). Two-matrix models and biorthogonal polynomials. In *The Oxford Handbook of Random Matrix Theory*, pp. 310–28. Oxford University Press, Oxford.

Brézin, E., Itzykson, C., Parisi G., and Zuber, J. B. (1978). Planar diagrams. *Comm. Math. Phys.* **59**, 35–51.

Chatterjee, S. (2015). Rigorous solution of strongly coupled so(n) lattice gauge theory in the large n limit. arXiv 1502.07719.

Collins, B., Guionnet, A., and Maurel-Segala, E. (2009). Asymptotics of unitary and orthogonal matrix integrals. *Adv. Math.* **222**, 1, 172–215.

Di Francesco, P., Golinelli, O., and Guitter, E. (2000). Meanders exact asymptotics. *Nucl. Phys. B* **570**(3), 699–712.

Eynard, B., and Kristjansen, C. (1995). Exact solution of the $O(n)$ model on a random lattice. *Nucl. Phys. B* **455**(3), 577–618.

Eynard, B., and Orantin, N. (2009). Topological recursion in enumerative geometry and random matrices. *J. Phys. A* **42**(29), 117.

Feinberg, J., and Zee, A. (1997). Non-Gaussian non-Hermitian random matrix theory: phase transition and addition formalism. *Nucl. Phys. B* **501**(3), 643–69.

Girko, V. L. (1984). The circular law. *Teor. Veroyatnost. i Primenen.* **29**(4), 669–79.

Götze, F., and Tikhomirov, A. (2010). The circular law for random matrices. *Ann. Probab.* **38**(4), 1444–91.

Guionnet, A. (2009). *Large Random Matrices: Lectures on Macroscopic Asymptotics*, vol. 1957 of Lecture Notes in Mathematics. Springer-Verlag, Berlin.

Guionnet, A., Jones, V. F. R., and Shlyakhtenko, D. (2010). Random matrices, free probability, planar algebras and subfactors. In *Quanta of Maths*, vol. 11 of Clay Math. Proc., pp. 201–239. American Mathematical Society, Providence, RI.

Guionnet, A., Jones, V. F. R., Shlyakhtenko, D., and Zinn-Justin, P. (2012). Loop models, random matrices and planar algebras. *Comm. Math. Phys.* **316**(1), 45–97.

Guionnet, A., Krishnapur, M., and Zeitouni, O. (2011). The single ring theorem. *Ann. Math.* **174**(2), 1189–217.

Guionnet, A., and Maurel-Segala, E. (2007). Second order asymptotics for matrix models. *Ann. Probab.* **35**(6), 2160–212.

Guionnet, A., and Novak, J. Asymptotics of unitary multimatrix models: the Schwinger-Dyson lattice and topological recursion. *J. Funct. Anal.* **268**(10), 2851–905.

Guionnet, A., and Zeitouni, O. (2012). Support convergence in the single ring theorem. *Probab. Theory Related Fields* **154**(3-4), 661–75.

Haagerup, U., and Larsen, F. (2000). Brown's spectral distribution measure for *R*-diagonal elements in finite von Neumann algebras. *J. Funct. Anal.* **176**(2), 331–67.

Harer, J., and Zagier, D. (1986). The Euler characteristic of the moduli space of curves. *Invent. Math.* **85**(3), 457–85.

Jones, V. F. R. (2000). The planar algebra of a bipartite graph. In *Knots in Hellas '98 (Delphi)*, vol. 24 of Ser. Knots Everything, pp. 94–117. World Science, River Edge, NJ.

Kostov, I. K., and Staudacher, M. (1992). Multicritical phases of the O(*n*) model on a random lattice. *Nuclear Phys. B* **384**(3), 459–83.

Maurel-Segala, E. (2006). High order expansion of matrix models and enumeration of maps. arXiv:math/0608192.

Rudelson, M., and Vershynin, R. (2014). Invertibility of random matrices: unitary and orthogonal perturbations. *J. Amer. Math. Soc.* **27**(2), 293–338.

Shlyakhtenko, D. (1996). Random gaussian band matrices and freeness with amalgamation. *Int. Math. Res. Not.* **20**, 1013–25.

Tao, T. (2012). *Topics in Random Matrix Theory*, vol. 132 of Graduate Studies in Mathematics. American Mathematical Society, Providence, RI.

Tao, T., and Vu, V. (2015). Random matrices: universality of local spectral statistics of non-Hermitian matrices. *Ann. Probab.* **43**(2), 782–874.

Wigner, E. P. (1958). On the distribution of the roots of certain symmetric matrices. *Ann. of Math.* **67**(2), 325–7.

Zinn-Justin, P., and Zuber, J.B. (2011). Knot theory and matrix integrals. In *The Oxford Handbook of Random matrix Theory*, pp. 557–77. Oxford University, Oxford.

4

The Kardar–Parisi–Zhang equation: a statistical physics perspective

Herbert SPOHN

Zentrum Mathematik and Physik Department,
Technische Universität München,
Boltzmannstraße 3,
85747 Garching, Germany

Spohn, H., 'The Kardar-Parisis-Zhang Equation: A Statistical Physics Perspective; in *Stochastic Processes and Random Matrices*. Edited by: Grégory Schehr et al, Oxford University Press (2017).
© Oxford University Press 2017. DOI 10.1093/oso/9780198797319.003.0004

Chapter Contents

4.1   Stable-metastable interface dynamics

The groundbreaking contribution of Kardar et al. (1986), for short KPZ, is entitled 'Dynamic scaling of growing interfaces'. They study the dynamics of an interface arising from a stable bulk phase in contact with a metastable one. Assuming that the two bulk phases have no conservation laws and relax exponentially fast, KPZ argue that the motion of the interface is governed by the stochastic PDE

$$\partial_t h = \tfrac{1}{2}\lambda(\nabla_x h)^2 + \nu\Delta_x h + \sqrt{D}\xi\,, \tag{4.1}$$

where $h(\boldsymbol{x},t)$ denotes the height function over the substrate space $\boldsymbol{x} \in \mathbb{R}^d$ at time $t \geq 0$. The nonlinearity arises from the asymmetry between the two phases. At the interface a transition from metastable to stable is fast while the reverse process is strongly suppressed. The Laplacian reflects the interface tension and the space-time white noise, $\xi(\boldsymbol{x},t)$, models the randomness in transitions from metastable to stable. λ, ν, D are material parameters, following the original KPZ notation, $\nu > 0$, $D > 0$.

Over the past fifteen years we have witnessed spectacular advances in the case of a two-dimensional bulk, one-dimensional interface, both from the experimental and theoretical side; see the reviews (Johansson 2005, Spohn 2006, Sasamoto and Spohn 2011, Ferrari and Spohn 2011, Quastel 2011, Corwin 2012, Borodin and Gorin 2012, Borodin and Petrov 2014, Takeuchi 2014, Quastel and Spohn 2015). Therefore, we restrict our discussions immediately to $d = 1$, in which case Eq. (4.1) reads

$$\partial_t h = \tfrac{1}{2}\lambda(\partial_x h)^2 + \nu(\partial_x)^2 h + \sqrt{D}\xi\,. \tag{4.2}$$

In a certain sense as a trade-off, also much progress has been achieved on the less understood case of $2 + 1$ dimensions. The long-standing open theoretical problem of an upper critical dimension is in perspective again. I refer to the recent contribution by Halpin-Healy and Takeuchi (2015), which serves as a perfect trail head and provides instructive details along the path.

To have a concrete physical picture and a better understanding of the approximations underlying (4.2), it is illuminating to first consider the two-dimensional ferromagnetic Ising model with Glauber spin flip dynamics, as one of the most basic model systems of statistical mechanics. Its spin configurations are denoted by $\sigma = \{\sigma_j, \boldsymbol{j} \in \mathbb{Z}^2\}$ with $\sigma_j = \pm 1$. The Ising energy is

$$H(\sigma) = -\sum_{i,j\in\mathbb{Z}^2, |i-j|=1} \sigma_i\sigma_j - h\sum_{j\in\mathbb{Z}^2}\sigma_j\,, \qquad H_0(\sigma) = H(\sigma)|_{h=0}\,, \tag{4.3}$$

where the first sum is over nearest neighbor pairs and the spin coupling is used as energy scale. The flip rate from σ_j to $-\sigma_j$ is given by

$$c_j(\sigma) = \begin{cases} 1, & \text{if } \Delta_j H(\sigma) \leq 0\,, \\ e^{-\beta\Delta_j H(\sigma)}, & \text{if } \Delta_j H(\sigma) > 0\,. \end{cases}$$

Here $\beta > 0$ is the inverse temperature and ΔH_j the energy difference in a spin flip at $\boldsymbol{j}$, $\Delta H_j(\sigma) = H(\sigma^j) - H(\sigma)$, where σ^j equals σ with σ_j flipped to $-\sigma_j$.

First note that the bulk dynamics has no conservation law and, away from criticality, an exponentially fast relaxation. If there would be bulk conservation laws, the interface dynamics to be studied would have very different properties.

Considering H_0 and $\beta > \beta_c$, $1/\beta_c$ being the critical temperature, the Glauber dynamics has exactly two (extremal) invariant measures, denoted by $\mu_\pm$. The μ_+ phase is obtained through the infinite volume limit of $Z^{-1}e^{-\beta H_0}$ with $+$ boundary conditions and μ_- equals μ_+ after a global spin flip. μ_+ has a strictly positive spontaneous magnetization. Both phases are stable and have exactly the same free energy. One could start, however, from a nonstationary initial state. A much studied example is a low-temperature quench, for which the initial state is equilibrium at $\beta = 0$, while the dynamics runs at $\beta \gg \beta_c$. In our context we fix $\beta > \beta_c$ and consider a setup with two rather large, possibly macroscopic, disjoint domains $\Lambda_{+(-)}$ with smooth boundaries such that $\Lambda_+ \cup \Lambda_- = \mathbb{Z}^2$. In Λ_+ we choose the state μ_+ and in Λ_- the state μ_-, adopting some physically reasonable choice at the interface $\partial \Lambda_+ \cup \partial \Lambda_-$. A standard example would be the half-spaces $\Lambda_{+(-)} = \{x, \vec{n} \cdot x \geq (<) \, 0\}$ specified by the normal $\vec{n}$. Thereby, a stable-stable interface is imposed. The interface is initially flat and remains sharply localized in the course of time under the Glauber dynamics for $h = 0$, i.e., with flip rates derived from H_0, but develops fluctuations with an amplitude of size $t^{1/4}$. Away from the interface the bulk has a statistics which in good approximation is described by either μ_+ or μ_-.

KPZ raised the issue of how such interface motion is modified when the Glauber dynamics is run at a small $h > 0$. Then μ_+ remains stable, but μ_- has turned metastable. At the interface the Glauber dynamics easily flips a$-$ spin to a$+$ spin, while the reversed process is suppressed. Thereby, the $+$ phase expands into the $-$ phase and the stable–metastable interface acquires a nonzero drift velocity. In addition, inside the metastable $-$ domain a stable $+$ nucleus could be formed, either far out statically or through a dynamical fluctuation. Such an event is unlikely, but once it happens the stable nucleus will grow and possibly collide with the already present interface. Our description will be restricted to times before such a collision. In fact, in most models such extra nucleation events are suppressed entirely. In the course of time the stable–metastable interface remains well localized, but roughens on top of the systematic motion. Our goal is to understand the space-time statistical properties of this roughening process. Note that the effective interface dynamics is no longer invariant under time reversal, in contrast to the underlying Glauber dynamics. As an additional issue of great interest, we have naturally arrived at a stochastic field theory with a nonsymmetric generator.

A theoretical study of the Glauber dynamics with such initial conditions seems to be exceedingly difficult. Fortunately, in the limit of zero temperature one arrives at tractable models. As a general consensus, one expects that the large-scale properties of the interface will not change when heating up, of course always staying below the critical temperature.

More specifically let us start from an interface given through a down-right lattice path. Below that path all spins are up and above they are down. At zero temperature only flips with $\Delta_j H \leq 0$ are admissible. Under this constraint a $-$ spin flips to $+$ with rate p while the reversed transition occurs with rate q, $p + q = 1$ to fix the time

scale. $h = 0$ corresponds to $p = q = \frac{1}{2}$. The dynamics is stochastically reversible. On the other hand, for small $h > 0$ the stable-metastable flip rates differ, $p > q$, and the stochastic dynamics is nonreversible. The interface width is one lattice unit. The zero temperature Glauber dynamics never leaves the set of down-right paths. Thus, in the spirit of the KPZ equation, we have accomplished an autonomous interface dynamics.

The conventional scheme to define a height function, $h(j, t)$ with $j \in \mathbb{Z}$, is to choose the anti-diagonal as reference line. The height function satisfies the constraint $|h(j+1, t) - h(j, t)| = 1$. We draw $h(j, t)$ as a continuous broken line with slope ± 1, $\diagup, \diagdown$, such that $h(j, t)$ is at the lattice points $\mathbb{Z} + \frac{1}{2}$. Under the height dynamics, independently a local minimum of h, $\diagdown\diagup$, flips to $\diagup\diagdown$ with rate p and thus $h(j) \Rightarrow h(j) + 2$. Correspondingly, a local maximum, $\diagup\diagdown$, flips to $\diagdown\diagup$ with rate q and thus $h(j) \Rightarrow h(j) - 2$. Since the slope takes only values ± 1, this height dynamics goes under the label 'single-step'. $p = q$ is the symmetric dynamics, corresponding to a stable-stable interface, and $p \neq q$ is the asymmetric case, including the totally asymmetric limits $p = 1$, $q = 1$.

For an interface parallel to one of the lattice axes, according to our rules the interface cannot move, no flip is allowed. To arrive at a nontrivial dynamics the limit $\beta \to \infty$ must be taken differently. As initial configuration let us assume that all spins in the upper half plane are down and are up in the lower half plane. Then at low temperatures the slow processes are flips from $-$ to $+$ with $\Delta_j H = 2$. Once this has happened the allowed flips with $\Delta_j H = 0$ are fast. Under a suitable scaling one arrives at the polynuclear growth (PNG) model. The height function is $h(x, t)$ with space $x \in \mathbb{R}$ and time $t \geq 0$. $x \mapsto h(x, t)$ is piecewise constant and takes values in $\mathbb{Z}$ such that up-steps are of size 1 and down-steps of size -1. In approximation the lateral motion is deterministic; up-steps move with velocity -1 and down-steps with velocity 1. Steps annihilate at collisions. In addition pairs of adjacent up-/down-steps are created according to a space-time Poisson process with uniform intensity, which for convenience will be set equal to 2.

For all these models, on a macroscopic scale the height is governed by a Hamilton–Jacobi equation of the form

$$\partial_t h = \Phi(\partial_x h), \tag{4.4}$$

which expresses that the local change in height depends only on the local slope, $u = \partial_x h$. For the single-step model one finds that $\Phi(u) = \frac{1}{2}(p-q)(1-u^2)$ and for the PNG model $\Phi(u) = \sqrt{4 + u^2}$. Some aspects of the interface dynamics for the Ising model at low temperatures are discussed in Spohn (1993).

The statistical mechanics problem is to characterize the space-time fluctuations relative to the shape governed by (4.4). In general, this turns out to be a challenging task and much of our understanding of the KPZ universality class relies on simplified models as single-step and PNG. Johansson (2000) studied the single-step model with $p = 1$ and wedge initial conditions, $h(j, 0) = |j|$. He succeeded in determining the exact probability density function of $h(0, t)$ for large t, which constituted the starting point in the search for further integrable stochastic interface models and their universal properties.

4.2   Scaling properties, KPZ equation as weak drive limit

At first sight the KPZ equation seems to unrelated to the single-step model. To eluci-
date the connection we first study the scaling properties of the KPZ equation and will
use them to guess the limit in which the single-step model is well approximated by the
KPZ equation. But before we note that, by rescaling x, t, h, any value of the material
coefficients λ, ν, D can be achieved. Also flipping h to $-h$ is equivalent to flipping λ to
$-\lambda$. Thus, without loss of generality we set $\nu = \frac{1}{2}$, $D = 1$, which makes the formulas
less clumsy. Hence, the KPZ equation reads

$$\partial_t h = \tfrac{1}{2}\lambda(\partial_x h)^2 + \tfrac{1}{2}\partial_x^2 h + \xi \,, \tag{4.5}$$

keeping the dependence on the nonlinearity strength parameter λ.

Physically, we are interested in the large space, long-time behavior of the KPZ
equation, with the view that in this limit the microscopic details will be irrelevant.
First, we note one important building block. The linear equation, $\lambda = 0$, is a Gaussian
process and its time-stationary measure is easily computed to be given by

$$Z^{-1}\exp\left[-\tfrac{1}{2}\int \mathrm{d}x\big((\partial_x h(x))^2 - 2\mu\partial_x h(x)\big)\right], \tag{4.6}$$

where μ is the average slope. As a general experience, the nonlinear part of the drift
will modify the time-stationary measure. However, the KPZ equation (4.2) is very
special to have the time-stationary measure independent of λ. One only must observe
that under the evolution governed by

$$\partial_t h = \tfrac{1}{2}\lambda(\partial_x h)^2 \tag{4.7}$$

the time change of the action is

$$\frac{\mathrm{d}}{\mathrm{d}t}\int \mathrm{d}x(\partial_x h(x))^2 = 2\int \mathrm{d}x\partial_x h(x)\partial_x\partial_t h(x) = \lambda\int \mathrm{d}x\partial_x h(x)\partial_x(\partial_x h(x))^2 = 0\,. \tag{4.8}$$

In principle one should worry also about the Jacobian. But formally the vector field
in (4.7) is divergence free and the Jacobian equals 1. Our argument fails in higher
dimensions. The steady state of the KPZ equation (4.1) is not known.

We now transform to large scales by

$$x \rightsquigarrow \epsilon^{-1}x, \qquad t \rightsquigarrow \epsilon^{-z}t\,, \tag{4.9}$$

where ϵ is the dimensionless scale parameter, $\epsilon \ll 1$, and x, t on the right side are
independent of ϵ. z is the dynamical scaling exponent, which still must be determined.
The height field is transformed to

$$h_\epsilon(x, t) = \epsilon^b h(\epsilon^{-1}x, \epsilon^{-z}t) \tag{4.10}$$

with b the fluctuation exponent. Recall that white noise satisfies $(a_1 a_2)^{1/2}\xi(a_1 x, a_2 t) = \xi(x, t)$. Thus, inserting (4.10) in (4.2) one arrives at

$$\partial_t h_\epsilon = \epsilon^{2-z-b}\tfrac{1}{2}\lambda(\partial_x h_\epsilon)^2 + \epsilon^{2-z}\tfrac{1}{2}\partial_x^2 h_\epsilon + \epsilon^{b+(1-z)/2}\xi\,. \tag{4.11}$$

Since the time-stationary measure is given by Eq. (4.6), the fluctuation exponent equals

$$b = \tfrac{1}{2} .$$ (4.12)

To have in (4.11) the nonlinearity maintained implies then the dynamic exponent

$$z = \tfrac{3}{2} .$$ (4.13)

The KPZ equation has two well-separated and distinct noise scales. Locally the dynamics tries to maintain stationarity; i.e., $x \mapsto h(x,t)$ has the statistical properties of a Brownian motion at some constant drift, in other words some locally averaged slope, which is constant on small scales but still changing on coarser space-time scales. For large scales the nonlinearity dominates, but the evolution is still noisy. Its properties will have to be computed. Scaling by itself is certainly not enough. But the gross features can be guessed already from Eq. (4.11), setting $b = \tfrac{1}{2}$ and $z = \tfrac{3}{2}$. Then the ratios *height* : *space* : *time* are given by $\epsilon^{-1/2} : \epsilon^{-1} : \epsilon^{-3/2}$. Choosing $\epsilon^{-3/2}$ as time unit, then, for large t, the typical height fluctuations are of order $t^{1/3}$ and correlations in x are of order $t^{2/3}$. Put differently, if one chooses a reference point x_0 and $|x - x_0| \ll t^{2/3}$, then on that spatial interval the KPZ solution $x \mapsto h(x,t)$ is like a Brownian motion with constant drift. To understand the statistical properties for $|x - x_0| \simeq t^{2/3}$ requires further input.

With this background we can tackle the approximation through the single-step model. First, note that if in (4.5) the nonlinearity λ is assumed to be equal to $\epsilon^{1/2}$ and space is scaled to $\epsilon^{-1}x$, time to $\epsilon^{-2}t$, then the scaled height function, $h_\epsilon(x,t) = \epsilon^{1/2}h(\epsilon^{-1}x, \epsilon^{-2}t)$, satisfies

$$\partial_t h_\epsilon = \tfrac{1}{2}(\partial_x h_\epsilon)^2 + \tfrac{1}{2}\partial_x^2 h_\epsilon + \xi .$$ (4.14)

Thus the small nonlinearity is precisely balanced by large space-time.

Such a limit is meaningful also for the single-step model. To distinguish, the single-step height is denoted by $h^{\text{step}}(j,t)$. We adopt a lattice spacing ϵ, i.e., $h^{\text{step}}_\epsilon(x) = h^{\text{step}}(\lfloor \epsilon^{-1}x \rfloor)$ with $\lfloor \cdot \rfloor$ denoting the integer part and x being independent of ϵ. From our experience with much simpler equations we expect that in the limit of zero lattice spacing, with an appropriate simultaneous rescaling of h^{step} and t, one obtains some continuum equation. The scaling properties suggest choosing a weak asymmetry as $p = \tfrac{1}{2}(1 + \kappa\sqrt{\epsilon}), q = \tfrac{1}{2}(1 - \kappa\sqrt{\epsilon}), \kappa > 0$. With this choice the rescaled height is

$$h^{\text{step}}_\epsilon(x,t) = \epsilon^{1/2}h^{\text{step}}(\lfloor \epsilon^{-1}x \rfloor, \epsilon^{-2}t).$$ (4.15)

Indeed, there is a theorem by Bertini and Giacomin (1997), which states that

$$\lim_{\epsilon \to 0} \left(h^{\text{step}}_\epsilon(x,t) - \epsilon^{-1}\kappa t \right) = h(x, \kappa t) ,$$ (4.16)

where the right-hand side is the solution to the KPZ equation (4.5) with $\lambda = 1$. In essence, what is required is only that the initial height profile grows less than linearly at infinity.

The proof of the limit (4.16) is not at all obvious. In the case of the single-step model one relies on a transformation discovered by Gärtner (1988), which shifts the nonlinearity into the noise term. In fact, currently there are only a few models for which such a limit can be established. A major advance is the solution theory of Hairer (2013). In a related undertaking (Gubinelli and Perkowski 2017) the solution theory serves as a tool for proving that a discretized version of the KPZ equation, as proposed in Sasamoto and Spohn (2009), converges to the continuum equation (4.5).

In our previous discussion we did not anticipate that, according to (4.16), one must switch to a moving frame of reference, whose velocity diverges as ϵ^{-1} for $\epsilon \to 0$. When approximating a continuum theory by a lattice based model, counter terms must be subtracted. In our case there is just a single term, independent of x, which is in spirit very similar to an energy renormalization in quantum field theory.

To summarize, the KPZ equation becomes exact in the limit of weak asymmetry. In the specific case of the Glauber model weak asymmetry corresponds to a small magnetic field h and a simultaneous rescaling of space-time and height. In this respect the KPZ equation is similar to other effective equations based on the availability of a small parameter. More exceptional is the feature to have a nonlinear limit dynamics which is still noisy.

4.3   Eden-type growth models

Before proceeding to the analysis of the KPZ equation we discuss another class of growth processes, known as Eden models (Eden 1961, Barabasi and Stanley 1995, Krug 1997, Meakin 1998). This time the reason is not beautiful mathematics. Rather Eden models are currently the best laboratory realized systems. In the Eden model the ambient metastable phase is ignored entirely, in accord with single step and PNG. Usually one starts from a seed and provides a rule for potential growth sites. In a single update one of the growth sites is filled according to a uniform probability. To have an example with $\mathbb{Z}^2$ as underlying lattice, the origin is taken to be the seed. Given the connected cluster at time t, any site with distance 1 is a growth site. The cluster at time $t + 1$ is obtained by filling one of the growth sites at random. After a long time a deterministic shape emerges, which looks circular, but nevertheless is anisotropic because of the underlying lattice. For us the shape fluctuations are the main interest. They are described approximately by the KPZ equation, but only in a small segment, since the KPZ height is the graph of a function.

The anisotropy of the Eden model implies that the coefficients of the approximating KPZ equation depend on the particular radial direction. This makes numerical simulations more difficult, since angular averaging will distort the universal result and is hence not advisable. An isotropic Eden model would be preferred. One possibility is to have growth on $\mathbb{R}^2$, where the basic building blocks are disks of fixed diameter. The seed is a disk located at the origin. Growth sites are disk centers such that the corresponding disk touches the current cluster, avoiding, however, any overlap with disks already present. The subsequent disk is attached according to the normalized Lebesgue measure on the union of arcs formed by the growth sites. By construction the

limit shape is now a circle. In a more physical variant, to every disk currently present a further disk is attached, independently of all other disks, uniformly over all touching points, at constant rate, and subject to the constraint of no overlap (Takeuchi 2012).

A further variant of Eden type models is ballistic deposition. Along random rays orthogonal to the substrate, mass is transported and attached to the current cluster according to some prescribed rule. Early experiments on KPZ growth tried to realize such ballistic deposition. Unfortunately, it is difficult to control what precisely happens when a particle touches the growing surface. The more elegant realization is to have only a change of type at the interface. For example, in the smouldering paper experiment (Miettinen et al. 2005), the paper switches from unburned to burned at the flame front. No mass is transported. For $2 + 1$ dimensions, ballistic deposition has been revived by noting that larger molecules are more favorable building blocks (Almeida et al. 2014, Halpin-Healy and Palasantzas 2014).

Takeuchi and Sano (2012) had the ingenious idea to use turbulent liquid crystal for realizing a stable–metastable interface. In the actual experiment the liquid crystal film is 16×16 mm at a height of 12 μm. The rod-like molecules are on average aligned to be orthogonal to the confining plates. Hence the system is in-plane isotropic. There is an external electric field, uniform in space and oscillating in time, which makes the bulk phases turbulent, thus ensuring rapid relaxation. In fact these are nonequilibrium steady states. One carefully selects a point in the phase diagram, at which a stable (DSM2) and a metastable (DSM1) phase coexist. The two phases are easily distinguished through transmission of light. DSM2 is black, transmitting no light, while DSM1 appears in a gray color. The film is prepared in the metastable DSM1 phase and a seed of DSM2 is planted by a very sharp laser pointer. Alternatively, the laser may print a line seed, through which the dependence on initial conditions can be studied. The cluster grows to its maximal size in approximately 40–60 s. On the order of 5×10^3 repeats are carried out. Adding the angular directions, one achieves a large sample set. In Fig. 4.1 we show the histogram, on a logarithmic scale, for a point seed and a line seed. The respective theoretical predictions are discussed in Sect. 4.4. In fact, the transition from metastable to stable seems to be a complicated physical process and currently there is little understanding of the precise mechanism. But empirically the isotropic Eden model captures the main features of the growing interface. For further details, we refer to Takeuchi and Sano (2012).

4.4   The KPZ universality class

Starting in 2000 for a variety of growth models exact universal scaling properties have been obtained, including the KPZ equation itself. Some of the most important results will be listed. To disentangle, however, which result has been proved for which model is beyond the present scope and must be looked up in more specialized articles (Ferrari and Spohn 2011, Borodin and Gorin 2012, Barraquand 2015). In particular I recommend the review article by Corwin (2012), which covers the field up to 2011. For better readability, I list the results as if obtained for the KPZ equation. In some cases this is actually correct (Amir et al. 2011, Sasamoto and Spohn 2010, Borodin et al. 2015).

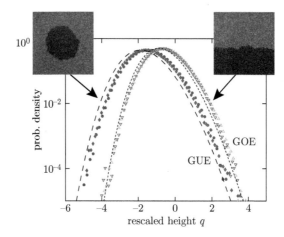

Fig. 4.1 Histogram of the rescaled local height q for the circular (solid symbols) and flat (open symbols) interfaces. The solid square and circle symbols display the histograms for the circular interfaces at $t = 10$ and 30 s, while the open up and down triangle symbols are for the flat interfaces at $t = 20$ and 60 s. The dashed and dotted curves show the densities of ξ_{cur} and ξ_{fla}, respectively, see (4.21) and (4.19). Note that the first moment is still shifting, while the higher cumulants have settled already to their asymptotic values.

In other cases the corresponding solution of the KPZ equation is widely open, but the asymptotics has been obtained using another model in the KPZ universality class. We are still at the stage at which very specific models are analyzed in considerable detail. To recall, the KPZ equation is written as

$$\partial_t h = \tfrac{1}{2}\lambda(\partial_x h)^2 + \tfrac{1}{2}\partial_x^2 h + \xi. \tag{4.17}$$

(i) *Initial conditions.* The long time asymptotics of the solution depends on the initial conditions. Three standard classes have been identified. They are (IC1) *flat*, $h(x,0) = 0$; (IC2) *curved*, e.g., $h(x,0) = -x^2/2$; (IC3) *stationary*, $x \mapsto h(x,0) = B(x)$, where $B(x)$ is a two-sided Brownian motion pinned as $\dot{B}(0) = 0$. Note that also $h(x,0) = ux + B(x)$ is stationary, but $\langle B(x)^2 \rangle = |x|$ is required by our choice of parameters. One can also consider domain walls formed by such initial data. For example, $h(x,0) = 0$ for $x \le 0$ and $x \mapsto h(x,0)$ a Brownian motion for $x \ge 0$. If the focus is far to the left, then one is in class (IC1) and far to the right in class (IC3). But near $x = 0$ novel cross-over statistical properties will be realized (Corwin 2012).

(ii) *Observables.* In statistical physics one learns that correlations, possibly higher order cumulants, are the central goal. KPZ is actually an area where full probability density functions are of considerable advantage. They seem to characterize more sharply the KPZ universality class than scaling exponents. In numerical simulations, and in experiments, the line shape often settles earlier than a definite power law.

4.4.1 One-point distributions

The most basic observable is the long time statistics of the height at one spatial reference point, which for the initial conditions just described can be taken as $x = 0$. As a generic result,

$$h(0,t) \simeq c_\diamond t + \sigma_\diamond (\Gamma_\diamond t)^{1/3} \xi_\diamond \,, \qquad \sigma_\diamond = \pm 1 \,, \tag{4.18}$$

valid for long times. The subscript $_\diamond$ stands for either 'fla', or 'cur', or 'sta', depending on the initial conditions. If one samples $h(0,t)$ at some large time t, then the distribution is shifted by $c_\diamond t$ and the fluctuations are of order $t^{1/3}$ with a random amplitude characterized by the random variable $\xi_\diamond$. The nonuniversal factor of the fluctuating term is best collected through the rate $\Gamma_\diamond > 0$. $\xi_\diamond$ is defined with a particular sign convention. But the random amplitude could be either $\xi_\diamond$ or $-\xi_\diamond$, depending on the particular situation. In Sect. 4.2 the exponent $\frac{1}{3}$ has been anticipated already on the basis of a simple scaling argument. But now we assert in addition the full probability density function. $c_\diamond$ and $\Gamma_\diamond$ are coefficients depending on the model. All other features are universal. For the KPZ equation one obtains $c_\diamond = -\frac{1}{24}\lambda^3 t$, $\sigma_\diamond = \mathrm{sgn}(\lambda)$, $\Gamma_{\mathrm{fla}} = \frac{1}{8}|\lambda|$, $\Gamma_{\mathrm{cur}} = \frac{1}{2}|\lambda|$, $\Gamma_{\mathrm{sta}} = \frac{1}{2}|\lambda|$. The nonuniversal coefficients are known also for a few other models. $c_\diamond$, $\sigma_\diamond$, and $\Gamma_\diamond$, may take different values in distinct equations.

For (IC1) the random amplitude ξ_{fla} is distributed as GOE Tracy–Widom, for (IC2) the amplitude ξ_{cur} is distributed as GUE Tracy–Widom, and for (IC3) the amplitude ξ_{sta} is distributed as Baik–Rains. These are non-Gaussian random variables and their distribution functions are written in terms of Fredholm determinants. More explicitly, for GOE Tracy–Widom (Tracy and Widom 1994)

$$\mathbb{P}(\xi_{\mathrm{fla}} \leq s) = \det(1 - K_{1,s})_{L^2(\mathbb{R}_+)} = F_1(s) \,, \tag{4.19}$$

where

$$K_{1,s}(x,y) = \mathrm{Ai}(x + y + s), \tag{4.20}$$

with Ai the standard Airy function; see Ferrari and Spohn (2005) for this particular representation. For GUE Tracy–Widom

$$\mathbb{P}(\xi_{\mathrm{cur}} \leq s) = \det(1 - K_{2,s})_{L^2(\mathbb{R}_+)} = F_2(s) \,, \tag{4.21}$$

where

$$K_{2,s}(x,y) = \int_0^\infty \mathrm{d}u \mathrm{Ai}(x + u + s)\mathrm{Ai}(y + u + s) \,. \tag{4.22}$$

The Baik–Rains distribution (Baik and Rains 2000) has a more complicated expression,

$$\mathbb{P}(\xi_{\mathrm{sta}} \leq s) = F_0(s) = \frac{\mathrm{d}}{\mathrm{d}s}\big(g(s)F_2(s)\big), \tag{4.23}$$

with

$$g(s) = s + \langle 1, (1 - K_{2,s})^{-1}(K_{1,s} - K_{2,s})1 \rangle_{L^2(\mathbb{R}_+)} . \tag{4.24}$$

All determinants are on the Hilbert space $L^2(\mathbb{R}_+)$ with inner product $\langle \cdot, \cdot \rangle_{L^2(\mathbb{R}_+)}$ and 1 is the constant function.

F_1 and F_2 have appeared before in the context of random matrix theory, where they characterize the fluctuations of the largest eigenvalue of GOE and GUE random matrices. The Baik–Rains distribution does not seem to have an obvious connection to random matrix theory.

Early numerical plots were based on the connection to the Hastings–McLeod solution of the Painlevé II differential equation. Since this solution is unstable, one must employ an ultra-precise shooting algorithm. Bornemann (2010) pointed out that a direct numerical evaluation of the suitably approximated Fredholm determinant is a more accessible approach and works equally well in cases when no connection to a differential equation is available.

Instead of the reference point $x = 0$ one can also consider x along the ray $\{x = vt\}$. Then, for long times,

$$h(vt, t) \simeq c_\diamond(v)t + \sigma_\diamond(\Gamma_\diamond(v)t)^{1/3}\xi_\diamond . \tag{4.25}$$

For flat initial conditions, $h(vt, t)$ is independent of v by translation invariance. For curved initial conditions the velocity v is arbitrary, within limits set by the model, but $c_{\mathrm{cur}}, \Gamma_{\mathrm{cur}}$ depend on v, in general. However, in the stationary case, there is only one specific velocity, v_0, for which anomalous fluctuations of order $t^{1/3}$ are observed. For all other rays the fluctuations are Gaussian of size $\sqrt{t}$. Physically, v_0 is the propagation velocity of a small localized perturbation in the slope. For the KPZ equation $v_0 = 0$ and the asymptotics (4.18) holds also for $\diamond = \mathrm{sta}$.

A particular case of curved initial conditions is the KPZ equation with sharp wedge initial data, i.e., $h(x, 0) = \lim_{\delta \to 0} -\delta^{-1}|x| - \log(2\delta)$. Then $h(x, t) = -(x^2/2\lambda t) - \frac{1}{24}\lambda^3 t + \eta(x, t)$ and $x \mapsto \eta(x, t)$ is stationary for fixed t. The exact probability density function of $\eta(0, t)$, denoted by $F_t'(s)$, can be written in terms of the difference of two Fredholm determinants (Sasamoto and Spohn 2010, Amir et al. 2011). In Fig. 4.2 we show a time sequence of such densities (Prolhac and Spohn 2011c), obtained using the Bornemann method. At early times the density is Gaussian with variance of order $t^{1/4}$, which then crosses over to the GUE Tracy–Widom density on scale $t^{1/3}$.

4.4.2 Multipoint distributions

Instead of the single reference point (vt, t), the joint distribution of the height at several space-time points could be considered. At such generality not much is known. For curved initial data, two times at the same space point, e.g., the joint distribution of $h(0, t), h(0, 2t)$, are recently studied by Johansson (2015). But otherwise all results refer to a single time, but with an arbitrary number of spatial points. As anticipated in Sect. 4.2, to obtain a nondegenerate universal limit the space points must be separated on the scale $t^{2/3}$. The asymptotics (4.18) generalizes to

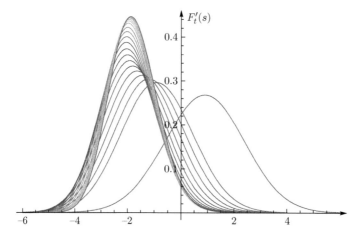

Fig. 4.2 Probability density function for time t from short times (lower curves) to long times (upper curves) for t ranging from 0.25 to 20, 000; see Prolhac and Spohn (2011c) for further details. For $t \to 0$, $F'_t(s)$ becomes a Gaussian (rightmost curve) and for $t \to \infty$, the density converges to the GUE Tracy–Widom distribution (upper black curve). Note that, in contrast to the experiment, the long time limit is approached from the left.

$$h(wt^{2/3}, t) \simeq c_\diamond t + \sigma_\diamond (\Gamma_\diamond t)^{1/3} \mathcal{A}_\diamond(w) \,, \qquad (4.26)$$

as a stochastic process in w, which means that the finite-dimensional distributions from the left, i.e., joint distributions for a finite number of reference points, converge to the one on the right. The limit process $\mathcal{A}_\diamond(w)$ is known as the Airy process. For each of the three classes of initial conditions there is a distinct Airy process.

4.4.3 Stationary covariance

With two-sided Brownian motion as initial conditions the height field is not stationary in the usual sense of the word. Rather, if $h(x, 0) = B(x)$, then $x \mapsto h(x, t) - h(0, t)$ is again two-sided Brownian motion. Note that the random shift $h(0, t) - h(0, 0)$ is correlated with $h(x, t) - h(0, t)$. Stationarity in the conventional sense is achieved by considering instead the slope $u(x, t) = \partial_x h(x, t)$, which is governed by the stochastic Burgers equation

$$\partial_t u - \partial_x \left(\tfrac{1}{2} \lambda u^2 + \tfrac{1}{2} \partial_x u + \xi \right) = 0 \,. \qquad (4.27)$$

In this case, the time-stationary measure is unit strength white noise in x and, for such initial conditions, $u(x, t)$ is a random field stationary in both space and time.

For the linear case, $\lambda = 0$, the covariance is easily computed to be given by

$$\langle u(x, t) u(x', t') \rangle = (2\pi |t - t'|)^{-1/2} \exp \left[-(x - x')^2 / 2|t - t'| \right] \,. \qquad (4.28)$$

A small perturbation in the slope spreads diffusively. For $\lambda \neq 0$ one must rely on the asymptotics in (4.26) with $\diamond = \text{sta}$. One first notes that

$$\partial_x^2 \langle (h(x,t) - h(0,0) - c_{\text{sta}}t)^2 \rangle = 2\langle u(x,t)u(0,0) \rangle \tag{4.29}$$

and, using (4.26), infers that

$$\langle (h(wt^{2/3}, t) - c_{\text{sta}}t)^2 \rangle = \langle ((\Gamma_{\text{sta}}t)^{1/3} \mathcal{A}_{\text{sta}}(w))^2 \rangle, \tag{4.30}$$

valid for large t. There is an explicit formula for the probability distribution of $\mathcal{A}_{\text{sta}}(w)$. Hence, one must compute its second moment and twice differentiate w.r.t. w to obtain the universal stationary scaling function. The result is a self-similar two-point function

$$\langle u(x,t)u(0,0) \rangle \simeq (\Gamma_{\text{sta}}|t|)^{-2/3} f_{\text{KPZ}}((\Gamma_{\text{sta}}|t|)^{-2/3}x), \quad \Gamma_{\text{sta}} = \sqrt{2}|\lambda|, \tag{4.31}$$

valid for large $|x|, |t|$. The function f_{KPZ} is tabulated in Prähofer (2005), denoted there by f. Its properties are $f_{\text{KPZ}} \geq 0$, $\int_{\mathbb{R}} dx f_{\text{KPZ}}(x) = 1$, $f_{\text{KPZ}}(x) = f_{\text{KPZ}}(-x)$, $\int_{\mathbb{R}} dx f_{\text{KPZ}}(x)x^2 \simeq 0.510523$. f_{KPZ} looks like a Gaussian with a large $|x|$ decay as $\exp[-0.295|x|^3]$. Plots are provided in Prähofer and Spohn (2004) and Prähofer (2005). As required by the conservation law, $\int_{\mathbb{R}} dx \langle u(x,t)u(0,0) \rangle = 1$.

4.5 Directed polymers in a random medium

One can rewrite the KPZ equation as a problem in equilibrium statistical mechanics of disordered systems. This step is a one-to-one map, no information is lost, and it offers a different intuition on the KPZ equation. Besides, new tools become available. As noted already by Hopf (1950) and Cole (1951) for the dissipative Burgers equation, (4.27) with zero noise, the transformation

$$Z(x,t) = e^{\lambda h(x,t)} \tag{4.32}$$

'linearizes' the KPZ equation as

$$\partial_t Z(x,t) = \tfrac{1}{2}\partial_x^2 Z(x,t) + \lambda \xi(x,t)Z(x,t). \tag{4.33}$$

Equation (4.33) is the heat equation with a space-time random potential, hence also called the stochastic heat equation. Following Feynman and Kac, its solution can be written as the expectation over an auxiliary Brownian motion, $b(t)$,

$$Z(x,t) = \mathbb{E}_{(x,t)}\left(\exp\left(\lambda \int_0^t ds\, \xi(b(s), s)\right) Z_0(b(t))\right). \tag{4.34}$$

$b(t)$ is the directed polymer, which starts at (x,t) and moves backward in time to end at $(b(t), 0)$. The directed polymer has an intrinsic elastic energy, implicit in the expectation $\mathbb{E}_{(x,t)}$, and a potential energy obtained by integrating the random potential $\xi(x,s)$ along its path. $Z_0(x) = e^{\lambda h(x,0)}$ is the initial condition. The partition function

$Z(x,t)$ is the sum over all paths weighted with the Boltzmann factor. Since the potential energy is random, the partition function is random and we arrived at a problem from the theory of disordered systems, which studies systems in thermal equilibrium for which the coupling constants appearing in the energy are random, but regarded as fixed for thermal averages. At the end, our interest is the random free energy

$$h(x,t) = \lambda^{-1} \log Z(x,t) \,. \tag{4.35}$$

It has a leading term linear in t, which is self-averaging, i.e.,

$$\lim_{t\to\infty} t^{-1}\lambda^{-1} \log Z(x,t) = v_0 \tag{4.36}$$

almost surely. For growing interfaces the key point are the fluctuations of the free energy, a not so well-studied quantity for disordered systems.

The directed polymer in (4.34) is called a continuum directed polymer. Since we are interested in large-scale properties, its local properties can be modelled fairly freely. A popular, and natural, choice is to replace the space-time continuum by the discrete lattice $\mathbb{Z}^2$ and the directed polymer by an up-right path w, starting at $(0,0)$ and ending at (N,N), say. The white noise is replaced by independent identically distributed random variables $\xi_{i,j}$, $(i,j) \in \mathbb{Z}^2$. The energy of a $2N$-step directed polymer is now

$$E(w) = \sum_{\ell=1}^{2N+1} \xi_{w(\ell)} \tag{4.37}$$

and the discretized version of the partition function (4.34) reads

$$Z_\beta(N,N) = \sum_{w:(0,0)\to(N,N)} e^{-\beta E(w)} \,, \tag{4.38}$$

where we introduced the more conventional inverse temperature β as parameter. In analogy, the height function is defined by

$$h_\beta(N,N) = -\beta^{-1} \log Z_\beta(N,N) \,. \tag{4.39}$$

This problem is called a point-to-point directed polymer and corresponds to a sharp wedge in the language of height functions. On the other hand, for flat initial conditions, $h(0,x) = 0$, implying that $Z_0(x) = 1$, which corresponds to summing over all directed polymers with only one end point fixed. This is called point-to-line directed polymer. From this perspective it is less surprising that flat and curved initial conditions have distinct fluctuation behavior.

In a discretized version, as in (4.38), one can take the limit $\beta \to \infty$. Then the log is traded against the exponential and the finite temperature problem turns into a ground state problem. Hence, the height function at a given point is

$$h_\infty(N,N) = \min_{w:(0,0)\to(N,N)} E(w) \,. \tag{4.40}$$

No surprise, one expects, and proves for a few very specific distributions of $\xi_{i,j}$ that

$$h_\infty(N,N) \simeq c_{\mathrm{DP}}N + (\Gamma_{\mathrm{DP}}N)^{1/3}\xi_{\mathrm{GUE}} \qquad (4.41)$$

for large N (Johansson 2000, Borodin et al. 2013, Georgiou et al. 2015). The PNG model is also included in the list as a shot noise limit. The random medium is now a homogeneous, two-dimensional Poisson point process. An admissible path is continuous, increasing in both coordinates, and consists of linear segments bordered by Poisson points. Each Poisson point carries a negative unit of energy and, as before, one studies the ground state energy. An optimal path is defined by transversing a maximal number of Poisson points.

In the representation through a directed polymer, one can ask how the continuum directed polymer is approximated through a discrete version. This is just like approximating the KPZ equation by a discrete growth model. In fact, the continuum directed polymer is obtained through a weak noise limit. We refer to Alberts et al. (2014), where a proof under fairly general assumptions is carried through.

4.6 Replica solutions

As noted already early on (Kardar 1987), there are closed evolution equations for the moments of the partition function $Z(x,t)$ as defined in (4.34). Let us consider the nth moment, now written with n auxiliary independent Brownian motions, called the replicas. Denoting the white noise average by $\langle \cdot \rangle$, one arrives at

$$\left\langle \prod_{j=1}^{n} Z(x_j,t) \right\rangle = \left\langle \prod_{j=1}^{n} \mathbb{E}_{(x_j,t)}\left(e^{\lambda \int_0^t ds\xi(b_j(s),s)} Z_0(b_j(t)) \right) \right\rangle . \qquad (4.42)$$

The white noise average can be carried out explicitly and is given by the exponential of

$$\tfrac{1}{2}\lambda^2 \sum_{i,j=1}^{n} \int_0^t \int_0^t ds_1 ds_2\, \delta(s_1 - s_2)\, \delta(b_i(s_1) - b_j(s_2)) . \qquad (4.43)$$

The summand with $i = j$ is defined only when smearing the δ-function. However, if for the stochastic integral in Eq. (4.34) one adopts the Itô discretization; then the diagonal term $i = j$ must be omitted. The double time integration reduces trivially to a single one. Hence, using the Feynman–Kac formula backwards, one obtains

$$\left\langle \prod_{j=1}^{n} Z(x_j,t) \right\rangle = \langle x_1,\ldots,x_n | e^{-H_n t} | (Z_0)^{\otimes n} \rangle . \qquad (4.44)$$

Here H_n is the n-particle Lieb–Liniger quantum Hamiltonian on the real line with attractive δ-interaction,

$$H_n = -\tfrac{1}{2}\sum_{j=1}^{n} \partial_{x_j}^2 - \tfrac{1}{2}\lambda^2 \sum_{i\neq j=1}^{n} \delta(x_i - x_j) . \qquad (4.45)$$

The quantum propagator acts on the initial product wave function $\prod_{j=1}^{n} Z_0(x_j)$, denoted by $(Z_0)^{\otimes n}$, and is evaluated at the point $(x_1, \ldots, x_n)$. Since $(Z_0)^{\otimes n}$ is symmetric, only the restriction of $\exp[-H_n t]$ to the subspace of permutation symmetric wave functions, the bosonic subspace, in $L^2(\mathbb{R}^n)$ is required. As a result the right-hand side of (4.44) is a symmetric function, as it should be.

For curved initial data, in the sharp wedge approximation $Z_0(x) = \delta(x)$, one obtains

$$\langle Z_{\mathrm{cur}}(0, t)^n \rangle = \langle 0 | e^{-H_n t} | 0 \rangle \tag{4.46}$$

with shorthand $|0\rangle = |0, \ldots, 0\rangle$. For flat initial conditions, $h(x, 0) = 0$, the nth moment of the partition function reads

$$\langle Z_{\mathrm{fla}}(0, t)^n \rangle = \int_{\mathbb{R}^n} dx_1 \ldots dx_n \langle 0 | e^{-H_n t} | x_1, \ldots, x_n \rangle . \tag{4.47}$$

Also for stationary initial data there is a concise formula, the initial wave function, however, being no longer of product form,

$$\langle Z_{\mathrm{sta}}(0, t)^n \rangle = \int_{\mathbb{R}^n} dx_1 \ldots dx_n \langle 0 | e^{-H_n t} | x_1, \ldots, x_n \rangle \exp \left[\tfrac{1}{2} \langle \big(\sum_{j=1}^{n} B(x_j) \big)^2 \rangle \right] , \tag{4.48}$$

where the right average is over the two-sided Brownian motion $B(x)$.

Of course, the general hope is to extract from the moments some information on the distribution of $\log Z(x, t)$. Unfortunately, the moments diverge as $\exp[n^3]$ and one is forced to fall back on formal resummation procedures. Even then there are prior difficulties. Firstly, from the Bethe ansatz solution of the Lieb–Liniger model, one must deduce a sufficiently concise formula for the particular matrix element of the propagator. It is not known how to proceed for general initial data, but for the three canonical initial conditions this step has been accomplished, at increasing complexity from wedge (Calabrese et al. 2010, Dotsenko 2010a, 2010b), to stationary (Imamura and Sasamoto 2012, Imamura and Sasamoto 2013), to flat (Calabrese and Le Doussal 2011, Le Doussal and Calabrese 2012). While for the curved and stationary cases there are corresponding rigorous results, the flat initial conditions are yet to be resolved (Ortmann et al. 2016a, 2016b). The second difficulty is part of working with a badly divergent series. One is not allowed to somehow cut, or otherwise approximate, the series at intermediate steps.

The current results all deal with a single reference point. For the joint distribution at several space points, one relies on an intermediate decoupling assumption (Prolhac and Spohn 2011a, 2011b). On a large scale the resulting expressions agree with the corresponding ones from lattice models, indicating that the decoupling is valid at least in approximation.

4.7   Statistical mechanics of line ensembles

There is a second mapping, which is more hidden and has been discovered only 15 years after the publication of the KPZ paper. In contrast to the Cole–Hopf transformation, the second mapping deals only with the data at some fixed time and thus provides partial information only. Nevertheless, it is this mapping through which many of the universal results were obtained first. Whether such mapping can be defined for the KPZ equation is not known at the moment and we turn instead to the PNG model (Prähofer and Spohn 2002). We consider droplet growth, which means that there is an initial ground layer expanding linearly in time up to $[-t, t]$ and all nucleation events outside this layer are suppressed. At time t we have the random height profile $x \mapsto h_{\mathrm{PNG}}(x, t)$. Now we claim that the statistics of $h_{\mathrm{PNG}}(x, t)$ at fixed t can be obtained through a direct construction, completely avoiding an explicit solution of the dynamics. We choose x as running parameter, $|x| \leq t$, and consider independent, time-continuous, symmetric, simple random walks $w_n(x)$, $n = 0, -1, \ldots$. w_n is already conditioned on $w_n(\pm t) = n$. Now we pick $M > 0$ and further condition on the event that the walks $w_n(x), n = 0, \ldots, -M$, do not intersect. Finally, we take the limit $M \to \infty$. This limit exists, since there is some smallest random index m such that $w_m(x) = m$ for all x. The such conditioned nonintersecting random walks are denoted again by $w_n(x)$. The theorem is that the distribution of $h_{\mathrm{PNG}}(x, t)$, t fixed, is identical to the one of the top random walk $w_0(x)$.

Because of entropic repulsion the typical shape of $w_0(x)$ is a droplet of the form of a semicircle, $h(x) = 2\sqrt{1 - x^2}$. The collection $\{w_n(x), n \in \mathbb{Z}_-\}$ is called a nonintersecting line ensemble, which this time is an object of equilibrium statistical mechanics. More generally, such an ensemble is defined through the Boltzmann weight

$$Z^{-1} \exp\left[-\beta \sum_{n=-M}^{-1} \int_{-t}^{t} \mathrm{d}y V(w_{n+1}(y) - w_n(y)) \right], \tag{4.49}$$

where V is a short range, strongly repulsive hard core potential. Equation (4.49) defines a random field $\{\zeta(x, j), x \in [-t, t], j \in \mathbb{Z}\}$, where $\zeta(x, j) = 1$ if a line passes through (x, j) and $\zeta(x, j) = 0$ otherwise. Of course, one still must specify the boundary conditions. In our example the lines are pinned at the border lines $\{|x| = \pm t\}$. Thereby, the equilibrium measure becomes inhomogeneous, in both x and j.

Our mapping comes with an additional powerful tool. In the limit of an infinitely strong point repulsion, which is equivalent to the nonintersection condition discussed earlier, the random walks $x \mapsto w_n(x)$ are the world lines of noninteracting fermions. x is the Euclidean time and j is space. The fermions start at $x = -t$ with the lattice $\mathbb{Z}$ half-filled from $-\infty$ to 0. They evolve in imaginary time by the standard symmetric nearest neighbor hopping. Thus, the one-particle Hamiltonian is the nearest neighbor Laplacian $-(\Delta f)_j = -f_{j+1} - f_{j-1} + 2f_j$. At time $x = t$ the fermions must return to their original positions. Such problem can be handled by free fermion techniques. To study $w_0(0)$ it is convenient to consider the full collection of points $\{w_n(0), n = 0, -1, \ldots\}$. This turns out to be the ground state of free fermions subject to a linear external potential j/t. Just like in the case of a gravitational potential there is a top

fermion. As $t \to \infty$, using that the slope of the linear potential decreases as $1/t$, the position of the top particle has the distribution

$$\omega_0(0) = h_{\mathrm{PNG}}(0,t) \simeq 2t + (\Gamma_{\mathrm{PNG}}t)^{1/3}\xi_{\mathrm{cur}}, \qquad (4.50)$$

valid for large t. Extending to several reference points, one concludes convergence to the full $\mathcal{A}_{\mathrm{cur}}(w)$ process.

A more complete discussion can be found in my write-up for the 2005 Summer School on 'Fundamental Problems in Statistical Mechanics' at Leuven (Spohn 2006).

4.8　Noisy local conservation laws

Any one of the topics mentioned so far deserves further explanations. But this would easily run oversize. Instead I will focus in much greater detail on one aspect, which is a recent development and covers physics yet different from growth processes.

On a purely formal level the generalization consists of replacing the scalar height $h(x,t)$ by an n-vector $\vec{h} = (h_1, \dots, h_n)$ (Ferrari et al. 2013, Spohn 2014). In our applications the relevant quantities will be the slopes $\partial_x \vec{h}$. Thus, we start from the KPZ equation in the slope form (4.27) and generalize it to

$$\partial_t u_\alpha + \partial_x \big((A\vec{u})_\alpha + \vec{u} \cdot (H^\alpha \vec{u}) + (D\vec{u})_\alpha + (B\vec{\xi})_\alpha \big) = 0. \qquad (4.51)$$

These are n coupled conservation laws, u_α, $\alpha = 1, \dots, n$, being the conserved fields. The currents have three pieces:

(i) *a nonlinear current.* We included terms only up to quadratic order, since by power counting higher orders are expected to be irrelevant. The linear part is specified by the $n \times n$ matrix A. For $n = 1$ it could be removed by switching to a frame moving with constant velocity. But for larger n this will not be possible unless all eigenvalues of A coincide. The quadratic part is specified by the symmetric Hessians H^α.

(ii) *dissipation.* This term is proportional to the gradients. The diffusion matrix, D, has positive eigenvalues. In principle D could depend also on $\vec{u}$, but this is regarded as a higher order effect.

(iii) *fluctuating currents.* ξ_α is Gaussian white noise with independent components, $\langle \xi_\alpha(x,t) \rangle = 0$, $\langle \xi_\alpha(x,t)\xi_{\alpha'}(x',t') \rangle = \delta_{\alpha\alpha'}\delta(x-x')\delta(t-t')$. The matrix B encodes possible correlations between the random currents.

The generalization (4.51) may look more natural after providing a few examples. First, we return to the single-step model. Its height has slopes ± 1. We regard -1 as the position of a particle and 1 as an empty site. Then under the single-step dynamics the particles form a stochastic system known as the asymmetric simple exclusion process (ASEP). Particles hop on the lattice $\mathbb{Z}$. Independently they jump to the right with rate p and to left with rate q. An attempted jump is suppressed in case it would lead to a double occupancy (the exclusion rule). 'Simple' refers to nearest neighbor jumps only. Clearly, the particle number is the only conserved field. The steady states are Bernoulli. Therefore, at density ρ, the average current, j, equals $j(\rho) = (p-q)\rho(1-\rho)$.

As explained in Sect. 4.2, on the mesoscopic scale the density is governed by the stochastic Burgers equation (4.27), at least for small $|p - q| \neq 0$. To generalize from one to n components we introduce particles of type $\alpha = 1, \ldots, n$. α particles jump under the exclusion rule on lane α, but the jump rate depends on the occupancy of the corresponding sites on the other lanes. In the limit of weak asymmetry this then leads to (4.51). Obviously, our theme allows for many variations. One particular version is to have two components with particles hoppping on $\mathbb{Z}$, subject to exclusion. If the exchange rates depend only on nearest neighbor occupations, one arrives at the well-studied AHR model (Arndt et al. 1999).

Also the PNG model can viewed as stochastic motion of particles. Now there are two types, $\pm$, already. A $+$ particle is located at a down-step moving to the right and a $-$ particle is located at an up-step moving to the left, both at unit speed. Particles annihilate at collisions. In addition, point-like $-|+$ pairs are generated at constant rate. We denote the two components by ρ_+, ρ_-. Only $\rho = \rho_+ - \rho_-$ is conserved, while $j = \rho_+ + \rho_-$ is not conserved. As anticipated by notation, j is the current for ρ. Using that the stationary states are uniform Poisson, one concludes that the current-density relation is $j(\rho) = \sqrt{4 + \rho^2}$. However, the PNG model has no natural asymmetry parameter. So there is no appropriate limit leading to the stochastic Burgers equation. Of course, multicomponent versions are still easily invented.

Even at that general level there is already a crucial distinction. The dominant term in (4.51) is the linear flow term $\partial_x A \vec{u}$. The anharmonic chains to be studied have three conservation laws, hence $n = 3$, and a matrix A with nondegenerate eigenvalues. There are early studies of multicomponent KPZ equations (Ertaş and Kardar 1993, Das et al. 2001), assuming, however, $A = 0$, which then leads to a scenario very different from the one discussed here.

For the remainder of this chapter we will discuss one-dimensional mechanical systems governed by Newton's equations of motion, either point particles or discrete nonlinear wave equations. Usually they have three conserved fields. But an example with $n = 2$ will also be considered. The methods to be developed can also be used for multicomponent ASEP or other stochastic models with several conservation laws; see Stoltz and Spohn (2015) and Popkov et al. (2015). We will spend some time defining the models, to argue how the coupled system (4.51) of conservation laws arises, and to explain how the model-dependent coefficients are computed. A separate task will be to extract out of a system of nonlinear stochastic conservation laws concrete predictions which may be checked through molecular dynamics simulations.

4.9   One-dimensional fluids and anharmonic chains

As microscopic model we focus on a classical fluid on the line consisting of particles with positions q_j and momenta p_j, $j = 1, \ldots, N$, $q_j, p_j \in \mathbb{R}$, possible boundary conditions to be delayed momentarily. We use units such that the mass of the particles equals 1. Then the Hamiltonian is of the standard form

$$H_N^{\text{fl}} = \sum_{j=1}^{N} \tfrac{1}{2} p_j^2 + \tfrac{1}{2} \sum_{i \neq j=1}^{N} V(q_i - q_j), \tag{4.52}$$

with pair potential $V(x) = V(-x)$. The potential may have a hard core and otherwise is assumed to be short ranged. The dynamics for long-range potentials is of independent interest (Miloshevich et al. 2014), but is not discussed here. The three conserved fields are density, momentum, and energy. One might want to add a periodic external potential. Then momentum is no more conserved and the dynamical properties will change dramatically.

A substantial simplification is achieved by assuming a hard core of diameter a, i.e., $V(x) = \infty$ for $|x| < a$, and restricting the range of the smooth part of the potential to at most $2a$. Then the particles maintain their order, $q_j \leq q_{j+1}$, and in addition only nearest neighbor particles interact. Hence H_N^{fl} simplifies to

$$H_N = \sum_{j=1}^{N} \tfrac{1}{2} p_j^2 + \sum_{j=1}^{N-1} V(q_{j+1} - q_j). \tag{4.53}$$

As a, at first sight very different, physical realization, we could interpret H_N as describing particles in one dimension coupled through anharmonic springs, which is then usually referred to as anharmonic chain.

In the second interpretation the spring potential can be more general than anticipated so far. No ordering constraint is required and the potential does not have to be even. To have well-defined thermodynamics the chain is pinned at both ends as $q_1 = 0$ and $q_{N+1} = \ell N$. It is convenient to introduce the stretch $r_j = q_{j+1} - q_j$. Then the boundary condition corresponds to the microcanonical constraint

$$\sum_{j=1}^{N} r_j = \ell N. \tag{4.54}$$

Switching to canonical equilibrium according to the standard rules, one then arrives at the obvious condition of a finite partition function

$$Z(P, \beta) = \int_{\mathbb{R}} \mathrm{d}x \, \mathrm{e}^{-\beta(V(x) + Px)} < \infty, \tag{4.55}$$

using the standard convention that the integral is over the entire real line. Here $\beta > 0$ is the inverse temperature and P is the thermodynamically conjugate variable to the stretch. By partial integration

$$P = -Z(P, \beta)^{-1} \int_{\mathbb{R}} \mathrm{d}x V'(x) \, \mathrm{e}^{-\beta(V(x) + Px)}, \tag{4.56}$$

implying that P is the average force in the spring between two adjacent particles, hence identified as thermodynamic pressure. To have a finite partition function, a natural condition on the potential is to be bounded from below and to have a one-sided linear bound as $V(x) \geq a_0 + b_0|x|$ for either $x > 0$ or $x < 0$ and $b_0 > 0$. Then there is a nonempty interval $I(\beta)$ such that $Z(P, \beta) < \infty$ for $P \in I(\beta)$. For the particular case of a hard-core fluid one must impose $P > 0$.

Note: The sign of P is chosen such that for a gas of hard-point particles one has the familiar ideal gas law $P = 1/\beta\ell$. The chain tension is $-P$.

Famous examples are the harmonic chain, $V_{\text{ha}}(x) = x^2$, the Fermi–Pasta–Ulam (FPU) chain, $V_{\text{FPU}}(x) = \frac{1}{2}x^2 + \frac{1}{3}\alpha x^3 + \frac{1}{4}\beta x^4$, in the historical notation (Fermi et al. 1965), and the Toda chain (Toda 1967), $V(x) = e^{-x}$, in which case $P > 0$ is required. The harmonic chain, the Toda chain, and the hard-core potential, $V_{\text{hc}}(x) = \infty$ for $|x| < a$ and $V_{\text{hc}}(x) = 0$ for $|x| \geq a$, are in fact integrable systems which have a very different correlation structure and will not be discussed here. Except for the harmonic chain, one simple way to break integrability is to assume alternating masses, say $m_j = m_0$ for even j and $m_j = m_1$ for odd j.

We will mostly deal with anharmonic chains described by the Hamiltonian (4.53), including one-dimensional hard-core fluids with a sufficiently small potential range. There are two good reasons. Firstly, amongst the large body of molecular dynamics simulations there is not a single one which deals with an 'honest' one-dimensional fluid. To be able to reach large system sizes all simulations are performed for anharmonic chains. Secondly, from a theoretical perspective, the equilibrium measures of anharmonic chains are particularly simple in being of product form in momentum and stretch variables. Thus, material parameters, as compressibility and sound speed, can be expressed in terms of one-dimensional integrals involving the Boltzmann factor $e^{-\beta(V(x)+Px)}$, $V(x)$, and x.

The dynamics of the anharmonic chain is governed by

$$\frac{d}{dt}q_j = p_j, \qquad \frac{d}{dt}p_j = V'(q_{j+1} - q_j) - V'(q_j - q_{j-1}). \qquad (4.57)$$

This can be viewed as the discretization of the nonlinear wave equation

$$\partial_t^2 u(x,t) = \partial_x V'(\partial_x u(x,t)), \qquad (4.58)$$

which for the harmonic potential reduces to the linear wave equation. In this physical interpretation $\{q_j, j = 1, \ldots, N\}$ is the discretized displacement field $u(x)$. Throughout we will stick to the lattice field theory point of view. For the initial conditions we choose a lattice cell of length N and require

$$q_{j+N} = q_j + \ell N, \qquad p_{j+N} = p_j \qquad (4.59)$$

for all $j \in \mathbb{Z}$. This property is preserved under the dynamics and thus properly mimics a system of finite length N. The stretches are then N-periodic, $r_{j+N} = r_j$, and the single-cell dynamics is given by

$$\frac{d}{dt}r_j = p_{j+1} - p_j, \qquad \frac{d}{dt}p_j = V'(r_j) - V'(r_{j-1}), \qquad (4.60)$$

$j = 1, \ldots, N$, together with the periodic boundary conditions $p_{1+N} = p_1$, $r_0 = r_N$ and the constraint (4.54). Through the stretch there is a coupling to the right neighbor and through the momentum a coupling to the left neighbor. The potential is defined only up to translations, since the dynamics does not change under a simultaneous shift

of $V(x)$ to $V(x-a)$ and r_j to r_j+a, in other words, the potential can be shifted by shifting the initial r-field. Note that our periodic boundary conditions are not identical to fluid particles moving on a ring, but they may become so for large system size when length fluctuations become negligible.

Before proceeding let us be more specific on the link between fluids and anharmonic chains. To avoid the issue of boundary conditions we start from the infinitely extended system. First of all, for a potential as $V(x) = x^2 + x^4$ the lattice field theory point of view is the natural option. Similarly, for $V(x) = 1, |x| < a$, and $V(x) = 0, |x| \geq a$, unlabeled particles moving on the real line is the obvious choice, q_j then being the physical position of the jth particle. So let us consider a potential, for which both fluid and solid picture are meaningful. Clearly, given (4.57), also Eq. (4.60) is satisfied. In reverse order, given the solution to (4.60), we still must fix the value of q_0, say. Then $q_0(t)$ follows from (4.57) and one thereby reconstructs all other positions. Thus, up to the choice of q_0 the two dynamics are identical. However, the observables for a hydrodynamic theory will be different. In case of a fluid, momentum and energy are attached to a particle. For example, the momentum field is defined by

$$u_\mathrm{fl}(x,t) = \sum_j \delta(x - q_j(t))p_j(t), \tag{4.61}$$

while for the lattice field theory it is merely $p_j(t)$. Both fields are locally conserved. The dynamical correlator for the fluid, $\langle u_\mathrm{fl}(x,t)u_\mathrm{fl}(0,0)\rangle$, differs from the corresponding lattice correlator, $\langle p_j(t)p_0(0)\rangle$, and there is no simple rule for transforming one into the other. On the other hand, the hydrodynamic theory to be developed could as well be carried through for fluids. Structurally both theories will be of the form of Eq. (4.51). By universality we thus expect to have the same behavior on large scales.

We return to anharmonic chains and note that Eqs. (4.60) are already of conservation type. Hence,

$$\frac{d}{dt}\sum_{j=1}^{N} r_j = 0, \quad \frac{d}{dt}\sum_{j=1}^{N} p_j = 0. \tag{4.62}$$

We define the local energy by

$$e_j = \tfrac{1}{2}p_j^2 + V(r_j). \tag{4.63}$$

Then its local conservation law reads

$$\frac{d}{dt}e_j = p_{j+1}V'(r_j) - p_j V'(r_{j-1}), \tag{4.64}$$

implying that

$$\frac{d}{dt}\sum_{j=1}^{N} e_j = 0. \tag{4.65}$$

The microcanonical equilibrium state is defined by the Lebesgue measure constrained to a particular value of the conserved fields as

$$\sum_{j=1}^{N} r_j = \ell N, \quad \sum_{j=1}^{N} p_j = uN, \quad \sum_{j=1}^{N} \left(\tfrac{1}{2}p_j^2 + V(r_j)\right) = \mathfrak{e}N, \tag{4.66}$$

with ℓ being the stretch, u the momentum, and $\mathfrak{e}$ the total energy per particle. In our context the equivalence of ensembles holds and computationally it is of advantage to switch to the canonical ensemble with respect to all three constraints. Then the dual variable for the stretch ℓ is the pressure P, for the momentum the average momentum, again denoted by u, and for the total energy $\mathfrak{e}$ the inverse temperature β. For the limit of infinite volume the symmetric choice $j \in [-N, \ldots, N]$ is more convenient. In the limit $N \to \infty$ either under the canonical equilibrium state, trivially, or under the microcanonical ensemble, by the equivalence of ensembles, the collection $(r_j, p_j)_{j \in \mathbb{Z}}$ are independent random variables. Their single site probability density is given by

$$Z(P, \beta)^{-1} e^{-\beta(V(r_j) + Pr_j)} (2\pi/\beta)^{-1/2} e^{-\frac{1}{2}\beta(p_j - u)^2}. \tag{4.67}$$

Averages with respect to (4.67) are denoted by $\langle \cdot \rangle_{P,\beta,u}$. The dependence on the average momentum can be removed by a Galilei transformation. Hence we mostly work with $u = 0$, in which case we merely drop the index u. We also introduce the internal energy, e, through $\mathfrak{e} = \frac{1}{2}u^2 + e$, which agrees with the total energy at $u = 0$. The canonical free energy, at $u = 0$, is defined by

$$G(P, \beta) = -\beta^{-1}\left(-\tfrac{1}{2}\log\beta + \log Z(P, \beta)\right). \tag{4.68}$$

Then

$$\ell = \langle r_0 \rangle_{P,\beta}, \quad e = \partial_\beta\big(\beta G(P, \beta)\big) - P\ell = \frac{1}{2\beta} + \langle V(r_0)\rangle_{P,\beta}. \tag{4.69}$$

The relation (4.69) defines $(P, \beta) \mapsto (\ell(P, \beta), e(P, \beta))$, thereby the inverse map $(\ell, e) \mapsto (P(\ell, e), \beta(\ell, e))$, and thus accomplishes the switch between the microcanonical thermodynamic variables ℓ, e and the canonical thermodynamic variables P, β.

It is convenient to collect the conserved fields as the 3-vector $\vec{g} = (g_1, g_2, g_3)$,

$$\vec{g}(j, t) = \big(r_j(t), p_j(t), e_j(t)\big), \tag{4.70}$$

$\vec{g}(j, 0) = \vec{g}(j)$. Then, the conservation laws are combined as

$$\frac{d}{dt}\vec{g}(j, t) + \vec{\mathcal{J}}(j + 1, t) - \vec{\mathcal{J}}(j, t) = 0, \tag{4.71}$$

with the local current functions

$$\vec{\mathcal{J}}(j) = \big(-p_j, -V'(r_{j-1}), -p_j V'(r_{j-1})\big). \tag{4.72}$$

Very roughly, our claim is that, for suitable random initial data, on the mesoscopic scale the conservation law (4.71) can be well approximated by a noisy conservation

law of the form (4.51) with $n = 3$. This leaves the physical setup widely unspecified. To be more concrete, also to have a well-defined control through molecular dynamics simulations and a quantity of physical importance, we consider time correlation functions in thermal equilibrium of the conserved fields. They are defined by

$$S_{\alpha\alpha'}(j,t) = \langle g_\alpha(j,t)g_{\alpha'}(0,0)\rangle_{P,\beta} - \langle g_\alpha(0,0)\rangle_{P,\beta}\langle g_{\alpha'}(0,0)\rangle_{P,\beta}, \qquad (4.73)$$

$\alpha, \alpha' = 1, 2, 3$. The infinite volume limit has been taken already and the average is with respect to thermal equilibrium at u = 0. It is known that such a limit exists (Bernardin and Olla 2016). Also the decay in j is exponentially fast, but with a correlation length increasing in time. In this context our central claim states that $S_{\alpha\alpha'}(j,t)$ can be well approximated by the stationary covariance of a Langevin equation of the same form as in Eq. (4.51) with coefficients which still must be computed.

Often it is convenient to regard $S(j,t)$, no indices, as a 3×3 matrix. In general, $S(j,t)$ has certain symmetries, the first set resulting from space-time stationarity and the second set from time reversal, even for $\alpha = 1, 3$, odd for $\alpha = 2$,

$$S_{\alpha\alpha'}(j,t) = S_{\alpha'\alpha}(-j,-t), \quad S_{\alpha\alpha'}(j,t) = (-1)^{\alpha+\alpha'}S_{\alpha\alpha'}(j,-t). \qquad (4.74)$$

At $t = 0$ the average (4.73) reduces to a static average, which is easily computed. In general the static susceptibility matrix is defined through

$$C = \sum_{j \in \mathbb{Z}} S(j,0). \qquad (4.75)$$

For anharmonic chains the fields are uncorrelated in j; hence,

$$S(j,0) = \delta_{j0}C \qquad (4.76)$$

with susceptibility matrix

$$C = \begin{pmatrix} \langle r_0; r_0 \rangle_{P,\beta} & 0 & \langle r_0; V_0 \rangle_{P,\beta} \\ 0 & \beta^{-1} & 0 \\ \langle r_0; V_0 \rangle_{P,\beta} & 0 & \frac{1}{2}\beta^{-2} + \langle V_0; V_0 \rangle_{P,\beta} \end{pmatrix}. \qquad (4.77)$$

Here, for X, Y arbitrary random variables, $\langle X; Y \rangle = \langle XY \rangle - \langle X \rangle \langle Y \rangle$ denotes the second cumulant and $V_0 = V(r_0)$, following the same notational convention as for e_0. Note that the conservation law implies the zeroth moment sum rule

$$\sum_{j \in \mathbb{Z}} S(j,t) = \sum_{j \in \mathbb{Z}} S(j,0) = C. \qquad (4.78)$$

Since our theoretical discussion will be somewhat lengthy, we indicate already now in which direction we are heading. Figure 4.3 displays a molecular dynamics simulation of a FPU chain with potential $V(x) = \frac{1}{2}x^2 + \frac{2}{3}x^3 + \frac{1}{4}x^4$ at inverse temperature $\beta = 2$ and pressure $P = 1$ (Das et al. 2014). The system size is $N = 8192$. Plotted is a time sequence of a generic matrix element of $j \mapsto S(j,t)$ with a normalization such that

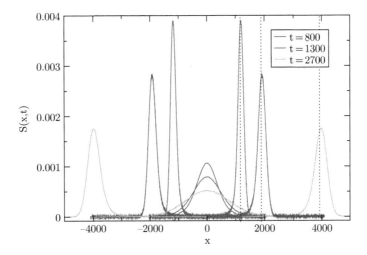

Fig. 4.3 Heat peak and two sound peaks, area normalized to 1, at times $t = 800, 1300, 2700$ for a FPU chain with $N = 8192$, $V(x) = \frac{1}{2}x^2 + \frac{2}{3}x^3 + \frac{1}{4}x^4$, pressure $P = 1$, and inverse temperature $\beta = 2$. The sound speed is $c = 1.45$. The heat peak has power law tails, which are cut off by the sound peaks. According to the theory, asymptotically the sound peaks are symmetric relative σct and have stretched exponential tails as $\exp[-|x|^3]$.

the area under each peak equals 1. There are two sound peaks, one moving to the right and its mirror image to the left with sound speed $c = 1.45$. In addition there is a peak standing still, which for thermodynamic reasons is called the heat peak. The peaks broaden in time as $t^{2/3}$ for sound and as $t^{3/5}$ for heat. The average is over 10^7 samples drawn from the canonical distribution (4.67).

4.10   Discrete nonlinear Schrödinger equation

In parallell to anharmonic chains, it is instructive to discuss as second microscopic model the nonlinear Schrödinger equation on the one-dimensional lattice (DNLS). As a novel feature, the large-scale dynamics at high temperatures is diffusive, while at low temperatures propagating modes appear because of an additional almost conserved field.

For DNLS the lattice field is $\psi_j \in \mathbb{C}$ and ψ, ψ^* are canonically conjugate fields, $*$ denoting complex conjugation. The Hamiltonian reads

$$H = \sum_{j=1}^{N} \left(\tfrac{1}{2}|\psi_{j+1} - \psi_j|^2 + \tfrac{1}{2}g|\psi_j|^4 \right), \tag{4.79}$$

with periodic boundary conditions. $g > 0$ is the coupling constant, a *defocusing* nonlinearity. The dynamics is defined through

$$\mathrm{i}\,\frac{\mathrm{d}}{\mathrm{d}t}\psi_j = \partial_{\psi_j^*} H \tag{4.80}$$

and thus

$$\mathrm{i}\,\frac{\mathrm{d}}{\mathrm{d}t}\psi_j = -\tfrac{1}{2}\Delta\psi_j + g\,|\psi_j|^2\,\psi_j, \tag{4.81}$$

with the lattice Laplacian $\Delta = -\partial^{\mathrm{T}}\partial$ and $\partial\psi_j = \psi_{j+1} - \psi_j$.

The DNLS has two obvious locally conserved fields, density and energy,

$$\rho_j = |\psi_j|^2\,, \qquad e_j = \tfrac{1}{2}|\psi_{j+1} - \psi_j|^2 + \tfrac{1}{2}\,g\,|\psi_j|^4\,. \tag{4.82}$$

According to the discussion in Ablowitz et al. (2004), the DNLS is nonintegrable and one expects density and energy to be the only locally conserved fields. They satisfy the conservation laws

$$\frac{\mathrm{d}}{\mathrm{d}t}\rho_j(t) + \mathcal{J}_{\rho,j+1}(t) - \mathcal{J}_{\rho,j}(t) = 0\,,$$

$$\frac{\mathrm{d}}{\mathrm{d}t}e_j(t) + \mathcal{J}_{e,j+1}(t) - \mathcal{J}_{e,j}(t) = 0\,, \tag{4.83}$$

with density current

$$\mathcal{J}_{\rho,j} = \tfrac{1}{2}\mathrm{i}\big(\psi_{j-1}\,\partial\psi_{j-1}^* - \psi_{j-1}^*\,\partial\psi_{j-1}\big) \tag{4.84}$$

and energy current

$$\mathcal{J}_{e,j} = \tfrac{1}{4}\mathrm{i}\big(\Delta\psi_j^*\,\partial\psi_{j-1} - \Delta\psi_j\,\partial\psi_{j-1}^*\big) + g|\psi_j|^2\mathcal{J}_{\rho,j}\,. \tag{4.85}$$

As a consequence the canonical equilibrium state is given by

$$Z^{-1}\mathrm{e}^{-\beta(H-\mu\mathsf{N})}\prod_{j=1}^{N}\mathrm{d}\psi_j\mathrm{d}\psi_j^*\,, \qquad \mathsf{N} = \sum_{j=1}^{N}|\psi_j|^2\,, \tag{4.86}$$

with chemical potential $\mu \in \mathbb{R}$. Here we assume $\beta > 0$. But also negative temperature states, in the microcanonical ensemble, have been studied (Iubini et al. 2013, 2014). Then the dynamics is dominated by a coarsening process mediated through breathers. In equilibrium, the ψ-field has high spikes at random locations embedded in a low noise background, which is very different from the positive temperature states considered here.

Canonically conjugate variables can also be introduced by splitting the wave function into its real and imaginary parts as

$$\psi_j = \tfrac{1}{\sqrt{2}}(q_j + \mathrm{i}p_j)\,. \tag{4.87}$$

In these variables, the Hamiltonian reads

$$H = \sum_{j=1}^{N}\Big(\tfrac{1}{4}\big((\partial q_j)^2 + (\partial p_j)^2\big) + \tfrac{1}{8}\,g\big(q_j^2 + p_j^2\big)^2\Big)\,. \tag{4.88}$$

The dynamics defined by Eq. (4.81) is then identical to the Hamiltonian system

$$\frac{\mathrm{d}}{\mathrm{d}t}q_j = \partial_{p_j} H\,, \quad \frac{\mathrm{d}}{\mathrm{d}t}p_j = -\partial_{q_j} H\,. \tag{4.89}$$

Note that H is symmetric under the interchange $q_j \leftrightarrow p_j$.

It will be convenient to make a canonical (symplectic) change of variables to polar coordinates as

$$\varphi_j = \arctan(p_j/q_j)\,, \quad \rho_j = \tfrac{1}{2}\left(p_j^2 + q_j^2\right)\,, \tag{4.90}$$

which is equivalent to the representation

$$\psi_j = \sqrt{\rho_j}\, e^{i\varphi_j}\,. \tag{4.91}$$

In the new variables the phase space becomes $(\rho_j, \varphi_j) \in \mathbb{R}_+ \times S^1$, with S^1 the unit circle. The corresponding Hamiltonian is given by

$$H = \sum_{j=1}^{N} \left(\tfrac{1}{2}\left(\sqrt{\rho_{j+1}\,\rho_j}\, 2\,(1 - \cos(\varphi_{j+1} - \varphi_j)) + (\sqrt{\rho_{j+1}} - \sqrt{\rho_j})^2\right) + \tfrac{1}{2}\, g\, \rho_j^2 \right)$$

$$= \sum_{j=1}^{N} \left(-\sqrt{\rho_{j+1}\,\rho_j}\, \cos(\varphi_{j+1} - \varphi_j) + \rho_j + \tfrac{1}{2}\, g\, \rho_j^2 \right)\,. \tag{4.92}$$

The equations of motion read then

$$\tfrac{\mathrm{d}}{\mathrm{d}t}\varphi_j = -\partial_{\rho_j} H\,, \quad \tfrac{\mathrm{d}}{\mathrm{d}t}\rho_j = \partial_{\varphi_j} H\,. \tag{4.93}$$

From the continuity of $\psi_j(t)$ when moving through the origin, one concludes that at $\rho_j(t) = 0$ the phase jumps from $\varphi_j(t)$ to $\varphi_j(t) + \pi$. The φ_j's are angles and therefore position-like variables, while the ρ_j's are actions and hence momentum-like variables. The Hamiltonian depends only on phase differences, which implies the invariance under the global shift $\varphi_j \mapsto \varphi_j + \phi$.

In (4.79) the kinetic energy is chosen such that in the limit of zero lattice spacing one arrives at the continuum nonlinear Schrödinger equation on $\mathbb{R}$. This is an integrable nonlinear wave equation, while the DNLS is nonintegrable. Thus, lattice and continuum version show distinct dynamical behavior.

4.11   Linearized Euler equations

Our goal is to predict the long-time behavior of the correlations of the conserved fields, compared with (4.73). $S_{\alpha\alpha'}(j, t)$ may be viewed as the response in the field α at (j, t) to equilibrium perturbed in the field α' at $(0, 0)$. One might hope to capture such a response on the basis of an evolution equation for the conserved fields when linearized at equilibrium. The most obvious macroscopic descriptions are the Euler equations which are a fairly direct consequence of the conservation laws. One starts the system

in a state of local equilibrium, which means to have the equilibrium parameters varying slowly on the scale of interparticle distances. If the dynamics is sufficiently chaotic, such a situation is expected to persist provided the parameters evolve according to the Euler equations. Their currents are thus obtained by averaging the microscopic currents in a local equilibrium state. In other words, the Euler currents are defined through static expectations. More specifically, in the case of anharmonic chains we use the microscopic currents (4.72). Then the average currents are

$$\langle \vec{\mathcal{J}}(j) \rangle_{\ell, \mathsf{u}, \mathsf{e}} = \left(-\mathsf{u}, P(\ell, \mathsf{e} - \tfrac{1}{2}\mathsf{u}^2), \mathsf{u}P(\ell, \mathsf{e} - \tfrac{1}{2}\mathsf{u}^2) \right) = \vec{\mathsf{j}}(\ell, \mathsf{u}, \mathsf{e}), \qquad (4.94)$$

with $P(\ell, \mathsf{e})$ defined implicitly through (4.69). On the macroscopic scale the difference becomes ∂_x and one arrives at the macroscopic Euler equations

$$\partial_t \ell - \partial_x \mathsf{u} = 0, \quad \partial_t \mathsf{u} + \partial_x P(\ell, \mathsf{e} - \tfrac{1}{2}\mathsf{u}^2) = 0, \quad \partial_t \mathsf{e} + \partial_x \left(\mathsf{u} P(\ell, \mathsf{e} - \tfrac{1}{2}\mathsf{u}^2) \right) = 0, \quad (4.95)$$

with the three conserved fields depending on x, t. We refer to a forthcoming monograph by Bernardin and Olla (2016), where the validity of the Euler equations is proved up to the first shock. Since, as emphasized already, it is difficult to deal with deterministic chaos, the authors add random velocity exchanges between neighboring particles which ensure that the dynamics locally enforces the microcanonical state.

For DNLS the situation is much simpler. The currents are symbolically of the from $i(z - z^*)$ and the equilibrium state is invariant under complex conjugation. Hence the average currents vanish. We will later see that at low temperatures an additional almost conserved field emerges. Then the Euler equations become nontrivial and have the same structure as in (4.95).

We are interested here only in small deviations from equilibrium and therefore linearize the Euler equations as $\ell + u_1(x)$, $0 + u_2(x)$, $\mathsf{e} + u_3(x)$ to linear order in the deviations $\vec{u}(x)$. This leads to the linear equation

$$\partial_t \vec{u}(x, t) + A \partial_x \vec{u}(x, t) = 0, \qquad (4.96)$$

with

$$A = \begin{pmatrix} 0 & -1 & 0 \\ \partial_\ell P & 0 & \partial_\mathsf{e} P \\ 0 & P & 0 \end{pmatrix}. \qquad (4.97)$$

Here, and in the following, the dependence of A, C, and similar quantities on the background values $\ell, \mathsf{u} = 0, \mathsf{e}$, hence on P, β, is suppressed from the notation. Beyond (4.78) there is the first moment sum rule which states that

$$\sum_{j \in \mathbb{Z}} j S(j, t) = AC\, t. \qquad (4.98)$$

A proof, which in essence uses only the conservation laws and space-time stationarity of the correlations, is given in (Spohn 2014), see also see Tóth and Valkó (2003) and

Grisi and Schütz (2011). Microscopic properties enter only minimally. However, since $C = C^{\mathrm{T}}$ and $S(j,t)^{\mathrm{T}} = S(-j,-t)$, Eq. (4.98) implies the important relation

$$AC = (AC)^{\mathrm{T}} = CA^{\mathrm{T}},\tag{4.99}$$

with ^{T} denoting transpose. Of course, (4.99) can be checked also directly from the definitions. Since $C > 0$, A is guaranteed to have real eigenvalues and a nondegenerate system of right and left eigenvectors. For A one obtains the three eigenvalues $0, \pm c$ with

$$c^2 = -\partial_\ell P + P\partial_e P > 0.\tag{4.100}$$

Thus the solution to the linearized equation has three modes, one standing still, one right moving with velocity c, and one left moving with velocity $-c$. Hence we have identified the adiabatic sound speed as being equal to c.

Equation (4.96) is a deterministic equation. But the initial data are random such that within our approximation

$$\langle u_\alpha(x,0)u_{\alpha'}(x',0)\rangle = C_{\alpha\alpha'}\delta(x - x').\tag{4.101}$$

To determine the correlator $S(x,t)$ with such initial conditions is most easily achieved by introducing the linear transformation R satisfying

$$RAR^{-1} = \mathrm{diag}(-c,0,c),\quad RCR^{\mathrm{T}} = 1.\tag{4.102}$$

Up to trivial phase factors, R is uniquely determined by these conditions. Explicit formulas are found in Spohn (2014). Setting $\vec{\phi} = A\vec{u}$, one concludes that

$$\partial\phi_\alpha + c_\alpha\partial_x\phi_\alpha = 0,\quad \alpha = -1,0,1,\tag{4.103}$$

with $\vec{c} = (-c,0,c)$. By construction, the random initial data have the correlator

$$\langle\phi_\alpha(x,0)\phi_{\alpha'}(x',0)\rangle = \delta_{\alpha\alpha'}\delta(x - x').\tag{4.104}$$

Hence

$$\langle\phi_\alpha(x,t)\phi_{\alpha'}(0,0)\rangle = \delta_{\alpha\alpha'}\delta(x - c_\alpha t).\tag{4.105}$$

We transform back to the physical fields. Then in the continuum approximation, at the linearized level,

$$S(x,t) = R^{-1}\mathrm{diag}\big(\delta(x + ct), \delta(x), \delta(x - ct)\big)R^{-\mathrm{T}},\tag{4.106}$$

with $R^{-\mathrm{T}} = (R^{-1})^{\mathrm{T}}$.

Rather easily we have gained a crucial insight: $S(j,t)$ has three peaks which separate linearly in time. For example, $S_{11}(j,t)$ has three sharp peaks moving with velocities $\pm c, 0$. The peak standing still, velocity 0, is called the heat peak and the two peaks moving with velocity $\pm c$ are called the sound peaks. Physically and as supported by

the molecular dynamics simulation displayed in Fig. 4.3, one expects such peaks not to be strictly sharp, but to broaden in the course of time because of dissipation. This issue will have to be explored in great detail. It follows from the zeroth moment sum rule that the area under each peak is preserved in time and thus determined through (4.106). Hence the weights can be computed from the matrix R^{-1}, usually called Landau–Plazcek ratios. A Landau–Placzek ratio could vanish, either accidentally or by a particular symmetry. An example is the momentum correlation $S_{22}(j, t)$. Since $(R^{-1})_{20} = 0$ always, its central peak is absent.

For integrable chains each conservation law generates a peak. Thus, e.g., $S_{11}(j, t)$ of the Toda chain is expected to have a broad spectrum expanding ballistically, rather than consisting of three sharp peaks.

4.12   Linear fluctuating hydrodynamics

The broadening of the peaks results from random fluctuations in the currents, which tend to be uncorrelated in space-time. Therefore, the crudest model would be to assume that the current statistics is space-time Gaussian white noise. In principle, the noise components could be correlated. But since the stretch current is itself conserved, its fluctuations will be taken care of by the momentum equation. Momentum and energy currents have different signature under time reversal; hence, their cross correlation vanishes. As a result, there is a fluctuating momentum current of strength σ_u and an independent energy current of strength σ_e. According to Onsager, noise is linked to dissipation as modeled by a diffusive term. Thus, the linearized equations (4.96) are extended to

$$\partial_t \vec{u}(x, t) + \partial_x \big(A\vec{u}(x, t) - \partial_x D\vec{u}(x, t) + B\vec{\xi}(x, t) \big) = 0 \,. \tag{4.107}$$

Here $\vec{\xi}(x, t)$ is standard white noise with covariance

$$\langle \xi_\alpha(x, t)\xi_{\alpha'}(x', t') \rangle = \delta_{\alpha\alpha'}\delta(x - x')\delta(t - t') \tag{4.108}$$

and, as argued, the noise strength matrix is diagonal as

$$B = \mathrm{diag}(0, \sigma_u, \sigma_e) \,. \tag{4.109}$$

To distinguish the linearized Euler equations (4.96) from the Langevin equations (4.107), we use $\vec{u} = (u_1, u_2, u_3)$ for the fluctuating fields.

The stationary measures for (4.107) are spatial white noise with arbitrary mean. Since small deviations from uniformity are considered, we always impose mean zero. Then the components are correlated as

$$\langle u_\alpha(x)u_{\alpha'}(x') \rangle = C_{\alpha\alpha'}\delta(x - x') \,. \tag{4.110}$$

Stationarity relates the linear drift and the noise strength through the steady state covariance as

$$- (AC - CA^{\mathrm{T}})\partial_x + (DC + CD^{\mathrm{T}})\partial_x^2 = BB^{\mathrm{T}}\partial_x^2 \,. \tag{4.111}$$

The first term vanishes by (4.99) and the diffusion matrix is uniquely determined as

$$D = \begin{pmatrix} 0 & 0 & 0 \\ 0 & D_u & 0 \\ \tilde{D}_e & 0 & D_e \end{pmatrix}, \qquad (4.112)$$

with $\tilde{D}_e = -\langle r_0; V_0 \rangle_{P,\beta} \langle r_0; r_0 \rangle_{P,\beta}^{-1} D_e$. Here $D_u > 0$ is the momentum and $D_e > 0$ the energy diffusion coefficient, which are related to the noise strength as

$$\sigma_u^2 = \langle p_0; p_0 \rangle_{P,\beta} D_u, \qquad \sigma_e^2 = \langle e_0; e_0 \rangle_{P,\beta} D_e. \qquad (4.113)$$

As an insert we return to the DNLS. To distinguish from A, B, C, D for an anharmonic chain, we use A, B, C, D as 2×2 matrices. There are two conserved fields, but the Euler currents vanish. Hence, linear fluctuating hydrodynamics takes the form

$$\partial_t \vec{u}(x,t) + \partial_x \big(-\partial_x D \vec{u}(x,t) + B \vec{\xi}(x,t) \big) = 0, \qquad (4.114)$$

with two components, index 1 for density and index 2 for energy. The matrices B, C satisfy the fluctuation relation $DC + CD = BB^{\mathrm{T}}$, the susceptibility matrix C being defined as in (4.75). Since the Hamiltonian in (4.92) contains nearest neighbor couplings, this matrix is not as readily computed as in the case of anharmonic chains. Physically there is a more important difference. One expects that DC can be written in terms of a time integral over the total current–current correlation matrix. While not computable in analytic form, D is thus a uniquely defined matrix. On the other hand in (4.107) the diffusion matrix D is a phenomenologically introduced coefficient. As will be argued, the true long-time behavior will depend only on C, which is uniquely defined, and not on B, D separately, thus circumventing the arbitrariness in D.

It is not difficult to solve (4.114). We note that D is in general not symmetric, but it has strictly positive eigenvalues. This leads to the stationary two-point function

$$\langle u_\alpha(x,t) u_{\alpha'}(0,0) \rangle = S_{\alpha\alpha'}(x,t) = \int_{\mathbb{R}} dk \, e^{-i2\pi kx} \big(e^{-(2\pi k)^2 Dt} C \big)_{\alpha\alpha'}, \qquad (4.115)$$

$\alpha, \alpha' = 1, 2$. The corresponding lattice correlator reads

$$S(j,t) = \begin{pmatrix} \langle \rho_j(t); \rho_0(0) \rangle & \langle \rho_j(t); e_0(0) \rangle \\ \langle e_j(t); \rho_0(0) \rangle & \langle e_j(t); e_0(0) \rangle \end{pmatrix}. \qquad (4.116)$$

Working out the Fourier transform, one arrives at the prediction

$$S(j,t) \simeq \frac{1}{\sqrt{4\pi D t}} \, e^{-j^2/(4Dt)} \, C. \qquad (4.117)$$

Such a prediction can be checked numerically, for which purpose we performed a molecular dynamics simulation with system size system $N = 4096$ and parameters $g = 1$, $\beta = 1$, and $\langle \rho_j \rangle = 1$ (Mendl and Spohn 2015b). The susceptibility and diffusion matrices are obtained as

$$C = \begin{pmatrix} 0.580 & 0.907 \\ 0.907 & 1.848 \end{pmatrix}, \qquad D = \begin{pmatrix} 3.079 & -0.350 \\ 2.298 & 0.897 \end{pmatrix}. \qquad (4.118)$$

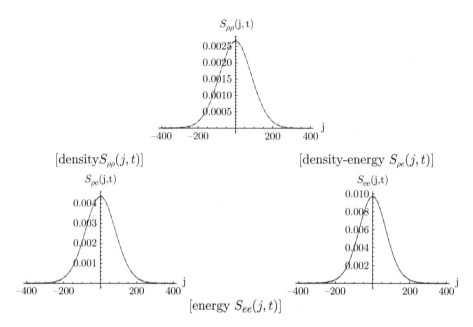

[density $S_{\rho\rho}(j,t)$]   [density-energy $S_{\rho e}(j,t)$]

[energy $S_{ee}(j,t)$]

Fig. 4.4 Equilibrium time correlations $S_{\alpha\alpha'}(j,t)$ of density and energy for the discrete NLS with $g = 1$. The initial states are drawn from the canonical ensemble (4.86) with inverse temperature $\beta = 1$ and $\langle\rho_j\rangle = 1$. The black dashed curves are the entries on the right of Eq. (4.117).

Both density and energy are even under time reversal, which explains that cross correlations are permitted. The Gaussian fit is essentially perfect, compare with Fig. 4.4. D is measured at the longest available time $t = 1536$. However, for $\beta = 15$ the behavior is drastically different, as confirmed by Fig. 4.5, which shows a right-moving sound peak broadening as $t^{2/3}$. The heat peak has in comparison a much smaller amplitude and is not resolved. For a theoretical explanation, we will have to wait until Sec. 4.15.

We return to anharmonic chains. Based on (4.107) one computes the stationary space-time covariance, which most easily is written in Fourier space,

$$S_{\alpha\alpha'}(x,t) = \langle u_\alpha(x,t)u_{\alpha'}(0,0)\rangle = \int_{\mathbb{R}} dk\, e^{i2\pi kx}\left(e^{-it2\pi kA-|t|(2\pi k)^2 D}C\right)_{\alpha\alpha'}. \quad (4.119)$$

To extract the long-time behavior it is convenient to transform to normal modes. But before, we must introduce a more systematic notation. We will use throughout the superscript $^\sharp$ for a normal mode quantity. Thus, for the anharmonic chain

$$S^\sharp(j,t) = RS(j,t)R^{\mathrm{T}}, \quad S^\sharp_{\alpha\alpha'}(j,t) = \langle(R\vec{g})_\alpha(j,t);(R\vec{g})_{\alpha'}(0,0)\rangle_{P,\beta}. \quad (4.120)$$

The hydrodynamic fluctuation fields are defined on the continuum, thus functions of x,t, and we write

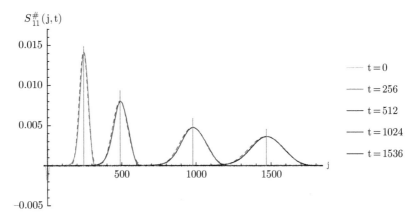

Fig. 4.5 Time sequence of the right moving sound peak, $S_{11}^{\sharp}(j,t)$, of the discrete NLS with parameters $N = 4096$, $g = 1$, $\beta = 15$, and $\langle \rho_j \rangle = 1$ corresponding to $\mu = 1.025$. The vertical lines show the theoretically predicted sound speed c and the dashed lines are the stationary KPZ scaling functions.

$$S_{\alpha\alpha'}(x,t) = \langle u_\alpha(x,t)u_{\alpha'}(0,0)\rangle, \quad S^{\sharp}(x,t) = RS(x,t)R^{\mathrm{T}}. \tag{4.121}$$

Correspondingly, $A^{\sharp} = RAR^{-1} = \mathrm{diag}(-c,0,c)$, $D^{\sharp} = RDR^{-1}$, $B^{\sharp} = RB$. Note that $\vec{u}(x,t)$ will change its meaning when switching from linear to nonlinear fluctuating hydrodynamics.

In normal mode representation Eq. (4.121) becomes

$$S^{\sharp}(x,t) = \int_{\mathbb{R}} \mathrm{d}k\, \mathrm{e}^{\mathrm{i}2\pi kx}\mathrm{e}^{-\mathrm{i}t2\pi kA^{\sharp}-|t|(2\pi k)^2 D^{\sharp}}. \tag{4.122}$$

The leading term, $\mathrm{i}t2\pi kA^{\sharp}$, is diagonal, while the diffusion matrix $D^{\sharp}$ couples the components. But for large t the peaks are far apart and the cross terms become small. More formally we split $D^{\sharp} = D_{\mathrm{dia}} + D_{\mathrm{off}}$ and regard the off-diagonal part D_{off} as perturbation. When expanding, one notes that the off-diagonal terms carry an oscillating factor with frequency $c_\alpha - c_{\alpha'}$, $\alpha \neq \alpha'$. Hence, these terms decay quickly and

$$S_{\alpha\alpha'}^{\sharp}(x,t) \simeq \delta_{\alpha\alpha'} \int_{\mathbb{R}} \mathrm{d}k\, \mathrm{e}^{\mathrm{i}2\pi kx}\mathrm{e}^{-\mathrm{i}t2\pi kc_\alpha-|t|(2\pi k)^2 D_{\alpha\alpha}^{\sharp}} \tag{4.123}$$

for large t. Each peak has a Gaussian shape function which broadens as $(D_{\alpha\alpha}^{\sharp}|t|)^{1/2}$.

Besides the peak structure, we have gained a second important insight. Since the peaks travel with distinct velocities, on the linearized level the three-component system decouples into three scalar equations, provided it is written in normal modes. The two sound peaks are mirror images of each other and broaden with $D_{11}^{\sharp} = D_{-1-1}^{\sharp}$ and the heat peak broadens with $D_{00}^{\sharp}$.

4.13 Second-order expansion, nonlinear fluctuating hydrodynamics

From the experience with the scalar case, we know that at least the second-order expansion of the Euler currents must be included. In this case the stationary measure is spatial white noise and the quadratic nonlinearity yields the dynamic exponent $z = \frac{3}{2}$. Thus, we retain dissipation and noise in (4.107), but expand the Euler currents of (4.94) beyond first order to now include second order in $\vec{u}$, which turns (4.107) into the equations of nonlinear fluctuating hydrodynamics,

$$\partial_t u_1 - \partial_x u_2 = 0, \tag{4.124}$$

$$\partial_t u_2 + \partial_x \big((\partial_\ell P) u_1 + (\partial_e P) u_3 + \tfrac{1}{2} (\partial_\ell^2 P) u_1^2$$
$$- \tfrac{1}{2} (\partial_e P) u_2^2 + \tfrac{1}{2} (\partial_e^2 P) u_3^2 + (\partial_\ell \partial_e P) u_1 u_3 - D_u \partial_x u_2 + \sigma_u \xi_2 \big) = 0,$$

$$\partial_t u_3 + \partial_x \big(P u_2 + (\partial_\ell P) u_1 u_2 + (\partial_e P) u_2 u_3 - \tilde{D}_e \partial_x u_1 - D_e \partial_x u_3 + \sigma_e \xi_3 \big) = 0.$$

To explore their consequences is a more demanding task than solving the linear Langevin equation and the results of the analysis will be more fragmentary.

To proceed further, it is convenient to write (4.124) in vector form,

$$\partial_t \vec{u}(x,t) + \partial_x \big(A\vec{u}(x,t) + \tfrac{1}{2} \langle \vec{u}, \vec{H}\vec{u} \rangle - \partial_x D\vec{u}(x,t) + B\vec{\xi}(x,t) \big) = 0, \tag{4.125}$$

where $\vec{H}$ is the vector consisting of the Hessians of the currents with derivatives evaluated at the background values $(\ell, 0, e)$,

$$H^\alpha_{\gamma\gamma'} = \partial_{u_\gamma} \partial_{u_{\gamma'}} j_\alpha, \qquad \langle \vec{u}, \vec{H}\vec{u} \rangle = \sum_{\gamma,\gamma'=1}^{3} \vec{H}_{\gamma\gamma'} u_\gamma u_{\gamma'}. \tag{4.126}$$

As for the linear Langevin equation we transform to normal modes through

$$\vec{\phi} = R\vec{u}. \tag{4.127}$$

Then

$$\partial_t \phi_\alpha + \partial_x \big(c_\alpha \phi_\alpha + \langle \vec{\phi}, G^\alpha \vec{\phi} \rangle - \partial_x (D^\sharp \vec{\phi})_\alpha + (B^\sharp \vec{\xi})_\alpha \big) = 0 \tag{4.128}$$

with $D^\sharp = RDR^{-1}$, $B^\sharp = RB$. By construction $B^\sharp B^{\sharp\mathrm{T}} = 2D^\sharp$. The nonlinear coupling constants, denoted by $\vec{G}$, are defined by

$$G^\alpha = \tfrac{1}{2} \sum_{\alpha'=1}^{3} R_{\alpha\alpha'} R^{-\mathrm{T}} H^{\alpha'} R^{-1}, \tag{4.129}$$

with the notation $R^{-\mathrm{T}} = (R^{-1})^{\mathrm{T}}$.

Since derived from a chain, the couplings are not completely arbitrary, but satisfy the symmetries

$$G^\alpha_{\beta\gamma} = G^\alpha_{\gamma\beta}, \qquad G^\sigma_{\alpha\beta} = -G^{-\sigma}_{-\alpha-\beta}, \qquad G^\sigma_{-10} = G^\sigma_{01},$$

$$G^0_{\sigma\sigma} = -G^0_{-\sigma-\sigma}, \quad G^0_{\alpha\beta} = 0 \quad \text{otherwise.} \qquad (4.130)$$

In particular note that

$$G^0_{00} = 0 \qquad (4.131)$$

always, while $G^1_{11} = -G^{-1}_{-1-1}$ are generically different from 0. This property signals that the heat peak will behave differently from the sound peaks. The $\vec{G}$-couplings are listed in Spohn (2014) and as a function of P, β expressed in cumulants up to third order in r_0, V_0. The algebra is somewhat messy. But there is a short MATHEMATICA program available (Mendl 2013) which, for given P, β, V, computes all coupling constants G, including the matrices C, A, R.

4.14   Coupling coefficients, dynamical phase diagram

The coupling coefficients, $\vec{G}$, determine the long-time behavior of the correlations of the conserved fields. A more quantitative treatment will follow, but I first want to discuss some general features independent of the specific underlying microscopic model. For a given class of models, one can change the model parameters, which are either thermodynamic, like pressure, or mechanical, like a coefficient in the potential. For the purpose of the discussion, I call both model parameters. The coupling coefficients are complicated functions of the model parameters. So dynamical phase diagrams mean the dynamical properties in dependence on model parameters as mediated through the $\vec{G}$-couplings.

We emphasize that the $\vec{G}$-couplings cannot distinguish whether the microscopic dynamics are integrable. For example, the $\vec{G}$-couplings of the Toda lattice are not much different from nearby nonintegrable chains. The assumption of chaotic microscopic dynamics is used already when writing down the Euler equations.

Our discussion assumes that the peak velocities are distinct, $c_\alpha \neq c_{\alpha'}$ for $\alpha \neq \alpha'$. The degenerate case still remains to be studied. One must distinguish three types of couplings:

— self-couplings: $G^\alpha_{\alpha\alpha}$,
— diagonal couplings, but not self: $G^\alpha_{\alpha'\alpha'}$, $\alpha \neq \alpha'$,
— off-diagonal couplings: $G^\alpha_{\alpha'\alpha''}$, $\alpha' \neq \alpha''$.

The off-diagonal couplings are irrelevant, in the sense that they do not show in the long-time behavior. Of course, they may strongly modify the short-term behavior. The rough argument is that for long times the peaks α' and α'' have spatially a very small overlap and therefore feed back little to the mode α. On the other hand, if $\alpha' = \alpha''$, then the two peaks are on top of each other, and the peak α can still interact with them through having a slow spatial decay.

The dynamical phase diagram is characterized by the vanishing of some relevant couplings $G^\alpha_{\alpha'\alpha'}$. For $n = 3$, there are 9 relevant couplings and hence 2^9 dynamical phases, at least in principle. Many of them will have coinciding properties and do not need to be distinguished. Within a given class of models generically not all phases are

actually realized. In fact, anharmonic chains have only three dynamical phases and the DNLS has only one, even at low temperatures. It may happen, within a given class of models, that $G^\alpha_{\alpha'\alpha'}$ either has a definite sign or vanishes identically. Thereby, the richness of the phase diagram is strongly reduced. From a theoretical perspective, the most interesting case is a coupling $G^\alpha_{\alpha'\alpha'}$ with no definite sign. Then the model parameters can be changed so to make $G^\alpha_{\alpha'\alpha'} = 0$. This could be so because of an additional symmetry. But it could simply be an accidental zero for a function on a high-dimensional parameter space. Once such a zero is located, the theory predicts that at this zero the dynamical properties change dramatically, so to speak out of the blue, because the parameters as such do not look very different from neighboring parameter values. Such a qualitative change is often more easily detectable, in comparison to a precise measurement of dynamical scaling exponents.

Anharmonic chains have special symmetries and not all possible couplings $\vec{G}$ can be realized. Listing the relevant parameters one has only four distinct parameters, $a_1, \ldots, a_4$,

$$\mathrm{diag}(G^{-1}) = (-a_3, -a_2, -a_1), \quad \mathrm{diag}(G^0) = (-a_4, 0, a_4), \quad \mathrm{diag}(G^1) = (a_1, a_2, a_3).$$
(4.132)

From the explicit expression of a_4 one concludes that $a_4 > 0$ and only three parameters remain. Their variation leads to three distinct phases,

(Phase 1) $G^1_{11} = -G^{-1}_{-1-1} \neq 0$. This is the generic phase. According to mode-coupling theory, to be discussed later, the two sound peaks broaden as KPZ, while the heat peak scales as symmetric Lévy $\frac{5}{3}$.

(Phase 2) $G^1_{11} = 0$ and $a_1 = 0, a_2 = 0$. Then the sound peaks decouple and broaden diffusively, while the heat peak scales as symmetric Lévy $\frac{3}{2}$.

(Phase 3) $G^1_{11} = 0$ and $a_1 \neq 0, a_2 \neq 0$ or $a_1 \neq 0, a_2 = 0$ or $a_1 = 0, a_2 \neq 0$. Since $G^0_{11} = -G^0_{-1-1} > 0$, all three peaks are cross-coupled. This leads to a heat peak which scales with symmetric Lévy κ, a left sound peak which scales as maximally left asymmetric Lévy κ, and a right sound peak which scales as maximally right asymmetric Lévy κ. Here κ is the golden mean, $\kappa = \frac{1}{2}(1 + \sqrt{5})$.

To realize Phase 2 one starts from the observation that the $\vec{G}$ coefficients are expressed through cumulants in r_0, V_0. If the integrands are antisymmetric under reflection, many terms vanish. The precise condition on the potential is to have some a_0, P_0 such that

$$V(x - a_0) + P_0 x = V(-x - a_0) - P_0 x$$
(4.133)

for all x. Then for $P = P_0$ and arbitrary β, one finds that

$$G^1_{11} = 0, \quad G^1_{-1-1} = -G^{-1}_{11} = 0, \quad G^1_{00} = -G^{-1}_{00} = 0,$$
(4.134)

while $\sigma G^0_{\sigma\sigma} > 0$. The standard examples for (4.133) to hold are a FPU chain with no cubic interaction term, the β-chain, and the square well potential with alternating masses, both at zero pressure.

Two examples of accidental zeros of G_{11}^1 have been found recently for the asymmetric FPU potential $V(x) = \frac{1}{3}ax^3 + \frac{1}{4}x^4$ (Lee-Dadswell 2015). The first example is $\beta = 1$, $P = 0.59$, $a = -2$, and the second one $\beta = 1$, $P = -0.5$, $a = 1.89$. In fact, the signature of the $\vec{G}$ matrices is identical to an even potential at $P = 0$. Note that, in contrast to the even potential, also the value of β is fixed.

The cross-coupling leading to Phase 3 is discussed in Spohn and Stoltz (2015) and Popkov et al. (2015). An explicit example has been discovered only during the Les Houches summer school (Mendl 2015) and reads

$$V(x) = \tfrac{1}{2}x^2 + \cos(\pi(x - \tfrac{1}{3})) + \tfrac{1}{8}x^4. \tag{4.135}$$

Then $G_{11}^1 = 10^{-8}$, $G_{-1-1}^1 = 0.164$, and $G_{00}^1 = 0.272$ at $P = 2.214$ and $\beta = 1$.

4.15   Low-temperature DNLS

We return to the DNLS introduced in Sect. 4.10. At high temperatures the conserved fields, density and energy, spread diffusively. However, at low temperatures the field of phase differences is almost conserved, which allows for propagating modes. According to the equilibrium measure, for $\beta \gg 1$ and $\langle \rho_j \rangle = 1$, the phase is similar to a discrete time random walk on S^1 with single-step variance of order β^{-1}, until the walk realizes the finite geometry and the variance crosses over to exponential decay. In one dimension there is no static phase transition, but in three and more dimensions a continuous symmetry can be broken. Then the field corresponding to such a broken symmetry must be added to the list of conserved fields (Forster 1975). In a similar spirit, for DNLS at low temperatures the phase difference must be included in the list of conserved fields.

We adjust the chemical potential μ such that the average density $\bar{\rho}$ is fixed, $\bar{\rho} > 0$. As one lowers the temperature, according to the equilibrium measure, the ρ_j's deviate from $\bar{\rho}$ by order $1/\sqrt{\beta}$ and also the phase difference $|\varphi_{j+1} - \varphi_j| = \mathcal{O}(1/\sqrt{\beta})$, while the phase itself is uniformly distributed over $[0, 2\pi]$. We introduce $\tilde{r}_j = \Theta(\varphi_{j+1} - \varphi_j)$, where Θ is 2π-periodic and $\Theta(x) = x$ for $|x| \leq \pi$. Since Θ has a jump discontinuity, $\tilde{r}_j$ is not conserved. In a more pictorial language, the event that $|\varphi_{j+1}(t) - \varphi_j(t)| = \pi$ is called an umklapp for phase difference $\tilde{r}_j$ or an umklapp process to emphasize its dynamical character. At low temperatures a jump of size π has a small probability of order $e^{-\beta\Delta V}$ with ΔV the height of a suitable potential barrier still to be determined. Hence, $\tilde{r}_j$ is locally conserved up to umklapp processes occurring with a very small frequency only; see Das and Dhar (2014) for a numerical validation.

How to incorporate an almost conserved field into fluctuation hydrodynamics is not so obvious. Therefore, we first construct an effective low-temperature Hamiltonian H_{lt}, in such a way that it has the same equilibrium measure and strictly conserves $\tilde{r}_j$. More concretely, we first parametrize the angles $\varphi_0, \ldots, \varphi_{N-1}$ through $r_j = \varphi_{j+1} - \varphi_j$ with $r_j \in [-\pi, \pi]$. To distinguish, we denote the angles in this particular parametrization by ϕ_j. (ϕ_j, ρ_j) is a pair of canonically conjugate variables. Umklapp is defined by $|r_j(t)| = \pi$. The dynamics governed by the DNLS Hamiltonian corresponds to periodic boundary conditions at $r_j = \pm\pi$. For an approximate low-temperature description we

impose instead specular reflection; i.e., if $r_j = \pm\pi$, then ρ_j, ρ_{j+1} are scattered to $\rho'_j = \rho_{j+1}$, $\rho'_{j+1} = \rho_j$. By fiat, the approximate low-temperature dynamics stricly conserves the phase difference r_j and is identical to the DNLS dynamics between two umklapp events. The corresponding Hamiltonian then reads

$$H_{\mathrm{lt}} = \sum_{j=0}^{N-1} \left(\sqrt{\rho_{j+1}\,\rho_j}\, U(\phi_{j+1} - \phi_j) + V(\rho_j) \right), \tag{4.136}$$

where

$$U(x) = -\cos(x) \;\; \text{for} \;\; |x| \le \pi, \qquad U(x) = \infty \;\; \text{for} \;\; |x| > \pi, \tag{4.137}$$

and

$$V(x) = x + \tfrac{1}{2} g\, x^2 \;\; \text{for} \;\; x \ge 0, \qquad V(x) = \infty \;\; \text{for} \;\; x < 0. \tag{4.138}$$

As required before, the Boltzmann weights are not modified, $\exp[-\beta H] = \exp[-\beta H_{\mathrm{lt}}]$. For some computations it will be convenient to replace the hard collision potentials U, V by a smooth variant, for which the infinite step is replaced by a rapidly diverging smooth potential. The dynamics is governed by

$$\frac{\mathrm{d}}{\mathrm{d}t}\phi_j = -\partial_{\rho_j} H_{\mathrm{lt}}, \qquad \frac{\mathrm{d}}{\mathrm{d}t}\rho_j = \partial_{\phi_j} H_{\mathrm{lt}}, \tag{4.139}$$

including the specular reflection of ρ_j at $\rho_j = 0$ and of r_j at $r_j = \pm\pi$. The solution trajectory agrees piecewise with the trajectory from the DNLS dynamics. The break points are the umklapp events, which are extremely rare at low temperatures.

There are two potential barriers, ΔU and ΔV. The minimum of $V(x) - x - \mu x$, $\mu > 0$, is at $\bar\rho = \mu/g$, hence $\Delta V = \tfrac{1}{2} g\bar\rho^2$. The minimum of U is at $\phi_{j+1} - \phi_j = 0$ and, setting $\rho_j = \bar\rho$, one arrives at $\Delta U = 2\bar\rho$. Thus the low-temperature regime is characterized by

$$\tfrac{1}{2}\beta g\bar\rho^2 \gtrsim 1, \qquad 2\beta\bar\rho \gtrsim 1. \tag{4.140}$$

In this parameter regime we expect the equilibrium time correlations based on H_{lt} to well approximate the time correlations of the exact DNLS.

The conserved fields are now ρ_j, r_j, and the energy

$$e_j = \sqrt{\rho_{j+1}\,\rho_j}\, U(r_j) + V(\rho_j). \tag{4.141}$$

This local energy differs from the one introduced in (4.83) by the term $\tfrac{1}{2}(\rho_{j+1} - \rho_j)$. In the expressions that follow such a difference term drops out and in the final result we could use either one. The local conservation laws and their currents read, for the density

$$\frac{\mathrm{d}}{\mathrm{d}t}\rho_j + \mathcal{J}_{\rho,j+1} - \mathcal{J}_{\rho,j} = 0, \tag{4.142}$$

with local density current

$$\mathcal{J}_{\rho,j} = \sqrt{\rho_{j-1}\,\rho_j}\,U'(r_{j-1})\,, \tag{4.143}$$

for the phase difference

$$\frac{\mathrm{d}}{\mathrm{d}t}r_j + \mathcal{J}_{r,j+1} - \mathcal{J}_{r,j} = 0, \tag{4.144}$$

with local phase difference current

$$\mathcal{J}_{r,j} = \tfrac{1}{2}\sqrt{\rho_{j+1}/\rho_j}\,U(r_j) + \tfrac{1}{2}\sqrt{\rho_{j-1}/\rho_j}\,U(r_{j-1}) + V'(\rho_j)\,, \tag{4.145}$$

and for the energy

$$\frac{\mathrm{d}}{\mathrm{d}t}e_j + \mathcal{J}_{e,j+1} - \mathcal{J}_{e,j} = 0, \tag{4.146}$$

with local energy current

$$\mathcal{J}_{e,j} = \tfrac{1}{2}\sqrt{\rho_{j-1}\,\rho_{j+1}}\left(U(r_{j-1})U'(r_j) + U'(r_{j-1})U(r_j)\right) + \sqrt{\rho_{j-1}\,\rho_j}\,U'(r_{j-1})V'(\rho_j)\,. \tag{4.147}$$

To shorten notation, we set $\vec{g}_j = (\rho_j, r_j, e_j)$ and $\vec{\mathcal{J}}_j = (\mathcal{J}_{\rho,j}, \mathcal{J}_{r,j}, \mathcal{J}_{e,j})$.

We follow the blueprint provided by the anharmonic chains. The Euler currents are determined as the thermal average of the microscopic currents. The canonical state has as parameters β, μ and, in addition, a second chemical potential, ν, which sets the average phase difference $\langle r_j \rangle$. Physically, the phase difference cannot be controlled; hence $\nu = 0$. However, for the second-order expansion one must first take derivatives and then set $\nu = 0$. At the first sight it looks difficult to draw any conclusions on the coupling matrices $\vec{G}$. There is, however, a surprising identity which comes for rescue,

$$\langle \vec{\mathcal{J}}_j \rangle_{\beta,\mu,\nu} = \langle (\mathcal{J}_{\rho,j}, \mathcal{J}_{r,j}, \mathcal{J}_{e,j}) \rangle_{\beta,\mu,\nu} = (\nu, \mu, \mu\,\nu) = \vec{\mathsf{j}}, \tag{4.148}$$

which should be compared with (4.94). A tricky computation is still ahead. But at the end one finds that $G^0_{00} = 0$ and $G^1_{11} > 0$, while G^1 equals $-G^{-1}$ transposed relative to the anti-diagonal.

Of course, the equations of nonlinear fluctuating hydrodynamics have the same structure as explained before. Thus, we conclude that the dynamical phase of low-temperature DNLS is identical to the generic Phase 1 of an anharmonic chain.

4.16   Mode-coupling theory

4.16.1   Decoupling hypothesis

For the linear Langevin equations the normal modes decouple for long times. As first argued by van Beijeren (2012) such decoupling persists when adding the quadratic nonlinearities. For the precise phrasing, we must be somewhat careful. We consider a

fixed component, α, in normal mode representation. It travels with velocity c_α, which is assumed to be distinct from all other mode velocities. If $G_{\alpha\alpha}^\alpha \neq 0$, then for the purpose of computing correlations of mode α at large scales, one can use the scalar conservation law

$$\partial_t \phi_\alpha + \partial_x \left(c_\alpha \phi_\alpha + G_{\alpha\alpha}^\alpha \phi_\alpha^2 - D_{\alpha\alpha}^\sharp \partial_x \phi_\alpha + B_{\alpha\alpha}^\sharp \xi_\alpha \right) = 0, \tag{4.149}$$

which coincides with the stochastic Burgers equation (4.27). If decoupling holds, one has the exact asymptotics as stated in (4.31) with $\lambda = 2G_{\alpha\alpha}^\alpha$.

As discussed in Sect. 4.14, for a generic anharmonic chain $G_{11}^1 = -G_{-1-1}^{-1} \neq 0$. Hence, if $G_{11}^1 \neq 0$, the decoupling hypothesis asserts that the *exact* scaling form of the sound peak is

$$S_{\sigma\sigma}^\sharp(x, t) \simeq 2^{-1}(\Gamma_s t)^{-2/3} f_{\text{KPZ}} \left(2^{-1}(\Gamma_s t)^{-2/3}(x - \sigma c t) \right), \quad \Gamma_s = |G_{\sigma\sigma}^\sigma|, \tag{4.150}$$

$\sigma = \pm 1$. On the other hand $G_{00}^0 = 0$ always. To find out about the scaling behavior of the heat peak, other methods must be developed.

4.16.2 One-loop, diagonal, and small overlap approximations

The Langevin equation (4.128) is slightly formal. To have a well-defined evolution, we discretize space by a lattice of N sites. The field $\vec{\phi}(x, t)$ then becomes $\vec{\phi}_j(t)$ with components $\phi_{j,\alpha}(t)$, $j = 1, \ldots, N$, $\alpha = 0, \pm 1$. The spatial finite difference operator is denoted by ∂_j, $\partial_j f_j = f_{j+1} - f_j$, with transpose $\partial_j^{\mathrm{T}} f_j = f_{j-1} - f_j$. Then the discretized equations of fluctuating hydrodynamics read

$$\partial_t \phi_{j,\alpha} + \partial_j \left(c_\alpha \phi_{j,\alpha} + \mathcal{N}_{j,\alpha} + \partial_j^{\mathrm{T}} D^\sharp \phi_{j,\alpha} + B^\sharp \xi_{j,\alpha} \right) = 0, \tag{4.151}$$

with $\vec{\phi}_j = \vec{\phi}_{N+j}$, $\vec{\xi}_0 = \vec{\xi}_N$, where $\xi_{j,\alpha}$ are independent Gaussian white noises with covariance

$$\langle \xi_{j,\alpha}(t) \xi_{j',\alpha'}(t') \rangle = \delta_{jj'} \delta_{\alpha\alpha'} \delta(t - t'). \tag{4.152}$$

The diffusion matrix $D^\sharp$ and noise strength $B^\sharp$ act on components, while the difference operator ∂_j acts on the lattice site index j.

$\mathcal{N}_{j,\alpha}$ is quadratic in ϕ. But let us first consider the case $\mathcal{N}_{j,\alpha} = 0$. Then $\phi_{j,\alpha}(t)$ is a Gaussian process. The noise strength has been chosen such that one invariant measure is the Gaussian

$$\prod_{j=1}^N \prod_{\alpha=0,\pm 1} \exp\left[-\tfrac{1}{2}\phi_{j,\alpha}^2\right](2\pi)^{-1/2} d\phi_{j,\alpha} = \rho_{\mathrm{G}}(\phi) \prod_{j=1}^N \prod_{\alpha=0,\pm 1} d\phi_{j,\alpha}. \tag{4.153}$$

Because of the conservation laws, the hyperplanes

$$\sum_{j=1}^N \phi_{j,\alpha} = N\rho_\alpha, \tag{4.154}$$

are invariant and on each hyperplane there is a Gaussian process with a unique invariant measure given by (4.153) conditioned on that hyperplane. For large N it would become independent Gaussians with mean ρ_α, our interest being the case of zero mean, $\rho_\alpha = 0$.

The generator of the diffusion process (4.151) with $\mathcal{N}_{j,\alpha} = 0$ is given by

$$
L_0 = \sum_{j=1}^{N} \left(- \sum_{\alpha=0,\pm 1} \partial_j \left(c_\alpha \phi_{j,\alpha} + \partial_j^{\mathrm{T}} D^\sharp \phi_{j,\alpha} \right) \partial_{\phi_{j,\alpha}} \right.
$$

$$
\left. + \sum_{\alpha,\alpha'=0,\pm 1} (B^\sharp B^{\sharp \mathrm{T}})_{\alpha\alpha'} \partial_j \partial_{\phi_{j,\alpha}} \partial_j \partial_{\phi_{j,\alpha'}} \right). \tag{4.155}
$$

The invariance of $\rho_{\mathrm{G}}(\phi)$ can be checked through

$$
L_0^* \rho_{\mathrm{G}}(\phi) = 0 \,, \tag{4.156}
$$

where * is the adjoint with respect to the flat volume measure. Furthermore, linear functions evolve to linear functions according to

$$
\mathrm{e}^{L_0 t} \phi_{j,\alpha} = \sum_{j'=1}^{N} \sum_{\alpha'=0,\pm 1} (\mathrm{e}^{At})_{j\alpha,j'\alpha'} \phi_{j',\alpha'} \,, \tag{4.157}
$$

where the matrix $\mathcal{A} = -\partial_j \otimes \mathrm{diag}(-c,0,c) - \partial_j \partial_j^{\mathrm{T}} \otimes D^\sharp$, the first factor acting on j and the second on α.

We now add the nonlinearity $\mathcal{N}_{j,\alpha}$. In general, this will modify the time-stationary measure and we have little control on how. Therefore, we propose to choose $\mathcal{N}_{j,\alpha}$ such that the corresponding vector field $\partial_j \mathcal{N}_{j,\alpha}$ is divergence free (Sasamoto and Spohn 2009). If $\mathcal{N}_{j,\alpha}$ depends only on the field at sites j and $j+1$, then the unique solution reads

$$
\mathcal{N}_{j,\alpha} = \tfrac{1}{3} \sum_{\gamma,\gamma'=0,\pm 1} G^\alpha_{\gamma\gamma'} \left(\phi_{j,\gamma} \phi_{j,\gamma'} + \phi_{j,\gamma} \phi_{j+1,\gamma'} + \phi_{j+1,\gamma} \phi_{j+1,\gamma'} \right). \tag{4.158}
$$

For ρ_{G} to remain invariant under the deterministic flow generated by the vector field $-\partial_j \mathcal{N}$ requires that

$$
L_1^* \rho_{\mathrm{G}} = 0 \,, \qquad L_1 = -\sum_{j=1}^{N} \sum_{\alpha=0,\pm 1} \partial_j \mathcal{N}_{j,\alpha} \partial_{\phi_{j,\alpha}} \,, \tag{4.159}
$$

which implies that

$$
\sum_{j=1}^{N} \sum_{\alpha=0,\pm 1} \phi_{j,\alpha} \partial_j \mathcal{N}_{j,\alpha} = 0 \tag{4.160}
$$

and thus the cyclicity constraints

$$
G^\alpha_{\beta\gamma} = G^\beta_{\alpha\gamma} \left(= G^\alpha_{\gamma\beta} \right) \tag{4.161}
$$

for all $\alpha, \beta, \gamma = 1, 2, 3$, where in brackets we added the symmetry which holds by definition. Denoting the generator of the Langevin equation (4.151) by

$$L = L_0 + L_1 , \qquad (4.162)$$

one concludes $L^* \rho_G = 0$, i.e., the time-invariance of ρ_G. The cyclicity condition also appears for coupled KPZ equations (Funaki 2015).

Unfortunately, the cyclicity condition fails for anharmonic chains. On the other hand, to derive the mode-coupling equations the explicit form of the stationary measure seems to be required. To avoid a deadlock we argue by universality. First, the relevant couplings $G^\alpha_{\alpha'\alpha'}$ are computed from the particular anharmonic chain. We then fix all other couplings by cyclicity. Of course, the true irrelevant couplings are different. But this will not show in the long-time behavior.

We return to the Langevin equation (4.151) and consider the mean zero, stationary $\phi_{j,\alpha}(t)$ process with ρ_G as $t = 0$ measure. The stationary covariance reads

$$S^\sharp_{\alpha\alpha'}(j, t) = \langle \phi_{j,\alpha}(t)\phi_{0,\alpha'}(0) \rangle = \langle \phi_{0,\alpha'} e^{Lt}\phi_{j,\alpha} \rangle_{\mathrm{eq}}, \quad t \geq 0 . \qquad (4.163)$$

On the left, $\langle \cdot \rangle$ denotes the average with respect to the stationary $\phi_{j,\alpha}(t)$ process and on the right $\langle \cdot \rangle_{\mathrm{eq}}$ refers to the average with respect to ρ_G. By construction

$$S^\sharp_{\alpha\alpha'}(j, 0) = \delta_{\alpha\alpha'}\delta_{j0} . \qquad (4.164)$$

The time derivative reads

$$\frac{\mathrm{d}}{\mathrm{d}t} S^\sharp_{\alpha\alpha'}(j, t) = \langle \phi_{0,\alpha'}(e^{Lt}L_0\phi_{j,\alpha}) \rangle_{\mathrm{eq}} + \langle \phi_{0,\alpha'}(e^{Lt}L_1\phi_{j,\alpha}) \rangle_{\mathrm{eq}} . \qquad (4.165)$$

We insert

$$e^{Lt} = e^{L_0 t} + \int_0^t \mathrm{d}s\, e^{L_0(t-s)} L_1 e^{Ls} \qquad (4.166)$$

in the second summand of (4.165). The term containing only $e^{L_0 t}$ is cubic in the time zero fields and hence its average vanishes. Therefore one arrives at

$$\frac{\mathrm{d}}{\mathrm{d}t} S^\sharp_{\alpha\alpha'}(j, t) = A S_{\alpha\alpha'}(j, t) + \int_0^t \mathrm{d}s \langle \phi_{0,\alpha'} e^{L_0(t-s)} L_1(e^{Ls} L_1 \phi_{j,\alpha}) \rangle_{\mathrm{eq}} . \qquad (4.167)$$

For the adjoint of $e^{L_0(t-s)}$ we use (4.157) and for the adjoint of L_1 we use

$$\langle \phi_{j,\alpha} L_1 F(\phi) \rangle_{\mathrm{eq}} = -\langle (L_1 \phi_{j,\alpha}) F(\phi) \rangle_{\mathrm{eq}} , \qquad (4.168)$$

which both rely on $\langle \cdot \rangle_{\mathrm{eq}}$ being the average with respect to ρ_G. Furthermore,

$$L_1 \phi_{j,\alpha} = -\partial_j \mathcal{N}_{j,\alpha} . \qquad (4.169)$$

Inserting in (4.167) one arrives at the identity

$$\frac{d}{dt}S_{\alpha\alpha'}^{\sharp}(j,t) = AS_{\alpha\alpha'}^{\sharp}(j,t) - \int_0^t ds\langle(e^{A^{\mathrm{T}}(t-s)}\partial_j\mathcal{N}_{0,\alpha'})(e^{Ls}\partial_j\mathcal{N}_{j,\alpha})\rangle_{\mathrm{eq}}. \tag{4.170}$$

To obtain a closed equation for $S^{\sharp}$ we note that the average

$$\langle\partial_{j'}\mathcal{N}_{j',\alpha'}e^{Ls}\partial_j\mathcal{N}_{j,\alpha}\rangle_{\mathrm{eq}} = \langle\partial_j\mathcal{N}_{j,\alpha}(s)\partial_{j'}\mathcal{N}_{j',\alpha'}(0)\rangle \tag{4.171}$$

is a four-point correlation. We invoke the Gaussian factorization as

$$\langle\phi(s)\phi(s)\phi(0)\phi(0)\rangle \cong \langle\phi(s)\phi(s)\rangle\langle\phi(0)\phi(0)\rangle + 2\langle\phi(s)\phi(0)\rangle\langle\phi(s)\phi(0)\rangle. \tag{4.172}$$

The first summand vanishes because of the difference operator ∂_j. Secondly, we replace the bare propagator $e^{A(t-s)}$ by the interacting propagator $S^{\sharp}(t-s)$, which corresponds to a partial resummation of the perturbation series in $\vec{G}$. Finally, we take a limit of zero lattice spacing. This step could be avoided, and is done so in our numerical scheme for the mode-coupling equations. We could also maintain the ring geometry which, for example, would make it possible to investigate collisions between the moving peaks. Universality is only expected for large j, t, hence in the limit of zero lattice spacing. The continuum limit of $S^{\sharp}(j,t)$ is denoted by $S^{\sharp}(x,t)$, $x \in \mathbb{R}$. With these steps we arrive at the mode-coupling equation

$$\partial_t S_{\alpha\beta}^{\sharp}(x,t) = \sum_{\alpha'=0,\pm1}\left(\left(-c_\alpha\delta_{\alpha\alpha'}\partial_x + D_{\alpha\alpha'}\partial_x^2\right)S_{\alpha'\beta}^{\sharp}(x,t)\right.$$

$$\left. + \int_0^t ds\int_{\mathbb{R}}dyM_{\alpha\alpha'}(y,s)\partial_x^2 S_{\alpha'\beta}^{\sharp}(x-y,t-s)\right) \tag{4.173}$$

with the memory kernel

$$M_{\alpha\alpha'}(x,t) = 2\sum_{\beta',\beta'',\gamma',\gamma''=0,\pm1}G_{\beta'\gamma'}^{\alpha}G_{\beta''\gamma''}^{\alpha'}S_{\beta'\beta''}^{\sharp}(x,t)S_{\gamma'\gamma''}^{\sharp}(x,t). \tag{4.174}$$

In numerical simulations of both, the mechanical model of anharmonic chains and the mode-coupling equations, it is consistently observed that $S_{\alpha\alpha'}^{\sharp}(j,t)$ becomes approximately diagonal fairly rapidly. To analyze the long-time asymptotics on the basis of (4.173) we, therefore, rely on the diagonal approximation

$$S_{\alpha\alpha'}^{\sharp}(x,t) \simeq \delta_{\alpha\alpha'}f_\alpha(x,t). \tag{4.175}$$

Then $f_\alpha(x,0) = \delta(x)$ and the f_α's satisfy

$$\partial_t f_\alpha(x,t) = (-c_\alpha\partial_x + D_{\alpha\alpha}^{\sharp}\partial_x^2)f_\alpha(x,t) + \int_0^t ds\int_{\mathbb{R}}dy\partial_x^2 f_\alpha(x-y,t-s)M_{\alpha\alpha}(y,s), \tag{4.176}$$

$\alpha = -1, 0, 1$, with memory kernel

$$M_{\alpha\alpha}(x,t) = 2 \sum_{\gamma,\gamma'=0,\pm 1} (G^{\alpha}_{\gamma\gamma'})^2 f_\gamma(x,t) f_{\gamma'}(x,t) \,. \tag{4.177}$$

The solution to (4.176) has two sound peaks centered at $\pm ct$ and the heat peak sitting at 0. All three peaks have a width much less than ct. But then, in case $\gamma \neq \gamma'$, the product $f_\gamma(x,t) f_{\gamma'}(x,t) \simeq 0$ for large t. Hence, for the memory kernel (4.177) we invoke a small overlap approximation as

$$M_{\alpha\alpha}(x,t) \simeq M^{\mathrm{dg}}_\alpha(x,t) = 2 \sum_{\gamma=0,\pm 1} (G^{\alpha}_{\gamma\gamma})^2 f_\gamma(x,t)^2 \,, \tag{4.178}$$

which is to be inserted in Eq. (4.176).

The decoupling hypothesis can also be applied to Eqs. (4.176) and (4.178). In fact, in the summer of 2012 C. Mendl implemented numerical solutions of the full mode-coupling equations (4.173) on fairly small lattices. The decoupling is seen very convincingly. One then arrives at the mode-coupling equation for a single component, written down already in van Beijeren et al. (1985). In the long-time limit the spreading is proportional to $t^{2/3}$ and the corresponding scaling function deviates only by approximately 5% from the exact scaling function (4.150).

We now have a tool available, by which also the heat peak can be handled, at least approximately. Setting $\alpha = 0$ in (4.176) with approximation (4.178), f_0 couples to $f_{-1}f_{-1}$ and $f_1 f_1$. But these are presumably close to f_{KPZ} of Eq. (4.150). In fact, the scaling exponent matters while the precise shape modifies only prefactors. Inserting f_{KPZ} in Eq. (4.176) and solving the resulting linear equation for f_0, one obtains that for long times the Fourier transform of f_0 is given by

$$\hat{f}_0(k,t) = \mathrm{e}^{-|k|^{5/3}|\Gamma_{\mathrm{h}}t|} \,. \tag{4.179}$$

There is a somewhat lengthy formula for Γ_{h}, see Eq. (4.12) of Spohn (2014). The right-hand side is the Fourier transform of the Lévy probability distribution with exponents $\alpha = \frac{5}{3}, \beta = 0$. The tail of the Lévy $\frac{5}{3}$ distribution decays $|x|^{-8/3}$, which has no second moment. But in fact the distribution is cut off at the sound peaks. There are no correlations propagating beyond the sound cone.

Besides the generic dynamical Phase 1, mode-coupling covers also the other dynamical phases. For Phase 2, the sound peaks are diffusive with scaling function

$$f_\sigma(x,t) = \frac{1}{\sqrt{4\pi D_{\mathrm{s}}t}} \mathrm{e}^{-(x-\sigma ct)^2/4D_{\mathrm{s}}t} \,. \tag{4.180}$$

D_{s} is a transport coefficient. It can be defined through a Green–Kubo formula, which also means that no reasonably explicit answer can be expected. The feedback of the sound peak to the central peak follows by the same steps as before, with the result

$$\hat{f}_0(k,t) = \mathrm{e}^{-|k|^{3/2}\Gamma_{\mathrm{h}}|t|} \,, \tag{4.181}$$

where

$$\Gamma_{\rm h} = (D_{\rm s})^{-1/2}(G^0_{\sigma\sigma})^2(4\pi)^2(2\pi c)^{-1/2}\int_0^\infty {\rm d}t\,t^{-1/2}\cos(t)(2\sqrt{\pi})^{-1}. \qquad (4.182)$$

Since $\frac{3}{2} < \frac{5}{3}$, the density $f_0(x,t)$ turns out to be broader than the Lévy $\frac{5}{3}$ from the Phase 1.

For Phase 3, one proceeds as before, but leaves the scaling exponent yet undetermined. Since modes are cross-coupled, one repeats the argument with another pair of modes. The only solution turns out to be Lévy with the golden mean $\alpha = \frac{1}{2}(1 + \sqrt{5})$. (In the theory of stable laws the two relevant parameters are usually denoted by α, β; see Uchaikin and Zolotarev (1999).) However, one also picks up phase factors. This then yields the asymmetry parameters $\beta = -1, 0, 1$, respectively for each peak, consistent with the physical principle of no correlations beyond the sound cone.

4.17   Molecular dynamics and other missed topics

Anomalous transport in one-dimensional chains is a fascinating topic with many contributions, including extensive molecular dynamics simulations, which in absence of accurate experiments is the only mean to check theoretical predictions. As with other items, I cannot provide any details here. Instead I refer to the article by Lepri et al. (2003) and the article by Dhar (2008). These reviews extensively cover the popular scheme of coupling the chain at its two ends to thermal reservoirs of different temperatures. While, in principle, nonlinear fluctuating hydrodynamics should cover also such boundary conditions, no progress has been achieved yet in this direction. Also I had to omit a discussion of the current–current correlations in equilibrium (Mendl and Spohn 2015a). There have been recent attempts to go beyond covariances and, in spirit of the progress on the KPZ equation, to uncover the full probability density function of the time-integrated currents. Over the past few years several molecular dynamics simulations have been performed with the goal to check the validity of nonlinear fluctuating hydrodynamics. A fairly complete discussion can be found in my contribution to a forthcoming volume of Springer Lecture Notes in Physics, edited by S. Lepri, on thermal transport in low dimensions (Spohn 2016).

Acknowledgements

In my explorations of the KPZ landscape I had the good luck of benefitting from outstanding collaborators, starting with Henk van Beijeren and Joachim Krug in the early days and more recently with Michael Prähofer, Patrik Ferrari, Tomohiro Sasamoto, Sylvain Prolhac, and Thomas Weiss. For nonlinear fluctuating hydrodynamics I highly appreciate the cooperation with Christian Mendl. I enjoyed tremendously the stimulating atmosphere at Les Houches.

References

Ablowitz, M. J., Prinari, B., and Trubatch, A. D. (2004). *Discrete and Continuous Nonlinear Schrödinger Systems*. Cambridge University Press, Cambridge.

Alberts, T., Khanin, K., and Quastel, J. (2014). The intermediate disorder regime for directed polymers in dimension $1 + 1$. *Ann. Probab.* **42**, 1212–56.

Almeida, R. A. L., Ferreira, S. O. T., Oliveira, J., and Aaró Reis, F. D. A. (2014). Universal fluctuations in the growth of semiconductor thin films. *Phys. Rev. B* **89**, 045309.

Amir, G., Corwin, I., and Quastel, J. (2011). Probability distribution of the free energy of the continuum directed random polymer in $1+1$ dimensions. *Comm. Pure Appl. Math.* **64**, 466–537.

Arndt, P. F., Heinzel, T., and Rittenberg, V. (1999). Spontaneous breaking of translational invariance and spatial condensation in stationary states on a ring. I. The neutral system *J. Stat. Phys.* **97**, 1–65.

Baik, J., and Rains, E. M. (2000). Limiting distributions for a polynuclear growth model with external sources. *J. Stat. Phys.* **100**, 523–54.

Barabasi, A. L., and Stanley, H. E. (1995). *Fractal Concepts in Surface Growth*. Cambridge University Press, Cambridge.

Barraquand, G. (2015). Some integrable models in the KPZ universality class. Ph.D. thesis, Université Paris Diderot.

van Beijeren, H. (2012). Exact results for anomalous transport in one-dimensional Hamiltonian systems. *Phys. Rev. Lett.* **108**, 180601.

van Beijeren, H., Kutner, R., and Spohn, H. (1985). Excess noise for driven diffusive systems. *Phys. Rev. Lett.* **54**, 2026–9.

Bernardin, C., and Olla, S. (2016). Thermodynamics and Non-equilibrium Macroscopic Dynamics of Chains of Anharmonic Oscillators (in progress). Available at `https:// www.ceremade.dauphine.fr/olla/`.

Bertini, L., and Giacomin, G. (1997). Stochastic Burgers and KPZ equations from particle systems. *Commun. Math. Phys.* **183**, 571–607.

Bornemann, F. (2010). On the numerical evaluation of Fredholm determinants. *Math. Comp.* **79**, 871–915.

Borodin, A., and Gorin, V. (2012). Lectures on integrable probability. arXiv:1212.3351.

Borodin, A., Corwin, I., and Remenik, D. (2013). Log-Gamma polymer free energy fluctuations via a Fredholm determinant identity. *Commun. Math. Phys.* **324**, 215–32.

Borodin, A., Corwin, I., Ferrari, P. L., and Vető, B. (2015). Height fluctuations for the stationary KPZ equation. *Math. Phys. Anal. Geom.* **18**, 20.

Borodin A., and Petrov, L. (2014). Integrable probability: from representation theory to Macdonald processes. *Probab. Surv.*, **11**, 1–58.

Calabrese, P., and Le Doussal, P. (2011). An exact solution for the KPZ equation with flat initial conditions. *Phys. Rev. Lett.* **106**, 250603.

Calabrese, P., Le Doussal, P., and Rosso, A. (2010). Free-energy distribution of the directed polymer at high temperature. *Europhys. Lett.* **90**, 200002.

Cole, J. D. (1951). On a quasi-linear parabolic equation occurring in aerodynamics. *Quart. Appl. Math.* **79**, 225–236.

Corwin, I. (2012). The Kardar-Parisi-Zhang equation and universality class. *Random Matrices Theory Appl.* **1**, 1.

Das, D., Basu, A., Barma, M., and Ramaswamy, S. (2001). Weak and strong dynamic scaling in a one-dimensional driven coupled-field model: effects of kinematic waves. *Phys. Rev. E* **64**, 021402.

Das, S. G., and Dhar, A. (2014). Role of conserved quantities in normal heat transport in one dimension. arXiv:1411.5247.

Das, S. G., Dhar, A., Saito, K., Mendl, C. B., and Spohn, H. (2014). Numerical test of hydrodynamic fluctuation theory in the Fermi-Pasta-Ulam chain. *Phys. Rev. E* **90**, 012124.

Dhar, A. (2008). Heat transport in low-dimensional systems. *Adv. Phys.* **57**, 457–537.

Dotsenko, V. (2010a). Bethe ansatz derivation of the Tracy-Widom distribution for one-dimensional directed polymers. *Europhys. Lett.* **90**, 200003.

Dotsenko, V. (2010b). Replica Bethe ansatz derivation of the Tracy-Widom distribution of the free energy fluctuations in one-dimensional directed polymers *J. Stat. Mech.* **2010**, P07010.

Le Doussal, P., and Calabrese, P. (2012). The KPZ equation with flat initial condition and the directed polymer with one free end. *J. Stat. Mech.* **2012**, P06001.

Eden, M. (1961). A two-dimensional growth process. In *Proceedings of the Fourth Berkeley Symposium on Mathematical Statistics and Probability*, Volume 4: *Contributions to Biology and Problems of Medicine*, pp 223–39, University of California Press, Los Angeles.

Ertaş, D., and Kardar, M. (1993). Dynamic relaxation of drifting polymers: phenomenological approach. *Phys. Rev. E* **48**, 1228–1245.

Fermi, E., Pasta, J., and Ulam, S. (1965). Studies of nonlinear problems. Los Alamos report LA-1940 (1955), published later in *Collected Papers of Enrico Fermi*, ed. E. Segré. University of Chicago Press, Chicago.

Ferrari, P. L., Sasamoto, S., and Spohn, H. (2013). Coupled Kardar-Parisi-Zhang equations in one dimension. *J. Stat. Phys.* **153**, 377–99.

Ferrari, P. L., and Spohn, H. (2005). A determinantal formula for the GOE Tracy-Widom distribution. *J. Phys. A, Math. Gen.* **38**, L557–L561.

Ferrari, P. L., and Spohn, H. (2011). Random growth models. In *The Oxford Handbook of Random Matrix Theory*, ed. G. Akemann, J. Baik and P. Di Francesco.

Forster, D. (1975). *Hydrodynamic Fluctuations, Broken Symmetry and Correlation Functions*. Benjamin, New York.

Funaki, T. (2015). Infinitesimal invariance for coupled KPZ equations. In *Memoriam Marc Yor—Séminaire de Probabilités XLVII*, Lecture Notes in Mathematics, Vol. 2137. Springer, Berlin.

Gärtner, J. (1988). Convergence towards Burgers' equation and propagation of chaos for weakly asymmetric exclusion processes. *Stochastic Processes Appl.* **27**, 233–60.

Georgiou, N., Rassoul-Agha, F., Seppäläinen, T., and Yilmaz, A. (2015). Ratios of partition functions for the log-gamma polymer *Ann. Probab.* **43**, 2282–331.

Grisi, R., and Schütz, G. (2011). Current symmetries for particle systems with several conservation laws. *J. Stat. Phys.* **145**, 1499–512.

Gubinelli, M., and Perkowski, N. (2017). KPZ reloaded. *Commun. Math. Phys.* **346**, 165–269.

Hairer, M. (2013). Solving the KPZ equation. *Ann. Math.* **178**, 559–664.

Halpin-Healy, T., and Palasantzas, G. (2014). Universal correlators and distributions as experimental signatures of (2 + 1)-dimensional Kardar-Parisi-Zhang growth. *Europhys. Lett.* **105**, 50001.

Halpin-Healy, T., and Takeuchi, K. A. (2015). A KPZ cocktail-shaken, not stirred: Toasting 30 years of kinetically roughened surfaces. *J. Stat. Phys.* **160**, 794–814.

Hopf, E. (1950). The partial differential equation $u_t + uu_x = \mu u_{xx}$. *Commun. Pure Appl. Math.* **3**, 201–230.

Imamura, T., and Sasamoto, T. (2012). Exact solution for the stationary KPZ equation. *Phys. Rev. Lett.* **108**, 190603.

Imamura, T., and Sasamoto, T. (2013). Stationary correlations for the 1D KPZ equation. *J. Stat. Phys.* **150**, 908–39.

Iubini, S., Lepri, S., Livi, R., and Politi, A. (2013). Off-equilibrium Langevin dynamics of the discrete nonlinear Schrödinger chain. *J. Stat. Mech.* **2013**, P08017.

Iubini, S., Politi, A., and Politi, P. (2014). Coarsening dynamics in a simplified DNLS model. *J. Stat. Phys.* **154**, 1057–73.

Johansson, K. (2000). Shape fluctuations and random matrices. *Commun. Math. Phys.* **209**, 437–76.

Johansson, K. (2006). Random matrices and determinantal processes. In: *Mathematical Statistical Physics, École d'été Physique*, Les Houches, session LXXXIII, arXiv:math-ph/0510038.

Johansson, K. (2017). Two time distribution in Brownian directed percolation. *Commun. Math. Phys.* **351**, 441–492.

Kardar, M. (1987). Replica Bethe ansatz studies of two-dimensional interfaces with quenched random impurities. *Nucl. Phys.* B **290**, 582–602.

Kardar, M., Parisi, G., and Zhang,Y. C. (1986). Dynamic scaling of growing interfaces. *Phys. Rev. Lett.* **56**, 889–92.

Krug, J. (1997). Origins of scale invariance in growth processes. *Adv. Phys.* **46**, 139–282.

Lee-Dadswell, G. R. (2015). Universality classes for thermal transport in one-dimensional oscillator chains. *Phys. Rev. E* **91**, 032102.

Lepri, S., Livi, R., and Politi, A. (2003). Thermal conduction in classical low-dimensional lattices. *Phys. Rep.* **377**,1–80.

Meakin, P. (1998). *Fractals, Scaling and Growth Far from Equilibrium*. Cambridge University Press, Cambridge.

Mendl, C. (2013). MATHEMATICA program available upon request.

Mendl, C. (2015). private communication.

Mendl, C., and Spohn, H. (2015a). Current fluctuations for anharmonic chains in thermal equilibrium. *J. Stat. Mech.*, **2015**, P03007.

Mendl, C., and Spohn, H. (2015b). Low temperature dynamics of the one-dimensional discrete nonlinear Schrödinger equation. *J. Stat. Mech.* **2015**, P08028.

Miettinen, L., Myllys, M., Merikoski, J., and Timonen, J. (2005). Experimental determination of KPZ height-fluctuation distributions, *Eur. Phys. Jour. B* **46**, 55–60.

Miloshevich, G., Nguenang, J.-P., Dauxois, T., Khomeriki, R., and Ruffo, S. (2014). Instabilities in long-range oscillator chains. *Phys. Rev. E* **91**, 032927.

Ortmann, J., Quastel, J., and Remenik, D. (2016). Exact formulas for random growth with half-flat initial data. *Ann. Appl. Probab.* **26**, 507–548.

Ortmann, J., Quastel, J., and Remenik, D. (2017). A Pfaffian representation for flat ASEP. *Comm. Pure Appl. Math.* **70**, 3–89.

Popkov, V., Schadschneider, A., Schmidt, J., and Schütz, G. M. (2015). Fibonacci family of dynamical universality classes. *PNAS* **112**, 12645–50.

Prähofer, M. (2005). Exact scaling functions for one-dimensional stationary KPZ growth. Available at http://www-m5.ma.tum.de/KPZ.

Prähofer, M., and Spohn, H. (2002). Scale invariance of the PNG droplet and the Airy process. *J. Stat. Phys.* **108**, 1071–106.

Prähofer, M., and Spohn, H. (2004). Exact scaling functions for one-dimensional stationary KPZ growth. *J. Stat. Phys.* **115**, 255–79.

Prolhac, S., and Spohn, H. (2011a). Two-point generating function of the free energy for a directed polymer in a random medium. *J. Stat. Mech.* **2011**, P01031.

Prolhac, S., and Spohn, H. (2011b). The one-dimensional KPZ equation and the Airy process. *J. Stat. Mech.* **2011**, P03020.

Prolhac, S., and Spohn, H. (2011c). The height distribution of the KPZ equation with sharp wedge initial condition: numerical evaluations. *Phys. Rev. E* **84**, 011119.

Quastel, J. (2011). Introduction to KPZ. *Curr. Dev. Math.* **2011**, 125–94.

Quastel, J., and Spohn, H. (2015). The one-dimensional KPZ equation and its universality class. *J. Stat. Phys.* **160**, 965–84.

Sasamoto, T., and Spohn, H. (2009). Superdiffusivity of the 1D lattice Kardar-Parisi-Zhang equation. *J. Stat. Phys.* **137**, 917–35.

Sasamoto T., and Spohn, H. (2010). Exact height distributions for the KPZ equation with narrow wedge initial condition. *Nucl. Phys. B* **834**, 523–42.

Sasamoto, T., and Spohn, H. (2011). The 1+1-dimensional Kardar-Parisi-Zhang equation and its universality class. Proceedings StatPhys 24, *J. Stat. Mech.* **2011**, P01031.

Spohn, H. (1993). Interface motion in models with stochastic dynamics, *J. Stat. Phys.* **71**, 1081–132.

Spohn, H. (2006). Exact solutions for KPZ-type growth processes, random matrices, and equilibrium shapes of crystals. *Physica A* **369**, 71 –99.

Spohn, H. (2014). Nonlinear fluctuating hydrodynamics for anharmonic chains. *J. Stat. Phys.* **154**, 1191–227.

Spohn, H. (2016). Fluctuating hydrodynamics approach to equilibrium time correlations for anharmonic chains. In *Thermal Transport in Low Dimensions: From Statistical Physics to Nanoscale Heat Transfer*, ed. S. Lepri, Lecture Notes

in Physics. Springer, Berlin. Springer Lecture Notes in Physics, Volume 921, pp. 107–158

Spohn, H., and Stoltz, G. (2015). Nonlinear fluctuating hydrodynamics in one dimension: the case of two conserved fields. *J. Stat. Phys.* **160**, 861–84.

Takeuchi, K. A. (2012). Statistics of circular interface fluctuations in an off-lattice Eden model. *J. Stat. Mech.* **2012**, P05007.

Takeuchi, K. A. (2014). Experimental approaches to universal out-of-equilibrium scaling laws: turbulent liquid crystal and other developments. Proceedings article for the StatPhys 25, *J. Stat. Mech.* **2014**, P01006.

Takeuchi, K. A., and Sano, M. (2012). Evidence for geometry-dependent universal fluctuations of the Kardar-Parisi-Zhang interfaces in liquid-crystal turbulence. *J. Stat. Phys.* **147**, 853–90.

Toda, M. (1967). Vibration of a chain with a non-linear interaction. *J. Phys. Soc. Jpn* **22**, 431–6.

Tóth, B., and Valkó, B. (2003). Onsager relations and Eulerian hydrodynamic limit for systems with several conservation laws. *J. Stat. Phys.* **112**, 497–521.

Tracy, C. A., and Widom, H. (1994). Level-spacing distributions and the Airy kernel. *Commun. Math. Phys.* **159**, 151–74.

Uchaikin, V., and Zolotarev, V. (1999). *Chance and Stability. Stable Distributions and Applications*. Modern Probability and Statistics Series, De Gruyter.

5

Random matrix theory and quantum chromodynamics

Gernot AKEMANN

Faculty of Physics, Bielefeld University, Postfach 100131,
D-33501 Bielefeld, Germany

Akemann, G., 'Random Matrix Theory and Quantum Chromodynamics' in *Stochastic Processes and Random Matrices*. Edited by: Grégory Schehr et al, Oxford University Press (2017). © Oxford University Press 2017. DOI 10.1093/oso/9780198797319.003.0005

Chapter Contents

This chapter is based on the lectures delivered at the Les Houches Summer School in July 2015. They are addressed at a mixed audience of physicists and mathematicians with some basic working knowledge of random matrix theory. The first part is devoted to the solution of the chiral Gaussian unitary ensemble in the presence of characteristic polynomials, using orthogonal polynomial techniques. This includes all eigenvalue density correlation functions, smallest eigenvalue distributions, and their microscopic limit at the origin. These quantities are relevant for the description of the Dirac operator spectrum in quantum chromodynamics with three colours in four Euclidean space-time dimensions. In the second part these two theories are related based on symmetries, and the random matrix approximation is explained. In the last part recent developments are covered, including the effect of finite chemical potential and finite space-time lattice spacing, and their corresponding orthogonal polynomials. We also give some open random matrix problems.

5.1   Introduction and motivation

In this short introduction we would like to introduce the two players in the title, random matrix theory (RMT) and quantum chromodynamics (QCD), on a superficial level. This gives a motivation why and where it will be beneficial to relate these two seemingly unrelated theories, and what the reader may expect to learn from these notes.

Let us begin with QCD, the theory of the strong interactions.[1] It is part of the standard model of elementary particles that also describes the weak and electromagnetic interactions among all elementary particles. The latter two interactions will, however, not play any role here. The particle content of QCD are the N_f quarks with masses m_q, $q = 1, \ldots, N_f$, and the gluons. In nature $N_f = 6$ flavours have been observed and named up, down, strange, charm, top, and bottom quark. In the following we will keep N_f as a free parameter, and often consider only the lightest up and down quark ($N_f = 2$) as they constitute the most common particles as proton (p), neutron (n), and pions ($\pi^{\pm,0}$). The quarks interact through the gluons, the carriers of force, with a coupling constant g_s. Both come in three colours and the interaction is described through a field strength and covariant derivative carrying an $SU(3)$ Lie group structure. QCD is a strongly interacting, relativistic quantum field theory (QFT), which is very difficult. Fortunately, we will only need to know a few features and some of its global symmetries that will be described in Section 5.3. For a standard textbook on perturbative QCD we refer for example to [1].

Roughly speaking QCD has two different phases that are schematically depicted in Fig. 5.1. They are characterized by an order parameter, the chiral condensate Σ. At high energies corresponding to high temperatures T, $\Sigma = 0$ and quarks and gluons form a plasma that has been observed in collision experiments, e.g., at the Relativistic Heavy Ion Collider RHIC in Brookhaven. This phase also existed in the early universe,

---

[1] I apologise for being very elementary here. My audience was not assumed to know particle or general physics.

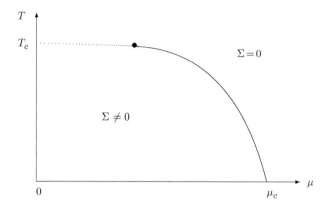

Fig. 5.1 Schematic phase diagram of QCD as a function of temperature T and quark chemical potential μ, for two massless quark flavours. The chemical potential is proportional to the Baryon density. For massless flavours $m_q = 0$ the dashed line corresponds to a second-order phase transition, merging with the full line representing a first-order transition in a tricritical point.

with the cooling down happening close to the temperature axis at low Baryon density parametrised by μ. In these situations as well as in single-particle collisions produced at collider experiments a perturbative expansion in powers of the coupling g_s typically applies, as described, e.g., in [1]. The second phase with $\Sigma \neq 0$ at low temperature and density is the one in which we live. Here quarks condense into colourless objects, that is into baryons made of three quarks of different colour adding up to white (like p or n), or mesons made of a quark and an anti-quark with its anti-colour (like the π's). Gluons are also confined to these objects, and a proof of confinement of quarks and gluons from QCD is still considered to be an open millennium problem. In QCD it happens that confinement goes along with the spontaneous breaking of chiral symmetry—hence the name of Σ. This global symmetry will be described in detail later, and eventually lead us to a RMT description.

In the phase with $\Sigma \neq 0$ perturbation theory breaks down and one has to apply other methods. If one wants to continue to work with first principles and the QCD action, one possibility is to study QCD on a finite space-time lattice of volume V and lattice spacing a, equipped with a Euclidean metric. The numerical solution of this theory nowadays reproduces the masses of particles that are composed of elementary ones to a very high precision, as testified in the particle data booklet [2]. To that aim two limits have to be taken, the continuum limit sending $a \to 0$ and the thermodynamical limit sending $V \to \infty$. For a standard textbook on lattice QCD we refer to [3]. A second possibility is to approximate QCD in the confined phase by effective theories that describe, for example, only the low energy excitation. One of these is chiral perturbation theory (chPT); see [4] for a review that describes the low momentum modes that appear after the spontaneous breaking of chiral symmetry, the so-called Goldstone Bosons. Another such theory is RMT. It should be clear by now that RMT will not solve QCD (nor chPT), in the continuum or on the lattice.

However, it will describe certain aspects in an analytic fashion, namely the spectral properties of the small eigenvalues of the QCD Dirac operator $\slashed{D}$. In a finite volume the density of eigenvalues of $\slashed{D}$ satisfies the Banks–Casher relation [5]

$$\rho_{\slashed{D}}(\lambda \approx 0) = \frac{1}{\pi}\Sigma V \,, \tag{5.1}$$

which relates it to the described setup. In particular RMT will predict the detailed dependence on the parameters m_q, μ, a, and V after being appropriately rescaled, as well as on the zero eigenvalues of $\slashed{D}$ which relate to a topological index ν. And most remarkably these predictions have been verified in comparison to lattice QCD by many groups. We will not repeat these findings here and refer to the literature at the end of each section.

The idea to apply RMT to QCD goes back to the seminal works [6, 7] that started from the simplest case with $N_f = 0$ which is called quenched approximation, at $\mu = 0 = a$. The field has since developed enormously, leading to the detailed analytical knowledge that we will describe. There exist a number of excellent reviews already, [8, 9], notably the lecture notes from the Les Houches session in 2004 by J. Verbaarschot [10]. This brings me to the main goals of these lecture notes, to be addressed in the order of the subsequent sections. What can we predict from RMT that can be compared to lattice QCD? In Section 5.2 we will start with the solution of the corresponding RMT using the theory of orthogonal polynomials. This section is rather mathematical and contains no further physics input. The second question, what the limit is, in which QCD reduces to RMT, is addressed in Section 5.3. Here the global symmetries of QCD are explained, leading to chPT and eventually to a RMT description. The more physics-inclined reader may jump to this section first. One of my personal motivations for adding another review on this topic is answered in Section 5.4, where the developments the past 10 years are reviewed, including some mathematical aspects. This covers the dependence on finite lattice spacing a and on chemical potential μ. For an earlier review on the latter, see [11].

Let us move to RMT, where I assume that the reader already has some working knowledge. RMT is a much older topic, beginning in the late 1920s with Wishart in mathematical statistics and in the late 1950s with Wigner and Dyson in nuclear physics. The number of its applications is huge and still increasing, and we refer to [12] for applications to quantum physics and to [13] for a recent compilation containing physics (including QCD in Chapter 32), mathematics, and more. In this chapter we will focus only on the theory of orthogonal polynomials. For the standard Gaussian ensembles they are covered in the 2004 edition of Mehta's classical book [14]; for more recent monographs, see [15, 16]. The symmetry class we will be interested in for the application to QCD is the chiral Gaussian unitary ensemble (chGUE), which is also known as the Wishart or Laguerre unitary ensemble, and extensions thereof. In the simplest case of $N_f = 0$ and when all the previous-mentioned parameters from QCD are absent it is defined by the probability space of complex $N \times N$ matrices W. All its matrix elements are independent and share the same complex normal distribution. The mean density $\rho(x)$ of the positive eigenvalues of $WW^\dagger$, where $\dagger$ denotes the Hermitian conjugate, is well known in the limit $N \to \infty$. After an appropriate rescaling, it is

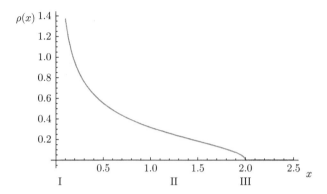

Fig. 5.2 The Marchenko–Pastur law $\rho(x) = \frac{1}{\pi}\sqrt{(2-x)/x}$ which is the limiting global spectral density of the chGUE. We also indicate the locations where different local statistics apply. They are given by (I) the Bessel kernel close to the hard edge, (II) the sine kernel in the bulk, and (III) the Airy kernel at the soft edge.

given by the Marchenko–Pastur law depicted in Fig. 5.2; cf. the lectures by Bouchaud where this was derived. The same law is obtained when taking real or quaternion-valued matrix elements, constituting the chiral Gaussian orthogonal and symplectic ensemble (chGOE and chGSE), respectively. However, the local statistics of the three ensembles differs, and depends on the location in the spectrum we consider. For finite N the eigenvalue statistics follows a determinantal (chGUE) or Pfaffian point process (chGOE, chGSE), and the limiting kernels in the various locations of the spectrum shown in Fig. 5.2 are given in terms of Bessel, Sine, or Airy functions. In view of the relation (5.1) we will be interested in the local statistics for the chGUE at the origin, also called hard edge. The other two symmetry classes at the hard edge will also find applications in QFT as pointed out in [17]; cf. [8], but will not be discussed here. In the limit where RMT applies the eigenvalues y_j of the QCD Dirac operator $\not{D}$ will be given by $y_j = \pm\sqrt{x_j}$ and thus come in pairs, where x_j are the eigenvalues of $WW^\dagger$ (the positive part $+\sqrt{x_j}$ are the singular values of W). After this change of variables the Marchenko–Pastur law becomes a semi-circle, which is why the global density of $\not{D}$ at the origin will become constant (and not divergent as in Fig. 5.2). For finite N the chGUE can be solved in terms of the classical Laguerre polynomials. It will be the content of Sections 5.2 and 5.4 to include more structure from QCD into this RMT, while maintaining its exact solvability.

5.2 Orthogonal polynomial approach to the Dirac operator spectrum

In this section we go directly to the RMT of QCD in an eigenvalue representation that generalizes the chGUE by including the mass terms of N_f quark flavours. In Subsection 5.2.1 we briefly recall the orthogonal polynomial (OP) formalism and define the quantities of our interest. We then attack the mathematical problem of computing

these by deriving properties of OP with a general weight, in particular including averages of characteristic polynomials in Subsection 5.2.2. This allows us to determine all quantities explicitly in terms of standard Laguerre polynomials in Subsection 5.2.3, which facilitates the large-N limit to be taken in Subsection 5.2.4.

5.2.1   The eigenvalue model and definition of its correlation functions

We begin by stating the RMT that we want to solve in its eigenvalue representation, defined by the partition function

$$
Z_N^{(\beta, N_f, \nu)} = \left(\prod_{j=1}^{N} \int_0^{\infty} \mathrm{d}x_j \; x_j^{\frac{\beta}{2}(\nu+1)-1} e^{-x_j} \prod_{f=1}^{N_f} (x_j + m_f^2) \right) |\Delta_N(\{x\})|^{\beta}
$$

$$
= \left(\prod_{j=1}^{N} \int_0^{\infty} \mathrm{d}x_j \right) \mathcal{P}_{jpdf}(x_1, \ldots, x_N) \, . \tag{5.2}
$$

It gives the normalization constant of the unnormalized joint probability density function (jpdf) $\mathcal{P}_{\mathrm{jpdf}}(x_1, \ldots, x_N)$ of all eigenvalues. Here we have included the dependence on the following parameters to be related to QCD later: $\nu = 0, 1, 2, \ldots$ called the topological index taking fixed values, the masses $m_f \in \mathbf{R}_+$, $f = 1, 2, \ldots, N_f$, with N_f counting the number of inserted characteristic polynomials. In the quenched theory with $N_f = 0$ their product is absent. These parameters are all collected in the weight function

$$
w(x) = x_j^{\frac{\beta}{2}(\nu+1)-1} e^{-x_j} \prod_{f=1}^{N_f} (x_j + m_f^2) \, , \tag{5.3}
$$

and finally we also have introduced the Vandermonde determinant

$$
\Delta_N(\{x\}) = \det_{1 \leq i,j \leq N} [x_i^{j-1}] = \prod_{1 \leq i < j \leq N} (x_j - x_i) \, . \tag{5.4}
$$

It depends on the set of all eigenvalues $\{x\} = \{x_i\}_{i=1,\ldots,N}$. The partition function Eq. (5.2) is relevant for QCD for $\beta = 2$, representing the chGUE in the presence of N_f characteristic polynomials (mass terms). For completeness and in order to state some open problems later we have also introduced the chGOE and chGSE with $\beta = 1$ and 4, respectively. All three RMT are also called Wishart–Laguerre ensembles and can be written in matrix representation,

$$
Z_N^{(\beta, N_f, \nu)} \sim \int [\mathrm{d}W] \prod_{f=1}^{N_f} \det \begin{bmatrix} m_f 1_N & iW \\ iW^{\dagger} & m_f 1_{N+\nu} \end{bmatrix} \exp[-\mathrm{Tr}(WW^{\dagger})] \, . \tag{5.5}
$$

Here W is an $N \times (N + \nu)$ matrix taking values $W_{ij} \in \mathbf{R}/\mathbf{C}/\mathbf{H}$ for $\beta = 1, 2, 4$, respectively, and $W^{\dagger}$ is its Hermitian conjugate. The integration $[\mathrm{d}W]$ is over the flat

Lebesgue measure of all independent matrix elements. The identity matrices in front of the scalar mass variables m_f of dimensions N and $N+\nu$ are denoted by 1_N and $1_{N+\nu}$, respectively. Here we see that ν counts the number of zero eigenvalues of the Wishart matrix $W^\dagger W$, and the x_j, $j = 1, \ldots, N$ denote its nonzero positive eigenvalues.[2] The change of variables from matrix elements to eigenvalues leads to the Vandermonde determinant to the power $\beta = 1, 2, 4$ times the ν-dependent part as the Jacobian, and we refer to standard textbooks on RMT for the derivation [14, 16]. In the lectures of J.-P. Bouchaud these ensembles were discussed for $N_f = 0$, in particular when ν is of the order of N. This matrix representation Eq. (5.5) will play an important role later in Section 5.3 when relating the RMT to QCD based on symmetries. In the following we will mainly focus on the chGUE with $\beta = 2$, which is the RMT relevant for QCD. The values $\beta = 1$ and 4 apply to other QFTs and are very briefly discussed at the end of Subsection 5.2.3.

What are the quantities we would like to calculate as functions of the masses m_f and ν, apart from the normalizing partition function? The jpdf represents a determinantal (or Pfaffian) point process introduced in the lectures by A. Borodin. Therefore we can determine all k-point eigenvalue density correlations defined for all β as

$$R_k(x_1, \ldots, x_k) = \frac{1}{Z_N^{(\beta, N_f, \nu)}} \frac{N!}{(N-k)!} \int_0^\infty dx_{k+1} \cdots \int_0^\infty dx_N \, \mathcal{P}_{jpdf}(x_1, \ldots, x_N) \quad (5.6)$$

$$= \prod_{j=1}^k w(x_j) \det_{1 \le i, j \le k} [K_N(x_i, x_j)], \quad \text{for } \beta = 2 . \quad (5.7)$$

In the second line valid for $\beta = 2$ only they can be expressed in terms of the determinant of the kernel

$$K_N(x, y) = \sum_{l=0}^{N-1} h_l^{-1} P_l(x) P_l(y) , \quad (5.8)$$

which does not contain the weights in our convention. It contains OP with respect to the weight Eq. (5.3),

$$\int_0^\infty dx \, w(x) P_k(x) P_l(x) = h_k \delta_{kl} . \quad (5.9)$$

For $\beta = 1, 4$ similar expressions exist in terms of Pfaffian determinants of a matrix kernel of OP with respect to a skew symmetric inner product replacing Eq. (5.9). These are called skew OP and we refer to [14, 16] for details. We choose the OP in Eq. (5.9) to be monic, $P_k(x) = x^k + O(x^{k-1})$, with squared norms $h_k = ||P_k||^2$. In particular Eq. (5.6) leads to the spectral density $R_1(x) = w(x) K_N(x, x)$ (normalized to N) and to the determinantal expression for the jpdf, $R_N(x_1, \ldots, x_N) =$

---

[2] For quaternionic matrix elements the eigenvalues are doubly degenerate (Kramer's degeneracy), which is why for $\beta = 4$ we always have an even number $2N_f$ of doubly degenerate masses in Eq. (5.2). We will come back to this at the end of Subsection 5.2.3.

$N!\mathcal{P}_{\mathrm{jpdf}}(x_1,\ldots,x_N)/Z_N^{(\beta,N_f,\nu)}$. Two further quantities that are of our interest can be defined. The kth gap probability defined for all β reads

$$E_k(s) = \frac{1}{Z_N^{(\beta,N_f,\nu)}}\frac{N!}{(N-k)!}$$

$$\times \int_0^s dx_1 \cdots \int_0^s dx_k \int_s^\infty dx_{k+1} \cdots \int_s^\infty dx_N \mathcal{P}_{\mathrm{jpdf}}(x_1,\ldots,x_N) ,$$

(5.10)

where $k = 0,1,\ldots,N$. It is the probability that k eigenvalues are in $[0,s]$ and $N-k$ in $[s,\infty)$. For example, $E_0(s)$ gives the gap probability that the interval $[0,s]$ is empty. If we order the eigenvalues $x_1 < x_2 < \ldots < x_N$ the kth eigenvalue distribution defined as

$$p_k(s) = \frac{k\binom{N}{k}}{Z_N^{(\beta,N_f,\nu)}}$$

$$\times \int_0^s dx_1 \cdots \int_0^s dx_{k-1} \int_s^\infty dx_{k+1} \cdots \int_s^\infty dx_N \mathcal{P}_{\mathrm{jpdf}}(x_1,\ldots,x_k = s,\ldots,x_N),$$

(5.11)

where $k = 1,\ldots,N$, gives the probability of finding the kth eigenvalue at $s = x_k$, $k-1$ eigenvalues in $[0,s]$ and $N-k$ in $[s,\infty)$. These individual eigenvalue distributions are normalized $\int_0^\infty ds\, p_k(s) = 1$, as can be easily checked (see, e.g., the appendix of [18]). We will be mostly interested in the example of the distribution of the smallest nonzero eigenvalue, $p_1(s)$. Equations (5.10) and (5.11) are not independent; in fact, the latter follow by differentiation:

$$\frac{\partial}{\partial s}E_0(s) = -p_1(s) , \qquad \frac{\partial}{\partial s}E_k(s) = k!(p_k(s) - p_{k+1}(s)) , \qquad k = 1,\ldots,N-1,$$

$$\frac{\partial}{\partial s}E_N(s) = N!p_N(s) ,$$

(5.12)

which can be solved for $p_k(s)$ as

$$p_k(s) = -\sum_{l=0}^{k-1}\frac{1}{l!}\frac{\partial}{\partial s}E_l(s) .$$

(5.13)

The gap probabilities can also be expressed in terms of the kernel $K_N(x,y)$ as a Fredholm determinant, cf. the lectures by A. Borodin and Eqs (5.48) and (5.49), but we will use a different route to compute them explicitly.

Our goal is now to determine all the quantities that we have introduced in this subsection, including the partition function. For $N_f = 0$ this would be an easy task as the polynomials orthogonal with respect to $w(x) = x^\nu \exp[-x]$ are the well-known Laguerre polynomials. It turns out that also for $N_f > 0$ we can express all quantities in terms of Laguerre polynomials, including the dependence on the parameter m_f and ν that we are after. This will be the goal of the following two subsections.

5.2.2 Properties of orthogonal polynomials with general weights

The first part of this subsection is standard material about OP on the real line and can be found in any standard reference, e.g., in [19], including also OP in the complex plane needed later. For the second part with weights containing characteristic polynomials we refer to [20] for more details.

Let us consider a general measurable weight function on the positive real half line with the condition that all moments exist, $M_k = \int_0^\infty dx\, w(x) x^k < \infty$. The monic polynomials $P_k(x)$ orthogonal with respect to $w(x)$ can then be recursively constructed using Gram–Schmidt, expressing them as the ratio of two determinants:

$$
P_k(x) = \begin{vmatrix} M_0 & M_1 & \dots & M_k \\ \vdots & \vdots & & \vdots \\ M_{k-1} & M_k & \dots & M_{2k-1} \\ 1 & x & \dots & x^k \end{vmatrix} \left(\det_{0 \le i,j \le k-1} [M_{i+j}] \right)^{-1} ; \tag{5.14}
$$

see [19] for details. It obviously satisfies $P_k(x) = x^k + \mathcal{O}(x^{k-1})$. These polynomials obey a three-step recurrence relation,

$$
x P_k(x) = P_{k+1}(x) + \alpha_k^k P_k(x) + \alpha_k^{k-1} P_{k-1}(x) , \tag{5.15}
$$

which can be seen as follows. With the $P_k(x)$ forming a complete set we can expand

$$
x P_k(x) = \sum_{l=0}^{k+1} \alpha_k^l P_l(x) , \quad \alpha_k^l = h_l^{-1} \int_0^\infty dx\, w(x) x P_k(x) P_l(x) . \tag{5.16}
$$

Using that in the last integral $x P_l(x)$ is a polynomial of degree $l+1$ and the orthogonality Eq. (5.9) it follows that $\alpha_k^l = 0$ for $l < k-1$ and $\alpha_k^{k-1} = h_k/h_{k-1}$. For the orthonormal polynomials $\hat{P}_k(x) = P_k(x)/\sqrt{h_k}$ the recursion Eq. (5.15) becomes more symmetric,

$$
x \hat{P}_k(x) = c_k \hat{P}_{k+1}(x) + \alpha_k^k \hat{P}_k(x) + c_{k-1} \hat{P}_{k-1}(x) , \quad c_{k-1} = \sqrt{h_k/h_{k-1}} , \quad k \ge 1 , \tag{5.17}
$$

and we obtain the Christoffel–Darboux formula for the kernel

$$
K_N(x,y) = \sum_{l=0}^{N-1} \hat{P}_l(x) \hat{P}_l(y) = c_{N-1} \frac{\hat{P}_N(x)\hat{P}_{N-1}(y) - \hat{P}_{N-1}(x)\hat{P}_N(y)}{x - y} , \quad x \ne y . \tag{5.18}
$$

This simply follows by multiplying the sum by $(x-y)$ with $x \ne y$, using the recursion and the fact that this is a telescopic sum. From l'Hôpital's rule for the kernel at equal arguments we see that in the asymptotic limit $N \to \infty$ we only need to evaluate the asymptotic of $P_N(x)$, $P_{N-1}(x)$, their derivatives, and the ratio of norms c_{N-1} to determine the spectral density:

$$
R_1(x) = w(x) K_N(x,x) = c_{N-1}(-\hat{P}_N(x)\hat{P}'_{N-1}(x) + \hat{P}_{N-1}(x)\hat{P}'_N(x))w(x). \tag{5.19}
$$

Let us give two examples. For the GUE with weight $w(x) = \exp[-x^2/2]$ on $\mathbf{R}$ instead of $\mathbf{R}_+$ we have

$$P_n(x) = \mathrm{He}_n(x) \ , \quad h_n = \sqrt{2\pi}\, n! \ , \tag{5.20}$$

the probabilist's Hermite polynomials. Second, for the Laguerre weight $w(x) = x^\nu \exp[-x]$, Eq. (5.3) with $N_f = 0$ on $\mathbf{R}_+$, the monic OP are given by the appropriately rescaled generalized Laguerre polynomials $L_n^\nu(x)$:

$$P_n(x) = (-1)^n n! L_n^\nu(x) \ , \quad h_n = n! \, \Gamma(n + \nu + 1) \ , \quad \nu > -1 \ , \tag{5.21}$$

where ν can also be real here. The asymptotic analysis of the kernel (and density) can now be made, and we will come back to this in Subsection 5.2.4.

In the following we will express the partition function, corresponding OPs, kernel, and gap probability $E_0(s)$ in terms of expectation values of characteristic polynomials in a first step, and then reduce them to determinants of Laguerre polynomials in a second step. In order to prepare this, the following exercise is very useful.

Consider the two Vandermode determinants, Eq. (5.4), inside the partition function (5.2) for $\beta = 2$. By adding columns inside the Vandermonde determinant we do not change its value. Starting with the last column of highest degree we can, thus, express $\Delta_N(\{x\})$ in terms of arbitrary monic polynomials $Q_k(x)$:

$$\Delta_N(\{x\}) = \det_{1 \le i,j \le N} [Q_{j-1}(x_i)] \ . \tag{5.22}$$

Choosing those monic polynomials $Q_k(x) = P_k(x)$ that precisely satisfy the orthogonality relation (5.9) we obtain for the partition function

$$
\begin{aligned}
Z_N^{(2,N_f,\nu)} &= \left(\prod_{l=1}^N \int_0^\infty \mathrm{d}x_l w(x_l) \right) \left(\det_{1 \le i,j \le N} [P_{j-1}(x_i)] \right)^2 \\
&= \sum_{\sigma,\sigma' \in S_N} (-1)^{\sigma + \sigma'} \prod_{l=1}^N \int_0^\infty \mathrm{d}x_l w(x_l) P_{\sigma(l)-1}(x_l) P_{\sigma'(l)-1}(x_l) \\
&= N! \prod_{l=1}^N h_{l-1} \ .
\end{aligned}
\tag{5.23}
$$

In the second step we have Laplace-expanded both determinants into sums over permutations $\sigma, \sigma' \in S_N$ and used the orthogonality (5.9). The computation of the partition function, thus, amounts to determining the squared norms of the OPs with respect to the weight Eq. (5.3).

Let us define the expectation value of an operator $\mathcal{O}(\{x\})$ that only depends on the set of eigenvalues $\{x\}$:

$$\langle \mathcal{O}(\{x\}) \rangle_N = \frac{1}{Z_N^{(\beta,N_f,\nu)}} \left(\prod_{l=1}^N \int_0^\infty \mathrm{d}x_l \right) P_{\mathrm{jpdf}}(x_1, \dots, x_N) \mathcal{O}(\{x\}) \ . \tag{5.24}$$

Then the following identity holds for the monic OPs:

$$P_L(x) = \left\langle \prod_{j=1}^{L} (x - x_j) \right\rangle_L . \tag{5.25}$$

It is given by the expectation value of a characteristic polynomial with respect to L eigenvalues for an arbitrary degree, where $0 \leq L \leq N$. It can also be expressed in terms of matrices of size L, choosing $\mathcal{O}(\{x\}) = \det[x - WW^{\dagger}]$. This relation is known as the Heine formula, dating back to the nineteenth century. The proof uses the simple identity for the Vandermonde determinant,

$$\prod_{j=1}^{L} (y - x_j) \Delta_L(\{x\}) = \Delta_{L+1}(\{x\}, y = x_{L+1}) = \det_{1 \leq i,j \leq L+1} [P_{j-1}(x_i)] . \tag{5.26}$$

Using this identity in the expectation value we arrive at the same situation as in Eq. (5.23), with one larger determinant:

$$\left\langle \prod_{j=1}^{L} (y - x_j) \right\rangle_L = \frac{1}{Z_L^{(2, N_f, \nu)}} \sum_{\substack{\sigma \in S_L \\ \sigma' \in S_{L+1}}} (-1)^{\sigma + \sigma'} \left(\prod_{l=1}^{L} \int_0^{\infty} dx_l w(x_l) P_{\sigma(l)-1}(x_l) P_{\sigma'(l)-1}(x_l) \right)$$

$$\times P_{\sigma'(L+1)-1}(y)$$

$$= P_L(y) . \tag{5.27}$$

Here the orthogonality of the polynomials enforces $\sigma'(L+1) = L+1$, and the remaining norms cancel due to Eq. (5.23). As a check also the leading coefficient of Eq. (5.25) can be compared, taking the asymptotic limit $x \gg 1$, which leads to $x^L + \mathcal{O}(x^{L-1})$ on both sides.

Note that Eq. (5.25) provides an L-fold integral representation of the $P_L(x)$ for an arbitrary weight function. Compared to the known single integral representation, e.g. for the Hermite polynomials in the example in Eq. (5.20), we thus have the duality relation

$$\mathrm{He}_n(x) = \frac{1}{\sqrt{2\pi}} \int_{-\infty}^{\infty} dt (x + it)^n \exp[-t^2/2] . \tag{5.28}$$

Here we take a single average over n copies of characteristic polynomials. The existence of such relations between 1- and n-fold averages of characteristic polynomials are known only for Gaussian weights. They can also be derived using the supersymmetric method as reviewed in Chapter 7 of [13] by Thomas Guhr; see also [21] for further dualities using OP techniques.

In the next step we will express the kernel itself as an expectation value of two characteristic polynomials, following the work of P. Zinn-Justin [22]. Namely, it holds that

$$K_{N+1}(x, y) = h_N^{-1} \left\langle \prod_{j=1}^{N} (x - x_j)(y - x_j) \right\rangle_N . \tag{5.29}$$

The proof uses the same idea as before, now including each product into a different Vandermonde determinant of size $N + 1$. We have

$$\left\langle \prod_{j=1}^{N}(x - x_j)(y - x_j) \right\rangle_N$$

$$= \frac{1}{Z_N^{(2,N_f,\nu)}} \sum_{\sigma,\sigma' \in S_{N+1}} (-1)^{\sigma+\sigma'} \left(\prod_{l=1}^{N} \int_0^\infty dx_l w(x_l) P_{\sigma(l)-1}(x_l) P_{\sigma'(l)-1}(x_l) \right)$$

$$\times P_{\sigma(N+1)-1}(x) P_{\sigma'(N+1)-1}(y)$$

$$= \frac{1}{Z_N^{(2,N_f,\nu)}} N! \sum_{\sigma(N+1)=\sigma'(N+1)=1}^{N+1} \prod_{j=1}^{N+1} h_{j-1} \frac{1}{h_{\sigma(N+1)-1}} P_{\sigma(N+1)-1}(x) P_{\sigma'(N+1)-1}(y)$$

$$= \frac{1}{Z_N^{(2,N_f,\nu)}} N! \prod_{j=1}^{N+1} h_{j-1} \sum_{l=0}^{N} \frac{P_l(x) P_l(y)}{h_l} = \frac{P_{N+1}(x) P_N(y) - P_N(x) P_{N+1}(y)}{x - y}. \quad (5.30)$$

Compared to eq. (5.27) the orthogonality only implies that $\sigma(N + 1) = \sigma'(N + 1)$, and its value can still run over all possible values from 1 to $N + 1$. In the last line we give the expression for the kernel $K_{N+1}(x, y)$ in terms of the Christoffel–Darboux formula (5.18), using monic polynomials instead. Once again the leading coefficients of the left- and right-hand sides can be seen to agree $\sim (xy)^N$ in the limit $x, y \gg 1$.

An immediate question arises: What is the expectation value for more than two products of characteristic polynomials? For that purpose let us introduce some more notation following Baik et al. [20], to where we refer for more details. Denote by $P_n^{[l]}(x)$ the OP with respect to weight $w^{[l]}(x) = (\prod_{j=1}^{l}(y_j - x))w(x)$ for $l = 1, 2, \ldots$, with $P_n^{[0]}(x) = P_n(x)$. The following relation, which is called the Christoffel formula, holds:

$$P_n^{[l]}(x) = \frac{1}{(x - y_1) \cdots (x - y_l)} \begin{vmatrix} P_n(y_1) & \cdots & P_{n+l}(y_1) \\ \vdots & & \vdots \\ P_n(y_l) & \cdots & P_{n+l}(y_l) \\ P_n(x) & \cdots & P_{n+l}(x) \end{vmatrix} \left(\det_{1 \leq i,j \leq l} [P_{n+j-1}(y_i)] \right)^{-1}.$$

$$(5.31)$$

As explained in [20] this can be seen as follows. It is clear that the determinant in the middle, which we call $q_n^{[l]}(x)$, is a polynomial of degree $n + l$ in x. It has zeros at $x = y_1, \ldots, y_l$, and thus $q_n^{[l]}(x)/(x - y_1) \cdots (x - y_l)$ is a polynomial of degree n in x. Thus, we have that

$$\int_0^\infty dx\, x^j \frac{q_n^{[l]}(x)}{(x - y_1) \cdots (x - y_l)} w^{[l]}(x) = 0 \text{ , for } j = 0, 1, \ldots, n - 1 \text{ ,} \quad (5.32)$$

due to $P_n(x)$ being OP with respect to weight $w(x)$, after cancelling the factors from $w^{[l]}(x)$. The last factor in Eq. (5.31) ensures that $P_n^{[l]}(x)$ is monic, as can bee seen from

taking the limit $x \gg 1$. The repeated product of the Christoffel formula (5.31) leads to the following theorem (see Theorem 2.3 [20]) for the average of products of $l+1$ characteristic polynomials with $l = 1, 2, \ldots$, which is due to Brézin and Hikami [23]:

$$\frac{1}{\Delta_{l+1}(\{y\})} \det_{1 \leq i,j \leq l+1} [P_{n+j-1}(y_i)] = \prod_{j=0}^{l} P_n^{[j]}(y_{j+1}) = \left\langle \prod_{i=1}^{n} \left(\prod_{j=1}^{l+1} (y_j - x_i) \right) \right\rangle_n. \quad (5.33)$$

For the last equality we have used Eq. (5.25) which leads to

$$P_n^{[j]}(y_{j+1}) = \frac{\left\langle \prod_{i=1}^{n} \left((y_{j+1} - x_i) \prod_{p=1}^{j} (y_p - x_i) \right) \right\rangle_n}{\left\langle \prod_{i=1}^{n} \left(\prod_{p=1}^{j} (y_p - x_i) \right) \right\rangle_n}. \quad (5.34)$$

In the product almost all factors cancel out. The case for a single characteristic polynomial with $l + 1 = 1$ in Eq. (5.33) was already stated in Eq. (5.25).

In [20] in Theorem 3.2 a further identity was derived for the expectation value of the product of an even number of characteristic polynomials, expressing it through a determinant of kernels divided by two Vandermonde determinants,

$$\frac{\prod_{l=N}^{N+K-1} h_l}{\Delta_K(\{\lambda\}) \Delta_K(\{\mu\})} \det_{1 \leq i,j \leq K} [K_{N+K}(\lambda_i, \mu_j)] = \left\langle \prod_{i=1}^{N} \left(\prod_{j=1}^{K} (\lambda_j - x_i)(\mu_j - x_i) \right) \right\rangle_N. \quad (5.35)$$

The simplest example for this relation was derived in Eq. (5.29). In [24] this set of identities was further generalized, expressing the expectation value of arbitrary products of characteristic polynomials by a determinant containing both kernels and polynomials divided by two Vandermonde determinants in many different and equivalent ways, cf. Eq. (5.133) later in this chapter. The simplest example with three products reads [24]

$$\frac{h_N}{v_2 - v_1} \begin{vmatrix} K_{N+1}(v_1, u) & P_{N+1}(v_1) \\ K_{N+1}(v_2, u) & P_{N+1}(v_2) \end{vmatrix} = \left\langle \prod_{i=1}^{n} (v_1 - x_i)(v_2 - x_i)(u - x_i) \right\rangle_N. \quad (5.36)$$

Let us mention that in [20] also expectation values of ratios of characteristic polynomials were determined. They can be expressed in terms of the Cauchy transforms $C_k(x)$ of the monic polynomials $P_k(x)$,

$$C_k(y) = \frac{1}{2\pi i} \int dt \frac{P_k(t)}{t - y} w(t) , \quad y \in \mathbf{C} \setminus \mathbf{R} , \quad (5.37)$$

as well as through mixed Chistoffel–Darboux kernels containing both Cauchy transforms and polynomials. The simplest example is given by

$$C_{L-1}(x) = \frac{-h_{L-1}}{2\pi i} \left\langle \frac{1}{\prod_{j=1}^{L}(x - x_j)} \right\rangle_L. \quad (5.38)$$

The fact that both sides agree $\sim x^{-L}$ for $x \gg 1$ can be easily seen from the definition (5.37), expanding the geometric series and using the orthogonality of the polynomial $P_k(t)$ to monic powers less than k.

A second particularly important example is the expectation value of the ratio of two characteristic polynomials, due to the following relation to the resolvent or Stieltjes transform $G(x)$:

$$G_N(x) = \left\langle \sum_{j=1}^{N} \frac{1}{x - x_j} \right\rangle_N = \frac{\partial}{\partial x} \left\langle \frac{\prod_{j=1}^{N}(x - x_j)}{\prod_{j=1}^{N}(y - x_j)} \right\rangle_N \Bigg|_{x=y} . \tag{5.39}$$

The resolvent can be used to obtain the spectral density through the relation

$$R_1(x) = \frac{-1}{2\pi i} \lim_{\epsilon \to 0^+} [G_N(x + i\epsilon) - G_N(x - i\epsilon)] . \tag{5.40}$$

There are many examples in RMT where the OP technique is not available, but where the expectation value of ratios of characteristic polynomials can be found by other means, e.g., by using supersymmetry, replicas, or loop equations. This then leads to an alternative way for determining the spectral density, or higher k-point correlation functions, by taking the average over k resolvents and then taking the imaginary parts with respect to each of the k arguments. We refer to Chapters 7, 8, and 16 in [13] for further details and references.

5.2.3 All correlation functions with masses in terms of Laguerre polynomials

Using the results from the previous subsection we are now ready to express the partition function, OPs, kernel, and consequently all k-point correlation functions with N_f characteristic polynomials, as well as the corresponding gap probabilities through expectation values of characteristic polynomials with respect to the quenched weight $w(x) = x^\nu \exp[-x]$ with $N_f = 0$. Because the OPs of this quenched weight are Laguerre polynomials, everything will be finally expressed through these, which yields explicit and exact expressions for any finite N and N_f. In the next Subsection 5.2.4 will use them to take the large-N limit based on the known asymptotic of the Laguerre polynomials.

We shall adopt the notation from the previous section, labelling all the previous listed quantities by superscript $[N_f]$ compared to the quenched weight without superscript. We have for the partition function Eq. (5.2) with masses

$$Z_N^{[N_f]}(\{m\}) = \left(\prod_{j=1}^{N} \int_0^\infty dx_j \, x_j^\nu e^{-x_j} \prod_{f=1}^{N_f} (x_j + m_f^2) \right) \Delta_N(\{x\})^2$$

$$= Z_N \left\langle \prod_{j=1}^{N} \prod_{f=1}^{N_f} (x_j + m_f^2) \right\rangle_N$$

$$= N!(-1)^{N_f(N+(N_f-1)/2)} \prod_{j=1}^{N} \Gamma(j+\nu)$$

$$\times \prod_{j=1}^{N+N_f} \Gamma(j) \frac{\det_{1\leq i,j\leq N_f} \left[L_{N+j-1}^{\nu}(-m_i^2)\right]}{\Delta_{N_f}(\{-m^2\})}. \tag{5.41}$$

This result simply follows from choosing $y_j = -m_j^2$ in Eq. (5.33), together with Eqs (5.23) and (5.21) for the quenched polynomials. It is clear that if some of the masses become zero, say L out of N_f, from looking at Eq. (5.2) for $\beta = 2$ without further calculation this leads to the shift $\nu \to \nu + L$ in the remaining determinant of size $N_f - L$. This property is called flavour-topology duality. In the case when all masses are degenerate, $m_f = m \ \forall f$, l'Hopital's rule eventually leads to [23]

$$Z_N^{[N_f]}(m) = \frac{(-1)^{NN_f}}{\prod_{l=0}^{N_h-1} l!} \det_{0\leq i,j\leq N_f-1} \left[(-1)^{N+i}(N+i)!(L_{N+i}^{\nu}(-m^2))^{(j)}\right], \tag{5.42}$$

where the superscript $^{(j)}$ denotes the jth derivative with respect to the argument $x = -m^2$.

The unquenched OP $P_n^{[N_f]}(x)$ follow directly from Eq. (5.34) at $j = N_f$, with $x = y_{j+1}$. For example, using Eq. (5.21) the OP for the weight $w^{[1]}(x) = (x+m^2)x^{\nu} \exp[-x]$ reads as follows (cf. [29]):

$$P_n^{[N_f=1]}(x) = \frac{h_n K_{n+1}(x, -m^2)}{P_n(-m^2)} = \frac{(-1)^{n+1}(n+1)!}{(x+m^2)L_n^{\nu}(-m^2)}$$
$$\times \left(L_{n+1}^{\nu}(x)L_n^{\nu}(-m^2) - L_{n+1}^{\nu}(-m^2)L_n^{\nu}(x)\right). \tag{5.43}$$

However, it is much simpler to directly use the kernel from Eq. (5.29), rather than expressing it through the polynomials $P_N^{[N_f]}(x)$ using the Christoffel–Darboux identity (5.18). Indeed the kernel is given by Eq. (5.29) as

$$K_N^{[N_f]}(x,y) = \frac{1}{h_{N-1}^{[N_f]}} \left\langle \prod_{i=1}^{N-1} (x-x_i)(y-x_i) \right\rangle_{N-1}^{[N_f]}$$

$$= \frac{1}{h_{N-1}} \frac{\left\langle \prod_{i=1}^{N-1} \left((x-x_i)(y-x_i) \prod_{f=1}^{N_f}(m_f^2+x_i)\right) \right\rangle_{N-1}}{\left\langle \prod_{i=1}^{N} \prod_{f=1}^{N_f}(m_f^2+x_i) \right\rangle_N}$$

$$= \frac{(-1)^{N_f-1}(N+N_f)!}{\Gamma(N+\nu)(y-x) \prod_{f=1}^{N_f}(y+m_f^2)(x+m_f^2)} \times \begin{vmatrix} L_{N-1}^{\nu}(-m_1^2) & \cdots & L_{N+N_f}^{\nu}(-m_1^2) \\ \vdots & & \vdots \\ L_{N-1}^{\nu}(-m_{N_f}^2) & \cdots & L_{N+N_f}^{\nu}(-m_{N_f}^2) \\ L_{N-1}^{\nu}(x) & \cdots & L_{N+N_f}^{\nu}(x) \\ L_{N-1}^{\nu}(y) & \cdots & L_{N+N_f}^{\nu}(y) \\ \det_{1\leq f,g\leq N_f}[L_{N+g-1}^{\nu}(-m_f^2)] \end{vmatrix}. \tag{5.44}$$

In the first step after inserting Eq. (5.29) we have included the mass-dependent norm $h^{[N_f]}_{N-1}$ into the massive partition function in the denominator. It is given by the product of its norms times $N!$. Thus, we have increasing the average from $N-1$ to N. In the final result in the second line we have already taken out all signs and factorials from the two determinants and cancelled them partly. Also the ratio of the two Vandermonde determinants, one of which contains the arguments x and y, has been simplified using the definition (5.4). The kernel Eq. (5.44) determines all k-point eigenvalue correlation functions from Eq. (5.7) including their mass dependence. In particular for $x = y$ we obtain the spectral density with N_f masses,

$$R^{[N_f]}_1(x) = x^\nu \exp[-x] \prod_{f=1}^{N_f} (x + m_f^2)\, K^{[N_f]}_N(x,x) \,, \tag{5.45}$$

after applying l'Hôpital's rule once. This leads to $L^\nu_{N+g-1}(x)'$ in the last row in the numerator. Note that the product from the weight in front of the kernel cancels part of its denominator.

Let us now turn to the gap probabilities and distribution of smallest eigenvalues, starting with $E_0(s)$ from Eq. (5.10). The following Andréief integral identity (cf. Borodin's lectures) can be used to get a first expression:

$$\int dx_1 \ldots \int dx_N \det_{1 \le i,j \le N}[\phi_i(x_j)] \det_{1 \le i,j \le N}[\psi_i(x_j)] = N! \det_{1 \le i,j \le N}\left[\int dx \phi_i(x)\psi_j(x)\right]. \tag{5.46}$$

The only condition to hold is that all integrals of the functions $\phi_j(x)$ and $\psi_j(x)$ exist. The proof merely uses the Laplace expansion of the left-hand side,

$$\sum_{\sigma,\sigma' \in S_N} (-1)^{\sigma+\sigma'} \prod_{j=1}^N \int dx_{\sigma'(j)} \phi_{\sigma^{-1}(\sigma'(j))}(x_{\sigma'(j)})\psi_j(x_{\sigma'(j)})$$

$$= N! \sum_{\sigma'' \in S_N} (-1)^{\sigma''} \prod_{j=1}^N \int dx \phi_{\sigma''(j)}(x)\psi_j(x) \,, \tag{5.47}$$

and integration over common arguments of $\phi_i(x_{\sigma(i)} = x_{\sigma'(j)})$ and $\psi_j(x_{\sigma'(j)})$. This implies that $i = \sigma^{-1}(\sigma'(j)) = \sigma''(j)$, which is yet another permutation. Following Eq. (5.10) we can thus write

$$E^{[N_f]}_0(s) = \frac{1}{N!} \prod_{p=1}^N \int_s^\infty dx_p \det_{1 \le i,j \le N}\left[\sqrt{w^{[N_f]}(x_j)}\hat{P}^{[N_f]}_{i-1}(x_j)\right]^2$$

$$= \det_{1 \le i,j \le N}\left[\delta_{ij} - \int_0^s dx w^{[N_f]}(x)\hat{P}^{[N_f]}_{i-1}(x)\hat{P}^{[N_f]}_{j-1}(x)\right], \tag{5.48}$$

which can be interpreted as a Fredholm determinant. In the first step we have re-placed the Vandermonde determinants by determinants of monic polynomials, and then included the square root of the weight functions and of the norms stemming from the normalizing partition function into the determinants, in order to make the polynomials orthonormal, $\hat{P}_{i-1}^{[N_f]}(x_j)$. In the next step we have used the Andréief for-mula and the orthonormality on $[0, \infty)$. Knowing both the monic polynomials from Eq. (5.34) and their squared norms from $h_{j-1}^{[N_f]} = Z_j^{[N_f]}/(jZ_{j-1}^{[N_f]})$ in terms of expect-ation values of Laguerre polynomials, Eq. (5.48) is a valid expression for the gap $E_0(s)$. However, we will use a more direct expression to be derived later. As a fur-ther remark we note that replacing $\int_s^\infty dx = \int_0^\infty dx - \int_0^s dx$ in the first equality in (5.48) we obtain an equivalent expansion of the Fredholm determinant in the second equation:

$$E_0^{[N_f]}(s) = 1 + \sum_{k=1}^{N} \frac{(-1)^k}{k!} \int_0^s dx_1 \dots \int_0^s dx_k R_k^{[N_f]}(x_1 \dots, x_k). \qquad (5.49)$$

In order to express the gap probability directly as an expectation value let us go back to its definition (5.10), which we state for general β:

$$E_0^{[N_f]}(s) = \frac{1}{Z_N^{(\beta,N_f,\nu)}} \left(\prod_{j=1}^{N} \int_s^\infty dx_j \, x_j^{\frac{\beta}{2}(\nu+1)-1} e^{-x_j} \prod_{f=1}^{N_f} (x_j + m_f^2) \right) |\Delta_N(\{x\})|^\beta$$

$$= \frac{e^{-Ns}}{Z_N^{(\beta,N_f,\nu)}} \left(\prod_{j=1}^{N} \int_0^\infty dy_j \, y_j^0(y_j + s)^{\frac{\beta}{2}(\nu+1)-1} e^{-y_j} \prod_{f=1}^{N_f} (y_j + s + m_f^2) \right)$$

$$\times |\Delta_N(\{y\})|^\beta. \qquad (5.50)$$

Here we have shifted all integration domains, substituting $x_j = y_j + s$, $j = 1, \dots, N$. The shift can be simply worked out; in particular, it leaves the Vandermonde de-terminant invariant, As a consequence the numerator is again given by a partition of $N_f + \frac{\beta}{2}(\nu+1) - 1$ masses with values $\sqrt{m_f^2 + s} = m_f'$ for the first N_f, and $\sqrt{s}$ for the remaining ones, with an effective topological charge ν_{eff} parameter sat-isfying $\frac{\beta}{2}(\nu_{\text{eff}} + 1) - 1 = 0$. This idea has been used in several papers [25–29] to give closed form expressions as an alternative to the Fredholm determinant above.

Let us first consider $\beta = 2$, with $\frac{\beta}{2}(\nu+1) - 1 = \nu$. Consequently Eq. (5.50) can be written as

$$\underline{\beta = 2}: \ E_0^{[N_f]}(s) = e^{-Ns} \frac{Z_N^{(2,N_f+\nu,0)}}{Z_N^{(2,N_f,\nu)}} \left\langle \prod_{i=1}^{N} \left((s + y_j)^\nu \prod_{f=1}^{N_f} (m_f^2 + s + y_j) \right) \right\rangle_{N,\nu_{eff}=0}.$$
$$(5.51)$$

The simplest examples with $N_f = 0$ and $\nu = 0, 1$ thus read

$$\nu = 0: \quad E_0(s) = e^{-Ns}, \tag{5.52}$$

$$\nu = 1: \quad E_0(s) = e^{-Ns} L_N^0(-s) \binom{N+\nu}{N}^{-1}. \tag{5.53}$$

In the second line we have replaced the expectation value of a single characteristic polynomial Eq. (5.25) by the Laguerre polynomial Eq. (5.21) and fixed the normalization by the requirement $E_0(s = 0) = 1$. The quenched gap probability ($N_f = 0$) with arbitrary ν can be computed from the massive partition functions Eq. (5.42) at complete degeneracy.

The smallest eigenvalue distribution $p_1(s)$ easily follows for the examples we have just given by differentiation of the gap probability; see Eq. (5.12). For higher gap probabilities and for computing $p_k(s)$ directly the same trick from Eq. (5.50) can be used. First, consider only the integrals $\int_s^\infty dx$ in the definitions (5.10) and (5.11). Then do the shift $x_j = y_j + s$, and perform the remaining integrals $\int_0^s dx$ over the obtained expectation value of characteristic polynomials in the end. We refer to [29] for more details.

Let us now briefly comment on $\beta = 1, 4$, mainly because of recent developments and open questions in these symmetry classes. We begin with $\beta = 1$. Here the effective topological charge is $\nu_{\text{eff}} = 1$ to satisfy $0 = (\nu_{\text{eff}} - 1)/2$ in Eq. (5.50). The additional mass terms originating from the shift by s in Eq. (5.50) appear with multiplicity $(\nu - 1)/2$:

$$\underline{\beta = 1}: \quad E_0^{[N_f]}(s) = e^{-Ns} \frac{Z_N^{(1, N_f + (\nu-1)/2, 1)}}{Z_N^{(1, N_f, \nu)}}$$

$$\times \left\langle \prod_{i=1}^{N} \left((s + y_j)^{(\nu-1)/2} \prod_{f=1}^{N_f} (m_f^2 + s + y_j) \right) \right\rangle_{N, \nu_{eff}=1}. \tag{5.54}$$

That is, for $\nu = 2l + 1$ odd, $l \in \mathbf{N}$ we have l extra masses $\sqrt{s}$ compared to the N_f shifted masses. Here the simplest case is $l = 0$ with $N_f = 0$, for which we obtain

$$\nu = 1: \quad E_0(s) = e^{-Ns} \tag{5.55}$$

for arbitrary N. It agrees with the result for $\beta = 2$ at $\nu = 0$, Eq. (5.52). For general odd ν the gap probabilities have been computed and we refer to [29] for results. However, for $\nu = 2l$ even we obtain a half-integer number $l - \frac{1}{2}$ of additional mass terms. This leads to the question of calculating expectation values including products of square roots of characteristic polynomials in Eq. (5.54). In [30] the expectation values needed to determine $E_0(s)$ for arbitrary even $\nu = 2l$ and general N_f have been computed for the ch-GOE, answering at least the question for the gap probability. The cases $\nu = 0$ [31] and

$\nu = 2$ [32] were previously known from different considerations. Namely, in [33] a recursive construction was made for the smallest eigenvalue for $\beta = 1$ valid for all ν. However, in this approach the Pfaffian structure appearing for $\nu \geq 4$ is not at all apparent.

Independently in [34] the question about expectation values of square roots of determinants has been asked and answered for special cases for the GOE in the context of scattering in chaotic quantum systems. It is an open problem if the most general expectation value of such products (or ratios) of square roots has a determinantal or Pfaffian structure as in Eq. (5.33).

Let us turn to $\beta = 4$. The effective topological charge we obtain is $\nu_{\text{eff}} = -\frac{1}{2}$ to satisfy $0 = 2\nu_{\text{eff}} + 1$ in Eq. (5.50). The fact that it is not an integer does not pose a problem as the index of the generalized Laguerre polynomials Eq. (5.21) in terms of which we have expressed expectation values of characteristic polynomials can be chosen accordingly. The more severe problem is the following. Because of Kramer's degeneracy, the eigenvalues x_j of a self-dual quaternion valued $N \times N$ matrix $H \in \mathbf{H}$ always come in pairs, implying that $\det[x - H] = \prod_{j=1}^{N}(x - x_j)^2$. Consequently, the mass terms generated from Eq. (5.5) always occur with an *even* power N_f of twofold degenerate masses in Eq. (5.2). In contrast in Eq. (5.50) we always need to evaluate an odd number of powers $(y_j + s)^{2\nu+3}$. Thus, as for $\beta = 1$ with even ν we need to evaluate square roots of determinants in order to compute the gap probability for $\beta = 4$. This is an open problem so far and only the gap probability (and smallest eigenvalue distribution) with $\nu = 0$ is known explicitly [31]. A Taylor series expansion exists though for $\nu > 0$, following from group integrals of Kaneko type [35].

5.2.4 The large-N limit at the hard edge

After having exhaustively presented the solution of the eigenvalues model Eq. (5.2) and its correlation functions for a finite number of eigenvalues N at arbitrary N_f for $\beta = 2$, we will now turn to the large-N limit. As it is true in general in RMT one has to distinguish between different large-N scaling regimes. The global spectral statistics, which includes the global macroscopic density given by the semi-circle law for the GUE, or the Marchenko–Pastur law for the chGUE (see Fig. 5.2), is concerned with correlations between eigenvalues that have many other eigenvalues (in fact, a finite fraction of all) in between them. For a discussion of this large-N limit alternative techniques are available, such as loop equations, and we refer to [36] for a standard reference where all correlation functions including subleading contributions are computed recursively for $\beta = 2$. In this limit the fluctuations of the eigenvalues on a local scale (of a few eigenvalues) are averaged out, and typically expectation values factorize.

In contrast when magnifying the fluctuations among eigenvalues at a distance of $1/N^\delta$ one speaks of microscopic limits. The value of δ and the form of the limiting kernel depend on the location in the spectrum; see Fig. 5.2. At the so-called soft edge of the Marchenko–Pastur law one finds the Airy kernel, whereas in the bulk of the spectrum the sine kernel is found; see e.g., [16]. Here we will be interested in the eigenvalues in the vicinity of the origin presenting a hard edge with $\delta = \frac{1}{2}$ and the limiting kernel to be computed below is the Bessel kernel. Why this limit is relevant

for the application of RMT to QCD has been indicated already in the Introduction after Eq. (5.1).

As a further consequence of this comparison to QCD we will change variables from Wishart eigenvalues x_j of $WW^\dagger$ to Dirac operator eigenvalues $y_j = \pm\sqrt{x_j}$ (more details follow in Section 5.3). Looking at the partition function Eq. (5.2) the Dirac eigenvalues and masses have to be rescaled with the same power in N. Otherwise, we would immediately lose the dependence on the masses. On the RMT side the hard-edge scaling limit zooming into the vicinity of the origin is conventionally defined, including factors by taking $N \to \infty$ and $y_j, m_f \to 0$, such that the following product remains finite:[3]

$$\tilde{y}_j = \lim_{\substack{N \to \infty \\ y_j \to 0}} 2\sqrt{N}\, y_j \,, \quad \tilde{m}_f = \lim_{\substack{N \to \infty \\ m_f \to 0}} 2\sqrt{N}\, m_f \,. \tag{5.56}$$

In the previous subsection we eventually expressed all quantities of interest in terms of the Laguerre polynomials of the quenched weight. Therefore, the only piece of information we need to take the asymptotic limit of all these quantities, is the well-known limit [37]:

$$\lim_{N\to\infty} N^{-\nu} L_N^\nu \left(\frac{x}{N}\right) = x^{-\nu/2} J_\nu \left(2\sqrt{x}\right) \,. \tag{5.57}$$

Consequently, all Laguerre polynomials of positive argument turn into J-Bessel functions. Those polynomials containing negative arguments will turn into I-Bessel functions, due to the relation $I_\nu(z) = i^{-\nu} J_\nu(iz)$ for integer ν. Note that both types of Bessel functions also appear together; see, e.g., in Eqs (5.43) and (5.44). The asymptotics of the Laguerre OP, Eq. (5.57), in fact holds for OP with respect to much more general weight functions. This phenomenon is called universality. After first results in [26, 38, 39] more sophisticated rigorous mathematical methods including the Riemann–Hilbert approach were developed; see Chapter 6 by A. Kuijlaars in [13] for a detailed discussion and for references.

In the following we will give a few examples for the hard edge limit, starting with the partition function. Looking at Eq. (5.41) it is clear that we have to normalize it differently, such that the limit exists and gives a function of the limiting masses from Eq. (5.56). For $N_f = 1$ this is very easy to do by dividing out the quenched partition function, and we define

$$\mathcal{Z}_N^{[N_f=1]}(m) = m^\nu N^{-\frac{\nu}{2}} \frac{Z_N^{[N_f=1]}(m)}{Z_N N!} = m^\nu N^{-\frac{\nu}{2}} L_N^\nu(-m^2) \,, \tag{5.58}$$

leading to the limit

$$\lim_{\substack{N \to \infty \\ m \to 0}} \mathcal{Z}_N^{[N_f=1]}(m) = \mathcal{Z}^{[N_f=1]}(\tilde{m}) = I_\nu(\tilde{m}) \,. \tag{5.59}$$

---

[3] Note that we consider weights that are N-independent here, in contrast to other conventions.

Note that our conventions are chosen such that for small mass the limiting partition function behaves as $Z_N^{[N_f=1]}(\tilde{m} \approx 0) \sim \tilde{m}^\nu$, rather than unity. The reason for this choice will also become transparent in the next section. For general N_f we do not spell out the normalization constant explicitly and directly give the limiting result for nondegenerate masses

$$\lim_{\substack{N \to \infty \\ m_f \to 0}} Z_N^{[N_f]}(\{m\}) = Z^{[N_f]}(\{\tilde{m}\}) = 2^{\frac{N_f(N_f-1)}{2}} \prod_{j=0}^{N_f-1} \Gamma(j+1)$$

$$\times \frac{\det_{1 \le f,g \le N_f} \left[\tilde{m}_f^{g-1} I_{\nu+g-1}(\tilde{m}_f) \right]}{\Delta_{N_f}(\{\tilde{m}^2\})} , \tag{5.60}$$

where we adopted the normalization from [40]. In the completely degenerate limit this leads to

$$Z^{[N_f]}(\tilde{m}) = \det_{1 \le f,g \le N_f} \left[I_{\nu+g-f}(\tilde{m}) \right] . \tag{5.61}$$

The direct limit $N \to \infty$ of Eq. (5.42) (or l'Hopital's) rule from Eq. (5.60)) leading to a determinant of derivatives of I-Bessel functions can be brought to this form of a Toeplitz determinant using identities for Bessel functions.

As the next step we will directly jump to the limiting kernel, the reason being twofold. First, it is shorter to directly construct all correlation functions from the kernel, Eq. (5.44), rather than constructing the kernel through the massive OPs first. Second, and more importantly, the limit of the individual OPs does not necessarily exist. While here in the hard edge limit this is not the case, the well-known sine and cosine asymptotic of the Hermite polynomials in the bulk of the spectrum is only achieved after multiplying them with the square root of the weight function. We obtain for the limiting kernel from Eq. (5.44), after appropriately normalizing and changing to Dirac eigenvalues,

$$K_s^{[N_f]}(\tilde{y}_1, \tilde{y}_2) \sim \lim_{\substack{N \to \infty \\ x_1, x_2, m_f \to 0}} (w(x_1)w(x_2))^{\frac{1}{2}} K_N^{[N_f]}(x_1, x_2)$$

$$= \frac{1}{(\tilde{y}_1^2 - \tilde{y}_2^2) \prod_{f=1}^{N_f} \sqrt{(\tilde{y}_1^2 + \tilde{m}_f^2)(\tilde{y}_2^2 + \tilde{m}_f^2)}}$$

$$\times \frac{\begin{vmatrix} I_\nu(\tilde{m}_1) & \cdots & \tilde{m}_1^{N_f+1} I_{N_f+\nu+1}(\tilde{m}_1) \\ \vdots & & \vdots \\ I_\nu(\tilde{m}_{N_f}) & \cdots & \tilde{m}_{N_f}^{N_f+1} I_{N_f+\nu+1}(\tilde{m}_{N_f}) \\ J_\nu(\tilde{y}_1) & \cdots & (-\tilde{y}_1)^{N_f+1} J_{\nu+N_f+1}(\tilde{y}_1) \\ J_\nu(\tilde{y}_2) & \cdots & (-\tilde{y}_2)^{N_f+1} J_{\nu+N_f+1}(\tilde{y}_2) \end{vmatrix}}{\det_{1 \le f,g \le N_f} [m_f^{g-1} I_{\nu+g-1}(\tilde{m}_f)]} . \tag{5.62}$$

At $N_f = 0$ it reduces to the standard Bessel kernel:

$$\mathcal{K}_s(\tilde{x}, \tilde{y}) = -\frac{J_\nu(\tilde{x})\tilde{y}J_{\nu+1}(\tilde{y}) - J_\nu(\tilde{y})\tilde{x}J_{\nu+1}(\tilde{x})}{\tilde{x}^2 - \tilde{y}^2} = \frac{J_\nu(\tilde{x})\tilde{y}J_{\nu-1}(\tilde{y}) - J_\nu(\tilde{y})\tilde{x}J_{\nu-1}(\tilde{x})}{\tilde{x}^2 - \tilde{y}^2} .$$

(5.63)

Here we have given two equivalent forms of the same kernel using an identity from [37] for J-Bessel functions, $2\nu J_\nu(x) = x(J_{\nu+1}(x) + J_{\nu-1}(x))$. Following Eq. (5.7) we have already included the weight function into the limiting kernel (5.62). While the exponential part cancels out due to $\exp[-x = -\tilde{y}^2/4N] \to 1$ the remaining ν-dependent part is contributing. We can, thus, immediately write down an example, the quenched microscopic density

$$\rho_s(\tilde{x}) = \mathcal{K}_s(\tilde{x}, \tilde{x}) = \frac{\tilde{x}}{2} \left(J_\nu(\tilde{x})^2 - J_{\nu-1}(\tilde{x})J_{\nu+1}(\tilde{x}) \right) .$$

(5.64)

We have omitted here the ν delta functions of the exact zero eigenvalues. They will enter the discussion later in Subsection 5.4.1 where they can be seen. The density is plotted in Fig. 5.3 for $\nu = 0$ and 1.

The next step is to obtain the limiting distribution of the smallest eigenvalue. Because we change from Wishart to Dirac eigenvalues $s \to x^2$, we obtain the following scaling limit for the examples of the gap probability (5.52) and (5.53) at $N_f = 0$, using the same scaling for the Dirac eigenvalues as in Eq. (5.56):

$$\nu = 0 : \quad \lim_{\substack{N \to \infty \\ x \to 0}} E_0(s = x^2 = \tilde{x}^2/(4N)) = \mathcal{E}_0(\tilde{x}) = e^{-\tilde{x}^2/4} ,$$

(5.65)

$$\nu = 1 : \quad \lim_{\substack{N \to \infty \\ x \to 0}} E_0(s = x^2 = \tilde{x}^2/(4N)) = \mathcal{E}_0(\tilde{x}) = e^{-\tilde{x}^2/4} I_0(\tilde{x}) .$$

(5.66)

Note that the extra power of N from the shift (5.50) leads to the fact that here the exponential is not vanishing. The liming quenched gap probability for general ν follows from the completely degenerate massive partition function (5.61):

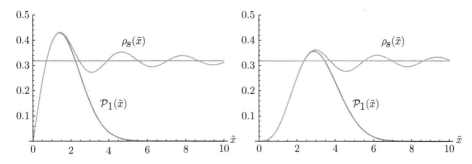

Fig. 5.3 The quenched microscopic spectral density from Eq. (5.64) and the distribution of the smallest eigenvalue for $\nu = 0$ (left) and $\nu = 1$ (right) using Eqs (5.68) and (5.69). It can be nicely seen how the distribution of the smallest eigenvalues follows the microscopic density for small $\tilde{x}$. The local maxima of the density further to the right mark the locations of the second, third, etc. eigenvalues; cf. [47]. The limiting value $\lim_{\tilde{x} \to \infty} \rho_s(\tilde{x}) = 1/\pi$ that equals the macroscopic density at the origin is marked as a horizontal line.

$$\mathcal{E}_0(\tilde{x}) = e^{-\tilde{x}^2/4} \det_{1 \le i,j \le \nu} [I_{i-j}(\tilde{x})] \;. \tag{5.67}$$

The limiting distribution of the corresponding smallest Dirac eigenvalue follows according to Eq. (5.12), after changing to Dirac eigenvalues,

$$\nu = 0: \quad \mathcal{P}_1(\tilde{x}) = -\partial_{\tilde{x}} \mathcal{E}_0(\tilde{x}) = \frac{\tilde{x}}{2} e^{-\tilde{x}^2/4} \;, \tag{5.68}$$

$$\nu = 1: \quad \mathcal{P}_1(\tilde{x}) = -\partial_{\tilde{x}} \mathcal{E}_0(\tilde{x}) = \frac{\tilde{x}}{2} e^{-\tilde{x}^2/4} I_2(\tilde{x}) \;, \tag{5.69}$$

where for the last step we have used an identity for Bessel functions. The last two formulas are compared to the quenched Bessel density, Eq. (5.64), in Fig. 5.3. It turns out that for general ν the derivative of the determinant in Eq. (5.67) can be written again as a determinant, following the ideas sketched after Eq. (5.53) that the distribution of the smallest eigenvalue can itself be expressed in term of an expectation value of characteristic polynomials,

$$\mathcal{P}_1(\tilde{x}) = \frac{\tilde{x}}{2} e^{-\tilde{x}^2/4} \det_{1 \le i,j \le \nu} [I_{i-j+2}(\tilde{x})] \;; \tag{5.70}$$

see [29] for a detailed derivation, including masses.

It should be mentioned that alternative representations exist for the gap probability Eq. (5.70) that are based on Fredholm determinant analysis [41]. They are given in terms of an exponential of an integral of the solution of a Painlevé V equation, which is valid for any real $\nu > -1$. For integer values of ν this expression reduces to the simpler forms that we have given here. Furthermore, in [42, 43] representations in terms of hypergeometric functions with matrix argument that are valid for all $\beta > 0$ were given. In [32] their equivalence to Eqs. (5.65)–(5.70) was proved for $\beta = 2$.

The distributions that we have computed so far, as for example the density and smallest eigenvalue in Fig. 5.3, have been extensively compared to numerical solutions from lattice QCD. For example, the quenched densitiy in Fig. 5.3 was matched with QCD in [44, 45], whereas the quenched distribution of the smallest eigenvalue also given in Fig. 5.3 was compared in [46] to QCD for different values of ν (and also to other QFTs corresponding to $\beta = 1, 4$). A matching to unquenched results for the smallest eigenvalues can be found in [47], and we refer to [8, 48] for a more exhaustive discussion of the lattice results.

5.3   Symmetries of QCD and its relation to RMT

Apart from giving nonspecialists some idea about the theory of quantum chromo-dynamics (QCD) we will have to answer three major questions in this section:

- How is the Dirac operator defined, the spectrum of which can be described by RMT?
- What are the global symmetries of QCD that determine the RMT to be used?
- What is the limit that leads from QCD to this RMT?

Let us give first short answers that also indicate how this section will be organized. The eigenvalues of the QCD Dirac operator will be partly described by RMT, actually only its smallest eigenvalues, and we will introduce this operator first in Subsection 5.3.1. The anti-Hermiticity properties of the Dirac operator and the concept of chiral symmetry will allow us to introduce some formal aspects and the definition of the QCD partition function. The breaking of chiral symmetry in Subsection 5.3.2 then leads us to the first approximation of QCD, to chiral perturbation theory that only contains the lightest excitations, the Goldstone Bosons. In a last step in Subsection 5.3.3 this interacting nonlinear, effective QFT is simplified in the epsilon regime, which agrees with RMT to leading order.

In what follows it is important that we talk about space-time coordinates $(x) = (\vec{x}, t) = (x_1, x_2, x_3, x_4)$ with components x_μ, $\mu = 1, 2, 3, 4$ equipped with the Euclidean metric $\delta_{\mu\nu}$ given by the Kronecker symbol.[4] This so-called Wick rotation from Minkowski to Euclidean metric is necessary for two reasons: only in this setting does the Dirac operator have definite anti-Hermiticity properties, and, second, only for QCD on a Euclidean space-time lattice, numerical solutions of QCD can be easily performed to which we can compare our RMT. We use Einstein's summation conventions, meaning that Greek indices appearing twice are automatically summed over from 1 to 4.

5.3.1   The Dirac operator and global symmetries of QCD

The QCD Dirac operator is a linear, matrix-valued differential operator defined as

$$\displaystyle{\not{D}} = \gamma_\mu {\not{D}}_\mu = \gamma_\mu \left(\partial_\mu + i g_s A_\mu(x) \right) = -{\not{D}}^\dagger , \tag{5.71}$$

where $\partial_\mu = \frac{\partial}{\partial x_\mu}$. The γ_μ are the Dirac gamma matrices in Euclidean space-time, satisfying the Clifford algebra

$$\{\gamma_\mu, \gamma_\nu\} = \gamma_\mu \gamma_\nu + \gamma_\nu \gamma_\mu = 2\delta_{\mu\nu}. \tag{5.72}$$

The curly brackets denote the anti-commutator. A standard representation as 4×4 matrices is given as follows in terms of the Pauli matrices σ_k, $k = 1, 2, 3$:

$$\gamma_k = \begin{pmatrix} 0 & i\sigma_k \\ -i\sigma_k & 0 \end{pmatrix}, \quad \gamma_4 = \begin{pmatrix} 0 & 1_2 \\ 1_2 & 0 \end{pmatrix}, \quad \gamma_5 = \gamma_1 \gamma_2 \gamma_3 \gamma_4 = \begin{pmatrix} 1_2 & 0 \\ 0 & -1_2 \end{pmatrix}. \tag{5.73}$$

All matrices are Hermitian, $\gamma_\mu^\dagger = \gamma_\mu$, including γ_5. This is only true in the Euclidean setting. Obviously, one can construct projection operators as

$$P_\pm = \frac{1}{2}(1_4 \pm \gamma_5), \quad P_\pm^2 = P_\pm, \quad P_+ P_- = P_- P_+ = 0, \quad P_+ + P_- = 1_4 . \tag{5.74}$$

The real parameter g_s in the Dirac operator is the coupling constant of the strong interactions; thus, $g_s = 0$ corresponds to the free Dirac operator without interactions.

---

[4] For this reason we don't need to distinguish between upper and lower indices as in Minkowski space-time.

It acts on 4-vectors $\psi(x)$ called spinors in Minkowski space, due to their transformation properties there, which we don't need to specify here. The real vector potential $A_\mu(x) \in SU_c(3)$ is an element of this Lie group for the three colours (the subscript c stands for colour). It is a scalar with respect to the γ_μ's, just as the partial derivatives in Eq. (5.71). The $\psi(x)$ come in three copies with colours blue b, green g, and red r, and so the interacting Dirac operator acts on the full vector $\Psi(x) = (\psi_b(x), \psi_g(x), \psi_r(x))$. For each quark flavour up, down, etc. we have such a vector $\Psi^q(x)$ with $q = 1, 2, \ldots N_f$. In the standard model of elementary particle physics we have $N_f = 6$, but often we will keep N_f fixed as a free parameter. We could have written the Dirac operator with tensor product notation to underline these structures, but the precise $SU_c(3)$ structure will not be crucial in the following.

We are now ready to state some important global symmetries of the Dirac operator. First, it follows from the algebra (5.72) that

$$0 = \{\slashed{D}, \gamma_5\} \quad \Rightarrow \quad \slashed{D} = \begin{pmatrix} 0 & i\mathcal{W} \\ i\mathcal{W}^\dagger & 0 \end{pmatrix}. \tag{5.75}$$

The Dirac operator is block off-diagonal also called chiral, where the symbol $\mathcal{W}$ still denotes a differential operator that depends on x and A_μ. Suppose we can find the eigenfunctions $\Phi_k(x)$ of the Dirac operator labelled by k

$$\slashed{D}\Phi_k(x) = i\lambda_k \Phi_k(x) , \tag{5.76}$$

in a suitably regularized setting (e.g., a finite box). Because of the anti-Hermiticity the eigenvalues $i\lambda_k \in i\mathbf{R}$ are purely imaginary. Furthermore, because of Eq. (5.75), multiplying Eq. (5.76) by γ_5 leads to different eigenfunctions $\gamma_5\Phi_k(x)$ with eigenvalues $-i\lambda_k$, provided that $\lambda_k \neq 0$. Thus, the nonzero eigenvalues of the Dirac operator come in pairs $\pm i\lambda_k$.

The Euclidean QCD partition function or path integral, as it is called in QFT, can be formally written down as

$$\mathcal{Z}^{\mathrm{QCD}} = \int [\mathrm{d}A_\mu] \int [\mathrm{d}\Psi] \exp\left[-\int \mathrm{d}^4x \sum_{q=1}^{N_f} \overline{\Psi}^q(x)(\slashed{D} + m_q)\Psi^q(x) - \int \mathrm{d}^4x \frac{1}{2}\mathrm{Tr}(F_{\mu\nu}F_{\mu\nu})\right]$$

$$= \int [\mathrm{d}A_\mu] \prod_{q=1}^{N_f} \det[\slashed{D} + m_q] \exp\left[-\int \mathrm{d}^4x \frac{1}{2}\mathrm{Tr}(F_{\mu\nu}F_{\mu\nu})\right] . \tag{5.77}$$

Here $\overline{\Psi}^q(x) = \Psi^{q\,\dagger}(x)\gamma_4$ is the Dirac conjugate and the $SU_c(3)$ field strength tensor is defined by the commutator $[\slashed{D}_\mu, \slashed{D}_\nu] = -ig_s F_{\mu\nu}$. The integration over $\mathrm{d}A_\mu$ and $\mathrm{d}\Psi$ are formal (path) integrals that we will not specify in the following; see standard textbooks on QCD for a discussion as [1]. In the second equation, however, we have formally integrated out the quarks due to the following observation. For a complex N-vector v and $N \times N$ matrix B we have

$$\int \mathrm{d}^{2N}v \exp[-v^\dagger Bv] \sim \frac{1}{\det[B]} , \quad \int \mathrm{d}^{2N}\psi \exp[-\psi^\dagger B\psi] \sim \det[B] , \tag{5.78}$$

whereas in the second identity we have integrated over a complex vector ψ of anti-commuting variables. Integration and differentiation over such fermionic variables can be defined in a precise way, and we refer to Chapter 7 in [13] for details. The reason for Fermions to be represented by anti-commuting variables is that in the canonical quantization of free Dirac fields one has to use anti-commutators for these fields.

Comparing Eq. (5.75) and the second line of Eq. (5.77) with Eq. (5.5) we get a first idea about the random matrix approximation of QCD, as was suggested initially in [6]: the Dirac operator from Eq. (5.75) is replaced by a constant matrix with the same block structure, $\mathcal{W} \to W$, and the average including the fields strength tensor is replaced by a Gaussian average over W. We further learn in this approximation that the eigenvalues $x_j = \lambda_k^2$ of the Gaussian random matrix $WW^\dagger$ correspond to the squared eigenvalues of the Dirac operator rotated to the real line, and that the number of eigenvalues N is proportional to the dimension of the QCD Dirac operator truncated in this way. However, as we will explain in the next subsection, the precise form of the limit from QCD to RMT was better understood later, representing a controlled approximation.

Furthermore, in the case when all masses vanish $m_q = 0$, the action in the exponent in the first line of Eq. (5.77) has an additional symmetry. Defining the vector $\Psi = (\Psi^1, \ldots \Psi^{N_f})$ as well as the projections $P_{\pm}\Psi = \Psi_{L/R}$ onto left (L) and right (R) chirality we can write the fermionic part of the action as

$$\sum_{q=1}^{N_f} \overline{\Psi}^q(x)\slashed{D}\Psi^q(x) = \overline{\Psi}_R(x)\slashed{D}\Psi_R(x) + \overline{\Psi}_L(x)\slashed{D}\Psi_L(x) . \tag{5.79}$$

It is invariant under the global rotations $\Psi_L(x) \to U_L\Psi_L(x)$ with $U_L \in U_L(N_f)$ and likewise for R. In contrast, even when all masses are equal, $m_q = m \; \forall q$, the mass term $m\overline{\Psi}\Psi = m(\overline{\Psi}_L\Psi_R + \overline{\Psi}_R\Psi_L)$ is only invariant under the diagonal transformation with $U_L = U_R$. In QCD the resulting global symmetry is $U_L(N_f) \times U_R(N_f) = U_V(1) \times U_A(1) \times SU_L(N_f) \times SU_R(N_f)$, with the $U(1)$-factors called $U_V(1)$ for vector and $U_A(1)$ for axial-vector split-off. The latter is broken through quantum effects also called an anomaly, whereas the former remains unbroken through the mass term. This leads to the explicit chiral symmetry breaking pattern in QCD:

$$SU_L(N_f) \times SU_R(N_f) \to SU(N_f) . \tag{5.80}$$

5.3.2   Chiral symmetry breaking and chiral perturbation theory

In this subsection we will describe the first step of simplifying QCD to what is called its low-energy effective theory. We begin with the following observation. It is found that the QCD vacuum $|0\rangle$, the ground state with the lowest energy, breaks chiral symmetry *spontaneously*, in precisely the same way as the theory with equal masses in Eq. (5.80) breaks it explicitly:

$$\Sigma = |\langle 0|\overline{\Psi}\Psi|0\rangle| = |\langle 0|\overline{\Psi}_L\Psi_R + \overline{\Psi}_R\Psi_L|0\rangle| \neq 0 . \tag{5.81}$$

The parameter Σ is called the chiral condensate and it is the order parameter for the spontaneous breaking of the chiral symmetry (5.80). It turns out that for QCD the low-energy phase with nonvanishing $\Sigma \neq 0$ coincides with the phase[5] where the quarks and the gluons, the carriers of the strong interaction represented by A_μ, are confined to colourless objects; see the phase diagram in Fig. 5.1. In the phase with $\Sigma \neq 0$ we need lattice QCD or other effective field theory descriptions (such as RMT) because a perturbative expansion in the coupling constant g_s will not work here, because the coupling is large.

An important consequence was drawn by Banks and Casher [5], relating the global continuum density $\rho_{\not{D}}(\lambda)$ of the Dirac operator eigenvalues in Eq. (5.76) at the origin, suitably regularized in a finite volume V, to Σ:

$$\rho_{\not{D}}(\lambda \approx 0) = \frac{1}{\pi}\Sigma V + |\lambda|\frac{\Sigma^2}{32\pi^2 F_\pi^4 N_f}(N_f^2 - 4) + o(\lambda) . \tag{5.82}$$

Here we added the second-order term to Eq. (5.1) derived by Smilga and Stern [49], containing the pion decay constant F_π as an additional parameter. The first term leads to the following interpretation: at the origin the global density is constant, and, thus, the average distance between eigenvalues there is $\lambda_k \sim 1/V$. This is very different from free particles in a box where $\lambda \sim 1/L$ with $V = L^4$ in 4 Euclidean dimensions. One can, thus, say that the smallest Dirac operator eigenvalues make an important contribution to build up the chiral condensate $\Sigma \neq 0$, by piling up very closely spaced at the origin. We are, thus, led to define rescaled, dimensionless eigenvalues and masses, as they appear on the same footing inside the determinant in Eq. (5.77), together with a rescaled microscopic density of the Dirac operator $\rho_s(\hat{\lambda})$:

$$\hat{\lambda}_k = \Sigma V \lambda_k, \quad \hat{m}_f = \Sigma V m_f, \quad \rho_s(\hat{\lambda}) = \lim_{V \to \infty} \frac{1}{\Sigma V}\rho_{\not{D}}(\hat{\lambda}/\Sigma V) . \tag{5.83}$$

This is called the microscopic limit of QCD. It can be compared, e.g., to Eq. (5.64) for $N_f = 0$, after identifying rescaled RMT eigenvalues $\tilde{y}_j = \hat{\lambda}_j$ and masses $\tilde{m}_f = \hat{m}_f$ from Eq. (5.56).

The breaking of a global continuous symmetry in any theory leads to Goldstone Bosons. This can be visualized by analogy in Fig. 5.4, showing the classical potential of a particle before (left) and after symmetry breaking (right). In the left picture with a convex potential, any excitation to make a particle move in the potential costs energy. After symmetry breaking the curvature at the origin has changed. Consequently, there now exists the possibility to excite a mode in angular direction along the valley of the potential, without costing potential energy. Modes that do need energy to be excited still exist in radial direction.

In our analogy particles that require energy to be excited correspond to massive particles. Consequently, after symmetry breaking there appear massless modes—the Goldstone Bosons—and modes that remain massive. If in QCD chiral symmetry was an exact symmetry, we would expect to observe exactly massless Goldstone Bosons

---

[5] In toy models of QCD, e.g., with two colours $SU_c(2)$ or different representations, this is not necessarily the case.

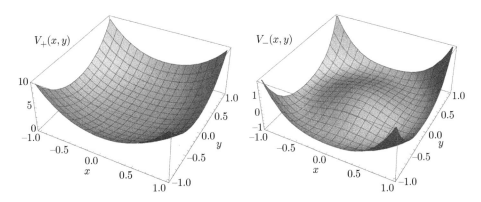

Fig. 5.4 The schematic potentials $V_\pm(x, y) = \pm 2(x^2 + y^2) + 1.5(x^2 + y^2)^2$ before symmetry breaking (V_+, left) and after breaking (V_-, right), obtained simply by switching the sign of the quadratic term.

in the phase with broken symmetry. However, it is only an approximate symmetry due to $m_q \neq 0$, leading to approximately massless Bosons after symmetry breaking. These are the three light pions observed in nature, $\pi^\pm$ and π^0, with masses of about 135 MeV, compared to the heavier non-Goldstone particles with masses of 800–1000 MeV. Recall that in the confined phase we only observe colourless bound states made from two quarks, mesons, which are Bosons, from three quarks, and baryons, which are Fermions, or more.[6] The pions are unstable and decay mostly into a muon and its anti-neutrino for π^+, the corresponding anti-particles for π^-, or two photons for π^0.

The number of Goldstone Bosons equals the dimension of the coset $SU_L(N_f) \times SU_R(N_f)/SU(N_f)$ of the groups before and after breaking, which is $N_f^2 - 1$, the dimension of $SU(N_f)$ in our case. To consider the pions as Goldstone Bosons thus means that we only keep the lightest quarks up and down, so that $N_f = 2$ in our theory. For many purposes at low energy this is a good approximation. Sometimes also the strange quark is included, with $N_f = 3$ leading to consider the eight lightest particles as Goldstone Bosons.

If we neglect all non-Goldstone Bosons in our theory—as they will not be relevant at low enough energies—and formally integrate them out, we arrive at the low-energy effective theory of the Goldstone Bosons,

$$\mathcal{Z}_{\text{chPT}} = \int [dU] \exp\left[-\int d^4x \frac{F_\pi^2}{4} \text{Tr} \left(\partial_\mu U(x) \partial_\mu U(x)^\dagger \right) + \int d^4x \frac{\Sigma}{2} \text{Tr} \left(M(U(x) + U^\dagger(x)) \right) \right],$$
(5.84)

where

$$U(x) = U_0 \exp\left[i \frac{\sqrt{2}}{F_\pi} \pi^b(x) t^b \right], \quad M = \text{diag}(m_1, \ldots, m_{N_f}).$$
(5.85)

[6] During the delivery of these lectures in July 2015 the discovery of the pentaquark was announced at CERN.

Here the implicit sum over b runs from 1 to $N_f^2 - 1$. For $N_f = 2$ the three fields $\pi^b(x)$ actually denote the pion fields, with $t^3 = \sigma_3$ and $t^\pm = \sigma^\pm = \sigma^1 \pm \sigma^2$. In Eq. (5.85) we have already split off the constant, x-independent modes U_0 of the fields $\pi^b(x)$ for later convenience. Usually at this stage U_0 is set to the identity. Equation (5.84) represents the partition function or path integral of chiral perturbation theory (chPT), which replaces that of QCD Eq. (5.77) in the low-energy regime. In Eq. (5.84) we only give the leading order (LO) Lagrangian, all higher order terms in powers of $U(x)$ and its derivatives that are invariant exist, and we will come back to the question when they contribute in the next section. For more details about chPT we refer to [4].

ChPT is an interacting, nonlinear QFT. When expanding the exponential in Eq. (5.85) in powers of $\pi^b(x)$ and considering only quadratic terms we obtain the standard Lagrangian for scalar fields that leads to the Klein–Gordon equation. From that expansion we can make the following identification for the masses M_π of the pion fields $\pi^b(x)$, the Gell-Mann–Oakes–Renner (GOR) relation:

$$F_\pi^2 M_\pi^2 = \Sigma(m_1 + \ldots + m_{N_f}) \,. \tag{5.86}$$

The QFT in Eq. (5.84) is not yet related to RMT, and we turn to this limit in the next subsection.

5.3.3   The limit to RMT: the epsilon regime

The so-called epsilon regime of chPT (echPT) was introduced by Gasser and Leutwyler [50] before a RMT of QCD was constructed. Their motivation was to introduce a regime where analytic computations in chPT were feasible, in particular to understand the mass dependence of the partition function and the role of topology. Leutwyler and Smilga later computed the partition function analytically in this regime and deduced so-called sum rules for the Dirac operator eigenvalues [51]. Working in a finite volume $V = L^4$ of linear size L, the authors of [50] introduced the scaling

$$\epsilon = L^{-1} \;\Rightarrow\; V \sim \epsilon^{-4}, \; \partial_\mu \sim \epsilon, \; \pi^b(x) \sim \epsilon \,. \tag{5.87}$$

This scaling implies, in particular, that

$$\Sigma V m_f = \mathcal{O}(1), \; \int d^4 x \, \partial_\mu \pi^b(x) \partial_\mu \pi^b(x) = \mathcal{O}(1), \tag{5.88}$$

where the former follows from the microscopic scaling limit of the Banks–Casher relation (5.83), and the latter ensures that the kinetic term of the LO action is dimensionless. The physical interpretation of the epsilon regime follows from the scaling of the pion mass, which is implied from multiplying the GOR relation Eq. (5.86) with V. It must hold that $M_\pi \sim L^{-2}$, such that both sides are of the order $\mathcal{O}(1)$. In other words, the Compton wavelength of the pion fields is much larger than the size of the box,

$$M_\pi^{-1} \sim L^2 \gg L \,, \tag{5.89}$$

which is clearly an unphysical regime. Normally, one would consider this to be un-desirable, because in order to observe physical, propagating pions one would choose a scaling such that the pions do fit into the box V, with $M_\pi^{-1} < L$. In the so-called p-regime both momenta and pion mass are scaled in the same way, ∂_μ, $M_\pi \sim \epsilon$, where such an inequality can be satisfied. We will come back to this issue very briefly later. In the epsilon regime introduced earlier we obtain the limit [50]

$$\lim_{\epsilon\text{-regime}} \mathcal{Z}_{\text{chPT}} = \int [\mathrm{d}\pi] \exp\left[-\int \mathrm{d}^4x \frac{1}{2}\partial_\mu \pi^b(x)\partial_\mu \pi^b(x)\right]$$

$$\times \int_{SU(N_f)} \mathrm{d}U_0 \exp\left[\frac{1}{2}\Sigma V \mathrm{Tr}(M(U_0 + U_0^\dagger)) + \mathcal{O}(\epsilon^2)\right]. \quad (5.90)$$

Here the propagating modes $\pi^b(x)$ completely factorize from the zero modes U_0 and contribute to the partition function as a free Gaussian field theory—which is just a constant factor. The reason that we do not expand in the constant modes of the pions as well, but keep them as a full matrix U_0 as in Eq. (5.85), is that their quantum fluctuations are of $\mathcal{O}(1) = V/M_\pi^2$; see [9] for a more detailed discussion. All higher order terms including those not given in the chiral Lagrangians (5.84) are subleading in this limit. It is in this limit that the mass dependence of the limiting partition function Eq. (5.90) and of RMT agree, and in fact also all Dirac operator eigenvalue correlation functions. Before we make this statement more precise let us add further comments. First, it is not only possible to make analytical computations in the epsilon regime, but also to compare them to lattice QCD. On the lattice the simulation parameters can be freely chosen, such that the epsilon regime is reached and a quantitative comparison is possible. Second, it is possible to compute so-called one or higher loop corrections to the partition function (and correlation functions) Eq. (5.90). For example writing down the $\mathcal{O}(\epsilon^2)$ term explicitly, expanding the exponential and using Wick's theorem with the finite volume propagator, corresponds to computing the one-loop corrections. Third, one may perform also perturbative computations in the physical p-regime, or even choose an extrapolation between the two regimes; cf. [9, 52] for a discussion and references.

Let us now make the map between RMT and echPT more precise. For that purpose it is convenient to introduce an extra term in the QCD Lagrangian, the so-called theta-vacuum term containing the dual field strength tensor $\tilde{F}_{\rho\sigma} \sim \varepsilon_{\rho\sigma\mu\nu}F^{\mu\nu}$. It is defined using the totally antisymmetric epsilon tensor $\varepsilon_{\rho\sigma\mu\nu}$ in four dimensions. The extra term in (5.77) reads in appropriate normalization:

$$-i\theta \int \mathrm{d}^4x \mathrm{Tr}\tilde{F}_{\rho\sigma}F_{\rho\sigma} = -i\nu\theta . \quad (5.91)$$

It is a topological term, and here ν is the number of left-handed (L) minus right-handed (R) zero modes of the Dirac operator Eq. (5.71), which so far has not made an appearance in this section about QCD. Due to the Atiah–Singer index theorem ν is equal to the winding number of the gauge fields A_μ, which is also a topological quantity. In the partition function Eq. (5.77) we have thus implicitly summed over all the topologies, which can now be written as

$$Z^{\mathrm{QCD}}(\theta) = \sum_{\nu=-\infty}^{\infty} e^{-i\nu\theta} Z_\nu^{\mathrm{QCD}} . \tag{5.92}$$

Here Z_ν^{QCD} denotes the QCD partition function at fixed topology. Inverting this equation it can be written as

$$Z_\nu^{\mathrm{QCD}} = \frac{1}{2\pi} \int_{-\pi}^{\pi} d\theta e^{+i\nu\theta} Z^{\mathrm{QCD}}(\theta) . \tag{5.93}$$

What is the advantage? In RMT in the previous section we always had a fixed $\nu \in \mathbf{N}$ — to which we should now compare for nonnegative values.[7] Furthermore, the unitary group integral over $SU(N_f)$ that we found in the epsilon regime Eq. (5.90) can now be promoted to an integral over the full unitary group, according to

$$\int_{-\pi}^{\pi} \frac{d\theta}{2\pi} e^{i\nu\theta} \int_{SU(N_f)} dU_0 = \int_{U(N_f)} dU_0 \det[U_0]^\nu . \tag{5.94}$$

Defining the rescaled quark masses in the epsilon regime as

$$\hat{M} = \mathrm{diag}(\hat{m}_1, \dots, \hat{m}_{N_f}) = \Sigma V M, \tag{5.95}$$

we obtain the limiting relation between the mass-dependent part of the QCD partition function in the epsilon regime at fixed topology and RMT:

$$\begin{aligned}
Z_{\mathrm{echPT}}^{[N_f]}(\{\hat{m}\}) &= \int_{U(N_f)} dU_0 \det[U_0]^\nu \exp\left[\frac{1}{2}\mathrm{Tr}(\hat{M}(U_0 + U_0^\dagger))\right] \\
&= 2^{\frac{N_f(N_f-1)}{2}} \prod_{j=0}^{N_f-1} \Gamma(j+1) \frac{\det_{1\le i,j\le N_f}[\hat{m}_i^{j-1} I_\nu(\hat{m}_i)]}{\Delta_{N_f}(\{\hat{m}^2\})} = Z^{[N_f]}(\{\hat{m}\}) .
\end{aligned} \tag{5.96}$$

In the last equality with the limiting RMT partition function from Eq. (5.60) we have identified rescaled random matrix and QCD quark masses $\tilde{m}_f = \hat{m}_f$ for all flavours. The computation of the integral over the appropriately normalized Haar measure of the unitary group $U(N_f)$ in Eq. (5.96) goes back to [53]; see also [54] and [40] for our normalization. A concise derivation in terms of group characters was presented in [40]. In the simplest case with $N_f = 1$ it is easy to see that

$$Z_{\mathrm{echPT}}^{[N_f=1]}(\hat{m}) = \int_{-\pi}^{\pi} \frac{d\theta}{2\pi} e^{i\nu\theta} \exp\left[\frac{1}{2}\hat{m}(e^{i\theta} + e^{-i\theta})\right] = I_\nu(\hat{m}) , \tag{5.97}$$

which is one of the standard definitions of the modified Bessel function. Furthermore, we can also verify the completely degenerate case $m_f = m$, $\forall\ f$, by applying the Andréief formula (5.46). Diagonalizing U_0 by a unitary transformation

---

[7] The QCD partition function is thought to be symmetric with respect to ν.

$U \in U(N)/U(1)^N$, $U_0 = U\text{diag}(e^{i\theta_1}, \ldots, e^{i\theta_{N_f}})U^\dagger$, we obtain again the squared Vandermonde determinant as a Jacobian:

$$\int_{U(N_f)} dU_0 \det[U_0]^\nu \exp\left[\frac{\hat{m}}{2}\text{Tr}(U_0 + U_0^\dagger)\right]$$

$$= const. \prod_{j=1}^{N_f} \left(\int_{-\pi}^\pi \frac{d\theta_j}{2\pi} \exp[i\nu\theta_j + \hat{m}\cos(\theta_j)]\right) |\Delta_{N_f}(\{e^{i\theta}\})|^2$$

$$= N_f! \, const. \det_{1 \leq j,k \leq N_f} \left[\int_{-\pi}^\pi \frac{d\theta}{2\pi} \exp[i(\nu + k - j)\theta + \hat{m}\cos(\theta)]\right]$$

$$= \det_{1 \leq j,k \leq N_f} [I_{\nu+k-j}(\hat{m})] . \tag{5.98}$$

It agrees with Eq. (5.61) from RMT, after appropriately normalizing. We observe that for both degenerate and nondegenerate masses the partition functions with N_f flavours Eqs (5.96) and (5.98) are determinants of single-flavour partition functions with $N_f = 1$.

The matching of the limiting RMT partition function and the group integral from the epsilon regime of chPT at fixed topology is only part of their equivalence. In particular this matching does not imply that all Dirac operator eigenvalue correlation functions agree. For example, at $N_f = 1$ the echPT partition function agrees for all three symmetry classes $\beta = 1, 2, 4$ [55], while their eigenvalue densities differ. At first sight it is not clear how to access the Dirac operator spectrum once we have moved from the quarks to the Goldstone Bosons as fundamental low-energy degrees of freedom. However, looking at the alternative construction of the spectral density via the resolvent Eq. (5.39) helps. Consider adding an auxiliary pair of quarks with masses x and y, respectively, to the QCD Lagrangian Eq. (5.77). If we arrange for the second quark to be bosonic, we obtain the following average with respect to the unperturbed QCD partition function, at fixed topology:

$$\lim_{\epsilon\text{-regime}} \left\langle \frac{\det[x - \slashed{D}]}{\det[y - \slashed{D}]} \right\rangle_\nu^{\text{QCD}} = \frac{\int_{U(N_f+1|1)} dU_0 \, \text{sdet}[U_0]^\nu \exp\left[\frac{1}{2}\text{sTr}(\hat{\mathcal{M}}(U_0 + U_0^\dagger))\right]}{\mathcal{Z}_{\text{echPT}}^{[N_f]}(\{\hat{m}\})} , \tag{5.99}$$

with $\hat{\mathcal{M}} = \text{diag}(\hat{m}_1, \ldots, \hat{m}_{N_f}, \hat{x}; \hat{y})$, in the limit of the epsilon regime. Because of the bosonic nature of the second auxiliary quark in the denominator the integral is now over the supergroup $U(N_f + 1|1)$ with the corresponding supersymmetric trace and determinant, sTr and sdet, respectively; see, e.g., Chapter 7 of [13]. Its evaluation and differentiation to obtain the resolvent must be done carefully. In [56], to where we refer for details, it was shown in this fashion that the microscopic density Eq. (5.64) and its extension to arbitrary N_f follows from Eq. (5.99), thus establishing the equivalence to RMT on the level of the microscopic density. In order to know the distribution of the smallest eigenvalues the knowledge of all density correlation functions is necessary, see Eq. (5.49), as was pointed out in [18]. The equivalence proof for all k-point density correlation functions was achieved in [57] by directly matching the k-point resolvent

generating functions with $2k$ additional ratios of quark determinants of both theories, using superbosonisation techniques.

A few comments regarding universality and corrections to the RMT regime are in place here. Already the fact that two apparently different theories as echPT and RMT yield the same Dirac operator correlation functions is a striking result which is based on symmetries and universality. It is thus surprising that even when taking into account one-loop corrections to echPT the same universal correlation functions were found in [58]. The only corrections to be made are subsumed in a modification $\Sigma \to \Sigma_{\rm eff} = \Sigma(1 + C_1/\sqrt{V})$, and $F_\pi \to F_{\rm eff} = F_\pi(1 + C_2/\sqrt{V})$, where C_1 and C_2 are constants that depend on N_f, F_π, and the geometry of the discretization; see, e.g., [59, 60] for details and references. So far F_π did not appear in eigenvalue correlation functions, but later when including a chemical potential it will. This modification means that the chiral condensate (and pion decay constant) and thus the rescaled masses in Eq. (5.95) and eigenvalues simply get renormalized accordingly. Only two-loop corrections within echPT contain nonuniversal terms that lead out of the RMT universality class [61].

However, even at the LO the agreement between the RMT and echPT rapidly breaks down when going to higher energies, meaning to higher rescaled Dirac operator eigenvalues. At some point the fluctuating modes will start to contribute, and eventually also the chPT approach breaks down as contributions from full QCD will appear. The scale where this happens has been called Thouless energy [62, 63] in analogy to applications in condensed matter physics, to where we refer for details. The point where this happens scales with $1/\sqrt{V}$ and F_π^2. Beyond this scale one sees a rise of the global spectral density as in Eq. (5.82), away from the RMT prediction. The scaling of the Thouless energy has been confirmed from lattice data in a different theory, in QFT with two colours [64]. For the same theory deep in the bulk regime the local statistics was seen to agree again with RMT bulk statistics of the GSE [65], which is the nonchiral $\beta = 4$ symmetry class.

Here we have already mentioned one of the two different symmetry classes corresponding to $\beta = 1$ and 4 in Eq. (5.2). They are also relevant but for different QFTs other than QCD, e.g., with only two colours $SU_c(2)$. These have chiral symmetry patterns different from that of Eq. (5.80) and we refer to [8] for a detailed discussion. The matching with the three chiral RMTs in Eq. (5.2) was initially proposed in [17]. While the agreement with lattice simulations of the respective theories (cf. [48]) leave little doubt that these RMT are equivalent to the epsilon regime of the corresponding chPT, such an equivalence has not been established even on the level of the partition function for general N_f. This is due to the more complicated structure of the corresponding group integral; see, however, [55] and [66] for a discussion.

Let us briefly comment also on the influence of the dimension. In three space-time dimensions there is no chiral symmetry, and the RMT for QCD is given by the GUE instead [67], including mass terms. In two dimensions, which again has chiral symmetry, the classification is much richer, and we refer to [68] for details. This makes contact with the symmetry classification of topological insulators [69] and the so-called Bott periodicity in any dimension.

5.4 Recent developments

In this section we will cover some recent developments in the application of RMT to QCD, where the detailed dependence of the partition function and Dirac operator eigenvalue correlation functions on finite lattice spacing a or chemical potential μ is computed. In Subsection 5.4.1 we will study the influence of a finite space-time lattice on the spectrum as it is expected to be seen in numerical simulation of QCD on a finite lattice far enough from the continuum. The corresponding RMT and effective field theory that we will consider go under the name of Wilson (W)RMT as introduced in [70], and Wilson (W)chPT; see [71] for a standard reference. We will only sketch the ideas beginning with the symmetries and WchPT, and then write down the joint density of eigenvalues of the Hermitian Wilson Dirac operator. Its solution and its non-Hermitian part will be referred to the literature.

The addition of a chemical potential term for the quarks to the Dirac operator renders its eigenvalues complex. The solution of the corresponding RMT using OP in the complex plane is sketched in Subsection 5.4.2, where we will give some details. The first application of such a RMT was already introduced by Stephanov in 1996 [72] to explain the difference between the quenched and unquenched theory. The development of the corresponding theory of OP in the complex plane started later in [73] and [74]. It is already partly covered in the Les Houches lecture notes from 2005 [10]; for a slightly more recent review on RMT of QCD with chemical potential, see [11]. Here we will comment on some recent developments relating to products of random matrices, see [75] for a detailed review (and Chapter 10 by Comtet and Tourigny in this volume).

5.4.1 RMT and QCD at finite lattice spacing—the Wilson Dirac operator

The discretization of derivatives on a space-time lattice is not a unique procedure, which also applies to the Dirac operator in Eq. (5.71). We would, thus, first like to discuss why in this subsection we will study the modification proposed by Wilson, the Wilson Dirac operator,

$$D_W = \slashed{D} + a\Delta \neq -D_W^\dagger , \qquad (5.100)$$

which is no longer anti-Hermitian. Here a is the lattice spacing and Δ denotes the Laplace operator, which is Hermitian. Of course, both operators still must be discretized. We cannot possibly give justice to the vast literature on this subject and will only focus on aspects relevant for the application of RMT. The reason Wilson proposed to add the Laplacian is the so-called doubler problem. In the continuum with $a = 0$ the relativistic energy-momentum relation for a particle with mass M, energy E, and 4-momentum k_μ reads

$$E^2 - M^2 = \sum_{\mu=1}^{4} k_\mu^2 . \qquad (5.101)$$

Here we have spelled out the sum explicitly. Particles satisfying this relation are called on-shell (on the energy shell). After standard discretization this relation becomes

$$E^2 - M^2 = \sum_{\mu=1}^{4} \frac{\sin(k_\mu a)^2}{a^2} , \tag{5.102}$$

which in the limit $a \to 0$ leads back to Eq. (5.101). In contrast to the continuum relation, here with any k_μ satisfying this equation also $(k_1 - \frac{\pi}{a}, k_2, k_3, k_4)$, $(k_1, k_2 - \frac{\pi}{a}, k_3, k_4)$, etc., fulfil Eq. (5.102). In total we have $2^4 = 16$ possibilities to add $-\pi/a$ to the components of k_μ, which all correspond to the same on-shell particle. This is the doubler problem. Wilson's modification to add the Laplacian to the Dirac operator amounts to adding the term $\gamma_\mu \sin(k_\mu a)$ to the right-hand side of Eq. (5.102). In the limit $a \to 0$ this makes the extra 15 particles heavy and thus removes the doublers. The modification comes with a price, as the Laplacian explicitly breaks chiral symmetry. We will not discuss other possibilities here, for example in [76] for a RMT of the so-called staggered Dirac operator, but rather stick to Wilson's choice. Despite the non-Hermiticity of D_W Eq. (5.100) it satisfies the so-called γ_5-Hermiticity:

$$D_W^\dagger = -\slashed{D} + a\Delta = \gamma_5 D_W \gamma_5 . \tag{5.103}$$

Here we have simply used that $(\gamma_5)^2 = 1_4$ and the anti-commutator from Eq. (5.75). Consequently one can define the following Hermitian operator called D_5,

$$D_5 = \gamma_5(D_W + m) = D_5^\dagger , \tag{5.104}$$

where traditionally the quark mass m is added to the definition. Looking at Eq. (5.75) for the Dirac operator and Eq. (5.73) for γ_5 we can immediately give the chiral block structure of D_W and D_5:

$$D_W = \begin{pmatrix} a\mathcal{A} & i\mathcal{W} \\ i\mathcal{W}^\dagger & a\mathcal{B} \end{pmatrix} , \quad D_5 = \begin{pmatrix} m & i\mathcal{W} \\ -i\mathcal{W}^\dagger & -m \end{pmatrix} + a \begin{pmatrix} \mathcal{A} & 0 \\ 0 & -\mathcal{B} \end{pmatrix}. \tag{5.105}$$

Here $\mathcal{A}$ and $\mathcal{B}$ are Hermitian operators. Note the change in sign in the mass term in the lower right block of D_5. As previously for $a = 0$ this immediately leads to a good Ansatz to make for a WRMT, as it was made in [70]. Before we turn to analyse this in some detail let us directly go to the epsilon regime of QCD in the Wilson formulation at fixed topology, WechPT. It turns out that at LO three extra terms appear in the chiral Lagrangian as was discussed in [71] (note the different sign convention).[8] They contain three new low-energy constants (LEC), W_6, W_7, and W_8, and all terms are proportionally to a^2:

$$\mathcal{Z}_{\text{WechPT}}^{[N_f]}(\{\hat{m}\}) = \int_{U(N_f)} dU_0 \det[U_0]^\nu \exp\left[+\frac{1}{2}\Sigma V \text{Tr}(M(U_0 + U_0^\dagger)) \right.$$
$$- a^2 V W_8 \text{Tr}(U_0^2 + U_0^{\dagger\,2}) - a^2 V W_6 (\text{Tr}(U_0 + U_0^\dagger))^2$$
$$\left. - a^2 V W_7 (\text{Tr}(U_0 - U_0^\dagger))^2 \right] . \tag{5.106}$$

---

[8] One may wonder how it is possible to describe QCD in the Wilson discretization scheme by a chiral Lagrangian in the continuum. Close to the continuum limit this is a good approximation; see [71] for further details.

It is clear that this implies the scaling $\hat{a}^2 = a^2 V$ or $a \sim \epsilon^2$, in addition to the scaling of the quark masses with the volume in Eq. (5.88). Let us discuss some simplifications first. Without fixing topology in the special case of $SU(2)$ the term $\mathrm{Tr}(U_0 - U_0^\dagger)$ vanishes, and the terms proportional to W_8 and W_6 are equivalent. In general, in the second line of Eq. (5.106) the two squares of the traces can be linearized by so-called Hubbart–Stratonovich transformations, at the expense of two extra Gaussian integrals. The linearized terms can be included into a shift of the mass for W_6, and into so-called axial mass terms $\frac{1}{2}\mathrm{Tr}(\hat{Z}(U_0 - U_0^\dagger))$ for W_7, see [77] for more details. Also for the latter term the corresponding group integral generalizing Eq. (5.96) is known; see [78]. Therefore, in the following we will set $W_6 = 0 = W_7$, having in mind that we need to perform two extra integrals for the full partition function (5.106) for nonzero values of these LECs.

Let us define $\hat{a}_8^2 = a^2 V W_8$ and compute the partition function only including this term. For $N_f = 1$ we obtain

$$
\begin{aligned}
\mathcal{Z}_{\mathrm{WechPT}}^{[N_f=1]}(\hat{m}) &= \int_{-\pi}^{\pi} \frac{\mathrm{d}\theta}{2\pi} e^{i\nu\theta + \hat{m}\cos(\theta) - 4\hat{a}_8^2 \cos(\theta)^2 + 2\hat{a}_8^2} \\
&= e^{2\hat{a}_8^2} \int_{-\infty}^{\infty} \frac{\mathrm{d}x}{\sqrt{\pi}} e^{-x^2} \left(\frac{\hat{m} - 4ix\hat{a}_8}{\hat{m} + 4ix\hat{a}_8} \right)^{-\frac{\nu}{2}} I_\nu \left(\sqrt{\hat{m}^2 + 16x^2 \hat{a}_8^2} \right). \quad (5.107)
\end{aligned}
$$

It is no longer elementary, but after linearizing the $\cos(\theta)^2$ term it can be written as a Gaussian integral over the one-flavour partition function $I_\nu(\hat{m})$ in the continuum at shifted mass. For degenerate masses and general N_f we can again diagonalize U_0 and use Andréief's integral formula, as in the derivation of Eq. (5.98), leading to

$$
\mathcal{Z}_{\mathrm{WechPT}}^{[N_f]}(\hat{m}) = \det_{1 \leq j,k \leq N_f} \left[e^{2\hat{a}_8^2} \int_{-\pi}^{\pi} \frac{\mathrm{d}\theta}{2\pi} \exp\left[i(\nu + k - j)\theta + \hat{m}\cos(\theta) - 4\hat{a}_8^2 \cos(\theta)^2 \right] \right].
$$
$$(5.108)$$

Once again it is given by the determinant of $N_f = 1$ flavour partition function. In order to determine the spectral density of D_5 or D_W one then must introduce additional auxiliary quark pairs as described in Eq. (5.99) (or use replicas). Prior to the RMT calculation this was done for the spectral density from WechPT in [70, 77], for both D_5 or D_W.

Let us now turn to the RMT side as introduced in [70]. For simplicity we will only consider the Hermitian operator D_5; for the discussion of the complex eigenvalue spectrum of D_W we refer to [77] and to [79] for its complete solution. Actually two slightly different RMT have been proposed, which have turned out to be equivalent on the level of the jpdf. In [70] the partition function (5.5) was generalized using Eq. (5.105) in an obvious way, by adding two Hermitian Gaussian matrices A and B of dimensions N and $N + \nu$, respectively:

$$Z_{WI,N}^{[N_f]} \sim \int [dW][dA][dB] \prod_{f=1}^{N_f} \det \begin{bmatrix} m_f 1_N + aA & iW \\ -iW^\dagger & -m_f 1_{N+\nu} - aB \end{bmatrix}$$

$$\times \exp\left[-\frac{1}{4} \mathrm{Tr}(A^2 + B^2 + 2WW^\dagger) \right]. \tag{5.109}$$

In [80] the following modification was made: rather than adding a block-diagonal Hermitian matrix to D_5 as the last term in Eq. (5.105), a full Hermitian matrix H was added to it, $D_5 = \gamma_5(\not{D}+m)+H$, filling the off-diagonal blocks with a rectangular complex matrix Ω (and trivially changing the sign of $B \to -B$):

$$Z_{WII,N}^{[N_f]} = \int [dW][dH] \prod_{f=1}^{N_f} \det \begin{bmatrix} m_f 1_N + A & iW + \Omega \\ -iW^\dagger + \Omega^\dagger & -m_f 1_{N+\nu} + B \end{bmatrix}$$

$$\times \exp\left[-\frac{1}{2(1-a^2)} \mathrm{Tr} WW^\dagger - \frac{1}{4a^2} \mathrm{Tr} H^2 \right],$$

$$\cdot \quad H = \begin{pmatrix} A & \Omega \\ \Omega^\dagger & B \end{pmatrix}, \quad a \in [0,1]. \tag{5.110}$$

The different rescaling of the variances with a was made in [80] to underline that on the level of RMT this corresponds to a two-matrix model that interpolates between the chGUE and the GUE, in the limits $a \to 0$ and $a \to 1$, respectively. It was shown in [70, 77] and [80] that in the limit $N \to \infty$, identifying $\hat{a}_8^2 = a^2 N/4$ in addition to the rescaled mass terms, the two RMT partition functions and Eq. (5.106) with $W_6 = W_7 = 0$ agree:

$$Z_{\mathrm{WchPT}}^{[N_f]} = \lim_{\substack{N \to \infty \\ a, m_f \to 0}} Z_{WI,N}^{[N_f]} = \lim_{\substack{N \to \infty \\ a, m_f \to 0}} Z_{WII,N}^{[N_f]}. \tag{5.111}$$

As we have said previously, the presence of the W_6 and W_7 terms can be achieved by two extra Gaussian integrals; see [77]. It is quite remarkable that RMT allows to include such a detailed structure of WchPT. The fact that we have 'only' considered RMTs for D_5 does not affect the equivalence argument of the partition functions. Because of Eq. (5.104) and $\det[\gamma_5] = 1$ the partition functions of D_W and D_5 agree.

Let us just state the jpdf of the eigenvalues d_j of D_5 for degenerate masses m without derivation; see [80] and [79] for details,

$$Z_{W,N}^{[N_f]}(m) \sim \exp\left[\frac{Nm^2}{2a^2(1-a^2)} \right] \int_{-\infty}^{\infty} \prod_{j=1}^{2n+\nu} dd_j d_j^{N_f} \exp\left[-\frac{d_j^2}{4a^2} \right] \Delta_{2n+\nu}(\{d\})$$

$$\times \mathrm{Pf}_{1 \le i,j \le 2n+\nu; 1 \le q \le \nu} \begin{bmatrix} F(d_j - d_i) & d_i^{q-1} \exp\left[-\frac{d_i m}{2a^2} \right] \\ -d_j^{q-1} \exp\left[-\frac{d_j m}{2a^2} \right] & 0 \end{bmatrix}, \tag{5.112}$$

where we have defined the antisymmetric weight

$$F(x) = \frac{4}{\sqrt{2\pi a^2(1 - a^2)}} \int_m^\infty du \, \exp\left[-\frac{u^2}{2a^2(1 - a^2)}\right] \sinh\left[\frac{xu}{2a^2}\right]. \tag{5.113}$$

This jpdf has two new features. First, there appears a Vandermonde times a Pfaffian (Pf) defined as $\mathrm{Pf}[A] = \sqrt{\det[A]}$ for antisymmetric matrices A of even dimension, see, e.g., [14] for a definition in terms of permutations. This is quite common for such interpolating two-matrix models, see [81] for classical works regarding the GUE–GOE and the GUE–GSE transitions. Second, and most importantly the solution for the correlation functions involves both skew OP and OP at the same time, where the OP are Hermite polynomials of a GUE of size $\nu \times \nu$. We refer to [79] for details, in particular that this substructure of a finite size GUE remains valid in the large-N limit. In [80] where only the cases $\nu = 0$ and 1 were solved this structure was not observed, because only Hermite polynomials of degree 0 and 1 occur.

Instead of giving any further details of the solution of WRMT we show plots in Fig. 5.5 for the limiting quenched microscopic density $\rho_s^{D_5}(\tilde{x})$ of D_5 for $\nu = 0, 1$, illustrating the effect of the rescaled finite lattice spacing $\hat{a}$. In order to compare to the results from the chGUE in Fig. 5.3 let us first discuss the difference between D_5 and $\slashed{D}$ at $a = 0$, where $D_5 = \gamma_5(\slashed{D}+m)$; see Eqs (5.100) and (5.104). The rotation with γ_5 merely makes the Dirac operator Hermitian rather than anti-Hermitian, which corresponds to rotating the eigenvalues from the imaginary to the real axis (as we had already done in writing the jpdf in Eq. (5.2) of real eigenvalues). The shift by $\gamma_5 m$, however, introduces a gap such that there are no eigenvalues of D_5 in $[-m, m]$. Furthermore, the ν exact zero eigenvalues not shown in Fig. 5.3 get moved away from the origin to one of the edges of this gap, which in our convention is at $-m$. Therefore, we now have to give the density on $\mathbf{R}$ instead of $\mathbf{R}_+$, Compared to (5.64) it holds for $N_f = 0$:

$$\lim_{\hat{a}_8 \to 0} \rho_s^{D_5}(\tilde{x}) = \frac{|\tilde{x}|}{2}\Theta(\tilde{x} - \tilde{m})\left(J_\nu(\sqrt{\tilde{x}^2 - \tilde{m}^2})^2 - J_{\nu-1}(\sqrt{\tilde{x}^2 - \tilde{m}^2})J_{\nu+1}(\sqrt{\tilde{x}^2 - \tilde{m}^2})\right)$$
$$+ \nu\delta(\tilde{x} + \tilde{m}) . \tag{5.114}$$

The effect of $\hat{a}_8 > 0$ is both to broaden the ν delta functions and to eventually fill the gap $[-\hat{m}, \hat{m}]$. Because eigenvalues repel each other the ν zero modes will now spread, and form approximately a GUE of finite size ν; see [77] for a further discussion. In order to make the comparison more transparent let us also choose $\hat{m} = 0$, so that we can plot the density on $\mathbf{R}_+$ only, as is shown in Fig. 5.5. The effect of $\hat{a}_8 > 0$ is particularly striking on the zero modes.

The density of the Hermitian Wilson Dirac operator D_5 has been compared to quenched lattice simulations in [82, 83]. Because this involves multiparameter fits several quantities have been proposed that are simple to measure [84]. A comparison to unquenched data remains a difficult question; see, however, [85]. The RMT approach presented here has also lead to insights about the sign and constraints amongst the LECs $W_{6,7,8}$ that were previously controversial [77, 86]. When the lattice artefacts become very strong, additional unphysical phases appear, and the possibility of accessing these as a function of the LEC and quark content N_f was clarified in [87]. The

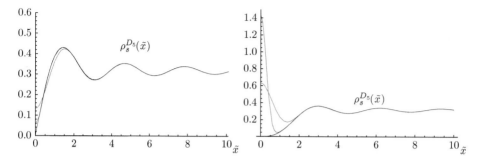

Fig. 5.5 The quenched microscopic spectral density $\rho_s^{D_5}(\tilde{x})$ for $\nu = 0$ (left) and $\nu = 1$ (right) as a function of $\hat{a}_8$, taken from [80]. Because of choosing $\hat{m} = 0$ it is still symmetric around the origin. For $\hat{a}_8 = 0$ we are back to the microscopic density of Bessel functions as in Fig. 5.3, whereas for $\hat{a}_8$ the density starts to spread into the region around the origin. For $\nu = 1$ the delta function of the one zero eigenvalue is broadened by increasing $\hat{a}_8 = 0.1$ (top curve) and 0.25 (middle curve), which is why this is plotted on a different scale compared to $\nu = 0$. At these small values of $\hat{a}_8$ the effect on the density is remarkably localized to the origin. Only in the limit $\hat{a}_8 \to \infty$ will we arrive at the microscopic density of the GUE [80], which is completely flat (and normalized to $1/\pi$ here).

derivation of all k-point correlation functions including $N_f \neq 0$ based on WechPT has been pushed forward in [88], leading to alternative representations compared to the OP approach [79]. For the density of D_5 in the p-regime see [89]. The computation of individual eigenvalue distribution is not known beyond the expansion of the Fredholm expansion proposed in [90].

5.4.2 RMT and QCD with chemical potential

In this subsection we will study the influence of a chemical potential μ on the Dirac operator spectrum. We will first state its global symmetry in QCD, then construct the RMT, and at the end link to the epsilon regime of the corresponding chiral Lagrangian. The addition of a chemical potential μ_q for each quark flavour to the QCD action in Eq. (5.77) amounts to adding the terms $\mu_q \Psi^{q\,\dagger}(x)\Psi^q(x) = \overline{\Psi}^q(x)\mu_q\gamma_4\Psi^q(x)$. In classical statistical mechanics the addition of a chemical potential allows the number of the corresponding particles to fluctuate. Here we will add the same chemical potential $\mu_q = \mu$ to all flavours (which makes it the Baryon chemical potential). In analogy to Eq. (5.75) the global symmetry of the Dirac operator plus $\mu\gamma_4$ changes to

$$\slashed{D}(\mu) = \begin{pmatrix} 0 & i\mathcal{W} + \mu \\ i\mathcal{W}^\dagger + \mu & 0 \end{pmatrix} \neq -\slashed{D}(\mu)^\dagger , \tag{5.115}$$

and thus becomes complex non-Hermitian. It holds that $-\slashed{D}(-\mu)^\dagger = \slashed{D}(\mu)$, and thus for purely imaginary chemical potential $\mu \to i\mu$ it is still anti-Hermitian. Because the operator remains chiral,

$$0 = \{\not{D}(\mu), \gamma_5\} , \quad \not{D}(\mu)\Phi_k(x) = iz_k\Phi_k(x) , \quad (5.116)$$

the nonzero complex eigenvalues continue to come in pairs $\pm iz_k$, using the same argument as after Eq. (5.76). The fact that the spectrum of the Dirac operator with chemical potential is complex (and thus the QCD action too) has severe repercussions for lattice QCD. Because the action can no longer be interpreted as a probability weight, the standard method on the lattice, weighting a given configuration by its action breaks down. Several methods have been invented to circumvent this so-called sign problem, and we refer to [91] for a review. In contrast the RMT for QCD with chemical potential remains analytically solvable with OP in the complex plane, as we will see.

In [72] a RMT was constructed by Stephanov where W was replaced by a complex Gaussian random matrix W_1; cf. Eq. (5.5) for $\beta = 2$ compared to (5.75). While this RMT was very useful for illuminating the difference between $N_f = 0$ and $N_f > 0$, which is more dramatic than for $\mu = 0$, this model was not analytically tractable with OP at finite N. Therefore, Osborn [74] introduced a two-matrix model where μ is multiplied by a second random matrix W_2, see Eq. (5.117), implying that the chemical potential term is not diagonal in that basis. While at first sight more complicated, this model turns out to be exactly solvable using OP in the complex plane, as we will demonstrate. A further equivalent representation was chosen in [92], multiplying $W_{1,2}$ by $\exp[\mp\mu]$ instead. This parametrization made it possible to understand why the sign problem in the numerical solution of this RMT for QCD with $\mu \neq 0$ can be circumvented [93]. Prior to introducing an RMT with μ as in Eq. (5.115) a similar model was introduced [94] describing the effect of temperature T. Such a model can only be solved using the supersymmetric method, see [95], with the effect of renormalizing $\Sigma \to \Sigma(T)$ to be temperature dependent, keeping the correlation functions otherwise unchanged. A further modification of Stephanov's model by including both parameters $\mu \to \mu + i\pi T$ in Eq. (5.115) was very successful as a schematic model for the phase diagram of QCD [96], predicting the qualitative features depicted in Fig. 5.1.

Let us turn to the RMT [74] defined as

$$Z_N(\mu) = \int [\mathrm{d}W_1][\mathrm{d}W_2] \prod_{f=1}^{N_f} \det \begin{bmatrix} m_f 1_N & iW_1 + \mu W_2 \\ iW_1^\dagger + \mu W_2^\dagger & m_f 1_{N+\nu} \end{bmatrix} \exp[-\mathrm{Tr}(W_1 W_1^\dagger + W_2 W_2^\dagger)]$$

$$\sim \int [\mathrm{d}X_1][\mathrm{d}X_2] \prod_{f=1}^{N_f} \det \begin{bmatrix} m_f 1_N & X_1 \\ X_2 & m_f 1_{N+\nu} \end{bmatrix} \exp\left[-a(\mu)\mathrm{Tr}(X_1 X_1^\dagger + X_2 X_2^\dagger)\right]$$

$$\times \exp\left[b(\mu)\mathrm{Tr}(X_1 X_2 + X_2^\dagger X_1^\dagger)\right] , \quad a(\mu) = \frac{1+\mu^2}{4\mu^2} , \quad b(\mu) = \frac{1-\mu^2}{4\mu^2} .$$

$$(5.117)$$

In the second step we have simply changed variables to the matrices $X_1 = iW_1 + \mu W_2$ and $X_2 = iW^\dagger + \mu W_2^\dagger$, which are now coupled. While at $\mu = 1$ the matrices X_1 and X_2 are again independent Gaussian random matrices—this is called maximal non-Hermiticity—at $\mu = 0$ we have $X_1^\dagger = -X_2$ with total correlation (and we are back

to the chGUE). A look at the characteristic equation for the massless RMT Dirac operator,

$$\det[z - \slashed{D}(\mu)] = \det[z^2 - X_1 X_2] , \qquad (5.118)$$

reveals that we are after the complex eigenvalues[9] of the product of the two correlated matrices $Y = X_1 X_2$. As in Eq. (5.5) we could have allowed for real or quaternion-valued matrix elements with $\beta = 1, 4$. However, the joint density of complex eigenvalues does not enjoy a closed form for all three β's as in Eq. (5.2). In particular, real matrices are special as they can have real eigenvalues and complex conjugated eigenvalues pairs, as the characteristic equation (5.118) is real. For the solution of the respective RMTs see [98] and [99], respectively.

The parametrization of a generic complex non-Hermitian matrix Y in terms of its complex eigenvalues is usually done using the Schur decomposition $Y = U(Z + T)U^\dagger$, where $U \in U(N)$ is unitary, $Z = \text{diag}(z_1, \dots, z_N)$ contains the complex eigenvalues, and T is a strictly upper triangular complex matrix (it can be chosen to be strictly lower triangular instead). For two matrices, however, in order to compute the Jacobian we have to use the generalized Schur decomposition, cf. [100]

$$X_1 = U(Z_1 + R)V , \quad X_2 = V^\dagger(Z_2 + S)U^\dagger \;\Rightarrow\; Y = X_1 X_2 = U(Z + T)U^\dagger , \quad (5.119)$$

with $Z = Z_1 Z_2$ and $T = R Z_2 + Z_1 S + R S$. Here $U, V \in U(N)$ are unitary, Z_1, Z_2 are diagonal matrices with complex entries (which are not the complex eigenvalues of X_1 and X_2), and R, S are strictly upper triangular. The complex eigenvalues of the product matrix Y are thus given by the elements of the diagonal matrix Z. Let us directly give the result for the joint density from [74] (see [101] for an alternative derivation):

$$Z_N(\mu)^{[N_f]} = \left(\prod_{j=1}^{N} \int d^2 z_j |z_j|^\nu K_\nu(a(\mu)|z_j|) \right.$$

$$\left. \times \exp\left[-\frac{b(\mu)}{2}(z_j + z_j^*) \right] \prod_{f=1}^{N_f}(z_j + m_f^2) \right) |\Delta_N(\{z\})|^2 \quad (5.120)$$

$$= \left(\prod_{j=1}^{N} \int d^2 z_j \right) \mathcal{P}_{jpdf}(z_1, \dots, z_N) . \quad (5.121)$$

Note that in view of Eq. (5.118) we should later take the square root of the eigenvalues z_j of $Y = X_1 X_2$ to change to Dirac operator eigenvalues. In the complex plane this is not unique and without loss of generality one can then restrict to the half-plane, also in view of the $\pm$ pairs of Dirac eigenvalues. Before we discuss the properties of Eq. (5.121) and its correlation functions let us try to understand why the Jacobian of

---

[9] For the singular values of $X_1 X_2$ in this setting see [97].

the transformation (5.119) does not only include the modulus-squared Vandermonde determinant (as in the complex Ginibre ensemble, see [14]) and the zero modes $|z_j|^\nu$, but also a modified K-Bessel function, even though we started from (coupled) Gaussian matrices. The same phenomenon happens when looking at products of complex Gaussian random variables $z_{1,2}$, and asking for the distribution $w(z)$ of the product random variable $z = z_1 z_2$:

$$w_2(z) = \int d^2 z_1 \int d^2 z_2 \exp[-|z_1|^2 - |z_1^2|]\delta^{(2)}(z - z_1 z_2)$$

$$= (2\pi)^2 \int_0^\infty \frac{dr\, r}{r^2} \exp[-r^2 - |z|^2/r^2] = 2\pi K_0(2|z|) \ . \tag{5.122}$$

Obviously, we are dealing with the analogue of quadratic matrices here. Multiplying $n \geq 2$ such random variables we obtain an $(n-1)$-fold integral representation for the distribution of z now given by a Meijer G-function,

$$w_n(z) = \prod_{j=1}^n \int d^2 z_j e^{-|z_j|^2} \delta^{(2)}(z - z_1 \cdots z_n)$$

$$= (2\pi)^{n-1} \prod_{j=1}^{n-1} \int_0^\infty \frac{dr_j}{r_j} e^{-r_j^2} e^{-\frac{|z|^2}{r_1^2 \cdots r_{n-1}^2}} = \pi^{n-1} G_{0\,n}^{n\,0}\left(\left. \begin{matrix} - \\ \vec{0} \end{matrix} \right| |z|^2 \right) ; \tag{5.123}$$

see [37] for its standard definition. These weights appear when studying the complex eigenvalues of the product of n independent Gaussian matrices $X_1, \ldots, X_n$ and we refer to [102] for the solution of this model.

The complex eigenvalue correlation functions of the model (5.121) are defined as in Eq. (5.6) and it holds that

$$R_k(z_1, \ldots, z_k) = \prod_{j=1}^k w(z_j) \det_{1 \leq i,j \leq k} [K_N(z_i, z_j^*)] \ . \tag{5.124}$$

They are again given in terms of the kernel

$$K_N(z, u^*) = \sum_{j=0}^{N-1} h_j^{-1} P_j(z) Q_j(u^*) \tag{5.125}$$

of polynomials $P_j(z)$ and $Q_j(z)$, which are biorthogonal in the complex plane with respect to the weight

$$w(z_j) = |z_j|^\nu K_\nu(a(\mu)|z_j|) \exp\left[-\frac{b(\mu)}{2}(z_j + z_j^*)\right] \prod_{f=1}^{N_f}(z_j + m_f^2) \ , \tag{5.126}$$

$$\int d^2 z\, w(z) P_k(z) Q_l(z) = h_k \delta_{kl} \ . \tag{5.127}$$

We will not discuss gap probabilities or individual eigenvalue distributions here and refer to [103]. Unlike in the real eigenvalue case they will now depend on the choice of the geometry of the region containing $k \geq 0$ eigenvalues, and only in the rotationally invariant case $\mu = 1$ a radial ordering of eigenvalues according to their modulus provides a natural choice.

Before we solve the RMT Eq. (5.120), a word of caution is in order. In standard OP theory in the complex plane only real positive weights are considered, which is only true for $N_f = 0$ in Eq. (5.126). In this case $P_k(z)^* = Q_k(z)$ and standard techniques such as Gram–Schmidt apply to construct these polynomials; see, e.g., [19]. When $N_f > 0$ in our case the weight is complex, and thus a priori the partition function is as well as the R_k are. Therefore, the density no longer has a probabilistic interpretation, it has a real and imaginary part, and we refer to [104] for a more detailed discussion.

We will first provide the solution of the simplest quenched case with $N_f = 0$. Then the partition function and general eigenvalue correlation functions including N_f mass terms can be obtained along the same lines as in Section 5.2, expressing them in terms of the OP at $N_f = 0$. Surprisingly, despite the complicated non-Gaussian weight function in the complex plane, the OP are again given by Laguerre polynomials of complex arguments [74].[10] The following holds, where we drop the argument μ of the parameters $a > b \geq 0$ in Eq. (5.117):

$$\int d^2z\,|z|^\nu K_\nu(a|z|)\exp[b\Re e(z)]L_j^\nu(cz)L_k^\nu(cz) = \delta_{jk}h_k \,, \quad \text{with} \quad c = \frac{a^2 - b^2}{2b} \,,$$

$$h_j = \frac{\pi(j+\nu)!}{j!\,a}\left(\frac{a}{b}\right)^{2j}\left(\frac{ac}{b}\right)^{\nu+1} \,. \tag{5.128}$$

Here $\nu = 0, 1, \ldots$ is an integer, and the Laguerre polynomials can be made monic as in Eq. (5.21). The proof we sketch follows [106] and uses induction of depth 2; for an earlier longer proof see [99]. Insert the following integral representations into the orthogonality relation (5.128) for $\nu = 0, 1$,

$$K_\nu(x) = \frac{x^\nu}{2^{\nu+1}}\int_0^\infty \frac{dt}{t^{\nu+1}}\exp[-t - x^2/(4t)] \,,$$

$$L_j^\nu(z) = \oint_\Gamma \frac{du}{2\pi}\frac{\exp\left[-\frac{zu}{1-u}\right]}{(1-u)^{\nu+1}u^{j+1}} \,, \tag{5.129}$$

where Γ is a contour enclosing the origin in positive direction, but not the point $z = 1$. Then the Gaussian integrals over the real and imaginary parts of $z = x + iy$ can be solved, and subsequently the integral over t from the K-Bessel function. The orthogonality then easily follows, applying the residue theorem twice. For the induction we use the three-step recurrence relation for the Laguerre polynomials and K-Bessel

---

[10] A first unsuccessful attempt with the incorrect weight (except for $\nu = \pm\frac{1}{2}$) was made in [105].

function [37]; see [106] for details. We thus have solved the quenched case as we know all correlation functions from the kernel

$$K_N(z, u^*) = \sum_{j=0}^{N-1} h_j^{-1} L_j^{\nu}(cz) L_j^{\nu}(cu^*) , \qquad (5.130)$$

in terms of the quantities in Eq. (5.128), by using Eq. (5.124). A similar solution exists [107] for the so-called elliptic complex Ginibre ensemble of a single complex matrix $W = H + iA$, where the Hermitian part $H = H^{\dagger}$ and anti-Hermitian part $A = A^{\dagger}$ of the matrix have different variances that depend on a parameter. The solution is given in term of Hermite polynomials in the complex plane and makes it possible to interpolate between the GUE and the complex Ginibre ensemble at maximal non-Hermiticity, where H and A have the same variance. We refer to [107] for a detailed discussion.

Let us now turn to the unquenched solution with $N_f > 0$ described in great detail in [104], expressing everything in terms of the Laguerre polynomials of the quenched solution from Eq. (5.128). Because we follow the same strategy as in Subsections 5.2.2 and 5.2.3 using expectation values of characteristic polynomials we can be very brief. The corresponding proofs for manipulating expectation values of characteristic polynomials and their complex conjugates defined as in Eq. (5.24) can be found in [24]. In particular for the monic OP in the complex plane satisfying Eq. (5.127) for a weight $w(z)$ that can be complex we have [24]

$$P_L(z) = \left\langle \prod_{j=1}^{L} (z - z_j) \right\rangle_L , \quad Q_L(u^*) = \left\langle \prod_{j=1}^{L} (u^* - z_j^*) \right\rangle_L . \qquad (5.131)$$

The proof goes exactly the same way as in Eq. (5.27), where for $P_L(z)$ we multiply the product into $\Delta_L(\{z\})$, increasing it to Δ_{L+1}. For $Q_L(u^*)$ we simply choose $\Delta_L(\{z\})^*$, increasing it to Δ_{L+1}^*. Likewise for the kernel we obtain

$$K_{N+1}(z, u^*) = h_N^{-1} \left\langle \prod_{j=1}^{N} (z - z_j)(u^* - z_j^*) \right\rangle_N = \sum_{j=0}^{N} h_j^{-1} P_j(z) Q_j(u^*) , \qquad (5.132)$$

modifying the proof in Eq. (5.30) accordingly by multiplying the first product into $\Delta_N(\{z\})$, and the second into $\Delta_N(\{z\})^*$. The only difference here is that in general no Christoffel–Darboux formula is available. This is despite the fact that the Laguerre polynomials in our example Eq. (5.128) still satisfy the standard three-step recurrence (which is not true for OP in the complex plane in general). The step in the proof of Eq. (5.18) that breaks down is the cancellation of the summands after multiplying Eq. (5.130) by $(z - u^*)$.

The most general result corresponding to Eq. (5.33) now contains both products of K characteristic polynomials and of L products of complex conjugated ones. Without loss of generality let us choose $K \geq L$ ($K < L$ can be obtained by complex conjugation), to state the result from [24]:

$$\left\langle \left(\prod_{l=1}^{N} \left(\prod_{i=1}^{K} (v_i - z_l) \right) \left(\prod_{j=1}^{L} (u_j^* - z_l^*) \right) \right) \right\rangle_N = \frac{\prod_{i=N}^{N+K-1} h_i^{\frac{1}{2}} \prod_{j=N}^{N+L-1} h_j^{\frac{1}{2}}}{\Delta_K(\{v\})\Delta_L(\{u\})^*}$$

$$\times \det_{1\leq l,m\leq K} \left[K_{N+L}(v_l, u_m^*) \frac{P_{N+m-1}(v_l)}{h_{N+m-1}^{\frac{1}{2}}} \right]. \tag{5.133}$$

Here the left block of kernels inside the determinant is of size $K \times L$. The proof is straightforward and uses induction. Using this result we are now ready to compute the partition function Eq. (5.120) for general $N_f \geq 0$. In the quenched case it is given again by $N!$ times the products of the norms from Eq. (5.128), as derived in Eq. (5.23). For $N_f > 0$ we can express it in terms of the Laguerre polynomials from Eq. (5.128), and obtain an almost identical answer to Eq. (5.41):

$$Z_N^{[N_f]}(\{m\}) = \frac{(-1)^{NN_f} N! \prod_{j=0}^{N-1} h_j}{\Delta_{N_f}(\{m\})} \det_{1\leq j,g\leq N_f} \left[(-1/c)^{N+g-1} (N+g-1)! L_{N+g-1}^{\nu}(-cm_f^2) \right]. \tag{5.134}$$

As in Eq. (5.41) it is given by a determinant of Laguerre polynomials of real arguments, which is real. The only difference to $\mu = 0$ is the scale factor $c = 1/(1-\mu^2)$ and the μ-dependence inside the norms h_j from Eq. (5.128). We will not give further details regarding the kernel $K_N^{[N_f]}(z, u^*)$ constituting the k-point correlation functions in Eq. (5.124). Following the first line of Eq. (5.44) it can be expressed through the quenched average of $N_f + 1$ characteristic polynomials times one conjugated characteristic polynomial, divided by the partition function from Eq. (5.134). The spectral density with N_f masses then follows as in Eq. (5.45), by multiplying $K_N^{[N_f]}(z, z^*)$ with the weight Eq. (5.126); see [74, 104] for more details. Clearly, it is not real in general.

We will now very briefly discuss the large-N limit and matching with the corresponding echPT. It turns out that for non-Hermitian matrices that are coupled as in Eq. (5.117) the possibility of taking such limits is much richer than in the Hermitian (or maximally non-Hermitian) case. In addition to the standard macroscopic and microscopic statistics a further class of limits exist, where the matrix $Y = X_1 X_2$ becomes almost Hermitian, taking the parameter $\mu \to 0$ at such a rate that that the products

$$\tilde{\mu}^2 = \lim_{\substack{N \to \infty \\ \mu \to 0}} 2N\mu^2 , \quad \tilde{z} = \lim_{\substack{N \to \infty \\ z \to 0}} 2\sqrt{N}z \tag{5.135}$$

are constant. Here the scaling of the complex eigenvalues (and masses) is unchanged compared to Eq. (5.56). This limit was first observed in the elliptic complex Ginibre ensemble already mentioned [107] and called weakly non-Hermitian [108]. This is in contrast to the limit at strong non-Hermiticity where $\mu \in (0, 1]$ is kept fixed and a different scaling of z is chosen. Here the statistics of the product of two independent complex matrices with $\mu = 1$ is found; see [104] and [103] for details. Looking at our result Eq. (5.134) it is clear that the limiting partition function at weak non-Hermiticity is the same as for $\mu = 0$ in Eq. (5.60) and is thus independent of the

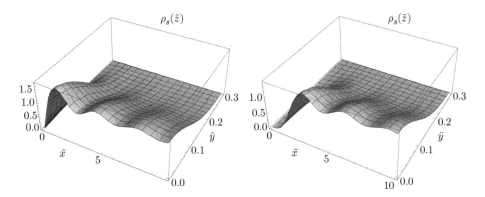

Fig. 5.6 The quenched microscopic spectral density from Eq. (5.136) as a function of $\tilde{z} = \tilde{x} + i\tilde{y}$ for $\nu = 0$ (left) and $\nu = 1$ (right), at $\tilde{\mu} = 0.1$. The density of real eigenvalues from Fig. 5.3 can still be recognized, it now spreads into the complex plane. Because in the limit $\tilde{\mu} \to 0$ it will become proportionally to a delta function $\delta(\tilde{y})$ times the Bessel density, the vertical scale is different.

rescaled $\tilde{\mu}$. In contrast already the limiting quenched density does depend on $\hat{\mu}$. It can be easily obtained from Eq. (5.130), replacing the sum by an integral and using Eq. (5.57) for the Laguerre polynomials,

$$\rho_s(\tilde{z}) = \frac{|\tilde{z}|^2}{2\pi\tilde{\mu}^2} K_\nu(|\tilde{z}|^2/4\tilde{\mu}^2) \exp[\Re e(\tilde{z}^2)/4\tilde{\mu}^2] \int_0^1 dt\, t \exp[-2t^2\tilde{\mu}^2]|J_\nu(\tilde{z}t)|^2 \ . \quad (5.136)$$

It is shown in Fig. 5.6. Here we have again changed to Dirac eigenvalues following [74]. This result was obtained earlier using replicas and the Toda equation in [109]. For $\tilde{\mu} = 0$ the integral becomes elementary and matches Eq. (5.64), while the prefactors turn into a delta function.

The fact that the partition function is $\tilde{\mu}$-independent, and the density is not, has been called the 'Silver blaze problem' [110], alluding to a novel by A. C. Doyle. How can we resolve this puzzle? Let us look again at chPT, the theory of Goldstone Bosons. Because these are mesons they don't carry baryon charge and thus do not feel the effect of adding μ to the action (mesons consist of a quark and anti-quark, whereas baryons, as the proton, consist of three quarks). We can still make the chPT partition function μ-dependent by adding an integer number N_f^* of quarks with the opposite chemical potential $-\mu$. As discussed after Eq. (5.115) this amounts to adding complex conjugated determinants of the Dirac operator to the action, leading to

$$Z_N^{[N_f + N_f^*]}(\{m\}, \{n\}) = Z_N \left\langle \prod_{l=1}^N \left(\prod_{i=1}^{N_f} (z_l + m_f) \right) \left(\prod_{j=1}^{N_f^*} (z_l^* + n_f^2) \right) \right\rangle_N \ , \quad (5.137)$$

which we can also compute using Eq. (5.133). The corresponding limiting partition function of echPT now depends on μ as follows:

$$\mathcal{Z}_{\text{echPT}}^{[N_f+N_f^*]}(\{m\},\{n\}) = \int_{U(N_f+N_f^*)} dU_0\, \det[U_0]^\nu$$

$$\times \exp\left[\frac{1}{2}\Sigma V \text{Tr}(M(U_0+U_0^\dagger)) - \frac{1}{4}\mu^2 V F_\pi^2[U_0,B][U_0,B]\right]. \tag{5.138}$$

Here $M = \text{diag}(m_1,\ldots,m_{N_f},n_1,\ldots,n_{N_f^*})$ contains both kinds of masses and $B = \text{diag}(1_{N_f},-1_{N_f^*})$ is the metric with the signature of the chemical potentials. The commutator comes from the fact that in the Dirac operator μ acts as an additional vector potential. It obviously vanishes at $N_f^* = 0$. Once again in the weak non-Hermiticity limit the partition functions from RMT and echPT agree, when identifying the rescaled masses and chemical potentials as $\hat{\mu}^2 = \mu^2 V F_\pi^2 = \tilde{\mu}^2$:

$$\lim_{\substack{N\to\infty\\ m_f,n_f,\mu\to 0}} \mathcal{Z}_N^{[N_f+N_f^*]}(\{m\},\{n\}) = \mathcal{Z}_{\text{echPT}}^{[N_f+N_f^*]}(\{\hat{m}\},\{\hat{n}\}). \tag{5.139}$$

They are given by the determinant from Eq. (5.133) in terms of the limiting kernel and the limiting polynomials, which are again $I_\nu(\hat{m})$. The corresponding group integral Eq. (5.138) was computed independently in [109] using an explicit parametrization of $U(N_f+N_f^*)$; see [111] for the general N_f case.

What about the equivalence of RMT and echPT for the density; in other words, how can we generate the spectral density Eq. (5.136) from a chiral Lagrangian? Although one may still define and generate the resolvent as in Eq. (5.39), the relation to the spectral density changes compared to Eq. (5.40),

$$R_1(z) = \frac{1}{\pi}\frac{\partial}{\partial z^*}G_N(z), \tag{5.140}$$

due to $\partial_{z^*}(1/z) = \delta^{(2)}(z)$ on **C**. While it is subtle to extract the non-holomorphic part from the expectation value of the ratio of characteristic polynomials two alternative ways exist for formulating the generation of a two-dimensional delta function. One way is to use so-called replicas, which amounts to introducing k extra pairs of complex conjugated quarks in the chiral Lagrangian. Applying the relation to the Toda lattice equation, the replica limit can be made well defined as was developed in [109]. The generating function for the density with $k = 1$, $\mathcal{Z}_{\text{echPT}}^{[N_f+1+1^*]}$ on the echPT side thus depends on $\tilde{\mu}$, which is why the resulting density also depends on $\tilde{\mu}$. The second possibility is to use supersymmetry, which will also involve pairs of complex conjugated Bosons [112],

$$R_1(z) = -\frac{1}{\pi}\lim_{\kappa\to 0}\partial_{z^*}\left(\partial_u\left\langle\frac{\det[(z-\slashed{D}(\mu))(z-\slashed{D}(\mu))^\dagger + \kappa^2]}{\det[(u-\slashed{D}(\mu))(u-\slashed{D}(\mu))^\dagger + \kappa^2]}\right\rangle_\nu^{\text{QCD}}\bigg|_{u=z}\right), \tag{5.141}$$

and both methods have been shown to be equivalent [113]. In either case the silver blaze puzzle is resolved. We would also like to mention the papers [114, 115] where averages of ratios of characteristic polynomials and their complex conjugates were

evaluated in a purely complex OP framework. The most prominent example for such an average is the distribution of the phase of $\mathbb{D}(\mu)$

$$\left\langle \prod_{j=1}^{N} \frac{z_j + m^2}{z_j^* + m^2} \right\rangle_N^{[N_f]}, \tag{5.142}$$

and we refer to [116] and [117] for a detailed discussion, including its physical interpretation.

When trying to compare the RMT predictions to QCD lattice simulations the presence of the chemical potential also has a virtue, despite the sign problem: it couples to the second LEC in the chiral Lagrangian, F_π, in Eq. (5.138). Just as for $\mu = 0 = a$, a fit of the density to lattice data helps to determine Σ, a two-parameter fit at $\tilde{\mu} \neq 0$ makes it possible to measure F_π in this way. Because for imaginary chemical potential $i\mu$ the Dirac operator spectrum remains real as discussed after Eq. (5.115), this was proposed to determine F_π using standard lattice simulations [118]. There is also a corresponding two-matrix, model [119], which in the large N with rescaled $i\mu$ as in Eq. (5.135) becomes equal to Eq. (5.138), with $\mu^2 \rightarrow -\mu^2$ (which is, of course, still convergent).

Turning back to real chemical potential, most comparisons were done within the quenched approximation, starting with [120]; cf. [121] for the influence of topology. Despite the difficulty of defining individual eigenvalue distributions these were compared within [122]. There are many issues that we have not discussed about the application to QCD with chemical potential, apart from methods different from the OP approach mentioned earlier. This includes the different origin of chiral symmetry breaking at $\mu \neq 0$ that is strongly related to the oscillatory behaviour of the unquenched density and we refer to [123] and to [124] for a review. Thanks to its analytical solution, the RMT Eq. (5.120) has been used as a testing ground for numerical ways to solve the sign problem. This has been successful for the RMT itself and is reported in [93].

Furthermore, the study of the singular values of the QCD Dirac operator with chemical potential has been proposed in [125], which could also be used to analyse different phases (not depicted in Fig. 5.1) for larger values of μ. Here both the lattice realization and the RMT have many open questions, and a first step in RMT has been taken in [97]. Here a different kernel is found at the origin, the so-called Meijer G-kernel [126] that generalizes the Bessel kernel. The mathematical question of universality in the limit of weak non-Hermiticity for non-Gaussian RMTs is still open at the moment, despite first heuristic attempts [127]. For a further discussion including all known limiting kernels in this weak non-Hermiticity limit resulting from Gaussian models, we refer to [128] and references therein.

5.5   Summary

Let us briefly summarize the lessons we have learned in these lectures. The theory of orthogonal polynomials on the real line and in the complex plane and their

properties has been extensively discussed. A particular focus was put on weights including characteristic polynomials, as this allowed us to determine the Dirac operator spectrum of QCD including the contribution of N_f quark flavours in a random matrix approach. Based on the global symmetries of QCD we have taken a limit to the low-energy regime that eventually leads to a random matrix description in a controlled approximation, to which corrections can be computed. Even though this is an unphysical limit it can be compared to numerical solutions of QCD on the lattice. This leads to a determination of the low-energy constants of the theory and tests of lattice algorithms to be used then in the physical regime. We have shown how the introduction of more details in the theory of QCD such as the effect of finite lattice spacing or of chemical potential, which make the eigenvalues complex, modify the spectrum. And we have mentioned several open problems relevant to RMT and to its application like the computation of new correlation functions or the proof of universality for existing results. They hopefully contribute to keep this an interesting and lively area of research for both mathematicians and theoretical physicists.

Acknowledgements

It has been a great pleasure to be back in Les Houches, now as a lecturer, and I would like to express my sincere gratitude to the organizers for inviting me. My understanding of this subject would not have been possible without the many collaborations on this topic, and I would like to thank all my coworkers for sharing their insights and compassion. I am particularly thankful to Jac Verbaarschot for many valuable comments on a first draft of this manuscript, and Ivan Parra for help in removing further typos. Furthermore I would like to thank the Simons Center for Geometry and Physics at Stony Brook University (programme on Statistical Mechanics and Combinatorics 2016) for the kind hospitality where part of this chapter was written up. This work was partly supported by DFG Grant AK 35/2-1.

References

1. T. Muta, *Foundations of Quantum Chromodynamics*, third ed. World Scientific, Singapore, 2010.
2. The Review of Particle Physics (2015), K. A. Olive et al. (Particle Data Group), *Chin. Phys. C*, 38, 090001 (2014) and 2015 update [http://pdg.lbl.gov].
3. I. Montvay and G. Münster, *Quantum Fields on a Lattice*. Cambridge University Press, Cambridge, 1997.
4. S. Scherer, *Adv. Nucl. Phys.* **27** (2003) 277 [hep-ph/0210398].
5. T. Banks and A. Casher, *Nucl. Phys. B* **169** (1980) 103.
6. E. V. Shuryak and J. J. M. Verbaarschot, *Nucl. Phys. A* **560** (1993) 306 [hep-th/9212088].
7. J. J. M. Verbaarschot and I. Zahed, *Phys. Rev. Lett.* **70** (1993) 3852 [hep-th/9303012].
8. J. J. M. Verbaarschot and T. Wettig, *Ann. Rev. Nucl. Part. Sci.* **50** (2000) 343 [hep-ph/0003017].

9. P. H. Damgaard, *J. Phys. Conf. Ser.* **287** (2011) 012004 [arXiv:1102.1295 [hep-ph]].

10. J. J. M. Verbaarschot, Lectures given at the Les Houches Summer School on Applications of Random Matrices in Physics, Les Houches, France, 6-25 Jun 2004, arXiv:hep-th/0502029.

11. G. Akemann, *Int. J. Mod. Phys. A* **22** (2007) 1077 [hep-th/0701175].

12. T. Guhr, A. Müller-Groeling, and H. A. Weidenmüller, *Phys. Rep.* **299** (1998) 190 [cond-mat/9707301].

13. G. Akemann, J. Baik, and P. Di Francesco (Eds.), *The Oxford Handbook of Random Matrix Theory*. Oxford University Press, Oxford, 2011.

14. M. L. Mehta, *Random Matrices*, third ed. Academic Press, London, 2004.

15. G. Anderson, A. Guionnet, and O. Zeitouni, *An Introduction to Random Matrices*. Cambridge Studies in Advanced Mathematics **118** (2009).

16. P. J. Forrester, *Log-Gases and Random Matrices*. London Mathematical Society Monographs Series, 34. Princeton University Press, Princeton, NJ, 2010.

17. J. J. M. Verbaarschot, *Phys. Rev. Lett.* **72** (1994) 2531 [hep-th/9401059].

18. G. Akemann and P. H. Damgaard, *Phys. Lett. B* **583** (2004) 199 [hep-th/0311171].

19. W. Van Assche, in *Special Functions, q-Series and Related Topics*, ed. M.E.H. Ismail et al., 211–45 Fields Institute Communications **14** (1997).

20. J. Baik, P. Deift, and E Strahov, *J. Math. Phys.* **44** (2003) 3657 [math-ph/0304016].

21. M. L. Mehta and J.-M. Normand, *J. Phys. A: Math. Gen.* **34** (22) (2001) 4627 [cond-mat/0101469].

22. P. Zinn-Justin, *Comm. Math. Phys.* **194** (1998) 631 [cond-mat/9705044].

23. E. Brézin and S. Hikami, *Comm. Math. Phys.* **214** (2000) 111 [math-ph/9910005].

24. G. Akemann and G. Vernizzi, *Nucl. Phys. B* **660** (2003) 532 [hep-th/0212051].

25. P. J. Forrester and T. D. Hughes, *J. Math. Phys.* **35** (1994) 6736.

26. T. Wilke, T. Guhr, and T. Wettig, *Phys. Rev. D* **57** (1998) 6486 [hep-th/9711057].

27. S. M. Nishigaki, P. H. Damgaard, and T. Wettig, *Phys. Rev. D* **58** (1998) 087704 [hep-th/9803007].

28. T. Nagao and P. J. Forrester, *Nucl. Phys. B* **509** (1998) 561.

29. P. H. Damgaard and S. M. Nishigaki, *Phys. Rev. D* **63** (2001) 045012 [hep-th/0006111].

30. G. Akemann, T. Guhr, M. Kieburg, R. Wegner, and T. Wirtz, *Phys. Rev. Lett.* **113** (2014) 250201 [arXiv:1409.0360 [math-ph]].

31. P. J. Forrester, *Nucl. Phys. B* **402** (1993) 709.

32. G. Akemann and P. Vivo, *J. Stat. Mech.* **1105** (2011) P05020 [arXiv:1103.5617 [math-ph]].

33. A. Edelman, *Lin. Alg. Appl.* **159** (1991) 55.

34. Y. V. Fyodorov and A. Nock, *J. Stat. Phys.* **159** (2015) 731 [arXiv:1410.5645 [math-ph]].

35. M. E. Berbenni-Bitsch, S. Meyer, and T. Wettig, *Phys. Rev. D* **58** (1998) 071502 [hep-lat/9804030].

36. J. Ambjorn, L. Chekhov, C. F. Kristjansen, and Y. Makeenko, *Nucl. Phys. B* **404** (1993) 127 [hep-th/9302014].
37. I. S. Gradshteyn, and I. M. Ryzhik, *Table of Integrals, Series, and Products*, ed. A. Jeffrey and D. Zwillinger, fifth ed. Academic Press, New York, 1994.
38. G. Akemann, P. H. Damgaard, U. Magnea, and S. Nishigaki, *Nucl. Phys. B* **487** (1997) 721 [hep-th/9609174].
39. P. H. Damgaard and S. M. Nishigaki, *Nucl. Phys. B* **518** (1998) 495 [hep-th/9711023].
40. A. B. Balantekin, *Phys. Rev. D* **62** (2000) 085017 [hep-th/0007161].
41. C. A. Tracy and H. Widom, *Comm. Math. Phys.* **161** (1994) 289 [hep-th/9304063].
42. P. J. Forrester, *J. Math. Phys.* **35** (1994) 2539.
43. Y. Chen, D.-Z. Liu, and D.-S. Zhou, *J. Phys. A: Math. Theor.* **43** (2010) 315303.
44. P. H. Damgaard, U. M. Heller, and A. Krasnitz, *Phys. Lett. B* **445** (1999) 366 [hep-lat/9810060].
45. M. Göckeler, H. Hehl, P. E. L. Rakow, A. Schäfer, and T. Wettig, *Phys. Rev. D* **59** (1999) 094503 [hep-lat/9811018].
46. R. G. Edwards, U. M. Heller, J. E. Kiskis, and R. Narayanan, *Phys. Rev. Lett.* **82** (1999) 4188 [hep-th/9902117].
47. P. H. Damgaard, U. M. Heller, R. Niclasen, and K. Rummukainen, *Phys. Lett. B* **495** (2000) 263 [hep-lat/0007041].
48. P. H. Damgaard, *Nucl. Phys. Proc. Suppl.* **106** (2002) 29 [hep-lat/0110192].
49. A. V. Smilga and J. Stern, *Phys. Lett. B* **318** (1993) 531.
50. J. Gasser and H. Leutwyler, *Phys. Lett. B* **184** (1987) 83.
51. H. Leutwyler and A. Smilga, *Phys. Rev. D* **46** (1992) 5607.
52. P. H. Damgaard and H. Fukaya, *JHEP* **0901** (2009) 052 [arXiv:0812.2797 [hep-lat]].
53. R. Brower, P. Rossi, and C.-I. Tan, *Nucl. Phys. B* **190** [FS3] (1981) 699.
54. A. D. Jackson, M. K. Sener, and J. J. M. Verbaarschot, *Phys. Lett. B* **387**, 355 (1996) [hep-th/9605183].
55. A. V. Smilga and J. J. M. Verbaarschot, *Phys. Rev. D* **51** (1995) 829 [hep-th/9404031].
56. P. H. Damgaard, J. C. Osborn, D. Toublan, and J. J. M. Verbaarschot, *Nucl. Phys. B* **547** (1999) 305 [hep-th/9811212].
57. F. Basile and G. Akemann, *JHEP* **0712** (2007) 043 [arXiv:0710.0376 [hep-th]].
58. C. Lehner and T. Wettig, *JHEP* **0911** (2009) 005 [arXiv:0909.1489 [hep-lat]].
59. G. Akemann, F. Basile, and L. Lellouch, *JHEP* **0812** (2008) 069 [arXiv:0804.3809 [hep-lat]].
60. P. H. Damgaard, T. DeGrand, and H. Fukaya, *JHEP* **0712** (2007) 060 [arXiv:0711.0167 [hep-lat]].
61. C. Lehner, S. Hashimoto, and T. Wettig, *JHEP* **1006** (2010) 028 [arXiv:1004.5584 [hep-lat]].
62. J. C. Osborn and J. J. M. Verbaarschot, *Phys. Rev. Lett.* **81** (1998) 268 [hep-ph/9807490].

63. R. A. Janik, M. A. Nowak, G. Papp, and I. Zahed, *Phys. Rev. Lett.* **81** (1998) 264 [hep-ph/9803289].

64. M. E. Berbenni-Bitsch et al., *Phys. Lett. B* **438** (1998) 14 [hep-ph/9804439].

65. T. Guhr, J. Z. Ma, S. Meyer, and T. Wilke, *Phys. Rev. D* **59** (1999) 054501 [hep-lat/9806003].

66. D. Toublan and J. J. M. Verbaarschot, *Nucl. Phys. B* **560** (1999) 259 [hep-th/9904199].

67. J. J. M. Verbaarschot and I. Zahed, *Phys. Rev. Lett.* **73** (1994) 2288 [hep-th/9405005].

68. M. Kieburg, J. J. M. Verbaarschot, and S. Zafeiropoulos, *Phys. Rev. D* **90** (2014) 085013 [arXiv:1405.0433 [hep-lat]].

69. A. P. Schnyder, S. Ryu, A. Furusaki, and A. W. W. Ludwig, *Phys. Rev. B* **78** (2008) 195125 [arXiv:0803.2786].

70. P. H. Damgaard, K. Splittorff, and J. J. M. Verbaarschot, *Phys. Rev. Lett.* **105** (2010) 162002 [arXiv:1001.2937].

71. S. R. Sharpe and R. L. Singleton, *Phys. Rev. D* **58** (1998) 074501 [arXiv:heplat/9804028].

72. M. A. Stephanov, *Phys. Rev. Lett.* **76** (1996) 4472 [arXiv:9604003 [hep-lat]].

73. G. Akemann, *Phys. Rev. D* **64** (2001) 114021 [hep-th/0106053].

74. J. C. Osborn, *Phys. Rev. Lett.* **93** (2004) 222001 [arXiv:0403131 [hep-th]].

75. G. Akemann and J. R. Ipsen, *Acta Phys. Polon. B* **46** (2015) 1747 [arXiv:1502.01667 [math-ph]].

76. J. C. Osborn, *Nucl. Phys. Proc. Suppl.* **129** (2004) 886 [hep-lat/0309123].

77. G. Akemann, P. H. Damgaard, K. Splittorff, and J. J. M. Verbaarschot, *Phys. Rev. D* **83** (2011) 085014 [arXiv:1012.0752].

78. B. Schlittgen and T. Wettig, *J. Phys. A* **36** (2003) 3195 [math-ph/0209030].

79. M. Kieburg, *J. Phys. A: Math. Theor.* **45** (2012) 205203 [arXiv:1202.1768].

80. G. Akemann and T. Nagao, JHEP **1110** (2011) 060 [arXiv:1108.3035 [math-ph]].

81. A. Pandey and M. L. Mehta, *Commun. Math. Phys.* **87** (1983) 449; M. L. Mehta and A. Pandey, *J. Phys. A* **16** (1983) 2655.

82. P. H. Damgaard, U. M. Heller, and K. Splittorff, *Phys. Rev. D* **85** (2012) 014505 [arXiv:1110.2851 [hep-lat]]; *Phys. Rev. D* **86** (2012) 094502 [arXiv:1206.4786 [hep-lat]].

83. A. Deuzeman, U. Wenger, and J. Wuilloud, *JHEP* **1112** (2011) 109 [arXiv:1110.4002 [hep-lat]].

84. M. Kieburg, J. J. M. Verbaarschot, and S. Zafeiropoulos, *Phys. Rev. Lett.* **108** (2012) 022001 [arXiv:1109.0656 [hep-lat]].

85. K. Cichy, E. Garcia-Ramos, K. Splittorff, and S. Zafeiropoulos, arXiv:1510.09169 [hep-lat].

86. G. Akemann, P. H. Damgaard, K. Splittorff, and J. Verbaarschot, *PoS LAT* **2010** (2010) 092 [arXiv:1011.5118 [hep-lat]].

87. M. Kieburg, K. Splittorff, and J. J. M. Verbaarschot, *Phys. Rev. D* **85** (2012) 094011 [arXiv:1202.0620 [hep-lat]].

88. K. Splittorff and J. J. M. Verbaarschot, *Phys. Rev. D* **84** (2011) 065031 [arXiv:1105.6229 [hep-lat]].

89. S. Necco and A. Shindler, *JHEP* **1104** (2011) 031 [arXiv:1101.1778 [hep-lat]].
90. G. Akemann and A. C. Ipsen, *JHEP* **1204** (2012) 102 [arXiv:1202.1241 [hep-lat]].
91. P. de Forcrand, *PoS LAT* **2009** (2009) 010 [arXiv:1005.0539 [hep-lat]].
92. J. Bloch, F. Bruckmann, M. Kieburg, K. Splittorff, and J. J. M. Verbaarschot, *Phys. Rev. D* **87** (2013) 034510 [arXiv:1211.3990 [hep-lat]].
93. J. Bloch, *Phys. Rev. Lett.* **107** (2011) 132002 [arXiv:1103.3467 [hep-lat]].
94. M. A. Stephanov, *Phys. Lett. B* **375** (1996) 249 [hep-lat/9601001].
95. B. Seif, T. Wettig, and T. Guhr, *Nucl. Phys. B* **548** (1999) 475 [hep-th/9811044].
96. A. M. Halasz, A. D. Jackson, R. E. Shrock, M. A. Stephanov, and J. J. M. Verbaarschot, *Phys. Rev. D* **58** (1998) 096007 [hep-ph/9804290].
97. G. Akemann and E. Strahov, *Commun. Math. Phys.* **354** (2016) 101 [arXiv:1504.02047 [math-ph]].
98. G. Akemann, M. J. Phillips, and H.-J. Sommers, *J. Phys. A* **43** (2010) 085211 [arXiv:0911.1276 [hep-th]].
99. G. Akemann, *Nucl. Phys. B* **730** (2005) 253 [hep-th/0507156].
100. G. H. Golub, and C. F. Van Loan, *Matrix Computations*, Vol. 3. JHU Press, Baltimore, 2012.
101. E. Kanzieper and N. Singh, *J. Math. Phys.* **51** (2010) 103510 [arXiv:1006.3096 [math-ph]].
102. G. Akemann and Z. Burda, *J. Phys. A: Math. Theo.* **45** (2012) 465201 [arXiv:1208.0187 [math-ph]].
103. G. Akemann, M. J. Phillips, and L. Shifrin, *J. Math. Phys.* **50** (2009) 063504 [arXiv:0901.0897 [math-ph]].
104. G. Akemann, J. C. Osborn, K. Splittorff, and J. J. M. Verbaarschot, *Nucl. Phys. B* **712** (2005) 287 [hep-th/0411030].
105. G. Akemann, *J. Phys. A* **36** (2003) 3363 [hep-th/0204246].
106. G. Akemann and M. Bender, *J. Math. Phys.* **51** (2010) 103524 [arXiv:1003.4222 [math-ph]].
107. Y. V. Fyodorov, B. A. Khoruzhenko, and H.-J. Sommers, *Ann. Inst. Henri Poincaré* **68** (1998) 449 [arXiv:chao-dyn/9802025].
108. Y. V. Fyodorov, B. A. Khoruzhenko, and H.-J. Sommers, *Phys. Lett. A* **226** (1997) 46 [arXiv:cond-mat/9606173]; *Phys. Rev. Lett.* **79** (1997) 557 [arXiv:cond-mat/9703152].
109. K. Splittorff and J. J. M. Verbaarschot, *Nucl. Phys. B* **683** (2004) 467 [hep-th/0310271].
110. T. D. Cohen, *Phys. Rev. Lett.* **91** (2003) 222001 [hep-ph/0307089].
111. G. Akemann, Y. V. Fyodorov, and G. Vernizzi, *Nucl. Phys. B* **694** (2004) 59 [hep-th/0404063].
112. H. Sompolinsky, A. Crisanti, and H.-J. Sommers, *Phys. Rev. Lett.* **61** (1988) 259.
113. K. Splittorff and J. J. M. Verbaarschot, *Nucl. Phys. B* **695** (2004) 84 [hep-th/0402177].
114. M. C. Bergere, hep-th/0404126.
115. G. Akemann and A. Pottier, *J. Phys. A* **37** (2004) L453 [math-ph/0404068].
116. M. P. Lombardo, K. Splittorff, and J. J. M. Verbaarschot, *Phys. Rev. D* **80** (2009) 054509 [arXiv:0904.2122 [hep-lat]].

117. J. C. R. Bloch and T. Wettig, *JHEP* **0903** (2009) 100 [arXiv:0812.0324 [hep-lat]].
118. P. H. Damgaard, U. M. Heller, K. Splittorff, and B. Svetitsky, *Phys. Rev. D* **72** (2005) 091501 [hep-lat/0508029].
119. G. Akemann, P. H. Damgaard, J. C. Osborn, and K. Splittorff, *Nucl. Phys. B* **766** (2007) 34; Erratum: [*Nucl. Phys. B* **800** (2008) 406] [hep-th/0609059].
120. G. Akemann and T. Wettig, *Phys. Rev. Lett.* **92** (2004) 102002; Erratum: [*Phys. Rev. Lett.* **96** (2006) 029902] [hep-lat/0308003].
121. J. C. R. Bloch and T. Wettig, *Phys. Rev. Lett.* **97** (2006) 012003 [hep-lat/0604020].
122. G. Akemann, J. C. R. Bloch, L. Shifrin, and T. Wettig, *Phys. Rev. Lett.* **100** (2008) 032002 [arXiv:0710.2865 [hep-lat]].
123. J. C. Osborn, K. Splittorff, and J. J. M. Verbaarschot, *Phys. Rev. Lett.* **94** (2005) 202001 [hep-th/0501210].
124. K. Splittorff, *PoS LAT* **2006** (2006) 023 [hep-lat/0610072].
125. T. Kanazawa and T. Wettig, *JHEP* **1410** (2014) 55 [arXiv:1406.6131 [hep-ph]].
126. A. B. J. Kuijlaars and L. Zhang, *Commun. Math Phys* **332** (2014) 759 [arXiv:1308.1003 [math-ph]].
127. G. Akemann, *Phys. Lett. B* **547** (2002) 100 [hep-th/0206086].
128. G. Akemann and M. J. Phillips, in *Random Matrices*, MSRI Publications ed. P. Deift and P. J. Forrester, Vol. 65 pp. 1–24. Cambridge University Press, Cambridge, 2014 [arXiv:1204.2740 [math-ph]].

6

Random matrix theory and (big) data analysis

Jean-Philippe BOUCHAUD

Capital Fund Management, 23 rue de l'Université, 75 007 Paris, France

Bouchaud, J.P., 'Random Matric Theory and (Big) Data Analysis' in *Stochastic Processes and Random Matrices*. Edited by: Grégory Schehr et al, Oxford University Press (2017). © Oxford University Press 2017. DOI 10.1093/oso/9780198797319.003.0006

Chapter Contents

In this chapter, we review methods from random matrix theory to extract information about a large signal matrix **C** (for example, a correlation matrix arising in big data problems), from its noisy observation matrix **M**. We show that the replica method can be used to obtain both the spectral density and the *overlaps* between noise corrupted eigenvectors and the true ones, for both additive and multiplicative noise. This allows one to construct optimal rotationally invariant estimators of **C** based on the observation of **M** alone. We also discuss the case of rectangular correlation matrices and the problem of random singular value decomposition. These notes are based on a common work with R. Allez, J. Bun, and M. Potters.

6.1 Introduction

In the present era of 'big data', new statistical methods are needed to decipher large dimensional data sets that are now routinely generated in almost all fields—physics, image analysis, genomics, epidemiology, engineering, economics, and finance, to quote only a few. It is, for example, very natural to try to identify common causes (or factors) that explain the joint dynamics of N quantities. These quantities might be daily returns of the different stocks of the S&P 500, temperature variations in different locations around the planet, velocities of individual grains in a packed granular medium, or different biological indicators (blood pressure, cholesterol, etc.) within a population, etc. The simplest mathematical object that quantifies the similarities between these observable is an $N \times N$ correlation matrix **C**. Its eigenvalues and eigenvectors can then be used to characterize the most important common dynamical modes—this is the well-known 'principal component analysis' (or PCA) method.

However, when N is large, a proper estimation of all the elements of **C** becomes problematic when the total number T of simultaneous observations of our N variables is not very large compared to N itself. In the example of stock returns, T is the total number of trading days in the sampled data; but in the biological example, T would be the size of the population sample, etc. Throughout the chapter, we will denote **M** the empirical correlation matrix, defined as

$$M_{ij} = \frac{1}{T} \sum_{t=1}^{T} R_i^t R_j^t \equiv \frac{1}{T} \left(\mathbf{R}^T \mathbf{R} \right)_{ij}, \tag{6.1}$$

where R_i^t is the realization of the ith observable ($i = 1, \ldots, N$) at 'time' t ($t = 1, \ldots, T$), which we be assume in the following to be demeaned and standardized. The empirical correlation matrix **M** (i.e., computed on a given realization) must be carefully distinguished from the 'true' correlation matrix **C** of the underlying statistical process (which might not even exist). In fact, the whole point of the present review is to characterize the difference between **M** and **C**, and discuss how well (or how badly) one may reconstruct **C** from the knowledge of **M**. Of course, if N is small (say $N = 4$) and the number of observations is huge (say $T = 10^6$), then one can intuitively expect that any observable computed using **M** will be very close to its 'true' value,

computed using $\mathbf{C}$. For example, $\mathrm{Tr}\mathbf{M}^{-1}$ is very close to being a consistent estimator of $\mathrm{Tr}\mathbf{C}^{-1}$ when T is large enough for a fixed N. This is the usual limit considered in statistics. However, in many applications where T is large, the number of observables N is also large, such that the ratio $q = N/T$ is not very small compared to unity. In this case, one finds that $\mathrm{Tr}\mathbf{M}^{-1} = \mathrm{Tr}\mathbf{C}^{-1}/(1-q)$; i.e., the estimator is strongly biased, and even diverges when $q \to 1$. Typical numbers in the case of stocks are $N = 500$ and $T = 2500$, corresponding to 10 years of daily data, already quite a long strand compared to the lifetime of stocks or the expected structural evolution time of markets, but correspond to $q = 0.2$. For macroeconomic indicators, say inflation, 20 years of monthly data produce a meager $T = 240$, whereas the number of sectors of activity for which inflation is recorded is around $N = 30$, such that $q = 0.125$. The relevant mathematical limit to focus on in these cases is $T \gg 1$, $N \gg 1$ but with $q = N/T = O(1)$.

The aim of this chapter is to review several random matrix theory (RMT) results that can be established in the above asymptotic limit. The best known result in the field is the Marčenko–Pastur 1967 theorem (Marchenko and Pastur 1967), which shows that the empirical density of eigenvalues (aka the 'spectral density') is strongly distorted when compared with the 'true' density (corresponding to $q \to 0$). However, when $T \to \infty$, $N \to \infty$, the spectrum has a large degree of universality with respect to the distribution of the R_i^t values; this makes RMT results particularly appealing.

While many of these RMT results (like the one of Marčenko–Pastur) are some decades old and quite well known, the field has witnessed a recent revival—with, in particular, financial time series in mind, although many other applications are concerned as well. Whereas many of the early RMT results only dealt with the statistics of *eigenvalues*, the most exciting, very recent developments relate to the *eigenvectors* and offer very interesting new tools to separate efficiently the wheat (the signal) from the chaff (the noise) in correlation matrices. Reviewing these new results is the aim of the present chapter.

6.2   Eigenvalue spectrum and eigenvector overlaps

Throughout this chapter, we will consider the signal matrix $\mathbf{C}$ to be a symmetric matrix of dimension N with N that goes to infinity. We denote by $c_1 \geq c_2 \geq \cdots \geq c_N$ its eigenvalues and by $|V_1\rangle, |V_2\rangle, \ldots, |V_N\rangle$ their corresponding eigenvectors. The corrupted, noisy matrix $\mathbf{M}$ will be assumed to be symmetric with eigenvalues denoted by $\mu_1 \geq \mu_2 \geq \cdots \geq \mu_N$ associated with the eigenvectors $|U_1\rangle, |U_2\rangle, \ldots, |U_N\rangle$. In the limit of large dimension, it is often more convenient to index the eigenvectors of both matrices by their corresponding eigenvalues, i.e., $|U_i\rangle \to |U_{\mu_i}\rangle$ and $|V_i\rangle \to |V_{c_i}\rangle$ for any integer $1 \leq i \leq N$, and this is the convention that we adopt henceforth. We will use the following notation for the (rescaled) *overlaps*:

$$O(\mu_i, c_j) := N\left[\langle U_i | V_j \rangle^2\right]. \tag{6.2}$$

6.2.1 Relation among resolvent, spectrum, and overlaps

A convenient way to work out the eigenvalue spectrum and eigenvector overlaps is to study the resolvent of the random matrix $\mathbf{M}$, defined as

$$\mathcal{G}_{\mathbf{M}}(z) := (z\mathbb{I}_N - \mathbf{M})^{-1}.$$

The claim is that for z not too close to the real axis, the matrix $\mathcal{G}_{\mathbf{M}}(z)$ is *self-averaging* in the large N limit so that its value is independent of the specific realization of $\mathbf{M}$ (at least for the examples studied in the following). More precisely, this means that $\mathcal{G}_{\mathbf{M}}(z)$ converges to a deterministic matrix for any fixed value (i.e., independent of N) of $z \in \mathbb{C} \setminus \mathbb{R}$ when $N \to \infty$.

The relation between the resolvent and the overlaps $O(\mu_j, c_i)$ is relatively straightforward. For $z = \mu - i\eta$ with $\mu \in \mathbb{R}$ and $\eta > 0$, we have

$$\mathcal{G}_{\mathbf{M}}(\mu - i\eta) = \sum_{k=1}^{N} \left[\frac{\mu}{(\mu - \mu_k)^2 + \eta^2} + i\frac{\eta}{(\mu - \mu_k)^2 + \eta^2} \right] |U_k\rangle\langle U_k|.$$

If we take the trace of this quantity, and take the limit $\eta \to 0$ (after $N \to \infty$), it is well known that one obtains the 'density of states' (or 'spectral density') $\rho_{\mathbf{M}}$:

$$\Im G_{\mathbf{M}}(\mu - i\eta) \equiv \Im \frac{1}{N}\mathrm{Tr}\mathcal{G}_{\mathbf{M}}(\mu - i\eta) = \pi \rho_{\mathbf{M}}(\mu) \tag{6.3}$$

(see Appendix A). Similarly, the elements of $\Im\mathcal{G}_{\mathbf{M}}(\mu - i\eta)$ can be written for $\eta > 0$ as

$$\langle V_i|\Im\mathcal{G}_{\mathbf{M}}(\mu - i\eta)|V_i\rangle = \sum_{k=1}^{N} \frac{\eta}{(\mu - \mu_k)^2 + \eta^2}\langle V_i|U_k\rangle^2. \tag{6.4}$$

This latter quantity is also self-averaging in the large N limit in the sense

$$\langle V_i|\Im\mathcal{G}_{\mathbf{M}}(\mu - i\eta)|V_i\rangle \xrightarrow[N\to\infty]{} \int_{\mathbb{R}} \frac{\eta}{(\mu - \mu')^2 + \eta^2}O(\mu', c_i)\rho_{\mathbf{M}}(\mu')\mathrm{d}\mu'.$$

where the overlap function $O(\mu', c_i)$ is extended (continuously) to arbitrary values of μ' inside the support of $\rho_{\mathbf{M}}$ in the large N limit. Sending $\eta \to 0$ (but after $N \to \infty$) in this latter equation, we finally obtain the following formula valid in the large N limit

$$\langle V_i|\Im\mathcal{G}_{\mathbf{M}}(\mu - i\eta)|V_i\rangle \approx \pi\rho_{\mathbf{M}}(\mu)O(\mu, c_i). \tag{6.5}$$

Equation (6.5) will thus enable us to investigate the overlaps $O(\mu, c_i)$ in great details through the calculation of the elements of the resolvent $\mathcal{G}_{\mathbf{M}}(z)$. This is what we aim for in the next subsections.

6.2.2 Free additive noise

The first model of noisy measurement that we consider is the case where the true signal $\mathbf{C}$ is corrupted by a free additive noise, that is to say

$$\mathbf{M} = \mathbf{C} + OBO^{\dagger}, \tag{6.6}$$

where $\mathbf{B}$ is a fixed matrix with eigenvalues $b_1 > b_2 > \cdots > b_N$ with limiting spectral density $\rho_{\mathbf{B}}$ and O is a random matrix chosen uniformly in the orthogonal group $O(N)$ (i.e., according to the Haar measure). A simple example is when the noisy matrix $OBO^{\dagger}$ is a symmetric Gaussian random matrix with independent and identically distributed (i.i.d.) entries, corresponding to the so-called Gaussian orthogonal ensemble (GOE). By construction, the eigenvectors of a GOE matrix are invariant under rotation.

It is now well known that the spectral density of $\mathbf{M}$ can be obtained from that of $\mathbf{C}$ and $\mathbf{B}$ using free addition; see Voiculescu et al. (1992) and, in the language of statistical physics, Zee (1996). The statistics of the eigenvalues of $\mathbf{M}$ has, therefore, been investigated in great detail; see Brézin and Zee (1995) and Brézin et al. (1995) for instance. However, the question of the eigenvectors has been much less studied, except recently in Allez and Bouchaud (2014), and Allez et al. (2013) in the special case where $OBO^{\dagger}$ belongs to the GOE (see Eq. (6.12)).

A way to obtain results for both the eigenvalue spectrum (and recover the free addition results) and the eigenvector overlaps is to use the replica method. (Other physical approaches, e.g., supersymmetric methods, can also be used to understand free addition in a more rigourous way: see, e.g., Mandt and Zirnbauer (2010).) For a general free additive noise, one finds that the estimate for the resolvent reads, in the large N limit—see Appendix B:

$$\mathcal{G}_{\mathbf{M}}(z) = \mathcal{G}_{\mathbf{C}}(Z(z)) \tag{6.7}$$

where the function $Z(z)$ is given by

$$Z(z) = z - R_{\mathbf{B}}(G_{\mathbf{M}}(z)), \tag{6.8}$$

and $R_{\mathbf{B}}$ is the so-called R-transform of $\mathbf{B}$ (see Appendix A for a reminder of the definition of the different useful spectral transforms).

Note that Eq. (6.7) is a matrix relation that simplifies when written on the basis where $\mathbf{C}$ is diagonal, since in this case $\mathcal{G}_{\mathbf{C}}(Z)$ is also diagonal. Therefore, the evaluation of the overlap $O(\mu, c)$ is straightforward using Eq. (6.5). Let us define the Hilbert transform $H_{\mathbf{M}}(\mu)$, which is simply the real part of the Stieltjes transform $G_{\mathbf{M}}(\mu - i\eta)$ in the limit $\eta \to 0$. Then the overlap for the free additive noise is given by

$$O(\mu, c) = \frac{\beta_1(\mu)}{(\mu - c - \alpha_1(\mu))^2 + \pi^2 \beta_1(\mu)^2 \rho_{\mathbf{M}}(\mu)^2}, \tag{6.9}$$

where c is the corresponding eigenvalue of the unperturbed matrix $\mathbf{C}$, and where we have defined

$$\begin{cases} \alpha_1(\mu) \equiv \Re[R_{\mathbf{B}}\left(H_{\mathbf{M}}(\mu) + i\pi \rho_{\mathbf{M}}(\mu)\right)], \\ \beta_1(\mu) \equiv \dfrac{\Im[R_{\mathbf{B}}(H_{\mathbf{M}}(\mu) + i\pi \rho_{\mathbf{M}}(\mu))]}{\pi \rho_{\mathbf{M}}(\mu)}. \end{cases} \tag{6.10}$$

As a first check of these results, let us consider the normalized trace of Eq. (6.7) and then set $u = G_{\mathbf{M}}(z) = G_{\mathbf{C}}(Z(z))$. One can find by using the blue transform that we indeed retrieve the free addition formula $R_{\mathbf{M}}(u) = R_{\mathbf{C}}(u) + R_{\mathbf{B}}(u)$ when $N \to \infty$, as it should be.

Deformed GOE

As a second verification, we specialize our result to the case where $OBO^{\dagger}$ is a GOE matrix such that the entries have a variance equal to σ^2/N. It is then well known that in this case $R_{\mathbf{B}}(z) = \sigma^2 z$, meaning that Eq. (6.8) simply becomes $Z(z) = z - \sigma^2 G_{\mathbf{M}}(z)$. This allows us to get a simpler expression for the resolvent of $\mathbf{M}$, i.e.,

$$\mathcal{G}_{\mathbf{M}}(z) = \mathcal{G}_{\mathbf{C}}(z - \sigma^2 \mathcal{G}_{\mathbf{M}}(z)). \tag{6.11}$$

This can also be interpreted as the solution of a Burgers' equation in the zero viscosity limit, see, e.g., Allez et al. (2013). For the overlaps, the expression also simplifies to

$$O(\mu, c) = \frac{\sigma^2}{(c - \mu + \sigma^2 H_{\mathbf{M}}(\mu))^2 + \sigma^4 \pi^2 \rho_{\mathbf{M}}(\mu)^2}, \tag{6.12}$$

which is exactly the result derived in Allez and Bouchaud (2014) and Allez et al. (2013) using other methods.

6.2.3 Free multiplicative noise and empirical covariance matrices

Our second model deals with *multiplicative* noise in the following sense: we consider that the noisy measurement matrix $\mathbf{M}$ can be written as

$$\mathbf{M} = \sqrt{\mathbf{C}} OBO^{\dagger} \sqrt{\mathbf{C}}, \tag{6.13}$$

where again $\mathbf{C}$ is the signal, $\mathbf{B}$ is a fixed matrix with eigenvalues $b_1 > b_2 > \cdots > b_N$ with limiting density $\rho_{\mathbf{B}}$, and O is a random matrix chosen in the orthogonal group $O(N)$ according to the Haar measure. Note that we implicitly require that $\mathbf{C}$ is positive definite with Eq. (6.13), so that the square root of $\mathbf{C}$ is well defined. The case of empirical correlation matrices, Eq. (6.1), corresponds to choosing $\mathbf{B}$ as a Wishart matrix; see below.

The Replica analysis leads to the following systems of equations (see Allez et al. (2016) for details) for the general problem of a free multiplicative noise, Eq. (6.13),

$$z\mathcal{G}_{\mathbf{M}}(z) = Z(z)\mathcal{G}_{\mathbf{C}}(Z(z)), \tag{6.14}$$

with

$$Z(z) = zS_{\mathbf{B}}(zG_{\mathbf{M}}(z) - 1), \tag{6.15}$$

where $S_{\mathbf{B}}$ is the so-called S-transform of $\mathbf{B}$ and $G_{\mathbf{M}}$ is the normalized trace of $\mathcal{G}_{\mathbf{M}}(z)$. The latter obeys, from Eqs. (6.14), the self-consistent equation

$$zG_{\mathbf{M}}(z) = Z(z)G_{\mathbf{C}}(Z(z)). \tag{6.16}$$

Again, Eq. (6.14) is a matrix relation that simplifies when written on the basis where $\mathbf{C}$ is diagonal. Note that Eqs (6.16) and (6.15) allow us to retrieve the usual free multiplicative convolution, that is to say,

$$S_{\mathbf{M}}(u) = S_{\mathbf{C}}(u)S_{\mathbf{B}}(u). \tag{6.17}$$

This result is thus the analog of our result (6.7) in the multiplicative case.

With the estimate for the resolvent given by Eqs. (6.14) and (6.15), we can obtain a general overlap formula for the free multiplicative noise case. Let us set $z = \mu - i\eta$ and take the limit $\eta \to 0$ as usual, and define the functions

$$\begin{cases} \alpha_2(\mu) \equiv \lim_{z \to \mu - i0^+} \Re\left[\dfrac{1}{S_{\mathbf{B}}(zG_{\mathbf{M}}(z)-1)} \right] \\[2ex] \beta_2(\mu) \equiv \lim_{z \to \mu - i0^+} \Im\left[\dfrac{1}{S_{\mathbf{B}}(zG_{\mathbf{M}}(z)-1)} \right] \dfrac{1}{\pi \rho_{\mathbf{M}}(\mu)}. \end{cases} \tag{6.18}$$

The overlap $O(\mu, c)$ between the eigenvectors of $\mathbf{C}$ and $\mathbf{M}$ are given by

$$O(\mu, c) = \frac{c\beta_2(\mu)}{(\mu - c\alpha_2(\mu))^2 + \pi^2 c^2 \beta_2(\mu)^2 \rho_{\mathbf{M}}(\mu)^2}. \tag{6.19}$$

In order to give more insights on our results, we will now specify these results to the well-known case of empirical correlation matrices.

Empirical covariance matrix

As mentioned previously, the most famous application of a model of the form (6.13) is given by the sample covariance estimator that we recall briefly. Let us define the $N \times T$ observation matrix $\mathbf{R}$ that comes from T independent and identically distributed samples $\mathbf{R}^t \equiv (\mathbf{R}_1^t, \dots, \mathbf{R}_N^t)$ with $t \in [1, T]$ and we assume that each sample has zero mean. The N elements of $\mathbf{R}^t$ generally display some degree of interdependence, that is often represented by the *true* (or also *population*) covariance matrix $\mathbf{C}$, defined as $\langle \mathbf{R}_i^t \mathbf{R}_j^{t'} \rangle = \mathbf{C}_{i,j} \delta_{t,t'}$, where $\delta_{t,t'}$ is the Kronecker symbol. As the signal $\mathbf{C}$ is unknown, the classical estimator for the covariance matrix is to compute the empirical (or sample) covariance matrix thanks to the Pearson estimator

$$\mathbf{M} = \frac{1}{T}\mathbf{R}\mathbf{R}^\dagger = \sqrt{\mathbf{C}}\frac{1}{T}\mathbf{X}\mathbf{X}^\dagger\sqrt{\mathbf{C}},$$

where $\mathbf{X}$ is a $N \times T$ matrix where all elements are i.i.d. random variables (i.e., their true covariance matrix is the identity matrix). So this model is a particular case of the model (6.13) with $\mathbf{B} := T^{-1}\mathbf{X}\mathbf{X}^{\dagger}$, i.e., a Wishart matrix. The S transform of $\mathbf{B} = T^{-1}\mathbf{X}\mathbf{X}^{\dagger}$ has an explicit form

$$S_{\mathbf{B}}(x) = \frac{1}{1 + qx}, \qquad q = \frac{N}{T}. \tag{6.20}$$

Using our general results Eqs (6.14) and (6.15), we obtain, on the basis where $\mathbf{C}$ is diagonal,

$$z\mathcal{G}_{\mathbf{M}}(z) = Z(z)\mathcal{G}_{\mathbf{C}}(Z(z)), \quad \text{with} \quad Z(z) = \frac{z}{1 - q + qz\mathcal{G}_{\mathbf{M}}(z)}, \tag{6.21}$$

which is exactly the result found in Burda et al. (2004) and also in Knowles and Yin (2014) at leading order. We can therefore recover the well-known Marčenko–Pastur equation (Marchenko and Pastur 1967), which gives a fixed point equation satisfied by the resolvent of $\mathbf{M}$ in term of the resolvent of the true matrix $\mathbf{C}$

$$z G_{\mathbf{M}}(z) = Z(z)G_{\mathbf{C}}(Z(z)), \quad \text{with} \quad Z(z) = \frac{z}{1 - q + qz G_{\mathbf{M}}(z)}, \tag{6.22}$$

from which one can express the spectral density of $\mathbf{M}$ in terms of that of $\mathbf{C}$. The most famous example is the case where $\mathbf{C} = \mathbb{I}_N$. In this case, we simply have $G_{\mathbf{C}}(z) = (z-1)^{-1}$ and solving for $G_{\mathbf{M}}(z)$ one finds that

$$G_{\mathbf{M}}(z) = \frac{z + q - 1 \pm \sqrt{(z+q-1)^2 - 4zq}}{2zq}, \tag{6.23}$$

which leads to the well-known Marčenko–Pastur spectral density

$$\rho_{\mathbf{M}}(\mu) = \frac{1}{2\pi q \mu} \sqrt{4q\mu - (\mu + q - 1)^2}, \quad \mu \in \left[(1 - \sqrt{q})^2, (1 + \sqrt{q})^2\right]. \tag{6.24}$$

The expression of the limiting overlaps can also be simplified in the particular case of empirical correlation matrices, and reads

$$O(\mu, c) = \frac{qc\mu}{(c(1-q) - \mu + qc\mu H_{\mathbf{M}}(\mu))^2 + q^2\mu^2 c^2 \pi^2 \rho_{\mathbf{M}}(\mu)^2}, \tag{6.25}$$

and we recover the result established in Ledoit and Péché (2011). As a conclusion, our result generalizes the standard Marčenko and Pastur formalism to an arbitrary multiplicative noise term $OBO^{\dagger}$.

6.3 Rotationally invariant estimators

6.3.1 The oracle estimator and the overlaps

We now attempt to construct an estimator $\widehat{\mathbf{C}}(\mathbf{M})$ of the true signal $\mathbf{C}$ that relies on the given dataset $\mathbf{M}$ at our disposal. It is well known that an estimator is optimal

with respect to a specific loss function (e.g., the distance) and a standard metric is to consider the (squared) Euclidean (or Frobenius) norm

$$\text{Tr}\left[(\mathbf{C} - \widehat{\mathbf{C}}(\mathbf{M}))^2\right].$$

The best estimator with respect to this loss function is the solution of the minimization problem

$$\widehat{\mathbf{C}}(\mathbf{M}) = \underset{\mathbf{C}(\mathbf{M})\in\mathcal{A}}{\text{argmin}} \sum_{i,j} [\mathbf{C}_{i,j} - \mathbf{C}(\mathbf{M})_{i,j}]^2, \tag{6.26}$$

considered over the set of all possible estimators $\mathbf{C}(\mathbf{M})$ of a certain class $\mathcal{A}$. In the following, we will assume that we have no prior on the structure of the eigenvectors of $\mathbf{C}$; i.e., we assume that $\mathbf{C}$ belongs to a rotationally invariant ensemble. In this case, the only possible unbiased choice for the eigenvectors of $\mathbf{C}$ is those of the empirical matrix $\mathbf{M}$. The corresponding estimator will be called a 'rotationally invariant estimator' (RIE), and the class $\mathcal{A}$ will therefore be the ensemble of symmetric matrices that have the same eigenvectors as $\mathbf{M}$.

In this case, the only free variables left in the constrained optimization problem (6.26) are the eigenvalues of $\mathbf{C}(\mathbf{M})$ and we can rewrite

$$\widehat{\mathbf{C}}(\mathbf{M}) = U\widehat{\mu}\,U^\dagger, \tag{6.27}$$

where the eigenvalues $\widehat{\mu}_1, \widehat{\mu}_2, \ldots, \widehat{\mu}_N$ are associated with the corresponding perturbed eigenvectors $|U_{\mu_1}\rangle, |U_{\mu_2}\rangle, \ldots, |U_{\mu_N}\rangle$. We seek for the $\widehat{\mu}_i$ that solve the optimization program

$$\widehat{\mu} = \underset{\{\widehat{\mu}_k\}_k\in\mathbb{R}^N}{\text{argmin}} \sum_{i,j=1}^{N} \left(\mathbf{C}_{i,j} - \sum_{k=1}^{N} U_{i,k}\widehat{\mu}_k U_{j,k}\right)^2, \tag{6.28}$$

where the $U_{i,k}$ denote the entries of the eigenvectors of $\mathbf{M}$. A simple computation leads to the following formulas for the $\widehat{\mu}_i$ in terms of the overlaps among the perturbed $|U_i\rangle$, the nonperturbed eigenvectors $|V_j\rangle$, and the eigenvalues of the true matrix $\mathbf{C}$:

$$\widehat{\mu}_i = \text{Tr}\left[|U_i\rangle\langle U_i|\mathbf{C}\right] \equiv \sum_{j=1}^{N} \langle U_i|V_j\rangle^2 c_j. \tag{6.29}$$

We will see that this estimator $\widehat{\mu}_i$ is self-averaging in the large N limit and can thus be approximated with its expected value

$$\widehat{\mu}_i \approx \sum_{j=1}^{N} \left[\langle U_i|V_j\rangle^2\right] c_j,$$

where $[\cdot]$ denotes the expected value with respect to the random eigenvectors $(|U_i\rangle)_i$ of the matrix $\mathbf{M}$. In statistics, the optimal RIE (6.29) is sometimes called the *oracle*

estimator because it depends explicitly on the knowledge of the true signal $\mathbf{C}$. The 'miracle' is that in the large N limit, and for a large class of problems, one can actually express the oracle estimator in terms of the (observable) spectral density of $\mathbf{M}$ only!

6.3.2 Optimal rotational invariant estimator

Equipped with the results of the previous section, we can now tackle the problem of the optimal RIE of the signal $\mathbf{C}$. Indeed, the high-dimensional limit $N \to \infty$, that allows one to reach some degree of universality. First, we rewrite the RIE (6.29) as

$$\widehat{\mu}_i \underset{N \to \infty}{=} \frac{1}{N} \sum_{j=1}^{N} O(\mu_i, c_j) c_j \approx \int \mathrm{d}c\, \rho_{\mathbf{C}}(c)\, O(\mu_i, c)c.$$

Quite remarkably, as we show in the following, the optimal RIE can be expressed as a function of the spectral measure of the observable (noisy) $\mathbf{M}$ only.

Free additive noise

We now specialize the RIE and we begin with the free additive noise case for which the noisy measurement is given by

$$\mathbf{M} = \mathbf{C} + OBO^\dagger.$$

It is easy to see from Eqs (6.5) and (6.7) that

$$\widehat{\mu}_i = \frac{1}{\pi \rho_{\mathbf{M}}(\mu_i)} \lim_{z \to \mu_i - i0^+} \Im \int \mathrm{d}c \frac{\rho_{\mathbf{C}}(c)c}{Z(z) - c} = \frac{1}{N \pi \rho_{\mathbf{M}}(\mu_i)} \lim_{z \to \mu_i - i0^+} \Im \mathrm{Tr}\left[\mathcal{G}_{\mathbf{M}}(z)\mathbf{C}\right],$$

$$(6.30)$$

where $Z(z)$ is given by Eq. (6.8). From Eq. (6.7) one also has $\mathrm{Tr}(\mathcal{G}_{\mathbf{M}}(z)\mathbf{C}) = N(Z(z)G_{\mathbf{M}}(z) - 1)$, and using Eq. (6.8) and (6.10), we end up with

$$\lim_{z \to \mu - i0^+} \Im \mathrm{Tr}\left[\mathcal{G}_{\mathbf{M}}(z)\mathbf{C}\right] = N \pi \rho_{M}(\mu)\left[\mu - \alpha(\mu) - \beta(\mu) H_{\mathbf{M}}(\mu)\right].$$

We therefore find the following optimal RIE nonlinear 'shrinkage' function F_1:

$$\widehat{\mu}_i = F_1(\mu_i); \qquad F_1(\mu) = \mu - \alpha_1(\mu) - \beta_1(\mu) H_{\mathbf{M}}(\mu), \qquad (6.31)$$

where α_1, β_1 are defined in, Eq. (6.10). This result states that if we consider a model where the signal $\mathbf{C}$ is perturbed with an additive noise (that is, free with respect to $\mathbf{C}$), the optimal way to 'clean' the eigenvalues of $\mathbf{M}$ in order to get $\widehat{\mathbf{C}}(\mathbf{M})$ is to keep the eigenvectors of $\mathbf{M}$ and apply the nonlinear shrinkage formula (6.31).

Let us, for example, consider the case where $OBO^\dagger$ is a GOE matrix. Using the definition of α_1 and β_1 given in Eq. (6.10), the nonlinear shrinkage function is given by

$$F_1(\mu) = \mu - 2\sigma^2 H_{\mathbf{M}}(\mu). \qquad (6.32)$$

Moreover, suppose that $\mathbf{C}$ is also a GOE matrix so that $\mathbf{M}$ is a also a GOE matrix with variance $\sigma_M^2 = \sigma_C^2 + \sigma^2$. As a consequence, the Hilbert transform of $\mathbf{M}$ can be computed straightforwardly from the Wigner semicircle law and we find that

$$H_{\mathbf{M}}(\mu) = \frac{\mu}{2\sigma_M^2}.$$

The optimal cleaning scheme to apply in this case is then given by

$$F_1(\mu) = \mu \left(\frac{\sigma_C^2}{\sigma_C^2 + \sigma^2} \right), \tag{6.33}$$

where one can see that the optimal cleaning is given by rescaling the empirical eigenvalues by the signal-to-noise ratio. This result is expected in the sense that we perturb a Gaussian signal by adding a Gaussian noise. We know in this case that the optimal estimator of the signal is given, element by element, by the Wiener filter (Wiener 1949), and this is exactly the result that we have obtained with (6.33). We can also note that the spectral density of the cleaned matrix is narrower than the true one. Indeed, let us define the signal-to-noise ratio $\mathrm{SNR} = \sigma_C^2/\sigma_M^2 \in [0, 1]$, and it is obvious from (6.33) that $\widehat{\mathbf{C}}(\mathbf{M})$ is a Wigner matrix with variance $\sigma_C^2 \times \mathrm{SNR}$ which leads to

$$\sigma_M^2 \geq \sigma_C^2 \geq \sigma_C^2 \times \mathrm{SNR}, \tag{6.34}$$

as it should be.

Free multiplicative noise

By proceeding in the same way as in the additive case, we can derive formally a nonlinear shrinkage estimator that depends on the observed eigenvalues μ of $\mathbf{M}$ defined by

$$\mathbf{M} = \sqrt{\mathbf{C}} \mathbf{O} \mathbf{B} \mathbf{O}^\dagger \sqrt{\mathbf{C}}.$$

Following the computations done earlier, we can find after some manipulations of the global law estimate (6.14):

$$\mathrm{Tr} \left(\mathcal{G}_{\mathbf{M}}(z)\mathbf{C} \right) = N(z G_{\mathbf{M}}(z) - 1) S_{\mathbf{B}}(z G_{\mathbf{M}}(z) - 1). \tag{6.35}$$

Using the analyticity of the S-transform, we define the function $\gamma_{\mathbf{B}}$ and $\omega_{\mathbf{B}}$ such that

$$\lim_{z \to \mu - i0^+} S_{\mathbf{B}}(z G_{\mathbf{M}}(z) - 1) := \gamma_{\mathbf{B}}(\mu) + i\pi \rho_{\mathbf{M}}(\mu)\omega_{\mathbf{B}}(\mu). \tag{6.36}$$

As a consequence, the optimal RIE (or nonlinear shrinkage formula) for the free multiplicative noise model (6.13) reads

$$\widehat{\mu}_i = F_2(\mu_i); \qquad F_2(\mu) = \mu\gamma_{\mathbf{B}}(\mu) + (\mu H_{\mathbf{M}}(\mu) - 1)\omega_{\mathbf{B}}(\mu), \tag{6.37}$$

and this is the analogue of the estimator (6.31) in the multiplicative case.

As an application of the general result Eq. (6.37), we reconsider the homogeneous Marčenko–Pastur setting where $\mathbf{B} = \frac{1}{T}\mathbf{X}\mathbf{X}^{\dagger}$. We trivially find from the definition of the S-transform that (6.36) yields in this case:

$$\gamma_{\mathbf{B}}(\mu) = \frac{1 - q + q\mu H_{\mathbf{M}}(\mu)}{|1 - q + q\mu \lim\limits_{z \to \mu - i0^+} G_{\mathbf{M}}(z)|^2} \quad \text{and} \quad \omega_{\mathbf{B}}(\mu) = -\frac{q\mu}{|1 - q + q\mu \lim\limits_{z \to \mu - i0^+} G_{\mathbf{M}}(z)|^2}.$$

(6.38)

The nonlinear shrinkage function F_2, thus, becomes

$$F_2(\mu) = \frac{\mu}{(1 - q + q\mu H_{\mathbf{M}}(\mu))^2 + q^2\mu^2\pi^2\rho_{\mathbf{M}}^2(\mu)},$$

(6.39)

which is precisely the Ledoit–Péché estimator derived in Ledoit and Péché (2011). Let us insist once again on the fact that this is the oracle estimator, but it can be computed without the knowledge of $\mathbf{C}$ itself, but only with its noisy version $\mathbf{M}$. This 'miracle' is, of course, only possible thanks to the $N \to \infty$ limit that allows the spectral properties of $\mathbf{M}$ and $\mathbf{C}$ to become deterministically related one to the other.

6.4 Rectangular correlation matrices

It is often interesting to consider nonsymmetrical, or even rectangular correlation matrices that measure the correlation between N 'input' variables X_i, $i = 1, ..., N$ and M 'output' variables Y_a, $a = 1, ..., M$. The X and the Y's may be completely different from one another (for example, X could be production indicators and Y inflation indexes), or, as in the earlier example the same set of observables but observed at different times: $N = M$, $X_i^t = R_i^t$ and $Y_a^t = R_a^{t+\tau}$. The cross-correlations between X's and Y's is characterized by a rectangular $N \times M$ matrix $\mathcal{C}$ defined as $\mathcal{C}_{ia} = \langle X_i Y_a \rangle$ (we assume that both X's and Y's have zero mean and variance unity). If there is a total of T observations, where both X_i^t and Y_a^t, $t = 1, ..., T$ are observed, the empirical estimate of $\mathcal{C}$ is, after standardizing X and Y:

$$\mathcal{M}_{ia} = \frac{1}{T} \sum_{t=1}^{T} X_i^t Y_a^t.$$

(6.40)

What is the generalization of eigenvalues and eigenvectors for rectangular, nonsymmetric correlation matrices? The singular value decomposition (SVD) of $\mathbf{C}$ answers the question in the following sense: what is the (normalized) linear combination of X's on the one hand, and of Y's on the other hand, that have the strongest mutual correlation? In other words, what is the best pair of predictor and predicted variables, given the data? The largest singular value $c_{\max}$ and its corresponding left and right eigenvectors answer precisely this question: the eigenvectors tell us how to construct these optimal linear combinations, and the associated singular value gives us the strength of the cross-correlation: $0 \leq c_{\max} \leq 1$. One can now restrict both the input and output spaces to the $N-1$ and $M-1$ dimensional subspaces orthogonal to the two eigenvectors, and repeat the operation. The list of singular values c_a gives the

prediction power, in decreasing order, of the corresponding linear combinations. This is called 'canonical component analysis' (CCA) in the literature (Johnstone 2008); surprisingly in view of its wide range of applications, this method of investigation has been somewhat neglected since it was first introduced in 1936 (Hotelling 1936).

What can be said about the singular value spectrum of $\mathbf{M}$ in the special limit $N, M, T \to \infty$, with $n = N/T$ and $m = M/T$ fixed? Whereas the natural null hypothesis for correlation matrices is $\mathbf{C} = 1$, that leads to the Marčenko–Pastur density, the null hypothesis for cross-correlations between *a priori* unrelated sets of input and output variables is $\mathcal{C} = \mathbf{0}$. However, in the general case, input and output variables can very well be correlated between themselves, for example if one chooses redundant input variables. In order to establish a universal result, one should therefore consider the *exact* normalized principal components for the sample variables X's and Y's,

$$\hat{X}_\alpha^t = \frac{1}{\sqrt{\lambda_\alpha}} \sum_i V_{\alpha,i} X_i^t; , \tag{6.41}$$

and similarly for the $\hat{Y}_a^t$. The λ_α and the $V_{\alpha,i}$ are the eigenvalues and eigenvectors of the sample correlation matrix $\mathbf{M}_X$ (or, respectively, $\mathbf{M}_Y$). We now define the normalized $M \times N$ cross-correlation matrix as $\hat{\mathbf{M}} = \hat{Y}\hat{X}^T$. One can then use the following tricks (Bouchaud et al. 2007):

- The nonzero eigenvalues of $\hat{\mathcal{M}}^T \hat{\mathcal{M}}$ are the same as those of $\hat{X}^T \hat{X} \hat{Y}^T \hat{Y}$.
- $\mathbf{A} = \hat{X}^T \hat{X}$ and $\mathbf{B} = \hat{Y}^T \hat{Y}$ are two mutually free $T \times T$ matrices, with N (M) eigenvalues exactly equal to 1 (due to the very construction of $\hat{X}$ and $\hat{Y}$), and $(T-N)^+$ $((T-M)^+)$ equal to 0.
- The S-transforms are multiplicative, allowing one to obtain the spectrum of $\mathbf{AB}$.

Due to the simplicity of the spectra of $\mathbf{A}$ and $\mathbf{B}$, the calculation of S-transforms is particularly easy (Bouchaud et al. 2007). The final result for the density of nonzero singular values (i.e., the square root of the eigenvalues of $\mathbf{AB}$) reads (see Wachter (1980) for an early derivation of this result; see also Johnstone (2008)):

$$\rho(c) = \max(m+n-1,0)\delta(c-1) + \Re\frac{\sqrt{(c^2 - \gamma_-)(\gamma_+ - c^2)}}{\pi c(1-c^2)}, \tag{6.42}$$

where $n = N/T$, $m = M/T$ and $\gamma_\pm$ are given by

$$\gamma_\pm = n + m - 2mn \pm 2\sqrt{mn(1-n)(1-m)}, \quad 0 \le \gamma_\pm \le 1. \tag{6.43}$$

The allowed c's are all between 0 and 1, as they should be since these singular values can be interpreted as correlation coefficients. In the limit $T \to \infty$ at fixed N, M, all singular values collapse to 0, as they should since there is no true correlations between X and Y; the allowed band in the limit $n, m \to 0$ becomes

$$c \in \left[\frac{|m-n|}{\sqrt{m} + \sqrt{n}}, \sqrt{m} + \sqrt{n} \right], \tag{6.44}$$

showing that for fixed N, M, the order of magnitude of allowed singular values decays as $T^{-1/2}$. As a simple case, imagine one has 20 input and 20 output economic variables, with a total time series of length $T = 400$ (this would correspond to a century of quarterly data). With $n = m = 0.05$, the upper edge of the spectrum is $c_{\max} \approx 2\sqrt{n} \approx 0.45$: there is a linear combination of input variables that is 45% correlated (empirically) to a linear combination of output variables, even if the two sets of variables are in fact completely uncorrelated! This is what is called in the literature the 'sunspot' effect, because the solar activity appears to be correlated to the stock market—which, of course, must be a statistical artefact.

Note that one could have considered a different benchmark ensemble, where one considers two independent vector time series X and Y with true correlation matrices C_X and C_Y equal to 1. The direct SVD spectrum in that case can also be computed as the S-convolution of two Marčenko–Pastur distributions with parameters m and n Bouchaud et al. (2007). This alternative benchmark is, however, not well suited in practice, since it mixes up the possibly nontrivial correlation structure of the input variables and of the output variables themselves with the *cross*-correlations between these variables.

6.5 Conclusion and (some) open problems

RMT is at the heart of many significant contributions when it comes to reconstructing a true signal matrix $\mathbf{C}$ of large dimension from a noisy measurement. In this chapter, we have focused on the overlap between the eigenvectors of the signal matrix with the corrupted ones, for the case of additive and multiplicative noise. This allows one to construct the so-called rotationally invariant oracle estimator which, miraculously, can be expressed *without any knowledge* of the signal $\mathbf{C}$ in the large N limit. This last observation, which generalizes the work of Ledoit and Péché (2011), should be of particular interest in practical cases.

All this analysis is well established for eigenvalues in the 'bulk' of the spectral density, but is a priori inadequate to treat isolated 'spikes'—which are nevertheless of great importance in many practical applications. In particular, the largest eigenvalue of financial correlation matrices is associated with the so-called market mode; i.e., the fact that all stocks tend to evolve in sync. Recent progress has been made in this direction, with the surprising result that the bulk result extends to spikes as well—see Bun and Knowles (2016) for details.

Note also that a RIE assumes no particular prior knowledge on the eigenvector structure of the true matrix $\mathbf{C}$, so that the only possible choice is those of the empirical matrix $\mathbf{M}$. However, it may happen in practice that one has a prior structure on the eigenvectors of $\mathbf{C}$ (for example, given by a factor model), and it would be interesting to see how one can extend the above analysis to a non-RI framework (for some recent work in this direction; see, e.g, Monasson and Villamaina (2015)).

Finally, the random SVD analysis of the previous section suggests many interesting, unexplored extensions. Can one construct RIEs in this case as well? Can one extend the analysis to the case of a nonnull, true underlying correlation structure $\mathcal{C} \neq \mathbf{0}$—much as

the Marčenko–Pastur spectral density Eq. (6.24) can be extended to an arbitrary 'true' correlation matrix $\mathbf{C}$? We leave all these problems to future investigations, maybe by some of the students of this Les Houches session.

6.6 Appendix A: Reminder on transforms in RMT

We give in this first appendix a short reminder on the different transforms that are useful in the study of the statistics of eigenvalues in RMT due to their link with free probability theory (see, e.g., Speicher (2011) or Burda (2013) for a review). We recall that the resolvent of $\mathbf{M}$ is defined by

$$\mathcal{G}_{\mathbf{M}}(z) := (z\mathbb{I}_N - \mathbf{M})^{-1}, \tag{6.45}$$

and the *Stieltjes* (or sometimes *Cauchy*) transform is the normalized trace of the resolvent:

$$G_{\mathbf{M}}(z) := \frac{1}{N}\mathrm{Tr}\mathcal{G}_{\mathbf{M}}(z) \quad = \quad \frac{1}{N}\sum_{k=1}^{N}\frac{1}{z - \mu_k},$$

$$\underset{N\to\infty}{\sim} \int \frac{\mathrm{d}\mu\rho_{\mathbf{M}}(\mu)}{z - \mu}. \tag{6.46}$$

The Stieltjes transform can be interpreted as the average law and is very convenient in order to describe the convergence of the eigenvalues density $\rho_{\mathbf{M}}$. If we set $z = \mu - i\eta$ and take the limit $\eta \to 0$, we have in the large N limit

$$G_{\mathbf{M}}(\mu_i - i\eta) = \mathrm{P.V.} \int \frac{\mathrm{d}\mu'\rho_{\mathbf{M}}(\mu')}{\mu - \mu'} + i\pi\rho_{\mathbf{M}}(\mu),$$

where the real part is often called the *Hilbert* transform $H_{\mathbf{M}}(\mu)$ and the imaginary part leads to the eigenvalues density.

When we consider the case of adding two random matrices that are (asymptotically) free with each other, it is suitable to introduce the functional inverse of the Stieltjes transform known as the *Blue* transform

$$B_{\mathbf{M}}(G_{\mathbf{M}}(z)) = z. \tag{6.47}$$

This allows us to define the so-called R-transform

$$R_{\mathbf{M}}(z) := B_{\mathbf{M}}(z) - \frac{1}{z}, \tag{6.48}$$

which can be seen as the analogue in RMT of the logarithm of the Fourier transform for free additive convolution. More precisely, if $\mathbf{A}$ and $\mathbf{B}$ are two $N \times N$ independent invariant symmetric random matrices, then in the large N limit, the spectral measure of $\mathbf{M} = \mathbf{A} + \mathbf{B}$ is given by

$$R_{\mathbf{M}}(z) = R_{\mathbf{A}}(z) + R_{\mathbf{B}}(z), \tag{6.49}$$

known as the free addition formula (Voiculescu et al. 1992). In this case, we note by $\rho_{A\boxplus B}$ the eigenvalues density of $\mathbf{M}$.

We can do the same for the free multiplicative convolution. In this case, we rather have to define the so-called T (or sometimes η (Tulino and Verdú 2004) transform given by

$$T_{\mathbf{M}}(z) = \int \frac{d\mu\rho_{\mathbf{M}}(\mu)\mu}{z - \mu} \equiv zG_{\mathbf{M}}(z) - 1, \tag{6.50}$$

which can be seen as the moment-generating function of $\mathbf{M}$. The S-transform of $\mathbf{M}$ is then defined as

$$S_{\mathbf{M}}(z) := \frac{z + 1}{zT_{\mathbf{M}}^{-1}(z)}, \tag{6.51}$$

where $T_{\mathbf{M}}^{-1}(z)$ is the functional inverse of the T-transform. Before showing why the S-transform is important in RMT, one must be careful about the notion of product of free matrices. Indeed, if we reconsider the two $N \times N$ independent symmetric random matrices $\mathbf{A}$ and $\mathbf{B}$, the product $\mathbf{AB}$ is in general not self-adjoint even if $\mathbf{A}$ and $\mathbf{B}$ are self-adjoint. However, if $\mathbf{A}$ is positive definite, then the product $\sqrt{\mathbf{A}}\mathbf{B}\sqrt{\mathbf{A}}$ makes sense and shares the same moments as the product $\mathbf{AB}$. We can thus study the spectral measure of $\mathbf{M} = \sqrt{\mathbf{A}}\mathbf{B}\sqrt{\mathbf{A}}$ in order to get the distribution of the free multiplicative convolution $\rho_{A\boxtimes B}$. The result, first obtained in Voiculescu et al. (1992), reads

$$S_{A\boxtimes B}(z) := S_{\mathbf{M}}(z) = S_{\mathbf{A}}(z)S_{\mathbf{B}}(z). \tag{6.52}$$

The S-transform is therefore the analogue of the Fourier transform for free multiplicative convolution.

6.7 Appendix B: Resolvents and replicas

6.7.1 The replica method

The starting point of our approach is to rewrite the entries of the resolvent $\mathcal{G}_{\mathbf{M}}(z)$ by the Gaussian integral representation of an inverse matrix

$$\mathcal{G}_{\mathbf{M}}(z)_{i,j} = \frac{\int \left(\prod_{k=1}^{N} d\eta_k\right) \eta_i\eta_j \exp\left\{-\frac{1}{2}\sum_{k,l=1}^{N} \eta_k(z\delta_{k,l} - \mathbf{M}_{k,l})\eta_l\right\}}{\int \left(\prod_{k=1}^{N} d\eta_k\right) \exp\left\{-\frac{1}{2}\sum_{k,l=1}^{N} \eta_k(z\delta_{k,l} - \mathbf{M}_{k,l})\eta_l\right\}}. \tag{6.53}$$

We recall that the claim is that for a complex z not too close to the real axis; we expect the resolvent to be self-averaging in the large N limit, that is to say independent of the specific realization of the matrix itself. Therefore, we can study the resolvent $\mathcal{G}_{\mathbf{M}}(z)$ through its ensemble average (denoted by $\langle \cdot \rangle$ in the following) given by

$$\langle \mathcal{G}_{\mathbf{M}}(z)_{i,j}\rangle = \left\langle \frac{1}{\mathcal{Z}} \int \left(\prod_{k=1}^{N} d\eta_k\right) \eta_i\eta_j \exp\left\{-\frac{1}{2}\sum_{k,l=1}^{N} \eta_k(z\delta_{k,l} - \mathbf{M}_{k,l})\eta_l\right\}\right\rangle, \tag{6.54}$$

where $\mathcal{Z}$ is the partition function, i.e., the denominator in Eq. (6.53). The computation of the average value is highly nontrivial in the general case. The replica method tells us that the expectation value can be handled thanks to the identity

$$
\langle \mathcal{G}_{\mathbf{M}}(z)_{i,j} \rangle = \lim_{n \to 0} \left\langle \mathcal{Z}^{n-1} \int \left(\prod_{k=1}^{N} d\eta_k \right) \eta_i \eta_j \exp \left\{ -\frac{1}{2} \sum_{k,l=1}^{N} \eta_k (z\delta_{k,l} - \mathbf{M}_{k,l}) \eta_l \right\} \right\rangle
$$

$$
= \lim_{n \to 0} \int \left(\prod_{k=1}^{N} \prod_{\alpha=1}^{n} d\eta_k^\alpha \right) \eta_i^1 \eta_j^1 \left\langle \exp \left\{ -\frac{1}{2} \sum_{\alpha=1}^{n} \sum_{k,l=1}^{N} \eta_k^\alpha (z\delta_{k,l} - \mathbf{M}_{k,l}) \eta_l^\alpha \right\} \right\rangle.
$$

$$(6.55)$$

We have thus transformed our problem to the computation of n replicas of the initial system (6.53). So when we have computed the average value in (6.54), it suffices to perform an analytical continuation of the result to real values of n and finally takes the limit $n \to 0$. The main concern of this nonrigorous approach is that we assume that the analytical continuation can be done with only n different set of points which could lead to uncontrolled approximation in some cases (Parisi 1980).

6.7.2   Free additive noise

We consider a model of the form

$$
\mathbf{M} = \mathbf{C} + OBO^\dagger,
$$

where $\mathbf{B}$ is a fixed matrix with eigenvalues $b_1 > b_2 > \cdots > b_N$ with spectral $\rho_{\mathbf{B}}$ and O is a random matrix chosen in the orthogonal group $O(N)$ according to the Haar measure. Clearly, the noise term is invariant under rotation so that we expect the resolvent of $\mathbf{M}$ to be on the same basis as $\mathbf{C}$. We therefore set without loss of generality that $\mathbf{C}$ is diagonal. In order to derive the global law estimate for the resolvent of the matrix $\mathbf{M}$, we must consider the ensemble average value of the resolvent over the Haar measure for the $O(N)$ group, which can be written as

$$
\langle \mathcal{G}_{\mathbf{M}}(z)_{i,j} \rangle = \int \left(\prod_{\alpha=1}^{n} \prod_{k=1}^{N} d\eta_k^\alpha \right) \eta_i^1 \eta_j^1 \prod_{\alpha=1}^{n} e^{-\frac{1}{2} \sum_{k=1}^{N} (\eta_k^\alpha)^2 (z - c_k)} \left\langle e^{-\frac{1}{2} \sum_{k,l=1}^{N} \eta_k^\alpha (OBO^\dagger)_{k,l} \eta_l^\alpha} \right\rangle_O.
$$

$$(6.56)$$

The evaluation of the later equation can be done straightforwardly if we set the measure dO to be a flat measure constrained to the fact that $OO^\dagger = \mathbb{I}_N$, or equivalently said,

$$
\mathcal{D}O \propto \prod_{i,j=1}^{N} dO_{i,j} \prod_{i,j=1}^{N} \delta \left(\sum_k O_{i,k} O_{j,k} - \delta_{i,j} \right)
$$

where $\delta(\cdot)$ is the Dirac delta function and $\delta_{i,j}$ is Kronecker delta. In the case where n is finite (and independent of N), one can note that Eq. (6.56) is the orthogonal low-rank version of the Harish–Chandra–Itzykson–Zuber integrals (Harish–Chandra 1957;

Itzykson and Zuber 1980). The result is known for all symmetry groups (Marinari et al. 1994 or Guionnet and Maida 2004 for a more rigorous derivation), and this reads for the rank-n case

$$\int \mathcal{D}O \exp\left[\mathrm{Tr}\left(\frac{1}{2}\sum_{\alpha=1}^{n}\eta^{\alpha}(\eta^{\alpha})^{\dagger}OBO^{\dagger}\right)\right] = \exp\left[\frac{N}{2}\sum_{\alpha=1}^{n}W_{\mathbf{B}}\left(\frac{1}{N}(\eta^{\alpha})^{\dagger}\eta^{\alpha}\right)\right], \quad (6.57)$$

with W_B being the primitive of the R-transform of B. The computation of the resolvent (6.56) becomes

$$\langle \mathcal{G}_{\mathbf{M}}(z)_{i,j}\rangle = \int\left(\prod_{k=1}^{N}\mathrm{d}\eta_k\right)\eta_i^1\eta_j^1\exp\left\{\frac{N}{2}\sum_{\alpha=1}^{n}\left[W_{\mathbf{B}}\left(\frac{1}{N}(\eta^{\alpha})^{\dagger}\eta^{\alpha}\right)-\frac{1}{2}\sum_{k=1}^{N}(\eta_k^{\alpha})^2(z-c_k)\right]\right\},$$

where we have introduced a Lagrange multiplier $p^{\alpha} = \frac{1}{N}(\eta^{\alpha})^{\dagger}\eta^{\alpha}$ which gives using Fourier transform (renaming $\zeta^{\alpha} = 2i\zeta^{\alpha}/N$)

$$\langle \mathcal{G}_{\mathbf{M}}(z)_{i,j}\rangle \propto \int\int\left(\prod_{\alpha=1}^{n}\mathrm{d}p^{\alpha}\mathrm{d}\zeta^{\alpha}\right)\exp\left\{\frac{N}{2}\sum_{\alpha=1}^{n}\left[W_{\mathbf{B}}(p^{\alpha})+p^{\alpha}\zeta^{\alpha}\right]\right\}$$

$$\times\int\left(\prod_{\alpha=1}^{n}\prod_{k=1}^{N}\mathrm{d}\eta_k^{\alpha}\right)\eta_i^1\eta_j^1\exp\left\{-\frac{1}{2}\sum_{\alpha=1}^{n}\sum_{k=1}^{N}(\eta_k^{\alpha})^2(z+\zeta^{\alpha}-c_k)\right\}.$$

This additional constraint allows one to retrieve a Gaussian integral over the $\{\eta_j\}$ which can be computed exactly. Ignoring normalization terms, we obtain

$$\langle \mathcal{G}_{\mathbf{M}}(z)_{i,j}\rangle \propto \int\int\left(\prod_{\alpha=1}^{n}\mathrm{d}p^{\alpha}\mathrm{d}\zeta^{\alpha}\right)\frac{\delta_{i,j}}{z+\zeta^1-c_i}\exp\left\{-\frac{Nn}{2}F_0(p^{\alpha},\zeta^{\alpha})\right\}, \quad (6.58)$$

where the 'free energy' F_0 is given by

$$F_0(p,\zeta) = \frac{1}{Nn}\sum_{\alpha=1}^{n}\left[\sum_{k=1}^{N}\log(z+\zeta^{\alpha}-c_k)-W_{\mathbf{B}}(p^{\alpha})-p^{\alpha}\zeta^{\alpha}\right]. \quad (6.59)$$

In the large N limit, the integral can be evaluated by considering the saddle-point of the free energy F_0 as the other term is obviously subleading. We now use the *replica symmetric* ansatz that tells us if the free energy is invariant under the action of the symmetry group $O(N)$, then we expect a saddle-point which is also invariant. This implies that we have at the saddle-point

$$p^{\alpha} = p \quad \text{and} \quad \zeta^{\alpha} = \zeta, \ \forall\alpha \in \{1,\dots,n\}, \quad (6.60)$$

and hence, we must solve the set of equations

$$\begin{cases} \zeta^* = -R_{\mathbf{B}}(p^*) \\ p^* = G_{\mathbf{C}}(z+\zeta^*). \end{cases}$$

The trick is to see that we can get rid of one variable by taking the normalized trace of the (average) resolvent which gives the following relation for the Stieltjes transform: $G_\mathbf{M}(z) = G_\mathbf{C}(z - R_\mathbf{B}(p^*)) = p^*$; that is to say,

$$p^* = G_\mathbf{M}(z) = G_\mathbf{C}\left(z - R_B\left(G_\mathbf{M}(z)\right)\right), \tag{6.61}$$

and therefore

$$\zeta^* = -R_\mathbf{B}(G_\mathbf{M}(z)). \tag{6.62}$$

In conclusion, by plugging (6.61) and (6.62) into (6.58) and then taking the limit $n \to 0$, we obtain the global law estimate

$$\langle \mathcal{G}_\mathbf{M}(z)_{i,j} \rangle = (Z(z)\mathbb{I}_N - \mathbf{C})_{i,i}^{-1} \delta_{i,j} \tag{6.63}$$

with

$$Z(z) = z - R_\mathbf{B}\left(G_\mathbf{M}(z)\right), \tag{6.64}$$

which are exactly the results stated in Eqs (6.7) and (6.8).

The case of free multiplicative noise when $\mathbf{M}$ reads

$$\mathbf{M} = \sqrt{\mathbf{C}} O B O^\dagger \sqrt{\mathbf{C}}, \tag{6.65}$$

where O is still a rotation matrix over the orthogonal group, $\mathbf{C}$ is a positive definite matrix, and $\mathbf{B}$ is such that $\mathrm{Tr}\mathbf{B} \neq 0$, can be treated very similarly. We refer to the appendix of Allez et al. (2016) for more details on this case.

References

Allez, R., and Bouchaud, J.-P. (2014). *Random Matrices: Theory and Applications* **03**, 1450010.

Allez, R.,, Bun, J., and Bouchaud, J.-P. (2013) arXiv preprint arXiv:1412.7108.

Allez, R., Bun, J., Bouchaud, J.-P. (2016). *IEEE Trans. Inform. Theory* **62**(12), 7475–90. and M. Potters, arXiv preprint arXiv:1502.06736.

Bouchaud, J.-P. Laloux, L. Miceli, M. A. and Potters, M. (2007). *Eur. Phys. J. B* **55**, 201.

Brézin, E., Hikami, S., and Zee, A. (1995). *Phys. Rev. E* **51**, 5442.

Brézin E., and Zee, A. (1995). *Nucl. Phys. B* **453**, 531.

Bun, J., and Knowles, A. (2016). In preparation.

Burda, Z. (2013). *J. Phys. Conf. Ser.* **473**(1). preprint arXiv:1309.2568.

Burda, Z., Görlich, A., Jarosz, A., and Jurkiewicz, J. (2004). *Phys. A: Stat. Mech. Appl.* **343**, 295.

Guionnet, A. and Maida, M. (2004). arXiv preprint math/0406121.

Harish-Chandra. (1957). *Ameri. J. Mathe.* 87–120.

Hotelling, H. (1936). *Biometrika* pp. 321–77.

Itzykson, C. and Zuber, J.-B. (1980). *J. Math. Phys.* **21**, 411.

Johnstone, I. M. (2008) *Ann. Stat.* **36**, 2638.

Knowles, A., and Yin, J. (2014). *Probab. Theory Related Fields* 1–96. arXiv 1502.06736.

Ledoit, O. and Péché, S. (2011). *Prob. Theory Rel. Fields* **151**, 233.

Mandt, S. and Zirnbauer, M. R., (2010). *J Phys A* **43**, 025201.

Marchenko V. A., and Pastur, L. A. (1967). *Matematicheskii Sbornik* **114**, 507.

Marinari, E., Parisi, G., and Ritort, F., (1994). *J. Phys. A: Math. Gen.* **27**, 7647.

Monasson, R., and Villamaina, D. (2015). *Europhys. Letter.* **i12**(5), 50001. arXiv:1503.00287.

Parisi, G. (1980). *J. Phys. A: Math. Gen.* **13**, L115.

Speicher, R., (2011). In *The Oxford Handbook of Random Matrix Theory*. Oxford University Press, Oxford.

Tanaka, T. (2008) *J. Phys. Conf. Ser.* **95**, 012002.

Tulino, A. M. and Verdú, S. (2004). *Communi. Inform. Theory* **1**, 1.

Voiculescu D.,, Dykema K.,, and Nica, A. 1992. *Free Random Variables*, 1. American Mathematical Soc., Washington, DC.

Wachter, K. W. (1980). *The Annals of Statistics*, pp. 937–57.

Wiener, N. (1949). *Extrapolation, Interpolation, and Smoothing of Stationary Time Series*, vol. 2. (MIT press, Cambridge, MA.

Zee, A. (1996). *Nucl. Phys. B* **474**, 726.

7

Random matrices and loop equations

Bertrand EYNARD

IPHT/CEA/Saclay, France, and
CRM, Montréal, Canada

Eynard, B., 'Random Matrices and Loop Equations' in *Stochastic Processes and Random Matrices*.
Edited by: Grégory Schehr et al, Oxford University Press (2017). © Oxford University Press 2017.
DOI 10.1093/oso/9780198797319.003.0007

Chapter Contents

7.1 Introduction

This chapter follows a series of three lectures given in the LesHouches school of 2015. Some good references for an introduction to matrix models can be found Mehta (2004) and the review Di Francesco et al. (1995). Here, our goal is to present the basics of algebraic methods in random matrix theory. In the first section, we will introduce random matrix ensembles, and see that going beyond the usual Wigner ensembles can be very useful, in particular by allowing eigenvalues to lie on some paths in the complex plane rather than on the real axis. Most often the model depends on a matrix size N, and on a scaling parameter $\hbar$. The interesting limit to look at is most often the 'semi-classical' limit $\hbar \to 0$. Most often we will be in a regime where the 't Hooft parameter $t = N\hbar$ remains finite, which means that the limit $\hbar \to 0$ is then the same as the large N limit.

As a detailed example, we will retrieve the Plancherel model (a discrete sum over partitions with the Plancherel measure) as a random matrix model.

In the second section, we will study the saddle-point approximation, also called the Coulomb gas method. Indeed, the eigenvalue integral resembles the quantum mechanics of particles with a Coulomb repulsive force. The classical limit, i.e., the equilibrium position of particles, leads to a system of algebraic equations that can be recognized as being the Bethe ansatz of a Gaudin model. We will study the space of solutions, and we will encounter an algebraic curve called the 'spectral curve' that will play an immense role, in the sense that the spectral curve determines all the large N expansion, of all observables, and in a geometric way.

In the third section, we will introduce the notion of 'loop equations'. These are, in fact, Schwinger-Dyson equations (invariance of an integral under change of variable) or also integration by parts. They are, in fact, very powerful. First, we will see that loop equations are linear, so their space of solutions is a vector space, and it coincides with the space of possible normal matrix models (i.e., any solution of loop equations corresponds to a matrix integral with eigenvalues on a certain complex domain and vice versa). Then, we shall see that it is possible to solve loop equations order by order in the semiclassical expansion $\hbar \to 0$. To each order, expectation values of the density and correlation functions are algebraic functions, and are determined in terms of the spectral curve. We will show that they can be computed recursively by a universal recursion: the 'topological recursion'. We will mention that the topological recursion goes well beyond matrix models, and plays an important role in algebraic geometry. We will give the famous example of the Mirzakhani' recursion.

7.1.1 Random matrices

Consider a random Hermitian matrix $M \in H_N$ of size N, with a probability measure

$$d\mu(M) = e^{-\frac{1}{\hbar} \operatorname{Tr} V(M)} \, dM \tag{7.1}$$

with dM the conjugation invariant Lebesgue measure

$$dM = \prod_i dM_{i,i} \prod_{i<j} d\Re M_{i,j} \wedge d\Im M_{i,j} = \frac{1}{(-2i)^{\frac{N(N-1)}{2}}} \prod_{i,j} dM_{i,j}, \tag{7.2}$$

and where $V(M)$ is a Boltzmann weight potential, for example $V(M) = \frac{M^2}{2}$ is the Gaussian law.

The last equality in (7.2), holds because for Hermitian matrices $M_{j,i} = \overline{M_{i,j}} = \Re M_{i,j} - i \Im M_{i,j}$, and if $z = x + iy$ we have

$$dz \wedge d\bar{z} = -2i \, dx \wedge dy. \tag{7.3}$$

Eigenvalues

Every Hermitian matrix can be diagonalized by a unitary transformation

$$M = U\Lambda U^\dagger \,, \qquad UU^\dagger = \mathrm{Id}\,, \quad \Lambda = \mathrm{diag}(\lambda_1, \dots, \lambda_N)\,, \lambda_i \in \mathbb{R}. \tag{7.4}$$

The decomposition is not unique; we may right multiply U by an arbitrary unitary diagonal matrix (thus in $U(1)^N$), and permute the eigenvalues. Therefore,

$$H_N = \frac{U(N)/U(1)^N \times \mathbb{R}^N}{\mathfrak{S}_N}. \tag{7.5}$$

The Jacobian in the change of variable is

$$dM = \frac{1}{(2\pi)^N N!} \Delta(\Lambda)^2 d\Lambda dU \,, \quad d\Lambda = \prod_{i=1}^{n} d\lambda_i \,, \quad dU = \text{Haar measure on } U(N), \tag{7.6}$$

where

$$\Delta(\Lambda) = \prod_{i<j} (\lambda_i - \lambda_j) \tag{7.7}$$

is called the Vandermonde determinant of Λ. It has the property that it is the smallest degree totally antisymmteric polynomial of N variables; it is also equal to

$$\Delta(\Lambda) = \det \lambda_i^{j-1}. \tag{7.8}$$

Proof The measure dM is invariant under $U(N)$ conjugations; therefore, it is enough to prove the equality at $U = \mathrm{Id}$. From $M = U\Lambda U^{-1}$, write the differential

$$dM = U d\Lambda U^{-1} + dU \Lambda U^{-1} - U\Lambda U^{-1} dU U^{-1}, \tag{7.9}$$

which at $U = \mathrm{Id}$ is

$$dM = d\Lambda + [dU, \Lambda] \tag{7.10}$$

and thus

$$dM_{i,j} = \delta_{i,j} d\lambda_i + (\lambda_j - \lambda_i) dU_{i,j}. \tag{7.11}$$

The $N^2 \times N^2$ Jacobian matrix $dM/(d\Lambda, dU)$ is diagonal, with N eigenvalues equal to 1, and $N(N-1)/2$ equal to $(\lambda_j - \lambda_i)$, appearing twice, for $M_{i,j}$ and $M_{j,i}$. Therefore the Jacobian is proportional to the $\Delta(\Lambda)^2$. The normalization constant $1/(2\pi)^N N!$ is the ratio between the Haar measure on $U(N)$ and on $U(N)/U(1)^N/\mathfrak{S}_N$.

Generalization 1: Normal random matrix with eigenvalues on a contour γ

A normal matrix is diagonalizable by a unitary transformation; we define the set of normal matrices with eigenvalues in a given domain $\gamma \subset \mathbb{C}$:

$$H_N(\gamma) = \{M = U\Lambda U^\dagger \,|\, U \in U(N),$$

$$\Lambda = \mathrm{diag}(\lambda_1, \ldots, \lambda_N), \lambda_i \in \gamma\} = \frac{U(N)/U(1)^N \times \gamma^N}{\mathfrak{S}_N}. \tag{7.12}$$

If γ is a Jordan arc, we put a measure (not necessarily real or normalized) on $H_N(\gamma)$:

$$\mathrm{d}M = \frac{1}{(2\pi)^N N!}\Delta(\Lambda)^2 \mathrm{d}\Lambda \mathrm{d}U, \quad \mathrm{d}\Lambda = \prod_{i=1}^{n} \mathrm{d}\lambda_i, \quad \mathrm{d}U = \text{Haar measure on } U(N),$$

$$\tag{7.13}$$

where now $\mathrm{d}\lambda_i$ means the curvilinear measure along γ: if $\gamma = \{f(s)\,|\,s \in [0,1]\}$, if $\lambda_i = f(s_i) \in \gamma$, we define $\mathrm{d}\lambda_i = f'(s_i)\mathrm{d}s_i$.

We may then also chose a Boltzmann weight measure on $H_N(\gamma)$:

$$\mathrm{d}\mu(M) = e^{-\frac{1}{\hbar} \operatorname{Tr} V(M)} \mathrm{d}M. \tag{7.14}$$

Let us give 2 examples:

- with $\gamma = \mathbb{R}$ we obviously have

$$H_N(\mathbb{R}) = H_N, \tag{7.15}$$

  and $\mathrm{d}M$ is indeed the standard Lebesgue measure on H_N.
- with $\gamma = S^1 =$ unit circle we have

$$H_N(S^1) = U(N), \tag{7.16}$$

  with our measure $\mathrm{d}M$ related to the usual Haar measure on $U(N)$ by

$$\mathrm{d}M \propto i^{N^2} \det M^N \mathrm{d}M_{\text{Haar}}. \tag{7.17}$$

Indeed, diagonalizing $M = U\mathrm{diag}(\lambda_1, \ldots, \lambda_N)U^{-1}$ with eigenvalues $\lambda_i \in S^1$, we have

$$\mathrm{d}M_{\text{Haar}} \propto \mathrm{d}U \prod_{i<j} |\lambda_i - \lambda_j|^2 \prod_i \frac{\mathrm{d}\lambda_i}{i\lambda_i}$$

$$\propto \mathrm{d}U \prod_{i<j}(\lambda_i - \lambda_j)(\lambda_i^{-1} - \lambda_j^{-1}) \prod_i \frac{\mathrm{d}\lambda_i}{i\lambda_i}$$

$$\propto i^{-N}\mathrm{d}U \prod_{i<j}(\lambda_i - \lambda_j)(\lambda_j - \lambda_i)(\lambda_i^{-1}\lambda_j^{-1}) \prod_i \frac{\mathrm{d}\lambda_i}{\lambda_i}$$

$$\propto i^{-N}(-1)^{N(N-1)/2}\mathrm{d}U \prod_{i<j}(\lambda_i - \lambda_j)^2 \prod_i \frac{\mathrm{d}\lambda_i}{\lambda_i^N}$$

$$\propto i^{-N^2} \frac{\mathrm{d}M}{\det M^N}. \tag{7.18}$$

This makes it possible to view both the Hermitian and circular ensemble as just a special case of $H_N(\gamma)$, and the methods we are going to describe work for any γ.

Remark 7.1.1. **saddle points are not always real** Let us mention that even if one is interested only in Hermitian ensembles, or circular, studying $H_N(\gamma)$ with any $\gamma \subset \mathbb{C}$ is not just a generalization, it is in fact necessary. Indeed, the saddle-point approximation consists in finding an extremum of the integrand, and deforming the integration contour to a steepest descent path going through the saddle-point. Very often, the saddle-point is in the complex plane and the steepest descent path is not real.

Remark 7.1.2. **dimension of the space of matrix integrals** Assume now that V is a polynomial (resp. V' is a rational fraction). Let γ be any contour going from ∞ to ∞ (resp. from poles of V' to poles of V') in sectors where $\Re V \to +\infty$. The integral

$$\int_{\gamma^N} \Delta(\lambda_1, \ldots, \lambda_N)^2 \prod_i e^{\frac{-1}{\hbar} V(\lambda_i)} d\lambda_i \tag{7.19}$$

is then absolutely convergent, and the integrand vanishes at the boundary of γ. Moreover, since the integrand is analytic, the integral is unchanged under homotopic deformations of γ. An homology class is a complex linear combination of homotopy classes.

We shall say that a homology class of manifold $\Gamma \subset \mathbb{C}^N \setminus \{\text{poles}\}$ is *admissible* for the measure $\Delta(\lambda_1, \ldots, \lambda_N)^2 \prod_i e^{\frac{-1}{\hbar} V(\lambda_i)} d\lambda_i$, if the integral on Γ is absolutely convergent, and if the integrand vanishes at the boundaries of Γ.

For a polynomial potential (resp. rational V') of degree $d + 1$ (resp. V' of total degree d), there are $d + 1$ asymptotic sectors in which $\Re V \to +\infty$, and that makes d homologically linearly independent choices for γ. We say that the dimension of the space of admissible homology classes γ for the 1-dimensional measure $e^{\frac{-1}{\hbar} V(\lambda)} d\lambda$ is

$$\dim H_1(e^{\frac{-1}{\hbar} V(\lambda)} d\lambda, \mathbb{C}) = d = \deg V'. \tag{7.20}$$

Let us chose a basis $\gamma_1, \ldots, \gamma_d$. Any $\Gamma = \gamma_1^{n_1} \times \cdots \times \gamma_d^{n_d}$ is then admissible, meaning that

$$\int_{\gamma_1^{n_1} \times \cdots \times \gamma_d^{n_d}} \Delta(\lambda_1, \ldots, \lambda_N)^2 \prod_i e^{\frac{-1}{\hbar} V(\lambda_i)} d\lambda_i, \quad \sum_{i=1}^{d} n_i = N, \tag{7.21}$$

is absolutely convergent, with the integrand vanishing at the boundary. We call this choice of Γ an eigenvalue model with fixed filling fractions n_i/N of eigenvalues on contour γ_i. We often used to write

$$\epsilon_i = \frac{\hbar n_i}{\alpha}, \quad \sum_{i=1}^{d} \epsilon_i = \hbar N. \tag{7.22}$$

Such fixed filling fraction domains $\Gamma = \gamma_1^{n_1} \times \cdots \times \gamma_d^{n_d}$ form a basis of admissible domains, and thus the dimension of the homology space of admissible manifolds for integrating our eigenvalues is the number of ways of partitioning N into d parts, i.e.,

$$\dim H_1 \left(\Delta(\lambda_1,\ldots,\lambda_N)^2 \prod_i e^{\frac{-1}{\hbar}V(\lambda_i)} d\lambda_i, \mathbb{C} \right) = \frac{(N+d-1)!}{N!(d-1)!}. \tag{7.23}$$

In other words this number is the number of linearly independent eigenvalue integrals with a given potential of degree $\deg V' = d$.

For example with the quadratic potential $V = \frac{x^2}{2}$, $d = \deg V' = 1$, there is a unique Gaussian integral.

Generalization 2: Non-Hermitian matrices

Let S_N the set of real symmetric $N \times N$ matrices $M = M^T$, with its $O(N)$ conjugations invariant Lebesgue measure

$$dM = \prod_{i \le j} dM_{ij}. \tag{7.24}$$

Any $M \in S_N$ can be diagonalized by an orthogonal transformation

$$M \in S_N \quad \to \quad M = U\Lambda U^T \quad U \in O(N), \; UU^T = \mathrm{Id}. \tag{7.25}$$

We have

$$S_N = \frac{O(N)/O(1)^N \times \mathbb{R}^N}{\mathfrak{S}_N}. \tag{7.26}$$

The Jacobian of the change of variable is

$$dM = \frac{1}{2^N N!} |\Delta(\Lambda)| \, dU \wedge d\Lambda. \tag{7.27}$$

Similarly, matrices in the set Q_N of $N \times N$ self–dual quaternion matrices (see Mehta's book (Mehta 2004)) can be diagonalized by a symplectic transformation and the Jacobian of the change of variable is

$$dM = \frac{1}{(2\pi^2)^N N!} \Delta(\Lambda)^4 \, dU \wedge d\Lambda. \tag{7.28}$$

In all three cases we have

$$dM = \frac{1}{V_{1,\beta}^N N!} \left(\Delta(\Lambda)^2 \right)^{\frac{\beta}{2}} dU \wedge d\Lambda, \tag{7.29}$$

with $\beta = 1$ for S_N, $\beta = 2$ for H_N, and $\beta = 4$ for Q_N, and $\beta = 0$ for the set of diagonal matrices. In each case the partition function is

$$Z = \frac{V_{N,\beta}}{V_{1,\beta}^N N!} \int_{\gamma^N} d\lambda_1 \ldots d\lambda_N \left(\Delta(\Lambda)^2 \right)^{\frac{\beta}{2}} \prod_{i=1}^{N} e^{-\frac{\beta}{2\hbar} V(\lambda_i)}. \tag{7.30}$$

Generalization 3: Lie algebras

Let $\mathfrak{g}$ be a Lie algebra of a Lie group $G = e^{\mathfrak{g}}$ (e.g., iH_N, the Lie algebra of the group $G = U(N)$). A Cartan subalgebra $\mathfrak{h}$ is a maximal Abelian subalgebra (e.g., $\mathfrak{h}$= set of imaginary diagonal $N \times N$ matrices $\subset iH_N$); any two Cartan subalgebras are related by an adjoint action of a group element (a conjugation). Any element $M \in \mathfrak{g}$ can be brought to a given Cartan subalgebra $\mathfrak{h}$ by an adjoint action of a group element (a conjugation),

$$M = \mathrm{Ad}_U(\Lambda) \qquad U \in G, \ \Lambda \in \mathfrak{h}, \tag{7.31}$$

which generalizes the diagonalization $M = U\Lambda U^{-1}$. Such a decomposition is not unique, U can be right multiplied by an element of the maximal torus $e^{\mathfrak{h}} \subset G$, and U and Λ can be 'permutted' by a Weil group element (the Weil group $\mathfrak{W}$ is the stabilizer of $e^{\mathfrak{h}}$ in G, modulo $e^{\mathfrak{h}}$). In the example of $G = U(N)$ and $e^{\mathfrak{h}} = U(1)^N$, the Weil group is $\mathfrak{W} = \mathfrak{S}_N$ the group of permutations. Since all $h \in \mathfrak{h}$ commute, their adjoint operators $\mathrm{Ad}_h : u \mapsto [h, u]$ must have a common basis of eigenvectors; call them u_i. The eigenvalue of Ad_h for the eigenvector u_i is called $r_i(h)$; it depends linearly on h and thus is a linear form $r_i \in \mathfrak{h}^*$. r_i is called a root; call E_{r_i} the eigenspace. We have

$$\mathfrak{g} = \mathfrak{h} \oplus \sum_i E_{r_i}, \tag{7.32}$$

and thus we see that the set of roots is finite with cardinal $\leq \dim \mathfrak{g} - \dim \mathfrak{h}$. The Jacobian of the change of variable is

$$dM \propto dU d\Lambda \prod_{\alpha = \text{roots}} (\alpha(\Lambda))^{m_\alpha/2}, \tag{7.33}$$

where m_α is the root's multiplicity. It generalizes the Vandermonde determinant.

Generalization 4: two-matrix models

A very commonly studied two-matrix ensemble is $H_N \times H_N$ with a Boltzmann weight probability measure

$$dM_1 dM_2 e^{-\frac{1}{\hbar} \mathrm{Tr} \, V_1(M_1) + V_2(M_2) - cM_1 M_2}. \tag{7.34}$$

Note that $H_N \times H_N$ is a submanifold of $M_N(\mathbb{C}) \times M_N(\mathbb{C})$ of half-dimension; it is an algebraic submanifold of equation $M_1^\dagger = M_1$, $M_2^\dagger = M_2$. This can be further generalized to $H_N(\gamma_1) \times H_N(\gamma_2)$, or more generally to any admissible half-dimension submanifold of $M_N(\mathbb{C}) \times M_N(\mathbb{C})$.

Another algebraic submanifold of $M_N(\mathbb{C}) \times M_N(\mathbb{C})$ of half-dimension has equation $M_1^\dagger = M_2$. The same probability measure would be in this one

$$dM_1 dM_1^\dagger e^{-\frac{1}{\hbar} \mathrm{Tr} \, V_1(M_1) + V_2(M_1^\dagger) - cM_1 M_1^\dagger}. \tag{7.35}$$

If one algebraic submanifold can be homotopically deformed into another one without encountering singularities of the integrand, then their integral is the same. This remark makes it possible to view the complex matrix model as a normal two-matrix model.

Generalization 4: multimatrix models Then one can invent anything with sub-manifolds of $M_N(\mathbb{C})^k$. This leads in particular to the notion of 'quivers', which is beyond the scope of this chapter. A particularly well-studied model is the chain of matrices (see, for instance, Di Francesco et al. (1995) or Mehta (2004))

$$\prod_{i=1}^{k} e^{\frac{-1}{\hbar} \operatorname{Tr} V_i(M_i)} \prod_{i=1}^{k-1} e^{\frac{c_i}{\hbar} \operatorname{Tr} M_i M_{i+1}} \prod_{i=1}^{k} dM_i. \tag{7.36}$$

Generalization 5: discrete matrix models Another application of this notion of normal matrix models with eigenvalues on a contour γ is to be able to view discrete sums as matrix integrals. Indeed, choose a Boltzmann weight $e^{\frac{-1}{\hbar} V(x)} dx$ that has simple poles at all positive integers (or, in fact, any prescribed set of points in $\mathbb{C}$), with residue at $n \in \mathbb{Z}_+$:

$$\operatorname{Res}_{n} e^{\frac{-1}{\hbar} V(x)} dx = R_n. \tag{7.37}$$

Taking a contour γ that surrounds all positive integers and no other singularity of V makes it possible to write the integral on γ as a sum of residues:

$$\int_{\gamma^N} \Delta(\lambda_1, \dots, \lambda_N)^2 \prod_i e^{\frac{-1}{\hbar} V(\lambda_i)} d\lambda_i = (2i\pi)^N \sum_{h_1, \dots, h_N \in \mathbb{Z}_+^N} \Delta(h_1, \dots, h_N)^2 \prod_i R_{h_i}. \tag{7.38}$$

Moreover, since the right-hand side vanishes at coinciding h_is and since it is symmetric, we may rewrite

$$\frac{1}{N!} \int_{\gamma^N} \Delta(\lambda_1, \dots, \lambda_N)^2 \prod_i e^{\frac{-1}{\hbar} V(\lambda_i)} d\lambda_i = (2i\pi)^N \sum_{h_1 > \dots > h_N \geq 0} \Delta(h_1, \dots, h_N)^2 \prod_i R_{h_i}. \tag{7.39}$$

Many discrete Dyson gases are of that form, and also many 'Nekrasov functions' and sums over Young tableaux are of that sort. This method makes it possible to rewrite them as matrix models. This is what we will illustrate now.

7.1.2 A detailed example: random partitions as random matrices

In this section, we show that some famous discrete models can be viewed as matrix models. This example will also illustrate the second section, 7.2.

Random partitions, Plancherel measure

Consider a partition $\mu = (\mu_1 \geq \mu_2 \geq \dots \geq \mu_N \geq 0)$. We call $|\mu| = \sum_i \mu_i$ the weight of μ, and $\ell(\mu) = \max(\{i, \mu_i > 0\})$ the length of μ.

Rotating by 45°, we define $h_i \in \mathbb{N}$ by

$$h_i = \mu_i - i + N \qquad , h_1 > h_2 > \dots > h_N \geq 0. \tag{7.40}$$

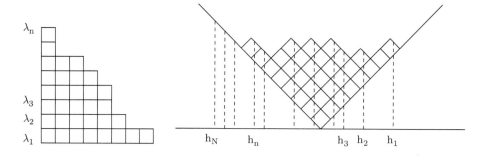

A partition corresponds to a representation of the unitary group, and the dimension of the representation is given by the formula

$$\dim \mu = \frac{\prod_{i<j}(h_i - h_j)}{\prod_i h_i!} |\mu|!, \tag{7.41}$$

which is independent of N as long as $N \geq \ell(\mu)$, and is always an integer.
Casimirs of μ are

$$C_k(\mu) = \frac{1}{k} \sum_{i=1}^{\ell(\mu)} \left(\mu_i - i + \frac{1}{2} \right)^k - \left(-i + \frac{1}{2} \right)^k = \frac{1}{k} \sum_{i=1}^{N} \left(h_i - N + \frac{1}{2} \right)^k - \left(-i + \frac{1}{2} \right)^k, \tag{7.42}$$

for instance,

$$C_1(\mu) = \sum_i \mu_i = |\mu| = \sum_i h_i - \frac{N(N-1)}{2} \tag{7.43}$$

$$C_2(\mu) = \frac{1}{2} \sum_i \mu_i(\mu_i - 2i + 1) = \frac{1}{2} \sum_i h_i^2 - (N - \frac{1}{2}) \sum_i h_i + \frac{N(N-1)(2N-1)}{6}. \tag{7.44}$$

The Plancherel measure is the following measure on the set of partitions of weight k:

$$P_k(\mu) = \frac{(\dim \mu)^2}{|\mu|!} \qquad |\mu| = k, \qquad \sum_{\mu, |\mu|=k} P_k(\mu) = 1. \tag{7.45}$$

Summing over all k with a factor $\frac{\hbar^{-2k}}{k!}$ gives the Poissonization (i.e., the grand canonical ensemble):

$$\sum_\mu P_{|\mu|}(\mu) \frac{\hbar^{-2|\mu|}}{|\mu|!} = e^{\hbar^{-2}}. \tag{7.46}$$

The expectation value of the partition weight is

$$\langle|\mu|\rangle = \frac{-e^{-\hbar^{-2}}}{2}\hbar\frac{d}{d\hbar}e^{\hbar^{-2}} = \hbar^{-2}e^{\hbar^{-2}}. \tag{7.47}$$

Letting $\hbar \to 0$ thus favours large partitions.

Interesting questions concern the statistics of partitions with the Plancherel measure, for instance expectation values of Casimirs $C_k(\mu)$s, or of characters $\chi_\nu(\mu)$s (Schur polynomials)

$$\langle C_k(\mu)\rangle, \ \langle\prod_i C_{k_i}(\mu)\rangle, \ \langle e^{\sum_k t_k C_k(\mu)}\rangle, \ \langle\chi_\nu(\mu)\rangle \ldots, \tag{7.48}$$

in particular in the $\hbar \to 0$ limit, i.e., for large partitions.

Statistics of jumping particles

Through the Robertson–Shengsted algorithm, the set of random partitions with the Plancherel measure, is equivalent to a statistics of N self-avoiding particles, located at $h_1, \ldots, h_N$. This is the 'Jeu de Taquin' model; see Romik and Śniady (2015). Each time a new box falls from the sky, exactly one of the h_is increases by 1, and h_is can never coincide; this is an 'exclusion process' (Fig. 7.1).

Rewriting as a matrix model

Let us rewrite the Plancherel measure as a matrix model (Eynard, 2008). Start from

$$Z_N = \sum_{h_1 > \cdots > h_N \geq 0} \Delta(h)^2 \frac{\prod_i \hbar^{-2h_i}}{\prod_i \Gamma(h_i + 1)^2}. \tag{7.49}$$

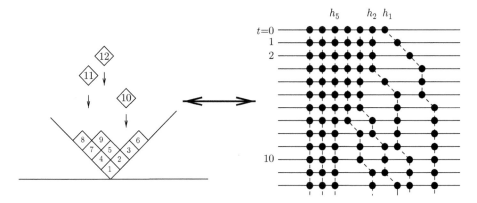

Fig. 7.1 On the left-hand side the number in the box indicates at what time it arrived. When a new box arrives, it must end at a place where it can't fall further, such that it reaches an inner corner. Adding a new box in an inner corner is equivalent to changing $h_i \to h_i + 1$. On the right-hand side, time is the vertical axis, oriented from top to bottom; the horizontal axis is the axis of the h_is.

First, observe that since $\Delta(h) = 0$ when some h_is coincide, we may extend the sum from strict inequalities $h_i > h_{i+1}$ to large inequalities $h_i \geq h_{i+1}$, and also the summand is symmetric, so up to a $1/N!$ factor, we may relax all constraints on the h_is, i.e.,

$$Z_N = \frac{1}{N!} \sum_{h_i \in \mathbf{N}} \Delta(h)^2 \frac{\prod_i \hbar^{-2h_i}}{\prod_i \Gamma(h_i + 1)^2}. \tag{7.50}$$

Then, note that a sum over integers can always be written as an integral by picking residues at all integers,

$$\sum_{\mathbf{N}} f(n) = \frac{1}{2\pi i} \int_\gamma dx \, \frac{\pi \, e^{i\pi x}}{\sin \pi x} f(x), \tag{7.51}$$

with γ a contour that surrounds all nonnegative integers and doesn't surround any singularity of f:

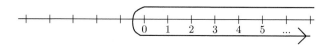

Therefore, we may rewrite

$$Z_N = \frac{1}{N!} \int_{\gamma^N} dx_1 \ldots dx_N \Delta(x)^2 \prod_{i=1}^{N} \frac{\hbar^{-2x_i}}{\Gamma(x_i + 1)^2} \frac{e^{i\pi x_i}}{2i \sin \pi x_i}. \tag{7.52}$$

Note that

$$\frac{\pi}{\Gamma(x+1) \sin \pi x} = \Gamma(-x). \tag{7.53}$$

Let us define the potential

$$\tilde{V}(x) = 2x \ln \hbar - i\pi x - \ln \frac{\Gamma(-x)}{2i\pi \Gamma(x+1)}, \tag{7.54}$$

so that

$$Z_N = \frac{1}{N!} \int_{\gamma^N} dx_1 \ldots dx_N \Delta(x)^2 \prod_{i=1}^{N} e^{-\tilde{V}(x_i)}. \tag{7.55}$$

In the $\hbar \to 0$ limit, let us rescale $x \to x/\hbar$ and define $V(x) = \hbar \tilde{V}(x/\hbar)$:

$$V(x) = 2x \ln \hbar - i\pi x - \hbar \ln \frac{\Gamma\left(\frac{-x}{\hbar}\right)}{2i\pi \, x\Gamma\left(\frac{x}{\hbar}\right)}. \tag{7.56}$$

$\tilde{V}$ is plotted here. V closely follows $x \mapsto 2x \ln x$:

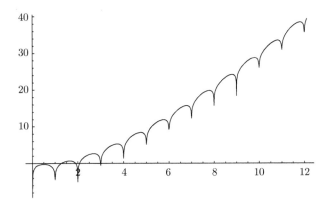

We thus obtain a matrix integral with the potential V

$$Z_N = \frac{\hbar^{-N^2}}{N!} \int_{\gamma^N} \mathrm{d}x_1 \ldots \mathrm{d}x_N \Delta(x)^2 \prod_{i=1}^{N} \mathrm{e}^{-\frac{1}{\hbar}V(x_i)} = \int_{H_N(\gamma)} \mathrm{d}M \mathrm{e}^{-\frac{1}{\hbar} \operatorname{Tr} V(M)}. \quad (7.57)$$

In the $\hbar \to 0$ limit, the Stirling approximation of the Γ function gives

$$\hbar \ln \Gamma(x/\hbar) \sim x \ln x - x \ln \hbar - \hbar \ln x + \hbar \ln \hbar - \frac{\hbar}{2} \ln 2\pi + \sum_{k=1}^{\infty} \frac{\mathcal{B}_k}{k(k-1)} \hbar^k x^{-k} \quad (7.58)$$

with $\mathcal{B}_k$ the kth Bernoulli number. Therefore, to leading order approximation we will be able to replace $V(x)$ with the function $x \mapsto 2x \ln x$, and

$$Z_N \sim \int_{H_N(\gamma)} \mathrm{d}M \mathrm{e}^{-\frac{2}{\hbar} \operatorname{Tr} M \ln M}. \quad (7.59)$$

Matrix models with potentials of the form $x \ln x$ are often called Eguchi–Kawai potentials from their appearance in gauge theories.

The goals of the next sections will be to study the $\hbar \to 0$ limit of the distribution of μ_is, or in other words the distributions of h_i's, which is the same as the distribution of eigenvalues of a random matrix with potential V, on the ensemble $H_N(\gamma)$. It is known that the $\hbar \to 0$ limit density of eigenvalues exists, and the partition μ converges almost surely to a fixed shape, first found by Vershik and Kerov (1977):

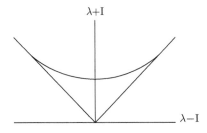

Our goal will be to retrieve it from random matrix methods.

7.2   Coulomb gas and saddle-point approximation

7.2.1   Coulomb gas of eigenvalues

For an arbitrary potential, the partition function

$$Z = \frac{\mathrm{Vol}(U(1)^N)}{\mathrm{Vol}(U(N))} \int_{H_N(\gamma)} dM e^{-\frac{1}{\hbar} \mathrm{Tr}\, V(M)} \tag{7.60}$$

can be rewritten in eigenvalues

$$Z = \frac{1}{N!} \int_{\gamma^N} d\lambda_1 \ldots d\lambda_N \prod_{i<j} (\lambda_i - \lambda_j)^2 \prod_i e^{-\frac{1}{\hbar} \mathrm{Tr}\, V(\lambda_i)}. \tag{7.61}$$

This can be reinterpreted as a 1-dimensional Coulomb gas of N particles at positions $\lambda_i \in \gamma$

$$Z = \frac{1}{N!} \int_{\gamma^N} d\lambda_1 \ldots d\lambda_N \, e^{-\frac{1}{\hbar^2} S(\lambda_1, \ldots, \lambda_N)} \tag{7.62}$$

and with an interaction energy

$$S(\lambda_1, \ldots, \lambda_N) = \hbar \sum_i V(\lambda_i) - 2\hbar^2 \sum_{i<j} \log(\lambda_i - \lambda_j), \tag{7.63}$$

made of a potential energy $\hbar V(\lambda_i)$ for each particle, and a logarithmic 2-body potential $-2\hbar^2 \ln(\lambda_i - \lambda_j)$, whose force—the derivative of the potential—is

$$\frac{-2\hbar^2}{\lambda_i - \lambda_j}, \tag{7.64}$$

i.e., a repulsive force proportional to the inverse of the distance, i.e., a Coulomb electrostatic force.

Here $\hbar^2$ can be intepreted as the temperature, and the fact that the Coulomb charge is $-2\hbar^2$ is very specific to normal Hermitian matrix models, this specific value of the charge enjoys special properties.

In most cases, for sufficiently regular potentials (for example, V polynomial), the interesting semi-classical regime is a limit where the temperature is small, $\hbar \to 0$, and at the same time the energy per particle remains finite, i.e., t remains finite, with

$$t = \hbar N. \tag{7.65}$$

t is called the ''t **Hooft**' parameter.

However, let us emphasize that for more exotic potentials Vs, the interesting semi–classical regime could be something else; this is the case for the Plancherel measure statistics introduced in section 7.1.2, for which the good 't Hooft parameter is computed with an 'effective size' $t = \hbar N_{\mathrm{eff}}(\hbar)$, as we will see in section 7.2.5. Indeed, the Plancherel measure is independent of N provided that N is large enough. The $\hbar \to 0$ limit is thus also independent of N provided that N is larger than the expected number of rows $\langle \ell(\mu) \rangle = N_{\mathrm{eff}}$.

7.2.2   Saddle-point approximation

The saddle-point approximation consists of finding the critical point of the energy, i.e., the equilibrium configuration $\partial S/\partial \lambda_i = 0$, in other words, the force vanishing:

$$\forall i, \qquad \frac{1}{\hbar}\frac{\partial S}{\partial \lambda_i} = 0 = V'(\lambda_i) - 2\hbar \sum_{j \neq i} \frac{1}{\lambda_i - \lambda_j}. \qquad (7.66)$$

For polynomial V (or more generally if V' is a rational function), this is a set of N coupled algebraic equations, and one can prove that the number of solutions is

$$\frac{(N + d - 1)!}{(d - 1)!}, \qquad (7.67)$$

where $d = \deg V'$ (if V' is rational, this is the sum of degrees at all poles, including the pole at ∞), and up to permutations:

$$\frac{(N + d - 1)!}{(d - 1)!N!}. \qquad (7.68)$$

Recognizing (7.23), we see that the number of saddle-points matches the number of possible independent matrix integrals. The way to interpret this is to say that a basis of the homology domains of integration is the basis of 'steepest descent' half-dimension manifolds going through saddle points.

Example. Gaussian potential. For the Gaussian potential $V(x) = \frac{x^2}{2}$, one has $d = \deg V' = 1$, and there is a unique solution up to permutations. As we shall see later, the unique solution of

$$\forall i = 1, \ldots, N, \qquad 2\hbar \sum_{j \neq i} \frac{1}{\lambda_i - \lambda_j} = \lambda_i \qquad (7.69)$$

is that $\frac{\lambda_1}{\sqrt{\hbar}}, \ldots, \frac{\lambda_N}{\sqrt{\hbar}}$ are the N zeroes of the Nth Hermite polynomial.

Discrete equilibrium equation, Bethe ansatz and Riccati Baxter equation

Readers with knowledge in integrable systems may recognize that Eqs (7.66) are the Bethe ansatz equations of the 'Gaudin model' (see Gaudin (1997)), in particular with the potential $V'(x) = \sum_{i=1}^{L} \frac{s_i}{x - \epsilon_i}$, with spins s_is and energies ϵ_is.

It is a general fact that the Bethe ansatz equations can be recast in the form of a Baxter equation. Let us see how it works here for this simple case.

Let us define the Baxter Q function

$$Q(x) = \prod_{i=1}^{N}(x - \lambda_i) \qquad (7.70)$$

where λ_i's are the Bethe roots, i.e., solutions of (7.66). Let us also define its logarithmic derivative

$$w(x) = \hbar\frac{Q'(x)}{Q(x)} = \hbar\sum_i \frac{1}{x - \lambda_i}. \tag{7.71}$$

Let us then compute

$$w(x)^2 + \hbar w'(x) = \hbar^2 \sum_{i,j} \frac{1}{(x - \lambda_i)(x - \lambda_j)} - \hbar^2 \sum_i \frac{1}{(x - \lambda_i)^2}$$

$$= \hbar^2 \sum_{i \neq j} \frac{1}{(x - \lambda_i)(x - \lambda_j)}$$

$$= \hbar^2 \sum_{i \neq j} \left(\frac{1}{x - \lambda_i} - \frac{1}{x - \lambda_j} \right) \frac{1}{\lambda_i - \lambda_j}$$

$$= 2\hbar^2 \sum_{i \neq j} \frac{1}{x - \lambda_i} \frac{1}{\lambda_i - \lambda_j}. \tag{7.72}$$

Now use (7.66); that gives

$$w(x)^2 + \hbar w'(x) = \hbar \sum_i \frac{V'(\lambda_i)}{x - \lambda_i}$$

$$= \hbar \sum_i \frac{V'(x) - (V'(x) - V'(\lambda_i))}{x - \lambda_i}$$

$$= V'(x)w(x) - \hbar \sum_i \frac{V'(x) - V'(\lambda_i)}{x - \lambda_i}. \tag{7.73}$$

Let us define

$$P(x) = \hbar \sum_i \frac{V'(x) - V'(\lambda_i)}{x - \lambda_i}. \tag{7.74}$$

If $V'(x)$ is polynomial (resp. rational) then this is a polynomial (resp. rational function) of x whose poles are the same as V' and whose degree at ∞ is $\leq \deg V''$ (resp. at a finite pole $\leq \deg V'$), and which behaves at ∞ like

$$P(x) \sim \hbar N \frac{V'(x)}{x} \tag{7.75}$$

(we recognize the 't Hooft parameter $t = \hbar N$).

If $\deg V' = d$ (sum of degrees of all poles, including the pole at ∞), then $P(x)$ is a polynomial (resp. rational function) characterized by $d - 1$ coefficients.

Let us assume for the moment that $P(x)$ is a known polynomial (resp. rational), then we have the 'Riccati' equation for $w(x)$:

$$w(x)^2 + \hbar w'(x) = V'(x)w(x) - P(x). \tag{7.76}$$

Writing that $w = \hbar \frac{Q'}{Q}$, we get a second-order differential equation for $Q(x)$

$$\hbar^2 Q''(x) = \hbar V'(x)Q'(x) - P(x)Q(x), \tag{7.77}$$

also written with a differential operator $T(x, \hbar d/dx)$

$$T(x, \hbar d/dx).Q(x) = 0, \quad T(x, \hbar d/dx) = \hbar^2 \frac{d^2}{dx^2} - V'(x)\hbar \frac{d}{dx} + P(x), \qquad (7.78)$$

which is called the 'T–Q Baxter equation'. This procedure of going from a Bethe ansatz to a Baxter equation or vice versa is very general: the Q Baxter function whose zeroes are the Bethe roots satisfies a linear equation of the form

$$T.Q(x) = 0, \qquad (7.79)$$

where T is a linear operator.

Here in our case of the Gaudin model, this linear operator is a second-order differential operator. In more general cases (multimatrix models, or nonrational potentials), it can be a higher order differential operator, or also a finite difference operator, an integral operator, or, in fact, any possible linear operator).

Rewriting $Q(x) = \psi(x)e^{\frac{V(x)}{2\hbar}}$ we get a Schrödinger equation for $\psi(x)$:

$$\hbar^2 \psi''(x) = U(x)\psi(x) \quad , \quad U(x) = \frac{V'(x)^2}{4} - P(x). \qquad (7.80)$$

WKB approximation

In the $\hbar \to 0$ limit, the WKB approximation makes it possible to write

$$\psi(x) \sim \frac{c_+}{U^{\frac{1}{4}}} \, e^{\frac{\pm 1}{\hbar} \int_0^x \sqrt{U(x')}dx'} \, (1 + O(\hbar)), \qquad (7.81)$$

where $c_\pm$ is some constant that can be determined by the asymptotics of $Q(x) \sim x^N$ at large x. This gives

$$\omega(x) = \hbar \frac{Q'(x)}{Q(x)} \sim \frac{V'(x)}{2} + \hbar \frac{\psi'(x)}{\psi(x)} = \frac{1}{2} \left(V'(x) \pm \sqrt{V'(x)^2 - 4P(x)} \right). \qquad (7.82)$$

In fact, (7.81) should be a sum of the two possible terms

$$\psi(x) \sim \frac{c_+}{U^{\frac{1}{4}}} \, e^{\frac{1}{\hbar} \int_0^x \sqrt{U(x')}dx'} \, (1 + O(\hbar)) + \frac{c_+}{U^{\frac{1}{4}}} \, e^{\frac{-1}{\hbar} \int_0^x \sqrt{U(x')}dx'} \, (1 + O(\hbar)). \qquad (7.83)$$

For most values of x, we have $\Re \int_0^x \sqrt{U(x')}dx' \neq 0$, and one of the two exponentials is dominant, so, unless the corresponding coefficients $c_\pm$ vanishes, for most values of x, only one of the two exponentials gives the leading behavior, the one with the highest real part $\pm \Re \int_0^x \sqrt{U(x')}dx' > 0$.

The lines where this real part vanishes are called 'Stokes lines':

$$\Re \int_0^x \sqrt{U(x')}dx' = 0. \qquad (7.84)$$

They are the lines where the two exponentials have the same order of magnitude. Therefore, they are also the only places where ψ can vanish, and asymptotically, the zeroes of ψ are thus the places where

$$x_n \text{ solution of } \quad \Im \int_0^{x_n} \sqrt{U(x')}dx' + \frac{\hbar}{2} \Im \ln \frac{c_+}{c_-} = n\pi, \quad n \in \mathbb{Z}. \quad (7.85)$$

The x_n's satisfying this equation for integer n are (in the semi-classical limit $\hbar \to 0$) the zeroes of ψ, i.e., the Bethe roots $x_n = \lambda_n$. This equation is the Bohr Sommerfeld quantization condition, or also the asymptotic Bethe ansatz equation.

The condition that ψ is a polynomial, and has a finite number N of zeroes, can be seen by requiring that ψ doesn't acquire a phase after going around any contour, and in particular this gives a nontrivial condition if we make a loop around a cut of $\sqrt{U(x)}$, i.e., going around a contour $\mathcal{A}$ that surrounds a pair of zeroes of $\sqrt{U(x)}$. This implies that

$$\forall \text{ cycle } \mathcal{A}, \quad \Re \oint_{\mathcal{A}} \sqrt{U(x)}dx = 0, \quad (7.86)$$

where $\mathcal{A}$ is any contour that surrounds a pair of zeroes of $U(x)$.

Moreover, $\omega(x)$ has poles, and thus is not analytic in the regions that contain some λ_j's. If a contour $\mathcal{A}$ surrounds n roots, one should have

$$\frac{1}{2\pi i} \oint_{\mathcal{A}} \omega(x)dx = \hbar n. \quad (7.87)$$

In the $\hbar \to 0$ limit, given the WKB asymptotic of ψ, i.e., the asymptotic (7.82), the only contours that can asymptotically enclose roots can be those surrounding cuts of $\sqrt{U(x)}$, and the number of roots contained along a cut is thus

$$\frac{1}{4\pi i} \oint_{\mathcal{A}_i} \sqrt{U(x)}\,dx = \hbar n_i, \quad (7.88)$$

and we must have $n_i \geq 0$, and $\sum_i n_i = N$.

We define the 'filling fractions' to be

$$\epsilon_i = \hbar n_i; \quad (7.89)$$

their sum is the 't Hooft parameter

$$\sum_i \epsilon_i = t. \quad (7.90)$$

Determining the polynomial P(x)

For a polynomial potential such that $\deg V' = d$, $U(x)$ has $2d$ zeroes; i.e. there are d pairs of zeroes, and thus one can gather them by pairs into d arcs surrounded by d noncrossing contours $\mathcal{A}_1, \ldots, \mathcal{A}_d$.

The system of equations

$$\begin{cases} \forall\, i = 1, \ldots, d\,, & \frac{1}{4\pi i \hbar} \oint_{\mathcal{A}_i} \sqrt{U(x)}\, \mathrm{d}x = n_i \in \mathbb{Z}_+ \\ \sum_{i=1}^{d} n_i = N \end{cases} \tag{7.91}$$

gives $d-1$ independent equations, which fully determine the $d-1$ unknown coefficients of the polynomial $P(x)$ (and determine at the same time the contours $\mathcal{A}_i$'s up to homotopy).

Proving that these equations have a unique solution for each given d-uple $(n_1, \ldots, n_d)$ involves highly nontrivial algebraic geometry, beyond the scope of this chapter.

The number of d-uples $(n_1, \ldots, n_d)$ with $n_i \geq 0$ and $\sum_i n_i = N$, is

$$\frac{(N + d - 1)!}{(d - 1)! N!}, \tag{7.92}$$

which matches the number of possible integrals (7.23) and the number of saddle-points (7.68).

7.2.3   Functional saddle-point

The Bethe ansatz approach is quite complicated; there is a much faster method for obtaining the same leading asymptotic result, which consists in taking a $\hbar \to 0$ limit before extremizing the energy S. Let us assume that in the $\hbar \to 0$ limit with $t = N\hbar$ finite, the λ_i's tend towards a smooth distribution with density $\rho(\lambda)\mathrm{d}\lambda$ on the contour γ. The energy can then be written as a function of the density:

$$S(\{\lambda_i\}) \to S(\rho) = t \int_\gamma V(x)\rho(x)\mathrm{d}x - t^2 \int_{\gamma \times \gamma} \rho(x)\mathrm{d}x\, \rho(x')\mathrm{d}x'\, \ln\,(x - x')^2. \tag{7.93}$$

It is a quadratic functional of ρ.

In the real eigenvalues case $\gamma = \mathbb{R}$, and assuming that $\rho(x)\mathrm{d}x$ is a real positive measure on $\mathbb{R}$, one can easily compute the Hessian (second derivative) of S, and see that it is positive definite (in fact, this is easier in Fourier transform $\rho(x) = \int \tilde{\rho}(k)\mathrm{d}k e^{ikx}$; we have $\int_{\gamma \times \gamma} \rho(x)\mathrm{d}x\, \rho(x')\mathrm{d}x'\, \ln\,(x - x')^2 = \int_0^\infty \frac{\tilde{\rho}(k)^2}{k}\mathrm{d}k$, so the Hessian is $1/k^2 > 0$). Therefore, $S(\rho)$ is a convex function on the set of real positive measures, and thus it has a unique minimum.

If $\gamma \neq \mathbb{R}$, and/or if $V(x)$ is not real, the situation is more complicated, $\Re S$ is a harmonic function, and it cannot have a unique minimum on the whole set of complex measures. It can have a minimum only on some subsets, which should be determined by 'free boundary problems' extremely difficult to solve. It is, in fact, a 'min-max' problem, where one must find the maximum (with respect to γ) of the minimum of $S(\rho)$ (with respect to a measure $\rho(x)\mathrm{d}x$ on γ).

To find an extremum of S we write the Euler–Lagrange equations. Before extremizing, we need to extremize with the constraint that $\rho(x)$ is a density with total mass 1,

i.e., $\int_\gamma \rho(x)dx = 1$. We may relax the constraint by introducing a Lagrange multiplier ℓ; i.e., we extremize $S - \ell(\int \rho(x)dx - 1)$. The Euler–Lagrange equations are then

$$\ell = \frac{\delta S}{\delta \rho(x)} = tV(x) - 2t^2 \fint \rho(x')dx' \ln(x - x'), \qquad (7.94)$$

where $\fint$ is the principal part. Taking a derivative with respect to x gives

$$0 = V'(x) - 2t \fint \rho(x')dx' \frac{1}{x - x'}. \qquad (7.95)$$

Let us define the 'Stieljes transform' of the density $\rho(x)dx$:

$$\omega(x) = t \int_{\text{supp}(\rho)} \rho(x')dx' \frac{1}{x - x'}. \qquad (7.96)$$

Note that this is the continuous limit of (7.71). The Stieljes transform of a distribution is analytic outside of the support, and has a discontinuity along the support:

$$\omega(x - i0) - \omega(x + i0) = 2\pi i \, t \, \rho(x). \qquad (7.97)$$

This means that we can recover the density as the discontinuity of ω.

The derivative of the Euler–Lagrange equation is then a 'Riemann–Hilbert' equation for the Stieljes transform:

$$\omega(x + i0) + \omega(x - i0) = V'(x) \qquad \forall x \in \text{supp}(\rho). \qquad (7.98)$$

This equation is the continuous version of the Bethe ansatz (7.66). The advantage is that it is a linear equation for ω.

Let us recover the Riccati version. Observe that

$$\forall x \in \mathbb{C} \qquad (\omega(x + i0) + \omega(x - i0) - V'(x))\,(\omega(x + i0) - \omega(x - i0)) = 0, \qquad (7.99)$$

and thus if we define

$$P(x) = V'(x)\omega(x) - \omega(x)^2, \qquad (7.100)$$

we have

$$P(x + i0) - P(x - i0) = 0. \qquad (7.101)$$

In other words $P(x)$ is analytic everywhere (except where V' is not analytic, typically it can have poles at the poles of V'), this is an entire function on $\mathbb{C}$−poles of V', and it behaves at ∞ like $\frac{V'(x)}{x}(1 + O(1/x))$, which implies that it must be a polynomial (resp. a rational fraction if V' is rational) with the same poles as V'. It gives the quadratic equation

$$\omega(x)^2 = V'(x)\omega(x) - P(x), \qquad (7.102)$$

which is the $\hbar \to 0$ limit of the Riccati equation (7.76), obtained by neglecting the $\hbar w'$ term.

The solution is again

$$w(x) = \frac{1}{2}\left(V'(x) - \sqrt{V'(x)^2 - 4P(x)}\right). \tag{7.103}$$

The $d - 1$ unknown coefficients of the polynomial $P(x)$ are determined by recovering the corresponding density ρ as the discontinuity (only the square root is discontinuous)

$$\rho(x) = \frac{1}{2\pi t}\sqrt{4P(x) - V'(x)^2}, \tag{7.104}$$

and we plug it into the full Euler–Lagrange equation (7.94). Since ℓ is the same for all connected components of the support, the integral between two connected components must vanish, i.e.,

$$\Re \int_{a_i}^{a_j} \sqrt{V'(x)^2 - 4P(x)}\,\mathrm{d}x = 0, \tag{7.105}$$

where a_i is on the ith connected component. On the other hand, we must have, as in section 7.2.2, that the following integrals are real positive for all $\mathcal{A}_i$ surrounding connected components of the supports

$$\forall\,\mathcal{A}_i \qquad \frac{1}{2\pi i}\oint_{\mathcal{A}_i} w(x)\,\mathrm{d}x = \epsilon_i > 0, \tag{7.106}$$

implying also that

$$\Re \oint_{\mathcal{A}_i} w(x)\,\mathrm{d}x = 0. \tag{7.107}$$

Therefore, real parts of integrals around connected components of the supports, and real parts of integrals from one component to another all vanish. This implies, in fact, that the real parts of all integrals of w around all possible contours must vanish. This is called the 'Boutroux' property.

In other words, we determine the polynomial $P(x)$ by requiring that $w(x)$ has the Boutroux property.

Remark (fixed filling fractions). If we wanted to have fixed filling fractions n_i/N on components γ_i of $\gamma = \sum_{i=1}^{d} c_i\gamma_i$, we would have to use d Lagrange multipliers ℓ_i for each γ_i, i.e., extremize

$$S(\rho) - \sum_{i=1}^{d} \ell_i\left(\int_{\gamma_i} \rho(x)\,\mathrm{d}x - \frac{n_i}{N}\right). \tag{7.108}$$

The Euler–Lagrange equations (7.94) would then be

$$\forall\,x \in \gamma_i: \qquad \ell_i = \frac{\delta S}{\delta\rho(x)} = tV(x) - 2t^2 \fint \rho(x')\,\mathrm{d}x'\,\ln(x - x'). \tag{7.109}$$

Taking the derivative with respect to x gives the same equation for $w(x)$, and the same solution (7.103). Since now ℓ_i is not the same on different connected components, the Boutroux property does not hold; instead, what determines the d coefficients of $P(x)$ are the d filling fraction equations:

$$\forall\,\mathcal{A}_i \qquad \frac{1}{2\pi i}\oint_{\mathcal{A}_i} w(x)\mathrm{d}x = \epsilon_i = \frac{tn_i}{N}. \qquad (7.110)$$

7.2.4   1-cut case, Joukovski transformation

Let us assume here that V is a polynomial of degree $d+1$.

A case that is particularly interesting is the 1-cut case, when $U(x) = \frac{V'(x)^2}{4} - P(x)$ has only one pair of simple zeroes, all the other zeroes being double zeroes. In that case we may write

$$V'(x)^2 - 4P(x) = M(x)^2\,(x-a)(x-b), \qquad (7.111)$$

where $M(x)$ is a polynomial of degree $d-1$ containing all the even zeroes. A convenient change of variable is to use instead of x a variable z related to x by

$$x = x(z) = \frac{a+b}{2} + \frac{a-b}{4}\left(z + \frac{1}{z}\right). \qquad (7.112)$$

The reason is that x is a rational fraction of z and so is

$$\sqrt{(x-a)(x-b)} = \frac{a-b}{4}\left(z - \frac{1}{z}\right), \qquad (7.113)$$

which means that $w(x)$ is a rational fraction in the variable z.

The change of variable from x to z maps the complex plane of x cut along $[a,b]$ to the exterior of the unit disc in the variable z.

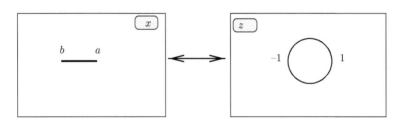

This conformal mapping was first considered by Joukovski, in aerodynamics, to conformally map a thin wing profile to a circular wing profile for which the equations of aerodynamics are easier to solve.

In the Joukovski variable $w(x(z))$ is a rational function of z, call it $y(z)$; it may have poles at $x = \infty$, i.e., at $z = 0$, and at $z = \infty$, we write it as a Laurent polynomial:

$$y(z) = w(x(z)) = \sum_{k\in\mathbb{Z}} u_k z^{-k}. \qquad (7.114)$$

Note that at large z, i.e., large x, we must have $w(x) \sim \frac{t}{x}$ and thus,

$$y(z) \sim \frac{t}{\gamma z} + O(z^{-2}) \qquad \gamma = \frac{a-b}{4}, \tag{7.115}$$

which implies that there is no negative k in (7.114). On the other hand, changing the sign of the square root amounts to $z \to 1/z$, and, thus, the saddle-point equation is

$$y(z) + y(z^{-1}) = V'(x(z)). \tag{7.116}$$

This implies that $y(z)$ is a polynomial of $1/z$ of degree $d = \deg V'$, taking the form

$$y(z) = \sum_{k=1}^{d} u_k z^{-k}, \quad u_1 = \frac{t}{\gamma}, \tag{7.117}$$

and we must have

$$V'(x(z)) = \sum_{k=1}^{d} u_k \left(z^k + z^{-k}\right). \tag{7.118}$$

These equations, together with $u_1 = t/\gamma$, and $u_0 = 0$ determine all the u_k's as well as a and b.

The density of eigenvalues is computed from the discontinuity of the Stieljes transform, i.e.,

$$\rho(x) = \frac{y(z^{-1}) - y(z)}{2t\pi i} = \frac{1}{\pi t} M(x) \sqrt{(a-x)(x-b)}. \tag{7.119}$$

The support of the density is the locus where $\rho(x)dx$ is real and positive.

7.2.5   Examples

Example 7.2.1.   *Gaussian potential* This corresponds to $V(x) = \frac{x^2}{2}$, i.e., $V'(x) = x$. Writing

$$x(z) = \alpha + \gamma(z + 1/z), \tag{7.120}$$

we have

$$V'(x(z)) = x(z) = \alpha + \gamma(z + 1/z) = \sum_k u_k(z^k + z^{-k}). \tag{7.121}$$

We read off: $u_0 = \frac{\alpha}{2}$, $u_1 = \gamma$, and $u_k = 0$ if $k \geq 2$.

The constraint $u_0 = 0$ fixes $\alpha = \frac{a+b}{2} = 0$, and the constraint $u_1 = \frac{t}{\gamma}$ fixes

$$u_1 = \gamma = \frac{t}{\gamma}, \tag{7.122}$$

i.e.,

$$\gamma = \frac{a-b}{4} = \sqrt{t}. \tag{7.123}$$

Finally, we have

$$a = -b = 2\sqrt{t}, \tag{7.124}$$

and

$$w(x) = y(z) = \frac{\sqrt{t}}{z} = \frac{1}{2}\left(x - \sqrt{x^2 - 4t}\right). \tag{7.125}$$

The corresponding density of eigenvalues is the famous 'Wigner's semi–circle'

$$\rho(x) = \frac{\sqrt{4t - x^2}}{\pi t}. \tag{7.126}$$

It is supported on the segment $[-2\sqrt{t}, 2\sqrt{t}]$.

Example 7.2.2. *Quartic potential* Let us chose $V(x) = -\frac{x^2}{2} + g\frac{x^4}{4}$, i.e., $V'(x) = -x + gx^3$. Writing

$$x(z) = \alpha + \gamma(z + 1/z), \tag{7.127}$$

we have

$$
\begin{aligned}
V'(x(z)) &= -x(z) + gx(z)^3 \\
&= -\alpha - \gamma(z + 1/z) + g(\alpha^3 + 3\alpha^2\gamma(z + 1/z) + 3\alpha\gamma^2(z^2 + 2 + z^{-2}) \\
&\quad + \gamma^3(z^3 + 3z + 3z^{-1} + z^{-3})) \\
&= \sum_k u_k(z^k + z^{-k}),
\end{aligned}
\tag{7.128}
$$

which gives

$$
\begin{aligned}
u_0 &= \alpha(-1 + g(\alpha^2 + 6\gamma^2)) \\
u_1 &= \gamma(-1 + g(3\alpha^2 + 3\gamma^2)) \\
u_2 &= 3g\alpha\gamma^2 \\
u_3 &= g\gamma^3.
\end{aligned}
\tag{7.129}
$$

The constraint $u_0 = 0$ gives (this is one solution, not the only one) $\alpha = 0$, and the constraint $u_1 = \frac{t}{\gamma}$ gives

$$t = -\gamma^2 + 3g\gamma^4, \tag{7.130}$$

i.e.,

$$\gamma^2 = \frac{1}{6g}\left(1 + \sqrt{1 + 12gt}\right). \tag{7.131}$$

Finally, we have

$$a = -b = 2\gamma, \tag{7.132}$$

and

$$w(x) = y(z) = \frac{t}{\gamma z} + \frac{g\gamma^3}{z^3} = \frac{1}{2}\left(-x + gx^3 - g\left(x^2 - 4\gamma^2 + \frac{\sqrt{1+12gt}}{g}\right)\sqrt{x^2 - 4\gamma^2}\right). \tag{7.133}$$

The corresponding density of eigenvalues is

$$\rho(x) = \frac{g}{\pi t}\left(x^2 - 4\gamma^2 + \frac{\sqrt{1-12gt}}{g}\right)\sqrt{4\gamma^2 - x^2}. \tag{7.134}$$

It is supported on the segment $[-2\gamma, 2\gamma]$, and is positive only if $t > \frac{1}{4g}$.

If $t < \frac{1}{4g}$, this means that the measure doesn't have one cut, but several cuts; the support is not connected. Let us look for 2-cut solutions, and, in fact, let us look for symmetric 2-cut $[b, a] \cup [-a, -b]$ solutions, and thus a double zero at 0, i.e.,

$$V'(x)^2 - 4P(x) = g^2 x^2 (x^2 - a^2)(x^2 - b^2). \tag{7.135}$$

This gives

$$w(x) = \frac{1}{2}\left(V'(x) - gx\sqrt{(x^2 - a^2)(x^2 - b^2)}\right)$$
$$\sim \frac{1}{2}\left(-x + gx^3 - gx^3\left(1 - \frac{a^2 + b^2}{2x^2} - \frac{(a^2 - b^2)^2}{8x^4}\right)\right) + O(x^{-3}). \tag{7.136}$$

The condition $w(x) \sim t/x$ at $x \to \infty$ determines

$$a^2 = \frac{1}{g} + 2\sqrt{t/g}, \quad b^2 = \frac{1}{g} - 2\sqrt{t/g}. \tag{7.137}$$

In this case we have 2 cuts, and since they are symmetric, the filling fractions are $n_i/N = \frac{1}{2}$.

Of course, other solutions with up to 3 cuts may exist, with any given filling fractions in the 3 cuts.

Example 7.2.3. *Plancherel measure* This is the follow-up of the example in section 7.1.2. Let us start by chosing the Stirling approximated potential $V(x) = 2x \ln x$, i.e., $V'(x) = 2(1 + \ln x)$. Assume that we are in a 1-cut case and thus use the Joukovski transformation

$$x(z) = \alpha + \gamma(z + 1/z). \tag{7.138}$$

$y(z)$ must be solution of

$$V'(x(z)) = 2(1 + \ln x(z)) = y(z) + y(z^{-1}). \tag{7.139}$$

This is not a polynomial problem; nevertheless, we can find a solution. Indeed, let us try

$$y(z) = 2 \ln \left(1 + \frac{c}{z} \right). \tag{7.140}$$

The equation for y gives

$$y(z) + y(z^{-1}) = 2 \ln \frac{c}{\gamma} + 2 \ln \gamma (c + c^{-1} + z + z^{-1}); \tag{7.141}$$

we thus require

$$\ln \frac{c}{\gamma} = 1, \qquad \alpha = \gamma(c + c^{-1}) \tag{7.142}$$

and the behavior $y \sim \frac{t}{\gamma z}$ at $z \to \infty$ implies that

$$c = \frac{t}{2\gamma}. \tag{7.143}$$

This implies that

$$\gamma = c = \sqrt{\frac{t}{2}}, \qquad \alpha = 1 + \frac{t}{2}. \tag{7.144}$$

and thus

$$w(x) = y(z) = 2 \ln \left(1 + \frac{\gamma}{z} \right) = \ln \frac{1}{2} \left(x + 1 - \frac{t}{2} - \sqrt{(x-1)^2 - t(x-1) - \frac{t^2}{4}} \right). \tag{7.145}$$

The support of the density is where $\Re(y(z) - y(z^{-1})) = 0$, and the density is worth

$$\rho(x) = \frac{y(z^{-1}) - y(z)}{2\pi i} = \frac{1}{\pi i} \ln \frac{1 + \gamma z}{1 + \gamma z^{-1}}. \tag{7.146}$$

Note that the density is positive only if $\gamma \leq 1$, i.e., if $t \leq 2$, or in other words $N \leq \frac{2}{\hbar}$. Chosing $t = 2$ we have $c = \gamma = 1$, $\alpha = 2$, and thus, writing $z = e^{i\phi}$

$$x(z) = z + 1/z + 2 = 2 + 2 \cos \phi, \quad y(z) = 2 \ln(1 + 1/z) = 2 \ln \left(1 + e^{-i\phi} \right), \tag{7.147}$$

$$\rho(x(z)) = \frac{y(z) - y(1/z)}{-2i\pi} = \frac{\phi}{\pi}. \tag{7.148}$$

The density $\rho(x)$ is the $\hbar \to 0$ limit of the distribution of eigenvalues $x_i = \hbar h_i$. The index i is the number of eigenvalues $\leq x_i$; i.e., it is the integral of the density up to x. In the $\hbar \to 0$ limit the index $i \mapsto x_i$ becomes, after rescaling $i \to i/N = \hbar i/2$, the function $x \mapsto I(x)$

$$I(x) = -\frac{1}{2}\int_{\alpha+2\gamma}^{x} \rho(x')\mathrm{d}x' = \int_0^\phi \frac{\phi'}{\pi}\sin\phi'\mathrm{d}\phi' = \frac{\sin\phi - \phi\cos\phi}{\pi}. \tag{7.149}$$

Also, since $h_i = \lambda_i - i + N$, in the continuum limit, $\frac{1}{N}\lambda_i$ becomes a function $x \mapsto \lambda(x) = \frac{x}{2}+I(x)-1$. The limit shape of a large partition $\frac{1}{N}(i, \lambda_i)$ tends to the plot of the parametric curve $(I(x), \frac{x}{2}+I(x)-1)$, or rotated by $45°$, the limit shape $\frac{1}{N}(2h_i, \lambda_i+i)$ becomes the plot of the function $x \mapsto \frac{x}{2}+2I(x)-1$. We recover the Vershik–Kerov limit shape of large partitions with the Plancherel measure:

$$\lambda + I = f(x) = \frac{\sqrt{1-x^2} - x\arccos x + x\pi/2}{\pi}. \tag{7.150}$$

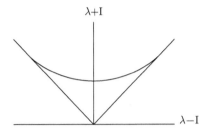

7.3   Loop equations

Loop equations is the name given to Schwinger–Dyson equations in the matrix model community (Migdal 1983; David 1990). They encode the reparametrization invariance under a change of variables, or equivalently, integration by parts.

In this section, we start from the partition function

$$Z = \int_{H_N(\gamma)} \mathrm{d}M e^{-\frac{1}{\hbar}\operatorname{Tr} V(M)} \tag{7.151}$$

and here we consider that V is a polynomial of degree $d+1$.

7.3.1   Basis of moments and observables

We shall consider all possible expectation values of symmetric polynomials of the eigenvalues. The power sums are products of traces of powers of the matrix: if $\mu = (\mu_1 \geq \mu_2 \geq \cdots \geq \mu_n > 0)$ is a partition, we denote by

$$p_\mu(M) = p_\mu(\lambda) = \prod_{i=1}^{\ell(\mu)}\left(\sum_{j=1}^{N}\lambda_j^{\mu_i}\right) = \prod_{i=1}^{\ell(\mu)} \operatorname{Tr} M^{\mu_i}. \tag{7.152}$$

A goal of random matrix theory is to compute their expectation values:

$$\hat{T}_{\mu_1,\dots,\mu_n} = Z\,\langle p_\mu(M)\rangle = \int_{H_N(\gamma)} \mathrm{d}M e^{-\frac{1}{\hbar}\operatorname{Tr} V(M)} \operatorname{Tr} M^{\mu_1}\dots\operatorname{Tr} M^{\mu_n} \tag{7.153}$$

and their cumulants (without the hat, and normalized, i.e., divided by Z):

$$T_{\mu_1,\dots,\mu_n} = \langle p_\mu(M) \rangle_c; \tag{7.154}$$

i.e., these are related by

$$\hat{T}_{\mu_1} = Z T_{\mu_1}$$
$$\hat{T}_{\mu_1,\mu_2} = Z \left(T_{\mu_1,\mu_2} + T_{\mu_1} T_{\mu_2} \right) \tag{7.155}$$

and so on,

$$\frac{\hat{T}_{\mu_1,\dots,\mu_n}}{Z} = \sum_{\nu \vdash \{1,\dots,n\}} \prod_{i=1}^{\ell(\nu)} T_{\mu_{\nu_i}}, \tag{7.156}$$

where the sum ν is over all ways of partitioning the μ_i's.

Power sums generate the vector space of all symmetric polynomials of a matrix M. But they are not independent. A basis of symmetric polynomials is provided by Schur polynomials:

$$s_\mu(\lambda) = \frac{1}{\Delta(\lambda)} \det \lambda_j^{\mu_i - i + N}. \tag{7.157}$$

The $s_\mu(\lambda)$'s can be expressed as combinations of power sums and vice versa. The Newton relations show that the p_μ's are not independent. An easy way to see it is to write the characteristic polynomial of M,

$$\det(1 - xM) = e^{\operatorname{Tr} \ln(1 - xM)}$$
$$= e^{-\sum_{k=1}^{\infty} \frac{x^k}{k} \operatorname{Tr} M^k}$$
$$= 1 + \sum_{n=1}^{\infty} \frac{(-1)^n}{n!} \sum_{\mu_1,\dots,\mu_n} \frac{x^{\sum_i \mu_i}}{\prod_i \mu_i} p_\mu(M), \tag{7.158}$$

writing that this is a polynomial of degree N means that the coefficient of x^k vanishes whenever $k > N$. We have relations

$$\forall k > N \qquad 0 = \sum_{\mu,\, |\mu| = k} \frac{(-1)^{\ell(\mu)}}{\prod_i n_i(\mu)! \; \prod_{i=1}^{\ell(\mu)} \mu_i} p_\mu, \tag{7.159}$$

where $n_i(\mu) = \#\{j \,|\, \mu_j = i\}$. It can be proved—for instance, by recursion on N—that, for a matrix of size N, a basis of symmetric polynomials is provided by the

$$p_\mu \qquad \text{with } \ell(\mu) \leq N. \tag{7.160}$$

The space of such symmetric polynomials is infinite dimensional. Indeed, the number of rows is bounded by N, but the size of each row is not bounded.

Generating series and resolvants

It is useful to also make generating series by introducing

$$\text{Tr}\,\frac{1}{x-M} = \sum_{k=0}^{\infty}\frac{\text{Tr}\,M^k}{x^{k+1}}, \tag{7.161}$$

the corresponding expectation values

$$W_1(x) = \left\langle \text{Tr}\,\frac{1}{x-M}\right\rangle = \frac{N}{x} + \sum_{k=1}^{\infty}\frac{T_k}{x^{k+1}}, \tag{7.162}$$

its disconnected part

$$\hat{W}_1(x) = ZW_1(x) = \frac{NZ}{x} + \sum_{k=1}^{\infty}\frac{\hat{T}_k}{x^{k+1}}, \tag{7.163}$$

and more generally

$$W_n(x_1,\ldots,x_n) = \sum_{\mu_1,\ldots,\mu_n}\frac{T_{\mu_1,\ldots,\mu_n}}{x_1^{\mu_1+1}\,x_2^{\mu_2+1}\,\ldots\,x_n^{\mu_n+1}}, \tag{7.164}$$

or equivalently for the noncumulants

$$\hat{W}_n(x_1,\ldots,x_n) = \sum_{\mu_1,\ldots,\mu_n}\frac{\hat{T}_{\mu_1,\ldots,\mu_n}}{x_1^{\mu_1+1}\,x_2^{\mu_2+1}\,\ldots\,x_n^{\mu_n+1}}. \tag{7.165}$$

$\text{Tr}\,\frac{1}{x-M}$ is called the 'resolvent', it is very useful because it is singular exactly on the spectrum of M.

7.3.2   Integrating by parts

Introduction

Let us warm up with an easy case; write that the integral of a total derivative vanishes,

$$\begin{aligned}
0 = \sum_i \int \mathrm{d}M\,\frac{\partial}{\partial M_{i,i}}\left((M^k)_{i,i}\;e^{-\frac{1}{\hbar}\,\text{Tr}\,V(M)}\right)\\
+\sum_{i<j}\int \mathrm{d}M\,\frac{\partial}{\partial\Re\,M_{i,j}}\left(((\Re\,M^k)_{i,j}\;e^{-\frac{1}{\hbar}\,\text{Tr}\,V(M)}\right)\\
+\sum_{i<j}\int \mathrm{d}M\,\frac{\partial}{\partial\Im\,M_{i,j}}\left(((\Im\,M^k)_{i,j}\;e^{-\frac{1}{\hbar}\,\text{Tr}\,V(M)}\right).
\end{aligned} \tag{7.166}$$

In complex coordinates $M_{i,j}$ and $M_{j,i} = \overline{M_{i,j}}$ rather than $\Re\,M_{i,j}$ and $\Im\,M_{i,j}$, this can be simply rewritten

$$0 = \sum_{i,j}\int \mathrm{d}M\,\frac{\partial}{\partial M_{i,j}}\left((M^k)_{i,j}\;e^{-\frac{1}{\hbar}\,\text{Tr}\,V(M)}\right). \tag{7.167}$$

Let us compute the derivatives in the right-hand side:

$$\frac{\partial}{\partial M_{i,j}} e^{-\frac{1}{\hbar}\operatorname{Tr}V(M)} = \frac{-1}{\hbar}\,(V'(M))_{j,i}\, e^{-\frac{1}{\hbar}\operatorname{Tr}V(M)} \tag{7.168}$$

and

$$\frac{\partial}{\partial M_{i,j}}(M^k)_{i,j} = \sum_{l=0}^{k-1}(M^l)_{i,i}\,(M^{k-1-l})_{j,j}. \tag{7.169}$$

This thus gives

$$0 = \sum_{l=0}^{k-1}\langle \operatorname{Tr} M^l \operatorname{Tr} M^{k-1-l}\rangle - \frac{1}{\hbar}\langle \operatorname{Tr} V'(M)M^k\rangle. \tag{7.170}$$

If we write that $V'(M) = \sum_{j=0}^{d} t_{j+1} M^j$, this says that

$$\sum_{j=0}^{d} t_{j+1}\langle p_{k+j}\rangle = \hbar \sum_{l=0}^{k-1}\langle p_l p_{k-1-l}\rangle \tag{7.171}$$

or also

$$\sum_{j=0}^{d} t_{j+1}\hat{T}_{k+j} = \hbar \sum_{l=0}^{k-1}\hat{T}_{l,k-1-l}, \tag{7.172}$$

and written with cumulants

$$\sum_{j=0}^{d} t_{j+1}T_{k+j} = \hbar \sum_{l=0}^{k-1}T_{l,k-1-l} + T_l\,T_{k-1-l}. \tag{7.173}$$

In other words, loop equations give linear relationships among the expectation values (and thus nonlinear relationships among cumulants).

When summing over k with $1/x^{k+1}$ this gives linear relationships for the generating series,

$$\left(V'(x)\hat{W}_1(x)\right)_- = \hbar\hat{W}_2(x,x), \tag{7.174}$$

or equivalently a quadratic relation for cumulants

$$(V'(x)W_1(x))_- = \hbar(W_1(x)^2 + W_2(x,x)). \tag{7.175}$$

where the negative part $(f(x))_-$ means keeping only negative powers of x in the large x expansion. The negative part is also

$$(f(x))_- = f(x) - (f(x))_+, \tag{7.176}$$

where the positive part $(f(x))_+$ means keeping only nonnegative powers of x in the large x expansion. Let us define the following polynomial of x of degree $d - 1$,

$$P_1(x) = (V'(x)W_1(x))_+ ; \qquad (7.177)$$

the loop equations can then be written

$$V'(x)W_1(x) - P_1(x) = \hbar\, W_1(x)^2 + \hbar W_2(x,x). \qquad (7.178)$$

This equation is similar to the quadratic equation (7.102) encountered in Sections 7.2.3.

Split and merge

In general, we are going to find loop equations by writing that the integral of a total derivative vanishes (if there is no boundary term, which is what we included in the definition of 'admissible domains' in Section 7.1.1). In all cases we will have

$$0 = \sum_{i,j} \int dM \frac{\partial}{\partial M_{i,j}} \left(f_{i,j}(M) e^{-\frac{1}{\hbar}\,\mathrm{Tr}\,V(M)} \right), \qquad (7.179)$$

where $f_{i,j}(M)$ is typically a polynomial of coefficients of M, or its generating function $\frac{1}{x-M}$. Depending on whether a power of M (i.e., a coefficient of the resolvent $\frac{1}{x-M}$) appears inside a trace or outside, two cases occur, as follows.

Let A and B be two arbitrary matrices; we then have the following.

- 'Split rule':

$$\sum_{i,j} \frac{\partial}{\partial M_{i,j}} (A\, M^k\, B)_{i,j} = \sum_{l=0}^{k-1} \mathrm{Tr}\, AM^l\ \mathrm{Tr}\, M^{k-1-l}B \qquad (7.180)$$

and in terms of generating series

$$\sum_{i,j} \frac{\partial}{\partial M_{i,j}} (A\, \frac{1}{x-M}\, B)_{i,j} = \mathrm{Tr}\, A\, \frac{1}{x-M}\ \mathrm{Tr}\, \frac{1}{x-M}\, B. \qquad (7.181)$$

- 'Merge rule':

$$\frac{\partial}{\partial M_{i,j}} A_{i,j}\ \mathrm{Tr}\,(M^k B) = \sum_{l=0}^{k-1} \mathrm{Tr}\, AM^l\, B\, M^{k-l-1} \qquad (7.182)$$

and in terms of generating series

$$\frac{\partial}{\partial M_{i,j}} A_{i,j}\ \mathrm{Tr}\, (\frac{1}{x-M}B) = \mathrm{Tr}\, A\frac{1}{x-M}\, B\, \frac{1}{x-M}. \qquad (7.183)$$

- 'Chain rule': these two rules are completed by the chain rule in the case where A or B are themselves functions of M.

    Those three rules are very useful for computing other loop equations, in particular for computing observables that are not symmetric in the eigenvalues, and also in matrix models that are not $U(N)$ invariant.

Hierarchy of loop equations and Virasoro

Let us now look at the general loop equations involving general expectation values. Let $\mu = (\mu_1, \ldots, \mu_n)$ (not necessarily ordered), and let $\mu' = (\mu_2, \ldots, \mu_n)$; we write that

$$0 = \sum_{i,j} \int dM \frac{\partial}{\partial M_{i,j}} \left((M^{\mu_1})_{i,j} \; e^{-\frac{1}{\hbar} \, \mathrm{Tr}\, V(M)} \, p_{\mu'}(M) \right). \tag{7.184}$$

The split-merge rules give (we write $p_\mu = p_\mu(M)$ for short)

$$\frac{1}{\hbar} \sum_{j=0}^{d} t_{j+1} \langle p_{\mu_1+j,\mu'} \rangle = \sum_{l=0}^{\mu_1-1} \langle p_{l,\mu_1-l-1,\mu'} \rangle + \sum_{j=2}^{n} \mu_j \langle p_{\mu_1+\mu_j-1,\mu'\setminus\{\mu_j\}} \rangle \tag{7.185}$$

i.e., a linear relationship for the expectation values

$$\frac{1}{\hbar} \sum_{j=0}^{d} t_{j+1} \hat{T}_{\mu_1+j,\mu'} = \sum_{l=0}^{\mu_1-1} \hat{T}_{l,\mu_1-l-1,\mu'} + \sum_{j=2}^{n} \mu_j \hat{T}_{\mu_1+\mu_j-1,\mu'\setminus\mu_j} \tag{7.186}$$

or a nonlinear relationship for the cumulants

$$\frac{1}{\hbar} \sum_{j=0}^{d} t_{j+1} T_{\mu_1+j,\mu'} = \sum_{l=0}^{\mu_1-1} T_{l,\mu_1-l-1,\mu'}$$

$$+ \sum_{l=0}^{\mu_1-1} \sum_{\nu_1 \uplus \nu_2 = \mu'} T_{l,\nu_1} \, T_{\mu_1-l-1,\nu_2}$$

$$+ \sum_{j=2}^{n} \mu_j T_{\mu_1+\mu_j-1,\mu'\setminus\mu_j}. \tag{7.187}$$

Equivalently in generating functions that gives a linear relationship

$$\Big(V'(x_1) \hat{W}_n(x_1, \ldots, x_n) \Big)_- = \hbar \hat{W}_{n+1}(x_1, x_1, x_2, \ldots, x_n)$$

$$+ \hbar \sum_{j=2}^{n} \frac{d}{dx_j} \frac{\hat{W}_{n-1}(x_1, \ldots, x_n \setminus x_j) - \hat{W}_n(x_2, \ldots, x_n)}{x_1 - x_j}, \tag{7.188}$$

and a quadratic relationship in terms of cumulants

$$(V'(x_1) W_n(x_1, \ldots, x_n))_- = \hbar W_{n+1}(x_1, x_1, x_2, \ldots, x_n)$$

$$+ \hbar \sum_{\nu_1 \uplus \nu_2 = \{x_2, \ldots, x_n\}} W_{1+|\nu_1|}(x_1, \nu_1) \, W_{1+|\nu_2|}(x_1, \nu_2)$$

$$+ \hbar \sum_{j=2}^{n} \frac{d}{dx_j} \frac{W_{n-1}(x_1, \ldots, x_n \setminus x_j) - W_n(x_2, \ldots, x_n)}{x_1 - x_j}. \tag{7.189}$$

Dimension of the space of solutions

The space of possible correlation functions is a linear space of infinite dimension, generated by the $\langle p_\mu(M)\rangle$'s of length $\ell(\mu) \leq N$. Loop equations are linear relationships; i.e., solutions of loop equations form a linear subspace of the space of possible correlations functions:

$$\mathcal{S} = \text{Solution of loop equations} \sim \frac{\{\langle p_\mu\rangle\}}{\text{loop equations} = 0}. \tag{7.190}$$

We will prove in the following that

$$\dim \mathcal{S} = \frac{(N+d-1)!}{(d-1)!\, N!}. \tag{7.191}$$

Indeed, let us prove that $\mathcal{S}$ is generated by the $\langle p_\mu\rangle$s with partition μ fitting in a rectangle of size $N \times (d-1)$, more precisely

$$\mathcal{S} \sim \{\langle p_\mu\rangle \mid \ell(\mu) \leq N \text{ and } \forall i,\, \mu_i \leq d-1\}. \tag{7.192}$$

We prove by recursion on the size $|\mu|$ that any $\langle p_\mu\rangle \in \mathcal{S}$ can be written as a linear combination of $\langle p_\nu\rangle$s with $\ell(\nu) \leq N$ and $\nu_i \leq d-1$. This is clearly true for $|\mu| = 0$. Assume that it holds for all $|\mu'| < |\mu|$. Let us show that it holds for $\langle p_\mu\rangle$. If $\ell(\mu) > N$, we rewrite p_μ as a linear combination of $p_{\mu'}$'s with $|\mu'| = |\mu|$ and $\ell(\mu') \leq N$, so we only need to prove the statement for the case where $\ell(\mu) \leq N$.

Then, assume that $\mu_1 > d-1$, then we can use loop equations to write $\langle p_\mu\rangle$ as a linear combinations of $p_{\mu'}$s with $|\mu'| < |\mu|$, for which the results hold by the recursion hypothesis.

Then, remark that the number of Young tableaux that fit in an $N \times (d-1)$ rectangle is

$$\frac{(N+d-1)!}{(d-1)!\, N!}. \tag{7.193}$$

That ends the proof.

Note that the dimension of solutions of loop equations matches the number of solutions to the saddle-point solutions, and matches the number of admissible eigenvalues integrals as in (7.23). This means that there is a 1:1 correspondence between an integration domain Γ of the eigenvalues, and a solution of loop equations: every solution of loop equation is an eigenvalue integral on an admissible domain and vice versa.

7.3.3   Perturbative topological expansion

We will now assume that there is a $\hbar \to 0$ expansion of the form

$$W_n(x_1, \ldots, x_n) \sim \sum_{g=0}^{\infty} \hbar^{2g-2+n}\, W_{g,n}(x_1, \ldots, x_n). \tag{7.194}$$

The existence of such an expansion, with leading term of order $\hbar^{n-2}$, is far from obvious, and in fact is not true for all potentials. It can hold only for some potentials, or alternatively, for some admissible integration domain of eigenvalues. Let us first solve the case where it holds, and we shall discuss other cases later on.

Leading order of W_1, spectral curve

Since we assume that $\exists \lim_{\hbar \to 0} \hbar W_1(x) = W_{0,1}(x)$ and $\exists \lim_{\hbar \to 0} W_2(x_1, x_2) = W_{0,2}(x_1, x_2)$, to leading order the loop equation (7.178) becomes in the limit

$$W_{0,1}(x)^2 = V'(x)W_{0,1}(x) - P_{0,1}(x), \qquad (7.195)$$

with $P_{0,1}(x) = \lim_{\hbar \to 0} \hbar P_1(x) = (V'(x)W_{0,1}(x))_+$. This equation is formally the same as (7.102), whose solution is

$$W_{0,1}(x) = \frac{1}{2}\left(V'(x) - \sqrt{V'(x)^2 - 4P_{0,1}(x)}\right), \qquad (7.196)$$

where $P_{0,1}(x)$ is a polynomial, found as in section 7.2.2 from Boutroux conditions or from fixed filling fractions.

We thus see that $y = W_{0,1}(x)$ is an algebraic function of x, which means that there is a polynomial relation between x and y:

$$E(x, y) = y^2 - V'(x)y + P_{0,1}(x) = 0. \qquad (7.197)$$

The locus of zeroes of an algebraic relation is a Riemann surface immersed in $\mathbb{C} \times \mathbb{C}$:

$$\Sigma = \{(x, y) \mid E(x, y) = 0\} \subset \mathbb{C} \times \mathbb{C}. \qquad (7.198)$$

It is indeed a complex codimension 1—i.e., a real codimension 2—submanifold of $\mathbb{C} \times \mathbb{C}$; it is thus a surface. It is also called a complex curve. Inheriting the complex structure of $\mathbb{C} \times \mathbb{C}$, it is a 'Riemann surface'.

Σ is called the 'spectral curve'. It can also be viewed as the immersion from an abstract compact Riemann surface $\mathcal{C}$ into $\mathbb{C} \times \mathbb{C}$, i.e., the data of a compact Riemann surface $\mathcal{C}$ and two functions $X : \mathcal{C} \to \mathbb{C}$ and $Y : \mathcal{C} \to \mathbb{C}$ that map $\mathcal{C}$ into $\mathbb{C} \times \mathbb{C}$:

$$\Sigma \leftrightarrow (\mathcal{C}, X, Y), \qquad \{(x, y) \mid E(x, y) = 0\} = \{(X(z), Y(z)) \mid z \in \mathcal{C}\}. \qquad (7.199)$$

Leading order of W_2

The 1st loop equation involving W_2 is

$$2W_1(x_1)W_2(x_1, x_2) + W_3(x_1, x_1, x_2) + \frac{\mathrm{d}}{\mathrm{d}x_2}\frac{W_1(x_1) - W_1(x_2)}{x_1 - x_2}$$

$$= \frac{1}{\hbar}\left(V'(x_1)W_2(x_1, x_2) - P_2(x_1, x_2)\right). \qquad (7.200)$$

To leading order in the topological expansion, W_3 disappears and we get

$$(V'(x_1) - 2W_{0,1}(x_1)) \, W_{0,2}(x_1, x_2) = \frac{\mathrm{d}}{\mathrm{d}x_2} \frac{W_{0,1}(x_1) - W_{0,1}(x_2)}{x_1 - x_2} + P_{0,2}(x_1, x_2). \quad (7.201)$$

Using that $W_{0,1}(x) = \frac{1}{2} \left(V'(x) - \sqrt{V'(x)^2 - 4P_{0,1}(x)} \right)$, this shows that

$$W_{0,2}(x_1, x_2) = \frac{-1}{2(x_1 - x_2)^2} + \frac{Q(x_1, x_2)}{2 \, (x_1 - x_2)^2 \, \sqrt{V'(x_1)^2 - 4P_{0,1}(x_1)} \, \sqrt{V'(x_2)^2 - 4P_{0,1}(x_2)}}, \quad (7.202)$$

where $Q(x_1, x_2)$ is some bivariate polynomial, which must satisfy the constraints

- symmetry: $Q(x_1, x_2) = Q(x_2, x_1)$,
- no pole at $x_1 = x_2 \Rightarrow Q(x, x) = V'(x)^2 - 4P_{0,1}(x)$,
- behavior at $x_1 \to \infty \Rightarrow \deg_{x_1} Q \le d$.

If, in addition, we require the filling fraction condition that

$$\forall x_1, \forall i, \qquad \oint_{x_2 \in \mathcal{A}_i} W_{0,2}(x_1, x_2) = 0, \qquad (7.203)$$

this determines completely $Q(x_1, x_2)$ and hence the polynomial $P_{0,2}(x_1, x_2)$.

In order to better write the final result, it is better to promote $W_{0,2}$ to a differential form, by multiplying it by $\mathrm{d}x_1 \, \mathrm{d}x_2$ (this means the tensor product $\mathrm{d}x_1 \otimes \mathrm{d}x_2$; i.e., it is a linear combination of 1-form in the variable x_1, whose coefficients are 1-forms of x_2. In particular, it is symmetric, not antisymmetric), and shifting by the trivial double pole:

$$B(x_1, x_2) = W_{0,2}(x_1, x_2) \mathrm{d}x_1 \mathrm{d}x_2 + \frac{\mathrm{d}x_1 \mathrm{d}x_2}{(x_1 - x_2)^2}. \qquad (7.204)$$

We thus observe that

- $B(x_1, x_2)$ is a symmetric bilinear differential form, with a double pole at $x_1 = x_2$ in the same sheet (i.e., with the same sign for the square roots):

$$B(x_1, x_2) = \frac{\mathrm{d}x_1 \mathrm{d}x_2}{(x_1 - x_2)^2} + \text{analytic}. \qquad (7.205)$$

- it has no other pole: it has no pole when $x_1 = x_2$ with different signs for the square roots, and it has no pole at $x_1 = \infty$, where it behaves like

$$B(x_1, x_2) \sim O(x_1^{-2} \mathrm{d}x_1) = \text{analytic at } \infty. \qquad (7.206)$$

- it is normalized on $\mathcal{A}$-cycles (the cycles defining the filling fractions as in (7.86)):

$$\forall x_1, \forall i, \qquad \oint_{x_2 \in \mathcal{A}_i} B(x_1, x_2) = 0. \qquad (7.207)$$

On every Riemann surface, there exists a unique symmetric bilinear differential having these properties. It is called the 'fundamental second kind differential', or also sometimes the 'Bergman–Schiffer kernel' (well explained in Fay's lectures (Fay, 1973), an introduction to the algebraic geometry of Riemann surfaces; see also Bergman (1953)).

The final result is, thus, that

$$W_{0,2}(x_1, x_2) dx_1 dx_2 = B(x_1, x_2) - \frac{dx_1 dx_2}{(x_1 - x_2)^2},$$

$$B = \text{fundamental second kind differential.} \tag{7.208}$$

Example. 1-cut case. In the 1-cut case, $V'(x)^2 - 4P_{0,1}(x)$ has only one pair of simple zeroes a and b, and we write

$$V'(x)^2 - 4P_{0,1}(x) = M(x)^2 (x - a)(x - b), \tag{7.209}$$

where $M(x)$ is some polynomial of degree $d - 1$ that contains all the double roots. We thus have

$$W_{0,2}(x_1, x_2) = \frac{-1}{2(x_1 - x_2)^2}$$

$$+ \frac{Q(x_1, x_2)}{2(x_1 - x_2)^2 M(x_1)M(x_2) \sqrt{(x_1 - a)(x_1 - b)} \sqrt{(x_2 - a)(x_2 - b)}}. \tag{7.210}$$

$\mathcal{A}$-cycles are contours surrounding a segment ending at a pair of zeroes of $V'^2 - 4P_{0,1}$, and thus some of them are circles encircling a double zero, i.e., a zero of $M(x)$, and whose contour integral gives a residue at the zero of $M(x)$. Thus, the contour integral condition implies that $Q(x_1, x_2)$ vanishes whenever x_1 (resp. x_2) is a zero of $M(x_1)$ (resp. $M(x_2)$). This implies that $Q(x_1, x_2)$ must be proportional to $M(x_1)M(x_2)$; i.e., there exists a bivariate symmetric polynomial $\tilde{Q}(x_1, x_2) = \frac{Q(x_1, x_2)}{M(x_1)M(x_2)}$, of degree 1 in each variable:

$$W_{0,2}(x_1, x_2) = \frac{-1}{2(x_1 - x_2)^2} + \frac{\tilde{Q}(x_1, x_2)}{2(x_1 - x_2)^2 \sqrt{(x_1 - a)(x_1 - b)} \sqrt{(x_2 - a)(x_2 - b)}}. \tag{7.211}$$

The condition that $\tilde{Q}(x, x) = (x - a)(x - b)$ then determines $\tilde{Q}$:

$$\tilde{Q}(x_1, x_2) = x_1 x_2 - \frac{a + b}{2} (x_1 + x_2) + ab. \tag{7.212}$$

In other words,

$$W_{0,2}(x_1, x_2) = \frac{-1}{2(x_1 - x_2)^2} + \frac{x_1 x_2 - \frac{a+b}{2} (x_1 + x_2) + ab}{2(x_1 - x_2)^2 \sqrt{(x_1 - a)(x_1 - b)} \sqrt{(x_2 - a)(x_2 - b)}}. \tag{7.213}$$

The result is better written using the Joukovski parametrization as in section 7.2.4, i.e., $x(z) = \alpha + \gamma(z + 1/z)$, and $\sqrt{(x - a)(x - b)} = \gamma(z - 1/z)$. Using $dx = x'(z)dz = \gamma(1 - z^{-2})dz$, and using (7.213), this gives in the z_1, z_2 variables

$$B = W_{0,2}(x_1, x_2)\mathrm{d}x_1\mathrm{d}x_2 + \frac{\mathrm{d}x_1\mathrm{d}x_2}{(x_1 - x_2)^2} = \frac{\mathrm{d}z_1\mathrm{d}z_2}{(z_1 - z_2)^2}. \tag{7.214}$$

In the 1-cut case, the fundamental second kind differential is thus simply the double pole in the Joukovski variables. This could have been found beforehand, since this is clearly the unique meromorphic bilinear symmetric differential having a double pole at $z_1 = z_2$ and nowhere else.

7.3.4 Higher orders

$W_{1,1}$

The loop equation (7.178), expanded to the next to leading order, gives

$$2W_{0,1}(x)W_{1,1}(x) + W_{0,2}(x, x) = V'(x)W_{1,1}(x) - P_{1,1}(x), \tag{7.215}$$

i.e.,

$$W_{1,1}(x) = \frac{W_{0,2}(x, x) + P_{1,1}(x)}{V'(x) - 2W_{0,1}(x)} = \frac{W_{0,2}(x, x) + P_{1,1}(x)}{\sqrt{V'(x)^2 - 4P_{0,1}(x)}}. \tag{7.216}$$

The polynomial $P_{1,1}(x)$ has degree $d - 2$. Its $d - 1$ coefficients can be completely fixed by requiring that

$$\forall i = 1, \ldots, d - 1, \qquad \oint_{\mathcal{A}_i} W_{1,1} = 0. \tag{7.217}$$

We will now see our first example of topological recursion by computing $W_{1,1}$. As a warm up, we will do it for the 1-cut case, and using the Joukovski variable.

First, as we have seen for $W_{0,2}$, it is more convenient to consider differential forms, and we define

$$\omega_{1,1}(z) = W_{1,1}(x)\mathrm{d}x, \tag{7.218}$$

where z is the Joukovski variable. Also, as in section 7.2.4, we have that

$$W_{0,1}(x) = y(z) = \frac{1}{2}\left(V'(x) - \sqrt{V'(x)^2 - 4P_{0,1}(x)}\right). \tag{7.219}$$

Observe that changing the sign of the square root is $z \to 1/z$, and thus

$$y(1/z) = \frac{1}{2}\left(V'(x) + \sqrt{V'(x)^2 - 4P_{0,1}(x)}\right), \tag{7.220}$$

and

$$y(1/z) - y(z) = \sqrt{V'(x)^2 - 4P_{0,1}(x)} = V'(x(z)) - 2W_{0,1}(x(z)). \tag{7.221}$$

We will also denote

$$\omega_{0,2}(z_1, z_2) = W_{0,2}(x(z_1), x(z_2))\, dx(z_1)\, dx(z_2) + \frac{dx(z_1)dx(z_2)}{(x(z_1) - x(z_2))^2} = B(z_1, z_2),$$

$$(7.222)$$

or in other words,

$$W_{0,2}(x(z_1), x(z_2))\, dx(z_1)\, dx(z_2) = B(z_1, z_2) - \frac{dx(z_1)dx(z_2)}{(x(z_1) - x(z_2))^2} = -B(z_1, 1/z_2).$$

$$(7.223)$$

The loop equation is thus

$$\omega_{1,1}(z) = \frac{-B(z, 1/z) + P_{1,1}(x(z))\, dx(z)^2}{(y(1/z) - y(z))\, dx(z)},$$

$$(7.224)$$

where $dx(z) = \gamma(1 - z^{-2})dz$. Remark that this is the ratio of a quadratic differential by a 1-form, so it is a 1-form.

We see that $\omega_{1,1}(z)$ may have poles at $z = \pm 1$, but also at all the zeroes of $M(x(z))$, i.e., the zeroes of $y(1/z) - y(z)$. However, assuming that

$$\forall i, \qquad \oint_{\mathcal{A}_i} W_{1,1} = 0$$

$$(7.225)$$

means that $P_{1,1}$ should be chosen so that the residues at the zeroes of $y(1/z) - y(z)$ vanish; i.e., $\omega_{1,1}$ has poles only at $z = \pm 1$.

Having understood that, we write the Cauchy residue theorem

$$\omega_{1,1}(z) = \operatorname*{Res}_{z' \to z} \frac{dz}{z' - z}\, \omega_{1,1}(z');$$

$$(7.226)$$

a residue is a contour integral around z, but moving the integration contour, we can surround all the other poles of $\omega_{1,1}$, i.e., $z' = \pm 1$. Therefore,

$$\omega_{1,1}(z) = - \operatorname*{Res}_{z' \to \pm 1} \frac{dz}{z' - z}\, \omega_{1,1}(z'),$$

$$(7.227)$$

then we use the loop equation, i.e.,

$$\omega_{1,1}(z) = \operatorname*{Res}_{z' \to \pm 1} \frac{dz}{z' - z} \frac{-B(z', 1/z') + P_{1,1}(x(z'))\, dx(z')^2}{(y(1/z') - y(z'))dx(z')}.$$

$$(7.228)$$

Note that the term with $P_{1,1}$ has no pole at $z' = \pm 1$, and $dx(z')$ vanishes at $z' = \pm 1$, so it does not contribute to the residue, and thus

$$\omega_{1,1}(z) = \operatorname*{Res}_{z' \to \pm 1} \frac{dz}{z - z'} \frac{B(z', 1/z')}{(y(1/z') - y(z'))dx(z')}.$$

$$(7.229)$$

Remarking also that we may change the variable $z' \to 1/z'$ we may symmetrize the integrand and write that

$$\omega_{1,1}(z) = \frac{1}{2} \operatorname*{Res}_{z' \to \pm 1} \left(\frac{\mathrm{d}z}{z - z'} - \frac{\mathrm{d}z}{z - \frac{1}{z'}} \right) \frac{B(z', 1/z')}{(y(1/z') - y(z')) \, \mathrm{d}x(z')}. \tag{7.230}$$

This formula makes it possible to compute $W_{1,1}$. This is an example of a more general formula called the 'topological recursion'. The quantity

$$K(z, z') = -\frac{1}{2} \left(\frac{\mathrm{d}z}{z - z'} - \frac{\mathrm{d}z}{z - \frac{1}{z'}} \right) \frac{1}{(y(z') - y(1/z')) \, \mathrm{d}x(z')} \tag{7.231}$$

is called the 'recursion kernel', it was introduced in Eynard (2005) and Chekhov et al. (2006).

General topological recursion, Joukovski case

In the previous section, we were able to compute $W_{1,1}$ (or rather $\omega_{1,1}$) from $W_{0,1}$ and $W_{0,2}$. The general case of $W_{g,n}$ is obtained by a similar reasoning. First, turn $W_{g,n}$ into a differential form,

$$\omega_{g,n}(z_1, \ldots, z_n) = W_{g,n}(x_1, \ldots, x_n) \, \mathrm{d}x_1 \ldots \mathrm{d}x_n + \delta_{g,0} \delta_{n,2} \frac{\mathrm{d}x_1 \mathrm{d}x_2}{(x_1 - x_2)^2}, \tag{7.232}$$

where $x_i = x(z_i)$ with z_i a point of the spectral curve. We also denote

$$W_{g,n}(z, z'; z_2, \ldots, z_n) = \omega_{g-1,n+1}(z, z', z_2, \ldots, z_n) \tag{7.233}$$

$$+ \sum_{g_1+g_2=g, \ I \uplus I'=\{z_2,\ldots,z_n\}}^{\prime} \omega_{g_1,1+|I|}(z, I) \, \omega_{g_2,1+|I'|}(z', I') \Big), \tag{7.234}$$

where $\sum'$ means that we exclude from the sum the terms where $(g_1, I_1) = (0, \emptyset)$ and $(g_2, I_2) = (0, \emptyset)$, $\uplus$ means disjoint union, and $|I|$ is the cardinal of $I \subset \{z_2, \ldots, z_n\}$. Completing the $\sum'$ into a full sum with no excluded terms, we recognize the left-hand side of loop equations (7.189), and thus

$$\omega_{0,1}(z_1)\omega_{g,n}(1/z_1, z_2, \ldots, z_n) + \omega_{0,1}(1/z_1)\omega_{g,n}(z_1, z_2, \ldots, z_n) + W_{g,n}(z_1, 1/z_1; z_2, \ldots, z_n)$$
$$= -P_{g,n}(x(z_1); z_2, \ldots, z_n)\mathrm{d}x(z_1)^2 \mathrm{d}x(z_2) \ldots \mathrm{d}x(z_n)$$
$$+ \mathrm{d}x(z_1)^2 \sum_{j=2}^{n} \mathrm{d}z_j \frac{\mathrm{d}}{\mathrm{d}z_j} \frac{\omega_{g,n-1}(z_2, \ldots, z_n)}{(x(z_1) - x(z_j))\mathrm{d}x(z_j)}. \tag{7.235}$$

By recursion on g and n (in fact by recursion on $2g + n$), these loop equations show that

$$\omega_{g,n}(z_1, \ldots, z_n)$$
$$= \frac{R_{g,n}(x(z_1); z_2, \ldots, z_n) + P_{g,n}(x(z_1); z_2, \ldots, z_n)}{(y(1/z_1) - y(z_1))} \, dx(z_1) \, dx(z_2) \ldots dx(z_n),$$

$$(7.236)$$

where $R_{g,n}$ is some rational fraction of the first variable, with poles only at the branch-points $z_1 = \pm 1$, and possibly at $z_1 = 0$ or $z_1 = \infty$. This shows in particular that $\omega_{g,n}$ is odd under $z_1 \to 1/z_1$:

$$\omega_{g,n}(1/z_1, \ldots, z_n) = -\omega_{g,n}(z_1, \ldots, z_n). \qquad (7.237)$$

Also, (7.236) shows that $\omega_{g,n}$ could possibly have poles at the zeroes of $y(z_1) - y(1/z_1)$, i.e., at the $d-1$ zeroes of $M(x(z_1))$. Again, using the requirement that contour integrals $\oint_{A_i} \omega_{g,n} = t\delta_{g,0}\delta_{n,1}$, we get that such poles have no residue, and since they are simple poles, they actually must be absent. The fact that $R_{g,n} + P_{g,n}$ vanishes at $d-1$ places determines—in principle—the polynomial $P_{g,n}$ of degree $d - 2$. However, this is not the method we will use.

We will rather, like in the previous section, write the Cauchy residue formula

$$\omega_{g,n}(z_1, \ldots, z_n) = \operatorname*{Res}_{z \to z_1} \frac{dz_1}{z - z_1} \omega_{g,n}(z, \ldots, z_n), \qquad (7.238)$$

and we move the integration contour to the only other poles $z = \pm 1$:

$$\omega_{g,n}(z_1, \ldots, z_n) = - \operatorname*{Res}_{z \to \pm 1} \frac{dz_1}{z - z_1} \omega_{g,n}(z, \ldots, z_n). \qquad (7.239)$$

We then use the loop equations (7.235) and note that the terms in the right-hand side of (7.235) have no poles at $z = \pm 1$; therefore,

$$\omega_{g,n}(z_1, \ldots, z_n) = \operatorname*{Res}_{z \to \pm 1} \frac{dz_1}{z - z_1} \frac{W_{g,n}(z, 1/z; z_2, \ldots, z_n)}{(y(z) - y(1/z))dx(z)}. \qquad (7.240)$$

Eventually, the topological recursion is the formula

$$\omega_{g,n+1}(z_1, \ldots, z_{n+1}) = \operatorname*{Res}_{z' \to \pm 1} K(z_1, z') \Big(\omega_{g-1,n+2}(z', 1/z', z_2, \ldots, z_{n+1}) \qquad (7.241)$$

$$+ \sum_{g_1+g_2=g} \sideset{}{'}\sum_{I \uplus I' = \{z_2, \ldots, z_{n+1}\}} \omega_{g_1, 1+|I|}(z', I) \, \omega_{g_2, 1+|I'|}(1/z', I') \Big)$$

$$(7.242)$$

with the same kernel $K(z_1, z)$ as in (7.231).

General topological recursion, general case

The topological recursion works not only for the Joukovski case; it works in every case. Indeed, it relies only on using the Cauchy residue formula

$$w_{g,n}(z_1, \ldots, z_n) = -\operatorname*{Res}_{z \to z_1} S(z_1, z) w_{g,n}(z, z_2, \ldots, z_n), \tag{7.243}$$

where $S(z_1, z)$ is a function of z on the Riemann surface on which the z_j's live, with a simple pole at $z = z_1$, with residu -1, and no other pole. An example of such a function is

$$S(z_1, z) = \int_{z'=o}^{z} B(z_1, z'), \tag{7.244}$$

where B is the fundamental second kind differential of (7.208), and o is an arbitrary base point on the curve. In the Joukovski case, chosing $o = \infty$, this was

$$S(z_1, z) = \frac{dz_1}{z_1 - z} = \int_{z'=\infty}^{z} \frac{dz_1 dz'}{(z_1 - z')^2}.$$

Then, we move the integration contour and proceed like in the Joukovski case.

We then get the same topological recursion for $w_{g,n}$ written in the geometry of the Riemann surface:

- We need to replace the residues at $z' = \pm 1$ by residues at the branchpoints, i.e., the zeroes of dx. Indeed in the Joukovski case, $dx(z') = \gamma(1 - z'^{-2})dz'$ vanishes at $z' = \pm 1$.
- Also the map $z' \mapsto 1/z'$ needs to be replaced by a map $z' \mapsto \sigma(z')$ satisfying

$$x(\sigma(z')) = x(z'). \tag{7.245}$$

  σ is called the local Galois involution; it is usually well defined only in the vicinity of branchpoints. It is the map that exchanges the two branches meeting at the branchpoint. In our examples, it changed the sign of the square root. In the Joukovski case, however, $\sigma(z') = 1/z'$ is well defined globally on $\mathbb{C}^*$.
- The fundamental second kind differential $B(z_1, z_2)$ is canonically defined on any Riemann surface with marked cycles $\mathcal{A}_i$s (see Fay (1973)). The 1-form $w_{0,1} = ydx$ is also well defined.
- The kernel is then

$$K(z_1, z) = \frac{1}{2} \frac{\int_{z'=\sigma(z)}^{z} B(z_1, z')}{w_{0,1}(z) - w_{0,1}(\sigma(z))}. \tag{7.246}$$

- The general topological recursion formula is then

$$w_{g,n+1}(z_1, \ldots, z_{n+1}) = \sum_{a} \operatorname*{Res}_{z' \to a} K_a(z_1, z') \left(w_{g-1,n+2}(z', \sigma_a(z'), z_2, \ldots, z_{n+1}) \right. \tag{7.247}$$

$$\left. + \sum_{g_1+g_2=g} \sum_{I \uplus I'=\{z_2,\ldots,z_{n+1}\}}^{'} w_{g_1,1+|I|}(z', I) w_{g_2,1+|I'|}(\sigma_a(z'), I') \right).$$

$$\tag{7.248}$$

This formula makes it possible to compute all $w_{g,n}$s by recursion on $2g + n$.

It solves the loop equations, not only of the 1-matrix model, but also of many other matrix models in fact, all generalizations mentioned in section 7.1.1:

- all normal matrix models $H_N(\gamma)$ with eigenvalues on some contours γ have their correlation function's expansion computed by the topological recursion; indeed since loop equations come from integration by parts, they are independent of the contour.
- many sorts of multimatrix models, including 2-matrix model, chain matrix model, $O(n)$ matrix model, and many others, have their correlation function's expansions computed by the topological recursion.
- non-Hermitian matrix models can also be computed by topological recursion, requiring only an adaptation of what B is.

Topological recursion beyond matrix models

In fact, many other things satisfy the topological recursion, in various fields of mathematics and physics. For instance, Gromov–Witten invariants, knot theory invariants (Jones polynomials) etc. satisfy the same topological recursion that solves matrix models. A famous example of topological recursion not coming from matrix models is Mirzakhani's recursion (she received the 2014 Fields Medal for it):

- **Mirzakhani's recursion** Let (g, n) be such that $2g - 2 + n > 0$. Let $\mathcal{M}_{g,n}(L_1, \ldots, L_n)$ be moduli space of hyperbolic surfaces (equipped with a metric of constant negative curvature -1), of genus g, and with n labeled geodesic boundaries, of respective lengths $L_1, \ldots, L_n$. A surface of genus g with n boundaries can be cut into $2g - 2 + n$ pairs of pants, along $3g - 3 + n$ closed and nonintersecting geodesics, of respective lengths $l_1, \ldots, l_{3g-3+n}$. Reciprocally, one can glue hyperbolic pairs of pants along their geodesic boundaries, provided that the glued geodesic lengths match, and in doing that one may rotate the $3g - 3 + n$ glued boundaries by some angles $\theta_1, \ldots, \theta_{3g-3+n}$. The $3g - 3 + n$ geodesic lengths and angles (l_i, θ_i) parametrize the gluing; they are a local coordinate system of $\mathcal{M}_{g,n}(L_1, \ldots, L_n)$. However, a cutting into pairs of pants is not unique, and those coordinates are only locally well defined, not globally.

However, the following volume form on $\mathcal{M}_{g,n}(L_1, \ldots, L_n)$ is well defined:

$$\Omega = \prod_{i=1}^{3g-3+n} dl_i \wedge d\theta_i. \tag{7.249}$$

It is independent of a choice of pants decomposition, and is called the Weil–Petersson volume form. This makes it possible to define the Weil–Petersson hyperbolic volumes

$$\mathcal{V}_{g,n}(L_1, \ldots, L_n) = \int_{\mathcal{M}_{g,n}(L_1,\ldots,L_n)} \Omega. \tag{7.250}$$

One can also equivalently be interested in their Laplace transforms

$$W_{g,n}(z_1,\ldots,z_n) = \int_0^\infty L_1 dL_1 \ldots \int_0^\infty L_n dL_n V_{g,n}(L_1,\ldots,L_n)\prod_{i=1}^n e^{-z_i L_i}, \quad (7.251)$$

where z_i are complex numbers, assumed to have $\Re z_i > 0$ for the integrals to converge. We also make differential forms by multiplying with $dz_1 \ldots dz_n$:

$$\omega_{g,n}(z_1,\ldots,z_n) = W_{g,n}(z_1,\ldots,z_n)dz_1 \ldots dz_n. \quad (7.252)$$

The challenge was to be able to compute these. Some cases are simple; for example when $(g,n) = (0,3)$, a surface in $\mathcal{M}_{0,3}(L_1,L_2,L_3)$ is already a pair of pants, it cannot be further decomposed, and one can prove that there exists a unique hyperbolic pair of pants of given boundary lengths. Therefore, $\mathcal{M}_{0,3}(L_1,L_2,L_3)$ is a set with a unique element, its volume is defined to be 1, and its Laplace transform is

$$V_{0,3}(L_1,L_2,L_3) = 1 \quad, \quad \omega_{0,3}(z_1,z_2,z_3) = \frac{dz_1 dz_2 dz_3}{z_1^2 z_2^2 z_3^3}. \quad (7.253)$$

In 2004, M. Mirzakhani discovered a recursion relation (on $2g - 2 + n$), making it possible to compute all volumes recursively. This recursion was rewritten in Eynard and Orantin (2007b) in terms of Laplace transform and is an instance of the general topological recursion

$$\omega_{g,n}(z_1,\ldots,z_n) = \operatorname*{Res}_{z\to 0} K(z_1,z)\Big(\omega_{g-1,n+1}(z,-z,z_2,\ldots,z_n)$$

$$+ \sum_{\substack{g_1+g-2=g,\, I_1\uplus I_2=\{z_2,\ldots,z_n\}}}' \omega_{g_1,1+|I_1|}(z,I_1)\omega_{g_2,1+|I_2|}(-z,I_2)\Big) \quad (7.254)$$

with

$$\omega_{0,2}(z_1,z_2) = \frac{dz_1 dz_2}{(z_1-z_2)^2}, \quad K(z_1,z) = \frac{\int_{z'=-z}^z \omega_{0,2}(z_1,z')}{\frac{-2z}{\pi}\sin(2\pi z)dz} = \frac{-\pi dz_1}{(z_1^2-z^2)\sin(2\pi z)dz}. \quad (7.255)$$

As an example let us compute $\omega_{1,1}(z_1)$ and $V_{1,1}(L_1)$:

$$\omega_{1,1}(z_1) = \operatorname*{Res}_{z\to 0} K(z_1,z)\omega_{0,2}(z,-z)$$

$$= \operatorname*{Res}_{z\to 0} K(z_1,z)\frac{-dz^2}{4z^2}$$

$$= \operatorname*{Res}_{z\to 0} \frac{dz_1}{z_1^2-z^2}\frac{1}{\frac{1}{\pi}\sin(2\pi z)dz}\frac{dz^2}{4z^2}$$

$$= \frac{dz_1}{8z_1^2}\operatorname*{Res}_{z\to 0}\frac{dz}{z^3}\left(1-\frac{z^2}{z_1^2}\right)^{-1}\left(\frac{\sin 2\pi z}{2\pi z}\right)^{-1}$$

$$= \frac{dz_1}{8z_1^2}\operatorname*{Res}_{z\to 0}\frac{dz}{z^3}\left(1+\frac{z^2}{z_1^2}+\frac{(2\pi z)^2}{6}+O(z^4)\right)$$

$$= \frac{\mathrm{d}z_1}{8z_1^2}\left(\frac{1}{z_1^2} + \frac{4\pi^2}{6}\right). \tag{7.256}$$

Its inverse Laplace transform gives

$$\mathcal{V}_{1,1}(L_1) = \frac{1}{24}\left(2\pi^2 + \frac{L_1^2}{2}\right). \tag{7.257}$$

There are many other applications of topological recursion, beyond the scope of this chapter, particularly toward algebraic geometry, cohomological field theories, and knot theory. See Eynard and Orantin (2008) for a review.

References

Bergman, S. and Schiffer, M. (1953). *Kernel Functions and Elliptic Differential Equations in Mathematical Physics*. Academic Press, New York.

Chekhov, L. Eynard, B. Orantin, N. (2006). Free energy topological expansion for the 2-matrix model, JHEP 0612 053, math-ph/0603003.

David, F. (1990). Loop equations and nonperturbative effects in two-dimensional quantum gravity. *Mod.Phys.Lett.* **A5**, 1019.

Di Francesco, P. Ginsparg, P., Zinn-Justin, J. (1995). 2D gravity and random matrices. *Phys. Rep.* **254**, 1.

Eynard, B. (2005). Topological expansion for the 1-Hermitian matrix model correlation functions. **JHEP** 112004. arXiv:hep-th/0407261.

Eynard, B. (2008). All orders asymptotic expansion of large partitions, math-ph: arxiv.0804.0381.

Eynard, B., and Orantin, N. (2007a). Invariants of algebraic curves and topological expansion, *Commun. Number Theory Phys.* **1**(2), 347–452. arXiv:math-ph/0702045.

Eynard, B., and Orantin, N. (2007b). Weil-Peterson volume of moduli space, Mirzakhani's recursion and matrix models. math.ph: arXiv:0705.3600v1.

Eynard, B., and Orantin, N. (2008). Algebraic methods in random matrices and enumerative geometry. arXiv:0811.3531.

Fay, J. D. (1973). *Theta functions on Riemann surfaces*. Springer Verlag, Berlin.

Gaudin, M. (1997). *La fonction d'onde de Bethe*. MASSON.

Mehta, M. L. (2004). *Random Matrices*, 3rd ed., Pure and Applied Mathematics, vol 142. Elsevier/Academic, Amsterdam.

Migdal, A.A. (1983). *Phys. Rep.* **102**, 199.

Romik, D., and Śniady, P. (2015). Jeu de taquin dynamics on infinite Young tableaux and second class particles. *Ann. Probab.* **43**, 682–737. arXiv1111.0575

Vershik, A. M., and Kerov, S. V. (1977). Asymptotics of the Plancherel measure of the symmetric group and the limiting form of Young tableau. *Dokl. AN SSSR* **233**(6), 1024–7; English translation *Sov. Math. Dokl.* **18**, 527–31.

8

Random matrices and number theory: some recent themes

Jon P. KEATING

School of Mathematics,
University of Bristol,
Bristol BS8 1TW, UK

Keating, J.P., 'Random Matrices and Number Theory: Some Recent Themes' in *Stochastic Processes and Random Matrices*. Edited by: Grégory Schehr et al, Oxford University Press (2017).
© Oxford University Press 2017. DOI 10.1093/oso/9780198797319.003.0008

Chapter Contents

The aim of this chapter is to motivate and describe some recent developments concerning the applications of random matrix theory to problems in number theory. The first section provides a brief and rather selective introduction to the theory of the Riemann zeta-function, in particular to those parts needed to understand the connections with random matrix theory. The second section focuses on the value distribution of the zeta function on its critical line, specifically on recent progress in understanding the extreme value statistics gained through a conjectural link to log-correlated Gaussian random fields and the statistical mechanics of glasses. The third section outlines some number-theoretic problems that can be resolved in function fields using random matrix methods. In this latter case, random matrix theory provides the only route we currently have for calculating certain important arithmetic statistics rigorously and unconditionally. These sections formed the basis for three lectures delivered at the Les Houches School on *Stochastic Processes and Random Matrices*. They, therefore, represent three connected narratives, rather than a systematic and comprehensive review of the field. They are written with the random matrix community in mind as the primary audience and are likely to be most profitably read in the context of previous overviews, for example, Keating (1993), Keating and Snaith (2003), Keating (2005), and Keating and Snaith (2011).

8.1   The Riemann zeta function

The Riemann zeta function provides a canonical example (historically, the first) of the use of random matrix theory in number theory. In this first section I review selected parts of the theory of the zeta function needed to understand this connection. For a more complete introduction, see, for example, Heath-Brown (2006) and Titchmarsh (1986).

8.1.1   Definition and basic properties

The Riemann zeta function is defined by the *Dirichlet series*

$$\zeta(s) = \sum_{n=1}^{\infty} \frac{1}{n^s}, \tag{8.1}$$

when $\mathrm{Re}(s) > 1$, where the series converges absolutely.

One reason why the zeta function is important is that it encodes information about the distribution of the primes p. This follows because one can also express it as a product over the primes (an *Euler product*)

$$\zeta(s) = \prod_{p} \left(1 - \frac{1}{p^s}\right)^{-1}, \tag{8.2}$$

which again converges when $\mathrm{Re}(s) > 1$. To see the connection with (8.1), observe that

$$\left(1 - \frac{1}{p^s}\right)^{-1} = 1 + \frac{1}{p^s} + \frac{1}{p^{2s}} + \frac{1}{p^{3s}} + \cdots, \tag{8.3}$$

and that the product then generates all multiplicative combinations of the primes and their powers, and so, by the fundamental theorem of arithmetic, all of the positive integers.

The zeta function has an analytic continuation to the rest of the complex s-plane, except for the point $s = 1$, where it has a simple pole. It satisfies a *functional equation* that is essentially a reflection symmetry about the line $\mathrm{Re}(s) = \frac{1}{2}$. Specifically,

$$\pi^{-s/2}\Gamma\left(\frac{s}{2}\right)\zeta(s) = \pi^{-(1-s)/2}\Gamma\left(\frac{1-s}{2}\right)\zeta(1-s). \tag{8.4}$$

It follows from the Euler product that the zeta function has no zeros with $\mathrm{Re}(s) > 1$; the product converges in this half-plane and none of its factors vanish there. It then follows from the functional equation that the only zeros with $\mathrm{Re}(s) < 0$ are at $s = -2, -4, -6, \ldots$ (coming from the poles of the Γ-function). These are known as the *trivial zeros*. Any other zeros must therefore lie in the *critical strip* $0 \leq \mathrm{Re}(s) \leq 1$. It turns out that there are infinitely many such *nontrivial zeros* (I will explain the idea behind one proof of this fact later in Section 8.1.4.)

We denote the nontrivial zeros by ρ_n, so $\zeta(\rho_n) = 0$ and $0 \leq \mathrm{Re}(\rho_n) \leq 1$. The *Riemann hypothesis* asserts that

$$\mathrm{Re}\rho_n = \frac{1}{2} \ \forall n. \tag{8.5}$$

Put another way, setting $\rho_n = \frac{1}{2} + it_n$, the Riemann hypothesis is that

$$t_n \in \mathbb{R} \ \forall n. \tag{8.6}$$

The Riemann hypothesis is known to be true for over 42% of the zeros (cf. Conrey (1989) and subsequent refinements), including the first 10^{13} of them and others in long ranges higher up on the critical strip.

8.1.2 Explicit formulae

One of the reasons the nontrivial zeros play a central role in number theory is that they govern the distribution of the primes. To see this, observe that, defining the *von Mangoldt function* by $\Lambda(n) = \log p$ if n is a power of the prime p (i.e., if $n = p^k$) and 0 otherwise,

$$\frac{\zeta'(s)}{\zeta(s)} = -\sum_{n=1}^{\infty} \frac{\Lambda(n)}{n^s} \tag{8.7}$$

when $\mathrm{Re}(s) > 1$. This follows from computing the logarithmic derivative of the Euler product (8.2). The left-hand side of this equation has poles at the pole of the zeta function at $s = 1$ and at the trivial and nontrivial zeros; the right-hand side is a sum over the primes.

One application of this formula is to count the primes in long intervals. To this end, consider

$$\psi(x) = \sum_{n \le x} \Lambda(n). \tag{8.8}$$

This is a counting function with jumps at values of x that are powers of primes. At the jumps the value is taken to be the average of the left-hand and right-hand limits. We have that for $c > 1$

$$\psi(x) = \frac{1}{2\pi i} \int_{c-i\infty}^{c+i\infty} \sum_{n=1}^{\infty} \frac{\Lambda(n)}{n^s} \frac{x^s}{s} ds. \tag{8.9}$$

This follows because on the line of integration the sum converges absolutely and so one can interchange the integral and the sum; the integral of the summand is $\Lambda(n)$ when $n < x$ (when the line of integration can be moved to the left, picking up a contribution from the pole at $s = 0$), 0 when $n > x$ (when the line of integration can be moved to the right, where there are no poles), and $\Lambda(n)/2$ when $n = x$. Now using (8.7), we have that

$$\psi(x) = -\frac{1}{2\pi i} \int_{c-i\infty}^{c+i\infty} \frac{\zeta'(s)}{\zeta(s)} \frac{x^s}{s} ds. \tag{8.10}$$

Shifting the contour to the left in this case, one picks up contributions from the poles at $s = 1$ (the pole of $\zeta(s)$), the nontrivial zeros, the pole at $s = 0$, and from the poles associated with the trivial zeros, giving

$$\psi(x) = x - \sum_{n} \frac{x^{\rho_n}}{\rho_n} - \frac{\zeta'(0)}{\zeta(0)} + \sum_{n=1}^{\infty} \frac{x^{-2n}}{2n}. \tag{8.11}$$

Hence, summing the final series,

$$\psi(x) = x - \sum_{n} \frac{x^{\rho_n}}{\rho_n} - \frac{\zeta'(0)}{\zeta(0)} - \frac{1}{2} \log \left(1 - x^{-2}\right). \tag{8.12}$$

The value of the constant term here may be calculated to be

$$-\frac{\zeta'(0)}{\zeta(0)} = -\log 2\pi. \tag{8.13}$$

We thus have an explicit formula for the prime sum $\psi(x)$.

The contribution from the trivial zeros is negligible when x is large. It can be shown that $0 < \mathrm{Re}\,\rho_n < 1 \ \forall n$, and that the density of the nontrivial zeros is not too large (see Section 8.1.4), so that

$$\sum_{n} \frac{x^{\rho_n}}{\rho_n} = o(x). \tag{8.14}$$

We thus have the *prime number theorem*

$$\psi(x) = x + o(x).\tag{8.15}$$

The Riemann hypothesis is then equivalent to

$$\psi(x) = x + O(x^{\frac{1}{2}+\epsilon})\tag{8.16}$$

for any $\epsilon > 0$.

8.1.3   *L-functions*

Just as the Riemann zeta function acts as a generating function for counting the primes in long intervals, other counting problems in number theory can be related to generalizations of the zeta function, known as *L-functions*. For example, *L-functions* can be used to count primes in arithmetic progressions, solutions of Diophantine equations, etc. These *L-functions* each have Dirichlet series representation, an Euler product representation, and a functional equation, and they each satisfy a Riemann hypothesis. Specifically, they are understood to satisfy a general set of axioms due to Selberg. Let S denote the Selberg class *L-functions*. For $F \in S$ primitive,

$$F(s) = \sum_{n=1}^{\infty} \frac{a_F(n)}{n^s},$$

let $m_F \geq 0$ be the order of the pole at $s = 1$ (for primitive *L-functions* other than the Riemann zeta function, $m_F = 0$),

$$\frac{F'}{F}(s) = -\sum_{n=1}^{\infty} \frac{\Lambda_F(n)}{n^s} \quad \text{and} \quad F(s)^{-1} = \sum_{n=1}^{\infty} \frac{\mu_F(n)}{n^s} \quad (\mathrm{Re}(s) > 1).$$

The function $F(s)$ has an Euler product

$$F(s) = \prod_{p} \exp\left(\sum_{l=1}^{\infty} \frac{b_F(p^l)}{p^{ls}} \right)\tag{8.17}$$

and satisfies a functional equation

$$\Phi(s) = \varepsilon_F \overline{\Phi}(1-s),$$

where

$$\Phi(s) = Q^s \left(\prod_{j=1}^{r} \Gamma(\lambda_j s + \mu_j) \right) F(s),$$

with some $Q > 0$, $\lambda_j > 0$, $\mathrm{Re}(\mu_j) \geq 0$, and $|\varepsilon_F| = 1$. Here $\overline{\Phi}(s) = \overline{\Phi(\bar{s})}$. The functional equation may also be written in the form

$$F(s) = X(s)\overline{F}(1-s),$$

where

$$X(s) = \varepsilon_F Q^{1-2s} \prod_{j=1}^{r} \frac{\Gamma(\lambda_j(1-s) + \overline{\mu_j})}{\Gamma(\lambda_j s + \mu_j)}.$$

The two important invariants of $F(s)$ are the degree d_F and the conductor $\mathfrak{q}_F$,

$$d_F = 2 \sum_{j=1}^{r} \lambda_j \quad \text{and} \quad \mathfrak{q}_F = (2\pi)^{d_F} Q^2 \prod_{j=1}^{r} \lambda_j^{2\lambda_j}.$$

The Riemann zeta function has $d_F = \mathfrak{q}_F = 1$. Elliptic curve L-functions have $d_F = 2$ and are then characterized by their conductor.

For $F \in \mathcal{S}$, it is expected that a generalized prime number theorem of the form

$$\psi_F(x) := \sum_{n \le x} \Lambda_F(n) = m_F x + o(x) \tag{8.18}$$

holds.

To give an example, let

$$\chi_d(p) = \left(\frac{d}{p}\right) = \begin{cases} +1 & \text{if } p \nmid d \text{ and } x^2 \equiv d \ (\text{mod } p) \ \text{ solvable} \\ 0 & \text{if } p|d \\ -1 & \text{if } p \nmid d \text{ and } x^2 \equiv d \ (\text{mod } p) \ \text{ not solvable} \end{cases} \tag{8.19}$$

denote the Legendre symbol. Then define

$$L_D(s, \chi_d) = \prod_{p} \left(1 - \frac{\chi_d(p)}{p^s}\right)^{-1}$$

$$= \sum_{n=1}^{\infty} \frac{\chi_d(n)}{n^s}, \tag{8.20}$$

where the product is over the prime numbers p. These functions form a family of L-functions parameterized by the integer index d. The Riemann zeta function is itself a member of this family $(d = 1)$.

8.1.4 Counting zeros

Let us denote the counting function of the nontrivial zeros by

$$N(T) = \#\{n : 0 < \ \text{Re}(t_n) \le T\}. \tag{8.21}$$

One can obtain a formula for $N(T)$ by integrating $\zeta'(s)/\zeta(s)$ around a rectangle containing the zeros with $0 < \ \text{Re}(t_n) \le T$. If this rectangle is reflection-symmetric around the critical line $\ \text{Re}(s) = \frac{1}{2}$, one can use the functional equation to relate the left half to the right half. The integral can then be evaluated straightforwardly, giving

$$N(T) = \bar{N}(T) + S(T) \tag{8.22}$$

with

$$\bar{N}(T) = 1 + \frac{1}{\pi} \operatorname{Im} \log \left\{ \pi^{-iT/2} \Gamma(\tfrac{1}{4} + \tfrac{1}{2}iT) \right\} \tag{8.23}$$

and

$$S(T) = \frac{1}{\pi} \operatorname{Im} \log \zeta(\tfrac{1}{2} + iT). \tag{8.24}$$

It follows from Stirling's formula that

$$\bar{N}(T) = \frac{T}{2\pi} \log \frac{T}{2\pi} - \frac{T}{2\pi} + \frac{7}{8} + O\left(\frac{1}{T}\right) \tag{8.25}$$

and it can be shown that

$$S(T) = O(\log T); \tag{8.26}$$

therefore,

$$N(T) \sim \frac{T}{2\pi} \log \frac{T}{2\pi} - \frac{T}{2\pi}. \tag{8.27}$$

Clearly, $N(T) \to \infty$ as $T \to \infty$, proving that there are infinitely many nontrivial zeros. However, their asymptotic density is sufficiently small to ensure the bound (8.14). Substituting the Euler product representation into (8.24) gives the explicit formula

$$N(T) = \bar{N}(T) - \frac{1}{\pi} \sum_{n=1}^{\infty} \frac{\Lambda(n)}{\log n \sqrt{n}} \sin(T \log n), \tag{8.28}$$

although this is to be interpreted formally because the product does not converge on the critical line. It can be made rigorous by integrating against a suitable test function, as we now describe.

8.1.5 General explicit formula

The explicit formulae of the previous two sections are examples of a general class of expressions that form the basis of the rigorous study of the statistics of the zeros. Let us set

$$\tilde{\Gamma}(s) = \pi^{-s/2} \Gamma(s/2), \tag{8.29}$$

and

$$\tilde{\zeta}(s) = \tilde{\Gamma}(s)\zeta(s). \tag{8.30}$$

Let $g \in C_c^{\infty}(\mathbb{R})$ be a compactly supported smooth test function, which we assume is even $g(-x) = g(x)$. Set

$$h(z) = \int_{-\infty}^{\infty} g(x) \exp(-izx) \mathrm{d}x. \tag{8.31}$$

Therefore, $h(z)$ is an entire function of z that decays exponentially for z real.

The general explicit formula is then as follows:

$$\sum_m h(t_m) - 2h\left(\tfrac{i}{2}\right) = \frac{1}{2\pi} \int_{-\infty}^{\infty} h(z) 2 \operatorname{Re} \frac{\tilde{\Gamma}'}{\tilde{\Gamma}}\left(\frac{1}{2} + iz\right) dz - 2 \sum_n \frac{\Lambda(n)}{\sqrt{n}} g(\log n). \quad (8.32)$$

To prove this formula one calculates

$$\frac{1}{2\pi i} \int_{c-i\infty}^{c+i\infty} H(s) \frac{\tilde{\zeta}'}{\tilde{\zeta}}(s) ds \quad (8.33)$$

for some $c > 1$, where

$$H(s) = \int_{-\infty}^{\infty} g(x) \exp\left((s - \tfrac{1}{2})x\right) dx, \quad (8.34)$$

i.e., $h(z) = H(\tfrac{1}{2} - iz)$. First, one uses

$$\frac{\tilde{\zeta}'}{\tilde{\zeta}}(s) = \frac{\tilde{\Gamma}'}{\tilde{\Gamma}}(s) - \sum_{n=1}^{\infty} \frac{\Lambda(n)}{n^s}, \quad (8.35)$$

interchanging the integral and the sum, and then shifting the contour to the critical line $\operatorname{Re}(s) = \tfrac{1}{2}$. This gives the terms on the right-hand side of (8.32). The terms on the left-hand side come from sifting the contour in (8.33) to $\operatorname{Re}(s) = 1 - c$, picking up contributions from the poles at $s = 0, 1$ and $s = \rho_m$, and then using the functional equation to relate the integral on the new contour to the original one.

A corresponding formula holds for other L-functions.

8.1.6   Connection with random matrices: Montgomery's pair-correlation conjecture

In this section I will, for ease of presentation, assume the Riemann hypothesis to be true. This is not strictly necessary—it simply makes some of the formulae more transparent.

It follows from (8.27) that the mean density of the nontrivial zeros increases logarithmically with height t up the critical line. Specifically, the *unfolded* zeros

$$w_n = t_n \frac{1}{2\pi} \log \frac{|t_n|}{2\pi} \quad (8.36)$$

satisfy

$$\lim_{W \to \infty} \frac{1}{W} \#\{w_n \in [0, W]\} = 1; \quad (8.37)$$

that is, the mean of $w_{n+1} - w_n$ is 1.

The connection between random matrix theory and number theory was first made by Montgomery (1973), who conjectured that

$$\lim_{W \to \infty} \frac{1}{W} \#\{w_n, w_m \in [0, W] : \alpha \le w_n - w_m < \beta\} = \int_\alpha^\beta \left(\delta(x) + 1 - \frac{\sin^2(\pi x)}{\pi^2 x^2} \right) dx.$$
(8.38)

This conjecture was motivated by a theorem Montgomery proved in the same paper which may be restated as

$$\lim_{N \to \infty} \frac{1}{N} \sum_{n,m \le N} f(w_n - w_m) = \int_{-\infty}^\infty f(x) \left(\delta(x) + 1 - \frac{\sin^2(\pi x)}{\pi^2 x^2} \right) dx \qquad (8.39)$$

for all test functions $f(x)$ whose Fourier transforms

$$\widehat{f}(\tau) = \int_{-\infty}^\infty f(x) \exp(2\pi i x \tau) dx \qquad (8.40)$$

have support in the range $(-1, 1)$ and are such that the sum and integral in (8.39) converge. The generalized form of the Montgomery conjecture is that (8.39) holds for all test functions such that the sum and integral converge, without any restriction on the support of $\widehat{f}(\tau)$. The form of the conjecture (8.38) then corresponds to the particular case in which $f(x)$ is taken to be the indicator function on the interval $[\alpha, \beta)$ (and so does not fall within the class of test functions covered by the theorem).

The connection to random matrix theory follows from the observation that the pair correlation of the nontrivial zeros conjectured by Montgomery coincides precisely with that which holds for the eigenvalues of random matrices taken from either the circular unitary ensemble (CUE) or the Gaussian unitary ensemble (GUE) of random matrices (i.e., random unitary or Hermitian matrices) in the limit of large matrix size (see, e.g., Dyson (1962) or Mehta (1991)). For example, let A be an $N \times N$ unitary matrix, so that $A(A^T)^* = AA^\dagger = I$. The eigenvalues of A lie on the unit circle; that is, they may be expressed in the form $e^{i\theta_n}$, $\theta_n \in \mathbb{R}$. Scaling the eigenphases θ_n so that they have unit mean spacing,

$$\phi_n = \theta_n \frac{N}{2\pi}, \qquad (8.41)$$

the two-point correlation function for a given matrix A may be defined as

$$R_2(A; x) = \frac{1}{N} \sum_{n=1}^N \sum_{m=1}^N \sum_{k=-\infty}^\infty \delta(x + kN - \phi_n + \phi_m), \qquad (8.42)$$

so that

$$\frac{1}{N} \sum_{n,m} f(\phi_n - \phi_m) = \int_0^N R_2(A; x) f(x) dx. \qquad (8.43)$$

$R_2(A; x)$ is clearly periodic in x, so can be expressed as a Fourier series:

$$R_2(A; x) = \frac{1}{N^2} \sum_{k=-\infty}^{\infty} |\operatorname{Tr}A^k|^2 e^{2\pi ikx/N}. \tag{8.44}$$

The CUE corresponds taking matrices from $U(N)$ with a probability measure given by the normalized Haar measure on the group. It follows from (8.44) that the CUE average of $R_2(A; x)$ may be evaluated by computing the corresponding average of the Fourier coefficients $|\operatorname{Tr}A^k|^2$. This was done in Dyson (1962):

$$\int_{U(N)} |\operatorname{Tr}A^k|^2 d\mu(A) = \begin{cases} N^2 & k = 0 \\ |k| & |k| \le N \\ N & |k| > N. \end{cases} \tag{8.45}$$

It follows that

$$\lim_{N\to\infty} \int_{U(N)} R_2(A; x)d\mu(A) = \delta(x) + 1 - \frac{\sin^2(\pi x)}{\pi^2 x^2}. \tag{8.46}$$

The idea behind the proof of Montgomery's theorem is to use the explicit formula to write the sum over pairs of zeros as a sum over pairs of primes and prime powers. The diagonal terms contribute a single sum over primes and their powers which can be evaluated using the prime number theorem, giving exactly the right-hand side of (8.39) for test functions satisfying the conditions of the theorem. It can then be shown that the off-diagonal terms do not contribute to leading asymptotic order, and so this establishes the result. For test functions such that $\hat{f}(\tau)$ has support outside (-1, 1), the off-diagonal terms do make a contribution at leading order. This contribution cannot be evaluated rigorously, but heuristically it can be calculated using a conjecture of Hardy and Littlewood (1923) concerning the pair correlation of the von Mangoldt function,

$$\sum_{n \le X} \Lambda(n)\Lambda(n + h) \sim XC_{\text{HL}}(h), \tag{8.47}$$

as $X \to \infty$, where $C_{\text{HL}}(h) = 0$ when h is odd, and when h is even

$$C_{\text{HL}}(h) = 2 \prod_{p>2} \left(1 - \frac{1}{(p-1)^2}\right) \prod_{q>2, q|h} \left(\frac{q-1}{q-2}\right). \tag{8.48}$$

These calculations extend to the the the n-point correlation functions of the Riemann zeros for all n (Bogomolny and Keating 1995, 1996a; Rudnick and Sarnak 1996). In each case the results agree with the corresponding random-matrix formulae. This leads to the general belief that all local statistics of the zeros (i.e., fluctuations on the scale of the mean spacing) coincide with those of the eigenvalues of random unitary matrices in the respective limits of large height up the critical line and large matrix size. There is extensive numerical evidence in support of this general belief (Odlyzko 1989).

It is worth emphasizing that zero statistics are expected to coincide with random-matrix eigenvalue statistics when correlations are measured over ranges that are fixed on the scale of the mean zero separation and in the limit as the height up the critical line tends to infinity. It is possible to use the explicit formula and the Hardy–Littlewood conjecture to derive a formula for the two-point correlation function that describes the correlations uniformly in the correlation scale and the height up the critical line: for f a suitable even test function

$$
\sum_{0<t_n,t_m\leq T} f(t_n - t_m) = \frac{f(0)}{2\pi}\int_0^T \log\frac{t}{2\pi}\,dt + \frac{1}{(2\pi)^2}\int_0^T\int_{-T}^T f(\eta)\left[\left(\log\frac{t}{2\pi}\right)^2 \right.
$$

$$
\left. +2\left(\left(\frac{\zeta'}{\zeta}\right)'(1+i\eta) + \left(\frac{t}{2\pi}\right)^{-i\eta} A(i\eta)\zeta(1-i\eta)\zeta(1+i\eta) - B(i\eta)\right)\right]d\eta\,dt + o(T),
$$

$$(8.49)$$

where

$$
A(r) = \prod_p \frac{(1-\frac{1}{p^{1+r}})(1-\frac{2}{p}+\frac{1}{p^{1+r}})}{(1-\frac{1}{p})^2}
$$

and

$$
B(r) = \sum_p \left(\frac{\log p}{p^{1+r}-1}\right)^2.
$$

Here the integral is to be regarded as a principal value near $\eta = 0$. This formula was originally obtained in Bogomolny and Keating (1996b) using the Hardy–Littlewood conjecture and was later shown by Conrey and Snaith (2007) to follow from a general conjecture, known as the *ratios conjecture*, for averages of ratios of values of the Riemann zeta function (Conrey et al. 2008b). Similar formulae can be written down for higher order correlations (Bogomolny and Keating 1996b, Conrey and Snaith 2008, Bogomolny and Keating 2013).

Montgomery's theorem and conjecture extend to the n-point correlations of other L-functions: when one takes a fixed L-function and averages along the critical line one expects the same result, corresponding to an average over $U(N)$, as for the Riemann zeta function. This can be analysed using the explicit formula for L-functions (Rudnick and Sarnak 1996). Alternatively, one can fix a point on the critical line and average over a family of L-functions. Then, depending on the family in question, one gets the unitary, orthogonal, or symplectic groups appearing (Katz and Sarnak 1999a, 1999b).

One of the most important applications of the Montgomery conjecture is to the statistics of the primes in short intervals. Goldston and Montgomery (1987) proved that the pair correlation conjecture (8.39) is equivalent to the following asymptotic formula for $X^\delta < H < X^{1-\delta}$ as $X \to \infty$,

$$
\frac{1}{X}\int_2^X \left|\sum_{n\in[x-\frac{H}{2},x+\frac{H}{2}]} \Lambda(n) - H\right|^2 dx \sim H\log(X/H),
$$

$$(8.50)$$

and using the Hardy–Littlewood conjecture, Montgomery and Soundararajan (2004) later extended this to

$$\frac{1}{X}\int_2^X \left|\sum_{n\in[x-\frac{H}{2},x+\frac{H}{2}]}\Lambda(n)-H\right|^2 dx \sim H\Big(\log(X/H)-(\gamma_E+\log 2\pi)\Big), \qquad (8.51)$$

where γ_E is the Euler–Mascheroni constant. This formula also follows from (8.49); i.e., the constant term is related to the corrections to the random matrix limit inherent in (8.49) (Bui et al. 2016).

For other L-functions, if we use

$$\frac{F'}{F}(s)=-\sum_{n=1}^\infty \frac{\Lambda_F(n)}{n^s}\qquad (\mathrm{Re}(s)>1),$$

the generalization (Bui et al. 2016) of (8.51) is that for $X^{1-1/d_F}<H<X^{1-\delta}$, as $X\to\infty$

$$\frac{1}{X}\int_2^X \left|\sum_{n\in[x-\frac{H}{2},x+\frac{H}{2}]}\Lambda_F(n)\right|^2 dx \sim H\Big(\log(X/H)+\log\mathfrak{q}_F-(\gamma_E-\log 2\pi)d_F\Big) \quad (8.52)$$

and for $X^\delta<H<X^{1-1/d_F}$, as $X\to\infty$

$$\frac{1}{X}\int_2^X \left|\sum_{n\in[x-\frac{H}{2},x+\frac{H}{2}]}\Lambda_F(n)\right|^2 dx \sim \frac{1}{6}H\Big(6\log X-(3+8\log 2)\Big). \qquad (8.53)$$

Clearly, when $d_F=1$ we have just one regime, and when $\mathfrak{q}_F=1$ we recover (8.51). When $d_F>1$ the new regime described by (8.53) becomes important.

Finally, even though we do not pursue this line of thought here, it is worth noting that the remarkable success of models for the statistical distribution of the Riemann zeros based on the eigenvalues of random matrices gives strong support for a spectral interpretation that might explain the Riemann hypothesis. See, for example, Berry and Keating (1999) for a review of ideas in this direction.

8.2   Value distribution of the zeta function on its critical line

In this second section 8.1, I will review how the connection with random matrix theory has informed our understanding of the distribution of the values taken by $\zeta(1/2+it)$.

8.2.1   Selberg's theorem

I will start by reviewing what is known about the value distribution of $\log\zeta(1/2+it)$. The most important general result, due originally to Selberg, is that this obeys a central limit theorem (see Titchmarsh (1986)): for any rectangle B in the complex plane,

$$\lim_{T \to \infty} \frac{1}{T} \text{meas.}\{T \le t \le 2T : \frac{\log \zeta(\frac{1}{2} + it)}{\sqrt{\frac{1}{2} \log \log \frac{t}{2\pi}}} \in B\} = \frac{1}{2\pi} \int \int_B e^{-\frac{1}{2}(x^2 + y^2)} dx dy.$$

(8.54)

The proof of this theorem relies on the fact that, loosely speaking, (8.2) suggests that the logarithm of $\zeta(1/2 + it)$ should be related to a sum over the primes, and the contributions from different primes should be statistically independent. The difficulties in making this rigorous are that the sum does not converge and that the contributions from very large primes are not independent. The resolution (cf. Titchmarsh 1986) was found by Selberg (1946) (with important later refinements by Soundararajan 2009), who showed, essentially, that for most values of t

$$\log \zeta\left(\frac{1}{2} + it\right) = \sum_{n \le t} \frac{\Lambda(n)}{\sqrt{n} \log n} \frac{1}{n^{it}} + O(1).$$

(8.55)

Specifically, this holds unless $\frac{1}{2} + it$ is close to a zero of the zeta function. The contributions to the sum in (8.55) may be treated as independent in a computation of the joint moments of the real and imaginary parts of $\log \zeta(1/2 + it)$, leading to the theorem stated earlier (Eq. (8.54)).

8.2.2 Moments of $\zeta\left(\frac{1}{2} + it\right)$

Next let us turn to the value distribution of $\zeta(1/2 + it)$ itself. Its moments satisfy the long-standing and important conjecture that as $T \to \infty$

$$\frac{1}{T} \int_0^T |\zeta(\frac{1}{2} + it)|^{2\lambda} dt \sim f_\zeta(\lambda) \prod_p \left[(1 - \frac{1}{p})^{\lambda^2} \sum_{m=0}^{\infty} \left(\frac{\Gamma(\lambda + m)}{m! \Gamma(\lambda)}\right)^2 p^{-m}\right] (\log \frac{T}{2\pi})^{\lambda^2}$$

(8.56)

for some function $f_\zeta(\lambda)$ (see Titchmarsh 1986).

This can be viewed in the following way. It asserts that the moments grow like $(\log \frac{T}{2\pi})^{\lambda^2}$ as $T \to \infty$. Treating the primes as being statistically independent of each other would give the right-hand side with $f_\zeta(\lambda) = 1$. $f_\zeta(\lambda)$ thus quantifies deviations from this simple-minded ansatz. Assuming that the moments do indeed grow like $(\log \frac{T}{2\pi})^{\lambda^2}$, the problem is then to determine $f_\zeta(\lambda)$.

The conjecture is known to be correct in only two nontrivial cases, when $\lambda = 1$ and $\lambda = 2$. It was shown by Hardy and Littlewood in 1918 that $f_\zeta(1) = 1$ and by Ingham in 1926 that $f_\zeta(2) = \frac{1}{12}$. On number-theoretical grounds, Conrey and Ghosh (1992) have conjectured that $f_\zeta(3) = \frac{42}{9!}$ and Conrey and Gonek (2001) that $f_\zeta(4) = \frac{24024}{16!}$. The methods leading to these vales involves the approximate functional equation (see Titchmarsh 1986)

$$\zeta(1/2 + it) = \sum_{n \le \sqrt{\frac{t}{2\pi}}} \frac{1}{n^{1/2+it}} + \pi^{it} \frac{\Gamma(1/4 - it/2)}{\Gamma(1/4 + it/2)} \sum_{m \le \sqrt{\frac{t}{2\pi}}} \frac{1}{m^{1/2-it}} + O(t^{-1/4}),$$

(8.57)

computing the moments directly in terms of the multiple sums that result. Unfortunately, this approach fails for $k \ge 5$. Only recently have new techniques made it

possible to go beyond $k = 4$ (Conrey and Keating 2017, 2016). It is not a priori obvious that $f_\zeta(k)$ should be a rational number, nor indeed that $(k^2)! f_\zeta(k)$ should be an integer, as the above values suggest.

8.2.3   Characteristic polynomials of random matrices

We will now look to random matrix theory to see what light, if any, it can shed on these issues. The idea is that since we believe that the zeros of the zeta function are distributed statistically like the eigenvalues of random unitary matrices $A \in U(N)$, it is natural to model the value distribution of $\zeta(1/2 + it)$ in terms of that of the characteristic polynomials

$$Z(A, \theta) = \det(I - Ae^{-i\theta})$$
$$= \prod_n (1 - e^{i(\theta_n - \theta)}), \tag{8.58}$$

where the eigenvalues of A are denoted $e^{i\theta_n}$.

We begin by considering the function

$$P_N(s, t) = \int_{U(N)} |Z(A, \theta)|^t e^{is \operatorname{Im} \log Z(A,\theta)} d\mu(A). \tag{8.59}$$

This is the moment generating function of $\log Z$: the joint moments of $\operatorname{Re} \log Z$ and $\operatorname{Im} \log Z$ are obtained from derivatives of P at $s = 0$ and $t = 0$, and

$$\int_{U(N)} \delta(x - \operatorname{Re} \log Z)\delta(y - \operatorname{Im} \log Z)d\mu(A) \tag{8.60}$$

$$= \frac{1}{4\pi^2} \int_{-\infty}^{\infty} \int_{-\infty}^{\infty} e^{-itx - isy} P(s, it)dsdt. \tag{8.61}$$

Written in terms of the eigenvalues,

$$P_N(s, t) = \int_{U(N)} \prod_{n=1}^{N} |1 - e^{i(\theta_n - \theta)}|^t e^{-is \sum_{m=1}^{\infty} \frac{\sin[(\theta_n - \theta)m]}{m}} d\mu(A). \tag{8.62}$$

Since the integrand is a class function, we can use Weyl's integration formula (Weyl 1946) to write

$$P_N(s, t) = \frac{1}{(2\pi)^N N!} \int_0^{2\pi} \cdots \int_0^{2\pi} \prod_{n=1}^{N} |1 - e^{i(\theta_n - \theta)}|^t.$$
$$\times e^{-is \sum_{m=1}^{\infty} \frac{\sin[(\theta_n - \theta)m]}{m}} \prod_{1 \leq j < k \leq N} |e^{i\theta_j} - e^{i\theta_k}|^2 d\theta_1 \cdots d\theta_N. \tag{8.63}$$

This integral can then be evaluated using a form of Selberg's integral described in Mehta (1991), giving (Keating and Snaith 2000a)

$$P_N(s,t) = \prod_{j=1}^{N} \frac{\Gamma(j)\Gamma(t+j)}{\Gamma(j+\frac{t}{2}+\frac{s}{2})\Gamma(j+\frac{t}{2}-\frac{s}{2})}. \tag{8.64}$$

Consider first the Taylor expansion

$$P_N(s,t) = e^{\alpha_{00}+\alpha_{10}t+\alpha_{01}s+\alpha_{20}t^2/2+\alpha_{11}ts+\alpha_{02}s^2/2+\cdots}. \tag{8.65}$$

The α_{m0} are the cumulants of $\mathrm{Re}\log Z$ and the α_{0n} are i^n times the cumulants of $\mathrm{Im}\log Z$. Expanding (8.64) gives

$$\alpha_{10} = \alpha_{01} = \alpha_{11} = 0; \tag{8.66}$$

$$\alpha_{20} = -\alpha_{02} = \frac{1}{2}\log N + \frac{1}{2}(\gamma+1) + O(\frac{1}{N^2}); \tag{8.67}$$

$$\alpha_{mn} = O(1) \quad \text{for} \quad m+n \geq 3; \tag{8.68}$$

and more specifically,

$$\alpha_{m0} = (-1)^m(1-\frac{1}{2^{m-1}})\Gamma(m)\zeta(m-1) + O(\frac{1}{N^{m-2}}), \quad \text{for } m \geq 3. \tag{8.69}$$

This leads to the following theorem (Keating and Snaith 2000a): for any rectangle B in the complex plane

$$\lim_{N\to\infty} \text{meas.} \left\{ A \in U(N) : \frac{\log Z(A,\theta)}{\sqrt{\frac{1}{2}\log N}} \in B \right\} = \frac{1}{2\pi} \int\int_B e^{-\frac{1}{2}(x^2+y^2)}dxdy. \tag{8.70}$$

Comparing this result to (8.54), one sees that $\log \zeta(1/2+it)$ and $\log Z$ both satisfy a central limit theorem when, respectively, $t \to \infty$ and $N \to \infty$. Note that the scalings in (8.54) and (8.70), corresponding to the asymptotic variances, are the same if we make the identification

$$N = \log \frac{t}{2\pi}. \tag{8.71}$$

This is the same as identifying the mean eigenvalue density with the mean zero density; cf. the unfolding factors in (8.41) and (8.36).

We now turn to the problem of the moments of $|\zeta(1/2+it)|$. It is natural to expect these moments to be related to those of the modulus of the characteristic polynomial Z, which are defined as

$$\int_{U(N)} |Z(A,\theta)|^{2\lambda} d\mu(A) = P(0,2\lambda)$$

$$= \prod_{j=1}^{N} \frac{\Gamma(j)\Gamma(j+2\lambda)}{(\Gamma(j+\lambda))^2} \tag{8.72}$$

$$= e^{\sum_{m=0}^{\infty} \alpha_{m0}(2\lambda)^n/n!}. \tag{8.73}$$

Therefore,

$$\lim_{N\to\infty} \frac{1}{N^{\lambda^2}} \int_{U(N)} |Z(A,\theta)|^{2\lambda} d\mu(A) \tag{8.74}$$

$$= e^{\lambda^2(\gamma+1)+\sum_{m=3}^{\infty}(-2\lambda)^m \frac{2^{m-1}-1}{2^{m-1}} \frac{\zeta(m-1)}{m}}, \tag{8.75}$$

for $|\lambda| < \frac{1}{2}$. Note that since we are identifying Z with $\zeta(1/2+it)$ and N with $\log \frac{t}{2\pi}$, the expression on the left-hand side of (8.75) corresponds precisely to (8.56).

We now recall some properties of the Barnes' G-function. This is an entire function of order 2 defined by

$$G(1+z) = (2\pi)^{z/2} e^{-[(1+\gamma)z^2+z]/2} \prod_{n=1}^{\infty} \left[(1+z/n)^n e^{-z+z^2/(2n)} \right]. \tag{8.76}$$

It satisfies

$$G(1) = 1, \tag{8.77}$$

$$G(z+1) = \Gamma(z)G(z) \tag{8.78}$$

and

$$\log G(1+z) = (\log 2\pi - 1)\frac{z}{2} - (1+\gamma)\frac{z^2}{2} + \sum_{n=3}^{\infty}(-1)^{n-1}\zeta(n-1)\frac{z^n}{n}. \tag{8.79}$$

Thus, we have that

$$\lim_{N\to\infty} \frac{1}{N^{\lambda^2}} \int_{U(N)} |Z(A,\theta)|^{2\lambda} d\mu(A) = f_U(\lambda), \tag{8.80}$$

with

$$f_U(\lambda) = \frac{G^2(1+\lambda)}{G(1+2\lambda)}. \tag{8.81}$$

Using (8.78) we further have that for positive integers k

$$f_U(k) = \prod_{j=0}^{k-1} \frac{j!}{(j+k)!}. \tag{8.82}$$

In particular, $f_U(1) = 1$, $f_U(2) = \frac{1}{12}$, $f_U(3) = \frac{42}{9!}$, and $f_U(4) = \frac{24024}{16!}$, which match the values of f_ζ listed after (8.56). This then motivates the conjecture (Keating and Snaith 2000a) that

$$f_\zeta(\lambda) = f_U(\lambda) \tag{8.83}$$

for all λ such that $\mathrm{Re}\lambda > -\frac{1}{2}$.

These calculations and those of the previous section extend to other L-functions (Conrey and Farmer 2000), as in those cases one also has a Weyl integration formula and the associated Selberg integrals (Keating and Snaith 2000b, Keating et al. 2003). They also extend to products and ratios of products of characteristic polynomials and L-functions, leading to conjectures for lower order terms in the moment asymptotics (Conrey et al. 2003, 2005, 2008a, b).

There are many applications of the moment formulae; for example, they lead to estimates for the number of elliptic curves with nonzero rank (Conrey et al. 2002, Bektemirov et al. 2007), and the statistical distribution of the zeros of elliptic curve L-functions (Dueñez et al. 2012).

8.2.4 Extreme values

The previous sections of this chapter have focused on the typical values of $\log \zeta(1/2+it)$ and of $\zeta(1/2+it)$, as captured by their moments and limit distributions. We now turn to the extreme large values taken by these functions. Selberg's central limit theorem implies that the typical size of $\log \zeta(1/2+it)$ is on the order of $\sqrt{\log \log t}$, the question is: what are the largest values one can expect it to take on intervals of a given length?

The Lindelöf hypothesis asserts that $|\zeta(1/2+it)| = o(t^\epsilon)$ for any $\epsilon > 0$; the Riemann hypothesis implies that

$$|\zeta(1/2+it)| = O\left(\exp\left(\frac{c_1 \log t}{\log \log t}\right)\right), \tag{8.84}$$

where c_1 is a constant (see Titchmarsh 1986), and, unconditionally, we know that (Bondarenko and Seip 2015)

$$\max_{t\in[T^{1/2},T]} |\zeta(1/2+it)| > \exp\left(c_2\sqrt{\frac{\log T \log \log \log T}{\log \log T}}\right), \tag{8.85}$$

for all $c_2 < 1/\sqrt{2}$. The exceptionally large values of $|\zeta(1/2+it)|$ thus lie in the range between (8.84) and (8.85). The problem of determining where precisely within this range they lie has attracted considerable attention in recent years, but it remains unresolved. The extreme values in question are so rare that extensive numerical computations have thus far failed to settle the matter.

Farmer et al. (2007) have conjectured, partly on the basis of the moments of the characteristic polynomials of random matrices, that

$$\max_{t\in[0,T]} |\zeta(1/2+it)| = \exp\left(\left(\frac{1}{\sqrt{2}} + o(1)\right)\sqrt{\log T \log \log T}\right). \tag{8.86}$$

Their analysis makes substantial use of a hybrid product formula for the Riemann zeta function, derived by Gonek et al. (2007), that interpolates between the Euler product over the primes and the Hadamard product over the zeros.

Our focus here will be on extreme values taken over shorter intervals of the critical line, specifically on

$$\zeta_{\max}(T) := \max_{T \leq t \leq T+2\pi} |\zeta(1/2 + it)|, \tag{8.87}$$

which we expect to be modelled by

$$Z_{\max}(A) := \max_{0 \leq \theta \leq 2\pi} |Z(A, \theta)|. \tag{8.88}$$

Note that both

$$V_Z(A, \theta) := -\log |Z(A, \theta)| \tag{8.89}$$

and

$$V_\zeta(t) := -\log |\zeta(1/2 + it)| \tag{8.90}$$

satisfy similar central limit theorems if we identify N and $\log(T/2\pi)$. Importantly, these functions are also correlated in the same way, namely,

$$\lim_{N \to \infty} \mathbb{E} V_Z(A, \theta) V_Z(A, \theta + x) \sim -\frac{1}{2} \log |x|, \tag{8.91}$$

where the expectation is computed with respect to $A \in U(N)$, and

$$\lim_{T \to \infty} \frac{1}{T} \int_T^{2T} V_\zeta(t) V_\zeta(t + x) dt \sim -\frac{1}{2} \log |x| \tag{8.92}$$

as $|x| \to 0$. This motivates the idea that both $\log |\zeta(1/2 + it)|$ and $\log |Z(A, \theta)|$ should be viewed in the context of the theory of log-correlated Gaussian random fields, such as the 2-D Gaussian free field. This idea is also suggested by the fact that (Hughes et al. 2001)

$$\log |Z(A, \theta)| = -\mathrm{Re} \sum_{n=1}^{\infty} \frac{1}{\sqrt{n}} \frac{\mathrm{Tr} A^n}{\sqrt{n}} e^{-in\theta} \tag{8.93}$$

and that the value distribution of the random variable $\frac{\mathrm{Tr} A^n}{\sqrt{n}}$ tends to a complex normal with mean zero and unit variance in the limit as $N \to \infty$ (Diaconis and Shahshahani 1994). $\log |Z(A, \theta)|$ thus resembles a one-dimensional cut through a 2-D Gaussian free field; more precisely, it resembles a random Fourier series associated with sampling a 2-D Gaussian free field along a circle of unit radius parametrized as

$z = e^{it}, t \in [0, 2\pi)$ (Fyodorov and Bouchaud 2008). Similarly, it follows from theorems of Selberg (1946) and of Soundararajan (2009) that, assuming the Riemann hypothesis,

$$\log |\zeta(1/2 + it)| = \operatorname{Re} \sum_{p < t/2\pi} \frac{1}{\sqrt{p}} e^{-it \log p} + O(1) \qquad (8.94)$$

for most values of t (specifically, away from the zeros of the zeta function). So the zeta function has a similar structure if the logarithms of the primes behave like independent random variables.

Given the formal similarity with log-correlated Gaussian random fields, one might conjecture that

$$\log Z_{\max}(A) = \log N - \frac{3}{4} \log \log N + x_A, \qquad (8.95)$$

where x_A is a random variable that is $O_{\mathbb{P}}(1)$ and where the probability density of values of x_A, $P(x)$, decays like $x \exp(-x)$ when $x \to \infty$; and correspondingly for the zeta function,

$$\log \zeta_{\max}(T) = \log \log T - \frac{3}{4} \log \log \log T + x_T, \qquad (8.96)$$

where x_T behaves like x_A (Fyodorov et al. 2012; Fyodorov and Keating 2014).

The factor $\frac{3}{4}$ in the second terms of (8.95) and (8.96) is characteristic of log-correlated Gaussian random fields, as is the form of the asymptotic decay of $P(x)$. Fields in which the correlations decay rapidly would have a factor $\frac{1}{4}$ and have fluctuations with a Gumbel-type distribution (see Fyodorov and Keating 2014).

A heuristic justification of (8.95) was developed in Fyodorov and Keating (2014) based on a heuristic calculation of the moments of the random variable

$$g(\beta, A) := \frac{1}{2\pi} \int_0^{2\pi} |Z(A, \theta)|^{2\beta} d\theta. \qquad (8.97)$$

Note that $g(\beta, A)$ resembles the partition function of a particle moving on the unit circle with energy given by $V_A(\theta)$, and that

$$Z_{\max}(A) = \lim_{\beta \to \infty} \frac{1}{2\beta} \log g(\beta, A). \qquad (8.98)$$

The calculation in Fyodorov and Keating (2014) proceeds via the multiple integral

$$\mathbb{E}[g(\beta, A)]^k = \frac{1}{(2\pi)^k} \int_0^{2\pi} \cdots \int_0^{2\pi} \mathbb{E} \prod_{j=1}^k |Z(A, \theta_j)|^{2\beta} d\theta_j. \qquad (8.99)$$

The integrand here may be computed, when $N \to \infty$ using the Fisher–Hartwig asymptotic formula (Widom 1973) and is then proportional to

$$\prod_{n < m} |e^{i\theta_n} - e^{i\theta_m}|^{-\beta^2}. \qquad (8.100)$$

The integrals over $\theta_1, \ldots, \theta_k$ may then be computed using the Selberg integral (Mehta 1991). The result is that for $k < 1/\beta^2$

$$\mathbb{E}[g(\beta, A)^k] \sim N^{k\beta^2} \left(\frac{(G(1+\beta))^2}{G(1+2\beta)\Gamma(1-\beta^2)} \right)^k \Gamma(1 - k\beta^2) \qquad (8.101)$$

and for $k > 1/\beta^2$

$$\mathbb{E}[g(\beta, A)^k] \sim c(\beta, k) N^{k^2\beta^2 + 1 - k} \qquad (8.102)$$

for some function $c(\beta, k)$. Inverting the moments for $\beta < 1$ and conjecturing a suitable analytical continuation of the resulting distribution to $\beta > 1$ leads to the prediction (8.95), with

$$P(x) = 2\exp(-x)\mathrm{K}_0\left(2\exp(-x/2)\right), \qquad (8.103)$$

where K_0 is a Bessel function. Note that as $x \to \infty$ this expression exhibits the tail asymptotic $P(x) \sim xe^{-x}$ expected to be universal for logarithmically correlated processes (Carpentier and Le Doussal 2001).

In Fig. 8.1 data are shown for the maxima of 10^6 random matrices of dimension $N = 50$ compared to the prediction (8.103).

For the Riemann zeta function, the calculation proceeds in the same way, using the formula of Conrey et al. (2005) in place of the Fisher–Hartwig asymptotic. This leads to the same predicted form for the fluctuations (8.103). Data are shown in Fig. 8.2 for the maxima of the zeta function over ranges of length 2π near height $T = 10^{28}$, compared to the prediction (8.103).

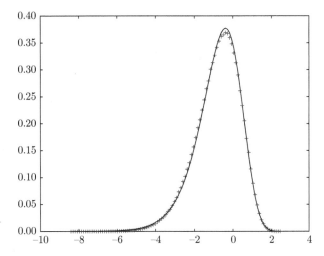

Fig. 8.1 The probability density for the fluctuations in $Z_{\max}(A)$ computed for 10^6 random matrices of dimension $N = 50$ compared to the prediction (8.103); data computed by Timothée Wintz.

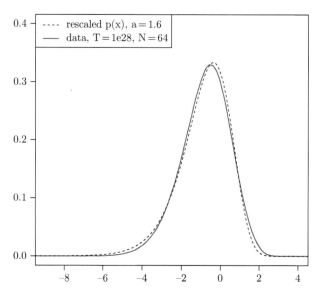

Fig. 8.2 Numerical computation of $\zeta_{\max}(T)$ (solid line) near height $T = 10^{28}$ compared to theoretical prediction (8.103) (dashed line) for $p(x)$. Data kindly provided by Professor Ghaith Hiary.

These predictions are supported by several recent theoretical developments. It was proved by Webb (2015) that the prediction for $\mathbb{E}[g(\beta, A)^k]$ holds when β is small, and by Claeys and Krasovsky (2015) that it holds when $k = 2$. Keating and Scott (2015) proved that $\mathbb{E}[g(\beta, A)^k]$ is a polynomial in N of order $k^2\beta^2 + 1 - k$ for $\beta = 1$, $k \in \{1, 2, 3, 4\}$, and for $\beta = 2$, $k \in \{1, 2\}$, where the polynomials in question are determined explicitly. Their calculation is based on using an exact formula for the integrand in (8.99) derived by Conrey et al. (2003).

More recently, Arguin et al. (2017a) have established rigorously the first term (i.e., the term proportional to $\log N$) in (8.95), and Paquette and Zeitouni (2016) have established the second (proportional to $\log \log N$). Furthermore, Arguin et al. (2017b) have established the first two terms in (8.96) for a random model of the zeta function introduced by Harper (2013) in which the term n^{it} in (8.55) is replaced by $\exp i\phi_n$, where the ϕ_n are i.i.d. random variables taken uniformly from $[0, 2\pi)$. These last developments have followed from establishing an interesting connection between the logarithms of the characteristic polynomials and the branching random walk, and similarly for the logarithm of the random model of the zeta function.

8.3 Function Fields

In this third section I will review the relationship between arithmetical fluctuations in function fields and random matrices. For further background, see Rudnick (2014) and Keating et al. (2015b).

8.3.1   Background

There is an interesting parallel between arithmetical problems in number fields and in function fields. Let $\mathbb{F}_q$ be a finite field of q elements and $\mathbb{F}_q[t]$ the ring of polynomials with coefficients in $\mathbb{F}_q$. For example, when q is a prime, $f \in \mathbb{F}_q[t]$ is a polynomial in the variable t with coefficients that are integers in $\{0, 1, \dots, q-1\}$, and arithmetical operations are performed modulo q. Let $\mathcal{P}_n = \{f \in \mathbb{F}_q[t] : \deg f = n\}$ be the set of polynomials of degree n and $\mathcal{M}_n \subset \mathcal{P}_n$ the subset of monic polynomials.

There is unique factorization in this context in terms of *irreducible polynomials*. The analogue of the von Mangoldt function is defined by $\Lambda(N) = \deg P$ if $N = cP^k$ with P an irreducible monic polynomial and $c \in \mathbb{F}_q^\times$, and by $\Lambda(N) = 0$ otherwise. The prime polynomial theorem in this setting is the identity

$$\sum_{f \in \mathcal{M}_n} \Lambda(f) = q^n . \tag{8.104}$$

This is analogous to the usual prime number theorem, which implies that the average value of the von Mangoldt function is 1, because the number of monic polynomials of degree n is precisely q^n. Comparing with (8.15), we are thus led to identify x with q^n.

The proof of (8.104) follows the same lines as in the number field setting. One can define a zeta function by

$$\zeta_q(s) = \prod_P \left(1 - \frac{1}{|P|^s}\right)^{-1} \tag{8.105}$$

when $\mathrm{Re}(s) > 1$, where the product runs over monic irreducible polynomials and the norm of a polynomial $f \in \mathbb{F}_q[t]$ is defined to be

$$|f| = q^{\deg f}. \tag{8.106}$$

Setting $u = q^{-s}$, the zeta function becomes

$$Z(u) := \prod_P (1 - u^{\deg P})^{-1} \tag{8.107}$$

The product in (8.107) may be expanded, using the fact that there is unique factorization, to give

$$Z(u) = \sum_f u^{\deg f} = \sum_{n=0}^{\infty} \sum_{f \in \mathcal{M}_n} u^{\deg f} = \sum_{n=0}^{\infty} q^n u^n \tag{8.108}$$

and so

$$Z(u) = \frac{1}{1 - qu}. \tag{8.109}$$

Hence $Z(u)$ has a pole at $u = 1/q$, but no zeros.

It follows from (8.107) that

$$\frac{Z'(u)}{Z(u)} = \sum_{n=1}^{\infty} \left(\sum_{f \in \mathcal{M}_n} \Lambda(f) \right) u^{n-1} \tag{8.110}$$

and from (8.109) that

$$\frac{Z'(u)}{Z(u)} = \sum_{n=1}^{\infty} q^n u^{n-1}; \tag{8.111}$$

equating powers of u in the preceding two equations then gives (8.104).

8.3.2   Zeta functions of curves

Let C be a smooth, projective, geometrically connected curve over $\mathbb{F}_q$, of genus g. Let N_n be the number of points on C in the extension field $\mathbb{F}_{q^n}$ of degree n. One defines the zeta function associated with C to be

$$Z_C(u) = \exp \left(\sum_{n=1}^{\infty} \frac{N_n u^n}{n} \right), \tag{8.112}$$

where the sum converges when $|u| < 1/q$.

It follows from deep results of Weil (see, for example, Katz and Sarnak 1999a) that

$$Z_C(u) = \frac{P_C(u)}{(1-u)(1-qu)}, \tag{8.113}$$

where $P_C(u)$ is a polynomial of degree $2g$. $P_C(u)$ satisfies the functional equation

$$P_C(u) = (qu^2)^g P_C \left(\frac{1}{qu} \right). \tag{8.114}$$

The roots of $P_C(u)$ can be proved to satisfy the analogue of the Riemann hypothesis in that if

$$P_C(u) = \prod_{j=1}^{2g} (1 - \alpha_j u), \tag{8.115}$$

then $|\alpha_j| = \sqrt{q} \ \forall j$. That is, all of the zeros of $P_C(u)$ lie on the circle $|u| = q^{-1/2}$.

To give an example, consider the case of hyperelliptic curves C_D, which take the form

$$y^2 = D(x), \qquad D(x) = x^{2g+1} + a_{2g} x^{2g} + \cdots + a_0. \tag{8.116}$$

Taking $D(x)$ to be square-free and of degree $2g + 1$ is equivalent to taking C_D to be smooth and of genus g. We then have that

$$Z_{C_D}(u) = \frac{\mathcal{L}(u, \chi_D)}{(1 - u)(1 - qu)} \tag{8.117}$$

with

$$\mathcal{L}(u, \chi_D) = \prod_P \left(1 - \chi_D(P)u^{\deg P}\right)^{-1} \tag{8.118}$$

where the product runs over monic irreducibles and

$$\chi_D(P) = \begin{cases} +1 & \text{if } P \nmid D \text{ and } f^2 \equiv D \ (\bmod\ P) \text{ solvable} \\ 0 & \text{if } P | D \\ -1 & \text{if } P \nmid D \text{ and } f^2 \equiv D \ (\bmod\ P) \text{ not solvable} \end{cases} \tag{8.119}$$

as in (8.19).

8.3.3   Spectral interpretation

It follows from the Riemann hypothesis in this context that we can write

$$P_C(u) = \det(I - uq^{1/2}\Theta_C), \tag{8.120}$$

where Θ_C is a unitary matrix representing the *unitarized Frobenius conjugacy class*. The zeros of $P_C(u)$ correspond to the eigenvalues of this matrix. We denote these eigenvalues by $e^{i\theta_n}$.

8.3.4   Spectral statistics

The spectral interpretation naturally leads to the question of spectral statistics: how are the eigenphases θ_n distributed as C varies? And what are the implications of any limiting distribution?

   To give a specific example, consider the *hyperelliptic ensemble*, defined in terms of the moduli space $H(g, q)$ of hyperelliptic curves of genus $g \geq 1$ over $\mathbb{F}_q$. The probability space in this context corresponds to picking $D(x)$ uniformly from square-free polynomials of the form given in (8.116), normalized using the fact that the number of such polynomials is precisely $(q - 1)q^{2g}$. So

$$\mathbb{E}F(D) = \frac{1}{(q - 1)q^{2g}} \sum_{D \in H(g,q)} F(D). \tag{8.121}$$

   There are two limits in which one can explore spectral statistics:

$g \to \infty$, *q-fixed*. This limit is similar to the usual number-field situation. It is con- jectured that the eigenvalues have a limiting distribution corresponding to that of random symplectic matrices of dimension $2g$ when $g \to \infty$. This is supported

by theorems and numerical data, as in the traditional number-field case. For example, one expects eigenvalue correlations and value distribution results for the L-functions corresponding to those described in the first two sections. See, for example, Andrade and Keating (2012, 2013, 2014), Rubinstein and Wu (2015), and Florea (2015a, b).

$q \to \infty$, g-fixed. This limit has no analogue in the usual number-field situation. To give an example, in the case of the hyperelliptic ensemble, it has been proved that Θ_C equidistributes when $q \to \infty$ in $USp(2g)$, and so the link between number theory and random matrix theory has a firm footing. Hence, correlations between the zeros and the value distribution of L-functions can be represented rigorously in terms of matrix integrals over $USp(2g)$.

8.3.5   Arithmetic statistics in function fields

To give an example of one application of these methods, let us now return to the problem of determining the variance of the number of primes in short intervals considered in Section 8.1. Specifically, we shall discuss recent progress in proving function-field analogies of the Goldston–Montgomery–Soundararajan conjecture (8.51) and its extension to other L-functions.

In the function-field setting the analogue of counting primes in short intervals is counting irreducible polynomials in subsets of $\mathcal{M}_n$ corresponding to short intervals. For $A \in \mathcal{P}_n$ of degree n, and $h < n$, we define these subsets as follows. Let

$$I(A; h) := \{f : |f - A| \le q^h\} = A + \mathcal{P}_{\le h} ,\tag{8.122}$$

where

$$\mathcal{P}_{\le h} = \{0\} \cup \bigcup_{0 \le m \le h} \mathcal{P}_m \tag{8.123}$$

is the space of polynomials of degree at most h (including 0). We have

$$\#I(A, h) = q^{h+1} .\tag{8.124}$$

Note that for $h < n$, if $|f - A| \le q^h$ then A monic if and only if f is monic. Hence for A monic, $I(A, h)$ consists of only monic polynomials and all monic f's of degree n are contained in one of the intervals $I(A, h)$ with A monic.

For $1 \le h < n$ and $A \in \mathcal{P}_n$, we define

$$\nu(A; h) = \sum_{f \in I(A,h), f(0) \ne 0} \Lambda(f).\tag{8.125}$$

It is not difficult to see that the mean value of $\nu(A; h)$, averaged over monic $A \in \mathcal{M}_n$, is

$$\langle \nu(\bullet; h) \rangle = q^{h+1}\left(1 - \frac{1}{q^n}\right).\tag{8.126}$$

The goal is to compute the variance of $\nu(A; h)$, in the limit $q \to \infty$.

One can calculate the sum of $\Lambda(f)$ over $\mathcal{M}_n$ in terms of the zeta function (8.107). This sum can be limited to an interval $I(A, h)$ using Fourier analysis, specifically, in this context, Dirichlet characters. This introduces characters into the zeta function, giving rise to L-functions, defined like those in Section 8.1.3. These L-functions can each be related to a matrix using the Riemann hypothesis. When the L-functions are summed over the characters involved, it follows from a theorem of N. Katz that the associated matrices equidistribute, resulting in a matrix integral which can be evaluated to prove (Keating and Rudnick 2014) that when $h < n - 3$,

$$\lim_{q \to \infty} \frac{1}{q^{h+1}} \operatorname{Var}(\nu(\bullet; h)) = \int_{U(n-h-2)} |\operatorname{Tr} A^n|^2 d\mu(A) = n - h - 2. \tag{8.127}$$

This corresponds precisely to the Goldston–Montgomery–Soundararajan conjecture (8.51), because n is the analogue of $\log X$ and $h + 1$ is the analogue of $\log H$. An extension to higher degree L-functions, similarly consistent with the general expression set out in Section 8.1, can be proved in the same way.

This approach extends in a similar way (i.e., via matrix integrals) to the variances of other important arithmetic functions, including the Möbius function (Keating and Rudnick 2016) and the generalized divisor functions (Keating et al. 2015a); the results either match previous number-theoretic conjectures, where these existed, or provide new conjectures.

The Möbius function $\mu(n)$ in the number-field setting is defined to be 0 if n is divisible by a square, and otherwise to be parity of the number of prime divisors of n. The prime number theorem is equivalent to

$$\sum_{n \leq X} \mu(n) = o(X). \tag{8.128}$$

It is a conjecture of Good and Churchhouse (1968) that for $X^\delta < H < X^{1-\delta}$, as $X \to \infty$

$$\frac{1}{X} \int_X^{2X} \left| \sum_{n \in [x - \frac{H}{2}, x + \frac{H}{2}]} \mu(n) \right|^2 dx \sim \frac{H}{\zeta(2)}. \tag{8.129}$$

In function fields one can define a Möbius function in exactly the same way. In this case we have for $n \geq 2$ that

$$\sum_{f \in \mathcal{M}_n} \mu(f) = 0, \tag{8.130}$$

which is the analogue of (8.104).

It was proved in Keating and Rudnick (2016) that if $0 \leq h \leq n-5$ then as $q \to \infty$, q odd,

$$\frac{1}{q^n} \sum_{A \in \mathcal{M}_n} \left| \sum_{f \in I(A;h)} \mu(f) \right|^2 \sim q^{h+1} \int_{U(n-h-2)} |\operatorname{Tr} \operatorname{Sym}^n A|^2 d\mu(A) = q^{h+1}. \tag{8.131}$$

This is consistent with the Good–Churchhouse conjecture if we write it as $H/\zeta_q(2)$, where $H = q^{h+1}$ and $\zeta_q(s) = \sum_{f \text{ monic}} \frac{1}{|f|^s}$, $\mathrm{Re}(s) > 1$, which tends to 1 as $q \to \infty$.

The kth divisor function $d_k(n)$ gives the number of ways of writing a (positive) integer as a product of k positive integers:

$$d_k(n) := \#\{(a_1, \ldots, a_k) : n = a_1 \cdot \ldots \cdot a_k, \quad a_1, \ldots a_k \geq 1\}. \tag{8.132}$$

The classical divisor function is $d(n) = d_2(n)$.

Dirichlet's divisor problem (Titchmarsh 1986) concerns the size of

$$\Delta_2(x) := \sum_{n \leq x} d_2(n) - x\left(\log x + (2\gamma_E - 1)\right). \tag{8.133}$$

For the higher divisor functions one defines a remainder term $\Delta_k(x)$ similarly as the difference between $\sum_{n \leq x} d_k(n)$ and $x P_{k-1}(\log x)$ where $P_{k-1}(u)$ is a certain polynomial of degree $k - 1$.

Let

$$\Delta_k(x; H) = \Delta_k(x + H) - \Delta_k(x) \tag{8.134}$$

be the remainder term for sums of d_k over short intervals $[x, x + H]$. It follows from results of Jutila (1984), Coppola and Salerno (2004), and Ivić (2009) that for $X^\epsilon < H < X^{1/2-\epsilon}$,

$$\frac{1}{X} \int_X^{2X} \left(\Delta_2(x, H)\right)^2 dx \sim H P_3(\log X - 2 \log H), \tag{8.135}$$

where $P_3(x)$ is a polynomial of degree 3 with positive leading coefficient.

For the higher divisor functions d_k, $k > 2$, essentially nothing is known about the mean square of $\Delta_k(x; H)$, other than Lester (2015) has shown that, assuming the Lindelöf hypothesis, if $h(x) = (\frac{x}{X})^{1-\frac{1}{k}} X^\delta$,

$$\frac{1}{X} \int_X^{2X} \left(\Delta_k(x, h(x))\right)^2 dx \sim a_k \frac{k^{k^2-1}}{\Gamma(k^2)} (1 - \frac{1}{k} - \delta)^{k^2-1} \frac{2^{2-\frac{1}{k}} - 1}{2 - \frac{1}{k}} X^\delta (\log X)^{k^2-1}, \tag{8.136}$$

provided $1 - \frac{1}{k-1} < \delta < 1 - \frac{1}{k}$, where

$$a_k = \prod_p \left\{(1 - \frac{1}{p})^{k^2} \sum_{j=0}^{\infty} \left(\frac{\Gamma(k+j)}{\Gamma(k)j!}\right)^2 \frac{1}{p^j}\right\}. \tag{8.137}$$

More generally, he argues that in the above range the variance is given in terms of a polynomial function of $\log X$ of degree $k^2 - 1$. For $k = 3$ and $\frac{7}{12} < \delta < \frac{2}{3}$, his results are unconditional.

It is natural to ask whether it is generally true that

$$\frac{1}{X} \int_X^{2X} \left(\Delta_k(x, H)\right)^2 dx \sim H P_{k^2-1}\left((1 - \frac{1}{k}) \log X - \log H\right) \tag{8.138}$$

in all ranges of interest.

In the function field setting one can define a generalized divisor function in exactly the same way as we've described here. It follows from a calculation similar to the one for the von Mangoldt function set out at the start of this section that

$$\frac{1}{q^n} \sum_{A \in \mathcal{M}_n} \sum_{f \in I(A;h)} d_k(f) = q^{h+1} \binom{n+k-1}{k-1}. \tag{8.139}$$

We therefore define

$$\Delta_k(A; h) = \sum_{f \in I(A;h)} d_k(f) - q^{h+1} \binom{n+k-1}{k-1}. \tag{8.140}$$

Using the methods outlined here, it is shown in Keating et al. (2015a) that for $n \geq 5$ and $h \leq \min(n - 5, (1 - \frac{1}{k})n - 2)$, as $q \to \infty$

$$\frac{1}{q^n} \sum_{A \in \mathcal{M}_n} |\Delta_k(A; h)|^2 \sim q^{h+1} I_k(n; n - h - 2), \tag{8.141}$$

where here, if we let $\Lambda^j : U(N) \to GL(\Lambda^j \mathbb{C}^N)$ be the exterior jth power representation $(0 \leq j \leq N)$

$$I_k(m; N) := \int_{U(N)} |\sum \mathrm{Tr} \Lambda^{j_1}(A) \ldots \mathrm{Tr} \Lambda^{j_k}(A)|^2 d\mu(A), \tag{8.142}$$

where the sum in the integrand is over $0 \leq j_1, \ldots, j_k \leq N$, subject to $j_1 + \cdots + j_k = m$.

By definition, $I_k(m; N) = 0$ for $m > kN$. It is nonzero for $m \leq kN$ satisfies a functional equation $I_k(m; N) = I_k(kN - m; N)$, and evaluates to

$$I_k(m; N) = \binom{m + k^2 - 1}{k^2 - 1}, \quad m \leq N. \tag{8.143}$$

In particular, if $h \leq n/2 - 1$ and $n \geq 8$ then

$$\frac{1}{q^n} \sum_{A \in \mathcal{M}_n} |\Delta_2(A; h)|^2 \sim H \frac{(n - 2h + 5)(n - 2h + 6)(n - 2h + 7)}{6}. \tag{8.144}$$

i.e., a cubic polynomial in $(n - 2h)$. This is the precise analogue of (8.135).

$I_k(m; N)$ can be shown to be equal to a count of lattice points $x = (x_i^{(j)}) \in (\mathbb{Z})^{k^2}$ satisfying certain relations. Furthermore, if we let $r := m/N$, then for $r \in [0, k]$,

$$I_k(m; N) = \gamma_k(r) N^{k^2 - 1} + O_k(N^{k^2 - 2}), \tag{8.145}$$

with

$$\gamma_k(r) = \int_{[0,1]^{k^2}} \delta_r(u_1^{(k)} + u_2^{(k-1)} + \cdots u_k^{(1)}) \mathbf{1}_{A_k}(u) \, d^{k^2} u. \tag{8.146}$$

Here $\delta_r(x) = \delta(x-r)$ is the delta distribution translated by r and $\mathbf{1}_{A_k}$ is the indicator function of the set A_k. Moreover,

$$\gamma_k(r) = \frac{1}{k!\, G(1+k)^2} \int_{[0,1]^k} \delta_r(w_1 + \cdot + w_k) \prod_{i<j}(w_i - w_j)^2 \, d^k w. \qquad (8.147)$$

Here G is the Barnes G-function, so that for positive integers k, $G(1+k) = 1! \cdot 2! \cdot 3! \cdots (k-1)!$.

From this one can prove that $\gamma_k(r)$ is a piecewise polynomial function of r, the polynomial changing when r passes through integer values, and that $\gamma_k(r) = \gamma_k(k-r)$. For example, when $0 < r < 1$ $\gamma_3(r) = \frac{r^8}{8!}$, when $2 < r < 3$ $\gamma_3(r) = \frac{(3-r)^8}{8!}$, and when $1 < r < 2$ $\gamma_3(r) = \frac{1}{8!}(-2r^8 + 24r^7 - 252r^6 + 1512r^5 - 4830r^4 + 8568r^3 - 8484r^2 + 4392r - 927)$.

This leads to the general conjecture (Keating et al. 2015a) that if $0 < \delta < 1 - \frac{1}{k}$ is fixed, then for $H = X^\delta$,

$$\frac{1}{X} \int_X^{2X} \Big(\Delta_k(x, H) \Big)^2 dx \sim a_k \mathcal{P}_k(\delta) H (\log X)^{k^2-1}, \quad X \to \infty, \qquad (8.148)$$

where $\mathcal{P}_k(\delta)$ is a piecewise polynomial function of δ, of degree $k^2 - 1$, given by

$$\mathcal{P}_k(\delta) = (1-\delta)^{k^2-1} \gamma_k\Big(\frac{1}{1-\delta}\Big). \qquad (8.149)$$

Acknowledgements

I am pleased to acknowledge support under EPSRC Programme Grant EP/K034383/1 LMF: L-Functions and Modular Forms. I am also grateful for the following additional support: a grant from the Leverhulme Trust, a Royal Society Wolfson Research Merit Award, a Royal Society Leverhulme Senior Research Fellowship, and a grant from the Air Force Office of Scientific Research, Air Force Material Command, USAF (number FA8655-10-1-3088). I further thank Mr. Timothée Wintz and Professor Ghaith Hiary for computing the data plotted in the figures in Section 8.2, and Ms. Emma Bailey and Professor Yan Fyodorov for their careful reading of these notes.

References

Andrade, J.C., and Keating, J.P. (2012). The mean value of $L(1/2, \chi)$ in the hyperelliptic ensemble. *J. Number Theory* **132**, 2793–816.

Andrade, J.C., and Keating, J.P. (2013). Mean value theorems for L-functions over prime polynomials for the rational function field. *Acta Arithmetica* **161**, 371–85.

Andrade, J.C., and Keating, J.P. (2014). Conjectures for the integral moments and ratios of L-functions over function fields. *J. Number Theory* **142**, 102–48.

Arguin, L.-P., Belius, D., and Bourgade, P. (2017a). Maximum of the characteristic polynomial of random unitary matrices, *Commun. Math. Phys.* **349**, 703–751.

Arguin, L.-P., Belius, D., and Harper, A.J. (2017b). Maxima of a randomized Riemann zeta function, and branching random walks. *Ann. Applied. Prob.* **27**, 178–215.

Bektemirov, B., Mazur, B., Stein, W., and M. Watkins (2007). Average ranks of elliptic curves: Tension between data and conjecture, *Bull. Am. Math. Soc* **44**, 233–254.

Berry, M.V., and Keating, J.P. (1999). The Riemann zeros and eigenvalue asymptotics. *SIAM Rev.* **41**, 236–66.

Bogomolny, E.B., and Keating, J.P. (1995). Random matrix theory and the Riemann zeros I: three- and four-point correlations. *Nonlinearity* **8**, 1115–31.

Bogomolny, E.B., and Keating, J.P. (1996a). Random matrix theory and the Riemann zeros II: n-point correlations. *Nonlinearity* **9**, 911–35.

Bogomolny, E.B., and Keating, J.P. (1996b). Gutzwiller's trace formula and spectral statistics: beyond the diagonal approximation. *Phys. Rev. Lett.* **77**, 1472–5.

Bogomolny, E.B., and Keating, J.P. (2013). A method for calculating spectral statistics based on random-matrix universality with an application to the three-point correlations of the Riemann zeros. *J. Phys. A* **46**, 305203

Bondarenko, A., and Seip, K. (2015). Large GCD sums and extreme values of the Riemann zeta function. arXiv:1507.05840.

Bui, H. M., Keating, J. P., and Smith, D. J. (2016). Sums of arithmetic functions over short intervals]On the variance of sums of arithmetic functions over primes in short intervals and pair correlation for L-functions in the Selberg class. *J. Lond. Math. Soc.* in press.

Carpentier, D., and Le Doussal, P. (2001). Glass transition of a particle in a random potential, front selection in nonlinear renormalization group, and entropic phenomena in Liouville and sinh-Gordon models. *Phys. Rev. E* **63**, 026110.

Claeys, T., and Krasovsky, I. (2015). Toeplitz determinants with merging singularities. *Duke Math. J.* **164**, 2897–987.

Conrey, J.B. (1989). More than 2/5 of the zeros of the Riemann zeta function are on the critical line. *J. Reine. Ang. Math.* **399**, 1–26.

Conrey, J.B., and Farmer, D.W. (2000). Mean values of L-functions and symmetry. *Int. Math. Res. Notices* **17**, 883–908.

Conrey, J.B., Farmer, D.W., Keating, J.P., Rubinstein, M.O., and N.C. Snaith (2003). Autocorrelation of random matrix polynomials. *Commun. Math. Phys.* **237**, 365–95.

Conrey, J.B., Farmer, D.W., Keating, J.P., Rubinstein, M.O., and Snaith, N.C. (2005). Integral moments of L-functions. *Proc. Lond. Math. Soc.* **91**, 33–104.

Conrey, J.B., Farmer, D.W., Keating, J.P., Rubinstein, M.O., and Snaith, N.C. (2008a). Lower order terms in the full moment conjecture for the Riemann zeta function. *J. Number Theory* **128**, 1516–54.

Conrey, J. B., Farmer, D. W., and Zirnbauer, M. R. (2008b). Autocorrelation of ratios of L-functions. *Commun. Number Theory Phys.* **2**, 593–636.

Conrey, J.B., and Ghosh, A. (1992). On mean values of the zeta-function. iii. *Proceedings of the Amalfi Conference on Analytic Number Theory, Università di Salerno.*

Conrey, J.B., and Gonek, S.M. (2001). High moments of the Riemann zeta-function. *Duke Math. J.* **107**, 577–604.

Conrey, J. B., and Keating, J. P. (2015a). Moments of zeta and correlations of divisor-sums: I. *Phil. Trans. R. Soc. A* **373**, 20140313; arXiv:1506.06842

Conrey, J. B., and Keating, J. P. (2015b). Moments of zeta and correlations of divisor-sums: II. In *Advances in the Theory of Numbers—Proceedings of the Thirteenth Conference of the Canadian Number Theory Association*, Fields Institute Communications, ed A. Alaca, S. Alaca, and K.S. Williams, 75–85. Springer, Berlin. arXiv:1506.06843

Conrey, J. B., and Keating, J. P. (2015c). Moments of zeta and correlations of divisor-sums: III. *Indagationes Mathematicae* **26**, 736–47. arXiv:1506.06844

Conrey, J. B., and Keating, J. P. (2016). Moments of zeta and correlations of divisor-sums: IV. *Res. Number Theory* 2:24

Conrey, J.B., Keating, J.P., Rubinstein, M.O., and Snaith, N.C. (2002). On the frequency of vanishing of quadratic twists of modular *L*-functions, In *Number Theory for the Millennium I: Proceedings of the Millennial Conference on Number Theory*; ed. M.A. Bennett et al., 301–15. A K Peters, Natick.

Conrey, J. B., and Snaith, N. C. (2007). Applications of the *L*-functions ratios conjectures, *Proc. Lond. Math. Soc.* **94**, 594–646.

Conrey, J. B., and Snaith, N. C. (2008). Correlations of eigenvalues and Riemann zeros. *Commun. Number Theory Phys.* **2**, 477–536.

Coppola, G., and Salerno, S. (2004). On the symmetry of the divisor function in almost all short intervals. *Acta Arith.* **113**, 189–201.

Diaconis, P., and Shahshahani, M. (1994). On the eigenvalues of random matrices. *J. Appl. Probab. A* **31**, 49–62.

Dueñez, E., Huynh, D.K., Keating, J.P., Miller, S.J., and Snaith, N.C. (2012). A random matrix model for elliptic curve L-functions of finite conductor. *J. Phys. A* **45**, 115207.

Dyson, F.J. (1962). Statistical theory of the energy levels of complex systems, i, ii and iii. *J. Math. Phys.* **3**, 140–75.

Farmer, D.W., Gonek, S.M., and Hughes, C.P. (2007). The maximum size of *L*-functions. *J. Reine Angew. Math (Crelle's Journal)* **609**, 215–36.

Florea, A. (2015a). Improving the error term in the mean value of $L(1/2, \chi)$ in the hyperelliptic ensemble. arXiv:1505.03094.

Florea, A. (2015b). The second and third moment of $L(1/2, \chi)$ in the hyperelliptic ensemble. arXiv:1507.02640.

Fyodorov, Y.V., and Bouchaud, J.P. (2008). Freezing and extreme-value statistics in a random energy model with logarithmically correlated potential. *J. Phys. A: Math. Theor.* **41**, 372001.

Fyodorov, Y.V., Hiary, G.A., and Keating, J.P. (2012). Freezing transition, characteristic polynomials of random matrices, and the Riemann zeta-function. *Phys. Rev. Lett.* **108**, 170601.

Fyodorov, Y.V., and Keating, J.P. (2014). Freezing transitions and extreme values: random matrix theory, $\zeta(1/2 + it)$, and disordered landscapes. *Phil. Trans. R. Soc. A* **372**, 20120503.

Goldston, D. A., and Montgomery, H. L. (1987). Pair correlation of zeros and primes in short intervals. *Progr. Math.* **70**, 183–203.

Gonek, S.M., Hughes, C.E., and Keating, J.P. (2007). A hybrid Euler-Hadamard product for the Riemann zeta function. *Duke Math. J.* **136**, 507–49.

Good, I. J., and Churchhouse, R. F. (1968). The Riemann hypothesis and pseudorandom features of the Möbius sequence. *Math. Comput.* **22**, 857–61.

Hardy, G.H., and Littlewood, J.E. (1918). Contributions to the theory of the Riemann zeta-function and the theory of the distribution of primes. *Acta Math.* **41**, 119–96.

Hardy, G.H., and Littlewood, J.E. (1923). Some problems in 'Partitio Numerorum' III: on the expression of a number as a sum of primes. *Acta Math.* **44**, 1–70.

Harper, A.J. (2013). A note on the maximum of the Riemann zeta function, and log-correlated random variables. arXiv:1304.0677.

Heath-Brown, D.R. (2006). Prime number theory and Riemann zeta function. In *Recent Perspectives in Random Matrix Theory and Number Theory*, ed. F. Mezzardi and N.C. Snaith, 1–30. London Mathematical Society Lecture Note Series, 322 Cambridge University Press, Cambridge.

Hughes, C.P., Keating, J.P., and O'Connell, N. (2001). On the characteristic polynomial of a random unitary matrix. *Commun. Math. Phys.* **220**, 429–51.

Ingham, A.E. (1926). Mean-value theorems in the theory of the Riemann zeta-function. *Proc. London Math. Soc. (2)* **27**, 273–300.

Ivić, A. (2009). On the mean square of the divisor function in short intervals. *J. Théor. Nombres Bordeaux* **21**, 251–61.

Jutila, M. (1984). On the divisor problem for short intervals, in *Studies in honour of Arto Kustaa Salomaa on the occasion of his fiftieth birthday, Ann. Univ. Turku. Ser. A I* **186**, 23–30.

Katz, N.M., and Sarnak, P. (1999a). *Random Matrices, Frobenius Eigenvalues and Monodromy.* American Mathematical Society Colloquium Publications, 45. American Mathematical Society, Providence, RI.

Katz, N.M., and Sarnak, P. (1999b). Zeros of zeta functions and symmetry. *Bull. Amer. Math. Soc.* **36**, 1–26.

Keating, J.P. (1993). The Riemann zeta function and quantum chaology. In *Quantum Chaos*, ed. G. Casati, I. Guarneri, and U. Smilansky, 145–85. North-Holland, Amsterdam.

Keating, J.P. (2005). *L*-functions and the characteristic polynomials of random matrices. In *Recent Perspectives in Random Matrix Theory and Number Theory*, ed. F. Mezzadri and N.C. Snaith, 251–78. London Mathematical Society Lecture Note Series 322. Cambridge University Press, Cambridge.

Keating, J.P., Linden, N., and Rudnick, Z. (2003). Random matrix theory, the exceptional Lie groups, and *L*-functions, *J. Phys. A-Math. Gen.* **36**, 2933–44.

Keating, J.P., Rodgers, B., Roditty-Gershon, E., and Rudnick, Z. (2017). Sums of divisor functions in $\mathbb{F}_q[t]$ and matrix integrals, *Math. Z.*, in press.

Keating, J.P., and Rudnick, Z. (2014). The variance of the number of prime polynomials in short intervals and in residue classes. *IMRN* **2014**, 259–88.

Keating, J.P., and Rudnick, Z. (2016). Squarefree polynomials and Mobius values in short intervals and arithmetic progressions. *Algebra and Number Theory* **10**, 375–420.

Keating, J.P., Rudnick, Z., and Wooley, T.D. (2015b). Number fields and function fields: coalescences, contrasts and emerging applications. *Phil. Trans. R. Soc. A* **373**.

Keating, J.P., and Scott, E. (2015). unpublished.

Keating, J.P., and Snaith, N.C. (2000a). Random matrix theory and $\zeta(1/2 + it)$. *Commun. Math. Phys.* **214**, 57–89.

Keating, J.P., and Snaith, N.C. (2000b). Random matrix theory and *L*-functions at $s = 1/2$. *Commun. Math. Phys* **214**, 91–110.

Keating, J.P., and Snaith, N.C. (2003). Random matrices and *L*-functions. *J. Phys. A* **36**, 2859–81.

Keating, J.P., and Snaith, N.C. (2011). Random matrix theory and number theory. In *The Handbook on Random Matrix Theory*, ed. G. Akemann, J. Baik, and P. Di Francesco. Oxford University Press, Oxford.

Lester, S. (2015). The variance of sums of divisor functions in short intervals. arXiv:1502.01170.

Mehta, M.L. (1991). *Random Matrices*, second edn Academic Press, London.

Montgomery, H.L. (1973). The pair correlation of zeros of the zeta function. *Proc. Symp. Pure Math.* **24**, 181–93.

Montgomery, H. L., and Soundararajan, K. (2004). Primes in short intervals. *Commun. Math. Phys.* **252**, 589–617.

Odlyzko, A.M. (1989). The 10^{20}th zero of the Riemann zeta function and 70 million of its neighbors. *Preprint*.

Paquette, E., and Zeitouni, O. (2016). The maximum of the CUE field. arXiv:1602.08875.

Rubinstein, M.O., and Wu, K. (2015). Moments of zeta functions associated to hyperelliptic curves over finite fields. *Phil. Trans. R. Soc. A* **373**, 20140307.

Rudnick, Z. (2014). Some problems in analytic number theory for polynomials over a finite field. arXiv:1501.01769.

Rudnick, Z., and Sarnak, P. (1996). Zeros of principal *L*-functions and random matrix theory. *Duke Math. J.* **81**, 269–322.

Selberg, A. (1946). Contribution to the theory of the Riemann zeta-function. *Arch. Math. Naturvid.* **48**, 89–155.

Soundararajan, K. (2009). Moments of the Riemann zeta function. *Ann. Math.* **170**, 981–93.

Titchmarsh, E.C. (1986). *The Theory of the Riemann Zeta Function*. Oxford University Press, Oxford.

Webb, C. (2015). The characteristic polynomial of a random unitary matrix and Gaussian multiplicative chaos—the L^2-phase. *Electronic J. Prob.* **20**, 1–21.

Weyl, H. (1946). *Classical Groups*. Princeton University Press, Princeton, NJ.

Widom, H. (1973). Toplitz determinants with singular generating functions. *Am. J. Math.* **95**, 333–83.

9

Modern telecommunications: a playground for physicists?

Aris L. MOUSTAKAS

Department of Physics, National and Kapodistrian University of Athens, Greece

Moustakas, A.L., 'Modern Telecommunications: A Playground for Physicists?' in *Stochastic Processes and Random Matrices*. Edited by: Grégory Schehr et al, Oxford University Press (2017).
© Oxford University Press 2017. DOI 10.1093/oso/9780198797319.003.0009

Chapter Contents

9.1   Introduction

Data traffic in wireless networks has been increasing exponentially for a long time and is expected to continue this trend. The emerging data-hungry applications, such as video-on-demand and cloud computing, as well as the exploding number of smart user devices demand the introduction of disruptive technologies. An analogous situation appears in the case of wireline (mostly fiber-optical) traffic, where the currently deployed infrastructure is expected to soon reach its limits, leading to the so-called 'capacity crunch' (Tkach 2010).

One way to counter this trend is the parallelization of information transmission in the spatial domain, thereby transmitting multiple data streams in parallel by using the same infrastructure (antennas) over the air in wireless communications or within the same optical fiber. The challenge is that, unlike the parallel use of orthogonal frequencies, the cross-talk between the different data streams can be significant, since there are no naturally occurring orthogonal modes due to the randomness of the medium. Foschini and Gans (1998) first developed an algorithm in the context of wireless communications for multiple antennas at the transmitter and receiver that could compensate this additional interference and promise unprecedented increases in data throughput. The acronym used for this system in the engineering community is 'MIMO', signifying multiple input and multiple output data streams.

Since it was first proposed, the technology has matured enough, at least in the context of wireless communications, so that current projections of what the next generation wireless systems will likely be envision massive (in terms of their number) antenna arrays transmitting parallel streams of data to many users nearby (hence called massive MIMO) (Andrews et al. 2014). Not surprisingly, similar projections have been made for the case of fiber-optical communications, where fibers with multiple cores have been proposed (Morioka et al. 2012).

It is therefore important to analyse the performance of such MIMO systems in the environments they are envisioned to operate. One very useful tool in this direction has been random matrix theory, with the help of which both exact and asymptotic expressions for various quantities of interest have been derived. After all in several of the occurring problems, such as massive MIMO mentioned in the previous paragraph, the asymptotic limit usually taken in random matrix theory is actually realistic. Therefore, such results are useful for performance prediction and network design, but also for providing intuition to system engineers on the way the network operates. This is so, because the obtained results show which system parameters are relevant, and which not. As a result, research in this field can be rewarding both for its scientific rigor but also for the direct applicability of its results.

The aim of this chapter is to introduce the physics and mathematics community to a number of relevant problems in communications research and the types of solutions that have been used to tackle them. In the process, interested readers may be able to further acquaint themselves with research in engineering bibliography cited herein.

9.1.1   Outline

After a brief introduction to basic metrics and quantities of interest in Section 9.2, Section 9.3 describes the solution to two problems in the context of wireless communications. More specifically, in Section 9.3.1 the statistics of information capacity in wireless MIMO systems are analysed, while Section 9.3.2 deals with the effects of macroscopic mobility of users. Section 9.4 provides two different ways to calculate the statistics of the mutual information in fiber-optical communications, all using various methods of random matrix theory. Generalizations, similar problems, shortcomings, and open problems are also mentioned in the text.

9.2   Information theory basics

In this section we introduce a few metrics that are relevant in information transmission, and will be used in further sections.

9.2.1   Information capacity

A key quantity in information theory is the mutual information between an input random variable X and an output random variable Y and is defined as

$$I(X,Y) = -\int dX \int dY \Pr(X,Y) \left[\log \Pr(Y) - \log \Pr(Y|X) \right], \qquad (9.1)$$

where the probability distribution $\Pr(Y|X)$ describes the type of noise the input X is subjected to, in order to produce the output Y. The maximum of this quantity with respect to the input distribution $\Pr(X)$, subject to certain constraints, such as maximum transmitted power, is called information capacity and represents the maximum number of nats (which are bits in the Neperian basis) that can be transmitted error-free per channel use. For a simple additive Gaussian-noise channel of the form

$$Y = \sqrt{\rho} X + Z, \qquad (9.2)$$

where $Z \sim \mathcal{CN}(0,1)$ is the noise, and ρ is the signal-to-noise ratio, the mutual information is maximized with a Gaussian input of unit variance $X \sim \mathcal{CN}(0,1)$. In this case the capacity can be expressed as (Cover and Thomas 1991)

$$I = \log \left(1 + \rho \right). \qquad (9.3)$$

The above analysis can be generalized in the case of N transmit and receive antennas with an average power constraint imposed at the transmitter. The corresponding channel equation can be expressed as

$$\mathbf{y} = \sqrt{\frac{\rho}{N}} \mathbf{G} \mathbf{x} + \mathbf{z}, \qquad (9.4)$$

where now $\mathbf{G}$ is the matrix of channel coefficients between the transmit and receive antennas and $\mathbf{y}$ and $\mathbf{z}$ are the N-dimensional output signal and noise vectors, respectively, with the latter assumed to be independent and complex Gaussian with unit

variance. In this case as well, the optimum input distribution is complex Gaussian. If
G is known at the transmitter the input covariance matrix $E[\mathbf{x}\mathbf{x}^\dagger]$ can be optimized
to take advantage of this knowledge. However, for simplicity, here we assume that
this information is not available at the receiver, in which case the covariance is unity,
i.e., $E[\mathbf{x}\mathbf{x}^\dagger] = \mathbf{I}_N$. When the channel is known at the receiver then the information
capacity (in nats) is

$$I_N = \log \det \left(\mathbf{I}_N + \rho GG^\dagger \right). \tag{9.5}$$

Note that for convenience, we have absorbed a factor of $N^{-1/2}$ in the definition of
G. This expression is also valid for channels where the transmitter has n_t antennas
available and the receiver has n_r antennas, i.e., when G is $n_t \times n_r$. In this case,
we need to replace $\mathbf{I}_N$ by $\mathbf{I}_{n_r}$. The capacity represents the maximum rate that can
be transmitted error-free for a given channel matrix G. Since the channel matrix
is randomly distributed the capacity itself is a random quantity. Its average $E[I_N]$
provides an estimate of what kind of throughput rate one should expect on average.
However, since G varies (albeit slowly) over time, the instantaneous rate must be fed
back to the transmitter to encode the data accordingly. If this is not possible, there
is always a finite probability that G will change in such a way that the encoded rate
is not supported in the transmission and errors will occur. In this case the outage
capacity is relevant, which is defined as the value R_{out} of the cumulative distribution
of I_N above for which the probability that $I_N < R$ is p_{out}, i.e.,

$$p_{\text{out}} = \Pr(I_N < R_{\text{out}}). \tag{9.6}$$

Therefore, the full distribution of I_N is important to characterize the transmission
performance.

9.2.2   Linear precoders

In the previous subsection we described the performance of a system of transmit
and receive antenna arrays, assuming that the received signal from the antennas can
be jointly processed. Often, however, the receive antennas are not collocated as they
correspond to different mobile users communicating with a multiantenna base-station.
This is the typical situation in a so-called massive MIMO system. In this case the
information capacity of each user takes the form of (9.3) with ρ substituted by an
appropriately defined signal-to-noise-ratio. However, in this case there is significant
interference between users. One way to counter this is to pre-multiply the signal vector
at the transmitter with an appropriately chosen matrix $\mathbf{V}$. Due to the linearity of
matrix multiplication, this approach is called linear precoding. As a result, the received
signal at user $k = 1, \ldots, n_r$ can be expressed as

$$y_k = \mathbf{g}_k^T \mathbf{V}\mathbf{x} + \sigma z_k, \tag{9.7}$$

where $\mathbf{g}_k^T$ is the kth row of the matrix G. Clearly, this only makes sense if the transmitter has some information about the G. There are several forms of precoding matrices,

one of which is the so-called 'zero-forcing' precoding matrix, which amounts to the pseudo-inverse of G, i.e.,

$$\mathbf{V} = G^\dagger \left(GG^\dagger\right)^{-1} \mathbf{P}^{1/2}, \tag{9.8}$$

where $\mathbf{P}$ is a diagonal matrix with elements the designated receive powers of each user p_k. Clearly, this matrix exists only if $n_t \leq n_r$. The benefit of using this precoding matrix is that the signal at each receiver is completely decoupled. Indeed plugging (9.8) into (9.7) results in the trivial

$$y_k = \sqrt{p_k} x_k + \sigma z_k, \tag{9.9}$$

and therefore the signal-to-noise ratio requirements are immediately met if $p_k = \rho_k * \sigma^2$, where ρ_k is the requested signal-to-noise ratio. The price for this is the increased transmitted power, which can be evaluated to be

$$P_{\text{tot}} = \frac{1}{n_t} \text{Tr} \left[\left(GG^\dagger\right)^{-1} \mathbf{P}\right]. \tag{9.10}$$

Additional precoding techniques exist in the literature (Sanguinetti et al. 2014), which tend to trade between interference cancellation at the receiver end and power consumption or channel information at the transmitter. When the receiver is equipped with multiple antennas, similar techniques can be applied there as well. However, in all cases one is left with an object, such as in (9.10), which depends on the channel randomness. Hence once again, random matrix theory can be of immediate help to get quantitative estimates.

9.3 Wireless communications: replicas and mobility

In this section we will provide two specific applications of random matrix theory in wireless communications.

Before moving ahead, it is important to introduce the statistics of the propagation channel matrix G. A good and reliable model for its elements $G_{i\alpha}$ is that they are complex Gaussian random variables due to multiple scattering. The correlations of the matrix elements can be evaluated in the diffusion approximation to be (Moustakas et al. 2000)

$$E\left[G_{i\alpha} G^*_{j\beta}\right] = \frac{\rho}{n_t} R_{ij} T_{\alpha\beta}. \tag{9.11}$$

In Eq. (9.11), R_{ij} and $T_{\alpha\beta}$ are the elements of the correlation matrices between the antennas at the receiver and transmitter arrays, respectively. R_{ij} can be expressed as (Moustakas et al. 2000)

$$R_{ij} = \ell\left(\mathbf{r}\right) \int d\Omega_{\mathbf{k}} \, \chi_i(\mathbf{k}) \, \chi_j(\mathbf{k}) \, e^{i\mathbf{k}\mathbf{d}_{ij}} \, w(\mathbf{k}), \tag{9.12}$$

where $\chi_i(\mathbf{k})$ is the response of the antenna i at incoming wavevector $\mathbf{k}$, $\mathbf{d}_{ij}$ the vector between antennas i and j, and $w(\mathbf{k})$ the weight of incoming power with a similar expression for $T_{\alpha\beta}$ (without the $\ell(\cdot)$-term). Thus, the further apart antennas are located in space, the less correlated they are, and the more evenly distributed over angles the incoming (or relevant outgoing) power is. Also, $\ell(\cdot)$ is the average power loss due to propagation and R_{tr} is the distance between receiver and transmitter array. A typical model for $\ell(x)$ is $\ell(x) = |x|^{-\beta}$, where the path-loss exponent is usually taken to be $\beta = 4 - 5$ (Calcev et al. 2007). Also, we have included the factor $1/n_t$ here that was absorbed into G earlier.

9.3.1 Capacity of correlated antennas

In this section we introduce a method based on replicas to obtain the asymptotic moments of the capacity distribution in the large antenna limit assuming the channel model discussed earlier. This methodology was first developed by Sengupta and Mitra (2006) and extended in Moustakas et al. (2000, 2003). While not rigorous it provides results in a few number of steps, which took a while to be established rigorously (Hachem et al. 2008).

The starting point is the moment-generating function

$$g(-\mu) = E\left[e^{-\mu I_N}\right] = E\left[\det\left(\mathbf{I}_{n_r} + \rho \mathsf{G}\,\mathsf{G}^{\dagger}\right)^{-\mu}\right]. \tag{9.13}$$

The key trick in the calculation is to express this determinant as a Gaussian complex integral, so that the matrices G will appear in the exponent and can then be averaged over. After some algebra we obtain

$$g(-\mu) = E\left[\int d\mathbf{X}\int d\mathbf{Y}\,e^{-\frac{1}{2}\mathrm{Tr}[\mathbf{X}^{\dagger}\mathbf{X}+\mathbf{Y}^{\dagger}\mathbf{Y}+\sqrt{\rho}\mathbf{X}^{\dagger}\mathsf{G}\mathbf{Y}-\sqrt{\rho}\mathbf{Y}^{\dagger}\mathsf{G}^{\dagger}\mathbf{X}]}\right], \tag{9.14}$$

where $\mathbf{X}$ and $\mathbf{Y}$ are $n_r \times \mu$ and $n_t \times \mu$ dimensional complex matrices with the appropriate integration measure $d\mathbf{X}$ and $d\mathbf{Y}$, respectively. After averaging over G, we obtain

$$g(-\mu) = \iint d\mathbf{X}d\mathbf{Y}\,e^{-\frac{1}{2}\mathrm{Tr}[\mathbf{X}^{\dagger}\mathbf{X}+\mathbf{Y}^{\dagger}\mathbf{Y}+\frac{\rho}{2n_t}\mathbf{X}^{\dagger}\mathsf{R}\mathbf{X}\mathbf{Y}^{\dagger}\mathsf{T}\mathbf{Y}]}. \tag{9.15}$$

The quartic term in the exponent cannot be integrated as such. However, in the large n_t limit we can treat it in a mean-field way. We now introduce the $\mu \times \mu$ matrices $\mathcal{T}$ and $\mathcal{R}$ through the identity

$$1 = \int D\mathcal{T}\,\delta\left(\mathcal{T} - \frac{\rho}{2\sqrt{n_t}}\mathbf{X}^{\dagger}\mathsf{R}\mathbf{X}\right) = \iint D\mathcal{T}D\mathcal{R}\,e^{\mathrm{Tr}\left[\mathcal{R}\mathcal{T} - \frac{\rho}{2\sqrt{n_t}}\mathcal{R}\mathbf{X}^{\dagger}\mathsf{R}\mathbf{X}\right]}, \tag{9.16}$$

where the δ-function appearing in is Eq. (9.16) shorthand notation for a product of δ-functions on all real and imaginary parts of the elements of the matrix $\mathcal{T}$. The integration of the elements of $\mathcal{T}$ is over the real axis, while that of the elements of $\mathcal{R}$

is over the imaginary axis, in agreement with Fourier integration. We then insert this identity into (9.15) getting

$$g(-\mu) = \iint D\mathcal{T} D\mathcal{R} \, \mathrm{e}^{\mathrm{Tr}(\mathcal{T}\mathcal{R})} \iint \mathrm{d}\mathbf{X} \mathrm{d}\mathbf{Y} \, \mathrm{e}^{-\frac{1}{2}\mathrm{Tr}\left[\mathbf{X}^{\dagger}\mathbf{X} + \mathbf{Y}^{\dagger}\mathbf{Y} + \frac{\rho}{\sqrt{n_t}}\mathbf{X}^{\dagger}\mathbf{R}\mathbf{X}\mathcal{R} + \frac{1}{\sqrt{n_t}}\mathbf{Y}^{\dagger}\mathbf{T}\mathbf{Y}\mathcal{T}\right]},$$

(9.17)

which, after integrating over $\mathbf{X}$, $\mathbf{Y}$, reduces to

$$g(-\mu) = \iint D\mathcal{T} D\mathcal{R} \, \mathrm{e}^{-\mathcal{S}}$$

(9.18)

$$\mathcal{S} = \log\det\left[\mathbf{I}_{n_t} \otimes \mathbf{I}_\mu + \frac{1}{\sqrt{n_t}}\mathbf{T} \otimes \mathcal{T}\right] + \log\det\left[\mathbf{I}_{n_r} \otimes \mathbf{I}_\mu + \frac{\rho}{\sqrt{n_t}}\mathbf{R} \otimes \mathcal{R}\right] - \mathrm{Tr}\left[\mathcal{T}\mathcal{R}\right].$$

The remaining integrals over the elements of the matrices $\mathcal{T}$, $\mathcal{R}$ will be performed using the saddle-point method. To do so, we need to 'guess' the structure of these matrices at the saddle-point. In the original spin-glass literature (Mézard et al. 1987) where the replica approach was introduced, the dynamic degrees of freedom are usually spin variables taking discrete values. Hence the corresponding correlation matrices at the replica symmetric saddle-point need to be invariant under the symmetric group, i.e., the set of all permutations of the replica indices. In contrast, here the dynamic variables, i.e, $\mathbf{X}$, $\mathbf{Y}$, are continuous and thus have $U(\mu)$ rotational symmetry in replica space. Therefore, at the replica-symmetric saddle-point, $\mathcal{T}$ and $\mathcal{R}$ need to be scalars, which we express them as

$$\mathcal{T} = t\sqrt{n_t}\,\mathbf{I}_\mu + \delta\mathcal{T}$$

(9.19)

$$\mathcal{R} = r\sqrt{n_t}\,\mathbf{I}_\mu + \delta\mathcal{R}.$$

Plugging these expressions into (9.18) we get to leading order

$$g(-\mu) \approx \mathrm{e}^{-\mathcal{S}_0}$$

$$\mathcal{S}_0 \equiv \mu\Gamma_0 = \mu\left(\log\det\left[\mathbf{I}_{n_t} + t\mathbf{T}\right] + \log\det\left[\mathbf{I}_{n_r} + \rho r\mathbf{R}\right] - n_t rt\right),$$

(9.20)

with r, t satisfying the saddle-point equations

$$r = \frac{1}{n_t}\mathrm{Tr}\left[\frac{\mathbf{T}}{\mathbf{I}_{n_t} + t\mathbf{T}}\right]$$

(9.21)

$$t = \frac{1}{n_t}\mathrm{Tr}\left[\frac{\rho\mathbf{R}}{\mathbf{I}_{n_r} + \rho r\mathbf{R}}\right].$$

Since $g'(0) = -E[I_N]$, we immediately see that to leading order in n_t, $E[I_{n_t}] = \Gamma_0$. To obtain higher moments of the distribution, we need to expand $\mathcal{S}$ in powers of $\delta\mathcal{R}$ and $\delta\mathcal{T}$. At the saddle-point, the linear terms vanish; hence, the leading term is the quadratic one,

$$\mathcal{S} = \mathcal{S}_0 - \frac{1}{2}\sum_{\mu,\nu}[\delta\mathcal{R}_{\mu\nu}, \delta\mathcal{T}_{\mu\nu}]\begin{pmatrix} r_2 & 1 \\ 1 & t_2 \end{pmatrix}\begin{bmatrix} \delta\mathcal{R}_{\nu\mu} \\ \delta\mathcal{T}_{\nu\mu} \end{bmatrix} + \mathcal{O}\left(\delta\mathcal{R}^3, \delta\mathcal{T}^3\right),$$

(9.22)

where

$$r_2 = \frac{1}{n_t} \text{Tr} \left[\frac{\mathbf{T}^2}{(\mathbf{I}_{n_t} + t\mathbf{T})^2} \right] \tag{9.23}$$

$$t_2 = \frac{1}{n_t} \text{Tr} \left[\frac{\rho^2 \mathbf{R}^2}{(\mathbf{I}_{n_r} + \rho r \mathbf{R})^2} \right].$$

Integrating over the quadratic term in the exponent by appropriately rotating the contour of integration close to the saddle-point, we obtain

$$g(-\mu) \approx e^{-\mu \Gamma_0 - \frac{\mu^2}{2} \log(1 - r_2 t_2)}. \tag{9.24}$$

As a result, the variance of the mutual information takes the simple form

$$\text{Var}(I_N) = -\log\left(1 - r_2 t_2\right). \tag{9.25}$$

It is worth contrasting this result with the standard central limit theorem for the sum of N random variables, in which the mean and the variance of the sum is $\mathcal{O}(N)$. Here the mean is $\mathcal{O}(N)$, while the variance is $\mathcal{O}(1)$. This vast reduction of fluctuations can be attributed to the fact that the underlying $\mathcal{O}(N)$ random degrees of freedom, i.e., the eigenvalues of the matrix $\mathsf{GG}^\dagger$, are highly correlated and (as we will see in Section 9.4.2) they are constrained to have positions very close to each other.

If we continue the perturbation expansion by including cubic and quartic terms in $\delta\mathcal{T}, \delta\mathcal{R}$, we obtain a $\mathcal{O}(1/n_t)$ correction term to $E[I_{n_t}]$ and a skewness of the same order (Moustakas et al. 2003). In fact, it can be established that all higher moments vanish when $n_t, n_r \to \infty$, thereby making the distribution asymptotically Gaussian. Interestingly, it can also be shown that the replica-symmetric saddle-point is stable (Moustakas and Simon 2007).

One reason these results are quite useful is that they are applicable not only for the case of very large antenna numbers, but also for just a few antennas. This can be seen explicitly in Fig. 9.1, where the agreement with simulations is remarkable even for $n_t = 3$. In conclusion, we have seen a first example where random matrix theory can provide useful results in wireless communications.

9.3.2   Effect of mobility on energy consumption

In addition to traffic growth, another related big challenge is the increasing energy consumption of cellular infrastructure equipment. As a result, energy consumption must be a key ingredient in the design of future cellular networks, especially in new rural regions of the developing world, where the electrical grid is unreliable or even nonexisting. In this section we will analyse the distribution of energy consumption for a particular case of linear precoding discussed in Section 9.2 when we take the mobility of users into account. Once again, the large system size will simplify the analysis considerably.

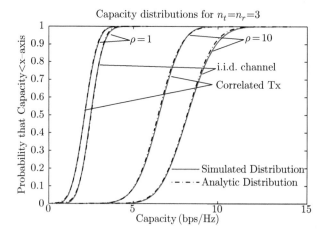

Fig. 9.1 Cumulative probability distributions of the capacity for a system with $n_t = n_r = 3$ antennas. One set of curves corresponds to uncorrelated antennas (denoted as 'i.i.d'), while the transmitter antennas of the other set are located on a line with distance between neighboring antennas fixed to $d = \lambda$ apart with a $\delta = 5°$ angle spread. Two different values of ρ are used. The theoretical curves are Gaussian distributions, with mean and variance calculated in 9.20 and 9.25, respectively. The agreement with the simulated curves is quite good.

We consider a base-station (BS) with n_t antennas serving n_r mobile single antenna users with $c \equiv n_r/n_t < 1$, in a square region centered at the BS with side length L. Here we focus in the downlink case, where the BS acts as a transmitter. In order to guarantee a certain signal-to-noise ratio ρ_k to user k, the transmitting array precodes the signal using the zero-forcing precoding matrix appearing in (9.8). As a result, the total transmitted power is given by (9.10). This power is time-dependent due to the movement of the users through the temporal variation of the channel coefficients. This, in turn, has two components. One originates from the relatively slow variation of the path loss due to the macroscopic movement of the users. The characteristic time for this is $\sim L/v$, where v is the typical velocity of the users. There is another much faster variation of G due to multiple (or Rayleigh) scattering. In the engineering literature, these fluctuations are called 'fast-fading'. The timescale here is much shorter, $\sim \lambda/v$. The analysis in the following will distinguish between these two processes.

For concreteness, we employ a simple mobility model for users, namely that of a Brownian motion, which is the continuous version of a simple random walk. Hence, $\Pr(\mathbf{x}, \mathbf{x}'; t - t')$, the probability of a user to be at position $\mathbf{x}$ at time t given that he was at x' at time t' satisfies the diffusion equation

$$\frac{\partial \Pr(\mathbf{x}, \mathbf{x}'; t - t')}{\partial t} = D\nabla^2 \Pr(\mathbf{x}, \mathbf{x}'; t - t'), \tag{9.26}$$

where the diffusion constant D characterizes the small scale mobility of the user. Further, we assume periodic boundary conditions at the borders of the square to mimic the existence of other users in neighboring cells entering the current cell.

The total energy consumed by the BS over time T is given by

$$E_T = \int_0^T P(t)\mathrm{d}t. \tag{9.27}$$

The aim of this section is to calculate the statistics of this quantity. We start by averaging the power $P(t)$ over fast-fading, keeping the positions of the users (roughly) fixed. This step can be done by using the results of the previous section. Starting from (9.10) we redefine the channel matrix $\mathbf{GP}^{-1/2} \to \mathbf{G}$, hence redefining the receiver correlation matrix $\mathbf{RP}^{-1} \to \mathbf{R}$. Then we observe that the expression in (9.10) can obtained from (9.20) by taking the $\rho \to \infty$ limit. More concretely,

$$P(t) = \lim_{\rho \to \infty} \left(-\rho^2 \frac{\mathrm{d}}{\mathrm{d}\rho} \left[\log \det \left[\mathbf{I}_{n_r} + \rho \mathbf{GG}^\dagger \right] - n_r \log \rho \right] \right). \tag{9.28}$$

Plugging in (9.20) into (9.28) (and re-introducing the matrix $\mathbf{P}$) gives

$$\begin{aligned}
\hat{P}(t) &\equiv E\left[P(t)|\{\mathbf{x}_k\}\right] \\
&= \frac{1}{n_t r} \mathrm{Tr}\left[\mathbf{R}^{-1}\mathbf{P}\right] \\
&= \frac{\sigma^2}{n_t r} \sum_{k=1}^{n_r} \frac{\rho_k}{\ell(\mathbf{x}_k(t))}.
\end{aligned} \tag{9.29}$$

The last line results from the fact that since the receive antennas are so far apart, they are uncorrelated, i.e., $R_{ij} = \ell(\mathbf{x})\delta_{ij}$. r can be found in this limit to be the solution of

$$\frac{1}{n_t} \sum_{j=1}^{n_t} \frac{T_j}{r + cT_j} = 1, \tag{9.30}$$

where T_j are the eigenvalues of the BS antenna correlation matrix $\mathbf{T}$. We see that when the transmitter antennas are uncorrelated $r = 1 - c$.

Now, $\hat{P}(t)$ is time dependent only due to the macroscopic movement of the mobile users $\mathbf{x}_k(t)$. Averaging over their movements as well, we obtain

$$\overline{P} = \frac{\sigma^2 c}{r} E\left[\ell^{-1}(\mathbf{x})\right] \frac{1}{n_r} \sum_{k=1}^{n_r} \rho_k. \tag{9.31}$$

Since the long-time spatial distribution of the Brownian motion is uniform the expectation above is over the whole square. As a result,

$$\overline{E}_T = \frac{\sigma^2 cT}{r} E\left[\ell^{-1}(\mathbf{x})\right] \frac{1}{n_r} \sum_{k=1}^{n_r} \rho_k. \tag{9.32}$$

To calculate the fluctuations of the consumed energy at the transmitter, we separate them into two parts according to their corresponding timescales as discussed earlier.

Hence,

$$P(t) - \overline{P} = \left(P(t) - \hat{P}(t)\right) + \left(\hat{P}(t) - \overline{P}\right). \tag{9.33}$$

The first part has fluctuations due to fast-fading, while the second due to user mobility. It can be observed from the relation with the previous section that the fluctuations of the first part scale as $1/n_r^2$. This has been rigorously established in Bai and Silverstein (2004). Hence since the decorrelation time is $\sim \lambda/v$ we conclude that the variance of the energy due to fast-fading will be $\sim \frac{T\lambda/v}{n_r^2}\overline{P}^2$. In contrast, as we will see, the fluctuations of $\hat{P}(t) - \overline{P}$ are of order $\mathcal{O}\left(n_r^{-1}\right)$. This is so because $\hat{P}(t)$, see (9.29), has n_r independent degrees of freedom (the positions of the n_r mobiles user). Since the decorrelation time of the Brownian motion is $\sim L^2/D$, the variance of this energy term will be of the order $\sim \frac{TL^2}{n_r D}\overline{P}^2$. Indeed, we can express the variance of the energy as

$$\mathrm{Var}(E_T) = \frac{1}{2}\int_0^T \mathrm{d}t \int_0^t \mathrm{d}t \int \mathrm{d}\mathbf{x} \int \mathrm{d}\mathbf{x}' \ell^{-1}(\mathbf{x})\ell^{-1}(\mathbf{x}') \tag{9.34}$$
$$\times \left(\mathrm{Pr}(\mathbf{x}, \mathbf{x}'; t - t') - \mathrm{Pr}(\mathbf{x}, \mathbf{x}'; t + t')\right),$$

for which, after expressing the diffusion probabilities in terms of the eigenfunctions of the diffusion equation $\Psi_{\mathbf{n}}(\mathbf{x}) = e^{i\mathbf{k}_{\mathbf{n}}\mathbf{x}}$ with eigenvalues $\mathbf{k}_{\mathbf{n}} = 2\pi\mathbf{n}/L$, where $\mathbf{n} \in \mathbb{Z}^2$, we obtain

$$\mathrm{Var}(E_T) = \frac{TL^2}{8\pi^2 D}\sum_{\mathbf{n}}\frac{\theta_{\mathbf{n}}^2}{|\mathbf{n}|^2} \tag{9.35}$$

$$\theta_{\mathbf{n}} = \int_{-1/2}^{1/2}\mathrm{d}x \int_{-1/2}^{1/2}\mathrm{d}y\, \ell^{-1}\left(\mathbf{x}L\right) e^{i2\pi\mathbf{n}\mathbf{x}}.$$

Since $\theta_{\mathbf{n}}$ falls off fast with increasing $|\mathbf{n}|$, the summation in Eq. (9.35) converges fast and only a few terms are necessary to evaluate it.

All higher moments of the energy can be easily shown to vanish faster in the large n_r limit. Hence, the energy consumption becomes a Gaussian variable with mean and variance calculated in Eqs. (9.32) and (9.35). This model can be used to approximate the probability that a battery-powered BS runs out of energy and also to design the cell radius for minimizing the energy consumption per unit area (Sanguinetti et al. 2014).

9.3.3   Discussion

In this section we briefly discuss various generalizations of the results of the previous section.

The results presented in Section 9.3.1 have been generalized to the calculation of the statistics of the capacity in cases where the channel matrix is not Gaussian. It turns out that only the second moment of the distribution is relevant (Couillet et al. 2011), at least for the mean capacity. Also, the methodology can be applied to

situations where the interference itself is a random matrix. In this case, the interference channel appears in two logarithms, which have different sign. In such a case, one needs to rely on supersymmetric methods, introducing integrals over Grassman variables (Moustakas et al. 2003, Taricco 2008). This results have still not been proved with more rigorous methods. Another generalization deals with the case, where the input distribution is binary rather than Gaussian (Müller 2003), in which case the replica approach is perhaps the only one that can provide an answer. The random matrix analysis of the behavior of precoders as in Section 9.3.2 is currently an active topic of research, due to their possible application in next-generation communications; see, for example, Wagner et al. (2012). Also, since precoders usually involve matrix inversions (9.8), a series of works has analysed the robust representation of precoders in terms of matrix polynomials (Müller and Verdú 2001). Finally, note that precoders do not necessarily need to be linear, and optimization over the nonlinear precoders has also been analysed using replicas and random matrix theory (Zaidel et al. 2013).

9.4   Optical communications: moments and tails

One way to increase the data throughput though optical fibers is to use more channels in each fiber. At a first level one can use more than one electromagnetic mode through existing fibers (Hsu et al. 2006, Tarighat et al. 2007). At a later stage, engineers envision a new generation of optical fibers, specially designed with multiple cores in each of them (Winzer and Foschini 2011, Morioka et al. 2012). Due to twisting, bending, and nonlinear coupling, these propagation channels mix strongly, especially when the fiber length extends over long distances. In contrast, backscattering can be assumed to be negligible. Another important difference from free space propagation in wireless communications is that fiber optical transmission is characterized by low loss. Hence the appropriate metric for describing the propagation is the transmission coefficients of the scattering matrix, since there is no reflection.

Although the total scattering matrix should be symmetric $\mathbf{S} = \mathbf{S}^T$ due to time reversal symmetry (Beenakker 1997), the transmission matrix $\mathbf{U}$ itself does not have any other symmetries or constraints, apart from the normalization condition $\mathbf{U}\mathbf{U}^\dagger = \mathbf{I}_N$, which is a direct consequence of the unitarity of $\mathbf{S}$. Therefore, in the strong mixing limit, we may neglect any bias between the various modes or cores or inhomogeneity in the mixing and assume that $\mathbf{U}$ is Haar random. It is convenient to define the channel matrix as $\mathsf{G} = \mathbf{T}^{1/2}\mathbf{U}\mathbf{T}^1/2$, where $\mathbf{R}$ and $\mathbf{T}$ are the correlations matrices of the transmitted and received signal, respectively. For example, in the optical fiber case they correspond to reflections and losses at the two edges of the link.

As a result, the corresponding MIMO channel for this system reads

$$\mathbf{y} = \mathbf{U}\mathbf{x} + \mathbf{z}, \tag{9.36}$$

with coherent detection and channel state information only at the receiver (Foschini and Gans 1998, Telatar 1999). $\mathbf{x}$, $\mathbf{y}$, and $\mathbf{z}$ are the N-dimensional input, output signal vectors and unit variance noise vector, respectively, all assumed for simplicity to be complex Gaussian. We also assume no differential delays between channels, which

effectively leads to frequency flat fading (Winzer and Foschini 2011) and no mode-dependent loss. As a result, the mutual information (in nats) can be expressed as

$$I_N(\mathbf{U}) = \log \det(I + \rho \mathsf{G}^\dagger \mathsf{G}). \tag{9.37}$$

As in the case of wireless communications, this expression can be generalized to cases where the number of active transmitters is $n_t \leq N$ and, correspondingly, the number of receiving elements is $n_r \leq N$. This can be done by making the matrices $\mathbf{T}$ and $\mathbf{R}$ be of rank n_t and n_r, respectively. This may correspond to the situation, where not all transmitting or receiving channels may be available to a given link.

9.4.1 Character expansions in communications

In this section we will calculate in closed form the moment-generating function of the mutual information

$$g(\mu) = E\left[\det\left(\mathbf{I}_N + \rho \mathsf{G}^\dagger \mathsf{G}\right)^\mu\right], \tag{9.38}$$

where the expectation is over the channel matrix G. To make progress, one could expand the quantity inside the expectation in (9.38) in terms of products of the matrices $\mathbf{U}$ and $\mathbf{U}^\dagger$ and then average the resulting products over the unitary group. These averages are, however, quite complicated and in most cases can only be treated in an asymptotic fashion (Brouwer and Beenakker 1996, Argaman and Zee 1996). Instead, here we employ a different approach first introduced by Balantekin (2000). Here, the expansion is performed using characters of the irreducible representations of the unitary group $U(N)$, the unitary matrices of size N. For completeness we summarize in the following some basic facts on representation theory of groups. The interested reader can refer to several textbooks, including Sternberg (1995).

A unitary representation V of a group G is a homomorphism from G to $U(N)$. An irreducible representation has no nontrivial invariant subspaces. The irreducible representations of the unitary group $U(N)$ (Hua 1963, Schlittgen and Wettig 2003), can be parameterized by an N-dimensional vector $\mathbf{m} = (m_1, m_2, \ldots, m_N)$, with integers $m_1 \geq m_2 \geq \ldots \geq m_N \geq 0$. The dimension $d_{\mathbf{m}}$ of an irreducible representation is the dimension of its invariant subspace. For the case of $U(N)$ expressions for $d_{\mathbf{m}}$ can be found in several books (Sternberg 1995). For the purposes of this analysis, a particular form of $d_{\mathbf{m}}$ will become handy (Simon et al. 2006), namely,

$$d_{\mathbf{m}} = \left[\prod_{i=1}^{N} \frac{1}{(N-i)!}\right] (-1)^{\frac{N(N-1)}{2}} \Delta(\mathbf{k}), \tag{9.39}$$

where $\Delta(\cdot)$ represents the Vandermonde determinant, defined as

$$\Delta(\mathbf{x}) \equiv \det\left(x_i^{j-1}\right) = \prod_{i>j}(x_i - x_j), \tag{9.40}$$

and the vector $\mathbf{k}$ has elements

$$k_i = m_i - i + N, \tag{9.41}$$

where $i = 1, \ldots, N$ and m_i are the elements of the representation vector $\mathbf{m}$. Now, the character $\chi(g)$ of a group element g in the representation V is equal to the trace of the corresponding matrix, i.e., $\chi(g) = Tr\,[V(g)]$. Thus, a character of a reducible representation can be written as a sum of characters of irreducible representations. Clearly, $\chi(g)$ depends only on the eigenvalues of $V(g)$. Calculating the characters of irreducible representations is greatly facilitated by Weyl's character formula (Weyl 1948, Hua 1963), which for $U(N)$ takes the form

$$\chi_{\mathbf{m}}(\mathbf{A}) = \frac{\det\left(a_i^{m_j + N - j}\right)}{\Delta(a_1, \ldots, a_N)}, \tag{9.42}$$

where the index $\mathbf{m}$ denotes the irreducible representation $(m_1, \ldots, m_N)$ and a_i, for $i = 1, \ldots, N$, are the eigenvalues of $\mathbf{A}$ in the fundamental (N-dimensional) representation. For example, the characters of the one-dimensional unitary group $U(1)$ are given by $\chi_n = e^{in\phi}$, where the character index n takes values $0, 1, \ldots$ and $e^{i\phi}$ is the (eigen)value of an arbitrary one-dimensional matrix in $U(1)$. Thus, a Fourier expansion can be seen as an expansion in the characters of $U(1)$ group. This suggests that the characters of a group can form a good basis of expanding functions, which are invariant under $U(N)$ group operations.

Balantekin (2000) showed that the product of any function $f(x)$ of the eigenvalues $\{a_i\}$ of an invertible matrix $\mathbf{A}$ can be expressed in the form

$$\prod_{i=1}^{N} f(a_i) = \sum_{\mathbf{m}} \alpha_{\mathbf{m}} \chi_{\mathbf{m}}(\mathbf{A}). \tag{9.43}$$

In this expression, $\chi_{\mathbf{m}}(\mathbf{A})$ is the character of $\mathbf{A}$ in the representation $\mathbf{m}$ and the sum is over all irreducible representations of $U(N)$ parameterized with the vector $\mathbf{m} = (m_1, m_2, \ldots, m_N)$, with integers $m_1 \geq m_2 \geq \ldots \geq m_N \geq 0$. In (9.43) the coefficient for each character, $\alpha_{\mathbf{m}}$, is given by

$$\alpha_{\mathbf{m}} = \det\left(f_{m_j + i - j}\right), \tag{9.44}$$

where f_n is the coefficient of the Taylor expansion of $f(x)$, i.e.,

$$f(x) = \sum_{n=0}^{\infty} f_n x^n. \tag{9.45}$$

In the particular case of (9.38) $\mathbf{A} = \mathbf{RUTU}^{\dagger}$ and $f(x) = (1 + \rho x)^{\mu}$ such that

$$f_n = \rho^n \frac{\Gamma(\mu + 1)}{\Gamma(\mu - n + 1)\Gamma(n + 1)}. \tag{9.46}$$

Note that, as expected, for integer μ, f_n vanishes for $n > \mu$. As a result, we have

$$g(\mu) = \sum_{\mathbf{m}} \alpha_{\mathbf{m}} E\left[\chi_{\mathbf{m}}\left(\mathbf{RUTU}^\dagger\right)\right]. \tag{9.47}$$

The expectation of the character $\chi_{\mathbf{m}}\left(\mathbf{RUTU}^\dagger\right) = \mathrm{Tr}\left[\mathbf{R}^{\mathbf{m}}\mathbf{U}^{\mathbf{m}}\mathbf{T}^{\mathbf{m}}\mathbf{U}^{\mathbf{m}\dagger}\right]$, where $\mathbf{U}^{\mathbf{m}}$, etc., is the group element of $\mathbf{U}$ in the representation $\mathbf{m}$, can be obtained using the identity (Sternberg 1995)

$$\int d\mathbf{U}\, U_{ij}^{(\mathbf{m})} U_{kl}^{(\mathbf{m}')*} = \frac{1}{d_{\mathbf{m}}} \delta_{\mathbf{mm}'} \delta_{ik}\delta_{jl}, \tag{9.48}$$

where $d\mathbf{U}$ is the standard Haar integration measure and $d_{\mathbf{m}}$ is the dimension of the representation $(\mathbf{m})$. As a result we have

$$g(\mu) = \sum_{\mathbf{m}} \frac{\alpha_{\mathbf{m}}}{d_{\mathbf{m}}} \chi_{\mathbf{m}}\left(\mathbf{R}\right) \chi_{\mathbf{m}}\left(\mathbf{T}\right). \tag{9.49}$$

This expression can become more transparent by using the characters of $\mathbf{R}$ and $\mathbf{T}$ through the Weyl character formula so that

$$g(\mu) = \sum_{\mathbf{m}} \frac{\alpha_{\mathbf{m}}}{d_{\mathbf{m}}} \frac{\det\left(t_i^{m_j+N-j}\right) \det\left(r_i^{m_j+N-j}\right)}{\Delta(\mathbf{t})\,\Delta(\mathbf{r})}, \tag{9.50}$$

where t_i and r_i are the eigenvalues of the matrices $\mathbf{T}$ and $\mathbf{R}$, respectively.

We now need to massage the expression of $\alpha_{\mathbf{m}}$. To do so we start by expressing it in the form

$$\alpha_{\mathbf{m}} = \prod_{i=1}^{N}\left(\frac{\Gamma(\mu+1)\rho^{k_i+i-N}}{\Gamma(\mu+N-k_i)\Gamma(k_i+1)}\right) \det\left[\frac{\Gamma(\mu-k_i+N)\Gamma(k_i+1)}{\Gamma(\mu-k_i+N-j)\Gamma(k_i+j-N+1)}\right], \tag{9.51}$$

where the indices $\mathbf{k}$ and $\mathbf{m}$ are related through (9.41). We observe that the (i,j) element of the matrix inside the square brackets in Eq. (9.51) is a $(N-1)$-degree polynomial of k_i, indexed by the row $j = 1,\dots,N$, expressed for compactness as $\pi_j(k_i)$. By performing linear operations on the columns of the matrix we can express the determinant of the matrix as $\Delta(\mathbf{k})$, up to an overall multiplicative factor $\overline{C}_N$, independent of k_i, which may be obtained in various ways, and will be discussed at the end. The key point of this calculation is that $\alpha_{\mathbf{m}}$ is proportional to $\Delta(\mathbf{k})$, which can then cancel the same factor appearing in $d_{\mathbf{m}}$. Thus, we have

$$g(\mu) = C_N \rho^{-\frac{N(N-1)}{2}} \sum_{\mathbf{k}} \prod_{i=1}^{N}\left(\frac{\Gamma(\mu+1)\rho^{k_i+i-N}}{\Gamma(\mu+N-k_i)\Gamma(k_i+1)}\right) \frac{\det\left(t_i^{k_j}\right) \det\left(r_i^{k_j}\right)}{\Delta(\mathbf{t})\,\Delta(\mathbf{r})}. \tag{9.52}$$

where we have absorbed all constant factors in C_N. The final step consists of using the Cauchy–Binet formula to sum over the indices k_i and one obtains

$$g(\mu) = C_N \rho^{-\frac{N(N-1)}{2}} \frac{\det\left[(1 + \rho t_i r_j)^{\mu+N-1}\right]}{\Delta(\mathbf{r})\Delta(\mathbf{t})}. \tag{9.53}$$

To obtain the constant C_N, we may take the limit of $r_j \to 0$, successively for $j = 1, \ldots, N$. Since both numerator and denominator vanish when we do this, we need to carefully apply the l'Hospital rule at each step (Simon et al. 2006). After some algebra we find that

$$C_N = \prod_{i=1}^{N-1} \left(\frac{\Gamma(i+1)\Gamma(\mu+N-i)}{\Gamma(\mu+N)} \right). \tag{9.54}$$

These expressions become singular when two or more of the eigenvalues t_i and/or r_j become equal, because then the Vandermonde determinants in the denominator vanish. However, since also the numerator determinant vanishes, one can repeatedly use the de l'Hospital rule to obtain finite answers. This becomes handy for the case discussed in the next section.

From the expression in (9.53), one can readily obtain the probability distribution of the mutual information $I_N(\rho)$, by Fourier transformation, i.e.,

$$\Pr(R) = \int_{-\infty}^{\infty} \frac{\mathrm{d}\mu}{2\pi} e^{-i\mu R} g(\mu). \tag{9.55}$$

In addition, the moments of the distribution can be evaluated by taking the appropriate derivatives with respect to μ. For example, the ergodic average of the mutual information can be expressed as

$$g'(0) = E[I_N] = \sum_{k=1}^{N-1} (\Psi(k) - \Psi(N)) \tag{9.56}$$

$$+ C_N^0 \rho^{-\frac{N(N-1)}{2}} \frac{\sum_{k=1}^{N} \det\left(\mathbf{Q}_k\right)}{\Delta(\mathbf{r})\Delta(\mathbf{t})},$$

where $\Psi(\cdot)$ is the digamma function (Gradshteyn and Ryzhik 1995), C_N^0 is the value of C_N in (9.54) evaluated at $\mu = 0$ and $\mathbf{Q}_k$ a set of matrices with elements given by

$$\mathbf{Q}_{ij,k} = \begin{cases} (1 + \rho t_i r_k)^{N-1} \ln(1 + \rho t_i r_k) & j = k \\ (1 + \rho t_i r_j)^{N-1} & j \neq k \end{cases}. \tag{9.57}$$

For $N = 2$, one obtains

$$g'(0) = E[I_N] = \frac{(\log(1 + \rho a_1 b_1) + \log(1 + \rho a_2 b_2))\,(1 + \rho a_1 b_1)(1 + \rho a_2 b_2)}{\rho(a_1 - a_2)(b_1 - b_2)} \tag{9.58}$$

$$- \frac{(\log(1 + \rho a_1 b_2) + \log(1 + \rho a_2 b_1))\,(1 + \rho a_1 b_2)(1 + \rho a_2 b_1)}{\rho(a_1 - a_2)(b_1 - b_2)} - 1.$$

Clearly, these expressions become unappealing for larger N and for higher moments due to the determinantal structure of $g(\mu)$. In the next section, we will show how the distribution of the mutual information can be evaluated asymptotically for large N for a special form of the matrices $\mathbf{R}$ and $\mathbf{T}$. A similar expression has been derived for correlated Gaussian channels (Simon and Moustakas 2004).

9.4.2 Tails of the mutual information

The previous section dealt with the exact calculation of the moment-generating function of the mutual information. However, in real communications systems a more important metric is the probability distribution of the mutual information itself and in particular its tails, which quantify the probability of error in decoding a packet that has been sent with too high a coding rate. This is particularly true in fiber-optical communications, where very low error rates are desirable, since no feedback is available to allow the transmitter to retransmit the packet, as is the case in wireless communications.

   In this section we will calculate the distribution of the mutual information of the optical MIMO channel in the large channel number N regime. By large N, we will signify that all n_t, n_r, N go to infinity, but with fixed ratios. To be able to do this calculation, the correlation matrices at the receiver and transmitter will be taken to have a simplified structure, namely $\mathbf{R} = \mathbf{P}_{n_r}$ and $\mathbf{T} = \mathbf{P}_{n_t}$ where $\mathbf{P}_n$ is the $N \times N$ projection matrix on an n-dimensional subspace. This simplification corresponds to idealized receiver and transmitter structures, with n_t transmitter channels, n_r receiver channels, and several untapped channels, which may be used by other transceivers or simply correspond to energy loss (Simon and Moustakas 2006). Since the exact nature of the subspaces will be irrelevant due to rotational symmetry, we take the matrices to be of the form $\mathbf{P}_n = \mathrm{diag}([1, \ldots, 1, 0, \ldots, 0])$, where the diagonal has n ones and $N - n$ zeros. In this case the joint probability distribution of the eigenvalues of the matrix $\mathbf{P}_{n_r} \mathbf{U} \mathbf{P}_{n_t} \mathbf{U}^\dagger \mathbf{P}_{n_r}$ has been shown to be (Simon and Moustakas 2006, Dar et al. 2013) for $n_t + n_r \leq N$ and $n_t > n_r$

$$\mathrm{Pr}_\lambda(\boldsymbol{\lambda}) \propto \Delta(\boldsymbol{\lambda})^2 \prod_{k=1}^{n_r} \lambda_k^{n_t - n_r} (1 - \lambda_k)^{N - n_t - n_r}, \tag{9.59}$$

while the remaining $N - n_r$ eigenvalues are 0, while for $n_t + n_r < N$ (Dar et al. 2013) there are $n_t + n_r - N$ eigenvalues equal to unity, $N - n_r$ zero eigenvalues, and the remaining eigenvalues have the density

$$\mathrm{Pr}_\lambda(\boldsymbol{\lambda}) \propto \Delta(\boldsymbol{\lambda})^2 \prod_{k=1}^{N - n_t} \lambda_k^{n_t - n_r} (1 - \lambda_k)^{n_t + n_r - N}. \tag{9.60}$$

The situation $n_t < n_r$ can be obtained directly from the above by interchanging n_t, n_r. For simplicity, we will now assume that $n_t + n_r < N$ and $n_t > n_r$.

   Now, based on the expressions of the probability distribution in (9.59) and (9.60), we can readily apply the methodology of the previous section to obtain the moment

generating function of the mutual information, which, in the case of $n_t + n_r < N$, can be expressed directly as

$$I_N = \sum_{k=1}^{n_r} \log(1 + \rho \lambda_k). \tag{9.61}$$

However, given the explicit expression of the probability distribution of the eigenvalues, more can be accomplished. We will follow closely the methodology by Majumdar (2006), Vivo et al. (2007a), Dean and Majumdar (2008) and Karadimitrakis et al. (2014) based on the analogy to a Coulomb gas pioneered by Dyson (1962). The key insight is that for large N the eigenvalues coalesce to a fluid that can be described as a density given by

$$n(x) = \frac{1}{n_r} \sum_{k=1}^{n_r} \delta(x - \lambda_k). \tag{9.62}$$

We start by conjecturing that the support of $n(x)$ is compact with borders $0 > a > b > 1$, which we will check in the end to be the case. Hence the logarithm of the distribution function in (9.59) can be expressed as

$$\frac{\log \mathrm{Pr}_\lambda(\{\lambda_k\})}{n_r^2} = \int_a^b n(x)\,(\beta \log(1-x) + \alpha \log(x))\, \mathrm{d}x \tag{9.63}$$

$$+ \iint_a^b n(x)n(y) \log|x-y| \mathrm{d}y \mathrm{d}x,$$

where $\alpha = (n_t - n_r)/n_r$ and $\beta = (N - n_t - n_r)/n_r$.

Note that $I_N(\rho)$ may now be written in terms of $n(\cdot)$ as

$$I_N(\rho) = n_r \int n(x) \log(1 + \rho x) \mathrm{d}x, \tag{9.64}$$

To evaluate the probability density of $r = I_N(\rho)/n_r$ we first express it in the form

$$\mathrm{Pr}_r(r) = n_r^2 E\left[\delta\left(n_r^2 \left\{r - \int n(x)\log(1+\rho x) \mathrm{d}x\right\}\right)\right] \tag{9.65}$$

$$= n_r^2 \int \frac{\mathrm{d}k}{2\pi i} E\left[e^{n_r^2 k(r - \int n(x)\log(1+\rho x))}\right].$$

To ensure the density $n(x)$ is properly normalized to unity, it is convenient to add another constraint in the form of a Fourier integral as in the case of the mutual information constraint in Eq. (9.65). As a result, we obtain

$$\mathrm{Pr}_r(r) \propto E\left[e^{-n_r^2 \mathcal{S}[n]}\right], \tag{9.66}$$

where the expectation is over all positive functions $n(x)$, k, and the corresponding Fourier integral for the normalization constraint, and

$$S[n] = -\int n(x)\left(\beta\log(1-x) + \alpha\log(x)\right)dx - \iint n(x)n(y)\log|x-y|dydx$$
$$- k\left(\int n(x)\log(1+\rho x)dx - r\right) - c\left(\int n(x)dx - 1\right). \tag{9.67}$$

In the large N limit, the path integral is dominated by the contribution around the saddle-point(s) of $S[n]$. Convexity arguments for $S[n]$ can assure that any solution will be unique (Ben Arous and Guionnet 1997, Hiai and Petz 1998). Taking the functional derivative on $S[n]$ with respect to $n(x)$ and setting the result to 0 we obtain

$$2\int_a^b n(y)\log|x-y|dy = -\beta\log(1-x) - \alpha\log(x) - k\log(1+\rho x) - c. \tag{9.68}$$

It is convenient to differentiate this expression with respect to x, which gives

$$2\mathcal{P}\int_a^b \frac{n(y)}{x-y}dy = -\frac{\beta}{1-x} - \frac{\alpha}{x} - \frac{k\rho}{1+\rho x}, \tag{9.69}$$

where $\mathcal{P}$ indicates the Cauchy principal value of the integral. Equation (9.69) has an appealing physical meaning, namely the balance of forces between the inter-eigenvalue (intercharge) repulsions and the forces imposed by external (one-body) potentials. Hence, the solution to this equation will provide the equilibrium (or most probable) density of eigenvalues consistent with rate r. Since both $\alpha, \beta > 0$, we expect an infinite force acting the charge density if either $a = 0$ or $b = 1$. Therefore, neither of this can be the case. Thankfully, this integral equation can be solved (see Tricomi (1957), Majumdar (2006)) with a general solution of the form

$$n(x) = \frac{\frac{\beta\sqrt{(1-a)(1-b)}}{1-x} - \frac{k\sqrt{(1+a\rho)(1+b\rho)}}{1+\rho x} - \frac{\alpha\sqrt{ab}}{x} + C}{2\pi\sqrt{(x-a)(b-x)}}, \tag{9.70}$$

where C is a constant. Assuming continuity at the boundary of the support, i.e., $n(a) = n(b) = 0$, we obtain

$$n^*(x) = \frac{\sqrt{(x-a)(b-x)}}{2\pi(1+\rho x)}\left(\frac{\beta(\rho+1)}{(1-x)\sqrt{(1-a)(1-b)}} + \frac{\alpha}{x\sqrt{ab}}\right), \tag{9.71}$$

with the additional constraint

$$\frac{\beta}{\sqrt{(1-a)(1-b)}} = \frac{\alpha}{\sqrt{ab}} + \frac{k\rho}{\sqrt{(1+\rho a)(1+\rho b)}}. \tag{9.72}$$

The parameters a, b, k can be evaluated uniquely from Eq. (9.72) in addition to the normalization constraint

$$\int_a^b n^*(x)dx = 1, \tag{9.73}$$

which demands that

$$\beta + \alpha + 2 + k = \frac{\alpha}{\sqrt{ab}} + \frac{k(1+\rho)}{\sqrt{(1+\rho a)(1+\rho b)}}, \tag{9.74}$$

and the rate constraint

$$r = \int n^*(x) \log(1 + \rho x) \tag{9.75}$$

$$= \log \Delta\rho + \frac{\beta}{2\sqrt{\bar{a}_c \bar{b}_c}} \left[G(\bar{a}_z, \bar{a}_z) - G(\bar{a}_z, -\bar{a}_c) \right]$$

$$+ \frac{\alpha}{2\sqrt{\bar{a}\bar{b}}} \left[G(\bar{a}_z, \bar{a}) - G(\bar{a}_z, \bar{a}_z) \right],$$

where $\Delta = b - a$, $z = \frac{1}{\rho}$. In Eq. (9.75) we have defined $\bar{a} = a/\Delta$, $a_c = 1 - a$, $\bar{a}_c = a_c/\Delta$, $\bar{a}_z = (a + z)/\Delta$, and $\bar{b} = b/\Delta$, $\bar{b}_c = b_c/\Delta = (1 - b)/\Delta$, $\bar{b}_z = (b + z)/\Delta$. The function $G(x, y)$ is given by (Karadimitrakis et al. 2014)

$$G(x, y) = \frac{1}{\pi} \int_0^1 \sqrt{t(1-t)} \frac{-\log(t + x)}{t + y} dt \tag{9.76}$$

$$= -2\mathrm{sgn}(y) \sqrt{|y(1 + y)|} \log \left[\frac{\sqrt{x|1 + y|} + \sqrt{|y|(1 + x)}}{\sqrt{|1 + y|} + \sqrt{|y|}} \right]$$

$$+ (1 + 2y) \log \left[\frac{\sqrt{1 + x} + \sqrt{x}}{2} \right] - \frac{1}{2} \left(\sqrt{1 + x} - \sqrt{x} \right)^2.$$

The probability density in (9.71) represents the most probable distribution of eigenvalues in the subspace where $I_N = n_r r$. We may now plug in the expression of $n^*(x)$ (Eq. (9.75)) into (9.67) and obtain an expression for $\mathcal{S}$ as follows:

$$\mathcal{S}^*(r) = \frac{k}{2}(r - \log(1 + b\rho)) - \frac{\log \Delta}{2}(\beta + \alpha + 2) - \frac{\beta}{2} \log b_c$$

$$- \frac{\beta^2}{4\sqrt{\bar{a}_c \bar{b}_c}} (G(\bar{b}_c, \bar{b}_c) - G(\bar{b}_c, -\bar{b}_z)) - \frac{\alpha}{2} \log b$$

$$+ \frac{\beta\alpha}{4\sqrt{\bar{a}\bar{b}}} (G(\bar{b}_c, -\bar{b}) - G(\bar{b}_c, -\bar{b}_z)) + \frac{\beta\alpha}{4\sqrt{\bar{a}_c \bar{b}_c}} (G(\bar{a}, -\bar{a}_c) - G(\bar{a}, \bar{a}_z))$$

$$- \frac{\alpha^2}{4\sqrt{\bar{a}\bar{b}}} (G(\bar{a}, \bar{a}) - G(\bar{a}, \bar{a}_z)) - \frac{\beta}{2\sqrt{\bar{a}_c \bar{b}_c}} (G(0, \bar{b}_c) - G(0, -\bar{b}_z))$$

$$+ \frac{\alpha}{2\sqrt{\bar{a}\bar{b}}} (G(0, -\bar{b}) - G(0, -\bar{b}_z)). \tag{9.77}$$

As a result, we have for large N

$$\mathrm{Pr}_r(r) \propto e^{-n_r^2 \mathcal{S}^*(r)}. \tag{9.78}$$

To find the normalization constant, we may just divide this expression in the absence of any constraints on $n(x)$ in (9.65), which corresponds to $k = 0$. In this case, the constraint for the mutual information is relaxed and the corresponding expression in (9.71) corresponds to the most probable distribution of eigenvalues in (9.59). After some work it is easy to see that the density of eigenvalues takes a similar form to the Marcenko–Pastur equation

$$n_0(x) = \frac{\sqrt{(x - a_0)(b_0 - x)}}{2\pi x(1 - x)}, \tag{9.79}$$

where

$$a_0, b_0 = \frac{\left(\sqrt{1 + \beta} \pm \sqrt{(\alpha + 1)(\beta + \alpha + 1)}\right)^2}{\beta + 2 + \alpha}, \tag{9.80}$$

which has been obtained using other methods in Simon and Moustakas (2006) and Debbah et al. (2003). The corresponding value of $\mathcal{S}_0$ can obtained directly by using this expression to evaluate $\mathcal{S}$ in (9.78). Analysing the behavior of $\mathcal{S}(r)$ close to the value of $k \approx 0$ it can be shown that

$$\mathcal{S}^*(r) - \mathcal{S}_0 \approx \frac{(r - r_{\text{erg}})^2}{2v_{\text{erg}}}, \tag{9.81}$$

where r_{erg} is the rate obtained using $n_0(x)$ in (9.75), which corresponds to the average (ergodic) value of the rate in the large N limit, and

$$v_{\text{erg}} = \log \frac{(\sqrt{1 + \rho b_0} + \sqrt{1 + \rho a_0})^2}{4\sqrt{1 + \rho b_0}\sqrt{1 + \rho a_0}}, \tag{9.82}$$

where a_0, b_0 are given in (9.80). Since the bulk of the probability distribution will be around the value $r = r_{\text{erg}}$, the normalization is to leading order identical to a Gaussian distribution centered at r_{erg} with variance v_{erg}. Hence,

$$\Pr{}_r(r) \approx n_r \frac{e^{-n_r^2(\mathcal{S}^*(r) - \mathcal{S}_0)}}{\sqrt{2\pi v_{\text{erg}}}}. \tag{9.83}$$

A few remarks are in order for the calculations performed in this section. First, although the large N limit was taken here, the results are valid also for reasonably valued n_t, n_r, etc. Indeed, in Fig. 9.2, the cumulative probability density is plotted as a function of r for a number of representative values of n_t, n_r. As we see, the agreement between Monte Carlo simulations and this approach is pretty good.

Furthermore, it should be noted that here we only analysed the generic case, when $\alpha, \beta > 0$. When $\alpha = 0$, two possible solutions for the eigenvalue density $n^*(x)$ may occur, depending on the value of r. The first extends all the way to the border $x = 0$, with a square root singularity, while the second has a positive lower limit of its support, i.e., $a > 0$. However, for a given value of r only one solution is acceptable, since the

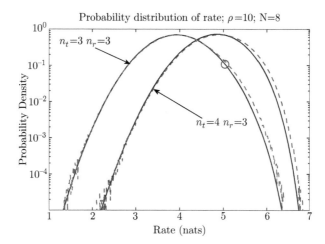

Fig. 9.2 Probability density of the mutual information for two different values of n_t. The agreement with Monte Carlo simulations is quite good. The circle represents the point where the minimum limit of the support of the optimum distribution becomes positive $a > 0$.

other becomes negative. At some critical value of the rate r there is a transition between these two solutions (see Fig. 9.2). Interestingly, only the third derivative of $\mathcal{S}^*(r)$ is discontinuous at this point. Similar behavior can be seen in the case when $\beta = 0$ at the upper limit of the support of $n^*(x)$. This behavior has been observed in other situations (Vivo et al. 2007b) and has been tied to the Tracy–Widom distribution (Majumdar and Schehr 2014). A similar analysis has been performed for complex Gaussian channels (Kazakopoulos et al. 2011).

9.4.3   Discussion

In this section we briefly discuss limitations of the results of the previous subsections.

One possible criticism of this analysis may be whether the expression of the mutual information used represents the true transmission rate for optical MIMO channels. After all, this expression assumes a complex Gaussian input signal, which at this point does not correspond to what is used in current fiber-optical communications systems. Nevertheless, complex Gaussian input is optimal when the noise is also Gaussian, as it happens to be (Cover and Thomas 1991). In addition, current modulation techniques are currently not too far from those used in wireless communications systems and therefore the results described in the previous subsections can be taken as a figure of merit for the fiber-optical channel.

Other limitations of that methodology have to do with the channel model used. For example, the fiber-optical channel has nonuniform mixing between different modes, as well as mode-dependent loss, in which the attenuation of each channel is different and in fact random (Winzer and Foschini 2011). Such details can, in principle, be included in the channel, within the current random matrix framework, by appropriately generalizing the statistics of the random matrix.

Another important limitation of this analysis is the omission of nonlinearities, which are inherently present in fiber-optical communications, especially for large distance light propagation. Although some models have been applied in this context for single mode fibers (Mitra and Stark 2001), a coherent approach for the capacity of the nonlinear optical MIMO channel is currently missing. However, it is hoped that the inverse large number of channels will provide a small parameter for meaningful approximations in the problem.

9.5   Conclusions and outlook

Modern telecommunications systems and algorithms are becoming increasingly complex and there is a need for mathematical tools that can tackle this complexity. Random matrix theory continues to be successful, not only in providing answers relevant in design and performance predictions, but also in providing intuition on the relevant issues. It is hoped that this introduction to the applicability of random matrix theory in communications and the description of how it can be used to tackle real problems will further inspire the cross-fertilization between these fields.

It is worth mentioning that tools developed in spin-glasses have also seen many applications in communications, signal processing, and optimization (Sourlas 1989, Montanari and Sourlas 2000, Tanaka 2002, Krzakala et al. 2012, Zaidel et al. 2013). Nevertheless, both random matrix theory and the discussed earlier applications of spin-glasses are essentially mean-field based. Perhaps the 'last' frontier for physics applications in telecommunications will be in the description of spatial and temporal fluctuations in 2-dimensional wireless networks, where mean-field approaches do not hold. One recent example can be found in Moustakas et al. (2016).

Acknowledgments

It is a pleasure to thank all my collaborators in this field over the years. However, I would like to specifically mention A. M. Sengupta and S. H. Simon, for their contribution in the early parts of the work. In addition, I would like to thank M. Debbah for rekindling my interest in this field in more recent years. Finally, thanks are due to A. Altland, Y. Fyodorov, N. O'Connel, and G. Schehr for organizing a cohesive and timely summer school and inviting me to be part of it.

References

Andrews, J. G., Buzzi, S., Choi, W., Hanly, S. V., Lozano, A., Soong, A. C. K., and Zhang, J. C. (2014). What will 5G be? *IEEE J. Sele. Areas in Commun.* **32**(6), 1065–82.

Argaman, N., and Zee, A. (1996). Diagrammatic theory of random scattering matrices for normal-metal-superconducting mesoscopic junctions. *Phys. Rev. B* **54**(10), 7406–20.

Bai, Z. D., and Silverstein, J W. (2004). CLT for linear spectral statistics of large dimensional sample covariance matrices. *Ann. Probab.* **32**, 553–605.

Balantekin, A. B. (2000). Character expansions, Itzykson-Zuber integrals and the QCD partition action. *Phys. Rev D* **62**(8), 5017.

Beenakker, C. W. J. (1997). Random-matrix theory of quantum transport. *Rev. Mod. Phys.* **69**, 731–808.

Ben Arous, G., and Guionnet, A. (1997). Large deviations for the Wigner's law and Voiculescu's non-commutative entropy. *Prob. Theory Relat. Fields*, **108**, 517–542.

Brouwer, P. W., and Beenakker, C. W. J. (1996). Diagrammatic method of integration over the unitary group, with applications to quantum transport in mesoscopic systems. *J. Math. Phys.* **37**(0), 4904–33.

Calcev, G., Chizhik, D., Goeransson, B., Howard, S., Huang, H., Kogiantis, A., Molisch, A. F., Moustakas, A. L., Reed, D., and Xu, H. (2007). A wideband spatial channel model for system-wide simulations. *IEEE Trans. Veh. Technol.* **56**(2), 389.

Couillet, R., Debbah, M., and Silverstein, J. W. (2011). A deterministic equivalent for the analysis of correlated MIMO multiple access channels. *IEEE Trans. Inform. Theory*, **57**(6), 3493–514.

Cover, T. M., and Thomas, J. A. (1991). *Information Theory*. Wiley, New York.

Dar, R., Feder, M., and Shtaif, M. (2013). The Jacobi MIMO channel. *IEEE Trans. Inform. Theory*, **59**(4), 2426–41.

Dean, D. S., and Majumdar, S. N. (2008). Extreme value statistics of eigenvalues of Gaussian random matrices. *Phys. Rev E*, **77**, 041108.

Debbah, M., Hachem, W., Loubaton, P., and de Courville, M. (2003). MMSE analysis of certain large isometric random precoded systems. *IEEE Trans. Inform. Theory*, **49**(5), 1293.

Dyson, F. (1962). Statistical theory of the energy levels of complex systems. I. *J. Math. Phys.* **3**, 140.

Foschini, G. J., and Gans, M. J. (1998). On limits of wireless communications in a fading environment when using multiple antennas. *Wireless Personal Communi.* **6**, 311–35.

Gradshteyn, I. S., and Ryzhik, I. M. (1995). *Table of Integrals, Series and Products*. Academic Press, New York.

Guo, D., and Verdú, S. (2005). Randomly spread CDMA: Asymptotics via statistical physics. *IEEE Trans Inform. Theory*, **51**(6), 1982–2010.

Hachem, W., Khorunzhiy, O., Loubaton, P., Najim, J., and Pastur, L. (2008). A new approach for capacity analysis of large dimensional multi-antenna channels. *IEEE Trans. Inform. Theory* **54**, 3987–4004.

Hiai, F., and Petz, D. (1998). Eigenvalue density of the wishart matrix and large deviations. *Infinite Dimensional Anal. Quantum Prob.* **1**, 633–46.

Hsu, R. C. J., Tarighat, A., Shah, A., Sayed, A. H., and Jalali, B. (2006). Capacity enhancement in coherent optical MIMO (COMIMO) multimode fiber links. *IEEE Comm. Lett.* **10**, 195–7.

Hua, L. K. (1963). *Harmonic Analysis of Functions of Several Complex Variables in the Classical Domains*. American Mathematical Society, Providence, RI.

Karadimitrakis, A., Moustakas, A. L., and Vivo, P. (2014). Outage capacity for the optical MIMO channel. *IEEE Trans. Inform. Theory* **60**(7), 4370.

Kazakopoulos, P., Mertikopoulos, P., Moustakas, A. L., and Caire, G. (2011). Living at the edge: a large deviations approach to the outage MIMO capacity. *IEEE Trans. Inform. Theory* **57**(4), 1984 –2007.

Krzakala, F., Mèzard, M., Sausset, F., Sun, Y. F., and Zdeborovà, L. (2011). Statistical-physics-based reconstruction in compressed sensing. *Phys. Rev X* **2**(2), 021005 (2012).

Majumdar, S. N. (2006). *Random Matrices, the Ulam Problem, Directed Polymers and Growth Models, and Sequence Matching.* Volume Complex Systems, Les Houches. Elsevier, Berline.

Majumdar, S. N., and Schehr G. (2014). Top eigenvalue of a random matrix: large deviations and third order phase transition. *J. Stat. Mech.* P01012.

Mézard, M., Parisi, G., and Virasoro, M. A. (1987). *Spin Glass Theory and Beyond.* World Scientific, Singapore.

Mitra, P. P., and Stark, J. B. (2001). Nonlinear limits to the information capacity of optical fibre communications. *Nature* **411**, 1027–30.

Montanari, A., and Sourlas, N. (2000). Statistical mechanics of turbo codes. *Eur. Phys. J. B*, **18**, 107–19.

Morioka, T., Awaji, Y., Ryf, R., Winzer, P., Richardson, D., and Poletti, F. (2012). Enhancing optical communications with brand new fibers. *Communi. Mag. IEEE* **50**(2), s31–s42.

Moustakas, A. L., Baranger, H. U., Balents, L., Sengupta, A. M., and Simon, S. H. (2000). Communication through a diffusive medium: coherence and capacity. *Science* **287**, 287–90.

Moustakas, A. L., Simon, S. H., and Sengupta, A. M. (2003). MIMO capacity through correlated channels in the presence of correlated interferers and noise: a (not so) large N analysis. *IEEE Trans. Inform. Theory* **49**(10), 2545–61.

Moustakas, A. L., and Simon, S. H. (2007). On the outage capacity of correlated multiple-path MIMO channels. *IEEE Trans. Inform. Theory*, **53**(11), 3887–903.

Moustakas, A. L., Mertikopoulos, P., and Bambos, N. (2016). Power optimization in random wireless Networks. *IEEE Trans. Inform. Theory* **62**(9), 5030–58.

Müller, R. R. (2003). Channel capacity and minimum probability of error in large dual antenna array systems with binary modulation. *IEEE Trans. Signal Processs.* **51**(11), 2821.

Müller, R. R., and Verdú, S. (2001). Design and analysis of low-complexity interference mitigation on vector channels. *IEEE J. Sel. Areas Commun.* **19**(8), 1429.

Sanguinetti, L., Moustakas, A.L., Bjornson, E., and Debbah, M. (2014). Large system analysis of the energy consumption distribution in multi-user MIMO systems with mobility. *Wireless Commun. IEEE Trans.* **PP**(99), 1–1.

Schlittgen, B., and Wettig, T. (2003). Generalizations of some integrals over unitary groups. *J. Phys. A: Math Gen* **36**, 3195–201.

Sengupta, A. M., and Mitra, P. P. (2006). Capacity of multivariate channels with multiplicative noise: Random matrix techniques and large-N expansions. *J. Stat. Phys.* **125**(5), 1223–42.

Simon, S. H., and Moustakas, A. L. (2004). Eigenvalue density of correlated random Wishart matrices. *Phys. Rev. E* **69**, 065101(R).

Simon, S. H., and Moustakas, A. L. (2006). Crossover from conserving to lossy in circular random matrix ensembles. *Phy. Rev. Lett.* **96**(13), 136805.

Simon, S. H., Moustakas, A. L., and Marinelli, L. (2006). Capacity and character expansions: moment generating function and other exact results for MIMO correlated channels. *IEEE Trans. Inform. Theory* **52**(12), 5336.

Sourlas, N. (1989). Spin glass models as error correcting codes. *Nature* **339**, (6227), 693–5.

Sternberg, S. (1995). *Group Theory and Physics*. Cambridge University Press, Cambridge, UK.

Tanaka, T. (2002). A statistical-mechanics approach to large-system analysis of CDMA multiuser detectors. *IEEE Trans. Inform. Theory* **48**(11), 2888–910.

Taricco, G. (2008). Asymptotic mutual information statistics of separately-correlated MIMO Rician fading channels. *IEEE Trans. Inform. Theory* **54**(8), 3490.

Tarighat, A., Hsu, R.C.J., Shah, A., Sayed, A.H., and Jalali, B. (2007). Fundamentals and challenges of optical multiple-input multiple-output multimode fiber links [topics in optical communications]. *Commun. Mag. IEEE* **45**(5), 57 –63.

Telatar, I. E. (1999). Capacity of multi-antenna Gaussian channels. *Europ. Trans. Telecommun. Related Technol.* **10**(6), 585–96.

Tkach, R. W. (2010). Scaling optical communications for the next decade and beyond. *Bell Labs Techn. J.* **14**(4), 3–9.

Tricomi, F. G. (1957). *Integral Equations*. Pure Appl. Math V. Interscience, London.

Vivo, P., Majumdar, S. N., and Bohigas, O. (2007a). Large deviations of the maximum eigenvalue in Wishart random matrices. *J. Phys. A* **40**, 4317–37.

Vivo, P., Majumdar, S. N., and Bohigas, O. (2007b). Large deviations of the maximum eigenvalue in Wishart random matrices. *J. Phys. A: Math. Theoret.* **40**(16), 4317–37.

Wagner, S., Couillet, R., Debbah, M., and Slock, D. T. M. (2012). Large system analysis of linear precoding in correlated MISO broadcast channels under limited feedback. *IEEE Trans. Inform. Theory* **58**(7), 4509–37.

Weyl, H. (1948). *The Classical Groups*. Princeton University Press, Princeton, NJ.

Winzer, P. J., and Foschini, G. J. (2011). MIMO capacities and outage probabilities in spatially multiplexed optical transport systems. *Opt. Express* **19**(17), 16680–96.

Zaidel, B. M., Müller, R. R., Moustakas, A. L., and de Miguel, R. (2013). Vector precoding for Gaussian MIMO broadcast channels: Impact of replica symmetry breaking. *IEEE Trans. Inform. Theory* **58**(3), 1413.

10

Random matrix approaches to open quantum systems

Henning SCHOMERUS

Department of Physics, Lancaster University, Lancaster, LA1 4YB, UK

Schomerus, H., 'Random Matrix Approaches to Open Quantum Systems' in *Stochastic Processes and Random Matrices*. Edited by: Grégory Schehr et al, Oxford University Press (2017). © Oxford University Press 2017. DOI 10.1093/oso/9780198797319.003.0010

Chapter Contents

10.1 Introduction

10.1.1 Welcome

Open quantum systems come in two variants. The first variant (on which we will focus more) are scattering systems in which the dynamics allow particles to enter and leave (Newton 2002; Messiah 2014). One then normally defines a scattering region, outside of which particles move free of any external forces or interactions. This situation is realized (at least to some level of approximation) in many decay or radiation processes (Weidenmüller and Mitchell 2009), but is also useful for describing phase-coherent transport in mesoscopic devices (Datta 1997; Beenakker 1997; Blanter and Büttiker 2000; Nazarov and Blanter 2009) or photonic structures (Cao and Wiersig 2015). The second variant (which we will encounter only briefly) are interacting systems in which the studied dynamical degrees of freedom are influenced by other degrees of freedom in the environment (Breuer and Petruccione 2002). This situation spans from the quantum-statistical foundations of thermodynamics (Gemmer et al. 2010) to the description of decoherence (Weiss 2008), with ample applications to quantum optics (Carmichael 2009), quantum-critical phenomena (Sachdev 1999), and quantum information processing (Nielsen and Chuang 2010).

While these two scenarios of openness are in many ways quite distinct, they have some important features in common—in particular, in both scenarios we are led to restrict our attention to a subsystem, while the processes that are involved often are very complex (meaning that we have no realistic handles to describe them in detail), be it due to underlying classical chaos, disorder, or uncontrolled interactions. Taken together, these features lay the foundations for a statistical description where individual systems are replaced by an appropriate ensemble. These ensembles are typically formulated in terms of effective models, e.g., for the Hamiltonian, the scattering matrix, or the density matrix, in which only the fundamental symmetries and the most essential time and energy scales are retained. Quantitative predictions then follow from explicit calculations and often turn out to be universal, i.e., applicable to generic representatives of the ensemble.

Over the past decades, a great body of theoretical and mathematical work has been devoted to these random-matrix descriptions (Beenakker 1997; Guhr et al. 1998; Mehta 2004; Stöckmann 2006; Haake 2010; Forrester 2010; Akemann et al. 2011; Pastur and Shcherbina 2011; Beenakker 2015). In this chapter we review the physical origins and mathematical structures of the underlying models, and collect key predictions which give insight into the typical system behaviour. In particular, we aim to give an idea how the different features are interlinked. This includes a detour to interacting systems, which we motivate by the overarching question of ergodicity. With this selection of topics, we hope to provide a useful bridge to the many excellent advanced sources, including the monographs and reviews mentioned earlier, which contain detailed expositions of the random-matrix calculations and further applications not covered here. In the remainder of this Introduction, we provide some basic background.

10.1.2   Primer

This chapter is based on lectures that were delivered to a mixed audience of mathematicians and physicists. To establish some common language, let us first review some basic notions of quantum mechanics (Peres 2002). This also gives us the opportunity to pinpoint the fundamental origins of the mathematical concepts and physical phenomena that we will encounter throughout this chapter—and further explain what this chapter is really about.

Let us recall, then, that quantum mechanics describes the physical states of a system in terms of vectors $|\psi\rangle$, $|\phi\rangle$, ... in a complex Hilbert space $\mathcal{H}$. The superposition principle means that the vectors can be freely combined to yield new physical states $\alpha|\psi\rangle + \beta|\phi\rangle$, $\alpha, \beta \in \mathbb{C}$. All vectors $\alpha|\psi\rangle$ that differ only by a multiplicative factor $\alpha \neq 0$ describe the same physical state, which is often exploited to impose the convenient normalization $\langle\psi|\psi\rangle = 1$. Following physics convention, we here use (what we term) the scalar product with $\langle\phi|(\alpha\psi + \beta\chi)\rangle = \alpha\langle\phi|\psi\rangle + \beta\langle\phi|\chi\rangle$, $\langle\psi|\phi\rangle = \langle\phi|\psi\rangle^*$, $\langle\psi|\psi\rangle > 0$ unless $|\psi\rangle = 0$, where * denotes complex conjugation. Two states with $\langle\phi|\psi\rangle = 0$ are called orthogonal, and a discrete basis with $\langle n|m\rangle = \delta_{nm}$ is called orthonormal. For a continuous basis, this is replaced by $\langle x|x'\rangle = \delta(x - x')$ with Dirac's delta function. In a given basis, states can be expanded as $|\psi\rangle = \sum_n \psi_n|n\rangle$ where $\psi_n = \langle n|\psi\rangle$, with the sum replaced by an integral when the basis is continuous.

Observables are represented by Hermitian linear operators $\hat{A}$, with $\hat{A}|\psi\rangle \equiv |\hat{A}\psi\rangle \in \mathcal{H}$ such that $\langle\phi|\hat{A}\psi\rangle = \langle\hat{A}\phi|\psi\rangle$. According to the measurement axiom, these operators predict physical observations via the expectation values $\mathcal{E}_\psi(A) = \langle\psi|\hat{A}\psi\rangle/\langle\psi|\psi\rangle$, which in reality are obtained by averaging the outcomes of experiments on systems in the same quantum state. The associated uncertainty (variance) is obtained from $\Delta A = [\mathcal{E}_\psi(A^2) - \mathcal{E}_\psi^2(A)]^{1/2}$, which in general is finite. Denoting by $E_a = \sum_n |\psi_{a,n}\rangle\langle\psi_{a,n}|$ the projector onto states that guarantee an outcome a with vanishing uncertainty $\Delta A = 0$, one finds that these are eigenstates with $\hat{A}|\psi_{a,n}\rangle = a|\psi_{a,n}\rangle$. In a general state, the probability of these outcomes $|\psi\rangle$ is then $P(a) = \langle\psi|E_a|\psi\rangle/\langle\psi|\psi\rangle$; no outcomes other than the associated eigenvalues are allowed. Beyond this probabilistic description, outcomes of individual experiments are unpredictable. Finally, the measurement axiom stipulates that right after the measurement with an outcome a, the quantum system acquires the state $E_a|\psi\rangle$.

Adopting the conventional Schrödinger picture, the time dependence of the quantum state arises from the Schrödinger equation

$$i\hbar\frac{\mathrm{d}}{\mathrm{d}t}|\psi(t)\rangle = \hat{H}(t)|\psi(t)\rangle. \tag{10.1}$$

Here $\hat{H}$ is the Hamiltonian, a Hermitian operator which represents energy, while $\hbar = h/2\pi$ is the reduced Planck's constant. The general solution can be written as $|\psi(t)\rangle = \hat{U}(t,t')|\psi(t')\rangle$, where $\hat{U}(t,t')$ is a unitary operator ($\hat{U}$ is unitary if always $\langle\hat{U}\phi|\hat{U}\psi\rangle = \langle\phi|\psi\rangle$). If $\hat{H}$ is independent of time, we can separate variables as $|\psi(t)\rangle = \exp(-iEt/\hbar)|\phi\rangle$ and arrive at the stationary Schrödinger equation $E|\phi\rangle = \hat{H}|\phi\rangle$. In this case, $\hat{U}(t,t') = \exp(-i\hat{H}(t-t')/\hbar)$.

In order to describe the incoherent mixture of normalized quantum states $|\psi_n\rangle$ one introduces the density matrix (statistical operator) $\hat{\rho} = \sum_n p_n |\psi_n\rangle\langle\psi_n|$ with positive weights p_n summing to $\sum p_n = 1$, so that $\mathrm{tr}\,\hat{\rho} = 1$. The expectation values $\mathcal{E}_\rho(A) = \mathrm{tr}\,(\hat{A}\hat{\rho}) = \sum_n p_n \mathcal{E}_{\psi_n}(A)$ are a combination of the quantum-mechanical average in each quantum state and the classical average over the weights p_n. The density operator is Hermitian and positive semidefinite, and for a pure state (with only one finite $p_n = 1$) becomes a projector, $\hat{\rho}^2 = \hat{\rho}$. To capture the departure from this situation one can consider the purity $\mathcal{P} = \mathrm{tr}\,\hat{\rho}^2$, which equals unity only for a pure state, as well as the von Neumann entropy $\mathcal{S} = -\mathrm{tr}\,\hat{\rho}\ln\hat{\rho}$, which vanishes for a pure state.

10.1.3 Open systems

The superposition principle mentioned in Section 10.1.2 is the origin of wave-like interference effects, the complexity of which we will aim to capture in a statistical description. To provide the states with some structure, we can often think of the state space being divided into sectors (which we here call regions), $\mathcal{H} = \mathcal{H}_1 \oplus \mathcal{H}_2$. We then can start to talk about local and non-local processes, within or between the regions, and introduce basic notions such as the exchange of particles or energy (see Fig. 10.1). An additional layer of complexity is added when we can view the system as being composed of separate degrees of freedom (which we here call parts). The Hilbert space then takes the form of a tensor product $\mathcal{H} = \mathcal{H}_1 \otimes \mathcal{H}_2$, with proper symmetrization or antisymmetrization if the parts are, in a physical sense, indistinguishable (e.g., when they describe identical bosonic or fermionic particles). Separable states are of the form $|\phi\rangle \otimes |\chi\rangle$, while superpositions of such states lead to quantum correlations (entanglement) that deeply enrich the behaviour of interacting systems. Based on these elements of structure, let us now agree, within the confines of this chapter, on two notions of open quantum systems. These are systems in which we can naturally focus on some region or part $\mathcal{H}_1$, which is either locally confined (as in $\mathcal{H} = \mathcal{H}_1 \oplus$

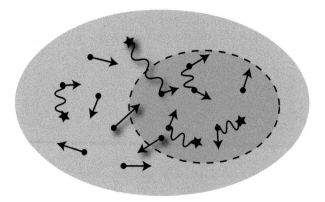

Fig. 10.1 Quantum systems couple to their environment by the exchange of particles and energy, and thereby by processes connected to the kinetic freedom of motion and the interactions of the various components.

$\mathcal{H}_2$) or constrained to some of the degrees of freedoms (as in $\mathcal{H} = \mathcal{H}_1 \otimes \mathcal{H}_2$). We are then naturally led down two roads: scattering-like scenarios, which describe the exchange of particles between a confined region and its surrounding environment, and scenarios dominated by the interactions, which often concern the exchange of energy and creation of entanglement. In order for this division to make some sense, the environment must be sufficiently structureless—either because the dynamics are simple and predictable (typically, the point of view taken in the case of scattering) or because they are so complex that they can be described in a simple statistical picture (typically, the point of view taken in the case of interactions). Physically, this requires that the rest of the system is large, and of a nature where incoming and outgoing particles are only simply correlated, and so is the energy flowing in or out of the system.

10.1.4   Preview

With these concepts at hand, we can now define our mission—to provide a statistical description of open quantum systems in terms of random matrices. This succeeds in situations where we can apply statistical considerations also to the complex dynamics in the region or part of interest, with constraints only arising from fundamental symmetries. We describe both settings in their purest incarnation.

(i) Our main focus is the elastic scattering of a noninteracting particle, which can undergo complex dynamics in the region of interest but enters and leaves in predictable ways. This is quantified in terms of the amplitudes of the incoming and outgoing waves, which are linearly related by a unitary scattering matrix S. As this pure setting is stationary, we can work in the energy domain, while time scales follow when we consider variations in energy. This setting also covers decay processes, where we initially confine the particle within the region of interest— effectively, this is described by a non-Hermitian Hamiltonian, with eigenvalues that coincide with the poles of the scattering matrix.

(ii) In a small detour at the end of this chapter, we consider purely interacting systems, with localized degrees of freedom that cannot move but evolve under the influence of their mutual environment. We quantify this in terms of a reduced density matrix, a Hermitian, positive semidefinite matrix which represents the quantum state when one ignores the other degrees of freedom. We again assume complex internal dynamics, and consider entropies that quantify entanglement.

As indicated, we will encounter, amongst others, random Hermitian Hamiltonians, unitary scattering matrices, positive semidefinite density and time-delay matrices, and non-Hermitian effective Hamiltonians. These are all naturally linked to canonical random-matrix ensembles (not surprisingly, as many of these ensembles were developed with such applications in mind), which we review in Section 10.2. In Section 10.3 we formulate effective scattering models that link these ensembles to physical effects. Section 10.4 provides an overview of key results concerning the decay, dynamics, and transport, where we focus on systems with fully random internal dynamics. Section 10.5 describes how localizing effects in low dimensions let systems depart from this ergodic behaviour, first for noninteracting systems and then in the context of interacting

systems. Section 10.6 gives a brief outlook, while the Appendix collects some simple derivations of relevant eigenvalue densities.

10.2   Foundations of random-matrix theory

In this section we review a range of classical random-matrix ensembles against the backdrop of closed-system behaviour, which informs the subsequent applications to open systems.

10.2.1   Random Hamiltonians and Gaussian Hermitian ensembles

Random-matrix descriptions in quantum mechanics naturally start out with considerations of closed systems. In this setting, the main object of interest is the Hamiltonian $\hat{H}$, whose eigenvalues give the energy levels. The energy spectrum can be characterized very neatly if one manages to identify a number of conserved quantities that commute with the Hamiltonian and amongst each other; considering joint eigenstates of these quantities helps to bring some order to the spectrum. In sufficiently complex systems, however, effects such as chaotic or diffractive scattering and interactions eliminate all conserved quantities, and the energy spectrum lacks any apparent regularities. It is natural to compare the resulting features with the case where the Hamiltonian can be considered as random. This was first proposed in the 1950s by Wigner (1956), who sought ways to analyse resonances in heavy nuclei. The idea is to focus on a suitable energy range, where the local spectral properties can then be studied by replacing the full Hamiltonian with a randomly chosen $M \times M$-dimensional Hermitian matrix (the limit $M \to \infty$ can be imposed later on).

The quality of this descriptions depends on the identification of a suitable random-matrix ensemble. To achieve this task we are allowed to incorporate any general feature of the system. These are, in particular, fundamental symmetries, rough geometric constraints such as dimensionality, as well as natural time and energy scales. The consideration of fundamental symmetries leads to ten symmetry classes (Zirnbauer 1996, 2011; Beenakker, 2015). These comprise the three Wigner–Dyson classes based on time-reversal symmetry (Dyson 1962a; Porter 1965; Mehta 2004; Guhr et al. 1998; Haake 2010), three corresponding classes with chiral symmetry (Verbaarschot 1994; Verbaarschot and Wettig 2000), and four classes based on a charge-conjugation symmetry (Altland and Zirnbauer 1997). These classes are developed in the present section, and listed in Table 10.1. We also describe the corresponding Hermitian matrix ensembles for the simplest situation, geometrically featureless systems in which the only relevant energy scale is the mean level spacing Δ. This reasonably applies when all system-specific information becomes indiscernible after a short time T_{erg}, which in particularly is much shorter than the Heisenberg time $T_H = 2\pi\hbar/\Delta$ (the minimal observation time at which individual energy levels can be resolved). Examples where this is realized are sufficiently featureless disordered (Efetov 1996) or classically chaotic systems (Stöckmann 2006; Haake 2010). The short-ranged level statistics then becomes universal, and can be captured by ensembles with Gaussian statistics of the matrix elements (Guhr et al. 1998; Mehta 2004; Haake 2010; Forrester 2010).

Table 10.1 Fundamental symmetries of hermitian random-matrix ensembles

Symmetries	Constraints	Realization $(H^*_{mn} = H_{nm})$
no symmetries	none besides $H = H^\dagger$	$H_{nm} \in \mathbb{C}$
$\mathcal{T} = K$	$H^* = H$	$H_{nm} \in \mathbb{R}$
$\mathcal{T} = \Omega K$	$H^* = \Omega H \Omega^{-1}$	$H_{nm} \in \mathbb{H}$
$\mathcal{C} = K$	$H^* = -H$	$H_{nm} \in i\mathbb{R}$
$\mathcal{C} = K, \mathcal{T} = \Omega K$	$H^* = -H = \Omega H \Omega^{-1}$	$H_{nm} = -\sigma_y H_{nm} \sigma_y \in \mathbb{H}$
$\mathcal{C} = \Omega K$	$H^* = -\Omega H \Omega^{-1}$	$H_{nm} \in i\mathbb{H}$
$\mathcal{C} = \Omega K, \mathcal{T} = K$	$H^* = H = -\Omega H \Omega^{-1}$	$H_{nm} = -\sigma_y H_{nm} \sigma_y \in i\mathbb{H}$
$\mathcal{X} = \tau_z \equiv \mathrm{diag}\,(1_{M_1}, -1_{M_2})$	$H = -\tau_z H \tau_z$	$H = \begin{pmatrix} 0 & A \\ A^\dagger & 0 \end{pmatrix}, A_{nm} \in \mathbb{C}$
$\mathcal{X} = \tau_z, \mathcal{C} = K \ (\mathcal{T} = \mathcal{X}\mathcal{C})$	$H = -\tau_z H \tau_z = -H^*$	$H = \begin{pmatrix} 0 & A \\ A^\dagger & 0 \end{pmatrix}, A_{nm} \in i\mathbb{R}$
$\mathcal{X} = \tau_z, \mathcal{C} = \Omega K \ (\mathcal{T} = \mathcal{X}\mathcal{C})$	$H = -\tau_z H \tau_z = -\Omega H^* \Omega^{-1}$	$H = \begin{pmatrix} 0 & A \\ A^\dagger & 0 \end{pmatrix}, A_{nm} \in i\mathbb{H}$

10.2.2   Time-reversal symmetry and the Wigner–Dyson ensembles

We start by considering the role of time reversal (Dyson 1962a; Haake 2010), instituted by an anti-unitary operator $\mathcal{T}$ fulfilling $\langle \mathcal{T}\phi | \mathcal{T}\psi \rangle = \langle \psi | \phi \rangle = \langle \phi | \psi \rangle^*$, which consequently may square to $\mathcal{T}^2 = 1$ or $\mathcal{T}^2 = -1$.

If the Hamiltonian obeys a time-reversal symmetry $\mathcal{T} H \mathcal{T}^{-1} = H$ with $\mathcal{T}^2 = 1$, we can adopt an invariant basis $|n\rangle$ in which $\langle \mathcal{T} n | \psi \rangle = \langle n | \psi \rangle$ for any $|\psi\rangle$. This implies that $\langle n | \mathcal{T}\psi \rangle = \langle n | \psi \rangle^*$, so that the time-reversal operation $\mathcal{T} = K$ amounts to the complex conjugation of the expansion coefficients $\psi_n = \langle n | \psi \rangle$ of any state. In this basis the matrix elements $H_{lm} = \langle l | \hat{H} | m \rangle = \langle \mathcal{T} l | \hat{H} \mathcal{T} | m \rangle = H^*_{lm}$ are real, while Hermiticity implies that the matrix is symmetric, $H_{ml} = H_{lm}$. This is known as the *orthogonal symmetry class* (OE), to which we associate the symmetry index $\beta = 1$.

In absence of any time-reversal symmetry, matrix elements of the Hamiltonian are in general complex, with $H_{lm} = H^*_{ml}$ because of Hermiticity, which defines the *unitary symmetry class* (UE) with symmetry index $\beta = 2$.

If we have a time-reversal symmetry $\mathcal{T} H \mathcal{T}^{-1} = H$ with $\mathcal{T}^2 = -1$ (*symplectic symmetry class* (SE) with symmetry index $\beta = 4$), we can adopt a basis arranged in pairs $|n\rangle = \mathcal{T} |\bar{n}\rangle$, so that the Hilbert space dimension $2M$ must be even. In this basis, $\mathcal{T} = \Omega K$ where $\Omega = i\sigma_y \otimes 1_M$, while the blocks $\begin{pmatrix} H_{lm} & H_{l\bar{m}} \\ H_{\bar{l}m} & H_{\bar{l}\bar{m}} \end{pmatrix} = a_{lm}1 + ib_{lm}\sigma_x + ic_{lm}\sigma_y + id_{lm}\sigma_z \in \mathbb{H}$ can be reinterpreted as quaternions, with real coefficients $a_{lm}, b_{lm}, c_{lm}, d_{lm}$ and Pauli matrices σ_r. Hermiticity requires that $a_{lm} = a_{ml}$ forms a symmetric matrix while $b_{lm} = -b_{ml}, c_{lm} = -c_{ml}, d_{lm} = -d_{ml}$ are antisymmetric. Expressed as an $M \times M$-dimensional matrix of quaternions, $H = \overline{H}$ is then seen to be quaternion self-conjugate, where by definition $(\overline{H})_{lm} = \overline{H}_{ml} = a_{ml}1 - ib_{ml}\sigma_x - ic_{ml}\sigma_y - id_{ml}\sigma_z$. For such a matrix, all energy levels appear in degenerate pairs, a phenomenon known

as Kramers degeneracy; in all the following considerations we count each pair as a single level. In keeping with this, the quaternion trace is defined as $\operatorname{tr} H = \sum_n a_{nn}$ (so differs by a factor of 2 from the conventional trace), and the quaternion determinant is similarly modified to maintain the relation $\det \exp A = \exp \operatorname{tr} A$, which makes it equivalent to a Pfaffian (Dyson 1970).

The symmetry index $\beta = 1, 2, 4$ mentioned earlier counts the real degrees of freedom in the matrix elements. The corresponding notions of orthogonal, unitary, and symplectic symmetry classes refer to the transformations $H = U D U^\dagger$, $D = \operatorname{diag}(E_n)$ that diagonalize these Hamiltonians. For $\beta = 1$ the matrix U is orthogonal, $UU^T = 1$, and hence belongs to the group $\mathrm{O}(M)$; for $\beta = 2$ $U \in \mathrm{U}(M)$ is a unitary matrix with $UU^\dagger = 1$, and for $\beta = 4$ the matrix is unitary symplectic, $U \in \mathrm{Sp}(2M)$ with $U\overline{U} = 1$. This 'threefold way' can be further justified within representation theory (Dyson 1962c).

Within these three Wigner–Dyson classes, the universal spectral features encountered in ergodic systems are captured by the Gaussian orthogonal, unitary, or symplectic ensemble (GOE, GUE, GSE), where the Hamiltonian obeys a probability density of the form $P(H) \propto \exp(-c_\beta \operatorname{tr} H^2)$ with $c_\beta = \beta \pi^2 / 4M\Delta^2$. The spectral statistics can then be determined from the joint probability distribution

$$P(\{E_n\}) \propto \prod_{n<m} |E_n - E_m|^\beta \prod_k \exp(-c_\beta E_k^2), \qquad (10.2)$$

which follows by a change of variables from the Hamiltonian to its eigenvalues and eigenvectors. This result can be obtained by sophisticated methods in the language of differential geometry (Forrester 2010), but in this specific incarnation also follows from elementary means and then acquires a simple geometric meaning. Given that $dH = dU\, D U^\dagger + U dD U^\dagger - U D U^\dagger dU U^\dagger$, consider the squared line element

$$\sum_{lm} |dH_{lm}|^2 = \operatorname{tr}(dH dH) = \operatorname{tr}(dXD - DdX)^\dagger(dXD - DdX) + \sum_m (dE_m)^2, \quad (10.3)$$

where D contains M real parameters (the eigenvalues) while $dX = -iU^\dagger dU$ depends on $M(M-1)/2$ real, complex, or quaternion parameters in the set of eigenvectors. The latter parameters can be associated with the rotations $R^{(nm)}$ in the nm plane of the diagonalized system, spanned by the eigenvectors with eigenvalues E_n and E_m. Each of these rotations then translates into a line element in the space of Hamiltonians of length $\propto |E_n - E_m|^\beta$, where the power arises from the fact that the rotation is parameterized by β real variables. (In particular, if we rotate the basis in a degenerate subspace the Hamiltonian does not change.) Hence $d\mu(H) \propto \left(\prod_{n<m} |E_n - E_m|^\beta\right) \left(\prod_k dE_k\right) d\mu(U)$, where $\mu(U)$ is the Haar measure arising from the form dX in the corresponding group of transformations. This measure is uniquely defined by the requirement that it is invariant under $U \to V'UV$ for any fixed V, V' from the same group.

The main characteristics of (10.2) is a universal degree of level repulsion $P(s) \sim s^\beta$ for small level spacings $s = |E_n - E_m|$ in the bulk of the spectrum. This feature was

first realized by Wigner, who put forward the famous surmise $P(s) \sim s^\beta \exp(-cs^2)$ with a suitable scale factor c (Porter 1965). As it turned out, this surmise is exact only for $M = 2$, but provides a very accurate estimate for any M. The exact result can be established by the method of orthogonal polynomials (here based on Hermite polynomials), which provides the complete set of correlation functions (Mehta 2004). When applied to a particular system, these correlations describe the short-ranged statistics in the bulk, i.e., over sufficiently small spectral ranges where the mean level spacing Δ is well defined (possibly, after some unfolding of the spectrum). In particular, the amount of level repulsion is considered as a prime indicator of whether a system displays the required ergodic dynamics, as further discussed in Section 10.5.

The mean level spacing itself is not universal; in real systems it varies systematically with energy, but for any comparison we wish to have it well defined in any given ensemble. In the Gaussian ensembles, this is guaranteed by the form of the eigenvalue density, which for large matrix dimensions $M \to \infty$ approaches the famous Wigner semicircle law (Wigner 1958)

$$\rho(E) = \frac{1}{\Delta}\sqrt{1 - E^2/E_0^2} \quad \text{for } |E| < E_0 = 2M\Delta/\pi. \tag{10.4}$$

A derivation of this classical result is given in the Appendix. It reveals that $\Delta = 1/\rho(0)$ is the mean level spacing at $E = 0$, defining the middle of the bulk around which we then determine the universal spectral features. Universal level statistics are also encountered around the spectral edges $\pm E_0$, whose actual positions are again system specific.

10.2.3   Chiral symmetry

Additional positions within the spectrum deserve dedicated attention when further symmetries come into play. In particular, this is encoutered when the Hamiltonian is antisymmetric under a suitable unitary or antiunitary transformation, an effect which often occurs in single-particle descriptions of fermions. Energy levels then appear in pairs E_n, $E_{\tilde{n}} = -E_n$, with the possible exception of levels pinned to the spectral symmetry point $E = 0$.

If $\mathcal{X}H\mathcal{X} = -H$ with a unitary involution $\mathcal{X}$ (such that $\mathcal{X}^2 = 1$), we talk of a chiral symmetry (Verbaarschot 1994; Verbaarschot and Wettig 2000). For a finite system of dimension $M = M_1 + M_2$, we can choose $\mathcal{X} = \text{diag}(1_{M_1}, -1_{M_2}) \equiv \tau_z$, so that the Hamiltonian takes a block form

$$H = \begin{pmatrix} 0 & A \\ A^\dagger & 0 \end{pmatrix}, \tag{10.5}$$

where A is an $M_1 \times M_2$-dimensional rectangular matrix.

The chiral symmetry arises in elementary particle physics (Verbaarschot and Wettig 2000; Akemann 2017), but can also be realized as an effective symmetry in electronic (Brouwer et al. 2002), superconducting (Fu and Kane 2008), and photonic systems (Schomerus and Halpern 2013; Lu et al. 2014; Poli et al. 2015). Given the structure (10.5), the symmetry generally applies to systems with two sublattices,

termed A and B, when the couplings within each isolated sublattice vanish (Sutherland 1986). The mentioned electronic and photonic implementations naturally extend this idea to suitably coupled subsystems.

An interesting aspect of these classes is the appearance of topological invariants, associated with the number of eigenenergies pinned to the symmetry point (Lieb 1989; Verbaarschot 1994; Brouwer et al. 2002). For a Hamiltonian of the form (10.5) with some finite $\nu = M_2 - M_1$ (so that A is not square), there are at least $|\nu|$ such zero modes. If $\nu < 0$ the associated eigenstates are of the form $\psi = (\psi_A, 0)^T$ with $A^\dagger \psi_A = 0$, while for $\nu > 0$ we have $\psi = (0, \psi_B)^T$ with $A\psi_B = 0$. The remaining paired levels with finite energy can be determined from the positive definite matrix $A^\dagger A$ or $AA^\dagger$, whose eigenvalues are given by E_n^2.

In combination with considerations of time-reversal symmetry one can now define *chiral orthogonal, unitary,* or *symplectic symmetry classes* (chOE, chUE, chSE) (Verbaarschot 1994; Verbaarschot and Wettig 2000; Akemann 2017), which are again associated with a symmetry index $\beta = 1, 2, 4$. Taking A as a random matrix with real, complex, or quaternion entries and $P(A) \propto \exp(-c_\beta \operatorname{tr} A^\dagger A)$ then leads to the Gaussian chiral ensembles (chGOE, chGUE, and chGSE), for which the positive energy levels in each pair follow the joint distribution

$$P(\{E_n\}) \propto \prod_{n<m, E_{n,m}>0} |E_n^2 - E_m^2|^\beta \prod_{k, E_k>0} E_k^{(|\nu|+1)\beta - 1} \exp(-c_\beta E_k^2). \tag{10.6}$$

The terms $E_n^2 - E_m^2 = (E_n - E_m)(E_n + E_m)$ include the repulsion from the negative-energy levels, while $E_k^{(|\nu|+1)\beta - 1}$ includes the repulsion from the mirror level at $E_{\bar{k}} = -E_k$ and from the zero modes. This modified repulsion follows again from the geometric argument described earlier, where the subspace to be explored by the rotations $R^{(nm)}$ corresponds to the case $M_1 = |\nu| + 1$, $M_2 = 1$. In this space, A becomes a vector and the eigenvalues and the squared eigenvalues $E_n^2 = E_{\bar{n}}^2 = |A|^2$ obey a χ^2 distribution.

These modifications affect the eigenvalue density around $E = 0$ over a range of a few level spacings,

$$\rho(E) - |\nu|\delta(E) \propto |E|^{(|\nu|+1)\beta - 1} \quad \text{for small } |E|, \tag{10.7}$$

which now becomes a universal spectral characteristic of the system. For a macroscopic number of zero modes with $M_2 \gg M_1 \gg 1$, the repulsion yields a hard gap around the symmetry point, corresponding to the mean density

$$\rho(E) = \frac{\pi}{M_1 \Delta^2 E} \sqrt{(E^2 - E_-^2)(E_+^2 - E^2)}, \quad E_\pm = \frac{M_1 \Delta}{\pi} (\sqrt{M_2/M_1} \pm 1) \tag{10.8}$$

for the M_1 positive eigenvalues (this expression follows from the Marchenko–Pastur law derived in the Appendix). For $M_1 = M_2 \gg 1$ this eigenvalue density reverts to a Wigner semicircle law (10.4), normalized to $2M_1$ eigenvalues in the whole energy range (the level repulsion (10.7) is not resolved as in this limit $\Delta \to 0$).

10.2.4 Charge-conjugation symmetry

If we admit for an antisymmetry $\mathcal{C}\mathcal{H}\mathcal{C}^{-1} = -\mathcal{H}$ with an antiunitary operator $\mathcal{C}$ we encounter four additional cases (Altland and Zirnbauer 1997). Two of these arise from the choices $\mathcal{C}^2 = \pm 1$, while the other two arise from an additional time-reversal symmetry with $\mathcal{T}^2 = -\mathcal{C}^2$.

If the antisymmetry is $\mathcal{C} = K$ ($\beta = \beta' = 2$), the Hamiltonian is imaginary and antisymmetric, $H = -H^* = -H^T$, and can be written in terms of matrix elements $H_{nm} \in i\mathbb{R}$. It is useful to denote this as the *real symmetry class* (RE) (Beenakker 2015). If we have in addition a time-reversal symmetry $\mathcal{T} = \Omega K$ ($\beta = 4$, $\beta' = 3$) we can write the Hamiltonian in the block form

$$H = \begin{pmatrix} A & B \\ B & -A \end{pmatrix}, \tag{10.9}$$

where $A = -A^T$, $B = -B^T$ are antisymmetric and $A_{nm}, B_{nm} \in i\mathbb{R}$. This can be usefully denoted as the *time-invariant real symmetry class* (T-RE).

For the antisymmetry $\mathcal{C} = \Omega K$ ($\beta = 2$, $\beta' = 0$) the Hamiltonian $\overline{H} = -H$ is anti-self-conjugate, and thus can be written in terms of matrix elements $H_{nm} \in i\mathbb{H}$. If, in addition, we also have the time-reversal symmetry $\mathcal{T} = K$ ($\beta = 1$, $\beta' = 0$), the Hamiltonian takes the block form (10.9) with symmetric matrices $A = A^T$, $B = B^T$ and elements $A_{nm}, B_{nm} \in \mathbb{R}$. The two cases define the *quaternion symmetry class* (QE) and the *time-invariant quaternion symmetry class* (T-QE).

In the two classes with $\mathcal{C}^2 = 1$, where the Hamiltonian can be made antisymmetric by an appropriate basis choice, a topologically protected zero mode exists if M is odd (when we have an additional time-reversal symmetry with $\mathcal{T}^2 = -1$ this mode is Kramers degenerate). The topological invariant counting such modes is then set to $\nu = 1$, while for even M we set $\nu = 0$. No such symmetry-protected zero modes exist in the two classes with $\mathcal{C}^2 = -1$.

Adopting again a Gaussian distribution $P(H) \propto \exp[-(c_\beta/2) \operatorname{tr} H^2]$ of matrix elements, these symmetry classes provide the joint probability density

$$P(\{E_n\}) \propto \prod_{n<m, E_{n,m}>0} |E_n^2 - E_m^2|^\beta \prod_{k, E_k>0} E_k^{(|\nu|+1)\beta - \beta'} \exp(-c_\beta E_k^2), \tag{10.10}$$

where β' modifies the repulsion from the mirror level as specified earlier (this follows again from the geometric argument in the small subspaces spanned by a level pair and any zero modes). As in the chiral classes, the spectral symmetry and the zero mode thus directly affect the level statistics in the closed system.

The symmetry associated with $\mathcal{C}$ is known as a charge-conjugation or particle–hole symmetry, and arises naturally in the context of superconducting systems. In a mean-field description, excitations are described as quasiparticles that obey the Boguliubov–de Gennes Hamiltonian

$$\mathcal{H} = \begin{pmatrix} H_0 - E_F & -i\sigma_y \otimes \Delta \\ i\sigma_y \otimes \Delta^* & E_F - H_0^* \end{pmatrix}, \tag{10.11}$$

where the blocks refer to the electron-like and hole-like degrees of freedom (addressed by Pauli matrices τ_i), the Pauli matrix σ_y acts in spin space, and $\Delta = \Delta^T$ is the s-wave pair potential. The charge conjugation is of the form $\mathcal{C} = \tau_x K$ and squares to $\mathcal{C}^2 = 1$. If $H_0 = H_+ \oplus H_-$ and $\Delta = \Delta_+ \oplus \Delta_-$ preserve the spin we can rearrange the Hamiltonian into two systems with

$$\mathcal{H}_\pm = \begin{pmatrix} H_\pm - E_F & \mp \Delta_\pm \\ \mp \Delta_\pm^* & E_F - H_\pm^* \end{pmatrix},$$

for which the charge-conjugation symmetry $\mathcal{C} = \Omega K$ with $\Omega = i\tau_y$ squares to $\mathcal{C}^2 = -1$.

In this setting, the zero modes in the classes with $\mathcal{C}^2 = 1$ are associated with Majorana fermions (Alicea 2012; Leijnse and Flensberg 2012; Beenakker 2013), previously elusive quasi-particles with possible applications for topological quantum computation (Nayak et al. 2008). These concepts can be generalized to surface and interface states in systems of specified spatial dimensions (Kitaev 2009; Teo and Kane 2010; Ryu et al. 2010), which are encountered in topological insulators and superconductors (Hasan and Kane 2010; Qi and Zhang 2011).

10.2.5 Random time-evolution operators and circular ensembles

To prepare how these considerations about the Hamiltonian translate to open systems, it is useful to turn to the dynamics and identify the corresponding symmetry classes of unitary matrices that exemplify the time evolution in the system. Of particular interest is the time evolution over a fixed time interval T_0, which also admits situations in which the Hamiltonian is itself time-dependent with that period. With a nod to the notion of a Floquet operator in the latter setting, we denote this stroboscopic time-evolution operator over a fixed time interval as F. Its eigenvalues $z_n = \exp(-i\varepsilon_n)$ lie on the unit circle, where the phases ε_n can be interpreted as quasi-energies. Similar considerations apply to quantum maps (Haake 2010) and quantum walks (Kitagawa et al. 2010).

As the time evolution is generated by the Schrödinger equation (10.1), we can symbolically write $F = \exp(-iHT_0/\hbar)$ with a suitable effective Hamiltonian H. The symmetries of F then follow from the symmetries of H, and thus comply with the ten symmetry classes described earlier (Zirnbauer 1996). In the resulting spaces of unitary matrices, some segments are smoothly connected to the identity, while others form disconnected pieces. This once more provides scope for topological invariants (Fulga et al. 2011; Beenakker 2015), which we specify in the following explicit constructions.

For the time-evolution operator, time-reversal symmetry implies that $\mathcal{T}F\mathcal{T}^{-1} = F^{-1}$. Given a time-reversal symmetry with $\mathcal{T}^2 = 1$ (orthogonal symmetry class with $\beta = 1$) and adopting a canonical basis where this is represented by $\mathcal{T} = K$, we find that F is symmetric under transposition, $F = F^T$. In absence of any symmetries (unitary symmetry class with $\beta = 2$), F is only constrained by $F^{-1} = F^\dagger$, so a member of the unitary group $U(M)$. For time-reversal symmetry with $\mathcal{T}^2 = -1$ (symplectic symmetry class with $\beta = 4$), the choice $\mathcal{T} = \Omega K$ implies that $F = \overline{F}$ is quaternion self-conjugate. The matrix $F_\Omega = \Omega F$ with elements $F_{\Omega,nm} = i\sigma_y F_{nm}$, written as a normal $2M \times 2M$ matrix, is then antisymmetric, $F_\Omega^T = -F_\Omega$. Notably, in the two classes arising from time-reversal symmetry, even though denoted as orthogonal and

symplectic, the spaces of matrices differ from the groups of orthogonal and symplectic matrices encountered in the diagonalization of the corresponding Hamiltonians. Only in the case of broken time-reversal symmetry the space remains associated with the unitary group.

In each of these three spaces we can again determine a Haar measure $\mu(F)$. This is uniquely defined by the requirement that the measure is invariant under transformations $F \to U'FU$, but now with unitary matrices U, U' that are subject to the constraints $U' = U^T$ in the orthogonal symmetry class, and $U' = \overline{U}$ in the symplectic symmetry class. Equipped with this measure, the corresponding ensembles are known as the circular ensembles (COE, CUE, and CSE) (Dyson 1962a). The joint distributions of phases φ_n in the unimodular eigenvalues $z_n = e^{i\varphi_n}$ is given by

$$P(\{\varphi_n\}) \propto \prod_{n<m} |e^{i\varphi_n} - e^{i\varphi_m}|^\beta, \tag{10.12}$$

and their density is uniform.

Chiral symmetry implies that $\mathcal{X}F\mathcal{X} = F^\dagger$, so that $F_X = \mathcal{X}F$ is Hermitian and only has eigenvalues ± 1. A topological invariant can then be defined as $\nu' = \frac{1}{2}\mathrm{tr}\,(F_X) = (M_+ - M_-)/2$, where $M_\pm$ counts the eigenvalues of either sign. One can again introduce a Haar measure, which in combination with the possible constraints from time-reversal symmetry leads to three chiral circular ensembles (chCOE, chCUE, and chCSE).

A charge-conjugation symmetry implies that $\mathcal{C}F\mathcal{C}^{-1} = F$. When we express $\mathcal{C} = K$ with $\mathcal{C}^2 = 1$ this implies that $F = F^*$ is real, and thus an element of the orthogonal group $O(M)$ (as the label OE is already taken, this justifies the notion of the real symmetry class, RE). We then have the invariant $\nu' = \det F$, where $\nu' = 1$ accounts for matrices from $SO(M)$. If in addition we have a time-reversal symmetry with $\mathcal{T} = K\Omega$ (class T-RE), such an invariant can be formulated with help of the Pfaffian $\nu' = \mathrm{pf}F_\Omega$ of the real antisymmetric matrix $F_\Omega = \Omega F$. For $\mathcal{C} = \Omega K$ with $\mathcal{C}^2 = -1$, the constraint can be written as $F^T\Omega F = \Omega$, which identifies F as symplectic (in quaternion language, $F\overline{F} = 1$, which justifies the notion of the quanternion universality class QE). If, in addition, we have a time-reversal symmetry with $\mathcal{T} = K$ (class T-QE), the matrix is furthermore constrained to be symmetric. Equipped with a Haar measure, the corresponding real and quaternion circular ensembles are denoted as CRE, T-CRE, CQE, and T-CQE (Beenakker 2015).

In a specific mathematical sense, it can now be argued that these ten classes provide a complete classification of random-matrix ensembles (Zirnbauer 1996; Caselle and Magnea 2004; Zirnbauer 2011)—they arise from the groups of unitary, orthogonal, and symplectic matrices and the associated compact symmetric Riemannian spaces, as classified by Cartan and summarized in Table 10.2. The three Wigner–Dyson classes with unitary, orthogonal, and symplectic symmetry (UE, OE, and SE) are labelled A, AI, AII; the corresponding chiral classes (chUE, chOE, and chSE) are labelled AIII, BDI, CII; the classes with charge-conjugation symmetry $\mathcal{C}^2 = 1$ and topological invariants (RE and T-RE) are labelled D and DIII, while the remaining two classes with $\mathcal{C}^2 = -1$ (QE and T-QE) are labelled C and CI.

Table 10.2 Classification of unitary matrix ensembles

Symmetries	Unitary matrices	Space	Cartan label
no symmetries	$F^{-1} = F^\dagger$	$U(M)$	A
$\mathcal{T} = K$	$F^{-1} = F^*$	$U(M)/O(M)$	AI
$\mathcal{T} = \Omega K$	$F^{-1} = \Omega F^* \Omega^{-1}$	$U(2M)/Sp(2M)$	AII
$\mathcal{C} = K$	$F^{-1} = F^T$	$O(M)$	D
$\mathcal{C} = K, \mathcal{T} = \Omega K$	$F^{-1} = F^T = \Omega F \Omega^{-1}$	$O(2M)/U(M)$	DIII
$\mathcal{C} = \Omega K$	$F^{-1} = \Omega F^T \Omega^{-1}$	$Sp(2M)$	C
$\mathcal{C} = \Omega K, \mathcal{T} = K$	$F^{-1} = F^* = \Omega F \Omega^{-1}$	$Sp(2M)/U(M)$	CI
$\mathcal{X} = \tau_z$	$(\mathcal{X}F) = (\mathcal{X}F)^\dagger$	$U(M_1 + M_2)/U(M_1) \otimes U(M_2)$	AIII
$\mathcal{X} = \tau_z, \mathcal{C} = K$	$(\mathcal{X}F) = (\mathcal{X}F)^T$ $= (\mathcal{X}F)^*$	$O(M_1 + M_2)/O(M_1) \otimes O(M_2)$	BDI
$\mathcal{X} = \tau_z, \mathcal{C} = \Omega K$	$(\mathcal{X}F) = (\mathcal{X}F)^\dagger$ $= \Omega(\mathcal{X}F)^* \Omega^{-1}$	$Sp(2M_1 + 2M_2)/Sp(2M_1) \otimes$ $Sp(2M_2)$	CII

10.2.6 Positive-definite matrices and Wishart–Laguerre ensembles

As we have seen in the construction of the ten Hamiltonian ensembles, it is often useful to study the blocks of a matrix, and compose new matrices out from them. This leads to natural extensions of the ensembles encountered so far, which can be justified via their connection to orthogonal polynomials (Mehta 2004; Forrester 2010). From this perspective, the Gaussian Hermitian matrix ensembles in the Wigner–Dyson classes are related to Hermite polynomials, while the other ensembles are related to Laguerre polynomials. As mentioned for the chiral symmetry classes, these ensembles are naturally related to positive semidefinite matrices $W = X^\dagger X$, where X is an $M' \times M$-dimensional matrix. It suffices to consider the case $M \leq M'$, as otherwise we can simply study $W = XX^\dagger$.

We again use the symmetry index $\beta = 1, 2, 4$ to distinguish settings where the matrix elements X_{lm} are real, complex, or quaternion. A Gaussian distribution

$$P(X) \propto \exp(-c'_\beta \operatorname{tr} X^\dagger X) \tag{10.13}$$

then defines the Wishart–Laguerre ensemble for W, where we set $c'_\beta = \beta/2\sigma^2$. This ensemble was first introduced by Wishart (1928) in the context of multivariate statistics, which marks the historical beginnings of random-matrix applications. The joint probability density of the eigenvalues λ of W is given by

$$P(\{\lambda_n\}) \propto \prod_{n<m} |\lambda_n - \lambda_m|^\beta \prod_k \lambda_k^{\beta(1+M'-M)/2-1} \exp(-c'_\beta \lambda_k), \tag{10.14}$$

which relates to the previously encountered eigenvalue distributions by the substitution $\lambda_n = E_n^2$. As mentioned earlier, the resulting eigenvalue correlations can be expressed in terms of Laguerre polynomials.

For large matrix dimensions the eigenvalue density approaches the Marchenko–Pastur law (Marčenko and Pastur 1967)

$$\rho(\lambda) = \frac{MT_0}{2\pi\lambda}\sqrt{(\lambda - \lambda_-)(\lambda_+ - \lambda)} \quad \text{for } \lambda_- < \lambda < \lambda_+, \tag{10.15}$$

where $\lambda_\pm = (\sqrt{M'} \pm \sqrt{M})^2/\sigma^2$ defines the range where this density is finite. This expression is derived in the Appendix.

10.2.7   Jacobi ensembles

A third class of classical orthogonal polynomials appearing in random-matrix problems are the Jacobi polynomials. These are associated with joint probability distributions of the form (Forrester 2010)

$$P(\{\mu_n\}) \propto \prod_{n<m} |\mu_n - \mu_m|^\beta \prod_k (1 - \mu_k)^{a\beta/2}(1 + \mu_k)^{b\beta/2}, \tag{10.16}$$

where $\mu_m \in [-1, 1]$, $m = 1, 2, 3, \ldots, M$.

Such distributions arise, for instance, when one considers the singular values of an $M' \times M$ dimensional off-diagonal block t of a suitable $N \times N$ dimensional unitary matrix F (Beenakker 1997, 2015). In particular, setting $N = M + M'$ with $M' \geq M$ and generating F from the three standard circular ensembles (COE, CUE, or CSE), the eigenvalues $T_n = (1 - \mu_n)/2 \in [0, 1]$ of $t^\dagger t$ obey a Jacobi ensemble with $a = M' - M + 1 - 2/\beta$, $b = 0$; similarly, if F is taken from $O(M + M')$ or $Sp(2M + 2M')$ (symmetry class D or C) one finds the same a but $b = 1 - 2/\beta$; the complete picture is presented in Section 10.4.4.

Alternatively (Forrester 2010), the quantities T_n can be interpreted as the eigenvalues of a so-called MANOVA matrix $(X^\dagger X + Y^\dagger Y)^{-1}X^\dagger X$, where X and Y are matrices of dimensions $M_x \times M$ and $M_y \times M$, distributed as Gaussians with equal variance σ according to Eq. (10.13). In this case, $a = M_x - M + 1 - 2/\beta$, $b = M_y - M + 1 - 2/\beta$. As shown based on this realization in the Appendix, in the limit of a large matrix dimension M with fixed $c_x = M_x/M$, $c_y = M_y/M$ the eigenvalue density approaches

$$\rho(T) = \frac{M(c_x + c_y)\sqrt{(T - T_-)(T_+ - T)}}{2\pi T(1 - T)}, \tag{10.17}$$

where

$$T_\pm = \frac{1}{1 + \lambda_\mp}, \quad \lambda_\pm = \left(\frac{\sqrt{c_x c_y} \pm \sqrt{c_x + c_y - 1}}{c_x - 1}\right)^2 \tag{10.18}$$

determines the range where the density is finite. In terms of the variables μ_n, this takes the form

$$\rho(\mu) = \frac{M(c_x + c_y)}{2\pi} \frac{\sqrt{(\mu - \mu_-)(\mu_+ - \mu)}}{1 - \mu^2}, \tag{10.19}$$

within the boundaries given by $\mu_\pm = (\lambda_\pm - 1)/(\lambda_\pm + 1)$.

10.2.8 Non-Hermitian matrices

The eigenvalues λ_n in the Wishart matrix $W = X^\dagger X$ are the squared singular values of the matrix X. For a square matrix of dimensions $M \times M$ we can also study the complex eigenvalues z_n of X, obtained from $X\mathbf{v}_n = z_n \mathbf{v}_n$ with eigenvectors $\mathbf{v}_n$. This leads to entirely different classes of random matrices (Ginibre 1965; Khoruzhenko and Sommers 2011). Since X is in general not normal (in particular, neither Hermitian nor unitary), there is no direct relation between the real singular values and the complex eigenvalues z_n. This key difference is intimately related to the fact that the eigenvectors $\mathbf{v}_n$ are not orthogonal to each other, so that the spectral decomposition $X = VDV^{-1}$ with $D = \text{diag}(z_n)$ involves a non-unitary matrix V. We therefore need to distinguish the right eigenvectors $\mathbf{v}_n$, which form the columns of V, from the left eigenvectors $\mathbf{w}_n$, which are obtained from $\mathbf{w}_n X = s_n \mathbf{w}_n$. Imposing the biorthogonality condition $\mathbf{w}_m \mathbf{v}_n = \delta_{nm}$, the left eigenvectors form the rows of V^{-1}.

This biorthogonal set of eigenvectors is in general no longer normalized. The extent of mode nonorthogonality can thus be quantified by the condition numbers (Chalker and Mehlig 1998; Janik et al. 1999; Schomerus et al. 2000)

$$O_{mn} = \frac{(\mathbf{v}_m^\dagger \mathbf{v}_n)(\mathbf{w}_n \mathbf{w}_m^\dagger)}{(\mathbf{v}_m^\dagger \mathbf{w}_m^\dagger)(\mathbf{w}_n \mathbf{v}_n)}, \tag{10.20}$$

which we have written in a way that does not rely on the chosen normalization condition. In terms of the matrix V,

$$O_{mn} = (V^\dagger V)_{mn}(V^{-1}V^{-1\dagger})_{nm}. \tag{10.21}$$

The diagonal elements $K_m = O_{mm}$ are real and obey $K_m \geq 1$, with $K_m = 1$ for all m only if V is unitary. These quantities become large in particular when two eigenvalues approach each other closely, and indeed diverge at eigenvalue degeneracies, so-called exceptional points (Berry 2004; Heiss 2012). Close to such a degeneracy with a coalescing pair $z_{n+1} = z_n$, X cannot be diagonalized but only be brought into a form involving Jordan blocks

$$\begin{pmatrix} z_n & 1 \\ 0 & z_n \end{pmatrix}. \tag{10.22}$$

This means that the eigenvectors of the modes become identical, in sharp contrast to Hermitian systems where the eigenvectors remain orthogonal as one approaches a degeneracy.

The probability distribution (10.13) for $M \times M$-dimensional square matrices X defines the Ginibre ensemble (Ginibre 1965; Khoruzhenko and Sommers 2011). For the complex Ginibre ensemble ($\beta = 2$), the joint distribution of eigenvalues is

$$P(\{z_n\}) \propto \prod_{n<m} |z_n - z_m|^2 \prod_k \exp(-c'_\beta z_k^2). \tag{10.23}$$

In the quaternion case $\beta = 4$ eigenvalues come in conjugate pairs, and the joint distribution of eigenvalues in the upper half of the complex plane

$$P(\{z_n\}) \propto \prod_{n<m} |z_n - z_m|^2 |z_n - z_m^*|^2 \prod_k |z_k - z_k^*|^2 \exp(-c'_\beta z_k^2) \tag{10.24}$$

contains the expected self-repulsion terms. For the real case $\beta = 1$, much more complicated expressions arise due to the accumulation of $O(\sqrt{M})$ eigenvalues on the real axis (Lehmann and Sommers 1991; Forrester and Nagao 2007). What is common to all three cases are the local spectral correlations of eigenvalues well inside the complex support (away from the boundaries and spectral symmetry lines), which irrespective of β are determined by the factors $|z_n - z_m|^2$. This yields a cubic level repulsion $P(s) \propto s^3$ for small spacings $s = |z_n - z_m|$, where one power of s arises from the area element in the complex plane.

As shown in the Appendix for the complex Ginibre ensemble, for a variance scaled to $\sigma^2 = 1/M$ and $M \to \infty$ the eigenvalue density in the complex plane approaches Ginibre's circular law $\rho(z) = \frac{M}{\pi}\Theta(1 - |z|)$, where Θ denotes the unit step function. As a side product of the calculation presented there (Janik et al. 1999), the condition number $\overline{K_m}|_{z_m=z} \sim M(1 - |z|^2)$ turns out to be large, unless one approaches the boundaries of the eigenvalue support.

From the general perspective of commutation and anticommutation with unitary and antiunitary symmetries, non-Hermitian matrices admit a very large number of symmetry classes (Magnea 2008). For a physical setting that illustrates this richness, we can consider photonic systems with absorption and amplification (Cao and Wiersig 2015). Without further constraints we may model these as a complex Ginibre ensemble ($\beta = 2$) with different weights of the Hermitian and non-Hermitian contributions, where the eigenvalue support becomes elliptic (Girko 1986). Time-reversal symmetry in optics (reciprocity) makes the matrix complex symmetric, $H = H^T \neq H^*$, which modifies the statistics but does not entail any spectral constraints. As a template for the real Ginibre ensemble ($\beta = 1$), we can take a system with balanced amplification and absorption, situated in regions that are mapped onto each other by a reflection or inversion P (Makris et al. 2008; Rüter et al. 2010). We then obtain a non-Hermitian PT-symmetric system with $PHP = H^* \neq H^T$ (Bender 2007), which in a suitable basis is represented by a real asymmetric matrix. In combination with magneto-optical effects, we can similarly construct PTT'-symmetric systems with $PHP = H^\dagger \neq H$ (Schomerus 2013a). The spectrum remains symmetric about the real axis, and a random-matrix analysis reveals a close connection to the real Ginibre ensemble, including the same accumulation of $O(\sqrt{M})$ eigenvalues on the real axis (Birchall and Schomerus 2012). Further examples can be constructed by modifying the role of P. In an optical system where P represents a chiral symmetry, we can realize the case $H = -PH^*P$ in which eigenvalues are symmetric with respect to the imaginary axis (Schomerus and Halpern 2013; Schomerus 2013b; Poli et al. 2015), as well as the case $H = H^* = -PHP$ in which eigenvalues are symmetric with respect

to both the real and the imaginary axes (Malzard et al. 2015). For a symmetry with $P^2 = -1$ (hence $P = -P^T$, assuming P is real), two interesting cases are the so-called Hamiltonian ensembles with $PHP = H^T$, as well as the skew-Hamiltonian ensembles with $PHP = -H^T$ (these notions relate to the symplectic structure of classical Hamiltonians, generated by an antisymmetric involution such as P; Beenakker et al. 2013). For a real Hamiltonian matrix with Gaussian statistics, $O(\sqrt{M})$ eigenvalues accumulate both on the real and on the imaginary axes; for a real skew-Hamiltonian matrix, all eigenvalues are twofold degenerate and $O(\sqrt{M})$ of these pairs accumulate on the real axis.

In the next section we will see that non-Hermitian matrices play a crucial role in the description of open scattering systems, where additional constraints arise from the physical implications of unitarity and causality.

10.3 The scattering matrix

In this chapter we develop effective models for the scattering matrix and use these to identify the associated random-matrix ensembles.

10.3.1 Points of interest

Consider a particle moving through a scattering region with a spatially varying potential energy V, as sketched for a simple one-dimensional setting in Fig. 10.2. The corresponding Hamiltonian is $\hat{H} = \hat{T} + \hat{V}$, where $\hat{T}$ represents the kinetic energy. Here are some natural phenomena that we may wish to consider: Decay, where we address the escape rate of a particle inserted into the scattering region; transport, where we

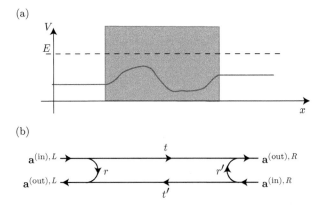

Fig. 10.2 (a) Sketch of a scattering region with a varying potential $V(x)$ in a one-dimensional system, with ideal leads attached to either side. Note that the potential does not need to be identical in both leads. (b) Scattering processes relating the amplitudes of propagating waves in the leads.

address the probability for an incident particle to be transmitted or reflected; dynamics, where we ask how long the particle engages with the scattering region and how many internal states it explores. We may also wish to identify system-specific details beyond the fundamental symmetries, such as regarding the role of different scattering subregions or the role of the contacts. All of these questions (and many more) can be addressed with the help of a single object, the scattering matrix $S(E)$.

10.3.2   Definition of the scattering matrix

To define the scattering matrix (Newton 2002; Messiah 2014) we stipulate that the motion outside the scattering region is ballistic. At any energy E, we then have access to a complete set of propagating scattering states $|\psi_n^{(\text{in})}\rangle$ in which the particle is approaching the scattering region (incoming channels), and a corresponding set of propagating states $|\psi_n^{(\text{out})}\rangle$ where the particle is moving away from the region (outgoing channels). These states are taken to be normalized to a unit probability flux through any closed surface surrounding the scattering region. We may also encounter a set of nonpropagating (evanescent) states $|\psi_m^{(\text{ev})}\rangle$ which decay away from the scattering region and do not carry any flux. Outside the scattering region, we then can write a state with a given energy as

$$|\psi\rangle = \sum_{n=1}^{N} a_n^{(\text{in})}|\psi_n^{(\text{in})}\rangle + \sum_{n=1}^{N} a_n^{(\text{out})}|\psi_n^{(\text{out})}\rangle + \sum_{l} a_l^{(\text{ev})}|\psi_m^{(\text{ev})}\rangle, \tag{10.25}$$

where N fixes the number of scattering channels. We collect the expansion coefficients into vectors $\mathbf{a}^{(\text{in})}$, $\mathbf{a}^{(\text{out})}$, and $\mathbf{a}^{(\text{ev})}$.

Inside the scattering region, we may expand the state in terms of any suitable complete set of modes, $|\psi\rangle = \sum_m b_m|\chi_m\rangle$ with a coefficient vector $\mathbf{b}$. With help of the stationary Schrödinger equation (10.1), the states inside and outside the scattering region are uniquely related. In particular, if we fix $\mathbf{a}^{(\text{in})}$ then the solution of the Schrödinger equation uniquely fixes $\mathbf{a}^{(\text{out})}$, $\mathbf{a}^{(\text{ev})}$, and $\mathbf{b}$, up to effectively decoupled parts that can be treated as a separate system. These relations must be linear, so that

$$\mathbf{a}^{(\text{out})} = S(E)\mathbf{a}^{(\text{in})}. \tag{10.26}$$

This defines the scattering matrix. Flux normalization ensures that for real energies $S(E)$ is unitary; hence, $S(E) \in U(N)$. Causality ensures that the poles E_l of S at complex energies are all confined to the lower half of the complex plane, $\text{Im} E_l < 0$. The number of propagating scattering channels N may change at certain energies, which gives rise to branch cuts in the complex-energy plane.

10.3.3   Preliminary answers

The scattering matrix addresses the phenomena listed at the beginning of this section in the following ways.

Decay—The complex poles $E_l = E'_l - i\hbar\gamma_l/2$ of the scattering matrix provide solutions where $\mathbf{a}^{(\text{out})}$ is finite while $\mathbf{a}^{(\text{in})} = 0$. These quasibound states provide a fundamental description of decay and resonant scattering (Guhr et al. 1998; Weidenmüller and Mitchell 2009; Moiseyev 2011). The time dependence of the quasibound states follows from the amplitude factor $A(t) = \exp(-itE_l/\hbar) = \exp(-itE'_l/\hbar)\exp(-t\gamma_l/2)$, so that the corresponding intensity $|A(t)|^2 = \exp(-t\gamma_l)$ decays with rate γ_l. For a particle prepared in this state at $t = 0$, the Fourier signal

$$A(\omega) = \int_0^\infty A(t)e^{i\omega t}\,dt = i[(\omega - E'_l/\hbar) + i\gamma_l/2]^{-1} \tag{10.27}$$

delivers the resonance-like frequency-resolved signal

$$|A(\omega)|^2 = \frac{1}{(\omega - E'_l/\hbar)^2 + \gamma_l^2/4}, \tag{10.28}$$

a Lorentzian centred at $E'_l/\hbar$ with full width at half-maximum γ_l. When the particle is prepared in a superposition of quasibound states, the resulting decay for long times depends on the characteristic decay rate $\gamma_0 = \inf \gamma_l$, defined such that $\gamma_l \geq \gamma_0$ for all contributing states. If $\gamma_0 > 0$ the decay becomes exponential, while for $\gamma_0 = 0$ one typically encounters a power law.

Transport—For a particle incoming in channel n, the probability to scatter into the outgoing channel n' is given by $|S_{n'n}|^2$. The unitarity of the scattering matrix guarantees that the sums of probabilities $\sum_n |S_{n'n}|^2 = \sum_{n'} |S_{n'n}|^2 = 1$ are normalized. This normalization also holds for an incident particle in any superposition of incoming modes, $|\mathbf{a}^{(\text{out})}|^2 = |\mathbf{a}^{(\text{in})}|^2$. These features are at the heart of the scattering approach to transport (Beenakker 1997; Blanter and Büttiker 2000; Nazarov and Blanter 2009).

In many settings, we are allowed to group the scattering amplitudes into subcomponents $\mathbf{a}^{(\text{in}),s}$, $\mathbf{a}^{(\text{out}),s}$, where s labels different asymptotic regions (leads). The scattering matrix is then formed of blocks $S_{s's}$ describing transmission from lead s to lead s', and reflections back into lead s if $s' = s$. The associated transmission probability is quantified by the dimensionless conductance $g_{s's} = \text{tr}\,(S_{s's}^\dagger S_{s's})$. In the case of two leads, designated as a left lead $s = L$ with N_L channels and a right lead $s = R$ with N_R channels, we write the blocks as

$$S = \begin{pmatrix} r & t' \\ t & r' \end{pmatrix}, \tag{10.29}$$

where r and t describe the reflection and transmission of particles arriving from the left, while r' and t' describe these processes for particles arriving from the right. This designation is illustrated in Fig. 10.2b. The dimensionless conductance is then given by $g = \text{tr}\,t^\dagger t = \text{tr}\,t'^\dagger t' = N_L - \text{tr}\,r^\dagger r = N_R - \text{tr}\,r'^\dagger r'$, where the stated identities follow from unitarity.

The eigenvalues $T_n \in [0, 1]$ of the Hermitian matrix $t^\dagger t$ are known as the transmission eigenvalues, and determine the dimensionless conductance via $g = \sum_n T_n$. The quantities $\sqrt{T_n}$ can be interpreted as the singular values of t, which generalizes to the polar decomposition of the scattering matrix,

$$S = \begin{pmatrix} V & 0 \\ 0 & V' \end{pmatrix} \begin{pmatrix} \sqrt{1-T} & \sqrt{T} \\ \sqrt{T} & -\sqrt{1-T} \end{pmatrix} \begin{pmatrix} V'' & 0 \\ 0 & V''' \end{pmatrix}, \quad T = \mathrm{diag}\,(T_n) \qquad (10.30)$$

with unitary matrices V, V', V'', and V'''.

The transmission eigenvalues determine many other transport properties, including the full counting statistics of electrons at low temperatures (Levitov and Lesovik 1993), with the shot noise characterized by the second binomial cumulant (Büttiker 1990; Blanter and Büttiker 2000)

$$\sum_n T_n(1 - T_n). \qquad (10.31)$$

Another example is the charge transport through a normal conductor into a conventional superconducting lead (Beenakker 1992, 1997), for which the dimensionless conductance at vanishing magnetic fields is given by

$$g_{NS} = \sum_n T_n^2/(2 - T_n)^2. \qquad (10.32)$$

Dynamics—Complementing the information about the scattering probabilities, the phase φ of a scattering amplitude $S_{n'n} = |S_{n'n}|e^{i\varphi}$ provides insight into the dynamics (de Carvalho and Nussenzveig 2002; Texier 2016). For instance, for ballistic propagation through a region of length L at a constant momentum $p(E)$, the particle picks up the dynamical phase $\varphi = pL/\hbar$. The energy sensitivity $\hbar d\varphi/dE = L/v = \tau$ of the phase, therefore, gives an indication of the travel time. In a semiclassical description of scattering from a slowly varying potential, we have $\varphi = S_{cl}/\hbar$, where the classical action S_{cl} again obeys $dS_{cl}/dE = \tau$.

These observations lead to the formal definition of the delay time of a particle that passes through the scattering region. For injection and extraction in individual channels, the delay time can be isolated by the logarithmic derivative $\mathrm{Im}\, S_{n'n}^{-1} dS_{n'n}/dE$. For multichannel scattering this is generalized by the Wigner–Smith time-delay matrix (Wigner 1955; Smith 1960)

$$Q = -i\hbar S^\dagger dS/dE. \qquad (10.33)$$

The unitarity of S at any energy ensures that $Q = Q^\dagger$ is Hermitian, while causality ensures that Q is positive semi-definite. Therefore, the eigenvalues τ_n of Q are real and positive. These eigenvalues are known as the proper delay times.

Noting that $v^{-1} = dp/dE$ also appears in semiclassical estimates of the accessible phase-space volume, the delay times are intimately related to the density of states. Indeed, the Wigner–Smith matrix directly quantifies the global density of states in the system, in terms of the Birman–Krein formula (Birman and Krein 1962)

$$\rho(E) = \frac{1}{2\pi\hbar}\operatorname{tr} Q. \tag{10.34}$$

Replacing the derivative $\mathrm{d}/\mathrm{d}E$ by a local variation of the potential $\partial/\partial V(x)$, this approach can be extended to obtain the local density of states (Gasparian et al. 1996). Analogous variations with respect to other parameters deliver a wide range of response functions, which can, for instance, be used to study adiabatic transport and quantum pumping (Büttiker et al. 1994; Brouwer 1998).

System-specific details—When we separate the scattering region into subregions, we can build up the total scattering matrix from the scattering problems of the subregions (Datta 1997; Beenakker 1997; Nazarov and Blanter 2009). This can be done exactly if we extend the scattering matrix to include evanescent states, and often still very reliably if we only account for the propagating states. The simple idea is to inspect each interface and identify the amplitudes of outgoing states from one region with the amplitudes of incoming states into the adjacent region.

For the case of two adjacent regions with scattering matrices S_1, S_2 of the form (10.29), the wave matching of propagating states leads to the composition law

$$S_{1\oplus2} = \begin{pmatrix} r_1 + t_1'r_2(1 - r_1'r_2)^{-1}t_1 & t_1'(1 - r_2r_1')^{-1}t_2' \\ t_2(1 - r_1'r_2)^{-1}t_1 & r_2' + t_2r_1'(1 - r_2r_1')^{-1}t_2' \end{pmatrix}. \tag{10.35}$$

This rule can be reformulated as a simple matrix multiplication $M = M_2M_1$ for the transfer matrix

$$M = \begin{pmatrix} t^{\dagger-1} & r't'^{-1} \\ r'^{\dagger}t^{\dagger-1} & t'^{-1} \end{pmatrix}, \tag{10.36}$$

which relates modes on the left and right according to

$$\begin{pmatrix} \mathbf{a}^{\mathrm{out},R} \\ \mathbf{a}^{\mathrm{in},R} \end{pmatrix} = M \begin{pmatrix} \mathbf{a}^{\mathrm{in},L} \\ \mathbf{a}^{\mathrm{out},L} \end{pmatrix}. \tag{10.37}$$

Flux conservation translates to the property $M^{\dagger}\sigma_z M = \sigma_z$, so that M is complex symplectic. The eigenvalues of $M^{\dagger}M$ and $(M^{\dagger}M)^{-1} = \sigma_z M^{\dagger}M\sigma_z$ are thus identical and appear in reciprocal pairs, which are given by $(\sqrt{1/T_n} \pm \sqrt{-1 + 1/T_n})^2$.

We note that in the composed system, according to Eq. (10.35) poles from the multiple scattering across the interface arise from

$$\det[1 - r_2(E)r_1'(E)] = 0. \tag{10.38}$$

Similarly, the role of a contact can be studied by inserting a static tunnel barrier at the corresponding boundary of the scattering region (Brouwer 1995; Beenakker 1997). For example, the scattering matrix

$$S_B = \begin{pmatrix} \sqrt{1 - \Gamma} & \sqrt{\Gamma} \\ \sqrt{\Gamma} & -\sqrt{1 - \Gamma} \end{pmatrix} \tag{10.39}$$

describes a barrier with uniform transparency $\Gamma \in [0, 1]$ in all channels. If we send $\Gamma \to 0$ for all contacts the system becomes closed. Poles approaching the real axis become the energy levels of the closed system, while poles moving deep into the complex plane are associated with direct reflection processes from the outside.

We can also artificially separate a closed system into two open systems joined by an interface. For a left and a right region, this is described by scattering matrices $S_1 = r_1'$ and $S_2 = r_2$, both only composed of a reflection block back to the interface. The quantization condition (10.38) can then be rewritten as

$$\det (S_1(E)S_2(E) - 1) = 0, \tag{10.40}$$

which determines the energies of the closed systems. This scattering quantization approach becomes exact when one includes the evanescent modes into the scattering description (Doron and Smilansky 1992; Bäcker 2003), and can be extended, e.g., to superconducting systems (Beenakker 2005) and non-Hermitian photonic systems (Schomerus 2013a).

10.3.4   Effective scattering models

In practice, many methods are available for calculating the scattering matrix in specific settings. This includes wave matching, Green function methods, and the boundary integral method, as well as iterative procedures based on the composition rule (10.35) of scattering matrices, and analogous rules for the Green function (Datta 1997). For the purpose of a statistical description, however, we require a generic model that captures the essential features of the internal dynamics and the coupling to the leads. This is delivered by the Mahaux–Weidenmüller formula (Mahaux and Weidenmüller 1969; Livsic 1973; Verbaarschot et al. 1985; Guhr et al. 1998; Bohigas and Weidenmüller 2017)

$$S(E) = \frac{i\pi W^{\dagger}(E - H)^{-1}W - 1}{i\pi W^{\dagger}(E - H)^{-1}W + 1}, \tag{10.41}$$

where H is an effective internal Hamiltonian of dimension $M \times M$ while W is a suitable $M \times N$-dimensional coupling matrix, specified fully in Eq. (10.64).

We provide a motivation of this formula via a detour to the stroboscopic scattering problem (Fyodorov and Sommers 2000; Tworzydło et al. 2003), which leads to its close cousin

$$S(\varepsilon) = \frac{K\mathcal{A}K^T - 1}{K\mathcal{A}K^T + 1}, \quad \mathcal{A} = \frac{1 + e^{i\varepsilon}F}{1 - e^{i\varepsilon}F} = -\mathcal{A}^{\dagger}. \tag{10.42}$$

Here F is an effective internal time-evolution operator over a fixed time period T_0, ε is the associated quasienergy, and the coupling matrix K is fully specified in (10.58). The Mahaux–Weidenmüller formula then follows in the continuum limit $T_0 \to 0$. We present this construction because it gives rather direct intuitive insight into scattering and decay problems, and also helps to isolate and justify the general features of the scattering matrix described in the previous section.

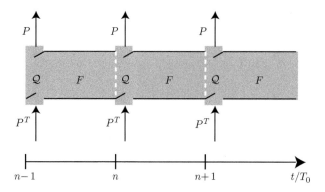

Fig. 10.3 Illustration of the stroboscopic scattering approach, in which particles are injected and collected at regular intervals.

10.3.5 Stroboscopic scattering approach

Stroboscopic ballistic decay

Our starting point is a simple, highly idealized scenario, which nonetheless can be easily extended to capture a large range of other cases. Consider a situation where the coupling of the scattering region to the outside occurs stroboscopically, at periodically spaced, discrete times $t = nT_0 \equiv t_n$, $n = 0, 1, 2, 3, \ldots$ (see Fig 10.3). Let us denote the state within the system just before these times as $|\psi_n\rangle = |\psi(t_n^-)\rangle$. This state evolves stroboscopically according to

$$|\psi_n\rangle = FQ|\psi_{n-1}\rangle = (FQ)^n|\psi_0\rangle, \tag{10.43}$$

where F is the unitary operator that describes the time evolution when the system is closed, while Q is a projector that describes what remains in the system when the system is open. In other words, in each time interval, we lose some internal wave amplitude according to the complementary projector $\mathcal{P} = 1 - Q$, while the remaining amplitude is propagated by the unitary time evolution operator F. As we assume that F and Q are independent of the time index n, we require that the details of the coupling are otherwise time independent and the internal dynamics are autonomous, or at least themselves time periodic with period T_0. The fact that we take Q as a projector means that the coupling is *ballistic*—the opening is fully transparent, without any partial reflection of the passing wave. This is also called an ideal lead.

According to Eq. (10.43), the decay of the amplitude within this system is described by the nonunitary operator FQ. In a basis where Q is diagonal this corresponds to *truncating* the unitary operator F. Let us specify this for a system with a finite internal Hilbert space of dimension M, coupled to N external channels such that rank $Q = M - N$. In the basis where $Q = \mathrm{diag}(0, 0, 0, \ldots, 0, 1, 1, \ldots, 1)$ (N zeros followed by $M - N$ ones), FQ is then obtained from F by setting the first N columns to 0.

In this setting, the quasibound states $|\phi_m\rangle$ are obtained from the eigenvalue problem

$$F\mathcal{Q}|\phi_m\rangle = z_m|\phi_m\rangle, \quad = 1, 2, \ldots, M. \tag{10.44}$$

Due to the projective nature of $\mathcal{Q}$, there will by N vanishing eigenvalues, while the remaining eigenvalues are in general complex and finite, with $|z_m| < 1$. Each eigenvalue describes the exponential stroboscopic decay of the associated quasibound state—if the initial state is $|\psi_0\rangle = |\phi_m\rangle$, the intensity within the system decays as

$$\langle\psi_n|\psi_n\rangle = |z_m|^{2n}\langle\psi_0|\psi_0\rangle. \tag{10.45}$$

Writing $z_m = \exp[-i(\varepsilon_m - i\gamma_m/2)]$, the decay constant over a period T_0 is given by γ_m. As indicated, this decay constant is best viewed as arising from the imaginary part of a complex quasienergy $\varepsilon_m^\star = \varepsilon_m - i\gamma_m/2$, where the real part is defined modulo 2π.

Stroboscopic scattering with ideal contacts

We now turn the stroboscopic decay problem into a stroboscopic scattering problem. This requires defining how the escape from the system translates into particles detected outside, as well as how to feed particles into the system. In other words, we need to define objects that connect the state within the system (residing in the internal Hilbert space in which F and $\mathcal{Q}$ operate) to the amplitudes of the N incoming modes (states $|\psi_n^{(\mathrm{in})}\rangle$) and the N outgoing modes (states $|\psi_n^{(\mathrm{out})}\rangle$) outside the system.

In the case of ballistic coupling that we study thus far, the outgoing state can be taken of the simple form

$$|\psi_n^{(\mathrm{out})}\rangle = P|\psi_n\rangle, \tag{10.46}$$

with P such that $\mathcal{P} = P^T P = 1 - \mathcal{Q}$ recovers the rank-N projector that complements $\mathcal{Q}$ in the internal Hilbert space. It follows that $PP^T = 1$ is the identity in the space of the external scattering channels (the rank does not change under the reordering and the resulting object is still a projector). Recall that the internal state refers to the instance just before we open the system. Therefore, the incoming particle injected in the previous step modifies this state according to

$$|\psi_n\rangle = F\mathcal{Q}|\psi_{n-1}\rangle + FP^T|\psi_{n-1}^{(\mathrm{in})}\rangle$$

$$= (F\mathcal{Q})^n|\psi_0\rangle + \sum_{l=0}^{n-1}(F\mathcal{Q})^l FP^T|\psi_{n-l-1}^{(\mathrm{in})}\rangle, \tag{10.47}$$

which replaces Eq. (10.43). Combining these expressions, we find that

$$|\psi_n^{(\mathrm{out})}\rangle = P(F\mathcal{Q})^n|\psi_0\rangle + P\sum_{l=0}^{n-1}(F\mathcal{Q})^l FP^T|\psi_{n-l-1}^{(\mathrm{in})}\rangle. \tag{10.48}$$

The first part recovers the decay of the initial state, while the remaining part describes the scattering. The pure decay problem is characterized by the absence of the incoming state, while the pure scattering problem is characterized by the absence of the initial state.

Both these problems now turn out to be intimately related. For this, we revert back to a continuous time variable, $|\psi^{(\text{out})}(t)\rangle = \sum_n \delta(t - nT_0)|\psi_n^{(\text{out})}\rangle$, and perform a Fourier decomposition of the scattered signal,

$$|\psi^{(\text{out})}(\varepsilon)\rangle = \sum_{n=0}^{\infty} e^{i\varepsilon n} |\psi_n^{(\text{out})}\rangle \tag{10.49}$$

$$= \sum_{l=0}^{\infty} \sum_{n=l+1}^{\infty} e^{i\varepsilon l} P(FQ)^l e^{i\varepsilon} F P^T e^{i\varepsilon(n-l-1)} |\psi_{n-l-1}^{(\text{in})}\rangle, \tag{10.50}$$

hence

$$|\psi^{(\text{out})}(\varepsilon)\rangle = S(\varepsilon)|\psi^{(\text{in})}(\varepsilon)\rangle \tag{10.51}$$

with the stroboscopic scattering matrix

$$S(\varepsilon) = P \sum_{l=0}^{\infty} [e^{i\varepsilon} FQ]^l e^{i\varepsilon} F P^T = P \frac{1}{1 - e^{i\varepsilon} FQ} e^{i\varepsilon} F P^T. \tag{10.52}$$

We now observe that the poles of the scattering matrix coincide with the complex quasienergies ε_m^*, as determined by the eigenvalue problem (10.44).

It is convenient to bring the scattering matrix (10.52) into the equivalent form

$$S(\varepsilon) = \frac{PAP^T - 1}{PAP^T + 1}, \quad A = \frac{1 + e^{i\varepsilon} F}{1 - e^{i\varepsilon} F} = -A^\dagger. \tag{10.53}$$

We then see that the scattering matrix is indeed unitary. Furthermore, this expression nicely generalizes to the case of nonideal contacts, which we address next.

Stroboscopic scattering with nonideal contacts

To account for nonideal coupling we insert an energy-independent scatterer at the place of the contact. The contact can be viewed as a region with N channels coupled to the outside and N channels coupled to the inside, and thus is described by a $2N \times 2N$-dimensional unitary scattering matrix

$$S_B = \begin{pmatrix} r_B & t_B' \\ t_B & r_B' \end{pmatrix}. \tag{10.54}$$

The blocks r_B and r_B' describe the partial reflection in the external and internal channels, while t_B and t_B' describe the transmission into and out of the system. This matrix is assumed to be energy independent, meaning that the reflection and transmission processes from the contact are instantaneous. The return of the particle to the contact is described by the ballistic scattering matrix S_0.

We can now match the waves at the contact according to Eq. (10.35), which results in the total scattering matrix

$$S = r_B + t'_B S_0 (1 - r'_B S_0)^{-1} t_B. \tag{10.55}$$

This expression has a simple interpretation: The incident wave either is directly reflected according to r_B, or enters into the system according to t_B. Once in the system, the wave undergoes a sequence of l events, each consisting of an internal scattering round trip S_0 followed by a partial reflection r'_B, until after another return S_0 to the contact it escapes according to t'_B. Equation (10.55) follows by summing over l, which is of the form of a geometric series.

Inserting for S_0 the stroboscopic scattering matrix (10.52) for ideal contacts, we find that this can be written more directly as

$$S = r_B + t'_B P \frac{1}{1 - e^{i\varepsilon} F(Q + P^T r'_B P)} e^{i\varepsilon} F P^T t_B. \tag{10.56}$$

To further simplify this expression we choose an appropriate basis for the internal state, as well as for the incoming and the outgoing state. This follows from the polar decomposition (10.30), which we need to adopt in the slightly more general form

$$S_B = \begin{pmatrix} V & 0 \\ 0 & V' \end{pmatrix} \begin{pmatrix} -\Sigma\sqrt{1-\Gamma} & \sqrt{\Gamma} \\ \sqrt{\Gamma} & \Sigma\sqrt{1-\Gamma} \end{pmatrix} \begin{pmatrix} V'' & 0 \\ 0 & V''' \end{pmatrix}, \quad \begin{cases} \Gamma = \mathrm{diag}\,(\Gamma_n) \\ \Sigma = \mathrm{diag}\,(\sigma_n) \end{cases}. \tag{10.57}$$

Here $\Gamma_n \in [0,1]$ are the transmission eigenvalues of the contact, while $\sigma_n = \pm 1$ discriminates two distinct ways to close a channel. The unitary matrices V, V', V'', and V''' can all be absorbed into the basis choice, which means that S_B is block diagonal and real. Starting from (10.56), this basis choice results in the desired generalization of Eq. (10.53),

$$S = \frac{KAK^T - 1}{KAK^T + 1}, \quad A = \frac{1 + e^{i\varepsilon}F}{1 - e^{i\varepsilon}F} = -A^\dagger, \quad K = \mathrm{diag}\,(\kappa_n^{\sigma_n})P, \tag{10.58}$$

where the contact is now characterized by the coupling coefficients

$$\kappa_n = \Gamma_n^{-1/2}(1 - \sqrt{1 - \Gamma_n}). \tag{10.59}$$

These coefficients take the value $\kappa_n = 1$ for $\Gamma_n = 1$ and $\kappa_n \approx \sqrt{\Gamma_n}/2$ for $\Gamma_n \ll 1$. As they enter the matrix K to the power σ_n, a semitransparent contact can be achieved both by decreasing the coupling ($\sigma_n = 1$) and by increasing the coupling ($\sigma_n = -1$). This completes the derivation of the stroboscopic scattering matrix (10.42).

10.3.6   Continuous-time scattering theory

To realize the time-continuous limit of the stroboscopic scattering theory, we set $\varepsilon = ET_0/\hbar$, $F = \exp(-iT_0 H/\hbar)$, and equate $T_0 \equiv 2\pi\hbar/M\Delta = T_H/M$ to the dwell time in

a continuous system with M channels and mean level spacing Δ (this is the mean time for a round trip F in the system). In the leading orders of T_0, we can approximate

$$e^{i\varepsilon} F \approx \frac{1 - iT_0(H - E)/2\hbar}{1 + iT_0(H - E)/2\hbar}, \tag{10.60}$$

so that

$$\mathcal{A} = \frac{1 + e^{i\varepsilon} F}{1 - e^{i\varepsilon} F} \approx \frac{2i\hbar}{T_0} G(E), \quad G(E) = \frac{1}{E - H}, \tag{10.61}$$

where $G(E)$ is the Green function (or resolvent) of the closed system. For the ideal case with scattering matrix (10.53), we then have

$$S = \frac{\frac{2i\hbar}{T_0} P(E - H)^{-1} P^T - 1}{\frac{2i\hbar}{T_0} P(E - H)^{-1} P^T + 1}, \tag{10.62}$$

while for nonideal leads P is replaced by K. Inserting T_0 completes the derivation of the Mahaux–Weidenmüller formula (10.41),

$$S(E) = \frac{i\pi W^{\dagger}(E - H)^{-1} W - 1}{i\pi W^{\dagger}(E - H)^{-1} W + 1}, \quad W = \frac{\sqrt{M\Delta}}{\pi} K^{\dagger}, \tag{10.63}$$

where the $M \times M$-dimensional Hermitian matrix H represents the Hamiltonian of the closed systems, while the $M \times N$-dimensional matrix W describes the coupling to the N scattering channels. With our basis choice, W is diagonal, with elements

$$W_{nn} = \frac{\sqrt{M\Delta}}{\pi} \kappa_n^{\sigma_n} \tag{10.64}$$

specified according to Eq. (10.59). The form of W in the nonideal case can also be obtained by starting with the scattering matrix (10.62) for ideal contacts and adding barriers by the construction (10.55).

Equation (10.63) can be rewritten in the equivalent form

$$S(E) = -1 + 2\pi i W^{\dagger}(E - H + i\pi WW^{\dagger})^{-1} W. \tag{10.65}$$

According to this, the poles of the scattering matrix are given by the eigenvalues of the effective non-Hermitian Hamiltonian $H - i\pi WW^{\dagger}$. The poles all lie in the lower half of the complex plane, as required by causality. Furthermore, the Wigner–Smith time-delay matrix $Q = -i\hbar S^{\dagger} dS/dE$ takes the form

$$Q = 2\pi \hbar W^{\dagger}(E - H - i\pi WW^{\dagger})^{-1}(E - H + i\pi WW^{\dagger})^{-1} W, \tag{10.66}$$

which is explicitly positive semi-definite, as again required by causality.

10.3.7   Merits

Via the stroboscopic model (10.58), the orthogonal, unitary, or symplectic symmetry of F in the three Wigner–Dyson classes with different forms of time-reversal symmetry translates directly into a corresponding symmetry of S. Via the continuous model (10.65), one finds that this also agrees with the corresponding symmetry class for H. In the symmetry classes with chiral or charge-conjugation symmetry, this translation holds when the scattering matrix is evaluated at the spectral symmetry points $E = 0$ or $\varepsilon = 0, \pi$ (away from these points, the symmetry reduces to the three Wigner–Dyson classes). Thus, the ten symmetry classes listed in Table 10.2 directly apply to the scattering matrix, with energy fixed to the symmetry point where required (Beenakker 2015).

It is instructive to verify these statements directly within the scattering picture (see Fig. 10.4). For this, consider that the time-reversal operation $\mathcal{T}$ transforms incoming modes into outgoing modes. If this is a symmetry of the Hamiltonian then the correspondingly transformed scattering state must be described by the original scattering matrix. For $\mathcal{T} = K$ this delivers

$$\mathbf{a}^{(\text{in})*} = S(E)\mathbf{a}^{(\text{out})*} = S(E)S^*(E)\mathbf{a}^{(\text{in})*}, \qquad (10.67)$$

such that $S^T(E) = S(E)$, as anticipated. Analogously, a time-reversal symmetry with $\mathcal{T} = \Omega K$ implies that $S^T(E) = \Omega S(E)\Omega^{-1}$, hence $[\Omega S(E)]^T = -\Omega S(E)$. For a chiral symmetry $\mathcal{X}$, we transform a solution at energy E into a solution at energy $-E$. This inverts the group velocity of the propagating modes, thus again transforming incoming modes into outgoing modes. It follows that $\mathcal{X}S(E)\mathcal{X} = S^\dagger(-E)$, and hence $[\mathcal{X}S(-E)]^\dagger = \mathcal{X}S(E)$. For a charge-conjugation symmetry $\mathcal{C}$, both effects on the propagation direction cancel such that $S(-E) = S^*(E)$ if $\mathcal{C} = K$, while $S(-E) = \Omega S^*(E)\Omega^{-1}$ if $\mathcal{C} = \Omega K$. This recovers all constraints in Table 10.2.

Based on this correspondence, the effective scattering models deliver an independent view on the topological quantum numbers associated with the Hamiltonian (Fulga et al. 2011; Beenakker 2015; Schomerus et al. 2015). In systems with a

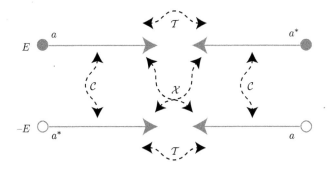

Fig. 10.4 Fundamental symmetries relate various states of motion, which constrains the scattering matrix in accordance to the ten universality classes for unitary matrix.

chiral symmetry, the matrix $S_{X0} = \mathcal{X}S(0)$ is unitary and Hermitian, so that the trace $\nu_0 = \frac{1}{2}\mathrm{tr}\,S_X$ quantifies the difference between eigenvalues ± 1. According to Eq. (10.65) with a chiral Hamiltonian of the form (10.5), this topological quantum number can then be expressed as $\nu_0 = [\nu + (N_A - N_B)/2]_{|\nu_0|\leq N/2}$, where N_A and N_B count the number of channels coupled to the two different chiral sectors; as indicated by the brackets, this saturates at $|\nu_0| = N/2$ where $N = N_A + N_B$. In systems with a charge-conjugation symmetry, where the Hamiltonian can be made antisymmetric by an appropriate basis choice and displays a zero mode if M is odd (modulo possible Kramers degeneracy), $\nu_0 = \det S(0) = \nu$ (class D) and $\nu_0 = \mathrm{pf}\,\Omega S(0) = \nu$ (class DIII) remain directly related to the internal topological quantum number.

Beyond the pure symmetry classification, and perhaps even more importantly, the effective scattering models also determine the appropriate statistical ensembles for the scattering matrix for ergodic internal wave propagation (Brouwer 1995). For ideal contacts, the circular ensembles for F translate via Eq. (10.52) into the corresponding circular ensembles for the ballistic scattering matrix S, with energy again fixed to the symmetry point where required. In the presence of a tunnel barrier, the Haar measure is deformed according to Eq. (10.58). In the three Wigner–Dyson classes this takes the form of a Poisson kernel

$$P(S) \propto |\det (1 - \overline{S}^\dagger S)|^{-\beta N - 2 + \beta}, \tag{10.68}$$

where the nonideal contacts are encoded in the average scattering matrix $\overline{S} = (1 - KK^T)/(1 + KK^T)$. In the additional symmetry classes with chiral or charge-conjugation symmetry, the analogue of the Poisson kernel can be constructed based on Eq. (10.55) (Béri 2009; Marciani et al. 2016), which we briefly illustrate in Section 10.4.3.

By carrying out the continuum limit for large M, one furthermore finds that the internal Hamiltonian H in the Mahaux–Weidenmüller formula (10.63) complies with the corresponding Gaussian ensemble (see again Brouwer 1995). In the three Wigner–Dyson ensembles, the Cayley transform (10.60) implies that at $E = 0$

$$F^\dagger dF = -i\Sigma dH\Sigma^\dagger, \quad \Sigma = \frac{1}{1 + iHT_0/2\hbar}, \tag{10.69}$$

which makes it possible to calculate the Jacobian for the transformation from F to H. This leads to a Cauchy distribution

$$P(H) \propto \det (1 + H^2 T_0^2/4\hbar^2)^{-(\beta M + 2 - \beta)/2}, \tag{10.70}$$

which for large M shares all leading p-point correlations functions with the corresponding Gaussian ensemble.

These considerations provide a solid link between the random-matrix models for closed and open systems with ergodic internal dynamics. For ideal leads, the stationary scattering at fixed energy is described by a unitary scattering matrix from a circular ensemble, while the related Poisson kernel applies when the contacts are

nonideal. Based on the appropriate Gaussian ensemble for H, the effective scattering model can also be employed to study the energy dependence, including the crossover between symmetry classes as the energy is steered away from a spectral symmetry point. Guided by the list of questions posed at the beginning of this chapter, we can now set out to describe scattering and decay from a random-matrix perspective.

10.4   Decay, dynamics, and transport

We now turn to the random-matrix description of the physical phenomena outlined in Section 10.3.3. We first cover decay processes described by complex eigenvalues and nonorthogonal eigenvectors (for an illustration of the general phenomenology see Fig. 10.5), then turn to decay processes (see Fig. 10.6), and finally consider transport (see Fig. 10.7).

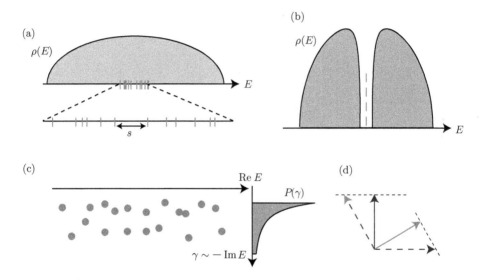

Fig. 10.5 (a) In a closed system, energy levels are constrained to be real, and random-matrix theory focuses on the spectral fluctuations, e.g., of the level spacings s. These occur against the nonuniversal backdrop of the mean density of states $\rho(E)$, here illustrated as the Wigner semi-circle law (10.4). (b) Fundamental symmetries can introduce spectral symmetries which induce universal aspects into the mean density of states. At the symmetry point, topologically protected zero modes can appear. This is here illustrated for the case of the chiral symmetry, with the mean density of states given by Eq. (10.8). (c) In an open system, the corresponding energies are complex and attention shifts to the decay rates γ of the states, here given in accordance to Eq. (10.80). (d) The states become nonorthogonal, which requires introducing a biorthogonal system as here illustrated for a pair of states.

10.4.1 Scattering poles

According to the Mahaux–Weidenmüller formula (10.65), the complex energies of the quasibound states (poles of the scattering matrix) are obtained from the eigenvalue problem

$$E_m|\phi_m\rangle = H_{\text{eff}}|\phi_m\rangle, \tag{10.71}$$

where the $M \times M$ dimensional effective non-Hermitian Hamiltonian is of the form $H_{\text{eff}} = H - i\pi WW^\dagger$ (Fyodorov and Sommers 1997, 2003; Fyodorov and Savin 2011). This consists of a Hermitian part H, which represents the dynamics in the closed system, and an anti-Hermitian part involving a positive semi-definite matrix $WW^\dagger$ of rank N. The eigenvalues are therefore confined to the lower half of the complex plane, where $\text{Im}\, E_m = -\hbar\gamma_m/2$ encodes the positive decay rates γ_m. Analogously, the poles of the stroboscopic scattering matrix can be read off Eq. (10.56), according to which they are obtained from the eigenvalue problem

$$z_m|\phi_m\rangle = F(\mathcal{Q} + P^T r'_B P)|\phi_m\rangle, \tag{10.72}$$

with $z_m = \exp(-i\varepsilon_m)$ confined by $|z_m| \leq 1$. The two problems are then related by identifying $\varepsilon_m = E_m T_0/\hbar$ with $T_0 = 2\pi\hbar/M\Delta$; see our discussion in Section 10.3.6. In a random-matrix description with large matrix dimension M, one typically finds that the eigenvalues populate a well-defined region, with universal statistics in the bulk (Fyodorov and Khoruzhenko 1999; Forrester 2010; Khoruzhenko and Sommers 2011). In particular, well inside the eigenvalue support the level repulsion is typically captured by a factor $\prod_{n<m}|E_n - E_m|^2$, as already encountered for the Ginibre ensembles, which then yields cubic level repulsion. For many physical applications, however, we are mainly interested in the properties of the longest living modes in a given energy range, which approach the real axis closest from below, and are automatically situated at the boundary of the spectral support. These modes determine the noticeable resonance patterns that one observes, e.g., in the scattering and decay of nuclei (Weidenmüller and Mitchell 2009) or in the emission properties of optical microresonators (Cao and Wiersig 2015). To determine their properties we need to work directly with the effective scattering models.

Particularly compact expression for the distribution of decay rates can be obtained for the stroboscopic model (10.52) with ideal leads (Zyczkowski and Sommers 2000). The quasibound states are then obtained from the eigenvalue problem (10.44) for the truncated time-evolution operator $F\mathcal{Q}$. We assume that $F \in U(M)$ is a random unitary matrix of dimension $M \times M$, distributed according to the Haar measure $\mu(F)$, which places us in the circular unitary ensemble (CUE) for systems without any further symmetries. Averaging over this ensemble, it is then possible to determine the density of eigenvalues z_m in the complex plane. In a first step, one finds the joint distribution of the nontrivial eigenvalues $z_m \neq 0$, to which we assign the indices $m = 1, 2, \ldots, M - N$. This joint distribution is given by

$$P(\{z_m\}) \propto \prod_{i<j}^{M-N} |z_i - z_j|^2 \prod_{k=1}^{M-N} (1 - |z_k|^2)^{N-1}, \tag{10.73}$$

where the first term signifies the expected level repulsion. The density of the eigenvalues in the complex plane follows by integrating out all but one eigenvalue, which gives

$$\rho(z) \propto (1 - |z|^2)^{N-1} \sum_{l=1}^{M-N} \frac{(N + l - 1)!}{(l-1)!} |z|^{2l-2} \quad \text{for } |z| < 1. \tag{10.74}$$

This density has several interesting limits. For $M, N \to \infty$ at fixed $N/M = 1 - \mu$, the modulus $r = |z|$ obeys

$$P(r) = (\mu^{-1} - 1) \frac{2r}{(1 - r^2)^2} \Theta(\mu - r^2), \tag{10.75}$$

while for $M \to \infty$ at fixed N we have, setting $(1 - r)/T_0 \to \gamma/2$,

$$P(\gamma) = \frac{\gamma^{N-1}}{(N-1)!} \left(\frac{-d}{d\gamma} \right)^N \frac{1 - e^{-\gamma T_H}}{\gamma T_H}, \tag{10.76}$$

where $T_H = 2\pi\hbar/\Delta$ is the Heisenberg time.

According to Eq. (10.75), in the considered limit all poles are confined to the region $r < \sqrt{\mu}$, and thus do not approach the unit circle closely. Such a hard gap is also obtained from large-N limit of equation (10.76) (thus, $1 \ll N \ll M$), in which

$$P(\gamma) = \frac{\gamma_0}{\gamma^2} \quad \text{if } \gamma > \gamma_0, \quad 0 \text{ otherwise.} \tag{10.77}$$

Here $\gamma_0 = N\Delta/2\pi\hbar = 1/T_D$ coincides with the classical decay rate out of a system with dwell time $T_D = T_H/N$. The corresponding energy scale $E_{\text{Th}} = \hbar/T_D = N\Delta/2\pi$ is known as the Thouless energy.

These results recover the main features obtained earlier by a direct analysis of the non-Hermitian eigenvalue problem (10.71). The most comprehensive insight is obtained using supersymmetric integration techniques, which predict Eq. (10.76) for ideal coupling and extend it to nonideal leads (Fyodorov and Sommers 1996, 1997, 2003). The result is

$$P(\gamma) = \frac{\hbar\pi}{\Delta} \mathcal{F}_1 \left(\frac{\hbar\pi}{\Delta} \gamma \right) \mathcal{F}_2 \left(\frac{\hbar\pi}{\Delta} \gamma \right),$$

$$\mathcal{F}_1(y) = \frac{1}{2\pi} \int_{-\infty}^{\infty} dx\, e^{-ixy} \prod_{n=1}^{N} \frac{1}{x_n - ix}, \quad \mathcal{F}_2(y) = \frac{1}{2} \int_{-1}^{1} dx\, e^{-xy} \prod_{n=1}^{N} (x_n + x),$$

$$\tag{10.78}$$

where $x_n = -1 + 2/\Gamma_n$ encodes the transparency of the contact. For a barrier with uniform transparency Γ (hence, dimensionless conductance $g_c = \Gamma N$), the distribution function can be written compactly as

$$P(\gamma) = \frac{\Delta}{2\pi\hbar\gamma^2(N-1)!} \int_{N(1-\Gamma)\gamma/\gamma_0}^{N\gamma/\gamma_0} \mathrm{d}x\, x^N \mathrm{e}^{-x}, \tag{10.79}$$

where now $\gamma_0 = \Gamma N\Delta/2\pi\hbar$. The large-$N$ limit (10.77) is then replaced by

$$P(\gamma) = \frac{\gamma_0}{\Gamma\gamma^2} \quad \text{if } \gamma_0 < \gamma < \gamma_0/(1-\Gamma), \tag{10.80}$$

so that the decay rates are reduced according to the increased classical dwell time $T_D = T_H/(\Gamma N)$.

The random-matrix results for the unitary symmetry class can be extended to the other symmetry classes. As with the Ginibre ensembles, many of the common characteristics remain unchanged, with the main modifications arising from spectral symmetries. In particular, in systems with time-reversal symmetry (orthogonal and symplectic symmetry class) no further spectral symmetries arise (these cases are therefore quite distinct from the real and symplectic Ginibre ensemble, which lends further justification to their careful construction). The main modifications arise from the altered level repulsion in the closed limit, which is felt by the longest living states (Sommers et al. 1999; Fyodorov and Savin 2011). At large matrix dimensions N and M, these modifications do not matter and a hard gap of order γ_0 again emerges for the decay rates (Haake et al. 1992; Lehmann et al. 1995; Janik et al. 1997). This induces the emergence of classical exponential decay in the time domain (Savin and Sokolov 1997).

In the classes with chiral or charge-conjugation symmetries, all poles come in pairs $E_l, -E_l^*$ which are symmetrically arranged with respect to the imaginary axis $\mathrm{Re}\, E = 0$. The exception is unpaired modes pinned to the imaginary axis, $\mathrm{Re}\, E_l = 0$, that arise from the zero modes in the closed setting, and add a topological feature to the complex spectrum (Pikulin and Nazarov 2012, 2013). These symmetry-respecting poles can only depart from the imaginary axis in pairs, involving an exceptional point where two poles meet, as described in Section 10.2.8. Thus, for an odd number of zero modes at least one such pole is always confined to the imaginary axis. For a superconducting system these poles describe Majorana zero modes that seep out of the system (Pikulin and Nazarov 2012, 2013; San-Jose et al. 2016), while in a photonic setting they can be employed for selective amplification (Schomerus and Halpern 2013; Schomerus 2013b; Poli et al. 2015). Within random-matrix theory, we describe the consequences for the density of states in Section 10.4.3.

In the construction of the effective scattering models we noted that channels can also be closed by increasing the coupling beyond a certain threshold ($\sigma_n = -1$ in Eq. (10.58) or Eq. (10.64)). Physically, this should again result in a reduced decay rate γ_0 of the longest living modes. The spectral decomposition of the effective Hamiltonian, on the other hand, implies the sum rule

$$\mathrm{Im\, tr}\, (H - i\pi WW^\dagger) = -\pi \mathrm{tr}\, WW^\dagger = \sum_m \mathrm{Im}\, E_m, \tag{10.81}$$

so that the sum of all decay rates must grow. These two expectations can be reconciled in a careful analysis which shows that N' strongly coupled channels result in a corresponding number of poles with very short life time (Haake et al. 1992). These poles are then well separated from the poles describing the long-living states, which retain a typical decay rate $\gamma_0 = \Gamma N \Delta / 2\pi \hbar$. This nontrivial reorganization of the complex spectrum is known as resonance trapping (Rotter 2009). In the symmetry classes with charge-conjugation symmetry, it can affect the Majorana pole pinned to the imaginary axis, which justifies identifying the case of ideal coupling as a topological phase transition (Akhmerov et al. 2011; Marciani et al. 2016).

The appearance of the classical decay rate in these considerations indicates that random-matrix theory is only applicable if the system-specific details become indiscernible before the classical dwell time $T_D = T_H/(\Gamma N)$. For a contact with dimensionless conductance $g_c = \Gamma N \gg 1$, this condition is more stringent than the requirement in the closed system, where T_D is replaced by T_H. A common occurrence where this condition is mildly violated are systems with ballistic decay routes, which result in additional short-living states that often form interweaving bands deep in the complex plane (Weich et al. 2014). In a classically chaotic systems, these routes apply to trajectories that escape before the Ehrenfest time $T_{\mathrm{Ehr}} \approx \lambda^{-1} \ln N$, where λ is the Lyapunov exponent (Berman and Zaslavsky 1978; Aleiner and Larkin 1996; Schomerus and Jacquod 2005). In the limit of large N and M, the fraction of long-living modes is then reduced by a factor $\exp(-T_{\mathrm{Ehr}}/T_D) = N^{-1/(\lambda T_D)}$ (Schomerus and Tworzydło 2004), a power law which agrees with a picture where these states are confined to the classical repeller (Lu et al. 2003; Keating et al. 2006). This modification due to ballistic chaotic decay is known as the fractal Weyl law (Nonnenmacher and Zworski 2005). In practice, random-matrix theory still provides a good description of the remaining long-living modes (Schomerus et al. 2009). Furthermore, partial reflections at the contacts and disorder are very effective mechanisms to remove the ballistic decay routes.

10.4.2   Mode nonorthogonality

Since the effective Hamiltonian $H_{\mathrm{eff}} = H - i\pi W W^\dagger$ is non-Hermitian, the quasibound states $|\phi_m\rangle$ from the eigenvalue problem (10.71) do not form an orthonormal basis. In a given basis, we thus have a spectral decompositions $H_{\mathrm{eff}} = V D V^{-1}$, $D = \mathrm{diag}\,(E_m)$ where the matrix V is not unitary. The extent of mode nonorthogonality is then quantified by the condition numbers O_{mn} introduced in Eq. (10.21).

In order to get insight into the significance of these objects we consider the divergent part

$$\mathrm{tr}\, S^\dagger S \approx \mathrm{tr}\, 2\pi W^\dagger (E - H - i\pi W W^\dagger)^{-1} 2\pi W^\dagger W (E - H + i\pi W W^\dagger)^{-1} W \equiv \sigma(E) \quad (10.82)$$

of the scattering strength for a complex energy close to a pole, $E \to E_n$ (Schomerus et al. 2000). Using the spectral decomposition for the effective Hamiltonian we find that

$$\sigma(E) = \sum_{nm} \frac{-(E_n - E_m^*)^2}{(E - E_n)(E - E_m^*)} O_{mn}, \tag{10.83}$$

where we used $2\pi W W^\dagger = iH_{\text{eff}} - iH_{\text{eff}}^\dagger = iVDV^{-1} - iV^{-1\dagger}D^*V^\dagger$. Very close to the pole, $\sigma(E) \approx \frac{(\hbar\gamma_n)^2}{|E-E_n|^2} K_n$ describes a Breit–Wigner resonance with peak height proportional to $K_n = O_{nn}$. Thus, the factors K_n are directly related to the scattering strengths of the quasibound states.

Energies in the complex plane are effectively probed in amplifying photonic systems, which can be described in a scattering approach that is amended to account for radiation created within the medium (Beenakker 1998; Schomerus et al. 2000; Schomerus 2009). Under ideal conditions, an active medium with amplification rate γ_a can generate spontaneously amplified radiation with frequency-resolved intensity

$$I(\omega) \approx (2\pi)^{-1} \operatorname{tr} (S^\dagger S - 1)|_{E=\hbar\omega - i\hbar\gamma_a/2}. \tag{10.84}$$

Close to the laser threshold, a single pole $E_m = \hbar\omega_m$ lies close to the real axis, producing a well-isolated Lorentzian emission line

$$I(\omega) \approx \frac{K_m}{2\pi} \frac{\gamma_n^2}{(\omega - \omega_m)^2 + (\gamma_m - \gamma_a)^2/4}. \tag{10.85}$$

In this context, K_m is know as the Petermann factor and signifies excess noise (Petermann 1979).

For lasers we can ignore magneto-optical effects, and thus are concerned with the orthogonal symmetry class where the effective Hamiltonian inherits the symmetry $H_{\text{eff}} = H_{\text{eff}}^T$. In this case we can normalize the right and left eigenstates so that $V^{-1} = V^T$ and find that

$$K_m = |(V^\dagger V)_{mm}|^2. \tag{10.86}$$

As described in Section 10.2.8 for the Ginibre ensemble, the Petermann factor of modes in the bulk of the complex spectrum should be large. For the effective Hamiltonian H_{eff} with $N, M \gg 1$, this can be verified in the free-probability approach (Janik et al. 1997), according to which

$$\overline{K_m}|_{\gamma_m=\gamma} \approx N \left(\frac{\gamma}{\gamma_0} - 1 \right) \left(1 - \frac{(1 - \Gamma)\gamma}{\gamma_0} \right) \tag{10.87}$$

for decay rates well within the range $\gamma_0 < \gamma < \gamma_0/(1 - \Gamma)$. However, this result breaks down close to the edges of the spectrum, where it violates the constraint $K_m \geq 1$, and hence does not apply to the long-living states that become the lasing modes.

These restrictions can be circumvented by the same supersymmetric techniques that address the poles (Schomerus et al. 2000). Equation (10.78) is then supplemented by

$$\overline{K_m}|_{\gamma_m=\gamma} = 1 + \frac{2\pi\hbar}{\Delta} \frac{S(\pi\hbar\gamma/\Delta)}{P(\gamma)}, \qquad S(y) = -\int_0^y dy' \, \mathcal{F}_1(y') \frac{\partial}{\partial y'} \mathcal{F}_2(y'), \tag{10.88}$$

which for identical transparencies $\Gamma_n = \Gamma$ can be brought into a compact form using

$$S(\pi\gamma/\hbar\Delta) = \frac{\Delta^2}{(2\pi\hbar\gamma)^2(N-1)!} \int_{N(1-\Gamma)\gamma/\gamma_0}^{N\gamma/\gamma_0} \mathrm{d}x\, x^{N-1} e^{-x} \left(\frac{N(1-\Gamma)\gamma}{\gamma_0} - x \right) \left(x - \frac{N\gamma}{\gamma_0} \right).$$
(10.89)

For large N, where we can apply a saddle-point approximation, it follows that the Petermann factor $\overline{K_m}\,|_{\gamma_m=\gamma_0} \sim \Gamma(\sqrt{2N/\pi} + 4\pi/3)$ of the long-living modes can still be parametrically large in N. When a large number L of such modes compete for the gain, the large-deviation tail of the decay-rate distribution (10.76) is probed, which reduces K_m by a factor $\sim 1/\sqrt{\ln L}$. For $\Gamma N \ll 1$, on the other hand, the system is almost closed, and $K_m \sim 1$ as mode-orthogonality is restored. Similarly, for $N = 1$ the typical Petermann factor $K_m \sim 1 + \Gamma\hbar\gamma_m/\Delta$ is also close to unity.

In all these cases, the Petermann factors of individual states can be much larger than the typical values quoted in the preceding paragraph. This is the case because K_m diverges if two complex eigenvalues become degenerate, thus, as one approaches an exceptional point. The cubic level repulsion makes such approaches rare, but long power tails still emerge in the probability distribution of K_m.

We mentioned that the Petermann factor signifies an enhanced sensitivity to noise generated by spontaneous emission. Similar considerations apply when external parameters are changed. A perturbative treatment then reveals an enhanced response compared to systems with orthogonal modes, which is again quantified by the mode nonorthogonality matrix (Fyodorov and Savin 2012). Close to an exceptional point, where the eigenvectors become degenerate and K_m diverges, the significantly enhanced response can be exploited for sensors (Wiersig 2014). This enhanced sensitivity also applies to the topological spectral transitions in non-Hermitian systems with a chiral or charge-conjugation symmetry (where they occur on the imaginary axis), or non-Hermitian systems with a parity-time symmetry (where they occur on the real axis). The radiation emitted from a parity-time symmetric photonic system indeed diverges when one closes the system (Schomerus 2010). For an open system close to an exceptional point, on the other hand, the formal divergence of the Petermann factor signifies a change of the line shape from the Lorentzian (10.84) to a squared Lorentzian (Yoo et al. 2011).

10.4.3   Delay times

We now turn to the Wigner–Smith time-delay matrix $Q = -i\hbar S^\dagger \mathrm{d}S/\mathrm{d}E$, which according to the Mahaux–Weidenmüller formula (10.65) can be written in the form (10.66),

$$Q = 2\pi\hbar W^\dagger (E - H - i\pi WW^\dagger)^{-1}(E - H + i\pi WW^\dagger)^{-1}W. \tag{10.90}$$

This matrix is manifestly Hermitian and positive-definite, as required by causality. According to the Birman–Krein formula (10.34), the density of states is then given by

$$\rho(E) = \mathrm{tr}\, W^\dagger (E - H - i\pi WW^\dagger)^{-1}(E - H + i\pi WW^\dagger)^{-1}W, \tag{10.91}$$

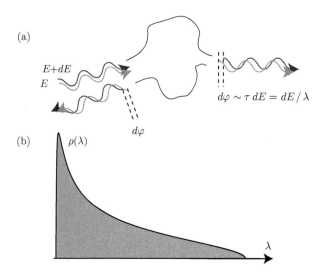

Fig. 10.6 (a) The Wigner–Smith delay times τ extract dynamical information by considering the energy sensitivity of the scattering phase in stationary scattering states. (b) Distribution of rates $\lambda = \tau^{-1}$ from random scattering, as predicted by the Marchenko–Pastur law (10.15) for the Wishart–Laguerre ensemble.

which is of a similar form as the scattering strength $\sigma(E)$ in Eq. (10.82). Using the spectral decomposition for the effective Hamiltonian we find that

$$\rho(E) = \frac{1}{2\pi} \sum_{nm} i \frac{(E_n - E_m^*)}{(E - E_n)(E - E_m^*)} O_{mn} = -\frac{1}{\pi} \text{Im} \sum_n \frac{1}{(E - E_n)}, \qquad (10.92)$$

where we used $\sum_n O_{nm} = \sum_m O_{nm} = 1$. Close to an isolated resonance $E \approx E_m$ this approaches

$$\rho(E) \approx \frac{1}{\pi} \frac{2\,\text{Im}\,E_n}{|E - E_n|^2},$$

which is a Lorentzian normalized to 1.

More direct insight into this problem is obtained from the proper delay times τ_n, defined as the eigenvalues of Q, which are all real and nonnegative. We first consider the case of ballistic coupling. In the three standard classes (Brouwer et al. 1997), it is useful to consider the matrix $Q_S = S^{1/2}QS^{-1/2}$, which has the same eigenvalues but whose statistical distribution is independent of S itself, so that $P(S, Q_S) = P(S)P(Q_S)$. Perturbation theory around the point where $S = -1$ then shows that the positive-definite rate matrix Q_S^{-1} follows the distribution

$$P(Q_S^{-1}) \propto (\det Q_S^{-1})^{N\beta/2} \exp[-(\beta T_H/2)\text{tr}\,Q_S^{-1}], \qquad (10.93)$$

with the Heisenberg time $T_H = 2\pi\hbar/\Delta$. This resembles a Wishart–Laguerre ensemble (10.13), but is directly expressed for Q_S and supplemented with a determinantal factor. The joint distribution of rates $\lambda_n = 1/\tau_n$ is given by

$$P(\{\lambda_n\}) \propto \prod_{n<m} |\lambda_n - \lambda_m|^\beta \prod_k \lambda_k^{N\beta/2} \exp(-\beta\lambda_k T_H/2), \qquad (10.94)$$

which indeed looks formally identical to the eigenvalue distribution (10.14) of a Wishart matrix, albeit with half-integer dimensions if $\beta = 4$. This still constitutes a Wishart–Laguerre ensemble.

The same independence of S and Q_S also occurs in the four classes with charge-conjugation symmetry at the symmetry point $E = 0$ (Marciani et al. 2014), where

$$P(\{\lambda_n\}) \propto \prod_{n<m} |\lambda_n - \lambda_m|^{\beta_T} \prod_k \lambda_k^{\beta_T' + N\beta_T/2} \exp(-\beta_T'' \lambda_k T_H/2) \qquad (10.95)$$

with $\beta_T = 1, 2, 4, 2$, $\beta_T' = -1, -1, 2, 1$, $\beta_T'' = 1, 2, 2, 1$ in the symmetry classes D, DIII, C, CI. In the classes C and CI all delay times occur in degenerate pairs, which in Eq. (10.95) are only accounted for once.

In contrast, the chiral symmetry condition $S(E) = \mathcal{X}S^\dagger(-E)S^\dagger$ implies that the Hermitian unitary matrix $S_{X0} = \mathcal{X}S(0)$ commutes with $Q(0)$, so that both matrices share a common structure (Schomerus et al. 2015). Recall that S_{X0} has eigenvalues ± 1, whose frequency is captured by the topological quantum number $\nu_0 = \frac{1}{2}\mathrm{tr}\, S_{X0} = [\nu + (N_A - N_B)/2]_{|\nu_0| \leq N/2}$ (see Section 10.3.7). Correspondingly, the delay times can be grouped into two sets, made of $N_+ = N/2 + \nu_0$ delay times $\tau_n^+ = 1/\lambda_n^+$ associated with the subspace where the eigenvalues of S_{X0} are 1, and $N_- = N/2 - \nu_0$ delay times $\tau_n^- = 1/\lambda_n^-$ associated with the subspace where the eigenvalues of S_{X0} are -1. These two sectors can be made manifest by considering the reordered matrix

$$\tilde{Q}(E) = 2\pi\hbar(E - H + i\pi WW^\dagger)^{-1} WW^\dagger (E - H - i\pi WW^\dagger)^{-1}, \qquad (10.96)$$

which has the same nonvanishing eigenvalues as Q. Inserting here the chiral Hamiltonian (10.5) and splitting the coupling matrix analogously into blocks $W = \mathrm{diag}\,(W_A, W_B)$ describing N_A and N_B open channels, respectively, this reordered matrix becomes block diagonal,

$$\tilde{Q}(0) = 2\pi\hbar\, \mathrm{diag}\,(\Lambda_-^{-1}, \Lambda_+^{-1}), \qquad (10.97)$$

where

$$\Lambda_- = \pi^2 W_A W_A^\dagger + A(W_B W_B^\dagger + 0^+)^{-1} A^\dagger, \qquad (10.98)$$

$$\Lambda_+ = \pi^2 W_B W_B^\dagger + A^\dagger(W_A W_A^\dagger + 0^+)^{-1} A. \qquad (10.99)$$

In the subspaces where these two matrices are finite, we can write $\Lambda_\pm = X_\pm^\dagger X_\pm$ with an $N \times N_\pm$ dimensional matrix X. For large M, the matrix X tends to a random Gaussian

matrix, so that the two sets of decay rates are both obtained from a Wishart–Laguerre ensemble,

$$P_\pm(\{\lambda_n^\pm\}) = \prod_{n<m} |\lambda_n^\pm - \lambda_m^\pm|^\beta \prod_k \lambda_k^{\beta/2-1+(\beta/4)|N\mp2\nu\pm N_B \mp N_A|} e^{-\beta\lambda_k T_H/4}. \quad (10.100)$$

The two sets are independent of each other, whereby the full joint distribution factorizes according to $P(\{\lambda_n^+, \lambda_n^-\}) = P_+(\{\lambda_n^+\})P_-(\{\lambda_n^-\})$.

We note that the joint distribution (10.95) does not involve the topological quantum number ν defined in classes D and DIII. In the chiral ensembles, on the other hand, the topological zero modes directly affect the joint distribution (10.100). This dependence also transfers to the mean density of states, which is given by

$$\overline{\rho} = \frac{1}{\Delta} \frac{N/2(N/2+1-2/\beta)+\nu_0^2}{(N/2+1-2/\beta)^2 - \nu_0^2} \quad \text{for } |\nu_0| < N/2, \quad (10.101)$$

$$\overline{\rho} = \frac{1}{\Delta} \frac{N/2}{|\nu-\nu_0+(N_A-N_B)/2|+1-2/\beta} \quad \text{for } |\nu_0| = N/2, \quad (10.102)$$

with the exceptions $|\nu - N_B| \leq 1$ or $|\nu + N_A| \leq 1$ (for $\beta = 1$) and $\nu = N_B$ or $\nu = -N_A$ (for $\beta = 2$) where the ensemble-average diverges.

These considerations can be extended to nonideal leads (Marciani et al. 2016), where one relates the scattering matrix S via Eq. (10.55) to the scattering matrix S_0 for ballistic coupling. The time-delay matrix then changes from Q_{S0} to $Q_S = \Sigma Q_{S0}\Sigma^\dagger$, where $\Sigma = (1-S^\dagger r_B)^{-1}t_B$. The transformation of the probability measure follows from the analogous relation $S^\dagger dS = \Sigma(S_0^\dagger dS_0)\Sigma^\dagger$. For the standard symmetry classes, the factorized distribution $P(S_0, Q_{S0}^{-1}) = P(S_0)P(Q_{S0}^{-1})$ transforms into

$$P(S, Q_S^{-1}) = (\det \Sigma\Sigma^\dagger)^{N\beta/2}(\det Q_S^{-1})^{N\beta/2} \exp[-(\beta T_H/2) \operatorname{tr} \Sigma^\dagger Q_S^{-1}\Sigma], \quad (10.103)$$

while in the classes with charge-conjugation symmetry this takes the form

$$P(S, Q_S^{-1}) = (\det \Sigma\Sigma^\dagger)^{N\beta_T/2}(\det Q_S^{-1})^{N\beta_T/2+\beta_T'} \exp[-(\beta_T'' T_H/2) \operatorname{tr} \Sigma^\dagger Q_S^{-1}\Sigma]. \quad (10.104)$$

The density of states $\rho = 2\pi\hbar^{-1} \operatorname{tr} \Sigma Q_{S0}\Sigma^\dagger$ can then be analysed directly using the independence of S (appearing in Σ) and Q_{S0}. By definition, $\operatorname{tr}\overline{Q_{S0}} = 2\pi\hbar\rho_0$ is given by the density of states for ideal coupling, while the scattering matrix itself follows the Poisson kernel distribution $P(S) \propto |\det(1-r_B^\dagger S)|^{-\beta_T N-2+\beta_T-2\beta_T'}$ (recovering the result of Béri (2009)). In class D, a barrier with mode-independent transparency Γ then yields the mean density of states

$$\overline{\rho} = \frac{N}{(N-2)\Delta}\left(1 - \frac{2}{N\Gamma}[\Gamma - 1 + (-1)^\nu(1-\Gamma)^{N/2}]\right), \quad (10.105)$$

which now depends on ν. More generally, in classes D and DIII a topological zero mode remains visible as long as none of the couplings are fully ballistic.

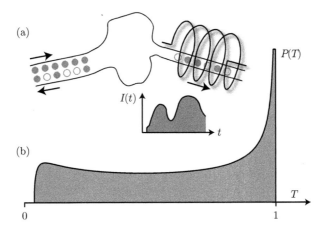

Fig. 10.7 (a) Phase-coherent electronic transport is characterized by partition noise, generated by the transmission of charge carriers with probability T. This noise can be detected in the current fluctuations $I(t)$. (b) Mean density of transmission probabilities T from Eq. (10.107), in accordance with the Jacobi ensemble of random-matrix theory.

10.4.4   Transport

Some of the best tested applications of random scattering matrices arise when one considers the low-temperature transport of electrons through a mesoscopic device in response to a small bias voltage V_b. These applications have been covered in two comprehensive reviews considering the standard ensembles (Beenakker 1997) and the additional ensembles with chiral or charge-conjugation symmetry (Beenakker 2015), supplemented by a detailed review on shot noise (Blanter and Büttiker 2000), and we refer to these sources throughout the section. In keeping with the rest of this chapter we remain focused on situations where the details of the geometry do not matter (this ignores the effects of Anderson localization, which we briefly pick up in the next Section 10.5). For the scattering at a fixed energy we are then directly led to the circular ensembles. This was first utilized by Blümel and Smilansky (1990), who found that the statistics of phase shifts from chaotic scattering agree with Eq. (10.12), while the applications to transport were pioneered by Baranger and Mello (1994) and Jalabert et al. (1994).

In the scattering approach to transport, the device is modelled as a scattering region attached to a left and a right lead, so that the scattering matrix is of the form (10.29). The quantities of interest are the transmission eigenvalues $T_n \in [0,1]$ of $t^\dagger t$, with the dimensionless conductance given by $g = \sum_n T_n$. We will assume that the number of channels $N_R \geq N_L$ so as to avoid $N_L - N_R$ vanishing eigenvalues (otherwise, we can simply study the eigenvalues of $tt^\dagger$). In the three standard circular ensembles (COE, CUE, and CSE), the joint distribution of the transmission eigenvalues is then given by

$$P(\{T_n\}) \propto \sum_{n<m} |T_n - T_m|^\beta \prod_k T_k^{-1+\beta(1+|N_L-N_R|)/2}, \qquad (10.106)$$

which can be interpreted as a Jacobi ensemble for variables $\mu_n = 1 - 2T_n$.

For large number of channels $N_L, N_R \gg 1$, the mean density of eigenvalues converges to

$$\rho(T) = \frac{N_L + N_R}{2\pi T} \left(\frac{T - T_c}{1 - T}\right)^{1/2} \qquad (10.107)$$

for $1 > T > T_c = (N_L - N_R)^2/(N_L + N_R)^2$; for $N_L = N_R$ this takes the form

$$\rho(T) = \frac{N_L}{\pi} \frac{1}{\sqrt{T(1-T)}}. \qquad (10.108)$$

In leading order of $N_L, N_R \gg 1$, Eq. (10.107) gives the ensemble-averaged dimensionless conductance $\bar{g} = N_L N_R/(N_L + N_R)$, so that $\bar{g}^{-1} = N_L^{-1} + N_R^{-1}$ resembles the series addition of two resistances. The exact result for finite N_L and N_R is

$$\bar{g} = \frac{N_L N_R}{N_L + N_R - 1 + 2/\beta}, \qquad (10.109)$$

so that the next-to-leading order reads

$$\bar{g} - \frac{N_L N_R}{N_L + N_R} \approx (1 - 2/\beta)\frac{N_L N_R}{(N_L + N_R)^2} \quad \text{for } N_L, N_R \gg 1. \qquad (10.110)$$

This ensemble-dependent correction, known as weak localization (for $\beta = 1$) and as weak antilocalization (for $\beta = 4$), can be related to the factors $T_k^{-1+\beta/2(1+|N_L-N_R|)}$ in the joint distribution (10.106), which induce a bias of the transmission eigenvalues to small or large values. The joint distribution also determines the variance of the conductance within the ensemble,

$$\text{var } g \approx \frac{(N_L N_R)^2}{\beta(N_L + N_R)^4} \quad \text{for } N_L, N_R \gg 1. \qquad (10.111)$$

Due to the repulsion $\sim |T_n - T_m|^\beta$ of the eigenvalues this variance is small, but depends on the symmetry class already in leading nonvanishing order.

In a more general picture, the transmission eigenvalues determine the full counting statistics of the electrons that pass through the system. Let $Q(s)$ be the accumulated charge over a time interval s, via the arrival of electrons with elementary charge **e**. In each eigenchannel, an incoming electron is transmitted with probability T_n, so that the counting statistics are given by a Bernoulli process. Noting that these transmission events occur with an attempt rate eV_b/h (with $h = 2\pi\hbar$), this process is described by the cumulant-generating function (Levitov and Lesovik 1993)

$$\ln\langle \exp(pQ(s)/\mathbf{e})\rangle = \sum_{k=1}^{\infty} \langle\langle Q(s)\rangle\rangle \frac{p^k}{\mathbf{e}^k k!} = s(eV_b/h) \sum_n \ln[1 + T_n(e^p - 1)]. \qquad (10.112)$$

The average current follows from $I = \lim_{s \to \infty} s^{-1} e \langle N(s) \rangle = (e^2 V_b/h) g$, while the shot-noise power is $P = \lim_{s \to \infty} 2s^{-1} e^2 \langle \langle N(s)^2 \rangle \rangle = (2e^3 V_b/h) \sum_n T_n (1 - T_n)$. If all transmission eigenvalues are small, the shot-noise power is $P = 2eI \equiv P_0$, while in general $P = f P_0$ with the so-called Fano factor $f = \sum_n T_n (1 - T_n) / \sum_n T_n \in [0, 1]$. For $N_L = N_R \gg 1$ we can calculate the cumulant-generating function exactly (Blanter et al. 2001),

$$\overline{\ln \langle \exp(pQ(s)/e) \rangle} = s \frac{eV}{h} \int dT \rho(T) \ln[1 + T(e^p - 1)] = 4sg \frac{eV}{h} \ln[\frac{1 + e^{p/2}}{2}]. \quad (10.113)$$

The Fano factor is then given by $\overline{f} = 1/4$. For $N_L, N_R \gg 1$ not necessarily equal, one finds that

$$\overline{f} \approx \frac{N_L N_R}{(N_L + N_R)^2} - (1 - 2/\beta) \frac{(N_L - N_R)^2}{(N_L + N_R)^3}, \quad (10.114)$$

where the weak-localization correction is seen to vanish if $N_L = N_R$.

As for the decay problem, these transport properties are modified by ballistic transport routes. A wavepacket injected into the opening can leave without any noticeable diffraction until the transport Ehrenfest time $T'_{\text{Ehr}} = \lambda^{-1} \ln N^2/M$ (Silvestrov et al. 2003), which results in transmission eigenvalues T_n close to 0 and 1. In particular, these processes can yield a noticeable suppression of shot noise (Tworzydło et al. 2003).

In the chiral symmetry classes, the statistics of the transmission eigenvalues is most conveniently expressed via $T_n = \sqrt{1 - r_n^2}$, where r_n are the eigenvalues of the Hermitian matrix $R_z = \tau_z r$ (Macedo-Junior and Macêdo 2002). We only consider the case $N_L = N_R$ with balanced coupling to both chiral subspaces ($N_A = N_B$), so that $\nu_0 = \text{tr} R_z$ determines the number of zero modes in the closed system. In this case one encounters $|\nu_0|$ closed transmission channels with $r_n^2 = 1$, while the remaining eigenvalues obey the joined distribution

$$P(\{r_n\}) \propto \prod_{n<m} |r_n - r_m|^\beta \prod_k (1 - r_k^2)^{-1+(|\nu_0|+1)\beta/2}. \quad (10.115)$$

In the symmetry classes with a charge-conjugation symmetry (Dahlhaus et al. 2010),

$$P(\{T_n\}) \propto \prod_{n<m} |T_n - T_m|^{\beta_T} \prod_k T_k^{-1+\beta_T(1+|N_L-N_R|)/2} (1 - T_k)^{\beta'_T/2}, \quad (10.116)$$

where the parameters $\beta_T = 1, 2, 4, 2$, $\beta'_T = -1, -1, 2, 1$ (classes D, DIII, C, CI) are the same as those encountered for the delay times. In the large-N limit, the eigenvalue density becomes again ensemble independent and approaches (10.107).

Note that the topological quantum number ν_0 only appears in the joint distribution (10.115) for chiral symmetry, but not in the joint distribution (10.116), so that any zero modes due to charge-conjugation symmetry cannot be detected in the transport with ideal leads—the same situation that we encountered for the density of states. This provides an incentive to consider the role of superconductivity and tunnel barriers in such systems, which we here will discuss for the classes D and BDI

(Pikulin et al. 2012). Instead of applying the Poisson kernel, we consider the experimentally relevant situation (Mourik et al. 2012) where the tunnel barrier is placed into a normal-conducting region, which is then interfaced with a superconductor.

In the context of such superconducting systems, the dimensionless conductance g relates to the particle (or heat) transport, while the charge transport is modified by the fact that holes carry an opposite charge. If a normal metallic region from the orthogonal symmetry class is attached to a conventional superconductor, the dimensionless conductance for charge transport is given by Eq. (10.32), which applies to systems with no magnetic fields and no spin–orbit scattering. The symmetry classes D arise in the presence of spin–orbit coupling and broken time-reversal symmetry, where only the charge-conjugation symmetry with $\mathcal{C}^2 = 1$ remains. The class BDI emerges from an additional chiral symmetry $\mathcal{X}$ that commutes with $\mathcal{C}$, which then also implies a time-reversal symmetry $\mathcal{T} = \mathcal{X}\mathcal{C}$ with $\mathcal{T}^2 = 1$. In these classes, the dimensionless conductance at the Fermi level can be written as

$$g_{NS} = \operatorname{tr} \Gamma(1 - U^*\sqrt{1 - \Gamma}U\sqrt{1 - \Gamma})^{-1}\Gamma(1 - U^\dagger\sqrt{1 - \Gamma}U^T\sqrt{1 - \Gamma})^{-1}, \qquad (10.117)$$

where $\Gamma = \operatorname{diag}(T_n)$ while the $N \times N$-dimensional unitary matrix U accounts for the mode-mixing from the spin–orbit scattering. For a large tunnel barrier in the normal region, we can assume that all transmission eigenvalues are identical, $T_n \equiv T$, so that

$$g_{NS} = T^2 \operatorname{tr} \frac{1}{1 - (1 - T)X} \frac{1}{1 - (1 - T)X^\dagger} = \sum_n \frac{T^2}{|1 - (1 - T)x_n|^2} \qquad (10.118)$$

is determined by the eigenvalues x_n of $X = U^*U$.

The structure of X implies that all eigenvalues occur in complex conjugated pairs $x_n, x_{\bar{n}} = x_n^*$, with the exception of possible eigenvalues pinned to 1. In class D, where U is only constrained to be unitary, such an eigenvalue occurs if N is odd; we thus have a topological index $\nu = N \bmod 2$. The paired eigenvalues can be specified by the quantities $\mu_n = (x_n + x_{\bar{n}})/2 = \operatorname{Re} x_n$. Sampling U from the circular unitary ensemble, these quantities then follow the distribution

$$P(\{\mu_n\}) \propto \prod_{n<m} (\mu_n - \mu_m)^2 \prod_k \frac{1 + \mu_k}{\sqrt{1 - \mu_k^2}} \quad \text{if } \nu = 0, \qquad (10.119)$$

$$P(\{\mu_n\}) \propto \prod_{n<m} (\mu_n - \mu_m)^2 \prod_k \sqrt{1 - \mu_k^2} \quad \text{if } \nu = 1. \qquad (10.120)$$

In class BDI, U is unitary and Hermitian, and the number of pinned eigenvalues $|\nu|$ follows from $\nu = \operatorname{tr} U$. Setting $U = V^\dagger \operatorname{diag}(1, 1, 1, \ldots, -1, -1, -1, \ldots)V$ with V again following the circular unitary ensemble, the paired eigenvalues are then described by the distribution

$$P(\{\mu_n\}) \propto \prod_{n<m} |\mu_n - \mu_m| \prod_k (1 - \mu_k)^{(|\nu|-1)/2}. \qquad (10.121)$$

The probability distributions (10.119) and (10.121) are both of the form of a Jacobi ensemble (10.16). Including the next-to-leading order in the large-N limit, the density

$$\rho(\mu) = \frac{N}{\pi} \frac{1}{\sqrt{1 - \mu^2}} + \frac{1}{2}\delta(\mu - 1) - \frac{1}{2}\delta(\mu + 1) \qquad (10.122)$$

becomes independent of the symmetry class and the topological indices. In leading orders, the ensemble-averaged dimensionless conductance is then given by

$$\overline{g_{NS}} = \frac{NT}{2 - T} + \frac{2(1 - T)}{(2 - T)^2}, \qquad (10.123)$$

again irrespective of the symmetry class.

Note that the joint distribution Eq. (10.115) of reflection coefficients in the classes with chiral symmetry can also be interpreted as a Jacobi distribution, while the joint distributions (10.106) and (10.116) can be brought into this form by a suitable shift $T_n = (1 - \mu_n)/2$ of the transmission coefficient. We take the appearance of this final class of classical random-matrix ensembles as our cue to wrap up the discussion of fully ergodic elastic scattering. Much more is known, in terms of both technical details and practical applications. This includes the full physical implications of superconductivity, such as Andreev reflection and Josephson currents (Beenakker 1997, 2015), as well as the interpretation of zero modes in terms of Majorana fermions (Alicea 2012; Leijnse and Flensberg 2012; Beenakker 2013). Another important aspect is the role of physical dimensions, which enters the full classification of topologically protected states (Kitaev 2009; Teo and Kane 2010; Ryu et al. 2010; Hasan and Kane 2010; Qi and Zhang 2011). In the following Section 10.5, we turn to one specific aspect of low-dimensional physics, the phenomenon of Anderson localization which prevents the full exploration of phase space. We also take this as an opportunity for a short detour into interacting systems, for which the related question of thermalization can be addressed by the density matrix.

10.5   Localization, thermalization, and entanglement

To round off this chapter we discuss a setting for random-matrix applications which has significance also for interacting systems. This brings us back to the origins of the field, which concerned the energy levels of heavy nuclei (Wigner 1956; Porter 1965; Weidenmüller and Mitchell 2009). There, interactions are sufficient for effectively coupling a large number of many-body states, thus resulting in a random Hamiltonian. Quite generally, statistical methods find broad applications to interacting systems, where the dynamics becomes particularly interesting when one considers low dimensions (Cardy 1996; Sachdev 1999). An interesting question is how such systems thermalize (Anderson 1958; Deutsch 1991; Srednicki 1994; Gemmer et al. 2010). Even in absence of interactions, the spread of energy and the propagation of particles can be inhibited by the same wave-interference effects that we so far have taken as the very justification for the application of random-matrix theory. These localization effects, first recognized by Anderson (1958), arise from the sparsity of the underlying matrices, be

it due to a reduced coordination number on a lattice, resulting in localization in real space (Kramer and MacKinnon 1993; Evers and Mirlin 2008), or due to the presence of interactions that only involve some few-body operators, resulting in localization in Fock space (Basko et al. 2006; Altman and Vosk 2015; Nandkishore and Huse 2015). The main question is when this sparsity can be felt, and how.

We first discuss this question briefly for noninteracting systems, where it can be addressed by the impact on the transport properties described in Section 10.4.4. We then turn to the many-body setting, where we focus on aspects of thermalization and entanglement.

10.5.1 Anderson localization

In random-matrix theory, systems with fully chaotic wave scattering are traditionally termed zero-dimensional systems. This is because in practice these systems are often realized by shrinking the size of two- or three-dimensional systems, as, e.g., in a planar quantum dot or a metallic grain. From a different perspective, such systems could be termed infinite-dimensional, as their main feature is the efficient dynamical coupling of states in the accessible Hilbert space, which is also observed in lattices or graphs with a fixed number of vertices and increasing coordination number. In properly scaled units, we can then assume that all of Hilbert space is instantly explored (ergodic time $T_{\mathrm{erg}} = 0$), so that only one relevant dynamical time scale remains—the dwell time T_D, which characterizes how long a particle will reside within the scattering region.

This approach reaches its limit when the internal transport within the system becomes inefficient. Consider a system made of L random scattering regions placed in series, with contacts carrying $N \gg 1$ channels (Iida et al. 1990). While the dimensionless conductance $g \sim N/2$ of each individual region may be large, the overall conductance $g_L \sim N/(1+L)$ of the composed system shrinks when L is increased. Once $g \lesssim 1$, one finds that the conductance decays exponentially with L, $\ln g_L \sim -2L/\xi$ where $\xi \sim \beta N$ is termed the localization length. The decay arises from a similar exponential decay of the wave functions in the closed system. This phenomenon is known as Anderson localization (Anderson 1958; Kramer and MacKinnon 1993; Evers and Mirlin 2008). Among its many signatures, it results in a significant reduction of the level repulsion, as energy levels of wave functions localized far apart can approach each other closely. We describe the underlying mechanism in the quasi one-dimensional setting, which is realized in a long and narrow disordered quantum wire or a disordered wave guide. Anderson localization then occurs at any strength of uncorrelated disorder.

For the detailed statistical description, one composes the system from slices of length L_0 that efficiently scramble all the modes according to a mean free path l, but are small enough so that the effect of each slice can be obtained in perturbation theory (Dorokhov 1982; Mello et al. 1988; Beenakker 1997; Nazarov and Blanter 2009). For $\beta = 2$, one obtains for each step

$$c_n = \frac{l}{L_0}\overline{\delta T_n} = -T_n^2 + 2T_n(1 - T_n)\sum_{m \neq n}\frac{T_m}{T_n - T_m}, \quad d_n = \frac{l}{L_0}\overline{(\delta T_n)^2} = 2T_n^2(1 - T_n).$$

$$(10.124)$$

The result can be fed into a Fokker–Planck equation

$$Nl\frac{\partial}{\partial L}P(\{T_n\}) = \sum_n \frac{\partial}{\partial T_n}\left(-c_n + \frac{1}{2}\frac{\partial}{\partial T_n}d_n\right)P(\{T_n\}) \qquad (10.125)$$

for the joint distribution of transmission eigenvalues, then known (up to a change of variables) as the Dorokhov–Mello–Pereira–Kumar (DMPK) equation. For this particular symmetry class, the joint distribution can be found exactly by a mapping to a Schrödinger equation describing free fermions (Beenakker and Rejaei 1993). DMPK equations can also be formulated for the other symmetry classes, where they reveal delocalizing effects near the spectral symmetry points (Brouwer et al. 2002, 2005). In the many-channel limit $N \gg 1$, these equations make equivalent predictions to nonlinear sigma models (Brouwer and Frahm 1996; Efetov 1996), both in the diffusive regime $l \ll L \ll \xi$ and in the localized regime $L \gg \xi$. This convergence of models indicates a large degree of universality, which we describe next (Beenakker 1997).

In the diffusive regime $l \ll L \ll \xi$ of a system with $N \gg 1$ channels, the density of transmission eigenvalues becomes independent of the symmetry class and is given by

$$\rho(T) = \frac{Nl}{2L}\frac{1}{T\sqrt{1-T}} \quad \text{for } T_- < T < 1, \quad T_- \sim 4\exp(-2L/l). \qquad (10.126)$$

Including the next-order corrections, the ensemble-averaged dimensionless conductance is $\bar{g} = Nl/L + \frac{1}{3}(1 - 2/\beta)$ and the variance is $\operatorname{var} g = 2/(15\beta)$. Furthermore, $\overline{T(1-T)} = \overline{T}/3$ so that the shot-noise Fano factor is $f = \frac{1}{3}$.

To capture the universal aspects of Anderson localization for $L \gg \xi$, it is useful to recall that the transfer matrix (10.36) of a composed system follows by multiplication of the transfer matrices of the components, $M = \prod_{l=1}^{L} M_l$. These aspects are therefore linked to products of random matrices (Crisanti et al. 1993). For $L \to \infty$, the eigenvalues $x_n > 1$ of $M^\dagger M$ display an exponential dependence, $\ln x_n \sim 2L/\xi_n$ with Lyapunov exponents ξ_n. The scaled exponents $(\ln x_n)/L$ exhibit diminishing fluctuations, which are captured by a log-normal distribution for x_n.

These general features directly translate to the transmission eigenvalues $T_n = 4/(x_n + 2 + 1/x_n)$. The details follow from the DMPK equation, which recovers the log-normal behaviour of $\ln T_n$ in the localized regime. When ordered by magnitude, the transmission eigenvalues fall into a pattern $1 \gg T_1 \gg T_2 \gg \ldots \gg T_N$, with $\ln T_n \sim -2L(1 + \beta n - \beta)/\xi$ and $\operatorname{var}\ln T \sim 4L/\xi$. The dimensionless conductance is dominated by the largest transmission eigenvalue, and also obeys a log-normal distribution.

These results indicate that the variance $\operatorname{var}\ln g = -2\overline{\ln g}$ in the localized regime is universal. This relation can also be obtained in a diagrammatic approach, where it results from the random-phase approximation (Anderson et al. 1980), and only breaks down when one approaches the band edges, while small corrections are observed near spectral symmetry points in the clean system (Schomerus and Titov 2003). The strong universality of the log-normal distribution underpins qualitative descriptions based on renormalization arguments, which extend the considerations to higher dimensions (Abrahams et al. 1979). For three dimensions, these arguments predict that

(a)

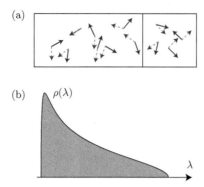

(b) $\rho(\lambda)$

λ

Fig. 10.8 (a) Bipartite entanglement concerns the quantum correlations between a subsystem and its complement. This information is captured in the eigenvalues λ of the reduced density matrix. (b) Up to a small correction accounting for normalization, the eigenvalues of a random reduced density matrix follow the Marchenko–Pastur law (10.15) for the Wishart–Laguerre ensemble.

localization sets in at a finite disorder strength, which is well supported by numerical investigations (Kramer and MacKinnon 1993; Evers and Mirlin 2008). However, an accurate statistical description is still lacking.

10.5.2 Thermalization and localization in many-body systems

In a low-dimensional many-body system, the localizing properties of disorder can be overcome by interactions. The paradigm is provided by thermal energy fluctuations that can liberate a particle from a trapped state. Such processes are also facilitated by the fact that the many-body level spacing is much smaller than the single-particle level spacing—in fact, with increasing system size the number of available states proliferates exponentially, which can be characterized by an entropy. On the other hand, this proliferation also makes it harder to establish ergodic dynamics (Anderson 1958; Basko et al. 2006; Altman and Vosk 2015; Nandkishore and Huse 2015). For the description of complex interacting systems it is therefore desirable to make contact with quantum statistical mechanics, where the posed questions tie to the concepts of ergodicity, entropy, and entanglement (Peres 2002; Gemmer et al. 2010).

Within the framework of quantum statistical mechanics, thermal equilibrium with a heat bath at temperature T is described by the canonical ensemble with density matrix $\rho = Z^{-1}\exp(-H/T)$, where $Z = \operatorname{tr}\exp(-H/T)$ is the partition function. The additional exchange of particles leads to the grandcanonical ensemble with density matrix $\rho = Z^{-1}\exp((\mu N - H)/T)$, where μ is the chemical potential and N the fluctuating particle number. In this thermodynamic setting, the associated entropy $S = -\operatorname{tr}\hat{\rho}\ln\hat{\rho}$ is an extensive quantity, which scales linearly with the volume, $S \propto V_S = O(L^d)$ for a system of size L in d dimensions. This implies that an exponential number $\sim \exp(cV_S)$ of states are populated. Deviations from these predictions occur

when one departs from equilibrium. This includes systems in which thermalization is inhibited, with the most notable example found in glasses.

Intriguingly, quantum statistical mechanics also covers the case of closed systems with a fixed energy and particle number, which allows us to focus on the intrinsic quantum-mechanical properties. These systems are described by the microcanonical ensemble, where the density matrix $\rho = M^{-1} \sum_{|E_n - E| < \delta E/2} |\psi_n\rangle\langle\psi_n|$ gives equal weight to $M \gg 1$ eigenstates residing in a classically small energy window δE around a fixed energy E. The expectation that the microcanonical entropy $S = \ln M$ is extensive indicates again that this involves an exponential number of available states.

The applicability of this description is intimately related to the question of thermalization in closed system, which in turn reveals whether the internal dynamics are ergodic (Deutsch 1991; Srednicki 1994). These links become apparent when we ask whether the microcanonical ensemble provides a good description of individual time-dependent quantum states. More precisely, we form a generic superposition of the states within the energy window and ask whether the time-averaged expectation values of some well-behaved, preselected observables $\hat{A}_n$ agree with their ensemble averages. As it turns out, a good agreement occurs when the matrix elements of the observables in the basis of participating eigenstates are sufficiently random. The ensuing self-averaging leads to an approximate state independence of the expectation values—a phenomenon known as eigenstate thermalization (Srednicki 1994; Polkovnikov et al. 2011; Nandkishore and Huse 2015). Deviations from these predictions then serve as an efficient tool for detecting insufficient coupling within the system.

In a useful picture, the state of a system becomes mixed because it is entangled with the environment (see Fig 10.8). Given a pure state $|\psi\rangle = \sum_{sb} x_{sb}|s\rangle \otimes |b\rangle \in \mathcal{H}_S \otimes \mathcal{H}_B$, with s labelling basis states of the system and b labelling basis states of the environment, we define the reduced density matrix of the system as $\hat{\rho}_S = \sum_{bss'} x_{sb} x_{s'b}^* |s\rangle\langle s'|$ (Peres 2002). For an observable $\hat{A} = \hat{A}_S \otimes 1$ that only depends on the state of the system, the expectation value follows from $\mathcal{E}_\psi(A) = \text{tr}(\hat{\rho}_S \hat{A}_S)$. The information loss from ignoring the environment can be quantified by the entanglement entropy

$$S_S = -\text{tr}\,\hat{\rho}_S \ln \hat{\rho}_S = -\sum_k \lambda_k \ln \lambda_k, \tag{10.127}$$

where λ_k are the positive eigenvalues of ρ_S. As indicated, this entropy measures the entanglement between the system and the environment; it vanishes when ρ_S describes a pure state, which requires that $|\psi\rangle = |\psi_S\rangle \otimes |\psi_B\rangle$ is separable.

To apply these concepts to the microcanonical setting of a closed system, we select a subsystem with Hilbert space dimension M and consider the remainder of (still finite) dimension M' as the environment, where for convenience we assume $M' \geq M$. For each normalized eigenstate $|\psi_n\rangle = \sum_{sb} x_{sb}|s\rangle \otimes |b\rangle$, we consider the coefficients x_{sb} as the elements of an $M \times M'$-dimensional matrix x, so that in this basis the reduced density matrix takes the form $\rho_S = xx^\dagger$, while $\rho_B = x^\dagger x$. Both are Hermitian, positive semidefinite matrices normalized to $\text{tr}\,\rho_S = \text{tr}\,\rho_B = 1$. The bipartite entanglement

entropy follows from Eq. (10.127). As we started out with a pure state for the total system we have $\mathcal{S}_S = \mathcal{S}_B$. Indeed, the nonvanishing eigenvalues of ρ_S and ρ_B are identical, so that we can pair each eigenstate $|\psi_{k,S}\rangle$ of ρ_S with an eigenstate $|\psi_{k,B}\rangle$ of ρ_B. This determines the Schmidt decomposition $|\psi_n\rangle = \sum_k \sqrt{\lambda_k}|\psi_{k,S}\rangle \otimes |\psi_{k,B}\rangle$, which reconstructs the underlying pure eigenstate. The bipartite entanglement entropy is a useful characteristic if the interactions in the system are local, and plays a central role in a broad range of physical situations, including quantum information (Nielsen and Chuang 2010), critical phenomena (Calabrese and Cardy 2009), and quantum gravity (Nishioka et al. 2009). In the ground state of a many-body system with local interactions, the entanglement entropy is found to be small, scaling with the surface area $\mathcal{S}_S \propto A_S = O(L^{d-1})$ instead of the volume V_S of the subsystem. This is termed an area law of entanglement. At phase transitions, the entanglement entropy in the ground state increases, and in 1D is often found to display a logarithmic dependence $\mathcal{S}_S \sim (c/3)\ln(L)$, where c can be interpreted as the central charge in a conformal field theory (Calabrese and Cardy 2009).

These considerations can be naturally informed by random-matrix theory, now applied directly to the structure of the eigenstates at a fixed energy. The simplest case arises when we assume that the system displays eigenstate thermalization. This can be modelled by a random reduced density matrix $\rho_S = XX^\dagger/Z$, $Z = \text{tr}\,XX^\dagger$, where X is distributed according to the Gaussian distribution (10.13) (Page 1993). The density matrix can then be interpreted as a Wishart matrix with posterior normalization (Zyczkowski and Sommers 2001; Majumdar 2011). The joint probability density of the eigenvalues λ_n follows directly by constraining the Wishart–Laguerre ensemble (10.14) to a normalized trace,

$$P(\{\lambda_n\}) \propto \delta\left(1 - \sum_k \lambda_k\right) \prod_{n<m} |\lambda_n - \lambda_m|^\beta \prod_k \lambda_k^{\beta(1+M'-M)/2-1}. \tag{10.128}$$

For $1 \ll M \leq M'$, the trace $\sum_k \lambda_k$ is self-averaging. The eigenvalue density then approaches the Marchenko–Pastur law (10.15) with $\overline{\lambda} = 1/M$,

$$\rho(\lambda) = \frac{1}{2\pi\lambda}\sqrt{4MM' - (M + M' - MM'\lambda)^2} \tag{10.129}$$

for $(\sqrt{M'} - \sqrt{M})^2 < MM'\lambda < (\sqrt{M'} + \sqrt{M})^2$. The average entanglement entropy follows as (Page 1993)

$$\overline{\mathcal{S}_B} = -\int d\lambda\, \rho(\lambda)\lambda \ln \lambda = \ln M - (M/2M'), \tag{10.130}$$

independent of the ensemble. This result signifies near-maximal entanglement. As the Hilbert space dimension of a many-body system grows exponentially with system size, the leading term corresponds to a volume law, $\mathcal{S}_S \propto V_S$, while the subleading term vanishes in the thermodynamic limit $M' \to \infty$. For highly excited states in an ergodic system obeying eigenstate thermalization, we therefore recover the expected thermodynamic behaviour.

Deviations from the eigenstate thermalization conditions should reduce the entanglement entropy. The expectation is that one recovers an area law when the disorder is increased beyond a certain threshold, a phenomenon termed many-body localization (Anderson 1958; Basko et al. 2006). This transition is indeed confirmed in numerical studies, which also detect a significant reduction of the levels repulsion (Altman and Vosk 2015; Nandkishore and Huse 2015). As for the noninteracting case, a complete statistical description of this transition is still missing.

Beyond this setting, many-body systems offer numerous deep applications of random-matrix theory. For instance, the logarithmic scaling in critical one-dimensional systems can be recovered from group integrals over unitary, orthogonal, or symplectic matrices equipped with the Haar measure (Keating and Mezzadri 2005). The ubiquitous appearance of such group integrals in field theories and other settings nicely leads us away from the theory of open quantum systems—see other chapters in this book—so we close here.

10.6   Conclusions

As this chapter illustrates, the applications of random matrices to open quantum systems are very diverse. Indeed, one can reasonably expect that this setting provides natural physical applications for (almost?) any notable random-matrix result. After all, matrices appear naturally in quantum mechanics, while openness liberates us from some of the algebraic constraints otherwise encountered. The relevance comes from the richness of complex dynamics, which helps to justify the approach for generic disorder (Efetov 1996) or underlying classical chaos (Haake 2010), and confronts us with a large number of interesting questions about the physical behaviour.

This richness already appeared in the two pure settings covered here—elastic single-particle scattering, and purely interacting systems. The latter topic we only covered briefly, and both effects can of course be combined. This is the subject of much ongoing research—e.g., regarding many-body localization and the topological protection in interacting fermionic systems, to mention just two examples. Furthermore, by combining various effects, links can be established to many other areas that enjoy random-matrix applications, as mentioned at various places in the text. As an example we recall the case of photonic systems with amplification and absorption, for which we can set up effectively non-Hermitian descriptions of the wave dynamics. When driven to the laser threshold, these systems provide means for directly probing the poles and residues of the scattering matrix, which gives a physical meaning to the Petermann factor. We can also define new, genuinely non-Hermitian symmetries, including the mentioned PT symmetry as well as non-Hermitian variants of the chiral and charge-conjugation symmetry, which all provide interesting topological effects. Such systems also display nonlinear phenomena, for which entirely new descriptions need to be developed.

It is of course important to consider where the predictive power of random-matrix descriptions may end. Take the design of small quantum devices. While their dedicated functionality is beyond the scope of this chapter, random-matrix theory can still help to

determine how well they may work—as is illustrated by our discussion of entanglement. Our system may also be insufficiently ergodic. For instance, localization effects in low dimensions lead to the search for new ensembles, a search that has not been completed. More subtle effects can arise from ballistic dynamics. These are the short-time signatures of classically deterministic motion captured by the fractal Weyl law, and dynamical constraints as encountered in a classically mixed (partially regular and chaotic) phase space. Given some suitable questions, random-matrix theory can often still be adapted to such situations, and otherwise serves as a useful benchmark for quantifying the system-specific behaviour. In general, deviations from random-matrix predictions can indicate exciting novel physics, leading to an endeavour that is nowhere near to end.

10.7 Appendix A: Eigenvalue densities of matrices with large dimensions

In the limit of large matrix dimensions M, eigenvalue distributions can be obtained very efficiently by applications of potential theory, which are based on the analogy of eigenvalues with fictitious particles in a Coulomb gas (Dyson 1962b, 1972; Beenakker 1997, Forrester 2010). In the case of the Gaussian ensembles, the leading order can also be obtained from self-consistent equations for the Green function (or resolvent) G (Pastur and Shcherbina 2011). The latter approach links neatly to the theory of free probability (Janik et al. 1997, 1999; Guionnet 2017), as we exploit in the following.

10.7.1 Gaussian Hermitian ensembles

For the Gaussian ensembles of Hermitian matrices H, we expand the Green function $G(E) = (E - H + i\varepsilon)^{-1}$ in a geometric series

$$G = E^{-1} \sum_{n=0}^{\infty} (HE^{-1})^n \tag{10.131}$$

and average using Wick's theorem, but only retaining noncrossing contractions,

$$\overline{G} = E^{-1} + E^{-1} \overline{\dot{H} E^{-1} \sum_{n=0}^{\infty} (HE^{-1})^n \dot{H} E^{-1} \sum_{n=0}^{\infty} (HE^{-1})^n}, \tag{10.132}$$

where the dot denotes terms that remain to be contracted. Denoting the variance $\overline{|H_{lm}|^2} = \sigma^2$ this gives in leading order $\overline{G} = E^{-1} + E^{-1}\sigma^2 (\operatorname{tr} \overline{G})\overline{G}$. In terms of the trace $g = \operatorname{tr} \overline{G}$, this leads to Pastur's equation

$$E = \sigma^2 g + M/g. \tag{10.133}$$

The solution $g = (1/2\sigma^2)\sqrt{E^2 - 4M\sigma^2}$ determines the density of states via

$$\rho(E) = -\lim_{\varepsilon \to 0^+} \frac{1}{\pi} \operatorname{Im} g(E + i\varepsilon) = \frac{2M}{\pi E_0^2} \sqrt{E_0^2 - E^2} \tag{10.134}$$

for $|E| < E_0$, where we identified $\sigma^2 = M\Delta^2/\pi^2 = E_0^2/4M$. This is the semicircle law (10.4). In the language of free probability, Eq. (10.133) leads to the notion of a Blue function $B_r(z) = \sigma_r^2 g + M/g$, where for later reference we equipped the variance with an index.

10.7.2 Appendix A.1: Wishart–Laguerre ensembles

For the Wishart-Laguerre ensembles, we analogously write that

$$G(\lambda) = (\lambda - X^\dagger X)^{-1} = \lambda^{-1} \sum_{n=0}^{\infty} (X^\dagger X \lambda^{-1})^n \qquad (10.135)$$

and express the noncrossing contractions as

$$\overline{G} = \lambda^{-1} + \dot{X}^\dagger \overline{(\lambda - XX^\dagger)^{-1} \dot{X} G} = \lambda^{-1} + \sigma^2 \overline{\mathrm{tr}(\lambda - XX^\dagger)^{-1} G} \qquad (10.136)$$

with $\overline{|X_{lm}|^2} = \sigma^2$. As $XX^\dagger$ differs from $X^\dagger X$ by $M' - M$ vanishing eigenvalues, we obtain that

$$g = \lambda^{-1} M + \sigma^2 (\lambda^{-1}(M' - M) + g)g, \qquad (10.137)$$

where again $g = \mathrm{tr}\,\overline{G}$. The solution

$$g = \frac{1}{2\sigma^2} - \frac{M' - M}{2\lambda} \pm \frac{1}{2\lambda\sigma^2} \sqrt{\lambda^2 - 2\lambda\sigma^2(M + M') + \sigma^4(M - M')^2} \qquad (10.138)$$

gives the Marchenko–Pastur law (10.15) via $\rho(\lambda) = -\frac{1}{\pi} \mathrm{Im}\, g$.

Note that in the chiral ensembles with Hamiltonian (10.5), the eigenvalues E_n^2 can be obtained from a Wishart matrix $AA^\dagger$; the joint distributions (10.14) and (10.6) are thus related by a change of variables $\lambda_n \propto E_n^2$, and so are the densities (10.15) and (10.8).

10.7.3 Appendix A.2: Jacobi ensembles

For the Jacobi ensembles, we base the considerations on the matrix $(X^\dagger X)^{-1} Y^\dagger Y$, whose eigenvalues λ_n determine the eigenvalues of the MANOVA matrix $(X^\dagger X + Y^\dagger Y)^{-1} X^\dagger X$ by $T_n = 1/(1 + \lambda_n)$. Consider the Green function

$$G = \begin{pmatrix} \lambda & Y \\ Y^\dagger & XX^\dagger - \lambda' \end{pmatrix}^{-1} = \begin{pmatrix} G_{11} & G_{12} \\ G_{21} & G_{22} \end{pmatrix}, \quad g = \begin{pmatrix} \mathrm{tr}\,G_{11} & \mathrm{tr}\,G_{12} \\ \mathrm{tr}\,G_{21} & \mathrm{tr}\,G_{22} \end{pmatrix}. \qquad (10.139)$$

This has matrix elements

$$G_{11} = [\lambda - Y(X^\dagger X - \lambda')^{-1} Y^\dagger]^{-1}, \quad G_{22} = (X^\dagger X - \lambda' - Y^\dagger Y/\lambda)^{-1}, \qquad (10.140)$$

with traces

$$g_{11} = (M_y - M)/\lambda + g_0, \quad \mathrm{tr}\, X G_{22} X^\dagger = \lambda' g_{22} + \lambda g_0, \qquad (10.141)$$

where

$$g_0 = \text{tr}[\lambda - (X^\dagger X - \lambda')^{-1} Y^\dagger Y]^{-1}. \qquad (10.142)$$

The eigenvalue density can then be obtained from $\rho(\lambda) = -\pi^{-1} \text{Im} \, \overline{g}_0|_{\lambda'=0}$.
    The noncrossing contractions give the relation

$$\overline{G} = \begin{pmatrix} 1/\lambda & 0 \\ 0 & -1/\lambda' \end{pmatrix} + \sigma^2 \begin{pmatrix} \text{tr} \, \overline{G}_{22}/\lambda & 0 \\ 0 & -\text{tr} \, \overline{G}_{11}/\lambda' \end{pmatrix} \overline{G} + (\sigma^2/\lambda') \begin{pmatrix} 0 & 0 \\ 0 & M_x - \text{tr} \, \overline{XG_{22}X^\dagger} \end{pmatrix} \overline{G},$$
$$(10.143)$$

while on average $\overline{G}_{12} = \overline{G}_{21} = 0$. For $\lambda' \to 0$, we therefore have

$$\overline{g}_{11} = M_y/\lambda + \sigma^2 \overline{g}_{11} \, \overline{g}_{22}/\lambda, \quad 0 = M + \sigma^2[(1+\lambda)\overline{g}_{11} + M - M_x - M_y]\overline{g}_{22}, \quad (10.144)$$

which determines

$$\overline{g}_0 = \frac{(M + M_x)\lambda + M - M_y - \sqrt{[(M - M_x)\lambda + M - M_y]^2 - 4M_x M_y \lambda}}{2\pi\lambda(1+\lambda)}. \qquad (10.145)$$

Denoting $c_x = M_x/M$, $c_y = M_y/M$, the eigenvalue density is thus given by

$$\rho(\lambda) = \frac{M(c_x - 1)\sqrt{(\lambda - \lambda_-)(\lambda_+ - \lambda)}}{2\pi\lambda(1+\lambda)}, \qquad (10.146)$$

where

$$\lambda_\pm = \left(\frac{\sqrt{c_x c_y} \pm \sqrt{c_x + c_y - 1}}{c_x - 1}\right)^2 = \left(\frac{c_y - 1}{\sqrt{c_x c_y} \mp \sqrt{c_x + c_y - 1}}\right)^2 \qquad (10.147)$$

determines the range where the density is finite. In terms of the variables T_n, the density is then given by Eq. (10.17).

10.7.4   Appendix A.3: Ginibre ensembles

As an example of non-hermitian matrix ensembles we consider the complex Ginibre ensemble, defined by (10.13) with $\beta = 2$. The Green function now must be extended to the block form (Janik et al. 1999)

$$G(z, z^*) = \begin{pmatrix} z - X & i\lambda \\ i\lambda & z^* - X^\dagger \end{pmatrix}^{-1} = \begin{pmatrix} G_{11} & G_{12} \\ G_{21} & G_{22} \end{pmatrix}, \quad g(z, z^*) = \lim_{\lambda \to 0^+} \begin{pmatrix} \text{tr} \, G_{11} & \text{tr} \, G_{12} \\ \text{tr} \, G_{21} & \text{tr} \, G_{22} \end{pmatrix},$$
$$(10.148)$$

which delivers the density of complex eigenvalues z_m via

$$\frac{1}{\pi} \frac{\partial g_{11}}{\partial z^*} = \rho(z) = \sum_m \delta(z - z_m), \qquad (10.149)$$

while the Petermann factors K_m are encoded in

$$-\frac{1}{\pi}g_{12}g_{21} = O(z) = \sum_m K_m \delta(z - z_m). \tag{10.150}$$

We denote again $\overline{|X_{lm}|^2} = \sigma^2$ and employ the expansion

$$G = \mathfrak{Z}^{-1} \sum_{n=0}^{\infty} (\mathfrak{X}\mathfrak{Z}^{-1})^n, \quad \mathfrak{Z} = \begin{pmatrix} z & i\lambda \\ i\lambda & z^* \end{pmatrix}, \quad \mathfrak{X} = \begin{pmatrix} X & 0 \\ 0 & X^\dagger \end{pmatrix}, \tag{10.151}$$

followed by the noncrossing approximation

$$\overline{G} = \mathfrak{Z}^{-1} + \mathfrak{Z}^{-1}\overline{\dot{\mathfrak{X}}\mathfrak{Z}^{-1}\sum_{n=0}^{\infty}(\mathfrak{X}\mathfrak{Z}^{-1})^n\dot{\mathfrak{X}}\mathfrak{Z}^{-1}\sum_{n=0}^{\infty}(\mathfrak{X}\mathfrak{Z}^{-1})^n} \tag{10.152}$$

$$= \mathfrak{Z}^{-1} + \mathfrak{Z}^{-1}\sigma^2 \begin{pmatrix} 0 & \overline{g}_{12} \\ \overline{g}_{21} & 0 \end{pmatrix} \overline{G}, \tag{10.153}$$

where the dot denotes elements to be Wick-contracted. The trace gives

$$\mathfrak{Z} = M/\overline{g} + \frac{\sigma^2}{2}(\overline{g} + \tilde{g}), \quad \tilde{g} = \begin{pmatrix} 1 & 0 \\ 0 & -1 \end{pmatrix} \overline{g} \begin{pmatrix} -1 & 0 \\ 0 & 1 \end{pmatrix}. \tag{10.154}$$

This agrees with the rules from free probability, according to which the Blue functions of the real and imaginary parts are $B_r(A) = \sigma_r^2 A + M/A$, $B_i(A) = \sigma_i^2\tilde{A} + M/A$, while the composition law reads $\mathfrak{Z} = B_r(g) + B_i(g) - M/g$; here $\sigma_r^2 = \sigma_i^2 = \sigma^2/2$. Let us set $\sigma^2 = 1/M$. For $|z| < 1$ the solution is then given by

$$\overline{g} = M \left(\frac{z^*}{\sqrt{|z^2| - 1}} \frac{\sqrt{|z^2| - 1}}{z} \right). \tag{10.155}$$

From this we recover Ginibre's circular law $\rho(z) = \frac{M}{\pi}\Theta(1 - |z|)$, while $O(z) = M^2(1 - |z|^2)/\pi$. According to Eq. (10.150), the ratio $O(z)/\rho(z) = M(1 - |z|^2) \sim \overline{K_m}|_{z_m=z}$ gives the average Petermann factor within the support of the spectrum.

Acknowledgements

Random-matrix theory has been developed by a large community of dedicated and inspired researchers, many of whom I had the pleasure to interact with over the years. My own thinking has been greatly influenced by Carlo Beenakker and Fritz Haake, to mention just two, and I am happy to acknowledge that this also informed the selection of topics covered here. I also thank the organizers (all themselves influential contributors to the field) for inviting me to first lecture, and now write about some aspects of this wonderful topic.

References

Abrahams, E., Anderson, P. W., Licciardello, D. C., and Ramakrishnan, T. V. (1979). Scaling theory of localization: absence of quantum diffusion in two dimensions. *Phys. Rev. Lett.* **42**, 673–6.

Akemann, G. (2017). Random matrix theory and quantum chromodynamics. In this volume.

Akemann, G., Baik, J., and Di Francesco, P. (eds.) (2011). *The Oxford Handbook of Random Matrix Theory*. Oxford University Press, Oxford.

Akhmerov, A. R., Dahlhaus, J. P., Hassler, F., Wimmer, M., and Beenakker, C. W. J. (2011). Quantized conductance at the Majorana phase transition in a disordered superconducting wire. *Phys. Rev. Lett.* **106**, 057001.

Aleiner, I. L., and Larkin, A. I. (1996). Divergence of classical trajectories and weak localization. *Phys. Rev. B* **54**, 14423–44.

Alicea, J. (2012). New directions in the pursuit of Majorana fermions in solid state systems. *Rep. Prog. Phys.* **75**, 076501.

Altland, A., and Zirnbauer, M. R. (1997). Nonstandard symmetry classes in mesoscopic normal-superconducting hybrid structures. *Phys. Rev. B* **55**, 1142–61.

Altman, E., and Vosk, R. (2015). Universal dynamics and renormalization in many-body-localized systems. *Annu. Rev. Condens. Matter Phys.* **6**, 383–409.

Anderson, P. W. (1958). Absence of diffusion in certain random lattices. *Phys. Rev.* **109**, 1492–505.

Anderson, P. W., Thouless, D. J., Abrahams, E., and Fisher, D. S. (1980). New method for a scaling theory of localization. *Phys. Rev. B* **22**, 3519–26.

Bäcker, A. (2003). Numerical aspects of eigenvalue and eigenfunction computations for chaotic quantum systems. In *The Mathematical Aspects of Quantum Maps*, ed. M. D. Esposito and S. Graffi, vol. 618, Lecture Notes in Physics, p. 91. Springer, Berlin.

Baranger, H. U., and Mello, P. A. (1994). Mesoscopic transport through chaotic cavities: A random S-matrix theory approach. *Phys. Rev. Lett.* **73**, 142–5.

Basko, D. M., Aleiner, I. L., and Altshuler, B. L. (2006). Metal-insulator transition in a weakly interacting many-electron system with localized single-particle states. *Ann. Phys. (N.Y.)* **321**, 1126–205.

Beenakker, C. W. J. (1992). Quantum transport in semiconductor-superconductor microjunctions. *Phys. Rev. B* **46**, 12841–4.

Beenakker, C. W. J. (1997). Random-matrix theory of quantum transport. *Rev. Mod. Phys.* **69**, 731–808.

Beenakker, C. W. J. (1998). Thermal radiation and amplified spontaneous emission from a random medium. *Phys. Rev. Lett.* **81**, 1829–32.

Beenakker, C. W. J. (2005). Andreev billiards. *Lecture Notes in Physics* **667**, 131–74. arXiv:cond-mat/0406018.

Beenakker, C. W. J. (2013). Search for Majorana fermions in superconductors. *Annu. Rev. Condens. Matter Phys.* **4**, 113–36.

Beenakker, C. W. J. (2015). Random-matrix theory of Majorana fermions and topological superconductors. *Rev. Mod. Phys.* **87**, 1037–66.

Beenakker, C. W. J., Edge, J. M., Dahlhaus, J. P., Pikulin, D. I., Mi, Shuo, and Wimmer, M. (2013). Wigner-poisson statistics of topological transitions in a Josephson junction. *Phys. Rev. Lett.* **111**, 037001.

Beenakker, C. W. J., and Rejaei, B. (1993). Nonlogarithmic repulsion of transmission eigenvalues in a disordered wire. *Phys. Rev. Lett.* **71**, 3689–92.

Bender, C. M. (2007). Making sense of non-Hermitian Hamiltonians. *Rep. Prog. Phys.* **70**, 947.

Béri, B. (2009). Random scattering matrices for Andreev quantum dots with nonideal leads. *Phys. Rev. B* **79**, 214506.

Berman, G. P., and Zaslavsky, G. M. (1978). Condition of stochasticity in quantum nonlinear systems. *Physica A* **91**, 450–60.

Berry, M. V. (2004). Physics of nonhermitian degeneracies. *Czech. J. Phys.* **54**, 1039–47.

Birchall, C., and Schomerus, H. (2012). Random-matrix theory of amplifying and absorbing resonators with $\mathcal{PT}$ or $\mathcal{PTT}'$ symmetry. *J. Phys. A.* **45**, 444006.

Birman, M. S., and Krein, M. G. (1962). On the theory of wave operators and scattering operators. *Sov. Math. Dokl.* **3**, 740–4.

Blanter, Ya. M., and Büttiker, M. (2000). Shot noise in mesoscopic conductors. *Phys. Rep.* **336**, 1–166.

Blanter, Ya. M., Schomerus, H., and Beenakker, C. W. J. (2001). Effect of dephasing on charge-counting statistics in chaotic cavities. *Physica E* **11**, 1–7.

Blümel, R., and Smilansky, U. (1990). Random-matrix description of chaotic scattering: semiclassical approach. *Phys. Rev. Lett.* **64**, 241–4.

Bohigas, O. and Weidenmüller, H.A. (2017). History—an overview. In this volume.

Breuer, H.-P., and Petruccione, F. (2002). *The Theory of Open Quantum Systems*. Oxford University Press, Oxford.

Brouwer, P. W. (1995). Generalized circular ensemble of scattering matrices for a chaotic cavity with nonideal leads. *Phys. Rev. B* **51**, 16878–84.

Brouwer, P. W. (1998). Scattering approach to parametric pumping. *Phys. Rev. B* **58**, R10135–8.

Brouwer, P. W., and Frahm, K. (1996). Quantum transport in disordered wires: Equivalence of the one-dimensional σ model and the Dorokhov-Mello-Pereyra-Kumar equation. *Phys. Rev. B* **53**, 1490–501.

Brouwer, P. W., Frahm, K. M., and Beenakker, C. W. J. (1997). Quantum mechanical time-delay matrix in chaotic scattering. *Phys. Rev. Lett.* **78**, 4737–40.

Brouwer, P. W., Furusaki, A., Mudry, C., and Ryu, S. (2005). Disorder-induced critical phenomena–new universality classes in Anderson localization. eprint arXiv:cond-mat/0511622.

Brouwer, P. W., Racine, E., Furusaki, A., Hatsugai, Y., Morita, Y., and Mudry, C. (2002). Zero modes in the random hopping model. *Phys. Rev. B* **66**, 014204.

Büttiker, M. (1990). Scattering theory of thermal and excess noise in open conductors. *Phys. Rev. Lett.* **65**, 2901–4.

Büttiker, M., Thomas, H., and Prêtre, A. (1994). Current partition in multiprobe conductors in the presence of slowly oscillating external potentials. *Z. Phys. B* **94**, 133–7.

Calabrese, P., and Cardy, J. (2009). Entanglement entropy and conformal field theory. *J. Phys. A.* **42**, 504005.

Cao, H., and Wiersig, J. (2015). Dielectric microcavities: Model systems for wave chaos and non-Hermitian physics. *Rev. Mod. Phys.* **87**, 61–111.

Cardy, J. (1996). *Scaling and Renormalization in Statistical Physics*. Cambridge Lecture Notes in Physics. Cambridge University Press, Cambridge, UK.

Carmichael, H. (2009). *An Open Systems Approach to Quantum Optics*. Lecture Notes in Physics Monographs. Springer, Berlin.

Caselle, M., and Magnea, U. (2004). Random matrix theory and symmetric spaces. *Phys. Rep.* **394**, 41–156.

Chalker, J. T., and Mehlig, B. (1998). Eigenvector statistics in non-hermitian random matrix ensembles. *Phys. Rev. Lett.* **81**, 3367–70.

Crisanti, A., Paladin, G., and Vulpiani, A. (1993). *Products of Random Matrices*. vol. 104, Springer Series in Solid-State Sciences. Springer, Berlin.

Dahlhaus, J. P., Béri, B., and Beenakker, C. W. J. (2010). Random-matrix theory of thermal conduction in superconducting quantum dots. *Phys. Rev. B* **82**, 014536.

Datta, S. (1997). *Electronic Transport in Mesoscopic Systems*. Cambridge University Press, Cambridge, UK.

de Carvalho, C. A. A., and Nussenzveig, H. M. (2002). Time delay. *Phys. Rep.* **364**, 83–174.

Deutsch, J. M. (1991). Quantum statistical mechanics in a closed system. *Phys. Rev. A* **43**, 2046–9.

Dorokhov, O. N. (1982). Transmission coefficient and the localization length of an electron in n bound disordered chains. *JETP Lett.* **36**, 318–21.

Doron, E., and Smilansky, U. (1992). Semiclassical quantization of chaotic billiards: a scattering theory approach. *Nonlinearity* **5**, 1055.

Dyson, F. J. (1962a). Statistical theory of the energy levels of complex systems. I. *J. Math. Phys.* **3**, 140–156.

Dyson, F. J. (1962b). Statistical theory of the energy levels of complex systems. II. *J. Math. Phys.* **3**, 157–65.

Dyson, F. J. (1962c). The threefold way. algebraic structure of symmetry groups and ensembles in quantum mechanics. *J. Math. Phys.* **3**, 1199–215.

Dyson, F. J. (1970). Correlations between eigenvalues of a random matrix. *Comm. Math. Phys.* **19**, 235–50.

Dyson, F. J. (1972). A class of matrix ensembles. *J. Math. Phys.* **13**, 90–7.

Efetov, K. (1996). *Supersymmetry in Disorder and Chaos*. Cambridge University Press, Cambridge UK.

Evers, F., and Mirlin, A. D. (2008). Anderson transitions. *Rev. Mod. Phys.* **80**, 1355–417.

Forrester, P. J. (2010). *Log-Gases and Random Matrices*. Princeton University Press, Princeton, NJ.

Forrester, P. J., and Nagao, T. (2007). Eigenvalue statistics of the real Ginibre ensemble. *Phys. Rev. Lett.* **99**, 050603.

Fu, L., and Kane, C. L. (2008). Superconducting proximity effect and Majorana fermions at the surface of a topological insulator. *Phys. Rev. Lett.* **100**, 096407.

Fulga, I. C., Hassler, F., Akhmerov, A. R., and Beenakker, C. W. J. (2011). Scattering formula for the topological quantum number of a disordered multimode wire. *Phys. Rev. B* **83**, 155429.

Fyodorov, Y. and Savin, D. (2011). Resonance scattering in chaotic systems. In *The Oxford Handbook of Random Matrix Theory*, ed. G. Akemann, J. Baik, and P. Di Franceseo. Oxford University Press, Oxford.

Fyodorov, Y. V., and Khoruzhenko, B. A. (1999). Systematic analytical approach to correlation functions of resonances in quantum chaotic scattering. *Phys. Rev. Lett.* **83**, 65–8.

Fyodorov, Y. V., and Savin, D. V. (2012). Statistics of resonance width shifts as a signature of eigenfunction nonorthogonality. *Phys. Rev. Lett.* **108**, 184101.

Fyodorov, Y. V., and Sommers, H.-J. (1996). Statistics of S-matrix poles in few-channel chaotic scattering: Crossover from isolated to overlapping resonances. *JETP Lett.* **63**, 1026–30.

Fyodorov, Y. V., and Sommers, H.-J. (1997). Statistics of resonance poles, phase shifts and time delays in quantum chaotic scattering: Random matrix approach for systems with broken time-reversal invariance. *J. Math. Phys.* **38**, 1918–81.

Fyodorov, Y. V., and Sommers, H.-J. (2000). Spectra of random contractions and scattering theory for discrete-time systems. *JETP Lett.* **72**, 422–6.

Fyodorov, Y. V., and Sommers, H.-J. (2003). Random matrices close to Hermitian or unitary: overview of methods and results. *J. Phys. A.* **36**, 3303–47.

Gasparian, V., Christen, T., and Büttiker, M. (1996). Partial densities of states, scattering matrices, and Green's functions. *Phys. Rev. A* **54**, 4022–31.

Gemmer, J., Michel, M., and Mahler, G. (2010). *Quantum Thermodynamics*, vol. 784, Lecture Notes in Physics. Springer, Berlin.

Ginibre, J. (1965). Statistical ensembles of complex, quaternion, and real matrices. *J. Math. Phys.* **6**, 440.

Girko, V. L. (1986). Elliptic law. *Theor. Prob. Appl.* **30**, 677–90.

Guhr, T., Müller-Groeling, A., and Weidenmüller, H. A. (1998). Random-matrix theories in quantum physics: common concepts. *Phys. Rep.* **299**, 189–425.

Guionnet, A. (2017). Free probability. In this volume.

Haake, F. (2010). *Quantum Signatures of Chaos*. Vol. 54, Springer Series in Synergetics. Springer, Berlin.

Haake, F., Izrailev, F., Lehmann, N., Saher, D., and Sommers, H.-J. (1992). Statistics of complex levels of random matrices for decaying systems. *Z. Phys. B* **88**, 359–370.

Hasan, M. Z., and Kane, C. L. (2010). *Colloquium*: topological insulators. *Rev. Mod. Phys.* **82**, 3045–67.

Heiss, W. D. (2012). The physics of exceptional points. *J. Phys. A: Math. Theor.* **45**, 444016.

Iida, S., Weidenmüller, H. A., and Zuk, J. A. (1990). Statistical scattering theory, the supersymmetry method and universal conductance fluctuations. *Ann. Phys. (N.Y.)* **200**, 219–70.

Jalabert, R. A., Pichard, J.-L., and Beenakker, C. W. J. (1994). Universal quantum signatures of chaos in ballistic transport. *Europhys. Lett.* **27**, 255.

Janik, R. A., Nörenberg, W., Nowak, M. A., Papp, G., and Zahed, I. (1999). Correlations of eigenvectors for non-Hermitian random-matrix models. *Phys. Rev. E* **60**, 2699–705.

Janik, R. A., Nowak, M. A., Papp, G., Wambach, J., and Zahed, I. (1997). Non-Hermitian random matrix models: free random variable approach. *Phys. Rev. E* **55**, 4100–6.

Keating, J. P., and Mezzadri, F. (2005). Entanglement in quantum spin chains, symmetry classes of random matrices, and conformal field theory. *Phys. Rev. Lett.* **94**, 050501.

Keating, J. P., Novaes, M., Prado, S. D., and Sieber, M. (2006). Semiclassical structure of chaotic resonance eigenfunctions. *Phys. Rev. Lett.* **97**, 150406.

Khoruzhenko, B. and Sommers, H.-J. (2011). Non-Hermitian ensembles. In *The Oxford Handbook of Random Matrix Theory*, ed. G. Akemann, J. Baik, and P. Di Francesco. Oxford University Press, Oxford.

Kitaev, A. (2009). Periodic table for topological insulators and superconductors. *AIP Conf. Proc.* **1134**, 22–30.

Kitagawa, T., Berg, E., Rudner, M., and Demler, E. (2010). Topological characterization of periodically driven quantum systems. *Phys. Rev. B* **82**, 235114.

Kramer, B. and MacKinnon, A. (1993). Localization: theory and experiment. *Rep. Prog. Phys.* **56**, 1469.

Lehmann, N., Saher, D., Sokolov, V. V., and Sommers, H.-J. (1995). Chaotic scattering: the supersymmetry method for large number of channels. *Nucl. Phys. A* **582**, 223–56.

Lehmann, N., and Sommers, H.-J. (1991). Eigenvalue statistics of random real matrices. *Phys. Rev. Lett.* **67**, 941–4.

Leijnse, M., and Flensberg, K. (2012). Introduction to topological superconductivity and Majorana fermions. *Semicond. Sci. Technol.* **27**, 124003.

Levitov, L. S., and Lesovik, G. B. (1993). Charge distribution in quantum shot noise. *JETP Lett.* **58**, 230.

Lieb, E. H. (1989). Two theorems on the Hubbard model. *Phys. Rev. Lett.* **62**, 1201–4.

Livsic, M. S. (1973). *Operators, Oscillations, Waves.* American Mathematical Society, Providence, RI.

Lu, L., Joannopoulos, J. D., and Soljačić, M. (2014). Topological photonics. *Nat. Photon.* **8**, 821–9.

Lu, W. T., Sridhar, S., and Zworski, M. (2003). Fractal Weyl laws for chaotic open systems. *Phys. Rev. Lett.* **91**, 154101.

Macedo-Junior, A. F., and Macêdo, A. M. S. (2002). Universal transport properties of quantum dots with chiral symmetry. *Phys. Rev. B* **66**, 041307.

Magnea, U. (2008). Random matrices beyond the Cartan classification. *J. Phys. A* **41**, 045203.

Mahaux, C., and Weidenmüller, H. A. (1969). *Shell-Model Approach to Nuclear Reactions.* North-Holland, Amsterdam.

Majumdar, S. N. (2011). Extreme eigenvalues of Wishart matrices and entangled bipartite system. In *The Oxford Handbook of Random Matrix Theory*, ed. G. Akemann, J. Baik, and P. Di Francesco. Oxford University Press, Oxford.

Makris, K. G., El-Ganainy, R., Christodoulides, D. N., and Musslimani, Z. H. (2008). Beam dynamics in $\mathcal{PT}$ symmetric optical lattices. *Phys. Rev. Lett.* **100**, 103904.

Malzard, S., Poli, C., and Schomerus, H. (2015). Topologically protected defect states in open photonic systems with non-Hermitian charge-conjugation and parity-time symmetry. *Phys. Rev. Lett.* **115**, 200402.

Marčenko, V. A., and Pastur, L. A. (1967). Distribution of eigenvalues for some sets of random matrices. *Math. USSR-Sb.* **1**, 457.

Marciani, M., Brouwer, P. W., and Beenakker, C. W. J. (2014). Time-delay matrix, midgap spectral peak, and thermopower of an Andreev billiard. *Phys. Rev. B* **90**, 045403.

Marciani, M., Schomerus, H., and Beenakker, C. W. J. (2016). Effect of a tunnel barrier on the scattering from a Majorana bound state in an Andreev billiard. *Physica E* **77**, 54–64.

Mehta, M. L. (2004). *Random Matrices*. Pure and Applied Mathematics. Elsevier Science, Amsterdam.

Mello, P. A., Pereyra, P., and Kumar, N. (1988). Macroscopic approach to multichannel disordered conductors. *Ann. Phys. (N.Y.)* **181**, 290–317.

Messiah, A. (2014). *Quantum Mechanics*. Dover Books on Physics. Dover, Mineola, NY.

Moiseyev, N. (2011). *Non-Hermitian Quantum Mechanics*. Cambridge University Press, Cambridge.

Mourik, V., Zuo, K., Frolov, S. M., Plissard, S. R., Bakkers, E. P. a. M., and Kouwenhoven, L. P. (2012). Signatures of Majorana fermions in hybrid superconductor–semiconductor nanowire devices. *Science* **336**, 1003–7.

Nandkishore, R., and Huse, D. A. (2015). Many-body localization and thermalization in quantum statistical mechanics. *Annu. Rev. Condens. Matter Phys.* **6**, 15–38.

Nayak, C., Simon, S. H., Stern, A., Freedman, M., and Das Sarma, S. (2008). Non-abelian anyons and topological quantum computation. *Rev. Mod. Phys.* **80**, 1083–159.

Nazarov, Y. V., and Blanter, Ya. M. (2009). *Quantum Transport: Introduction to Nanoscience*. Cambridge University Press, Cambridge, UK.

Newton, R. G. (2002). *Scattering Theory of Waves and Particles*. Dover, Mineola, NY.

Nielsen, M. A., and Chuang, I. L. (2010). *Quantum Computation and Quantum Information*. Cambridge University Press, Cambridge, UK.

Nishioka, T., Ryu, S., and Takayanagi, T. (2009). Holographic entanglement entropy: an overview. *J. Phys. A.* **42**, 504008.

Nonnenmacher, S., and Zworski, M. (2005). Fractal Weyl laws in discrete models of chaotic scattering. *J. Phys. A.* **38**, 10682–3.

Page, D. N. (1993). Average entropy of a subsystem. *Phys. Rev. Lett.* **71**, 1291–4.

Pastur, L. A., and Shcherbina, M. (2011). *Eigenvalue Distribution of Large Random Matrices*. Mathematical surveys and monographs. American Mathematical Society, Providence, RI.

Peres, A. (2002). *Quantum Theory: Concepts and Methods*. Springer, Dordrecht.

Petermann, K. (1979). Calculated spontaneous emission factor for double-heterostructure injection lasers with gain-induced waveguiding. *IEEE J. Quantum Electron.* **15**, 566–70.

Pikulin, D. I., Dahlhaus, J. P., Wimmer, M., Schomerus, H., and Beenakker, C. W. J. (2012). A zero-voltage conductance peak from weak antilocalization in a Majorana nanowire. *New J. Phys.* **14**, 125011.

Pikulin, D. I., and Nazarov, Yu. V. (2012). Topological properties of superconducting junctions. *JETP Lett.* **94**, 693–7.

Pikulin, D. I. and Nazarov, Y. V. (2013). Two types of topological transitions in finite Majorana wires. *Phys. Rev. B* **87**, 235421.

Poli, C., Bellec, M., Kuhl, U., Mortessagne, F., and Schomerus, H. (2015). Selective enhancement of topologically induced interface states in a dielectric resonator chain. *Nature Commun.* **6**, 6710.

Polkovnikov, A., Sengupta, K., Silva, A., and Vengalattore, M. (2011). Colloquium: Nonequilibrium dynamics of closed interacting quantum systems. *Rev. Mod. Phys.* **83**, 863–83.

Porter, C. E. (1965). *Statistical Theory of Spectra: Fluctuations*. Academic Press, New York.

Qi, X.-L., and Zhang, S.-C. (2011). Topological insulators and superconductors. *Rev. Mod. Phys.* **83**, 1057–110.

Rotter, I. (2009). A non-Hermitian hamilton operator and the physics of open quantum systems. *J. Phys. A* **42**, 153001.

Rüter, C. E., Makris, K. G., El-Ganainy, R., Christodoulides, D. N., Segev, M., and Kip, D. (2010). Observation of parity-time symmetry in optics. *Nat. Phys.* **6**, 192–5.

Ryu, S., Schnyder, A. P., Furusaki, A., and Ludwig, A. W. W. (2010). Topological insulators and superconductors: tenfold way and dimensional hierarchy. *New J. Phys.* **12**, 065010.

Sachdev, S. (1999). *Quantum Phase Transitions*. Cambridge University Press, Cambridge, UK.

San-Jose, P., Cayao, J., Prada, E., and Aguado, R. (2016). Majorana bound states from exceptional points in non-topological superconductors. *Sci. Rep.* **6**, 21427.

Savin, D. V., and Sokolov, V. V. (1997). Quantum versus classical decay laws in open chaotic systems. *Phys. Rev. E* **56**, R4911–13.

Schomerus, H. (2009). Excess quantum noise due to mode nonorthogonality in dielectric microresonators. *Phys. Rev. A* **79**, 061801.

Schomerus, H. (2010). Quantum noise and self-sustained radiation of PT-symmetric systems. *Phys. Rev. Lett.* **104**, 233601.

Schomerus, H. (2013a). From scattering theory to complex wave dynamics in non-Hermitian PT-symmetric resonators. *Phil. Trans. R. Soc. A* **371**, 20120194.

Schomerus, H. (2013b). Topologically protected midgap states in complex photonic lattices. *Opt. Lett.* **38**, 1912.

Schomerus, H., Frahm, K. M., Patra, M., and Beenakker, C. W. J. (2000). Quantum limit of the laser line width in chaotic cavities and statistics of residues of scattering matrix poles. *Physica A* **278**, 469–96.

Schomerus, H. and Halpern, N. Y. (2013). Parity anomaly and Landau-level lasing in strained photonic honeycomb lattices. *Phys. Rev. Lett.* **110**, 013903.

Schomerus, H., and Jacquod, P. (2005). Quantum-to-classical correspondence in open chaotic systems. *J. Phys. A.* **38**, 10663.

Schomerus, H., Marciani, M., and Beenakker, C. W. J. (2015). Effect of chiral symmetry on chaotic scattering from Majorana zero modes. *Phys. Rev. Lett.* **114**, 166803.

Schomerus, H., and Titov, M. (2003). Band-center anomaly of the conductance distribution in one-dimensional anderson localization. *Phys. Rev. B* **67**, 100201.

Schomerus, H., and Tworzydło, J. (2004). Quantum-to-classical crossover of quasibound states in open quantum systems. *Phys. Rev. Lett.* **93**, 154102.

Schomerus, H., Wiersig, J., and Main, J. (2009). Lifetime statistics in chaotic dielectric microresonators. *Phys. Rev. A* **79**, 053806.

Silvestrov, P. G., Goorden, M. C., and Beenakker, C. W. J. (2003). Noiseless scattering states in a chaotic cavity. *Phys. Rev. B* **67**, 241301.

Smith, F. T. (1960). Lifetime matrix in collision theory. *Phys. Rev.* **118**, 349–56.

Sommers, H.-J., Fyodorov, Y. V., and Titov, M. (1999). S-matrix poles for chaotic quantum systems as eigenvalues of complex symmetric random matrices: from isolated to overlapping resonances. *J. Phys. A: Math. Gen.* **32**, L77.

Srednicki, M. (1994). Chaos and quantum thermalization. *Phys. Rev. E* **50**, 888–901.

Stöckmann, H.-J. (2006). *Quantum Chaos: An Introduction.* Cambridge University Press, Cambridge, UK.

Sutherland, B. (1986). Localization of electronic wave functions due to local topology. *Phys. Rev. B* **34**, 5208–11.

Teo, J. C. Y., and Kane, C. L. (2010). Topological defects and gapless modes in insulators and superconductors. *Phys. Rev. B* **82**, 115120.

Texier, C. (2016). Wigner time delay and related concepts: Application to transport in coherent conductors. *Physica E* **82**, 16–33.

Tworzydło, J., Tajic, A., Schomerus, H., and Beenakker, C. W. J. (2003). Dynamical model for the quantum-to-classical crossover of shot noise. *Phys. Rev. B* **68**, 115313.

Verbaarschot, J. J. M., Weidenmüller, H. A., and Zirnbauer, M. R. (1985). Grassmann integration in stochastic quantum physics: The case of compound-nucleus scattering. *Phys. Rep.* **129**, 367–438.

Verbaarschot, J. J. M., (1994). The spectrum of the QCD Dirac operator and chiral random matrix theory: the threefold way. *Phys. Rev. Lett.* **72**, 2531–3.

Verbaarschot, J. J. M., and Wettig, T. (2000). Random matrix theory and chiral symmetry in QCD. *Annu. Rev. Nucl. Part. Sci.* **50**, 343–410.

Weich, T., Barkhofen, S., Kuhl, U., Poli, C., and Schomerus, H. (2014). Formation and interaction of resonance chains in the open three-disk system. *New J. Phys.* **16**, 033029.

Weidenmüller, H. A., and Mitchell, G. E. (2009). Random matrices and chaos in nuclear physics: nuclear structure. *Rev. Mod. Phys.* **81**, 539–89.

Weiss, U. (2008). *Quantum Dissipative Systems*, 3rd edn. World Scientific, Singapore.

Wiersig, J. (2014). Enhancing the sensitivity of frequency and energy splitting detection by using exceptional points: application to microcavity sensors for single-particle detection. *Phys. Rev. Lett.* **112**, 203901.

Wigner, E. P. (1955). Lower limit for the energy derivative of the scattering phase shift. *Phys. Rev.* **98**, 145–7.

Wigner, E. P. (1956). Proceedings of the conference on neutron physics by time-of-flight. *ORNL Rep.* **2309**, 59–70.

Wigner, E. P. (1958). On the distribution of the roots of certain symmetric matrices. *Ann. Math.* **67**, 325–7.

Wishart, J. (1928). The generalised product moment distribution in samples from a normal multivariate population. *Biometrika* **20A**, 32–52.

Yoo, G., Sim, H.-S., and Schomerus, H. (2011). Quantum noise and mode nonorthogonality in non-Hermitian $\mathcal{PT}$-symmetric optical resonators. *Phys. Rev. A* **84**(6), 063833.

Zirnbauer, M. (2011). Symmetry classes. In *The Oxford Handbook of Random Matrix Theory*, ed. G. Akemann, J. Baik, and P. Di Francesco. Oxford University Press, Oxford.

Zirnbauer, M. R. (1996). Riemannian symmetric superspaces and their origin in random-matrix theory. *J. Math. Phys.* **37**, 4986–5018.

Zyczkowski, K., and Sommers, H.-J. (2000). Truncations of random unitary matrices. *J. Phys. A.*, **33**, 2045.

Zyczkowski, K., and Sommers, H.-J. (2001). Induced measures in the space of mixed quantum states. *J. Phys. A* **34**, 7111.

11

Impurity models and products of random matrices

Alain COMTET[1] and Yves TOURIGNY[2]

[1]LPTMS, Université Paris 11, 91400, Orsay, France, Sorbonne Universités, UPMC, Université Paris 06, 70005, Paris, France,
[2]School of Mathematics, University of Bristol, Bristol BS8 1TW, United Kingdom

Comtet, A. and Tourigny, Y., 'Impurity Models and Products of Random Matrices' in *Stochastic Processes and Random Matrices*. Edited by: Grégory Schehr et al, Oxford University Press (2017).
© Oxford University Press 2017. DOI 10.1093/oso/9780198797319.003.0011

Chapter Contents

This is an introduction to the theory of one-dimensional disordered systems and products of random matrices, confined to the 2×2 case. The notion of *impurity model*— that is, a system in which the interactions are highly localized—links the two themes and enables their study by elementary mathematical tools. After discussing the spectral theory of some impurity models, we state and illustrate Furstenberg's theorem, which gives sufficient conditions for the exponential growth of a product of independent, identically-distributed matrices.

11.1 Introduction

11.1.1 Product of matrices

Consider a sequence

$$A_1, A_2, A_3, \ldots$$

of invertible $d \times d$ matrices drawn independently from some distribution, say μ. In this chapter, we will be interested in the large-n behaviour of the product

$$\Pi_n := A_n A_{n-1} \cdots A_1.$$

More precisely, we will investigate conditions under which the product *grows exponentially* with n and seek to compute the growth rate.

The solution of this problem is of great interest in the physics of disordered systems. Such systems are often modelled in terms of linear difference or differential equations with random coefficients and, in order to understand the behaviour of the system, one must study the spectral problem associated with the equation. For the models we will consider, the general solution of the equation can be expressed in terms of the product Π_n for a suitable choice of the A_n.

11.1.2 Disordered systems

The existence of widely separated scales is a remarkable feature of the physical world. The fact that the corresponding degrees of freedom can—to a very good approximation—decouple makes it possible to construct effective theories of condensed matter physics, and even of particle physics. The same crucial simplification occurs in the physics of disordered systems: one can distinguish between fast variables, which evolve very quickly, and slow variables, which, in a real experiment, are frozen in a specific configuration. It is then legitimate to treat the slow variables as (static but) random variables, distributed according to a prescribed probability law. Such systems are usually modelled using the concept of *random operator*.

Consider, for instance, the quantum mechanics of an electron interacting with a collection of scatterers, and think of these scatterers as 'impurities' in an otherwise homogeneous medium. This problem is modelled by the Hamiltonian

$$H = \frac{\mathbf{p}^2}{2m} + \sum_{j=1}^{n} V(\mathbf{r} - \mathbf{r}_j),$$

where the random parameters include, among others, the positions $\mathbf{r}_j$ of the impurities and their number n. We could, for instance, assume that the positions of the scatterers are independent and distributed uniformly in space, and consider the thermodynamic limit for a fixed density. Although this model looks very simple, its analytical treatment presents substantial difficulties; even basic quantities of physical interest, such as the large-volume distribution of eigenvalues, can seldom be computed exactly.

In the one-dimensional case, however, the problem is more tractable: much progress can be made by considering *initial-value problems*. Techniques pioneered by Dyson (1953) and Schmidt (1957) that lead in a few instances to explicit formulae have been developed. Schmidt's justification for bothering with the one-dimensional case was the 'hope that it gives in some respects a qualitatively correct description of real three-dimensional [systems]'. Dyson, on the other hand, felt that 'interest for working on one-dimensional problems is merely that they are fun'. Our own view is that the value of exact calculations is in revealing possible connections between objects which, at first sight, may have seemed unrelated.

The primary purpose of these lectures is to elaborate the connection between the Dyson–Schmidt methodology, applied to a certain class of disordered systems, and one of this school's themes, namely products of random matrices. Historically, it may be said that the desire to understand the behaviour of disordered systems provided one of the main motivations for the development, by Furstenberg (1963) and others, of the theory of products of random matrices. The classic reference on the interplay between the two subjects is the book by Bougerol and Lacroix (1985). That book is divided into two parts: the first part gives a rigorous account of Furstenberg's theorem on the growth of products of matrices; the second part applies that theory to products obtained from Anderson's model, which uses a 'tight-binding' (i.e., a finite-difference) approximation of the Schrödinger equation with a random potential. Roughly speaking, the programme there is to apply Furstenberg's theory to some disordered systems and in so doing deduce localization properties of the disordered states.

11.1.3 Outline

In the present chapter, we revisit these connections, but the flow of ideas is in the opposite direction: we begin by considering a class of disordered systems which use the notion of *random point interaction*. We call these disordered systems *impurity models*. We explain how, via the well-known technique of separation of variables, these models lead to products of 2×2 matrices, and we undertake the study of their spectral properties; it will be seen that these are intimately linked with the growth of the solutions. We also briefly touch upon a number of related topics, including the scattering problem for a disordered sample. In applying and developing these ideas in the context of our impurity models, we encounter many of the objects that feature in Furstenberg's work. Our approach thus provides some physically motivated insights into the abstract mathematics of the Furstenberg theory.

11.1.4 Recommended reading

This chapter constitutes an extensive development of some of the ideas presented in the papers by Comtet et al. (2010, 2011, 2013a) and Comtet et al. (2013b); from a

mathematical point of view, they are more or less self-contained, in the sense that anyone who is familiar with the basic concepts of probability, differential equations, complex variables, and group theory should be able to learn something from them.

We make no attempt to provide a systematic survey of the literature, but merely refer to those papers and books that we have personally found helpful in developing our own understanding of the subject. For an alternative introduction to disordered systems, the reader is encouraged to consult Pastur's survey (Pastur 1973), or the more elaborate account in Lifshits et al. (1988). The first chapter of Luck's excellent but insufficiently known monograph is also highly recommended (Luck 1992).

The material leading to Furstenberg's theorem is based on his fundamental paper (Furstenberg 1963), and on the later monographs by Bougerol and Lacroix (1985) and Carmona and Lacroix (1990).

11.2 Some impurity models

This section presents some examples of impurity models and explains how products of matrices arise naturally from them.

11.2.1 The vibrating string and Dyson's random chain model

Consider an inextensible string of unit tension, tied at the ends of the interval $[0, L]$. Denote by $M(x)$ the total mass of the string segment corresponding to the interval $[0, x)$ and by $y(x, t)$ the vertical displacement of the string from its equilibrium position above the point x at time t. The Lagrangian associated with this mechanical system is

$$\frac{1}{2} \int_0^L \left[M'(x) \left(\frac{\partial y(x, t)}{\partial t} \right)^2 - \left(\frac{\partial y(x, t)}{\partial x} \right)^2 \right] dx .$$

It follows from Hamilton's principle of least action— see, for instance, Simmons (1972)— that $y(x, t)$ obeys the *wave equation*

$$M'(x) \frac{\partial^2 y(x, t)}{\partial t^2} - \frac{\partial^2 y(x, t)}{\partial x^2} = 0, \quad 0 < x < L, \ t > 0 .$$

This equation admits separable solutions of the form

$$y(x, t) = \psi(x) \, e^{i\omega t}$$

provided that ω is a characteristic frequency of the string, i.e., a number such that there exists a nontrivial solution of the two-point boundary-value problem

$$\psi''(x) + \omega^2 M'(x) \psi(x) = 0, \quad 0 < x < L, \quad \psi(0) = \psi(L) = 0 . \tag{11.1}$$

For an 'ideal' string with a uniform distribution, the characteristic frequencies may be calculated exactly (see Exercise 11.3.2). More realistically, however, the manufacturing process may cause variations in the thickness of the string, or the material of which the string is made may contain defects. The statistical study of the effect of such imperfections on the characteristic frequencies, in the large-L limit, was initiated

by Dyson (1953). He considered the particular case of a string consisting of point masses, i.e.,

$$M'(x) = \sum_{j=1}^{\infty} m_j \, \delta \, (x - x_j), \tag{11.2}$$

where the masses $m_j > 0$ and the positions

$$0 =: x_0 < x_1 < x_2 < \cdots \tag{11.3}$$

are *random*. It follows from the definition of the Dirac delta that the differential equation in (11.1) admits solutions that are continuous and piecewise linear with respect to the partition (11.3). More precisely, after setting

$$\lambda := \omega^2$$

we have the recurrence relations

$$\psi'(x_{j+1}-) - \psi'(x_j-) + \lambda m_j \psi(x_j) = 0 \tag{11.4}$$

for $j = 1, 2, \cdots,$ and

$$\psi'(x_{j+1}-) = \frac{\psi(x_{j+1}) - \psi(x_j)}{\ell_j} \tag{11.5}$$

for $j = 0, 1, \cdots,$ where

$$\ell_j := x_{j+1} - x_j. \tag{11.6}$$

Expressed in matrix form, these relations become

$$\begin{pmatrix} \psi'(x_{j+1}-) \\ \psi(x_{j+1}) \end{pmatrix} = A_j \begin{pmatrix} \psi'(x_j-) \\ \psi(x_j) \end{pmatrix}, \tag{11.7}$$

where

$$A_j := \begin{cases} \begin{pmatrix} 1 & 0 \\ \ell_0 & 1 \end{pmatrix} & \text{if } j = 0 \\ \\ \begin{pmatrix} 1 & 0 \\ \ell_j & 1 \end{pmatrix} \begin{pmatrix} 1 & -\lambda m_j \\ 0 & 1 \end{pmatrix} & \text{otherwise} \end{cases} . \tag{11.8}$$

Hence we may express the solution of the *initial-value problem* for the string equation as

$$\begin{pmatrix} \psi'(x_{n+1}-) \\ \psi(x_{n+1}) \end{pmatrix} = \Pi_n A_0 \begin{pmatrix} \psi'(0-) \\ \psi(0) \end{pmatrix}. \tag{11.9}$$

where Π_n is the product of matrices

$$\Pi_n := A_n \cdots A_2 A_1 . \tag{11.10}$$

We will return in section 11.3 to the spectral problem for this string.

Remark 11.2.1. Our notation differs from Dyson's; his string equation is

$$K_j \left(x_{j+1} - x_j \right) + K_{j-1} \left(x_{j-1} - x_j \right) = -m_j \omega^2 x_j .$$

In his notation, the x_j are the positions of particles coupled together by springs that obey Hooke's law, and K_j is the elastic modulus of the spring between the jth and $(j+1)$th particles. The correspondence between this and Eqs (11.4)–(11.5) is

$$x_j \sim \psi(x_j) \quad \text{and} \quad K_j \sim 1/\ell_j .$$

Exercise 11.2.2. For $\omega^2 = \lambda$, let $\psi(\cdot, \lambda)$ be the particular solution of Eq. (11.1) such that $\psi(0, \lambda) = \cos \alpha$ and $\psi'(0-, \lambda) = \sin \alpha$. Show by induction on n that, for the mass density (11.2), $\psi'(x_{n+1}-, \lambda)$ and $\psi(x_{n+1}, \lambda)$ are polynomials of degree n in $(-\lambda)$ and that the leading coefficient of $\psi(x_{n+1}, \lambda)$ is

$$c_n := (\cos \alpha + \ell_0 \sin \alpha) \prod_{j=1}^{n} (\ell_j m_j) .$$

Exercise 11.2.3. *Consider the slightly more general mass distribution*

$$M'(x) = \mu + \sum_{j=1}^{\infty} m_j \, \delta \left(x - x_j \right) , \quad \mu \geq 0 .$$

Show that the solution of the string equation (11.1) is again of the form (11.9) if the A_j are appropriately modified.

11.2.2 The Frisch–Lloyd model

The fundamental equation of quantum mechanics is the time-dependent Schrödinger equation; see, for instance, Texier (2011). For a single particle on the positive half-line, the method of separation of variables leads to the time-independent version

$$-\psi''(x) + V(x)\psi(x) = E\psi(x), \quad x > 0, \tag{11.11}$$

where V is the potential function and E is the energy of the particle.

Consider a potential of the form

$$V(x) = \sum_{j=1}^{\infty} v_j \, \delta(x - x_j) . \tag{11.12}$$

One interpretation for this choice of potential is as follows: there are impurities located at the points x_j of the partition (11.3), and v_j is the 'coupling constant' of the interaction at x_j (Frisch and Lloyd 1960). For such a potential, the general solution of Eq. (11.11) may again be constructed in a piecewise fashion: for $x_j < x < x_{j+1}$ and $E = k^2$ with $k > 0$,

$$\begin{pmatrix} \psi'(x) \\ \psi(x) \end{pmatrix} = \begin{pmatrix} \cos\left[k(x - x_j)\right] & -k\sin\left[k(x - x_j)\right] \\ \sin\left[k(x - x_j)\right]/k & \cos\left[k(x - x_j)\right] \end{pmatrix} \begin{pmatrix} 1 & v_j \\ 0 & 1 \end{pmatrix} \begin{pmatrix} \psi'(x_j-) \\ \psi(x_j-) \end{pmatrix}.$$

Iterating, we deduce that

$$\begin{pmatrix} \psi'(x_{n+1}-) \\ \psi(x_{n+1}-) \end{pmatrix} = \Pi_n A_0 \begin{pmatrix} \psi'(0-) \\ \psi(0) \end{pmatrix},$$

where the matrices in the product Π_n, defined by Eq. (11.10), are now given by

$$A_j := \begin{pmatrix} \cos(k\ell_j) & -k\sin(k\ell_j) \\ \sin(k\ell_j)/k & \cos(k\ell_j) \end{pmatrix} \begin{pmatrix} 1 & v_j \\ 0 & 1 \end{pmatrix} \tag{11.13}$$

and

$$A_0 := \begin{pmatrix} \cos(k\ell_0) & -k\sin(k\ell_0) \\ \sin(k\ell_0)/k & \cos(k\ell_0) \end{pmatrix}.$$

It is useful to comment briefly on the *deterministic case*

$$A_j = A := \begin{pmatrix} \cos(k\ell) & -k\sin(k\ell) \\ \sin(k\ell)/k & \cos(k\ell) \end{pmatrix} \begin{pmatrix} 1 & v \\ 0 & 1 \end{pmatrix} \quad \text{for every } j \in \mathbb{N}.$$

The Frisch–Lloyd model then reduces to the famous Kronig–Penney model which was introduced long ago to analyse the band structure of crystalline materials (Kronig and Penney 1931). The asymptotic behaviour of this product is readily determined by examining the eigenvalues and eigenvectors of A. Since A has unit determinant, the reciprocal of an eigenvalue of A is also an eigenvalue. In particular, if $|\mathrm{Tr}A| > 2$ then A has two distinct real eigenvalues—one of which must exceed unity—and the product grows exponentially with n. On the other hand, if $|\mathrm{Tr}A| < 2$, then the eigenvalues form a complex conjugate pair on the unit circle and there is no growth. In physical terms, this means that, for a periodic system without disorder, the allowed values of the energy are given by those ranges of k for which the inequality

$$|\cos(k\ell) + \sin(k\ell)\, v/(2k)| < 1$$

is satisfied. For other values of the energy, there are no travelling or Bloch-like solutions, so that forbidden gaps in the energy spectrum are formed.

Exercise 11.2.4. *Show that, if one considers instead the case $E = -k^2 < 0$, $k > 0$, then the general solution of the Frisch–Lloyd model is in terms of a product of matrices*

where

$$A_j := \begin{pmatrix} ch(k\ell_j) & k\,sh(k\ell_j) \\ sh(k\ell_j)/k & ch(k\ell_j) \end{pmatrix} \begin{pmatrix} 1 & v_j \\ 0 & 1 \end{pmatrix} . \tag{11.14}$$

11.2.3   The Anderson model

For the mathematical description of the *localization phenomenon*, P. W. Anderson used the following approximation of the stationary Schrödinger equation (Anderson 1958):

$$-\psi_{n+1} + V_n\psi_n - \psi_{n-1} = E\psi_n , \quad n \in \mathbb{N} .$$

This is, in fact, the model that is usually studied in introductions to the theory of disordered systems (Luck 1992). The concepts that underly its mathematical treatment are analogous to those used for our impurity models, but we shall not consider them in what follows (see, however, section 11.12). The principal benefit of restricting our attention to impurity models is that, since they are described by differential (as opposed to difference) equations, one can make full use of the tools of differential calculus.

11.2.4   Further motivations

Products of random matrices are encountered not only in quantum impurity models but also in the classical statistical physics of disordered systems. Consider, for instance, a one-dimensional Ising chain consisting of n spins embedded in an external inhomogeneous magnetic field. The Hamiltonian can be written in the form

$$H(\sigma) := -J \sum_{j=1}^{n} \sigma_j\sigma_{j+1} - \sum_{j=1}^{n} h_j\sigma_j .$$

In this expression, J is the coupling constant and h_j is the local magnetic field at site j which is linearly coupled to the spin $\sigma_j = \pm 1$ there. For a positive (negative) coupling constant, the interaction is ferromagnetic (antiferromagnetic); this favours the alignment or antialignment of neighbouring spins. In terms of the slow/fast dichotomy mentioned in the Introduction, we assume here that the spin variables are the fast variables, and that they are in thermal equilibrium in a specific frozen configuration $\{h_j\}$. The object is then to evaluate the canonical partition function of the frozen system at temperature $\beta = 1/(kT)$:

$$Z_n = \sum_{\sigma} e^{-\beta H(\sigma)} .$$

Physical observables are in principle extracted by taking the thermodynamic limit $n \to \infty$. A remarkable feature of disordered systems is that certain extensive quantities such as the free energy

$$-\frac{kT}{n} \ln Z_n$$

converge as $n \to \infty$ to a nonrandom limit with probability 1. Such quantities are called *self-averaging*; this means that a typical realization of a large enough system gives access to the thermodynamic quantities. The self-averaging property of the free energy is a consequence of the central limit theorem for products of random matrices, and it enables one to express the free energy of the random Ising chain in terms of the growth rate of a certain product.

The easiest way of calculating Z_n is to use the transfer matrix technique. The partition functions $Z_j^{\pm}$, conditioned on the last spin $\sigma_j = 1$ or $\sigma_j = -1$, obey the recurrence relation

$$\begin{pmatrix} Z_j^+ \\ Z_j^- \end{pmatrix} = \begin{pmatrix} e^{\beta(J+h_j)} & e^{\beta(-J+h_j)} \\ e^{\beta(-J-h_j)} & e^{\beta(J-h_j)} \end{pmatrix} \begin{pmatrix} Z_{j-1}^+ \\ Z_{j-1}^- \end{pmatrix}.$$

For a chain of n spins with periodic boundary conditions the partition function takes the form

$$Z_n = \operatorname{Tr} \Pi_n,$$

where the matrices in the product (11.10) are

$$A_j = \begin{pmatrix} e^{\beta(J+h_j)} & e^{\beta(-J+h_j)} \\ e^{\beta(-J-h_j)} & e^{\beta(J-h_j)} \end{pmatrix}.$$

11.3   The spectral problem

In the previous section, we used the method of separation of variables to reduce the solution of some time-dependent models to particular instances of the 'master' system

$$-\psi'' + Q'\psi = \lambda M'\psi, \quad x > 0 \tag{11.15}$$

subject to the condition

$$\sin \alpha \, \psi(0) - \cos \alpha \, \psi'(0) = 0. \tag{11.16}$$

Here $\{Q(x) : x \geq 0\}$ and $\{M(x) : x \geq 0\}$, with M nondecreasing, are two processes, λ is the spectral parameter, and $\alpha \in [0, \pi)$ is some fixed parameter independent of λ. We have chosen to introduce α in order to make explicit the dependence of the spectral quantities on the condition at $x = 0$. The case $\alpha = 0$ corresponds to a Neumann condition, and the case $\alpha = \pi/2$ to a Dirichlet condition.

Every student of physics and mathematics is familiar with the next stage of the method of separation of variables: the object is to find all the values of the spectral parameter for which a nontrivial solution exists, and then to expand the solution of the *time-dependent problem* in terms of these 'eigenfunctions'. In the simplest cases, such as the heat and wave equations with constant coefficients, the eigenfunctions are trigonometric functions; the expansion is then the familiar Fourier series or transform, and there exist corresponding inversion formulae which permit the recovery of the solution from its Fourier coefficients or transform.

Our purpose in this section is to outline the extension of these concepts to the more complicated equation (11.15). In the general case, there is of course no hope of having explicit formulae for the 'eigenvalues' and 'eigenfunctions' of the problem. Nevertheless, we shall see that the necessary information for the construction of Fourier-like expansions is contained in a so-called *spectral measure*, which is itself accessible via an important object called the *Weyl coefficient* of the spectral problem. This beautiful theory, developed by Weyl and Titchmarsh in the deterministic case, brings out the important part played by the asymptotic behaviour of the solutions for large x. Our account is based on Coddington and Levinson (1955), Chapter 9. Some care must be taken when this theory is applied to Dyson's impurity model, because in that case the coefficient M' is not smooth. We shall indicate briefly the necessary adjustments that have been worked out by M. G. Krein and his school (Kac and Krein 1974).

It will be useful to denote by $\psi(\cdot, \lambda)$ and $\varphi(\cdot, \lambda)$ the particular solutions of the differential equation that satisfy the initial conditions

$$\psi(0, \lambda) = -\varphi'(0, \lambda) = \cos \alpha \quad \text{and} \quad \psi'(0, \lambda) = \varphi(0, \lambda) = \sin \alpha. \tag{11.17}$$

In technical terms, the spectral problem is *singular* because the independent variable x runs over an infinite interval. In order to understand the essential features of the singular case, we begin by considering a truncated version: For $L > 0$,

$$- \psi'' + Q'\psi = \lambda M'\psi, \quad 0 < x < L, \tag{11.18}$$

subject to the conditions

$$\sin \alpha \, \psi(0) - \cos \alpha \, \psi'(0) = 0 \quad \text{and} \quad \psi'(L-) = z\,\psi(L). \tag{11.19}$$

The parameter z that is used to specify the boundary at $x = L$ can take any real value (independent of the spectral parameter λ), as well as the "value" ∞; the boundary condition at $x = L$ is then interpreted as $\psi(L) = 0$, i.e. Dirichlet's condition. We shall see that it is very instructive to consider the entire range of this parameter.

To discuss the spectral problem, we introduce the operator

$$\mathscr{L} := \frac{\mathrm{d}x}{\mathrm{d}M}\left[-\frac{\mathrm{d}^2}{\mathrm{d}x^2} + \frac{\mathrm{d}Q}{\mathrm{d}x}\right] \tag{11.20}$$

associated with the problem (11.18)–(11.19). When M is smooth, $\mathscr{L}$ is a classical differential operator. In particular, we then have *Green's identity*:

$$\int_0^L \mathscr{L}u(x)\,\overline{v(x)}\,\mathrm{d}M(x) - \int_0^L u(x)\,\overline{\mathscr{L}v(x)}\,\mathrm{d}M(x)$$
$$= \left[u(x)\,\overline{v'(x)} - u'(x)\,\overline{v(x)}\right]\Big|_{0+}^{L-} \tag{11.21}$$

for every sufficiently smooth complex-valued functions u and v defined on $[0, L]$. This key identity implies, among other things, that the Wronskian of the particular solutions

$\varphi(\cdot, \lambda)$ and $\psi(\cdot, \lambda)$ is identically equal to 1. More importantly, it also follows that $\mathscr{L}$ is *self-adjoint* with respect to the inner product

$$(u, v) := \int_0^L u(x)\, \overline{v(x)}\, dM(x) \tag{11.22}$$

in a suitable space of functions that satisfy the boundary conditions (11.19). We associate with this inner product the norm $\|\cdot\| := \sqrt{(\cdot, \cdot)}$.

For Dyson's model, however, M is merely piecewise constant with respect to the partition (11.3) and the interpretation of Eq. (11.20) requires some clarification. A thorough discussion may be found in Kac and Krein (1974). For our immediate purposes, it will be sufficient to remark that, if we accept (11.20) and proceed formally, then

$$\int_0^L \mathscr{L}u(x)\, \overline{v(x)}\, dM(x) = \int_0^L \overline{v(x)}\, [-du'(x) + u(x)\, dQ(x)] \ .$$

For instance, if u is continuous and piecewise linear with respect to the partition (11.3), the meaning of the integral on the right-hand side is clear; by using the 'integration by parts' formula

$$\int_0^L \overline{v(x)}\, du'(x) = u'(x)\, \overline{v(x)}\Big|_{0+}^{L-} - \int_0^L u'(x)\, \overline{v'(x)}\, dx$$

it may be shown that Green's identity remains valid for Dyson's model.

11.3.1 The spectral measure

The eigenvalues of $\mathscr{L}$ are the zeros of the function

$$\lambda \mapsto \psi'(L-, \lambda) - z\, \psi(L, \lambda)\,.$$

For $z \in \mathbb{R} \cup \{\infty\}$, they are real and simple and may be ordered:

$$-\infty < \lambda_1 < \lambda_2 < \cdots .$$

The eigenfunction corresponding to λ_j is a multiple of $\psi(x, \lambda_j)$. Furthermore, every function f whose norm is finite may be expressed as a 'Fourier' series:

$$f(x) = \sum_j f_j \psi(x, \lambda_j)\,, \quad f_j := \frac{(f, \psi(\cdot, \lambda_j))}{\|\psi(\cdot, \lambda_j)\|^2}\,.$$

This expansion may be expressed neatly in the form

$$f(x) = \int_{\mathbb{R}} \widehat{f}_L(\lambda)\, \psi(x, \lambda)\, d\sigma_L(\lambda), \tag{11.23}$$

where $\widehat{f}_L$ is the 'transform'

$$\widehat{f}_L(\lambda) := \int_0^L f(x)\,\psi(x,\lambda)\,\mathrm{d}M(x) \tag{11.24}$$

and

$$\sigma'_L(\lambda) = \sum_j \|\psi(\cdot,\lambda_j)\|^{-2}\delta(\lambda - \lambda_j)\,. \tag{11.25}$$

The measure σ_L is called the *spectral measure* associated with the truncated problem.

11.3.2   The integrated density of states per unit length

The spectral measure is a complicated object. For the models we have in mind, it is the eigenvalues that may be measured experimentally and, for this reason, physicists are often more interested in the function

$$N_L(\lambda) := \frac{\#\,\{n \in \mathbb{N} : \lambda_n < \lambda\}}{L}\,, \tag{11.26}$$

which simply counts the number of eigenvalues per unit length. Thus, N_L retains only 'half the information' contained in the spectral measure; the norm of the eigenfunctions $\psi(\cdot,\lambda_j)$ has been lost.

The measure N_L may have a weak limit as $L \to \infty$; that is, there may be a measure N such that, for every smooth function η with compact support in $\mathbb{R}$, there holds

$$\int_{-\infty}^{\infty} \eta(\lambda)\,\mathrm{d}N_L(\lambda) \xrightarrow[L\to\infty]{} \int_{-\infty}^{\infty} \eta(\lambda)\,\mathrm{d}N(\lambda)\,.$$

We call this limit measure, if it exists, the *integrated density of states* per unit length.

11.3.3   The Weyl coefficient

The spectral characteristics of the truncated problem may be accessed by considering another particular solution of the differential equation (11.15); it is defined as the linear combination

$$\chi(x,\lambda) = \varphi(x,\lambda) + w_L(\lambda)\psi(x,\lambda), \tag{11.27}$$

where the coefficient $w_L(\lambda)$ is chosen so that

$$\chi'_L(L-,\lambda) = z\,\chi_L(L,\lambda)\,. \tag{11.28}$$

It is then readily seen that

$$w_L(\lambda) = -\frac{\varphi'(L-,\lambda) - z\,\varphi(L,\lambda)}{\psi'(L-,\lambda) - z\,\psi(L,\lambda)}\,. \tag{11.29}$$

The function $\lambda \mapsto w_L(\lambda)$ is called the *Weyl coefficient* associated with the truncated spectral problem. It turns out that the spectral measure σ_L may be recovered from it.

Notation

For a function F of the complex variable $x + iy$, we write that

$$F(x \pm i0) := \lim_{0 < \varepsilon \to 0} F(x \pm i\varepsilon).$$

Theorem 11.3.1.

$$\sigma_L(\lambda) - \sigma_L(\lambda_0) = \frac{1}{\pi} \int_{\lambda_0}^{\lambda} d\zeta \, Im\, w_L(\zeta + i0)$$

at points of continuity λ and λ_0 of σ_L.

A proof may be found in Coddington and Levinson (1955), Section 9.3.

Exercise 11.3.2. *For the particular case $\alpha = \pi/2$, $Q = 0$, and $M' = 1$:*

(a) *Show that*

$$w_L(\lambda) = \frac{z\sqrt{\lambda}\cot\left(\sqrt{\lambda}L\right) + \lambda}{\sqrt{\lambda}\cot\left(\sqrt{\lambda}L\right) - z}.$$

The Weyl coefficient is thus a meromorphic function of λ, with poles at the eigenvalues λ_n.

(b) *For the case $z = \infty$, compute the residue at λ_n. Thus, show by comparison with Eq. (11.25) that Theorem 11.3.1 does indeed hold in this case. Help: In order to evaluate the integral of the imaginary part of Weyl's coefficient, use Cauchy's theorem with the rectangular contour of height $2\varepsilon > 0$ centred on the interval $0 < x < \lambda$. Then let $\varepsilon \to 0$.*

(c) *Show that N_L has a weak limit as $L \to \infty$ given by*

$$N(\lambda) = \frac{1}{\pi}\sqrt{\lambda}, \quad \text{for } \lambda > 0.$$

Exercise 11.3.3. *Show that*

$$w_L(\lambda) = \frac{\frac{\chi'_L(0,\lambda)}{\chi_L(0,\lambda)}\sin\alpha + \cos\alpha}{\sin\alpha - \frac{\chi'_L(0,\lambda)}{\chi_L(0,\lambda)}\cos\alpha}.$$

11.3.4 The Riccati equation

Let us pause for a moment in order to draw attention to the very simple form taken by the Weyl coefficient for our impurity models. Set

$$Z(x) = \frac{\chi'_L(x,\lambda)}{\chi_L(x,\lambda)}.$$

The foregoing exercise shows that the Weyl coefficient may be expressed in terms of $Z(0)$. In particular, we have

$$w_L(\lambda) = Z(0) = \chi'_L(0, \lambda) \text{ for } \alpha = \pi/2.$$

On the other hand, since $\chi_L(\cdot, \lambda)$ solves the truncated equation (11.18), Z is the particular solution of the Riccati equation

$$Z' = -Z^2 + Q' - \lambda M', \quad x > 0, \tag{11.30}$$

which satisfies the condition $Z(L-) = z$.

For our impurity models, we can construct this particular solution by proceeding as follows: we associate with

$$A = \begin{pmatrix} a & b \\ c & d \end{pmatrix} \in \mathrm{SL}(2, \mathbb{R})$$

the linear fractional transformation $\mathcal{A} : \mathbb{C} \cup \{\infty\} \to \mathbb{C} \cup \{\infty\}$ defined by

$$\mathcal{A}(z) = \begin{cases} a/c & \text{if } z = \infty \\ \frac{az+b}{cz+d} & \text{otherwise} \end{cases}. \tag{11.31}$$

These linear fractional transformations form a group for the operation of composition.

Exercise 11.3.4.   *Verify the following formulae:*

(a)
$$A = \begin{pmatrix} 1 & v \\ 0 & 1 \end{pmatrix} \implies \mathcal{A}(z) = v + z.$$

(b)
$$A = \begin{pmatrix} 1 & 0 \\ \ell & 1 \end{pmatrix} \implies \mathcal{A}(z) = \frac{1}{\ell + \dfrac{1}{z}}.$$

(c)
$$A = \begin{pmatrix} 1 & 0 \\ \ell & 1 \end{pmatrix} \begin{pmatrix} 1 & -\lambda m \\ 0 & 1 \end{pmatrix} \implies \mathcal{A}(z) = \frac{1}{\ell + \dfrac{1}{-\lambda m + z}}.$$

(d)
$$A = \begin{pmatrix} \cos(k\ell) & -k\sin(k\ell) \\ \sin(k\ell)/k & \cos(k\ell) \end{pmatrix} \implies \mathcal{A}(z) = \frac{1}{\tau + \dfrac{1 + k^2\tau^2}{-k^2\tau + z}},$$

where $\tau = \tan(k\ell)/k$.

(e)
$$A = \begin{pmatrix} \cos(k\ell) & -k\sin(k\ell) \\ \sin(k\ell)/k & \cos(k\ell) \end{pmatrix} \begin{pmatrix} 1 & v \\ 0 & 1 \end{pmatrix} \implies \mathcal{A}(z) = \frac{1}{\tau + \dfrac{1 + k^2\tau^2}{v - k^2\tau + z}}.$$

Exercise 11.3.5. *In Exercise 11.2.4, we expressed the general solution of the Frisch–Lloyd model for $E = -k^2 < 0$, with $k > 0$, in terms of a product of random matrices where A_j is given by Eq. (11.14). Derive the formula*

$$A_j^{-1}(z) = \frac{z\,\mathrm{ch}(k\ell_j) - k\,\mathrm{sh}(k\ell_j)}{\mathrm{ch}(k\ell_j) - z\,\mathrm{sh}(k\ell_j)/k} - v_j\,,$$

and show that

$$\frac{\partial}{\partial \ell_j} A_j^{-1}(z) = (z^2 - k^2)\frac{dA_j^{-1}(z)}{dz}\,.$$

In terms of the Riccati variable Z, we can therefore express

$$\begin{pmatrix} \psi'(x_{j+1}-) \\ \psi(x_{j+1}) \end{pmatrix} = A_j \begin{pmatrix} \psi'(x_j-) \\ \psi(x_j) \end{pmatrix}$$

as

$$Z(x_j-) = A_j^{-1}\left(Z(x_{j+1}-)\right) \quad \text{for } j \geq 0\,.$$

For the impurity model, the general solution of the Riccati equation is therefore given by

$$Z(x_{n+1}-) = A_n \circ \cdots \circ A_1 \circ A_0(Z(0-)). \tag{11.32}$$

In particular, if $L = x_{n+1}$, then

$$Z(0-) = A_0^{-1} \circ A_1^{-1} \circ \cdots \circ A_n^{-1}(z)\,. \tag{11.33}$$

The right-hand side in this last equation may be written as a finite continued fraction.

Exercise 11.3.6. *Set $L = x_{n+1}$. Show the following:*

(a) For Dyson's string,

$$\frac{\chi_L'(0,\lambda)}{\chi_L(0,\lambda)} = \cfrac{1}{-\ell_0 + \cfrac{1}{\lambda m_1 + \cfrac{1}{-\ell_1 + \cdots + \cfrac{1}{\lambda m_n + \cfrac{1}{-\ell_n + \cfrac{1}{z}}}}}}\,.$$

(b) For the Frisch–Lloyd model,

$$\frac{\chi_L'(0,\lambda)}{\chi_L(0,\lambda)} = \cfrac{1}{-\tau_0 + \cfrac{1 + k^2\tau_0^2}{k^2\tau_0 - v_1 + \cfrac{1}{-\tau_1 + \cdots + \cfrac{1 + k^2\tau_{n-1}^2}{k^2\tau_{n-1} - v_n + \cfrac{1}{-\tau_n + \cfrac{1}{z}}}}}},$$

where $\tau_j := \tan(k\ell_j)/k$.

Remark 11.3.7.   It is clear from the foregoing discussion that

$$\frac{\chi_L'(0,\lambda)}{\chi_L(0,\lambda)}$$

is *independent of* α. Thus, although the Weyl coefficient itself does depend on the boundary condition at $x = 0$, Exercise 11.3.3 shows that the dependence is trivial.

11.3.5   Classification in terms of limit-circle and limit-point types

In order to understand what happens to the Weyl coefficient of the truncated problem as we take the limit $L \to \infty$, it is helpful to allow the parameter z to assume *complex* values.

Notation

$$\mathbb{C}_+ := \{x + iy : x \in \mathbb{R} \text{ and } y > 0\}.$$

$$\overline{\mathbb{C}_+} := \{x + iy : x \in \mathbb{R} \text{ and } y \geq 0\} \cup \{\infty\}.$$

Although the problem consisting of Eq. (11.18) and the boundary conditions (11.19) is no longer self-adjoint when z has a nonzero imaginary part, the particular solutions $\psi(\cdot, \lambda)$, $\varphi(\cdot, \lambda)$ and $\chi_L(\cdot, \lambda)$ remain well defined. We set

$$w = w_L(\lambda) = -\frac{\varphi'(L-,\lambda) - z\,\varphi(L,\lambda)}{\psi'(L-,\lambda) - z\,\psi(L,\lambda)} \tag{11.34}$$

and view w as a function of the complex parameter z.

Exercise 11.3.8.   *Show that*

$$\operatorname{Im} \lambda \, \|\chi_L(\cdot,\lambda)\|^2 = \operatorname{Im} w - \operatorname{Im} z \, |\chi_L(L,\lambda)|^2 .$$

Help: *Use Green's identity.*

This last equation says

$$\mathrm{Im}\,\lambda \int_0^L |\varphi(x,\lambda) + w\,\psi(x,\lambda)|^2 \; \mathrm{d}M(x) - \mathrm{Im}\,w$$
$$= -\mathrm{Im}\,z \; |\varphi(L,\lambda) + w\,\psi(L,\lambda)|^2 \; . \tag{11.35}$$

In particular,

$$\mathrm{Im}\,\lambda \int_0^L |\varphi(x,\lambda) + w\,\psi(x,\lambda)|^2 \; \mathrm{d}M(x) - \mathrm{Im}\,w = 0 \;\; \text{for } z \in \mathbb{R} \cup \{\infty\}\,.$$

We claim that this is the equation of a *circle*, say ∂D_L, in the complex w-plane. To see this, we first rewrite Eq. (11.34) in the form

$$z = \frac{\varphi'(L-,\lambda) + w\,\psi'(L-,\lambda)}{\varphi(L,\lambda) + w\,\psi(L,\lambda)}\;.$$

Then the equation $\mathrm{Im}\,z = 0$ in the complex z-plane corresponds to the following equation in the complex w-plane

$$\mathrm{Im}\left[(\varphi' + w\,\psi')\,(\overline{\varphi} + \overline{w}\,\overline{\psi})\right] = 0,$$

where, for convenience, we momentarily omit to make explicit the dependence of φ and ψ on L and λ. An equivalent form of this equation is

$$(\varphi' + w\,\psi')\,(\overline{\varphi} + \overline{w}\,\overline{\psi}) = (\overline{\varphi'} + \overline{w}\,\overline{\psi'})\,(\varphi + w\,\psi)\,.$$

This, in turn, may be expressed as

$$|w - c_L|^2 = R_L^2,$$

where

$$c_L := \frac{\overline{\psi'(L-,\lambda)}\varphi(L,\lambda) - \varphi'(L-,\lambda)\overline{\psi(L,\lambda)}}{\psi'(L-,\lambda)\overline{\psi(L,\lambda)} - \overline{\psi'(L-,\lambda)}\psi(L,\lambda)} \tag{11.36}$$

and

$$R_L := \left|\frac{\psi'(L-,\lambda)\varphi(L,\lambda) - \varphi'(L-,\lambda)\psi(L,\lambda)}{\psi'(L-,\lambda)\overline{\psi(L,\lambda)} - \overline{\psi'(L-,\lambda)}\psi(L,\lambda)}\right|\,. \tag{11.37}$$

Hence our claim.

Exercise 11.3.9. *Show that*

$$\frac{1}{R_L} = 2\,|\mathrm{Im}\,\lambda| \int_0^L |\psi(x,\lambda)|^2 \; \mathrm{d}M(x)\,.$$

Help: *Use Green's identity to evaluate the numerator and the denominator on the right-hand side of Eq. (11.37).*

For definiteness, let us from now on assume that $\lambda \in \mathbb{C}_+$, and denote by D_L the open disk in the w-plane that has ∂D_L as its boundary. We remark that $w \in D_L$ if and only if

$$\operatorname{Im} \lambda \int_0^L |\varphi(x, \lambda) + w \, \psi(x, \lambda)|^2 \, \mathrm{d}M(x) - \operatorname{Im} w < 0 \,.$$

Let us show that

$$L' \ge L \implies D_{L'} \subseteq D_L \,.$$

Indeed, if $L' \ge L$ and $w \in D_{L'}$ then

$$0 > \operatorname{Im} \lambda \int_0^{L'} |\varphi(x, \lambda) + w \, \psi(x, \lambda)|^2 \, \mathrm{d}M(x) - \operatorname{Im} w$$

$$\ge \operatorname{Im} \lambda \int_0^L |\varphi(x, \lambda) + w \, \psi(x, \lambda)|^2 \, \mathrm{d}M(x) - \operatorname{Im} w \,.$$

The geometrical implication is that $\{D_L\}_{L>0}$ is a *nested family* of disks, so that the limits

$$R := \lim_{L \to \infty} R_L \quad \text{and} \quad \partial D := \lim_{L \to \infty} \partial D_L$$

are well defined.

Two possibilities therefore arise as $L \to \infty$:

1. ∂D is a circle of radius $R > 0$. In this case, we say that the point at infinity is of *limit-circle* type. This implies among other things that the solution

$$\varphi(\cdot, \lambda) + w \, \psi(\cdot, \lambda)$$

belongs to $L_M^2(0, \infty)$ for every $w \in D$. One can therefore assert that *every* solution belongs to $L_M^2(0, \infty)$.

2. $R = 0$ and ∂D consists of a single point:

$$w(\lambda) := \lim_{L \to \infty} w_L(\lambda) \,. \tag{11.38}$$

In this case, we say that the point at infinity is of *limit-point* type. Importantly, the limit (11.38) is then *independent of* $z \in \overline{\mathbb{C}}$. The function

$$\chi(\cdot, \lambda) := \varphi(\cdot, \lambda) + w(\lambda) \, \psi(\cdot, \lambda) \tag{11.39}$$

is, up to a factor, the *only* solution of Eq. (11.15) that belongs to $L_M^2(0, \infty)$.

Now that we have explained this classification, let us return to the spectral problem for the infinite interval. For our particular impurity models, we make the assumptions

$$\begin{cases} \lim_{j\to\infty} x_j = \infty \\ \\ \{m_j\}_{j\in\mathbb{N}} \subset \mathbb{R}_+ \text{ (Dyson) or } \{v_j\}_{j\in\mathbb{N}} \subset \mathbb{R}_+ \text{ (Frisch–Lloyd).} \end{cases} \tag{11.40}$$

Under this assumption, it is known that the point at infinity is of limit-point type when $\text{Im}\,\lambda > 0$; see Kac and Krein (1974), section 12, for Dyson's string, and Coddington and Levinson (1955), Chapter 9, Theorem 2.4, for the Frisch–Lloyd model. Whatever $z \in \mathbb{R} \cup \{\infty\}$ one chooses to specify the boundary condition at $x = L$ in the truncated problem, the large-L limit will correspond to imposing the integrability condition

$$\psi \in L^2_M(0, \infty). \tag{11.41}$$

In particular, there is a limit measure σ such that

$$\int_{\mathbb{R}} \eta(\lambda)\,d\sigma_L(\lambda) \xrightarrow[L\to\infty]{} \int_{\mathbb{R}} \eta(\lambda)\,d\sigma(\lambda)$$

for every smooth function η with compact support. The measure σ is called the spectral measure of the singular problem (11.15); we denote its support by Σ.

The spectral measure can, as in the truncated case, be recovered from the Weyl coefficient: Theorem 11.3.1 remains valid in the limit as $L \to \infty$ and so we have

$$\sigma(\lambda) - \sigma(\lambda_0) = \frac{1}{\pi} \int_{\lambda_0}^{\lambda} d\zeta\, \text{Im}\, w(\zeta + i0) \tag{11.42}$$

at points of continuity λ and λ_0 of σ.

Exercise 11.3.10. *Consider the case $Q' = 0$ and $M' = 1$.*

(a) For $\lambda = -k^2$ with $k > 0$, show that

$$w(\lambda) = \frac{\cos\alpha - k\sin\alpha}{k\cos\alpha + \sin\alpha}.$$

(b) Deduce that

$$\sigma'(\lambda) = \frac{1}{\pi} \frac{\sqrt{\lambda}}{\sin^2\alpha + \lambda\cos^2\alpha}, \quad \lambda > 0.$$

As the exercise illustrates, unlike σ_L, the spectral measure σ need not consist only of point masses. In general, one has the decomposition

$$\text{supp}\,\sigma =: \Sigma = \Sigma_p \cup \Sigma_a \cup \Sigma_s, \tag{11.43}$$

where the sets on the right-hand side are, respectively, the *point*, the *absolutely continuous*, and the *singular continuous* spectrum. The point spectrum contains the eigenvalues, i.e., the values of λ for which there is a nonzero solution of Eq. (11.15) that is normalizable. The absolutely continuous spectrum consists of the values of λ for which the derivative σ' exists in the usual sense. The singular continuous spectrum is what is left after Σ_p and Σ_a have been removed. It is generally difficult to prove that a measure is singularly continuous; an example of a measure which is conjectured to be such is given in section 11.12, Example 11.12.6.

One further important consequence of the foregoing discussion is that, for the impurity models we have considered, the limit-point case arises whenever a certain continued fraction (see Exercise 11.3.6) converges as $n \to \infty$ to a limit independent of $z \in \overline{\mathbb{C}}$. This yields 'explicit' formulae for the Weyl coefficient:

$$\frac{\chi'(0,\lambda)}{\chi(0,\lambda)} = \cfrac{1}{-\ell_0 + \cfrac{1}{\lambda m_1 + \cfrac{1}{-\ell_1 + \cdots}}} \quad \text{(Dyson)} \tag{11.44}$$

and

$$\frac{\chi'(0,\lambda)}{\chi(0,\lambda)} = \cfrac{1}{-\tau_0 + \cfrac{1+k^2\tau_0^2}{k^2\tau_0 - v_1 + \cfrac{1}{-\tau_1 + \cdots}}} \quad \text{(Frisch–Lloyd)}, \tag{11.45}$$

where $\tau_j := \tan(k\ell_j)/k$.

Remark 11.3.11.   The continued fraction (11.44) is essentially of the type studied by Stieltjes (1894, 1895) in connection with the solution of the moment problem for a measure supported on $\mathbb{R}_+$.

Remark 11.3.12.   The spectrum depends on M and Q, and so in the random case, it will depend on the particular realization of those processes. Nevertheless, for the disordered models that we consider here, it may be shown (see Pastur (1980)) that the spectrum is the same for almost every realization. We refer to this common, non-random spectrum, as *the* spectrum, and use the letter Σ to denote it.

Remark 11.3.13.   The Weyl coefficient w is an analytic function of λ in the upper half-plane. It may, however, have simple poles on the real line. For instance, in Example 11.3.10, w has a simple pole at $\lambda = -\tan^2 \alpha$ unless $\alpha = 0$ or $\pi/2$.

11.3.6   Dyson's disordered strings of Type I

Let us now give an example of a random string for which the distribution of the Weyl coefficient may be found explicitly.

In his paper, Dyson discusses two kinds of disorder, which he calls 'Type I' and 'Type II'. Strings of the latter type are obtained by choosing the m_j and ℓ_j independently; such strings will be studied later in this chapter. Strings of Type I are constructed as follows: let $c_0, c_1, \ldots$ be positive random variables that are independent and identically distributed. Set

$$\ell_0 := \frac{1}{c_0}, \quad m_1 := \frac{c_0}{c_1}, \quad \ell_1 := \frac{c_1}{c_0\,c_2}, \quad m_2 := \frac{c_0 c_2}{c_1 c_3}, \quad \text{etc.}$$

With this definition, the ℓ_j and the m_j are obviously *dependent*.

Exercise 11.3.14. *Set $\alpha = \pi/2$. Show that, for $\lambda \notin \mathbb{R}_+$, the Weyl coefficient of the Type I string is given by*

$$\frac{w(\lambda)}{\lambda} = Y\left(-1/\lambda\right),$$

where

$$Y(t) := \cfrac{c_0 t}{1 + \cfrac{c_1 t}{1 + \cfrac{c_2 t}{1 + \ddots}}}.$$

The method devised by Dyson for finding the integrated density of states of the random string uses this continued fraction. He found the distribution of $Y(t)$ when $t > 0$ and the c_j have a gamma distribution with a certain choice of the parameters.

Notation

We say that the positive random variable X has a Gamma(p, q) distribution if its probability density is of the form

$$\frac{1}{q^p \Gamma(p)} x^{p-1} \exp\left(-x/q\right) \mathbf{1}_{(0,\infty)}(x)$$

for some $p, q > 0$. We say that the random variable Y has a Kummer(p, q, r) distribution if its probability density is of the form

$$\frac{1}{\Gamma(p)\Psi(p, 1-q; r)} y^{p-1}(1+y)^{-p-q} e^{-ry} \mathbf{1}_{(0,\infty)}(y)$$

for some $p > 0$, $q \in \mathbb{R}$ and $r > 0$, where Ψ denotes the Kummer function. We use the $\sim$ symbol to indicate the law of a random variable, i.e.,

$$X \sim \text{Gamma}(p, q) \quad \text{and} \quad Y \sim \text{Kummer}(p, q, r).$$

We state here Letac's generalization of Dyson's result:

$$X \sim \text{Gamma}(p, 1/r) \text{ and } Y \sim \text{Kummer}(p, q, r)$$

$$\implies \frac{X}{1+Y} \sim \text{Kummer}(p+q, -q, r),$$

where it is assumed that X and Y are independent. A proof may be found in Letac (2009).

Notation

Let X and Y be two random variables. The notation

$$X \overset{\text{law}}{=} Y$$

means that X and Y have the same distribution.
    Letac's result therefore implies, in particular, that

$$X \sim \text{Gamma}(p, 1/r) \text{ and } Y \sim \text{Kummer}(p, 0, r) \implies Y \overset{\text{law}}{=} \frac{X}{1+Y}.$$

Hence,

$$c_j \sim \text{Gamma}(p, q) \implies \frac{w(\lambda)}{\lambda} \sim \text{Kummer}(p, 0, -\lambda/q). \tag{11.46}$$

11.4   The complex Lyapunov exponent

For $\text{Im}\,\lambda > 0$, the particular solutions $\psi(\cdot, \lambda)$ and $\chi(\cdot, \lambda)$ form a basis for the solution space of the differential equation (11.15), and so every solution $\psi(x)$ is a linear combination of them:

$$\psi(x) = a\,\psi(x, \lambda) + b\,\chi(x, \lambda).$$

In the limit point case, the only solutions that belong to $L_M^2(0, \infty)$ are those for which $a = 0$. The other solutions do not decay sufficiently quickly as $x \to \infty$ and, for them, we have

$$\psi(x) \sim a\,\psi(x, \lambda) \text{ as } x \to \infty.$$

In particular, if $|\psi(\cdot, \lambda)|$ grows exponentially, then $|\psi|$ will grow *at the same rate.*

Definition 11.4.1.   *The (real) Lyapunov exponent of the system (11.15) is*

$$\gamma(\lambda) := \lim_{L \to \infty} \mathbb{E}\left(\frac{\ln|\psi(L, \lambda)|}{L}\right), \quad \text{Im}\,\lambda > 0, \tag{11.47}$$

where the expectation is over Q and M.

In words, the Lyapunov exponent measures the expected rate of exponential growth of the particular solution $\psi(\cdot, \lambda)$, and hence of every solution not proportional to $\chi(\cdot, \lambda)$.

Remark 11.4.2. Although the particular solution $\psi(\cdot, \lambda)$ defined by (11.17) depends on the (real) parameter α which specifies the boundary condition at $x = 0$ for the spectral problem (11.15), it is clear from the foregoing discussion that the Lyapunov exponent itself is, for $\operatorname{Im} \lambda > 0$, independent of α.

Remark 11.4.3. For our disordered model, it will turn out that the limit

$$\lim_{L \to \infty} \frac{\ln |\psi(L, \lambda)|}{L},$$

whose expectation appears in the definition of the Lyapunov exponent, is the same for almost every realization of the disorder. Thus, we have here another example of a random variable that is, in the terminology of section 11.2.4, *self-averaging*. The Lyapunov exponent and the integrated density of states are self-averaging; *by contrast, the spectral measure is not*. This may be explained heuristically by the fact that, as pointed out earlier, the spectral measure of the truncated problem contains information about the norms of the eigenfunctions $\psi(\cdot, \lambda_j)$.

Remark 11.4.4. We emphasize that the Lyapunov exponent is *not defined in the spectrum*, where the function $\psi(\cdot, \lambda)$ has its zeros. Nevertheless, the limit $\gamma(\lambda + i0)$, $\lambda \in \Sigma$, is meaningful, and contains valuable spectral information. We state without proof the following fact: *intervals where $\gamma(\lambda + i0)$ does not vanish cannot belong to the absolutely continuous spectrum*. In particular, it may be shown (Kotani 1982) that

$$\forall \lambda \in \Sigma, \ \gamma(\lambda + i0) > 0 \implies \Sigma_{ac} = \emptyset.$$

Heuristically, if the absolutely continuous spectrum is empty, then one might expect the spectrum to be purely punctual; the eigenfunctions must decay sufficiently rapidly to belong to $L_M^2(0, \infty)$. More concretely, one may think of $\gamma(\lambda + i0)$ as the rate of (exponential) decay of the eigenfunction corresponding to λ, and of the reciprocal as a measure of the spatial extent of the eigenfunction's support. In the literature on disordered systems, one often refers to $1/\gamma$ as the *localization length*.

It will be helpful to work with the following 'complex version' of the Lyapunov exponent.

Definition 11.4.5. *The function*

$$\Omega(\lambda) := \lim_{L \to \infty} \mathbb{E} \left(\frac{\ln \psi(L, \lambda)}{L} \right) \tag{11.48}$$

is called the complex Lyapunov exponent *of the system (11.15).*

Obviously,

$$\gamma = \operatorname{Re} \Omega .$$

11.4.1   The imaginary part

Let us now discuss the significance of the imaginary part; for convenience, we set $\alpha = \pi/2$. From the result of Exercise 11.2.2, we deduce that, for Dyson's model,

$$\psi(x_{n+1}, \lambda) = c_n \, (\lambda_1 - \lambda) \cdots (\lambda_n - \lambda) . \tag{11.49}$$

The λ_j are, therefore, the eigenvalues of Dyson's problem truncated to the interval $[0, L]$ with $L = x_{n+1}$ and $z = \infty$.

Let $\lambda > 0$, and suppose that it does not equal any of the λ_j. By virtue of Eq. (11.49), we then have

$$\ln \psi(x_{n+1}, \lambda + i0) = \ln c_n + \sum_{\lambda_j < \lambda} (-i\pi + \ln |\lambda_j - \lambda|) + \sum_{\lambda_j > \lambda} \ln |\lambda_j - \lambda|$$

$$= \ln |\psi(x_{n+1}, \lambda)| - i\pi \sum_{\lambda_j < \lambda} 1 .$$

We deduce

$$\ln \psi(x_{n+1}, \lambda + i0) = -i\pi \, x_{n+1} N_{x_{n+1}}(\lambda) + \ln |\psi(x_{n+1}, \lambda)|$$

where $N_L(\lambda)$ is the counting measure defined by Eq. (11.26). Divide both sides by x_{n+1}, and take the limit as $n \to \infty$. By definition of the complex Lyapunov exponent and of the integrated density of states, we obtain

$$\operatorname{Im} \Omega(\lambda + i0) = -\pi N(\lambda) . \tag{11.50}$$

For λ 'just above' the spectrum, the imaginary part of the complex Lyapunov exponent therefore yields the integrated density of states that Dyson set out to calculate.

Formula (11.50) remains valid for the Frisch–Lloyd model, but the proof is more complicated. A justification will be given later in section 11.5.

Remark 11.4.6.   As will become manifest in section 11.5, the integrated density of states per unit length $N(\lambda)$ does not depend on the boundary condition at $x = 0$— and hence on the parameter α. This fact, together with our earlier comments, makes it clear that the spectral information contained in the complex Lyapunov exponent Ω is much coarser than that contained in the Weyl coefficient w.

11.4.2   Some deterministic examples

Ω has the advantage of being *analytic* outside the spectrum, except at points where the Weyl coefficient happens to have poles; this important observation, which goes back to Dyson himself (Dyson 1953) and has been exploited by others (see, for instance, Luck (1992)), can greatly facilitate its calculation.

Example 11.4.7. We illustrate this point by revisiting the simple but instructive case where $Q' = 0$ and $M' = 1$. In this (deterministic) case, $\Sigma \subset \mathbb{R}_+$, and so the complex Lyapunov exponent is analytic along the negative half-line. Let us first compute Ω there: setting $\lambda = -k^2 < 0$, with $k > 0$, in Eq. (11.15) gives

$$-\psi'' = -k^2\psi \,.$$

Therefore,

$$\psi(x, -k^2) = \cos\alpha \,\mathrm{ch}(kx) + \frac{\sin\alpha}{k}\,\mathrm{sh}(kx)$$

and so, for $k \neq -\tan\alpha$,

$$\Omega(-k^2) = \lim_{L\to\infty} \frac{\ln\psi(L, -k^2)}{L} = k \,.$$

By using

$$k = \sqrt{-\lambda}$$

we can continue this formula for the complex Lyapunov exponent to other values of the spectral parameter. In particular, for $\lambda > 0$, we find that

$$\Omega(\lambda + \mathrm{i}0) = -\mathrm{i}\sqrt{\lambda} \,.$$

Hence, with the help of Formula (11.50), we easily recover the result obtained previously in Exercise 11.3.2 (c). Furthermore,

$$\gamma(\lambda + \mathrm{i}0) = \mathrm{Re}\,\Omega(\lambda + \mathrm{i}0) = 0 \,.$$

The fact that, in this example, the real Lyapunov exponent vanishes along the spectrum is consistent with the fact that the spectral measure is absolutely continuous. Another less trivial case where this result can be verified is that of the Kronig–Penney model discussed in section 11.2.2. The band spectrum Σ is characterized by

$$\Sigma = \{\lambda \in \mathbb{R} : \gamma(\lambda + \mathrm{i}0) = 0\} = \{k^2 \in \mathbb{R} : |\cos(k\ell) + \sin(k\ell)\,v/(2k)| < 1\} \,.$$

We will see that, by contrast, in the *disordered* case, $\gamma(\lambda+\mathrm{i}0)$ is strictly positive everywhere in the spectrum; this is the mathematical manifestation, in the one-dimensional case, of the phenomenon known as *Anderson localization*.

Example 11.4.8. Consider a deterministic homogeneous string such that all the masses equal m and the spacing between two successive impurities equals ℓ. Let us calculate Ω explicitly for $\lambda < 0$.

In this case, the general solution ψ of Eq. (11.15) is a continuous function of x that is piecewise linear with respect to the partition

$$0 = x_0 < x_1 < \dots, \quad \text{where } x_j = j\ell \,.$$

The values taken at the x_j satisfy the recurrence relation

$$\psi(x_{j+1}) - 2\psi(x_j) + \psi(x_{j-1}) = -\lambda m\ell\, \psi(x_j)\,, \quad j = 1, 2, \ldots.$$

Hence,

$$\psi(x_j) = a\,\mathrm{ch}(\Omega x_j) + b\,\mathrm{sh}(\Omega x_j),$$

where

$$e^{\pm\Omega\ell} = 1 - \frac{\lambda m\ell}{2} \pm \sqrt{\left(1 - \frac{\lambda m\ell}{2}\right)^2 - 1}\,.$$

From this, we readily deduce exact expressions for the particular solutions $\psi(x, \lambda)$ and for $\chi(x, \lambda)$: for instance, in the particular case of a Dirichlet boundary condition at $x = 0$, i.e., $\alpha = \pi/2$, we have

$$\chi(x, \lambda) = \frac{x_j - x}{\ell}e^{-\Omega x_{j-1}} + \frac{x - x_{j-1}}{\ell}e^{-\Omega x_j} \quad \text{for } x_{j-1} \leq x \leq x_j$$

and

$$\psi(x, \lambda) = \frac{\ell}{\mathrm{sh}(\Omega\ell)}\left[\frac{x_j - x}{\ell}\mathrm{sh}(\Omega x_{j-1}) + \frac{x - x_{j-1}}{\ell}\mathrm{sh}(\Omega x_j)\right] \quad \text{for } x_{j-1} \leq x \leq x_j\,.$$

Therefore,

$$\lim_{x\to\infty} \frac{\ln\psi(x, \lambda)}{x} = \Omega(\lambda) = \frac{1}{\ell}\ln\left[1 - \frac{\lambda m\ell}{2} + \sqrt{\left(1 - \frac{\lambda m\ell}{2}\right)^2 - 1}\right]$$

and

$$w(\lambda) = \chi'(0, \lambda) = -\frac{1 - e^{-\Omega\ell}}{\ell}\,.$$

Exercise 11.4.9. *Consider the homogeneous string with $\alpha = \pi/2$.*

(a) Show that its spectral measure is supported on the interval $0 < \lambda < 4/(m\ell)$, where it is given by

$$\sigma'(\lambda) = \frac{1}{\pi\ell}\sqrt{1 - (1 - \lambda m\ell/2)^2}\,.$$

(b) Show that, for $0 < \lambda < 4/(m\ell)$,

$$N(\lambda) = \frac{1}{\pi\ell}\arccos\left(1 - \frac{\lambda m\ell}{2}\right).$$

11.4.3   A disordered example: Kotani's Type II string

We remark that, in Example 11.4.7, Ω is the negative of the Weyl coefficient corresponding to a Dirichlet condition at $x = 0$. Example 11.4.8 shows that this relationship between the complex Lyapunov exponent and the Weyl coefficient does not hold in general. Nevertheless, we proceed to give an example which suggests that a connection between these two objects does exist if the system is disordered.

Let

$$Q' = 0 \text{ and } M' = \sum_{j=1}^{\infty} m_j \delta(x - x_j).$$

Kotani (1976), in his Example 2, section 5, considered the case of a random string of Type II where the m_j and the ℓ_j are two sequences of independent random variables with

$$m_j \sim \text{Gamma}(1, m) \text{ and } \ell_j \sim \text{Gamma}(1, \ell). \tag{11.51}$$

By a completely rigorous calculation, he found that

$$N(\lambda) = \frac{m\lambda/\pi^2}{J_1^2\left(\frac{2}{\sqrt{m\lambda\ell}}\right) + Y_1^2\left(\frac{2}{\sqrt{m\lambda\ell}}\right)}. \tag{11.52}$$

Our purpose in what follows is to show that the right-hand side is also the expected value of the spectral density σ' for the particular choice $\alpha = \pi/2$.

As mentioned already, it is useful to work outside the spectrum. We let $\lambda < 0$ and $\alpha = \pi/2$; the continued fraction (11.44) for $-w(\lambda)$ is then clearly a *positive* random variable.

Notation

We say that the positive random variable X has a $\text{GIG}(p, q, r)$ (generalized inverse Gaussian) distribution if its probability density function is of the form

$$\frac{(q/r)^{p/2}}{2K_p(\sqrt{qr})} x^{p-1} \exp\left[-\frac{1}{2}(qx + r/x)\right] \mathbf{1}_{(0,\infty)}(x)$$

for some nonnegative numbers p, q, and r.

Letac and Seshadri (1983) showed the following: let a, b and p be positive numbers, and let

$$\Gamma_1 \sim \text{Gamma}(p, 2/b), \quad \Gamma_2 \sim \text{Gamma}(p, 2/a), \text{ and } X \sim \text{GIG}(-p, a, b)$$

be three mutually independent random variables. Then

$$X \stackrel{\text{law}}{=} \frac{1}{\Gamma_1 + \dfrac{1}{\Gamma_2 + X}}. \tag{11.53}$$

When we apply this result to the random string of Type II (11.51), we find that

$$-w(\lambda) \sim \mathrm{GIG}\,(-p, a, b) \quad \text{with } p = 1,\ a = \frac{-2}{m\lambda} \text{ and } b = \frac{2}{\ell}.$$

Exercise 11.4.10.   *Show that*

$$\mathbb{E}\left[-w(\lambda)\right] = \left(-\frac{m\lambda}{\ell}\right)^{\frac{1}{2}} \frac{K_0\left(\frac{2}{\sqrt{-m\lambda\ell}}\right)}{K_1\left(\frac{2}{\sqrt{-m\lambda\ell}}\right)} \quad \text{for } \lambda < 0.$$

Next, we use analytic continuation in λ: for $-\pi/2 \le \arg z \le \pi$,

$$K_\nu(z) = -\frac{1}{2}\pi \mathrm{i}\, e^{-\mathrm{i}\pi\nu/2} H_\nu^{(2)}\left(z e^{-\mathrm{i}\pi/2}\right).$$

Therefore,

$$\mathbb{E}\left[-w(\lambda + \mathrm{i}0)\right] = \left(\frac{m\lambda}{\ell}\right)^{\frac{1}{2}} \frac{H_0^{(2)}\left(\frac{2}{\sqrt{m\lambda\ell}}\right)}{H_1^{(2)}\left(\frac{2}{\sqrt{m\lambda\ell}}\right)} \quad \text{for } \lambda > 0.$$

Now,

$$2\,\mathrm{Im}\, \frac{H_0^{(2)}(z)}{H_1^{(2)}(z)} = 2\,\mathrm{Im}\, \frac{H_0^{(2)}(z)H_1^{(1)}(z)}{\left|H_1^{(1)}(z)\right|^2} = \mathrm{Im}\frac{H_0^{(2)}(z)H_1^{(1)}(z) - H_0^{(1)}(z)H_1^{(2)}(z)}{\left|H_1^{(1)}(z)\right|^2}.$$

Equation 10.5.5, of NIST's Digital Library of Mathematical Functions, says that

$$H_0^{(2)}(z)H_1^{(1)}(z) - H_0^{(1)}(z)H_1^{(2)}(z) = -\frac{4\mathrm{i}}{\pi z}.$$

Hence, after some re-arrangement, we obtain, for $\lambda > 0$,

$$\frac{1}{\pi}\mathbb{E}\left[w(\lambda + \mathrm{i}0)\right] = \frac{m\lambda}{\pi^2} \frac{1}{\left|H_1^{(1)}(z)\right|^2}.$$

The right-hand side is the same as in Kotani's formula (11.52). Hence, for $\lambda > 0$,

$$\mathbb{E}\left[w(\lambda + \mathrm{i}0)\right] = -\mathrm{Im}\,\Omega(\lambda + \mathrm{i}0) \quad \text{for } \lambda > 0.$$

Theorem 11.3.1 then implies that

$$\mathbb{E}\left[\sigma'(\lambda)\right] = N(\lambda). \tag{11.54}$$

We will, in section 11.4.5, show that this relationship holds more generally provided that certain conditions are satisfied.

Exercise 11.4.11. *Show that, for Kotani's string,*

$$N(\lambda) \sim \left(\frac{m}{\ell}\right)^{\frac{1}{2}} \frac{\sqrt{\lambda}}{\pi} \quad \text{as } \lambda \to 0+ .$$

This expresses the fact that, for a large wavelength, the density of states approaches the density of states corresponding to a homogeneous string (for which $M' > 0$ is independent of x).

11.4.4 Calculation of the Lyapunov exponent

Just as was the case for the Weyl coefficient, the calculation of the complex Lyapunov exponent makes use of the Riccati equation (11.30) introduced earlier. This time, however, we are interested in the particular solution

$$Z(x, \lambda) := \frac{\psi'(x, \lambda)}{\psi(x, \lambda)}, \tag{11.55}$$

which we will refer to as the *Riccati process*.

Suppose that the A_j, $j = 1, 2, \ldots$, are independent and with the same distribution μ.

By applying Formula (11.32) for the general solution of the Riccati equation, we see that the random variables $\{Z(x_j-, \lambda)\}_{j \in \mathbb{N}}$ form a Markov chain:

$$Z(x_{j+1}-, \lambda) = A_j \left(Z(x_j-, \lambda)\right) \quad \text{for } j = 1, 2, \ldots . \tag{11.56}$$

This Markov chain has a stationary distribution; we denote the stationary probability measure by ν. When λ is *real*, the support of ν is contained in $\mathbb{R}$; more generally, it is contained in the complex plane.

Exercise 11.4.12. *Show that, if ν has a density supported on the real line, say*

$$d\nu(z) = f(z) \, dz ,$$

then

$$f(z) = \mathbb{E}\left([f \circ A^{-1}](z) \frac{d A^{-1}}{dz}(z) \right), \tag{11.57}$$

where the expectation is over the μ-distributed matrix random variable A.

This equation for the stationary density $f(z)$ of the Riccati process is known in the physics literature as the *Dyson–Schmidt equation*. There is no systematic method for solving it. Nevertheless, there are cases where the solution has been obtained explicitly. It will be instructive in what follows to consider also its integrated version:

$$\mathbb{E}\left(\int_{z}^{A^{-1}(z)} dt \, f(t) \right) = J . \tag{11.58}$$

We will see in due course that the constant of integration J has a physical interpretation.

Returning to the calculation of the complex Lyapunov exponent, we have

$$\frac{1}{L}\ln\frac{\psi(L,\lambda)}{\psi(a,\lambda)} = \frac{1}{L}\int_a^L Z(x,\lambda)\,\mathrm{d}x,$$

where $a > 0$ is an arbitrary point.

We make the hypothesis that the Riccati process behaves ergodically.

Concretely, this means that we assume that

$$\Omega(\lambda) = \lim_{L\to\infty}\frac{1}{L}\int_a^L Z(x,\lambda)\,\mathrm{d}x = \int_{\mathrm{supp}\,\nu} z\,\mathrm{d}\nu(z)\quad\text{almost surely.}\qquad(11.59)$$

11.4.5 The relationship between $\Omega(\lambda)$ and $w(\lambda)$

In order to clarify the relationship between the complex Lyapunov exponent and the Weyl coefficient, we introduce, with Letac (1986), the so-called *backward iterates* $\{Z_{n+1}\}$ associated with the (forward) Markov chain $\{Z(x_n-,\lambda)\}_{n\in\mathbb{N}}$:

$$Z_{n+1} := \mathcal{A}_1 \circ \mathcal{A}_2 \circ \cdots \circ \mathcal{A}_n(Z(x_1-,\lambda)).\qquad(11.60)$$

Since, by assumption, the $\mathcal{A}_j$ are independent and μ-distributed, Formula (11.32) implies that

$$Z_{n+1} \overset{\text{law}}{=} \mathcal{A}_n \circ \mathcal{A}_{n-1} \circ \cdots \circ \mathcal{A}_1(Z(x_1-,\lambda)) = Z(x_{n+1}-,\lambda).\qquad(11.61)$$

We have for these backward iterates the following counterpart of Exercise 11.3.6.

Exercise 11.4.13. *Show the following:*

(a) For Dyson's string,

$$Z_{n+1} = \cfrac{1}{\ell_1 + \cfrac{1}{-\lambda m_1 + \cdots + \cfrac{1}{\ell_n + \cfrac{1}{-\lambda m_n + Z(x_1-,\lambda)}}}}.$$

(b) For the Frisch–Lloyd model,

$$Z_{n+1} = \cfrac{1}{\tau_1 + \cfrac{1+k^2\tau_1^2}{v_1 - k^2\tau_1 + \cdots + \cfrac{1}{\tau_n + \cfrac{1+k^2\tau_n^2}{v_n - k^2\tau_n + Z(x_1-,\lambda)}}}},$$

where $\tau_j := \tan(k\ell_j)/k$.

For every fixed n, the backward and forward iterates have the same law, but their large-n behaviours are very different; see Fig. 11.3 for an illustration. The $Z(x_n-, \lambda)$ behave ergodically. By contrast, the Z_n converge almost surely to a random limit:

$$Z_\infty := \lim_{n \to \infty} \mathcal{A}_1 \circ \mathcal{A}_2 \circ \cdots \circ \mathcal{A}_n(Z(x_1-, \lambda)).\tag{11.62}$$

This limit is an infinite continued fraction independent of $Z(x_1-, \lambda)$:

$$Z_\infty = \cfrac{1}{\ell_1 + \cfrac{1}{-\lambda m_1 + \cfrac{1}{\ell_2 + \cfrac{1}{-\lambda m_2 + \ddots}}}} \qquad \text{(Dyson)}\tag{11.63}$$

and

$$Z_\infty = \cfrac{1}{\tau_1 + \cfrac{1 + k^2 \tau_1^2}{v_1 - k^2 \tau_1 + \cfrac{1}{\tau_2 + \cfrac{1 + k^2 \tau_2^2}{v_2 - k^2 \tau_2 + \ddots}}}} \qquad \text{(Frisch–Lloyd)}.\tag{11.64}$$

Now, by Eq. (11.61), the law of Z_∞ is the stationary distribution ν of the Markov chain. Hence Formula (11.59) may be expressed alternatively as

$$\Omega = \mathbb{E}(Z_\infty).\tag{11.65}$$

Since the ℓ_j are independent and identically distributed, as are the m_j and the v_j, we deduce upon comparison with the continued fractions in Eqs (11.44) and (11.45) that

$$Z_\infty \overset{\text{law}}{=} -\frac{\chi'_L(0, \lambda)}{\chi_L(0, \lambda)}.\tag{11.66}$$

For both of these impurity models, we therefore have

$$\Omega(\lambda) = -\mathbb{E}\left[\frac{\chi'_L(0, \lambda)}{\chi_L(0, \lambda)}\right].\tag{11.67}$$

This takes a particularly simple form when $\alpha = \pi/2$; we may then use Theorem 11.3.1 and take the expectation to deduce that

$$\mathbb{E}(\sigma') = -\frac{1}{\pi} \text{Im}\,\Omega(\lambda + i0).$$

For Dyson's string, the right-hand side is the integrated density of states, and so Eq. (11.54), obtained in the context of a specific random string, holds more generally.

Remark 11.4.14. We emphasize that the equality in law (11.66) is a nontrivial fact. For our particular impurity models, it relies on the 'i.i.d.' nature of the coefficients that appear in the continued fraction expansions, and also on the hypothesis that the Riccati process enjoys the ergodic property (11.59).

11.4.6 Application to the Frisch–Lloyd model

For the Frisch–Lloyd model, we do not have 'ready-made' formulae for the distribution of the random continued fractions (11.45) and (11.64). We must, therefore, work with the Riccati process, taking as our starting point the Dyson–Schmidt equation (11.57) for the unknown probability density f. If we assume that the coupling constants v_j are positive then the spectrum is contained in $\mathbb{R}_+$, and it will again be convenient to work on the negative half-line $\lambda = -k^2 < 0$, where the Lyapunov exponent is analytic.

11.4.7 Notation

For the sake of economy, we will in what follows use the same symbols ℓ_j and v_j to denote random variables or variables of integration.

In view of the result obtained in Exercise 11.3.5, we can write the Dyson–Schmidt equation in the form

$$(z^2 - k^2) f(z) = \int_{\text{supp}\,\mu} d\mu\, f\left(\mathcal{A}_j^{-1}(z)\right) \frac{\partial \mathcal{A}_j^{-1}(z)}{\partial \ell_j}. \tag{11.68}$$

We now consider the special case where the ℓ_j are independent with a common exponential distribution with mean ℓ, i.e., for every measurable subset $S \subseteq \mathbb{R}_+$,

$$\mathbb{P}\left(\ell_j \in S\right) = \int_S \frac{1}{\ell} e^{-x/\ell}\, dx.$$

$1/\ell$ is therefore the mean density of the impurities. We suppose also that the v_j are mutually independent random variables, independent also of the ℓ_j, with a common distribution whose density we will denote by ρ.

Exercise 11.4.15. *Set*

$$\Phi(\ell_j, v_j, z) := \int_{z-v_j}^{\mathcal{A}_j^{-1}(z)} dt\, f(t).$$

Show the following:

(a)

$$(z^2 - k^2) f(z) = \int_{\text{supp}\rho} dv_j\, \rho(v_j) \int_0^\infty d\ell_j\, \frac{1}{\ell}\, e^{-\ell_j/\ell} \frac{\partial \Phi}{\partial \ell_j}.$$

(b)

$$(z^2 - k^2)f(z) = \frac{1}{\ell} \mathbb{E}\left[\Phi(\ell_j, v_j, z)\right].$$

(c) For every $z > 0$,

$$(z^2 - k^2)f(z) + \frac{1}{\ell} \int_{supp\rho} dv_j\, \rho(v_j) \int_z^{z - v_j} dt\, f(t) = \frac{J}{\ell}. \tag{11.69}$$

(d) Deduce that, if Z_∞ has a mean, then $J = 0$.

The upshot of the exercise is that we have reduced the Dyson–Schmidt equation to an equation that is *linear*. The presence of an integral term makes it difficult to solve directly, but Frisch and Lloyd (1960) observed that the problem simplifies if one works with the *Laplace transform*

$$F(p) := \int_0^\infty f(z)\, e^{-pz}\, dz. \tag{11.70}$$

Exercise 11.4.16. *Show that, for $\lambda = -k^2 < 0$,*

$$F'' - k^2 F + \frac{1}{\ell} \mathbb{E}\left(\frac{e^{-pv_j} - 1}{p}\right) F = 0. \tag{11.71}$$

In terms of F, we have

$$\Omega(-k^2) = -F'(0+)$$

and so the problem reduces to finding a solution of a *linear homogeneous differential equation* that is positive and decays to 0 as $p \to \infty$.

Example 11.4.17. Let the v_j be independent and exponentially-distributed with mean $v > 0$, i.e.,

$$\rho(x) = \frac{1}{v} e^{-x/v} \mathbf{1}_{(0,\infty)}(x).$$

In this case, Equation (11.4.16) becomes

$$F'' - k^2 F - \frac{1/\ell}{p + 1/v} F = 0.$$

The general solution may be expressed in terms of the Whittaker functions; see NIST's Digital Library of Mathematical Functions, section 13.14. The only decaying solutions are those proportional to

$$W_{\frac{-1}{2k\ell}, \frac{1}{2}}\left(2k(p + 1/v)\right).$$

Hence,

$$
\Omega(\lambda) = -2\sqrt{-\lambda}\,\frac{W'_{\frac{-1}{2\ell\sqrt{-\lambda}},\frac{1}{2}}\left(2/v\,\sqrt{-\lambda}\right)}{W_{\frac{-1}{2\ell\sqrt{-\lambda}},\frac{1}{2}}\left(2/v\,\sqrt{-\lambda}\right)} \quad \text{for } \lambda < 0.
$$

Analytic continuation then yields

$$
\Omega(\lambda + \mathrm{i}0) = 2\mathrm{i}\sqrt{\lambda}\,\frac{W'_{\frac{-\mathrm{i}}{2\ell\sqrt{\lambda}},\frac{1}{2}}\left(-2\mathrm{i}\sqrt{\lambda}/v\right)}{W_{\frac{-\mathrm{i}}{2\ell\sqrt{\lambda}},\frac{1}{2}}\left(-2\mathrm{i}\sqrt{\lambda}/v\right)} \quad \text{for } \lambda > 0. \tag{11.72}
$$

This formula for the characteristic function was discovered by Nieuwenhuizen (1983).

11.5 Further remarks on the Frisch–Lloyd model

For Dyson's string, we were able to show that the imaginary part of the complex Lyapunov exponent yields the integrated density of states, i.e.,

$$
-\frac{1}{\pi}\operatorname{Im}\Omega(\lambda + \mathrm{i}0) = N(\lambda) := \lim_{L\to\infty} N_L(\lambda),
$$

where N_L is the counting measure

$$
N_L(\lambda) := \frac{\#\{n \in \mathbb{N} : \lambda_n < \lambda\}}{L}.
$$

For the Frisch–Lloyd model, the function $\psi(L, \lambda)$ is *not* a polynomial in λ, and the elementary approach used for Dyson's string is inapplicable. We will instead proceed in three steps:

1. First, in section 11.5.1, we point out the close relationship between the zeros of $x \mapsto \psi(x, \lambda)$ and those of $\lambda \mapsto \psi(x, \lambda)$.
2. Secondly, in section 11.5.2, we relate the distribution of these zeros to the tail behaviour of the stationary probability density of the Riccati variable.
3. Finally, the connection with the complex Lyapunov exponent is made in section 11.5.3.

For the sake of brevity, we restrict our attention to the Frisch–Lloyd model, i.e., $Q' = V$ and $M' = 1$ and, for the truncated problem, will consider only the case where the boundary condition at $x = L$ is Dirichlet's, i.e., $z = \infty$.

11.5.1 Prüfer variables: the phase formalism

Let

$$
-\infty < \lambda_1 < \lambda_2 < \cdots
$$

denote the eigenvalues of the spectral problem for (11.18) subject to the boundary conditions (11.19) with $z = \infty$. For $\lambda \in \mathbb{R}$, define

$$\tilde{N}_L(\lambda) := \frac{\#\{x \in (0, L) : \psi(x, \lambda) = 0\}}{L}$$

so that $L\tilde{N}_L(\lambda)$ is the number of zeros of the function

$$x \mapsto \psi(x, \lambda)$$

in $(0, L)$. Then $\lambda \mapsto L\tilde{N}_L(\lambda)$ is a nondecreasing function and

$$L\tilde{N}_L(\lambda_n) = n.$$

These facts are particular applications of *Sturm's oscillation and comparison theorems*; see Theorems 1.2 and 2.1 in Coddington and Levinson (1955), Chapter 8.

Exercise 11.5.1. *Deduce from the theorem that, for every $\lambda \in \mathbb{R}$,*

$$\left| \tilde{N}_L(\lambda) - N_L(\lambda) \right| \leq \frac{1}{L}.$$

The upshot is that we can study the large-L limit of N_L by counting the zeros of the function $x \mapsto \psi(x, \lambda)$ for $\lambda = k^2$ positive. The evolution of this function between two neighbouring impurities has a very simple 'phase plane' interpretation:

- On the interval (x_j, x_{j+1}) the evolution is free and the vector $(\psi, k\psi')$ undergoes a rotation of random angle $k\ell_j$.
- Across an impurity, the wave function is continuous but its derivative makes a jump of random height $\psi'(x_j + 0, \lambda) - \psi'(x_j - 0, \lambda) = v_j \psi(x_j, \lambda)$.

A typical trajectory is shown in Fig. 11.1, which illustrates two important features of the case $\lambda > 0$: first, the trajectory winds around the origin and, second, it expands exponentially. This motivates the parametrization

$$\frac{1}{\sqrt{\lambda}} \psi'(x, \lambda) + i\,\psi(x, \lambda) = \varrho(x, \lambda)\, e^{i\theta(x, \lambda)} \tag{11.73}$$

of the solution $\psi(\cdot, \lambda)$ and its derivative. Mathematicians sometimes refer to the real variables ϱ and θ as the *Prüfer variables*; in terms of these variables, the differential equation for ψ, namely

$$-\psi'' + V(x)\psi = k^2\psi,$$

becomes

$$\theta' = k - \frac{V(x)}{k} \sin^2 \theta \tag{11.74}$$

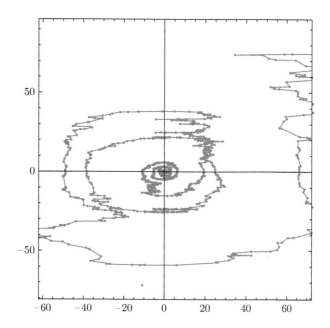

Fig. 11.1 Typical phase-plane trajectory for the Schrödinger equation with a white noise potential and $E = k^2 > 0$.

$$\frac{\varrho'}{\varrho} = \frac{V(x)}{2k} \sin(2\theta).$$ (11.75)

Benderskii and Pastur (1970) showed that, for a very large class of processes Q,

$$\frac{\theta(L, \lambda)}{L} \xrightarrow[L \to \infty]{} \pi N(\lambda)$$ (11.76)

almost surely.

The physical meaning of Eq. (11.74) is quite clear: it describes the motion of a classical rotator under the action of a constant drift term (depending on the energy) and a random term which reduces or increases the drift depending on the sign of the potential. This is consistent with the heuristic idea that valleys in the random potential will create new states and thus increase the density of states. Equation (11.75) expresses the local growth rate of the wave function. The same equations also arise in the context of dynamical systems perturbed by a noise—a context in which their physical interpretation is even clearer. The general problem is to ascertain the effect of a stochastic perturbation on a time-independent integrable Hamiltonian (Hansel and Luciani 1989, Tessieri and Izrailev 2000, Mallick and Marcq 2002). One of the simplest questions concerns the stability of the system: what will happen for large times? Let us address this question in the case of a random-frequency oscillator with the Hamiltonian

$$H(q, p, t) = \frac{1}{2}p^2 + \frac{1}{2}k^2 q^2 - \frac{1}{2}q^2 V(t) \,.$$

The classical equation of motion coincides with the Schrödinger equation after the substitution $(x, \psi) \to (t, q)$. By disregarding the terms that do not contribute to the large-time limit, it is easy to see that the Lyapunov exponent provides a measure of the asymptotic growth rate of the system's energy:

$$\gamma(k^2) = \lim_{t \to \infty} \frac{1}{t} \ln \left[p^2(t) + k^2 q^2(t) \right] \,.$$

Thus, positivity of the Lyapunov exponent implies that the total energy stored in the system grows exponentially. One can pursue the analysis a little further and view Eqs (11.74)–(11.75) as equations satisfied by the action-angle variables $I = \sqrt{\varrho}$ and θ of the system, where

$$\frac{p}{\sqrt{k}} := \sqrt{2I} \cos \theta \quad \text{and} \quad \sqrt{k} \, q := \sqrt{2I} \sin \theta \,.$$

In the analysis of this problem, one expects that what really matters is not the total phase but the reduced phase

$$\pi \left\{ \theta(x, \lambda) \right\} \in [0, \pi),$$

where $\{a\}$ denotes the fractional part of the number a.

11.5.2 Riccati analysis: a qualitative picture

It will be helpful in what follows to think of the spatial variable x as 'time', so that the zeros of $x \mapsto \psi(x, \lambda)$ then correspond to 'times' at which the Riccati variable

$$Z(x, \lambda) := \frac{\psi'(x, \lambda)}{\psi(x, \lambda)}$$

blows up. For the Frisch–Lloyd model, $Z(\cdot, \lambda)$ is a particular solution of

$$Z' = -Z^2 - \lambda + V \,.$$

Between impurities, $V = 0$, and the Riccati equation is an autonomous system describing the motion of a fictitious particle constrained to roll along the potential curve

$$U(z) = \frac{z^3}{3} + \lambda z$$

in such a way that its 'velocity' at 'time' x and 'position' Z is given by the slope $-U'(z)$ (see Fig. 11.2):

$$Z' = -U'(Z), \quad x \notin \{x_j\}_{j \in \mathbb{N}} \,. \tag{11.77}$$

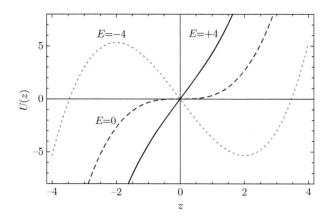

Fig. 11.2 The 'potential" $U(z)$ associated with the unperturbed Riccati equation $z' = -U'(z) = -(z^2 + E)$.

When the particle hits an impurity, say x_j, the Ricatti variable makes a jump

$$Z(x_j+) - Z(x_j-) = v_j .$$

We may regard the occurence of these jumps as a perturbation of the autonomous system (11.77), and the mean density of impurities as the perturbation parameter.

Let us consider first the unperturbed system. For $\lambda = k^2 > 0$, the system has no equilibrium point: the particle rolls down to $-\infty$, and reappears immediately at $+\infty$, reflecting the fact that the solution $\psi(\cdot, \lambda)$ of the corresponding Schrödinger equation has a zero at the 'time' x when the particle escapes to infinity. This behaviour of the Riccati variable indicates that every positive value of λ belongs to the spectrum.

Let us now turn to the case $\lambda = -k^2 < 0$, $k > 0$. In this case, the unperturbed system has an unstable equilibrium point at $-k$, and a stable equilibrium point at k. Unless the particle starts at the unstable equilibrium, it must tend asymptotically to the stable equilibrium point. The fact that the particle cannot reach infinity more than once indicates that the spectrum lies entirely in $\mathbb{R}_+$. The solution of the Frisch–Lloyd equation is

$$f = \delta(z - k) .$$

Let us now consider how the occurence of jumps can affect the system. For $\lambda > 0$, the jumps, as long as they are finite and infrequent (i.e., a small density of impurities), cannot prevent the particle from visiting $-\infty$ repeatedly; the system should therefore behave in much the same way as in the unperturbed case. The situation for $\lambda = -k^2 < 0$ is more complicated. Roughly speaking, *positive* jumps, i.e., discontinuous increases of z, enable the particle to make excursions to the right of the stable equilibrium point $z = k$, but the particle can never overcome the infinite barrier and so it rolls back down towards k. On the other hand, *negative* jumps, i.e., discontinuous decreases of z, enable the particle to make excursions to the left of k. If the jump is large enough,

the particle can overcome the potential barrier at $-k$ and escape to $-\infty$, raising the possibility that part of the spectrum of the Schrödinger operator lies in $\mathbb{R}_-$.

11.5.3 The Rice formula

There remains to relate the zeros of $x \mapsto \psi(x, \lambda)$ to the complex Lyapunov exponent.

In the absence of impurities, Eq. (11.77) gives the velocity of the fictitious particle as a function of its position. Hence the time, say τ, taken to go from $+\infty$ to $-\infty$ is

$$\tau = -\int_{+\infty}^{-\infty} \frac{dz}{z^2 + k^2} = \frac{\pi}{k}.$$

The fact that one 'particle' is transferred from ∞ to $-\infty$ in time τ means that there is a net current equal to k/π. Of course, τ is also the distance between two consecutive zeroes of $\psi(\cdot, \lambda)$. From the Sturm oscillation theorem, we deduce that the (free) integrated density of states per unit length is

$$N(\lambda) = \frac{1}{\tau} = \frac{k}{\pi}.$$

Now, the solution of the Frisch–Lloyd equation for $\lambda = k^2 > 0$ without impurities is the Cauchy density

$$f(z) = \frac{k/\pi}{z^2 + k^2}.$$

Hence,

$$N(\lambda) = \lim_{|z| \to \infty} z^2 f(z). \tag{11.78}$$

This relationship between the integrated density of states and the tail of the stationary density of the Riccati variable is a particular instance of the *Rice formula* (Rice 1944) and remains valid in the presence of impurities.

The Rice formula makes it clear that, in the spectrum, the stationary distribution of the Riccati variable does not have a mean. Nevertheless, the Cauchy principal value

$$\fint_{\mathbb{R}} zf(z)\,dz := \lim_{a \to \infty} \int_{-a}^{a} zf(z)\,dz$$

exists, and it can be shown that

$$\mathrm{Re}\,\Omega\,(\lambda + i0) = \fint_{\mathbb{R}} zf(z)\,dz, \quad \text{for } \lambda > 0. \tag{11.79}$$

Exercise 11.5.2. *Verify the validity of Eq. (11.79) in the absence of impurities.*

Let us revisit the concrete example discussed earlier in section 11.4.6, and show how to exploit the Rice formula to connect N with Ω when impurities are present.

For $\lambda = k^2 > 0$, Eq. (11.69) must be replaced by

$$(z^2 + k^2)f(z) + \frac{1}{\ell} \int_{\mathrm{supp}\rho} \mathrm{d}v_j\, \rho(v_j) \int_z^{z-v_j} \mathrm{d}t\, f(t) = \frac{J}{\ell},$$

where J is again the integration constant in Eq. (11.58). In particular, if we let $z \to \infty$, the Rice formula implies that

$$J = \ell N\,. \tag{11.80}$$

Hence J measures the current of 'Riccati particles'.

The heuristic reasoning that follows borrows heavily from Halperin (1965). Since, for $\lambda = k^2 > 0$, the density f is supported on the entire real line, we must now work with the *Fourier transform*

$$F(q) := \int_{\mathbb{R}} f(z)\, \mathrm{e}^{-iqz}\, \mathrm{d}z \tag{11.81}$$

of the unknown stationary density. For $\lambda \in \mathbb{C}_+$, f decays sufficiently fast as $|z| \to \infty$ to ensure that we can permute integration with respect to z and differentiation with respect to q; this gives

$$iF'(0) = \int_{\mathbb{R}} z f(z)\, \mathrm{d}z \quad \text{for } \lambda \in \mathbb{C}_+\,. \tag{11.82}$$

By contrast, for $\lambda = k^2$, $f(z)$ decays like $1/z^2$ and so F is not differentiable at $q = 0$. Nevertheless, the following calculation makes clear that $F'(0+)$ and $F'(0-)$ do exist. Indeed, we have for the unknown transform F a counterpart of Eq. (11.71):

$$\frac{\mathrm{d}^2 F}{\mathrm{d}q^2} + \frac{1}{\ell}\, \mathbb{E}\left(\frac{1 - \mathrm{e}^{-iqv_j}}{iq}\right) F = \lambda F - 2\pi N(\lambda)\, \delta(q)\,. \tag{11.83}$$

Integrate this equation over q from $-\varepsilon$ to ε and let $\varepsilon \to 0+$; there comes

$$F'(0+) - F'(0-) = -2\pi N(\lambda)\,. \tag{11.84}$$

Now, from

$$F(-q) = \overline{F(q)}\,, \quad q \in \mathbb{R}\,,$$

we deduce that

$$iF'(0-) = \overline{iF'(0+)}\,.$$

Reporting this in Eq. (11.84), we find that

$$\mathrm{Im}\left[iF'(0\pm)\right] = \mp \pi N(\lambda)\,. \tag{11.85}$$

On the other hand,

$$\mathrm{Re}\,[\mathrm{i}F'(0+)] = \mathrm{Re}\,[\mathrm{i}F'(0-)] = \oint_{\mathbb{R}} z f(z)\,\mathrm{d}z = \gamma(k^2)\,. \tag{11.86}$$

Combining Eqs (11.85) and (11.86) yields

$$\mathrm{i}F'(0\pm) = \gamma(k^2) \mp \mathrm{i}\pi N(k^2)$$

and when we compare this with Eq. (11.81), we see that the equality

$$\Omega(k^2 \pm \mathrm{i}0) = \mathrm{i}F'(0\pm) \tag{11.87}$$

is plausible. In particular,

$$-\frac{1}{\pi}\,\mathrm{Im}\,\Omega(k^2 + \mathrm{i}0) = N(k^2)\,.$$

11.6 The white noise limit

We have seen that the Frisch–Lloyd model can sometimes be 'solved' explicitly when the integral equation for the stationary density of the Riccati variable can be transformed into an ordinary differential equation. The question of the extent to which this can be done for an arbitrary discrete model has been the subject of recent investigations that we will not discuss here. Our goal is more modest: we will focus on a particular limit known as the *white noise limit*. It describes situations in which the impurities are densely packed along the real line whilst the variance of the coupling constants is small—a scenario which suggests a description in terms of a continuous model. We will show that the limit model is one in which the potential is a Gaussian process with delta correlations.

11.6.1 Lévy processes

Let us write, as before,

$$V(x) = Q'(x) := \sum_{j=1}^{\infty} v_j \delta(x - x_j)\,, \tag{11.88}$$

and consider the particular case where the spacings ℓ_j between successive impurities are exponentially distributed, independent random variables, and where the v_j are also identically distributed random variables with a common probability density ρ. This defines a random process

$$Q(x) := \int_0^x V(t)\,\mathrm{d}t\,. \tag{11.89}$$

The simplest case corresponds to taking all the v_j equal to 1. Then

$$Q(x) = n(x) := \#\{j \in \mathbb{N} : x_j < x\}$$

is the number of impurities contained in the interval $(0, x)$. For our choice of ℓ_j, this is the familiar *Poisson process* of intensity $1/\ell$, which has the following properties (Feller 1971):

1. The process starts at 0, i.e.,

$$Q(0) = 0 \,.$$

2. For every realization of the process, the function

$$x \mapsto Q(x)$$

   is right-continuous, and the limit

$$\lim_{y \to x-} Q(y)$$

   exists for every $x > 0$.
3. The increments are stationary, i.e., for every $x, y \geq 0$,

$$Q(x + y) - Q(x) \overset{\text{law}}{=} Q(y) \,.$$

4. The increments are independent, i.e., for every $0 \leq u < x \leq y < z$,

$$Q(x) - Q(u) \quad \text{and} \quad Q(z) - Q(y)$$

   are independent.

A process with these properties is called a *Lévy process*. For other distributions of the v_j, the impurity potential (11.88) can be expressed in the form

$$Q(x) = \sum_{j=1}^{n(x)} v_j \,. \tag{11.90}$$

Q is then called a *compound Poisson Process* and it retains Properties 1–4. There are other Lévy processes which are *not* of the form (11.90); see Applebaum (2004). The best-known example is *Brownian motion*: this is a continuous Gaussian stochastic process on the positive half-line such that, for every $x, y \geq 0$, there holds

$$\mathbb{E}\left[Q(x)\right] = 0 \quad \text{and} \quad \mathbb{E}\left[Q(x)Q(y)\right] = \min\{x, y\} \,.$$

We have thus embedded our impurity potential in a larger class which includes, in particular, the Gaussian white noise potential

$$V := \sqrt{\sigma} B' \tag{11.91}$$

with $\sigma > 0$. There remains to explain how, and in what sense, V may be approximated by potentials of the form (11.90).

By exploiting the properties satisfied by every Lévy process Q, it may be shown that

$$\mathbb{E}\left[e^{i\theta Q(x)}\right] = \exp\left[x\Lambda(\theta)\right],$$

where

$$\Lambda(\theta) := \lim_{h\to 0} \mathbb{E}\left[\frac{e^{i\theta Q(h)} - 1}{h}\right]. \tag{11.92}$$

Λ is called the *Lévy exponent* of Q and completely characterizes it.

Example 11.6.1. Let us compute the Lévy exponent of the compound Poisson process corresponding to the impurity potential (11.88) or, what is the same, (11.90). It is easy to show that

$$\mathbb{P}\left(n(x) = j\right) = e^{-x/\ell}\frac{(x/\ell)^j}{j!}.$$

Then

$$\mathbb{E}\left(e^{i\theta Q(x)}\right) = \sum_{j=0}^{\infty} \mathbb{E}\left(e^{i\theta Q(x)}\,\middle|\, n(x) = j\right)\mathbb{P}\left(n(x) = j\right)$$

$$= \sum_{j=0}^{\infty} \mathbb{E}\left(e^{i\theta \sum_{i=1}^{j} v_i}\right) e^{-x/\ell}\frac{(x/\ell)^j}{j!} = e^{-x/\ell}\sum_{j=0}^{\infty}\mathbb{E}\left(e^{i\theta v_1}\right)^j\frac{(x/\ell)^j}{j!}$$

$$= \exp\left\{x\frac{\mathbb{E}\left(e^{i\theta v_1}\right) - 1}{\ell}\right\}.$$

Hence

$$\Lambda(\theta) = \frac{\mathbb{E}\left(e^{i\theta v_1}\right) - 1}{\ell}.$$

Exercise 11.6.2. *Consider the special case where the probability density of the v_j is given by*

$$\rho(x) = \frac{a}{2}e^{-a|x|}, \quad a > 0. \tag{11.93}$$

(a) *What are the mean and the variance of the v_j?*
(b) *Work out the Lévy exponent.*
(c) *Let $\sigma > 0$. Show that*

$$\lim_{\substack{\ell\to 0,\,a\to\infty \\ \ell a^2 = 2/\sigma}} \Lambda(\theta) = -\frac{\sigma\theta^2}{2}. \tag{11.94}$$

(d) *Show that the right-hand side of (11.94) is the Lévy exponent of $\sqrt{\sigma}B$, where B is a standard Brownian motion.*

Help: *Use the fact that, for every $x > 0$, $B(x)$ is a Gaussian random variable of mean 0 and variance x.*

For the particular impurity model in this exercise, we thus have

$$V \rightharpoonup \sqrt{\sigma} B' \,. \tag{11.95}$$

This weak limit is in the sense of convergence of the Lévy exponent. Although in the exercise we considered a specific choice of distribution for the coupling constants, the conclusion holds for a large class of models; what really matters is the scaling condition that $\mathbb{E}(v_j^2)/\ell$ tend to a nonzero finite value.

11.6.2   The Lyapunov exponent

Let us now turn to the calculation of the complex Lyapunov exponent of the continuum limit. We begin by remarking that Eq. (11.83) for the Fourier transform of the invariant probability density $f(z)$ of the Riccati variable may be expressed in terms of the Lévy exponent as follows:

$$F'' - \frac{\Lambda(-q)}{iq} F = k^2 F - 2\pi N(k^2)\, \delta(q) \,.$$

In particular, for the white noise potential $V = \sqrt{\sigma} B'$, this becomes

$$F'' - i\frac{\sigma}{2} q F = k^2 F \,, \ q > 0 \,. \tag{11.96}$$

This equation can easily be reduced to the Airy equation (Abramowitz and Stegun 1964). The Fourier transform of f is proportional to the solution that decays to 0 as $q \to \infty$. Hence

$$F(q) = c \left[\mathrm{Ai} \left(-k^2/\xi^2 - i\xi q \right) - i\, \mathrm{Bi} \left(-k^2/\xi^2 - i\xi q \right) \right], \quad \xi := (\sigma/2)^{\frac{1}{3}} \,.$$

By the same reasoning as in section 11.5.3, we eventually obtain

$$\Omega(k^2 + i0) = \xi \, \frac{\mathrm{Ai}' \left(-k^2/\xi^2 \right) - i\, \mathrm{Bi}' \left(-k^2/\xi^2 \right)}{\mathrm{Ai} \left(-k^2/\xi^2 \right) - i\, \mathrm{Bi} \left(-k^2/\xi^2 \right)} \,. \tag{11.97}$$

This is a minor extension of a calculation first performed by Halperin (1965).

Exercise 11.6.3.   *Deduce that, for the white noise potential $V = \sqrt{\sigma} B'$, the integrated density of states is given by*

$$N(k^2) = \frac{\xi/\pi^2}{\mathrm{Ai}^2 \left(-k^2/\xi^2 \right) + \mathrm{Bi}^2 \left(-k^2/\xi^2 \right)} \,, \quad \xi := (\sigma/2)^{\frac{1}{3}} \,.$$

One can also find a more elementary expression for N by working with the stationary density of the Riccati variable. Indeed, applying the inverse Fourier transform to Eq. (11.96) yields

$$\frac{\sigma}{2} f'(z) + (z^2 + k^2) f(z) = N(k^2).$$

Hence,

$$f(z) = \frac{2N(k^2)}{\sigma} \exp\left[-\frac{2}{\sigma}\left(\frac{z^3}{3} + k^2 z\right)\right] \int_{-\infty}^{z} dt \exp\left[\frac{2}{\sigma}\left(\frac{t^3}{3} + k^2 t\right)\right]. \qquad (11.98)$$

Exercise 11.6.4. *Deduce the integral formula*

$$\frac{1}{N(k^2)} = \sqrt{2\pi/\sigma} \int_{0}^{\infty} \frac{du}{\sqrt{u}} \exp\left[-\frac{2}{\sigma}\left(\frac{u^3}{12} + k^2 u\right)\right]. \qquad (11.99)$$

Help: *Integrate Eq. (11.98) over z. This results in two nested integrals; by making a judicious substitution for one of the variables, one integral can be done explicitly.*

11.6.3   Riccati analysis

Let us indicate briefly how this result can also be derived by a Riccati analysis in the spirit of section 11.5.2. Sturm's oscillation theorem states that the density of states is obtained by counting the number of zeros of the wave function, which is also the number of times the process $\{Z(x)\}_{x\geq 0}$ goes to $-\infty$. Denote these random times by τ_j, so that

$$\tau_1 = \inf\{x > 0 : Z(x) = -\infty\}$$
$$\tau_2 = \inf\{x > \tau_1 : Z(x) = -\infty\}$$
$$\dots$$
$$\tau_n = \inf\{x > \tau_{n-1} : Z(x) = -\infty\}.$$

The integrated density of states is

$$N(E) = \lim_{L\to\infty} \frac{1}{L} \#\{j : \tau_j \leq L\} = \lim_{n\to\infty} \frac{n}{\tau_n}.$$

Therefore, if we assume that the process $\{Z(x)\}_{x\geq 0}$ is ergodic, one finds that

$$N(E) = \frac{1}{\mathbb{E}_{\infty}(\tau_1)}, \qquad (11.100)$$

where

$$\mathbb{E}_z(\tau_1) = \mathbb{E}\left(\tau_1 \,\middle|\, Z(0) = z\right).$$

It may be shown by using the tools of stochastic calculus that the Laplace transform

$$h_p(z) := \mathbb{E}_z\left(e^{-p\tau_1}\right)$$

is the recessive solution of the equation

$$\left[\frac{\sigma}{2}\frac{d^2}{dz^2} - (z^2 + k^2)\frac{d}{dz}\right] u = pu$$

that satisfies the condition $u(z) = 1$ as $z \to -\infty$. By using the obvious integrating factor, we can reformulate the differential equation as an integral equation and incorporate the condition at $-\infty$:

$$h_p(z) = 1 - \frac{2p}{\sigma}\int_{-\infty}^{z} dx\, e^{\frac{2}{\sigma}\left(\frac{x^3}{3}+k^2 x\right)}\int_{x}^{\infty} dy\, e^{-\frac{2}{\sigma}\left(\frac{y^3}{3}+k^2 y\right)} h_p(y). \qquad (11.101)$$

Exercise 11.6.5.   *Use Eq. (11.101) to recover Formula (11.99).*

11.7   Lifshitz tails and Lifshitz singularities

The density of states and the Lyapunov exponent often display interesting behaviour at the bottom of the spectrum. For instance, if the potential is a white noise and is therefore unbounded one can probe the region $E \to -\infty$ by a saddle approximation of (11.99):

$$\ln N(E) \underset{E \to -\infty}{\sim} -\frac{8}{3\sigma}|E|^{3/2}. \qquad (11.102)$$

This limit behaviour is called a *Lifshitz tail*. Because the behaviour is not analytic in E, one says that N has a *Lifshitz singularity* at $-\infty$. It may be explained by a very simple physical argument that involves the Riccati process (Jona–Lasinio 1983). As shown in section 11.5.2, for $E = -k^2 < 0$, the effective potential $U(z)$ develops a local minimum at $z = -k$ and a local maximum at $z = k$; this creates a potential barrier of height

$$\Delta U = U(-k) - U(k) = \frac{4k^3}{3}.$$

The expected hitting time of $-\infty$ is essentially given by the time that the particle takes to overcome the potential barrier. The mean first passage time, estimated by the Arrhenius formula, is

$$\mathbb{E}_\infty(\tau_1) = e^{\frac{2\Delta U}{\sigma}}$$

and so (11.102) follows. Lifshitz singularities occur in several disordered systems and have been extensively studied in the literature. Let us briefly review some interesting cases.

11.7.1 Short-range, repulsive potentials

Then

$$\ln N(E) \sim -\frac{c}{\sqrt{E}} \quad \text{as } E \to 0+ .$$

Again, this nonanalytic behaviour has a simple explanation, due to Lifshitz: low-energy modes may appear when there are large regions of space free of impurities, in which the particle can move freely. Such configurations have an exponentially small probability. For instance, if the impurities are uniformly distributed with mean spacing ℓ, then the probability of finding an impurity-free region of size L is $e^{-L/\ell}$. Since the lowest energy mode is of order $k = \pi/L$ this gives

$$\ln N(E) \sim -\frac{\pi}{\ell\sqrt{E}} \quad \text{as } E \to 0+ .$$

This argument generalizes to d dimensions; the exponent of E becomes $-d/2$—a result that may also be derived via instanton techniques.

11.7.2 Strings of Type I

For Dyson's strings, the behaviour at the bottom of the spectrum depends on the type.: The Type II string of Exercise 11.4.11 exhibits a square-root behaviour; by contrast, Dyson showed that the string of Type I discussed in section 11.3.6 is such that

$$\ln N(\lambda) \sim -2 \ln |\ln \lambda| \quad \text{as } \lambda \to 0+ .$$

Such singularities are called *Dyson singularities*.

Exercise 11.7.1. *Consider a random string of Type I, with a Dirichlet condition at $x = 0$ (i.e. $\alpha = \pi/2$), such that*

$$c_j \sim Gamma(p, q) .$$

Show that, for $\lambda < 0$,

$$\mathbb{E}\left[w(\lambda)\right] = -\lambda \frac{\Psi'(p, 1; -\lambda/q)}{\Psi(p, 1; -\lambda/q)} .$$

Let us use this to compute $\mathbb{E}(\sigma')$. Analytic continuation gives, for $\lambda > 0$,

$$\frac{\mathbb{E}\left[w(\lambda + i0)\right]}{\lambda} = -\frac{\Psi'(p, 1; -\lambda/q - i0)}{\Psi(p, 1; -\lambda/q - i0)} = -\frac{\Psi'(p, 1; -\lambda/q - i0)\Psi(p, 1; -\lambda/q + i0)}{|\Psi(p, 1; -\lambda/q - i0)|^2} .$$

Therefore,

$$- \frac{2\mathrm{i}\,\mathrm{Im}\,w(\lambda + \mathrm{i}0)}{\lambda}$$
$$= \frac{\Psi'(p, 1; -\lambda/q - \mathrm{i}0)\Psi(p, 1; -\lambda/q + \mathrm{i}0) - \Psi'(p, 1; -\lambda/q + \mathrm{i}0)\Psi(p, 1; -\lambda/q - \mathrm{i}0)}{|\Psi(p, 1; -\lambda/q - \mathrm{i}0)|^2}.$$

Ismail and Kelker (1979) worked out a simple expression for the numerator of the right-hand side; their Eq. (3.3) says that this numerator equals

$$-\frac{2\pi\mathrm{i}}{\lambda}\frac{q\,\mathrm{e}^{-\lambda/q}}{\Gamma(p)^2}.$$

Hence,

$$\mathbb{E}\left[\sigma'(\lambda)\right] = \frac{q\mathrm{e}^{-\lambda/q}}{\Gamma(p)^2\,|\Psi(p, 1; -\lambda/q - \mathrm{i}0)|^2}.$$

It follows, in particular, that

$$\mathbb{E}\left[\sigma'(\lambda)\right] \sim \frac{q}{\ln^2(\lambda/q)} \quad \text{as } \lambda \to 0+.$$

11.7.3 Supersymmetric potentials

A potential of the form

$$V(x) = W'(x) + W^2(x)$$

is called *supersymmetric* and $W(x)$ is called the *superpotential*. For such potentials, the spectral measure of the Schrödinger Hamiltonian is always supported on $[0, \infty)$. When $W(x)$ is random, the Lifshitz tails can be of the form

$$\ln N(E) \sim c\ln E \quad \text{as } E \to 0+$$

or feature a Dyson singularity. A physical picture that accounts for such behaviours is presented in Comtet et al. (1995). It turns out that the mechanism is just the opposite of the Lifshitz mechanism described earlier. Instead of being localized in impurity-free regions, the low-energy states are localized near the impurities. We note incidentally that the Schrödinger equation with a supersymmetric potential can be recast as the first-order system

$$-\psi' + W\psi = \sqrt{E}\,\phi \quad \text{and} \quad \phi' + W\phi = \sqrt{E}\,\psi.$$

These equations may be viewed as a degenerate form of the Dirac equation with the Hamiltonian

$$\mathscr{H} := \begin{pmatrix} \mathscr{O} & \mathscr{Q} \\ \mathscr{Q}^\dagger & \mathscr{O} \end{pmatrix},$$

where $\mathcal{O}$ denotes the zero operator and $\mathcal{Q}$ is a first-order differential operator. Hamiltonians with such a structure are called *chiral*. They obey the symmetry condition

$$\mathcal{H} = -\sigma_3 \mathcal{H} \sigma_3 \,.$$

It implies that positive and negative energy levels appear in pairs $\pm E$; if ψ_E is an eigenstate of energy E then $\sigma_3 \psi_E$ is an eigenstate of energy $-E$. The presence of this symmetry places severe constraints on the localization properties of the system. The one-dimensional supersymmetric Hamiltonian is thus an interesting toy model that illustrates the importance of symmetry considerations in disordered systems (Altland and Zirnbauer 1997).

11.7.4   Further reading

An introduction to supersymmetric quantum mechanics and a comprehensive review of its applications in quantum and statistical physics can be found in the book by Junker (1996). These models are relevant in several contexts of condensed matter physics, including one-dimensional disordered semiconductors (Ovchinnikov and Érikhman 1977), random spin chains, and organic conductors; see Comtet and Texier (1998) for a review. They have also found interesting applications in the context of diffusion in a random environment (Bouchaud et al. 1990, Le Doussal 2009, Grabsch et al. 2014). A table summarizing the low-energy behaviours of such models is given in Grabsch et al. (2014). Further information on their spectral properties can be found in Comtet et al. (2011).

11.8   Distribution of the ground state energy and statistics of energy levels

We have shown that the correct tail behaviour of the integrated density of states may be obtained by using a simple analogy with a physical activation process. The same analogy can also be used to investigate finer properties of the spectrum, such as the distribution of the ground state energy and the statistics of energy levels. In order to give a precise meaning to this statement, consider the truncated spectral problem

$$\mathcal{H}\psi = E\psi \,, \quad 0 < x < L \,,$$

with the Dirichlet boundary conditions

$$\psi(0) = \psi(L) = 0 \,.$$

In particular, we would like to compute the probability distribution of the ground state energy E_0. As before, this problem can be reduced to studying the zeros of the particular solution $\psi(\cdot, E)$ of the truncated problem that satisfies the initial conditions $\psi(0, E) = 0, \psi'(0, E) = 1$. Denoting by τ_1 the first strictly positive zero of $\psi(\cdot, E)$, Sturm's oscillation theorem implies that the event $E_0 > E$ is the same as the event $\tau_1 > L$. Hence,

$$\mathbb{P}(E_0 > E) = \mathbb{P}(\tau_1 > L)$$

and the problem of computing the statistics of E_0 amounts to solving a *first passage problem* for the Riccati process $\{Z(x)\}_{x \geq 0}$ started at infinity (Grenkova et al. Grenkova) .

Following Texier (2000), let us show how to compute the tail of this distribution by using the activation analogy. As discussed earlier, the expected exit time from the potential barrier is related to the integrated density of states via

$$\tau := \mathbb{E}_\infty(\tau_1) = \frac{1}{N(E)}$$

We require—not just the mean—but rather the whole distribution of the time of escape from the well. In the weak-disorder limit $E\sigma^{-2/3} \to -\infty$, the probability density, say p, of τ_1 is given by

$$p(t) = \frac{1}{\tau} e^{-\frac{t}{\tau}} .$$

Therefore, the ground state energy distribution is given by

$$\mathbb{P}(E_0 > E) = \mathbb{P}(\tau_1 > L) = \int_L^\infty p(t)\, \mathrm{d}t = e^{-LN(E)} . \tag{11.103}$$

When the size L of the system is very large, the ground state energy E_0—properly centred and rescaled—will have a limiting law. Take, for example, the case of the white noise potential $V = \sqrt{\sigma}B'$ and set for simplicity $\sigma = 1$ so that the tail of the density of states is given by

$$N(E) = \frac{\sqrt{-E}}{\pi} e^{-\frac{8}{3}(-E)^{3/2}} . \tag{11.104}$$

One expects that Eq. (11.103) will give the correct tail of the true probability distribution in the limit $L \to \infty$. After some algebra, one gets

$$\lim_{L \to \infty} \mathbb{P}\left[\frac{E_0 + (\frac{3}{8}\log\frac{L}{\pi})^{2/3}}{(24\log L)^{-1/3}} < x \right] = 1 - e^{-e^x} .$$

This result, derived for the case of a white noise potential in McKean (1994), is, in fact, discussed in greater generality in an earlier paper (Grenkova et al. Grenkova). As pointed out in Texier (2000), it expresses the fact that the distribution of E_0—properly centred and rescaled—is a *Gumbel law*. This distribution belongs to one of the three classes of extreme value statistics characterizing the minimum or the maximum of a set of independent, identically distributed random variables.

For the Frisch–Lloyd model where the x_n are the points of a Poisson process and the v_n independent, identically distributed random variables with an exponential distribution (Grenkova et al. Grenkova), and for its white noise limit (Texier 2000), the analysis can be pushed a little further and generalized to excited states. One finds that

$$\mathbb{P}[E_{n-1} < E < E_n] = \mathbb{P}[\tau_1 + \tau_2 + \ldots + \tau_n < L < \tau_1 + \tau_2 + \ldots + \tau_{n+1}],$$

where the τ_j are independent and have the same distribution as τ_1. Thus, for $n = 1, 2, \ldots$,

$$\mathbb{P}[E_{n-1} < E < E_n] = \frac{[LN(E)]^n}{n!} e^{-LN(E)}$$

and

$$\lim_{L \to \infty} \mathbb{P}\left[\frac{E_{n-1} - f_{n-1}(L)}{\sigma_{n-1}(L)} \in dx\right] = \frac{n^{n-1/2}}{(n-1)!} \exp\left(x\sqrt{n} - n e^{x/\sqrt{n}}\right) dx$$

for some functions f_n and σ_n that can be worked out explicitly. This result expresses the fact that the energy levels are uncorrelated (Molcanov 1981, Grenkova et al. Grenkova), and this is expected to hold whenever the quantum states are localized. When delocalization occurs at the bottom of the spectrum, as it does for the supersymmetric disordered Hamiltonian, the first energy levels are correlated and the energy levels do not obey Gumbel laws (Texier 2000, Texier and Hagendorf 2010).

11.9 Scattering and hyperbolic geometry

So far, our discussion of impurity models has focused on the spectral properties of the master equation (11.15) and of its truncated version (11.18). Another problem of physical interest concerns the effect of disorder on the transmission of a plane wave through a sample. This problem may be stated as follows: given $\lambda = k^2 > 0$ and $L > 0$, find complex numbers R_L and T_L such that there is a solution $\psi_L(x)$ of Eq. (11.18) satisfying

$$\psi_L(x) = \begin{cases} e^{ikx} + R_L\, e^{-ikx} & \text{if } x \leq 0 \\ T_L\, e^{ikx} & \text{if } x \geq L \end{cases}. \tag{11.105}$$

The numbers R_L and T_L are called the reflexion coefficient and the transmission coefficient, respectively.

Our aim in this section will be to determine how the transmission coefficient changes as the length L increases.

Remark 11.9.1. This problem makes sense for both the Frisch–Lloyd and Dyson models. For the string equation, however, it is more natural to consider the generalized version in Exercise 11.2.3 with $\mu = 1$.

Exercise 11.9.2. *Show that*

$$|R_L|^2 + |T_L|^2 = 1.$$

Help: *Use Green's identity.*

Recall that the general solution of the truncated equation (11.18) is

$$\begin{pmatrix} \psi'(x_{n+1}-) \\ \psi(x_{n+1}) \end{pmatrix} = \Pi_n A_0 \begin{pmatrix} \psi'(0-) \\ k\psi(0) \end{pmatrix},$$

where the matrices A_j are those given in section 11.2. Let now

$$L = x_{n+1} \quad \text{and} \quad k = 1.$$

(This restriction on k is merely for convenience and entails no real loss of generality.) By working with the Riccati variable

$$Z_L = \frac{\psi'_L}{\psi_L} \tag{11.106}$$

we deduce easily that

$$R_L = -\frac{\mathcal{X}_{n+1}(\mathrm{i}) - \mathrm{i}}{\mathcal{X}_{n+1}(\mathrm{i}) + \mathrm{i}}, \tag{11.107}$$

where

$$\mathcal{X}_{n+1} := A_0^{-1} \circ A_1^{-1} \circ \cdots \circ A_n^{-1}. \tag{11.108}$$

The sequence $\{\mathcal{X}_n\}_{n \in \mathbb{N}}$ defined by this last equation satisfies the recurrence relation

$$\mathcal{X}_{n+1} = \mathcal{X}_n \circ A_n^{-1}. \tag{11.109}$$

In particular, when the A_n are independent and identically distributed, the sequence is a random walk in the group of linear fractional transformations on the upper half-plane

$$\mathbb{H} := \{x + \mathrm{i}y : x \in \mathbb{R},\ y > 0\}$$

starting from the identity element.

Exercise 11.9.3.   *Let $A \in SL(2, \mathbb{R})$. Show that*

$$A(\mathbb{H}) \subseteq \mathbb{H}.$$

11.9.1   Hyperbolic geometry

Let us step back from impurity models for a moment and consider the equation satisfied by the Riccati variable (11.106):

$$Z' = -Z^2 + Q' - M', \quad 0 \le x < L, \tag{11.110}$$

where, for convenience, we assume that Q and M are differentiable in the usual sense. Put

$$Z = X + \mathrm{i}Y$$

so that

$$X' = Q' - M' + Y^2 - X^2 \text{ and } Y' = -2XY.$$

There are two possibilities: either Y is identically zero or it is always of the same sign. The case of interest for our purposes is obviously $Y > 0$. Then $\{Z(x) : x \geq 0\}$ is a curve contained in the upper half-plane. This is consistent with the result of Exercise 11.9.3.

Further insight into the nature of the Riccati flow may be gained by renaming the independent variable t and setting

$$q := X \text{ and } p := 1/Y.$$

Then the Riccati equation is equivalent to the real Hamiltonian system

$$\dot{p} = -\frac{\partial H}{\partial q} \text{ and } \dot{q} = \frac{\partial H}{\partial p},$$

where

$$H(q, p) := \left(Q' - M' - q^2\right) p - \frac{1}{p}. \tag{11.111}$$

In particular, by Liouville's theorem, the Hamiltonian flow in phase space is *incompressible*. In terms of the original Riccati variable, the volume element is

$$dp\,dq = \frac{dX\,dY}{Y^2} \tag{11.112}$$

and this is conserved by the Riccati flow.

It turns out that this is also precisely the infinitesimal volume associated with Poincaré's half-plane model of two-dimensional hyperbolic geometry. In this model, the length, say l_Z, of a path

$$Z = X + iY : [0, 1] \to \mathbb{H}$$

is defined by the formula

$$l_Z := \int_0^1 \frac{|\dot{Z}(t)|}{Y(t)}\,dt,$$

where the dot indicates differentiation with respect to the parameter t. This length defines a hyperbolic metric $\varrho(z_0, z_1) : \mathbb{H} \times \mathbb{H} \to \mathbb{R}_+$ via

$$\varrho(z_0, z_1) := \min_{\{Z : Z(0) = z_0,\ Z(1) = z_1\}} l_Z.$$

The hyperbolic distance between two points may be calculated explicitly; one finds that

$$\operatorname{ch}\left[\varrho(z_0, z_1)\right] = \frac{(x_0 - x_1)^2 + y_0^2 + y_1^2}{2y_0 y_1}.$$

Exercise 11.9.4.   *Show that, for every $z \in \mathbb{H}$,*

$$\text{th}\left[\frac{\varrho(z, i)}{2}\right] = \left|\frac{z - i}{z + i}\right|.$$

11.9.2   Decay of the transmission coefficient

Let us now return to our impurity model. Formula (11.108) for the reflexion coefficient is in terms of linear fractional transformations with *real* coefficients. Such transformations play a very important part in the analysis of Poincaré's half-plane model of hyperbolic geometry.

Exercise 11.9.5.   *Let $A \in SL(2, \mathbb{R})$ and $Z : [0, 1] \to \mathbb{H}$. Consider the transformed path*

$$\mathcal{A}(Z) := [0, 1] \to \mathcal{A}(Z(t)).$$

Show that

$$l_{\mathcal{A}(Z)} = l_Z.$$

In words, the length functional is invariant under the linear fractional transformation $\mathcal{A}$.

It follows from this exercise that the Poincaré metric ϱ is also invariant under the linear fractional transformations associated with $SL(2, \mathbb{R})$, i.e.,

$$\varrho\left(\mathcal{A}(z_0), \mathcal{A}(z_1)\right) = \varrho(z_0, z_1) \quad \text{for every } A \in SL(2, \mathbb{R}).$$

Exercise 11.9.6.   *Show that, for every*

$$A = \begin{pmatrix} a & b \\ c & d \end{pmatrix} \in SL(2, \mathbb{R}),$$

there holds

$$2 \, ch\left[\varrho\left(i, \mathcal{A}(i)\right)\right] = |A|^2 := a^2 + b^2 + c^2 + d^2.$$

We now have in place all the elements needed in order to relate the transmission coefficient to the the product of matrices: By combining the result of Exercise 11.9.4 with Formula (11.107), we deduce that

$$|R_L| = \text{th}\left[\frac{\varrho\left(\mathcal{X}_{n+1}(i), i\right)}{2}\right]$$

and hence, by the result of Exercise 11.9.2,

$$|T_L|^2 = \frac{1}{\text{ch}^2\left[\frac{\varrho(\mathcal{X}_{n+1}(i), i)}{2}\right]} = \frac{2}{1 + \text{ch}\left[\varrho\left(\mathcal{X}_{n+1}(i), i\right)\right]}.$$

Furthermore, by using Formula (11.108) and the result of Exercise 11.9.6, we obtain

$$|T_L|^2 = \frac{4}{2 + |\Pi_n|^2} \, . \tag{11.113}$$

In particular, if the Lyapunov exponent is positive, then

$$\lim_{L \to \infty} \frac{\ln |T_L|}{L} = -\gamma(k^2) \, . \tag{11.114}$$

This says that the coefficient of transmission through a disordered sample *decays exponentially with the sample length* and that the decay rate is precisely the Lyapunov exponent.

11.9.3 Distribution of the reflexion phase

It follows, in particular, that a plane wave incident on an infinite disordered sample is totally reflected:

$$\lim_{L \to \infty} |R_L| = 1 \, .$$

The distribution of the phase of R_L for large samples is an interesting observable which has received some attention in the physics literature (Sulem 1973, Barnes and Luck 1990).

For our impurity models, Eq. (11.107) expresses the reflexion coefficient for a sample of length $L = x_{n+1}$ in terms of the finite continued fraction

$$\mathcal{X}_{n+1}(i) = \mathcal{A}_0^{-1} \circ \mathcal{A}_1^{-1} \circ \cdots \circ \mathcal{A}_n^{-1}(i) \, . \tag{11.115}$$

For every n, its value is a point in the half-plane $\mathbb{H}$. However, since $\lambda > 0$, the matrices A_n are real, and so, if the continued fraction converges, its limit must be a *real* random variable, say

$$X_\infty := \lim_{n \to \infty} \mathcal{X}_n(i) \in \mathbb{R} \, . \tag{11.116}$$

This is illustrated in Fig. 11.3. This limit plays for the scattering problem the same part as the Weyl coefficient $w(\lambda)$ for the spectral problem. In particular, the probability density of $-X_\infty$ is the stationary probability density f of the Riccati process. Thus, if we write

$$\lim_{L \to \infty} R_L = e^{i\Theta_\infty} \, ,$$

then the probability density of the reflexion phase Θ_∞ of an infinite sample is

$$\frac{1}{1 + \cos\theta} f\left(\frac{\sin\theta}{1 + \cos\theta} \right) \, .$$

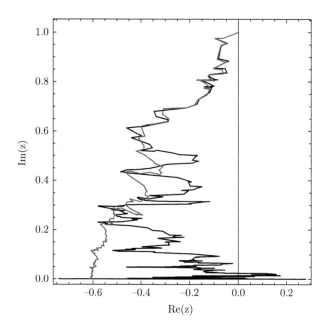

Fig. 11.3 The grey curve corresponds to a typical trajectory of the backward iteration $\{\mathcal{X}_n\}_{n\in\mathbb{N}}$ defined by (11.115). The black curve shows the trajectory of the corresponding forward iteration.

Explicit formulae for the density function f for the Frisch–Lloyd model of Example 11.4.17 and for the Dyson model of Exercise 11.4.10 may be found in Comtet et al. (2010) and Marklof et al. (2008), respectively.

The fact that a limiting distribution does exist is not entirely trivial. Although $\mathcal{X}_{n+1}$ is like the continued fraction for the Weyl coefficient of the truncated spectral problem that we studied in section 11.3, there is one significant difference between the two: for the scattering problem, $\lambda = k^2 = 1$, and so *the point at infinity is no longer of the limit-point type*. This means that the Weyl–Titchmarsh theory cannot be invoked to deduce the convergence of $\mathcal{X}_{n+1}$ as $n \to \infty$. A different approach is required, and Bougerol and Lacroix (1985), Chapter 2, show that convergence takes place if the Lyapunov exponent is strictly positive. Furstenberg's theorem, which will be discussed next, gives explicit conditions on the distribution of the A_n for this to be the case, and so the scattering problem underlines the practical significance of Furstenberg's theory.

11.10   The Lyapunov exponent of a product of random matrices

We now leave the subject of impurity models and turn our attention to the large-n behaviour of the product

$$\amalg_n := A_n \cdots A_2 A_1 \tag{11.117}$$

when the A_n are arbitrary random $d \times d$ invertible matrices, i.e., $A_n \in \mathrm{GL}(d, \mathbb{R})$. More precisely, we will be concerned with the limit

$$\lim_{n \to \infty} \frac{1}{n} \ln |\Pi_n \mathbf{x}|,$$

where $\mathbf{x}$ is some nonrandom, nonzero vector and $|\cdot|$ is the familiar Euclidean norm in $\mathbb{R}^d$, i.e.,

$$|\mathbf{x}| := \left(\sum_{j=1}^{d} x_j^2 \right)^{\frac{1}{2}}. \tag{11.118}$$

Also, we will restrict our attention to the case where the A_n are *independent and identically distributed*. We denote by μ the distribution from which they are drawn, i.e., for every set S of matrices,

$$\mathbb{P}(A_n \in S) = \int_S \mathrm{d}\mu(A).$$

In order to get a feel for the problem, let us begin by considering some simple particular cases.

Example 11.10.1. When $d = 1$, the A_n are just *numbers*. Then $|\cdot|$ is the absolute value and we have

$$\frac{1}{n} \ln |\Pi_n x| = \frac{1}{n} \sum_{j=1}^{n} \ln |A_j| + \frac{\ln |x|}{n} \xrightarrow[n \to \infty]{\text{a.s.}} \mathbb{E}(\ln |A|),$$

where A is μ-distributed. This result is nothing but the familiar *Law of Large Numbers*, which says that, if one draws numbers repeatedly and independently from the same distribution, then the average value after many draws approaches the mean of that distribution. In the 1×1 case, the growth of the product is therefore easily expressed in terms of the distribution μ, since

$$\mathbb{E}(\ln |A|) = \int_{\mathrm{GL}(1,\mathbb{R})} \ln |A| \, \mathrm{d}\mu(A).$$

This argument extends to the case where d is arbitrary but the A_n are *diagonalizable* and *commute*, i.e.

$$A_m A_n = A_n A_m \text{ for every } m, n \in \mathbb{N}.$$

Then, for every realization of the sequence $\{A_n\}_{n \in \mathbb{N}}$, the matrices A_n share the same eigenvectors. So one can find a an invertible matrix independent of n, say M, such that for every $n \in \mathbb{N}$

$$M A_n M^{-1} = \mathrm{diag}(\lambda_{n,j}),$$

where the $\lambda_{n,j}$, $1 \leq j \leq d$, are the eigenvalues of A_n.

Denote by $|A|$ the norm of the matrix A, i.e.,

$$|A| := \sup_{|\mathbf{x}| \leq 1} |A\mathbf{x}| . \tag{11.119}$$

For typical A, $B \in \mathrm{GL}(d, \mathbb{R})$, we have

$$|AB| \leq |A| |B|,$$

rather than the strict equality of the $d = 1$ case. So, unless the matrices in the product commute, the argument used in Example 11.10.1 breaks down. Nevertheless, as we will see, the result

$$\frac{1}{n} \ln |\Pi_n \mathbf{x}| \xrightarrow[n \to \infty]{\text{a.s.}} \gamma_\mu$$

holds under conditions that are quite natural. The number γ_μ is called the *Lyapunov exponent* of the product.

Remark 11.10.2. The norm (11.119) induced by the choice (11.118) differs from that used in section 11.9, Exercise 11.9.6. We point out, however, that in a finite-dimensional space, all norms are equivalent. Hence, the value of the Lyapunov exponent does not depend on the particular norm chosen; the choice is merely a matter of convenience.

Let $\mathbf{x}$ be a nonzero vector and set

$$Y_1 = \ln \left| A_1 \frac{\mathbf{x}}{|\mathbf{x}|} \right|, \quad Y_2 = \ln \left| A_2 \frac{A_1 \mathbf{x}}{|A_1 \mathbf{x}|} \right|, \quad \ldots, \quad Y_n = \ln \left| A_n \frac{A_{n-1} \cdots A_1 \mathbf{x}}{|A_{n-1} \cdots A_1 \mathbf{x}|} \right| .$$

Then we have the 'telescopic' formula

$$\frac{1}{n} \ln \frac{|A_n A_{n-1} \cdots A_1 \mathbf{x}|}{|\mathbf{x}|} = \frac{1}{n} \sum_{j=1}^{n} Y_j . \tag{11.120}$$

11.10.1 The Cohen–Newman example

Cohen and Newman (1984) remarked the following: suppose that μ, the distribution of A, is such that the random variable $|A\mathbf{u}|$ has the same distribution for every unit vector $\mathbf{u}$. Then the random variables Y_j defined earlier are identically distributed. They are also *independent*. Indeed, if the S_j are measurable sets, then

$$\mathbb{P}\left(Y_{n+1} \in S \mid Y_j \in S_j, 1 \leq j \leq n\right)$$

$$= \mathbb{P}\left(\ln \left| A_{n+1} \frac{A_n \cdots A_1 \mathbf{x}}{|A_n \cdots A_1 \mathbf{x}|} \right| \in S \mid Y_j \in S_j, 1 \leq j \leq n\right)$$

$$= \mathbb{P}\left(\ln|A_{n+1}\mathbf{u}| \in S \,\Big|\, Y_j \in S_j,\, 1 \le j \le n\right) = \mathbb{P}\left(\ln|A_{n+1}\mathbf{u}| \in S\right),$$

where $\mathbf{u}$ is any fixed nonrandom vector. Hence

$$\mathbb{P}\left(Y_{n+1} \in S \,\Big|\, Y_j \in S_j,\, 1 \le j \le n\right) = \mathbb{P}\left(Y_{n+1} \in S\right).$$

We may therefore use the Law of Large Numbers to conclude from Eq. (11.120) that

$$\gamma_\mu = \mathbb{E}\left(\ln\left|A\frac{\mathbf{x}}{|\mathbf{x}|}\right|\right).$$

Exercise 11.10.3. *Let*

$$A := \begin{pmatrix} \alpha & \beta \\ 0 & \frac{1}{\alpha} \end{pmatrix} \begin{pmatrix} \cos\theta & -\sin\theta \\ \sin\theta & \cos\theta \end{pmatrix},$$

where $\alpha \ne 0$ and β are fixed, nonrandom numbers, and θ is uniformly distributed over $[0, 2\pi)$.

(a) Show that

$$|A\mathbf{u}|$$

has the same law for every unit vector $\mathbf{u}$.

(b) Deduce that the Lyapunov exponent for the corresponding product is

$$\gamma_\mu = \ln\frac{(\alpha + 1/\alpha)^2 + \beta^2}{4}.$$

11.10.2 Application to the Frisch–Lloyd model

Let us apply the result of the previous exercise to a disordered system. Consider the Frisch–Lloyd model with impurities that are uniformly distributed in $(0, \infty)$ with mean spacing ℓ, but *fixed deterministic* coupling constants. In other words, for every $j \in \mathbb{N}$,

$$\mathbb{P}(\ell_{j-1} \in (a, b)) = \int_a^b dx\, \frac{1}{\ell}e^{-x/\ell} \quad \text{and} \quad v_j = v.$$

Exercise 11.10.4. *With these assumptions:*

(a) Show that if one uses

$$\begin{pmatrix} \psi'(x-) \\ k\psi(x) \end{pmatrix} \quad \text{instead of} \quad \begin{pmatrix} \psi'(x-) \\ \psi(x) \end{pmatrix},$$

then the solution of the Frisch–Lloyd model may be expressed as

$$\begin{pmatrix} \psi'(x_{n+1}-) \\ k\psi(x_{n+1}) \end{pmatrix} = \begin{pmatrix} \cos(k\ell_n) & -\sin(k\ell_n) \\ \sin(k\ell_n) & \cos(k\ell_n) \end{pmatrix} \Pi_n \begin{pmatrix} \psi'(0-) \\ k\psi(0) \end{pmatrix},$$

where

$$\Pi_n := A_n \cdots A_1$$

with

$$A_j := \begin{pmatrix} 1 & v/k \\ 0 & 1 \end{pmatrix} \begin{pmatrix} \cos(k\ell_{j-1}) & -\sin(k\ell_{j-1}) \\ \sin(k\ell_{j-1}) & \cos(k\ell_{j-1}) \end{pmatrix}.$$

(b) Introduce the reduced phase

$$\theta_j := 2\pi \left\{ \frac{k\ell_{j-1}}{2\pi} \right\} \in [0, 2\pi).$$

Equivalently,

$$\theta_j = k\ell_{j-1} \mod 2\pi.$$

Show that, for large $k\ell$—that is, when the mean spacing between successive impurities is large compared to the wavelength—θ_j is approximately uniformly distributed on $[0, 2\pi)$.

(c) Use the result of Exercise 11.10.3 to find the limit

$$\lim_{L\to\infty} \frac{\ln \sqrt{|\psi'(L-, k^2)|^2 + k^2 |\psi(L, k^2)|^2}}{L}.$$

Compare this with the result obtained by Bienaimé and Texier (2008).

11.11   Furstenberg's formula for the Lyapunov exponent

We stress that the telescopic property (11.120) always holds but that, in general the Y_j are *neither independent nor identically distributed* random variables. So the Law of Large Numbers cannot be used. Instead, as we shall see in this section, the Lyapunov exponent can be expressed, via the ergodic theorem, as a functional of a certain Markov chain. This fact will lead to a general formula for the Lyapunov exponent which is, in some sense, the 'product version' of the usual Law of Large Numbers. Importantly, the formula involves averaging— not only over the matrices in the product—but also over a certain space called the *projective space*.

11.11.1   The projective space

We say that two vectors $\mathbf{x}$ and $\mathbf{y}$ in $\mathbb{R}^d$ have the same *direction* if one is a scalar multiple of the other; i.e., there exists $c \in \mathbb{R}\backslash\{0\}$ such that

$$\mathbf{y} = c\mathbf{x}.$$

This defines an equivalence relation in $\mathbb{R}^d$; the set of all directions can be partitioned into equivalence classes, and each equivalence class can be identified with a straight line through the origin. The set of all such lines is called the *projective space* and is denoted $P\left(\mathbb{R}^d\right)$.

Notation

We will use $\overline{\mathbf{x}}$ to denote the direction of the vector $\mathbf{x} \in \mathbb{R}^d\backslash\{\mathbf{0}\}$. We will frequently abuse this notation by treating the map

$$\mathbf{x} \mapsto \overline{\mathbf{x}}$$

as though it were invertible. That is, we will use $\overline{\mathbf{x}}$ to denote also an arbitrary element of $P\left(\mathbb{R}^d\right)$ to which we then associate a vector $\mathbf{x} \in \mathbb{R}^d\backslash\{\mathbf{0}\}$ whose direction is $\overline{\mathbf{x}}$. This does no harm, as long as the result does not depend on the choice of the particular vector $\mathbf{x}$.

11.11.2   A Markov chain and its stationary distribution

Let $A \in \mathrm{SL}\left(d, \mathbb{R}\right)$ and $\overline{\mathbf{x}} \in P(\mathbb{R}^d)$. Since, by assumption, the determinant of A does not vanish, the vector $A\mathbf{x}$ is nonzero and so lies along a line in $\mathbb{R}^d$.

Notation

$$A \cdot \overline{\mathbf{x}} := \overline{A\mathbf{x}}.$$

Exercise 11.11.1.   *Show that, for every A_1, $A_2 \in SL(d, \mathbb{R})$ and every $\overline{x} \in P(\mathbb{R}^d)$,*

$$A_1 A_2 \cdot \overline{x} = A_1 \cdot (A_2 \cdot \overline{x}) .$$

One says that the group $SL(d, \mathbb{R})$ acts on the projective space.

Suppose now that the A_n are independent and μ-distributed. Let $\overline{\mathbf{x}} \in P(\mathbb{R}^d)$ and define the random sequence $\{\overline{\mathbf{x}}_n\}_{n \in \mathbb{N}}$ by recurrence:

$$\overline{\mathbf{x}}_1 := \overline{\mathbf{x}} \text{ and } \overline{\mathbf{x}}_{j+1} := A_j \cdot \overline{\mathbf{x}}_j \text{ for } j = 1, 2, \cdots.$$

It is easy to see that the sequence

$$\{(A_n, \overline{\mathbf{x}}_n)\}_{n \in \mathbb{N}}$$

is a Markov chain in the product space $\mathrm{SL}(d, \mathbb{R}) \times P\left(\mathbb{R}^d\right)$. Furthermore, the telescopic formula (11.120) may be expressed as

$$\frac{1}{n} \ln \frac{|A_n A_{n-1} \cdots A_1 \mathbf{x}|}{|\mathbf{x}|} = \frac{1}{n} \sum_{j=1}^{n} F\left(A_j, \overline{\mathbf{x}}_j\right), \tag{11.121}$$

where $\bar{\mathbf{x}}$ is the direction of $\mathbf{x}$ and

$$F\left(A,\bar{\mathbf{x}}\right) = \ln \left| A\frac{\mathbf{x}}{|\mathbf{x}|} \right| .$$

Definition 11.11.2.   *Let ν be a probability distribution on $P\left(\mathbb{R}^d\right)$. We say that ν is stationary for μ if, for every ν-distributed direction $\bar{x}$ and every independent μ-distributed matrix A, we have*

$$A \cdot \bar{x} \overset{law}{=} \bar{x} .$$

Now, suppose that there exists a unique μ-stationary probability measure ν on $P\left(\mathbb{R}^d\right)$ and that $\bar{\mathbf{x}}$ is ν-distributed. It is easy to see from the construction of the Markov chain that

$$(A_n, \bar{\mathbf{x}}_n) \overset{law}{=} (A_1, \bar{\mathbf{x}})$$

for every $n \in \mathbb{N}$. In other words, the product measure $\mu(A)\nu(\bar{\mathbf{x}})$ is the stationary distribution of the Markov chain. By the ergodic theorem, we therefore deduce from Eq. (11.121) that

$$\frac{1}{n} \ln \frac{|A_n A_{n-1} \cdots A_1 \mathbf{x}|}{|\mathbf{x}|} \xrightarrow[n \to \infty]{} \int_{SL(d,\mathbb{R})} d\mu(A) \int_{P(\mathbb{R}^d)} d\nu(\bar{\mathbf{x}})\, F(A,\bar{\mathbf{x}}) .$$

In other words,

$$\gamma_\mu = \int_{SL(d,\mathbb{R})} d\mu(A) \int_{P(\mathbb{R}^d)} d\nu(\bar{\mathbf{x}}) \ln \frac{|A\mathbf{x}|}{|\mathbf{x}|} . \tag{11.122}$$

This is Furstenberg's formula for the Lyapunov exponent of the product of random matrices. The essential difficulty in its practical application is that the μ-stationary measure ν is not known a priori.

11.11.3   The case $d = 2$

In this case, we can make the foregoing discussion much more concrete by relating it to familiar geometrical concepts. We speak of $P(\mathbb{R}^2)$ as the *projective line*. To specify a particular member of the projective line, we may use (the reciprocal of) its slope:

$$\mathbf{0} \neq \mathbf{x} = \begin{pmatrix} x_1 \\ x_2 \end{pmatrix} \implies \bar{\mathbf{x}} = z := x_1/x_2 . \tag{11.123}$$

The direction z may be finite or infinite. This defines a bijection between the projective line and the set of 'numbers' $\bar{\mathbb{R}} = \mathbb{R} \cup \{\infty\}$. With some abuse of notation, we will sometimes write

$$P\left(\mathbb{R}^2\right) = \bar{\mathbb{R}} .$$

The reader will immediately recognize that this variable z defined by Eq. (11.123) coincides with the Riccati variable introduced in the particular context of impurity models. The relevance of the projective space in the more general context may be understood intuitively as follows. Suppose for simplicity that all the matrices have *positive* determinant, so that there is no further loss of generality in assuming that their determinant is one. When $d = 2$, we can write the product in the column form

$$\prod_{j=1}^{n} A_j = (\mathbf{p}_n \ \mathbf{q}_n) \ .$$

Recall the geometrical interpretation of the determinant in the 2×2 case: its modulus is the area of the parallelogram spanned by the columns. We see that unimodularity implies that

$$|\mathbf{p}_n| \, |\mathbf{q}_n| \, |\sin \theta_n| = 1,$$

where θ_n is the *angle* between the columns. Hence, if we show that the columns tend to align along the same direction, then $\theta_n \to 0$ as $n \to \infty$ and at least one of $|\mathbf{p}_n|$ or $|\mathbf{q}_n|$ must grow.

Let

$$A := \begin{pmatrix} a & b \\ c & d \end{pmatrix} .$$

and

$$\mathbf{x} = \begin{pmatrix} x_1 \\ x_2 \end{pmatrix} \neq \mathbf{0} .$$

The reciprocal of the slope of this line is

$$\frac{ax_1 + bx_2}{cx_1 + dx_2} = \frac{az + b}{cz + d} = \mathcal{A}(z)$$

where $\mathcal{A}$ is the linear fractional transformation introduced in Eq. (11.31). The matrix A has 'acted' on the line of direction of $z = x_1/x_2$ and mapped it to another line whose direction is

$$A \cdot z = \mathcal{A}(z).$$

Returning to the formula (11.122) for the Lyapunov exponent, this is how one should read the right-hand side when $d = 2$: the number

$$\frac{|A\mathbf{x}|}{|\mathbf{x}|} = \left| A \frac{\mathbf{x}}{|\mathbf{x}|} \right|$$

depends only on A and on the *direction* $z \in \mathbb{R}$ of the nonzero vector $\mathbf{x}$. So we can write

$$\frac{|A\mathbf{x}|}{|\mathbf{x}|} = \frac{\left| A \begin{pmatrix} z \\ 1 \end{pmatrix} \right|}{\left| \begin{pmatrix} z \\ 1 \end{pmatrix} \right|}.$$

The formula for the Lyapunov exponent then takes the more readable form

$$\gamma_\mu = \int_{\mathbb{R}} \int_{\mathrm{SL}(2,\mathbb{R})} \ln \frac{\left| A \begin{pmatrix} z \\ 1 \end{pmatrix} \right|}{\left| \begin{pmatrix} z \\ 1 \end{pmatrix} \right|} \, \mathrm{d}\mu(A) \, \mathrm{d}\nu(z). \tag{11.124}$$

How can one find ν? In the case $d = 2$, if we assume that ν has a density f, then this density satisfies the Dyson–Schmidt equation (11.57):

$$f(z) = \mathbb{E}\left([f \circ \mathcal{A}^{-1}](z) \frac{\mathrm{d}\mathcal{A}^{-1}}{\mathrm{d}z}(z) \right).$$

The existence and uniqueness of a solution is intimately connected with the convergence of the continued fraction

$$\lim_{n \to \infty} \mathcal{A}_1 \circ \cdots \circ \mathcal{A}_n(z). \tag{11.125}$$

In particular, for the impurity models, we saw in section 11.4 that, under quite mild assumptions, the convergence of this continued fraction followed from the fact that the differential problem was in the limit-point case. We also saw that, when f exists, for $\lambda \in \mathbb{R}$,

$$\gamma(\lambda + \mathrm{i}0) = \operatorname{Re}\Omega(\lambda + \mathrm{i}0) = \int_{\mathbb{R}} z f(z) \, \mathrm{d}z.$$

On the other hand,

$$\gamma(\lambda + \mathrm{i}0) = \left(\lim_{n \to \infty} \frac{n}{x_n} \right) \gamma_\mu,$$

where γ_μ is the Lyapunov exponent—in the sense of Furstenberg—of the product of matrices, evaluated at $\lambda + \mathrm{i}0$, associated with the impurity model. It is striking that Formula (11.124) for γ_μ looks much more complicated than the formula for $\gamma(\lambda + \mathrm{i}0)$. The fact the *both formulae are correct* is proved in Comtet et al. (2010). We will not reproduce the proof here, but some of the calculations in later sections will illustrate the simplifications that can occur when applying Furstenberg's formula to specific products of 2×2 matrices.

11.12 Furstenberg's Theorem

We have so far emphasized the practical aspects of the *computation* of the Lyapunov exponent. It should be apparent by now that explicit calculations are possible only in exceptional cases. For theoretical purposes, it is often sufficient to determine whether the Lyapunov exponent is strictly positive. Furstenberg (1963) and others developed a general theory with the aim of addressing this question in the abstract context of random walks on groups. Here we apply it to the case where the group is $G :=$ SL $(2, \mathbb{R})$. For the proofs of the results stated, the reader should consult Bougerol and Lacroix (1985), and Carmona and Lacroix (1990).

Using the matrix norm introduced in section 11.10, we can speak of limits of sequences, and of bounded and closed sets in G.

Notation

Given a probability measure μ on G, we denote by G_μ the smallest closed subgroup of G containing the support of μ.

Definition 11.12.1. *A subgroup H of G is said to be* strongly irreducible *if there is no finite union*

$$S = \cup_{j=1}^m S_j$$

of one-dimensional subspaces S_j of $\mathbb{R}^2$ such that, for every $A \in H$,

$$A(S) = S.$$

By extension, we say that the measure μ itself is strongly irreducible if G_μ is strongly irreducible.

The following criterion will be useful.

Proposition 11.12.2. *If G_μ is unbounded, then G_μ is strongly irreducible if and only if the following holds: for every $z \in P\left(\mathbb{R}^2\right)$,*

$$\# \left\{ \mathcal{A}(z) \,:\, A \in G_\mu \right\} > 2.$$

Remark 11.12.3. Intuitively, in terms of the impurity models, the concept of strong irreducibility expresses the requirement that there should should be 'enough' disorder.

Theorem 11.12.4 (Furstenberg). Suppose that

$$\mathbb{E}\left(\ln |A|\right) < \infty.$$

Suppose also that G_μ is strongly irreducible and unbounded. Then the following statements are true:

1. There exists a unique measure ν on $P\left(\mathbb{R}^2\right)$ that is stationary for μ.
2. For every nonrandom, nonzero vector $\mathbf{x}$, we have

$$\frac{1}{n} \ln |\Pi_n \mathbf{x}| \xrightarrow[n\to\infty]{\text{a.s.}} \gamma_\mu,$$

where γ_μ is given by Formula (11.124).
1. $\gamma_\mu > 0$.

We will illustrate the use and the content of Furstenberg's theorem by means of elementary examples. Some of these examples have in common that the product of matrices arises from the solution of the difference equation

$$\psi_{n+1} = a_n \psi_n + b_n \psi_{n-1}, \quad n \in \mathbb{N}. \tag{11.126}$$

This may be viewed as a generalization of the Anderson model mentioned in section 11.2.3. Obviously, its general solution is given by

$$\begin{pmatrix} \psi_{n+1} \\ \psi_n \end{pmatrix} = \Pi_n \begin{pmatrix} \psi_1 \\ \psi_0 \end{pmatrix},$$

where the A_j in the product (11.10) are given by

$$A_j = \begin{pmatrix} a_j & b_j \\ 1 & 0 \end{pmatrix}.$$

Example 11.12.5.   Our first example is deterministic! The Fibonacci sequence satisfies the recurrence relation

$$\psi_{n+1} = \psi_n + \psi_{n-1}, \quad n \in \mathbb{N}, \tag{11.127}$$

with $\psi_0 = \psi_1 = 1$. Alternatively,

$$\begin{pmatrix} \psi_{n+1} \\ \psi_n \end{pmatrix} = \Pi_n \begin{pmatrix} 1 \\ 1 \end{pmatrix},$$

where

$$A_j = A := \begin{pmatrix} 1 & 1 \\ 1 & 0 \end{pmatrix} \quad \text{for } j \in \mathbb{N}.$$

We may think of Π_n as a product of 'random' matrices with a distribution μ whose mass is concentrated at A; we express this as

$$\mu = \delta_A.$$

This μ is *not* strongly irreducible: the matrix A has two eigenvectors, and the straight lines along which these eigenvectors lie are invariant under multiplication by A. So Furstenberg's theorem does not hold, and this manifests itself in the fact that the limit of $Z_n(z)$ in Eq. (11.125) depends on z. Indeed, let $z_\pm$ be the two roots of

$$z = \mathcal{A}(z) = \frac{z+1}{z}.$$

Then, for every $n \in \mathbb{N}$, we have

$$Z_n(z_\pm) = z_\pm,$$

and so there are *two* invariant measures, namely ν_+ and ν_-, concentrated, respectively, on z_+ and z_-.

Example 11.12.6. Consider the following randomized version of the previous example: let the A_j be drawn from

$$\operatorname{supp}\mu := \left\{ \begin{pmatrix} -1 & 1 \\ 1 & 0 \end{pmatrix}, \ \begin{pmatrix} 1 & 1 \\ 1 & 0 \end{pmatrix} \right\}$$

with equal probability.

G_μ is larger than in the previous (deterministic) example, and it follows from Proposition 11.12.2 that Furstenberg's theorem holds.

The following illustrates the kind of manipulations involved in the calculation of the Lyapunov exponent: We write that

$$A = \begin{pmatrix} a & 1 \\ 1 & 0 \end{pmatrix}.$$

Then

$$\gamma_\mu = \int_\mathbb{R} \int_{\mathrm{GL}(2,\mathbb{R})} \ln \frac{\left| A \begin{pmatrix} z \\ 1 \end{pmatrix} \right|}{\left| \begin{pmatrix} z \\ 1 \end{pmatrix} \right|} \, d\mu(A) \, d\nu(z)$$

$$= \frac{1}{2} \int_\mathbb{R} \int_{\mathrm{GL}(2,\mathbb{R})} \ln \frac{\left| A \begin{pmatrix} z \\ 1 \end{pmatrix} \right|^2}{\left| \begin{pmatrix} z \\ 1 \end{pmatrix} \right|^2} \, d\mu(A) \, d\nu(z)$$

$$= \frac{1}{2} \int_\mathbb{R} \int_{\mathrm{GL}(2,\mathbb{R})} \ln \frac{(az+1)^2 + z^2}{1+x^2} \, d\mu(A) \, d\nu(z)$$

$$= \frac{1}{2} \int_\mathbb{R} \int_{\mathrm{GL}(2,\mathbb{R})} \ln \left\{ \frac{z^2}{1+z^2} \left[1 + [\mathcal{A}(z)]^2 \right] \right\} \, d\mu(A) \, d\nu(z)$$

$$= \frac{1}{2} \int_\mathbb{R} \int_{\mathrm{GL}(2,\mathbb{R})} \ln \frac{z^2}{1+z^2} \, d\mu(A) \, d\nu(z)$$

$$+ \frac{1}{2} \int_\mathbb{R} \int_{\mathrm{GL}(2,\mathbb{R})} \ln \left[1 + [\mathcal{A}(z)]^2 \right] \, d\mu(A) \, d\nu(z).$$

At this point, we observe that, in the first of these integrals, the integrand is independent of A. Hence,

$$\frac{1}{2} \int_{\mathbb{R}} \int_{GL(2,\mathbb{R})} \ln \frac{z^2}{1+z^2} \, d\mu(A) \, d\nu(z) = \frac{1}{2} \int_{\mathbb{R}} \ln \frac{z^2}{1+z^2} \, d\nu(x) \,.$$

Furthermore, using the fact that ν is stationary for μ,

$$\frac{1}{2} \int_{\mathbb{R}} \int_{GL(2,\mathbb{R})} \ln \left(1 + [A(z)]^2\right) \, d\mu(A) \, d\nu(z) = \frac{1}{2} \int_{\mathbb{R}} \ln \left(1 + z^2\right) \, d\nu(z) \,.$$

Putting these results together, we obtain a much deflated formula for the Lyapunov exponent:

$$\gamma_\mu = \int_{\mathbb{R}} \ln |z| \, d\nu(z) \,.$$

Viswanath (2000) considered and solved the problem of finding the μ-stationary measure ν for this example. It turns out that ν is not a smooth measure; it is *singular continuous*. There is no explicit formula for it, but the measure of any real interval may be computed exactly to any desired accuracy by means of a recursion. Then

$$\gamma_\mu \in (0.1239755980, 0.1239755995) \,.$$

This value is, as it should be, smaller than the growth rate of the (deterministic) Fibonacci sequence, i.e.,

$$\ln \frac{\sqrt{5}+1}{2} = 0.481 \ldots \,.$$

Figure 11.4 shows a histogram of the the first 8000 terms of the sequence defined by Eq. (11.56) with a normal random variable as starting value. The sequence is ergodic and so the histogram may be thought of as the 'graph' of '$d\nu(z)$'.

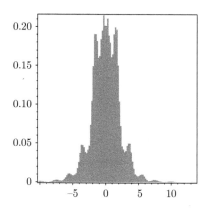

Fig. 11.4 A histogram of the first 8000 of the forward iterates for the random Fibonacci example.

Example 11.12.7. Let us give an example, taken from Bougerol and Lacroix (1985), of a product that does *not* grow. Let $\alpha > 0$ and set

$$D := \begin{pmatrix} \alpha & 0 \\ 0 & 1/\alpha \end{pmatrix} \quad \text{and} \quad R := \begin{pmatrix} 0 & -1 \\ 1 & 0 \end{pmatrix}.$$

Consider the distribution μ supported on $\{D, R\}$ such that

$$\mathbb{P}(A = D) = p \quad \text{and} \quad \mathbb{P}(A = R) = 1 - p, \quad \text{where } p \in [0, 1].$$

The matrix R is a rotation matrix; i.e., $R\mathbf{x}$ is the vector obtained after rotating $\mathbf{x}$ by the angle $\pi/2$. It is, therefore, obvious that if $p = 0$, then $\gamma_\mu = 0$. On the other hand, if $p = 1$ and $\alpha \neq 1$, one of the columns of the corresponding product of matrices will grow. What happens if $0 < p < 1$?

Figure 11.5 shows a plot of the quantity

$$\frac{1}{n} \ln |\Pi_n|$$

against n in the case where $p = \frac{1}{2}$. The suggestion is that $\gamma_\mu = 0$. Turning to the statement of Furstenberg's theorem, we note that

$$\mathbb{E}\left(|A_1|\right) = p \ln \alpha.$$

For $0 < p < 1$, G_μ contains D, R, D^{-1}, R^{-1} and every product of these. A little calculation shows that

$$G_\mu = \left\{ \begin{pmatrix} \beta^n & 0 \\ 0 & \beta^{-n} \end{pmatrix}, \begin{pmatrix} 0 & \beta^n \\ -\beta^{-n} & 0 \end{pmatrix} : \beta \in \{\pm\alpha, \pm1/\alpha\}, n \in \mathbb{Z} \right\}.$$

In particular, for every natural number n, $D^n \in G_\mu$ and so G_μ is unbounded.

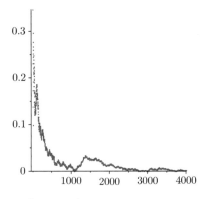

Fig. 11.5 Plot of $n^{-1} \ln \left| \prod_{j=1}^n A_j \right|$ against n for the Bougerol–Lacroix example.

There only remains to examine the strong irreducibility assumption. Let $z = 0$. Then we have

$$\begin{pmatrix} \beta^n & 0 \\ 0 & \beta^{-n} \end{pmatrix} \cdot z = 0 \text{ and } \begin{pmatrix} 0 & \beta^n \\ -\beta^{-n} & 0 \end{pmatrix} \cdot z = \infty.$$

We see that μ fails to satisfy the strong irreducibility criterion contained in Proposition 11.12.2.

11.13   Concluding remarks

In this chapter, we have presented a number of one-dimensional models of disordered systems whose study reduces to that of products of random matrices. For pedagogical reasons, we considered exclusively the case of 2×2 matrices drawn from the group $\mathrm{SL}(2, \mathbb{R})$. For this particular group, we have shown that the Iwasawa decomposition has a very natural interpretation in terms of an impurity model. The methodology developed in that context generalizes to a certain extent to groups of larger matrices associated with the multichannel systems of condensed matter physics. Important advances have been made in the study of these quasi-one-dimensional systems by exploiting the general classification of the groups of symmetries that these systems possess (Altland and Zirnbauer 1997). In the recent literature, approaches based on transfer matrices have proved highly useful in the discovery and analysis of unusual behaviour and of critical features that arise in the context of topological transitions (Brouwer et al. 2000). The reader interested in a more mathematical treatment of these ideas may also consult Ludwig et al. (2013), in which a perturbative calculation of the Lyapunov exponent for all 10 symmetry classes is performed. Apart from such perturbative studies, very few physical models have yielded to analytical treatment, although the Lyapunov exponent has been obtained for some examples of finite and infinite products of random matrices drawn from highly 'isotropic' distributions (Forrester 2013, Akemann et al. 2013). A particular multichannel disordered system for which the isotropy hypothesis can be relaxed has been the subject of a very recent study by Grabsch and Texier (2016). In that model, the Riccati process is matrix-valued, yet some calculations of the type that we have described in these notes can be done exactly. In particular, Grabsch and Texier managed to express the density of states as a determinant, and found for the Lyapunov spectrum interesting transitions which may be interpreted in topological terms.

Acknowledgements

It is our pleasure to thank Christophe Texier for kindly providing us with some of the figures used, and for making a number of suggestions which improved an earlier draft of these notes.

References

Abramowitz, M., and Stegun, I. (1964). *Handbook of Mathematical Functions with Formulas, Graphs and Mathematical Tables*. Dover, New York.

Akemann, G., Ipsen, J. R., and Kieburg, M. (2013). Products of rectangular random matrices: Singular values and progressive scattering. *Phys. Rev. E* **88**, 052118.

Altland, A., and Zirnbauer, M. R. (1997). Nonstandard symmetry classes in mesoscopic normal-superconducting hybrid structures. *Phys.Rev. B* **55(2)**, 1142–61.

Anderson, P. W. (1958). Absence of diffusion in certain random lattices. *Phys. Rev.* **109**, 1492–505.

Applebaum, D. (2004). Lévy processes–from probability to finance and quantum groups. *Notices Amer. Math. Soc.* **51**, 1336–47.

Barnes, C., and Luck, J. M. (1990). The distribution of the reflexion phase of disordered conductors. *J. Phys. A: Math. Theor.* **23**, 1717–34.

Benderskii, M. M., and Pastur, L. A. (1970). On the spectrum of the one-dimensional Schrödinger equation with a random potential. *Math. USSR Sbornik* **11**, 245–56.

Bienaimé, T., and Texier, C. (2008). Localization for one dimensional random potentials with large local fluctuations. *J. Phys. A: Math.Theor.*, **41** 475001.

Bouchaud, J. P., Comtet, A., Georges, A., and Le Doussal, P. (1990). Classical diffusion of a particle in a one-dimensional random force field. *Ann. Phys. (N.Y.)*, **201** 285–341.

Bougerol, P., and Lacroix, J. (1985). *Products of Random Matrices with Application to Random Schrödinger Operators*. Birkhaüser, Boston.

Brouwer, P. W., Furusaki, A., Gruzberg, I. A., and Mudry, C. (2000). Localization and delocalization in dirty superconducting wires, *Phys. Rev. Lett.* **85**, 1064–67.

Carmona, R., and Lacroix, J. (1990). *Spectral Theory of Random Schrödinger Operators*. Birkhaüser, Boston.

Coddington, E. A., and Levinson, N. (1955). *Theory of Ordinary Differential Equations*. McGraw–Hill, New–York.

Cohen, J. E., and Newman, C. M. (1984). The stability of large random matrices and their products. *Ann. Probab.* **12**, 283–310.

Comtet, A., Desbois, J., and Monthus, C. (1995). Localization properties of one dimensional disordered supersymmetric quantum mechanics. *Ann. Phys.* **239**, 312–50.

Comtet, A., Luck, J. M., Texier, C., and Tourigny, Y. (2013b). The Lyapunov exponent of a product of 2×2 matrices close to the identity. *J. Stat. Phys.* **150**, 13–65.

Comtet, A., and Texier, C. (1998). One dimensional disordered supersymmetric quantum mechanics: a brief survey. In *Supersymmetry and integrable models*, H. Aratyn, T.D.Imbo, W.Y.Keung, and U. Sukhatme, Lecture notes in physics, vol. 502, pp. 313–18. Springer, Berlin.

Comtet, A., Texier, C., and Tourigny, Y. (2010). Products of random matrices and generalised quantum point scatterers. *J. Stat. Phys.* **140**, 427–66.

Comtet, A., Texier, C., and Tourigny, Y. (2011). Supersymmetry quantum mechanics with Lévy disorder in one dimension. *J. Stat. Phys.* **145**, 1291–1323.

Comtet, A., Texier, C., and Tourigny, Y. (2013a). Lyapunov exponents, one-dimensional Anderson localization and products of random matrices. *J. Phys. A* **46**, 254003.

Dyson, F. J. (1953). The dynamics of a disordered linear chain. *Phys. Rev.* **92**, 1131–8.

Feller, W. (1971). *An Introduction to Probability Theory and Its Applications*, Vol. 2. Wiley, New York.

Forrester, P. J. (2013). Lyapunov exponents for products of complex Gaussian random matrices. *J. Stat. Phys.* **151**, 796-808.

Frisch, H. L., and Lloyd, S. P. (1960). Electron levels in a one-dimensional lattice. *Phys. Rev.* **120**, 1175–89.

Furstenberg, H. (1963). Noncommuting random products. *Trans. Amer. Math. Soc.* **108**, 377–428.

Grabsch, A., Texier, C., and Tourigny, Y. (2014). One-dimensional disordered quantum mechanics and Sinai diffusion with random absorbers. *J. Stat. Phys.* **155**, 237–76.

Grabsch, A., and Texier, C. (2016). Topological phase transitions and super universality in the 1d multichannel Dirac equation with random mass. `arXiv:1506.05322`.

Grenkova, L. N., Molcanov, S. A., and Sudarev, J. N. (1983). On the basic states of one-dimensional disordered structures. *Commun. Math. Phys.* **90**, 101–23.

Halperin, B. I. (1965). Green's functions for a particle in a one-dimensional random potential. *Phys. Rev.* **139**, A104–17.

Hansel, D., and Luciani, J. F. (1989). On diffusion equations for dynamical systems driven by noise. *J. Stat. Phys.* **54**, 971–95.

Ismail, M. E. H., and Kelker, D. H. (1979). Special functions, Stieltjes transforms and infinite divisibility, *SIAM J. Math. Anal.* **10**, 884–901.

Jona–Lasinio, G. (1983). Qualitative theory of stochastic differential equations and quantum mechanics of disordered systems. *Helv. Phys. Act.* **56**, 61–71.

Junker, G. (1996). *Supersymmetric Methods in Quantum and Statistical Physics.* Springer, Berlin.

Kac, I. S. and Krein, M. G. (1974). On the spectral functions of the string. *Amer. Math. Soc. Transl. Ser. 2*, **103**, 19–102.

Kotani, S. (1976). On asymptotic behaviour of the spectra of a one-dimensional Hamiltonian with a certain random coefficient. *Publ. RIMS, Kyoto Univ.* **12**, 447–92.

Kotani, S. (1982). Liapunov indices determine absolutely continuous spectra of stationary random one-dimensional Schrödinger operators. *Taniguchi Symp. SA*, Katata, 225–47.

Kronig, R. de L., and Penney, W. G. (1931). Quantum mechanics of electrons in crystal lattices. *Proc. Roy. Soc. London A*, **130**, 499–513.

Le Doussal, P. (2009). The Sinai model in the presence of dilute absorbers. *J. Stat. Mech.* P07032.

Letac, G. (1986). A contraction principle for certain Markov chains and its applications. *Contemp. Math.* 263–73.

Letac, G. (2009). *The random continued fractions of Dyson and their extensions.* Unpublished notes of a seminar given at Charles University, Prague, on November 25th.

Letac, G., and Seshadri, V. (1983). A characterisation of the generalised inverse Gaussian distribution by continued fractions. *Z. Wahrsch. Verw. Gebiete* **62**, 485–9.

Lifshits, I. M., Gredeskul, S. A., and Pastur, L. A. (1988). *Introduction to the Theory of Disordered Systems.* Wiley, New York.

Luck, J. M. (1992). *Systèmes Désordonnés Unidimensionels*. Aléa, Saclay.

Ludwig, A. W. W., Schulz–Baldes, H., and Stolz, M. (2013). Lyapunov spectra for all ten symmetry classes of quasi-one-dimensional disordered systems of non-interacting fermions, *J. Stat. Phys.* **152**, 275–304.

Mallick, K., and Marcq, P. (2002). Anomalous diffusion in non-linear oscillators with multiplicative noise. *Phys. Rev. E* **66**, 041113.

Marklof, J., Tourigny, Y., and Wołowski, L. (2008). Explicit invariant measures for products of random matrices. *Trans. Amer. Math. Soc.* **360**(7), 3391–427.

McKean, H. P. (1994). A limit law for the ground state of Hill's equation. *J. Stat. Phys.* **74**, 1227–32.

Molcanov, S. A. (1981). The local structure of the spectrum of the one-dimensional Schrdinger operator. *Commun. Math. Phys.* **78**, 429-46.

National Institute of Standards (NIST). *Digital Library of Mathematical Functions.* Available at http://dlmf.nist.gov/, Release 1.0.10 of 2015-08-07.

Nieuwenhuizen, T. M. (1983). Exact electronic spectra and inverse localization lengths in one-dimensional random systems: I. Random alloy, liquid metal and liquid alloy. *Physica* **120A**, 468–514.

Ovchinnikov, A. A., and Érikhman, N. S. (1977). Density of states in a one-dimensional random potential. *Sov. Phys. JETP* **46**, 340–6.

Pastur, L. A. (1973). Spectra of self-adjoint operators. *Russ. Math. Surv.* **28**, 1–67.

Pastur, L. A. (1980). Spectral properties of disordered systems in the one-body approximation. *Commun. Math. Phys.* **75**, 179–96.

Rice, S. O. (1944). Mathematical analysis of random noise. *Bell System Tech. J.* **23**, 282–332.

Schmidt, H. (1957). Disordered one-dimensional crystals. *Phys. Rev.* **105**, 425–41.

Simmons, G. F. (1972). *Differential Equations with Applications and Historical Notes*, McGraw–Hill, New York.

Stieltjes, T. J. (1894). Recherches sur les fractions continues. *Ann. Fac. Sc. Toulouse Ser. I*, **8**, J1–J122.

Stieltjes, T. J. (1895). Recherches sur les fractions continues [Suite et fin]. *Ann. Fac. Sc. Toulouse Ser. I*, **9**, A5–A47.

Sulem, P. L. (1973). Total reflexion of a plane wave from a semi-infinite, one dimensional random medium: distribution of the phase. *Physica* **70**, 190–208.

Tessieri, L., and Izrailev, F. M. (2000). Anderson localization as a parametric instability of the linear kicked oscillator. *Phys. Rev. E* **62**, 3090–5.

Texier, C. (2000). Individual energy levels distributions for one-dimensional diagonal and off-diagonal disorder. *J. Phys. A: Math. Theor.* **33**, 6095–128.

Texier, C. (2011). *Mécanique Quantique*. Dunod, Paris.

Texier, C., and Hagendorf, C. (2010). The effect of boundaries on the spectrum of a one-dimensional random mass Dirac Hamiltonian. *J. Phys. A: Math.Theor.* **43**, 025002.

Viswanath, D. (2000). Random Fibonacci sequences and the number 1.13198824 .... *Math. Comput.* **69**, 1131–55.

12

Gaussian multiplicative chaos and Liouville quantum gravity

Rémi RHODES[1] and Vincent VARGAS[2]

[1]Université Paris-Est Marne la Vallée, LAMA, Champs sur Marne, France,
and [2]ENS Ulm, DMA, 45 rue d'Ulm, 75005 Paris, France

Rhodes, R. and Vargas, V., 'Gaussian Multiplicative Chaos and Liouville Quantum Gravity' in
Stochastic Processes and Random Matrices. Edited by: Grégory Schehr et al, Oxford University
Press (2017). © Oxford University Press 2017. DOI 10.1093/oso/9780198797319.003.0012

Chapter Contents

12.1   Introduction

In 1985, Kahane laid the foundations of Gaussian multiplicative chaos theory (GMC). Roughly speaking, GMC is a theory which defines rigorously random measures with the formal definition

$$M_\gamma(\mathrm{d}x) = e^{\gamma X(x)}\sigma(\mathrm{d}x), \tag{12.1}$$

where σ is a Radon measure on some metric space D (equipped with a metric d), $\gamma > 0$ is a parameter, and $X : D \to \mathbb{R}$ is a centered Gaussian field. The definition (12.1) should be seen as formal since in the interesting cases the variable X does not live in the space of functions on D but rather in a space of distributions in the sense of Schwartz. In that case, $X(x)$ does not make sense pointwise. Of course, we could make the change of variables $X \to \gamma X$ and absorb the dependence in γ in the field X but we will not do so for reasons which will become clear in the sequel. In fact, Kahane's GMC theory is quite general and the metric space D need not be some subspace of $\mathbb{R}^d$; however, motivated by the study of 2d Liouville quantum gravity (LQG), we will consider in the sequel the very important subcase where D is some subdomain of $\mathbb{R}^d$, σ is a Radon measure on D, and X has a covariance kernel of log-type, namely

$$K(x, y) := \mathbb{E}[X(x)X(y)] = \ln_+ \frac{1}{|x - y|} + g(x, y), \tag{12.2}$$

where $\ln_+(x) = \max(\ln x, 0)$ and g is a bounded function over $D \times D$. In that case, one can show that X lives in the space of distributions: this just means that for all smooth function φ with compact support the integral $\int_D \varphi(x)X(x)\mathrm{d}x$ makes sense. In fact, even if we will not discuss this here, GMC measures associated with log-correlated X, namely with covariance (12.2), appear in many other fields among which are the following: mathematical finance (see Bacry et al. (2008) for a review), 3d turbulence (Chevillard et al. 2010), decaying Burgers turbulence (Fyodorov et al. 2010), the extremes of log-correlated Gaussian fields (Bramson et al. 2015, Biskup and Louidor 2016, Madaule 2015), the glassy phase of disordered systems (Carpentier and Le Doussal 2001, Fyodorov and Bouchaud 2008, Fyodorov et al. 2009, Madaule 2016) and the eigenvalues of Haar distributed random matrices (Webb 2014). However, we will focus in this chapter on applications to LQG.

   The purpose of this chapter is twofold: first, give a rigorous definition of measures of the type (12.1) and review some of their main properties. Emphasis will be put on explaining the main ideas and not on giving rigorous proofs. Second, we will show how to use these measures to define Polyakov's (1981) theory of LQG on the Riemann sphere; in this specific case, one can identify LQG with Liouville quantum field theory (LQFT). Here, emphasis will be put on explaining the construction of the so-called Liouville measures and explaining their (conjectured) relation with random planar maps: the construction will rely on section 12.2 on GMC.

12.1.1   Notations

We will denote by $|.|$ the standard Euclidean metric; i.e., $|x-y|$ will denote the distance between two points x and y. Also, if A is some set then $|A|$ will stand for the Euclidean volume of A. It should be clear from the context which convention is used for $|.|$. The Eucliden ball of center x and radius $r > 0$ will be denoted by $B(x,r)$. The standard Lebesgue measure will be $\mathrm{d}x$ in section 12.2; however, in section 12.3 on LQG, we will work exclusively in $2d$ so the Lebesgue measure will be denoted $\mathrm{d}z$.

In this chapter, we will only study the theory of GMC in the case where D is some subdomain of $\mathbb{R}^d$, σ is a Radon measure of the form $f(x)\mathrm{d}x$ with $\mathrm{d}x$ the Lebesgue measure, f a nonnegative $L^1(\mathrm{d}x)$ function, and X has a covariance kernel of log-type (12.2). The underlying probability space will be $(\Omega, \mathcal{F}, \mathbb{P})$ and we will denote $\mathbb{E}[.]$ the associated expectation. The vector space of p integrable random variables with $p \geq 1$ will be denoted L^p. We will call a function $\theta : \mathbb{R} \to \mathbb{R}$ a smooth mollifier if θ is C^∞ with compact support and such that $\int_{\mathbb{R}^d} \theta(x)\mathrm{d}x = 1$. We will use θ to regularize the field X by convolution; we will denote by $f * g$ the convolution between two distributions f and g. When θ is smooth, the convolution $X * \theta$ is in fact C^∞ and in particular the exponential of $X * \theta$ is well defined.

In section 12.2, we will also consider centered Gaussian fields Y, Z with continuous covariances kernels and which are almost surely continuous.

12.2   Gaussian multiplicative chaos

Before explaining the construction of the GMC measures, we first give a few reminders on Gaussian vectors and processes.

12.2.1   Reminder on Gaussian vectors and processes

Here, we recall basic properties of Gaussian vectors and processes that we will need in this chapter. The first one is the Girsanov transform.

Theorem 12.2.1 (Girsanov theorem).   Let $(Y(x))_{x\in D}$ be a smooth centered Gaussian field with covariance kernel K and W some Gaussian variable which belongs to the L^2 closure of the subspace spanned by $(Y(x))_{x\in D}$. Let F be some bounded function defined on the space of continuous functions. Then we have the identity

$$\mathbb{E}[e^{W-\frac{\mathbb{E}[W^2]}{2}}F((Y(x))_x)] = \mathbb{E}[F((Y(x) + E[WY(x)])_x)].$$

Though we state the Girsanov theorem under this form, it is usually stated in the following equivalent form: under the new probability measure $e^{W-\frac{\mathbb{E}[W^2]}{2}}\mathrm{d}\mathbb{P}$, the field $(Y(x))_{x\in D}$ has same law as the (shifted) field $(Y(x) + E[WY(x)])_{x\in D}$ under $\mathbb{P}$.

We will also need the following beautiful comparison principle first discovered by Kahane.

Theorem 12.2.2.   Convexity inequalities (Kahane 1985). Let $(Y(x))_{x \in D}$ and $(Z(x))_{x \in D}$ be continuous centered Gaussian fields such that

$$\mathbb{E}[Y(x)Y(y)] \leq \mathbb{E}[Z(x)Z(y)].$$

Then for all convex (resp. concave) functions $F : \mathbb{R}_+ \to \mathbb{R}$ with at most polynomial growth at infinity

$$\mathbb{E}\left[F\left(\int_D e^{Y(x) - \frac{\mathbb{E}[Y(x)^2]}{2}} \sigma(dx)\right)\right] \leq (\text{resp. } \geq) \, \mathbb{E}\left[F\left(\int_D e^{Z(x) - \frac{\mathbb{E}[Z(x)^2]}{2}} \sigma(dx)\right)\right].$$

$$(12.3)$$

12.2.2   Construction of the GMC measures

In this section, we will state a quite general theorem which will be used as definition of the GMC measure. The idea to construct a GMC measure is rather simple and standard: one defines the measure as the limit as ϵ goes to 0 of $c_\epsilon e^{\gamma X_\epsilon} \sigma(dx)$, where X_ϵ is a sequence which converges to X as ϵ goes to 0 and c_ϵ is some normalization sequence which ensures that the limit is nontrivial.

Theorem 12.2.3.   Let θ be a smooth mollifier. Set $X_\epsilon = X * \theta_\epsilon$ where X has a covariance kernel of log-type (12.2) and $\theta_\epsilon = \frac{1}{\epsilon^d}\theta(\frac{\cdot}{\epsilon})$. The random measures

$$M_{\epsilon,\gamma}(dx) = e^{\gamma X_\epsilon - \frac{\gamma^2 \mathbb{E}[X_\epsilon(x)^2]}{2}} \sigma(dx)$$

converge in probability in the space of Radon measures (equipped with the topology of weak convergence) toward a random measure M_γ. The random measure does not depend on the mollifier θ. If $\sigma(dx) = f(x)dx$ with $f > 0$, the measure M_γ is different from 0 if and only if $\gamma < \sqrt{2d}$.

Proof   For simplicity, we will prove Theorem 12.2.3 in the simple case where $\gamma < \sqrt{d}$, the so-called L^2 case. It is not restrictive to suppose $f = 1$ in the proof (the proof works the same with general f). Let θ be some smooth mollifier and $X_\epsilon = X * \theta_\epsilon$. For all compact A, we have by Fubini

$$\mathbb{E}[M_{\epsilon,\gamma}(A)] = \int_A \mathbb{E}[e^{\gamma X_\epsilon(x) - \frac{\gamma^2 \mathbb{E}[X_\epsilon(x)^2]}{2}}]dx = |A|.$$

Hence, we see that the average of $M_{\epsilon,\gamma}(A)$ is constant and equal to the Lebesgue volume of A: this explains the normalization term $\frac{\gamma^2 \mathbb{E}[X_\epsilon(x)^2]}{2}$ in the exponential. By a simple computation, one can show that for all $\epsilon' \leq \epsilon$ there exists global constants $c, C > 0$ such that

$$c + \ln \frac{1}{|y - x| + \epsilon} \leq \mathbb{E}[X_{\epsilon'}(x)X_\epsilon(y)] \leq C + \ln \frac{1}{|y - x| + \epsilon}. \qquad (12.4)$$

One can note that the bounds in (12.4) are independent of the smaller scale ϵ'. Hence, using Fubini, we get that for all compact A

$$\mathbb{E}[M_{\epsilon,\gamma}(A)^2] = \mathbb{E}\left[\left(\int_A e^{\gamma X_\epsilon(x) - \frac{\gamma^2 \mathbb{E}[X_\epsilon(x)^2]}{2}} dx\right)^2\right]$$

$$= \int_A \int_A \mathbb{E}\left[e^{\gamma(X_\epsilon(x) + X_\epsilon(y)) - \frac{\gamma^2 \mathbb{E}[X_\epsilon(x)^2]}{2} - \frac{\gamma^2 \mathbb{E}[X_\epsilon(y)^2]}{2}}\right] dx\, dy$$

$$= \int_A \int_A e^{\gamma^2 \mathbb{E}[X_\epsilon(x) X_\epsilon(y)]} dx\, dy$$

$$\underset{\epsilon \to 0}{\to} \int_A \int_A e^{\gamma^2 K(x,y)} dx\, dy,$$

where the last convergence is a consequence of the simple convergence of $\mathbb{E}[X_\epsilon(x) X_\epsilon(y)]$ toward K for $x \neq y$ and the dominated convergence theorem using (12.4) (the condition $\gamma^2 < d$ ensures the integrability of $e^{\gamma^2 K(x,y)}$).

Now, along the same lines (using Fubini), one can expand for $\epsilon' < \epsilon$ the quantity $\mathbb{E}[(M_{\epsilon,\gamma}(A) - M_{\epsilon',\gamma}(A))^2]$ and show that

$$\mathbb{E}[(M_{\epsilon,\gamma}(A) - M_{\epsilon',\gamma}(A))^2]$$

$$= \int_A \int_A e^{\gamma^2 \mathbb{E}[X_\epsilon(x) X_\epsilon(y)]} dx\, dy + \int_A \int_A e^{\gamma^2 \mathbb{E}[X_{\epsilon'}(x) X_{\epsilon'}(y)]} dx\, dy$$

$$- 2 \int_A \int_A e^{\gamma^2 \mathbb{E}[X_{\epsilon'}(x) X_\epsilon(y)]} dx\, dy$$

$$\underset{\epsilon',\epsilon \to 0}{\to} \int_A \int_A e^{\gamma^2 K(x,y)} dx\, dy + \int_A \int_A e^{\gamma^2 K(x,y)} dx\, dy - 2 \int_A \int_A e^{\gamma^2 K(x,y)} dx\, dy$$

$$= 0.$$

Hence $(M_{\epsilon,\gamma}(A))_{\epsilon > 0}$ is a Cauchy sequence.

Let $\bar{\theta}$ be another smooth mollifier and let $\bar{M}_{\epsilon,\gamma}(dx) = e^{\gamma \bar{X}_\epsilon - \frac{\gamma^2 \mathbb{E}[\bar{X}_\epsilon(x)^2]}{2}} dx$ with $\bar{X}_\epsilon = X * \bar{\theta}_\epsilon$. Along the same lines as previously, one can show that $M_{\epsilon,\gamma}(A) - \bar{M}_{\epsilon,\gamma}(A)$ converges to 0.

In conclusion, we have shown that for all compact A, the variable $M_{\epsilon,\gamma}(A)$ converges in L^2 to some random variable $Z(A)$ of mean $|A|$, and the limit $Z(A)$ does not depend on the smooth mollifier θ. Using standard results of the theory of random measures (see Daley and Vere-Jones (2007)), one can show that there exists a random measure version M_γ of the variables $Z(A)$ such that, in fact, $M_{\epsilon,\gamma}$ converges in probability in the space of random measures (equipped with the weak topology) towards M_γ. Of course, M_γ is nontrivial since for all compact A we have $\mathbb{E}[M_\gamma(A)] = |A|$.

Now for the case $\gamma \in [\sqrt{d}, \sqrt{2d}[$, the earlier L^2 computations no longer converge and one must use more refined techniques to show convergence: we refer to Berestycki's (2015) approach for a simple proof in that case.

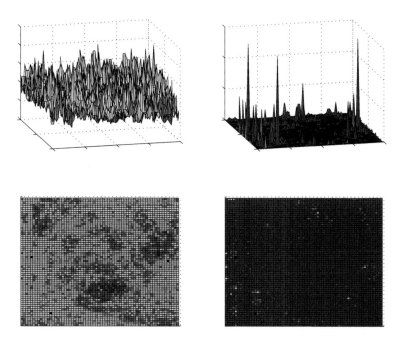

Fig. 12.1   Two examples of GMC measures. Left: weak parameter γ. Right: γ close to 2.

A brief historic on the construction of the GMC measures

In fact, the above convergence result could be strengthened to more general cut-off approximations X_ϵ of the field X. However, for the sake of simplicity, we only stated the theorem with approximations X_ϵ of the form $X * \theta_\epsilon$. Before stating important properties of the measures M_γ, let us briefly review the historics of Theorem 12.2.3. In his 1985 founding paper, Kahane defined the GMC measures by using a sequence of discrete approximations X_n to X: he considered the simplified assumption that the random functions $(X_{n+1}(\cdot) - X_n(\cdot))_n$ are independent.[1] Within this framework, he defined the GMC measure as the almost sure limit of $M_{n,\gamma}(\mathrm{d}x) = e^{\gamma X_n(x) - \frac{\gamma^2 \mathbb{E}[X_n(x)^2]}{2}} \sigma(\mathrm{d}x)$ and showed that the law of the limiting measure does not depend on the sequence X_n. Around 20 years later, Robert and Vargas (2010) proved a weak form of Theorem 12.2.3 by showing convergence in law of $M_{\epsilon,\gamma}(\mathrm{d}x)$. Duplantier–Sheffield (2011) proved Theorem 12.2.3 in the special case where X is the GFF and X_ϵ is a circle average[2] (this work was followed by the work of Chen–Jakobson (2014) where the authors adapted arguments from Duplantier–Sheffield (2011) to the $4d$ case). Recently, the convergence in law proved in Robert

---

[1] Kahane's motivation was the rigorous construction of Mandelbrot's limit lognormal model in turbulence defined in Mandelbrot (1972). Part of Mandelbrot's work is rigorous; Hoegh–Krohn (1971) also proved similar results around the same time.

[2] Duplantier-Sheffield call this specific GMC measure the Liouville measure; in this chapter, we choose a different convention for the terminology Liouville measure.

and Vargas (2010) was reinforced to a convergence in probability by Shamov (2014); the work of Shamov (2014), which relies on abstract Gaussian space theory, is, in fact, quite general and does not concern just log-correlated X. Finally, let us mention that other works have now also established theorem 12.2.3 by rather elementary methods: see Berestycki (2015) and Junnila–Saksman (2015) (this work is also interesting because it extends the theory to the critical case $\gamma = \sqrt{2d}$ where one can define a modified GMC theory; however, we will not consider the critical case $\gamma = \sqrt{2d}$ in this chapter). Berestycki (2015) is probably a very good starting point for someone who wants to learn GMC theory.

12.2.3 Main properties of the GMC measures

Now, we turn to some important properties of the GMC measures which we will need in our study of LQG.

Existence of moments and multifractality

Theorem 12.2.4. For $\gamma < \sqrt{2d}$, let M_γ be a GMC measure associated with a log-correlated field X with covariance (12.2) and $\sigma(\mathrm{d}x) = f(x)\mathrm{d}x$ with bounded f. Then, for $O \subset D$ an open ball we have

$$\mathbb{E}[M_\gamma(O)^q] < \infty$$

if and only if $q \in] -\infty, \frac{2d}{\gamma^2}[$.

We will not prove this theorem here: we refer to Robert and Vargas (2010) for a proof. Now, we turn to the multifractal scaling of the measure. This is the content of the following.

Proposition 12.2.5. For $\gamma < \sqrt{2d}$, let M_γ be a GMC measure associated with a log-correlated field X with covariance (12.2) and $\sigma(\mathrm{d}x) = f(x)\mathrm{d}x$ with bounded continuous f. Then for all x and all $q \in] -\infty, \frac{2d}{\gamma^2}[$, there exists some constant $C_x > 0$ (which depends also on f, q, and the exact form of the kernel K in (12.2)) such that

$$\mathbb{E}[M_\gamma(B(x,r))^q] \underset{r \to 0}{\sim} C_x r^{\zeta(q)}, \tag{12.5}$$

where $\zeta(q) = (d + \frac{\gamma^2}{2})q - \frac{\gamma^2 q^2}{2}$ is called the structure function of M_γ.

Proposition 12.5 implies that the GMC measure associated with a log-correlated field X exhibits multifractal behaviour; i.e., the measure is not scale invariant but rather is locally Hölder around each point. The Hölder exponent depends on the point (for more on the so-called multifractal formalism, see the next subsection). More generally, one can take as a definition that a random measure satisfying (12.5) where ζ is a strictly concave function is a multifractal measure.

Multifractal formalism

Now, we turn to the multifractal formalism of the measures M_γ. The measures M_γ are multifractal in the sense that the regularity of the measure around a point $x \in D$ depends on the point x: this can easily be seen in Fig. 12.1. Multifractal formalism is a general theory for studying the regularity of measures like M_γ around each point: for more background on this see section 4 in Rhodes and Vargas (2014).

For $\gamma^2 < 2d$ and $q \in]0, \frac{\sqrt{2d}}{\gamma}[$, we consider the set

$$K_{\gamma,q} = \left\{ x \in D; \lim_{\epsilon \to 0} \frac{\ln M_\gamma(B(x,\epsilon))}{\ln \epsilon} = d + (\frac{1}{2} - q)\gamma^2 \right\}.$$

In words, the set $K_{\gamma,q}$ is made of the points x such that $M_\gamma(B(x,r)) \underset{r \to 0}{\approx} r^{d+(\frac{1}{2}-q)\gamma^2}$. We can state the following theorem.

Theorem 12.2.6.   The set $K_{\gamma,q}$ has Hausdorff dimension $d - \frac{\gamma^2 q^2}{2}$.

In fact, the same theorem holds with the set $\bar{K}_{\gamma,q}$ defined by

$$\bar{K}_{\gamma,q} = \left\{ x \in D; \lim_{\epsilon \to 0} \frac{X_\epsilon(x)}{-\ln \epsilon} = \gamma q \right\},$$

where $X_\epsilon = X * \theta_\epsilon$ with θ any smooth mollifier. The reason is that it is useful to have in mind the approximation

$$M_\gamma(B(x,r)) \underset{r \to 0}{\approx} r^d e^{\gamma X_r(x) - \frac{\gamma^2 \mathbb{E}[X_r(x)^2]}{2}}, \tag{12.6}$$

where here $a_r \approx b_r$ means that the ratio a/b is a (random) constant C_r of order 1; i.e., $\mathbb{E}[C_r]$ belongs to an interval $[c, C]$ with $c, C > 0$ independent of r. The main difficulty in our context is that the random constant $C_{x,r}$ for the ratio of both sides in (12.6) really also depends on x so the approximation (12.6) cannot be used directly but is rather a guideline to get intuition on the behavior of M_γ. In our case, if we assume $C_{x,r} = 1$, then the sets $K_{\gamma,q} = \bar{K}_{\gamma,q}$ are the same (however, we stress that rigorously these two sets are not the same). Finally, following the terminology of Hu et al. (2010), a point x which belongs to $\bar{K}_{\gamma,1}$ is nowadays called a γ-thick point.

Now, among the sets $K_{\gamma,q}$ (and $\bar{K}_{\gamma,q}$), the set $K_{\gamma,1}$ (resp. $\bar{K}_{\gamma,1}$) is of particular importance for M_γ since it is the set on which the measure M_γ 'lives'. More specifically, we have

$$M_\gamma(^c K_{\gamma,1} \cup {}^c \bar{K}_{\gamma,1}) = 0. \tag{12.7}$$

In the modern terminology of Hu et al. (2010), one says that M_γ lives on the γ-thick points of X. This property was proved by Kahane in his seminal paper (Kahane 1985). Here, we will show a slightly weaker result, namely that

$$M_\gamma \left({}^c \left\{ x \in D; \lim_{n \to \infty} \frac{X_{\frac{1}{2^n}}(x)}{n \ln 2} = \gamma \right\} \right) = 0. \tag{12.8}$$

The only difference with $\bar{K}_{\gamma,1}$ is that we restrict the limit in $\bar{K}_{\gamma,1}$ to a dyadic sequence (in fact, with little effort, one can reinforce (12.8) to prove (12.7)).

Proof of (12.8) We introduce $\eta > 0$ and a compact set A. We have for all $n \leq p$ and by using the Girsanov Theorem 12.2.1 that

$$
\mathbb{E}\Big[\int_A \mathbb{1}_{\left\{x \in D;\ \frac{X_{2-n}(x)}{n \ln 2} \in {}^c[\gamma-\eta,\gamma+\eta]\right\}} e^{\gamma X_{2-p}(x) - \frac{\gamma^2}{2}\mathbb{E}[X_{2-p}(x)^2]} dx\Big]
$$

$$
= \int_A \mathbb{E}[\mathbb{1}_{\left\{x \in D;\ \frac{X_{2-n}(x)}{n \ln 2} \in {}^c[\gamma-\eta,\gamma+\eta]\right\}} e^{\gamma X_{2-p}(x) - \frac{\gamma^2}{2}\mathbb{E}[X_{2-p}(x)^2]}] dx
$$

$$
= \int_A \mathbb{E}[\mathbb{1}_{\left\{x \in D;\ \frac{X_{2-n}(x) + \gamma \mathbb{E}[X_{2-n}(x) X_{2-p}(x)]}{n \ln 2} \in {}^c[\gamma-\eta,\gamma+\eta]\right\}}] dx
$$

$$
\approx \int_A \mathbb{P}\left(\frac{X_{2-n}(x)}{n \ln 2} \in {}^c[-\eta,\eta]\right) dx.
$$

Now, since $X_{2-n}(x)$ is a Gaussian of variance roughly equal to $n \ln 2$ by (12.4), we get that

$$
\mathbb{P}\left(\frac{X_{2-n}(x)}{n \ln 2} \in {}^c[-\eta,\eta]\right) \leq 2e^{-n\eta^2 \frac{(\ln 2)^2}{2}}.
$$

Therefore, by taking the limit $p \to \infty$ in the above considerations, we get that there exists $C > 0$

$$
M_\gamma\left(\left\{x \in D;\ \frac{X_{2-n}(x)}{n \ln 2} \in {}^c[\gamma-\eta,\gamma+\eta]\right\}\right) \leq Ce^{-n\eta^2 \frac{(\ln 2)^2}{2}}.
$$

One can easily deduce from this by a Borell–Cantelli type argument that

$$
M_\gamma\left(\cap_N \cup_{n \geq N} \left\{x \in D;\ \frac{X_{2-n}(x)}{n \ln 2} \in {}^c[\gamma-\eta,\gamma+\eta]\right\}\right) = 0.
$$

Since the result is valid for all $\eta > 0$, we get (12.8). □

The first Seiberg bound

In this subsection, we state and prove a theorem we will need to define LQG: indeed, we will see that it corresponds to the so-called Seiberg bound in LQG. We have the following.

Lemma 12.2.7. *Let $\alpha \in \mathbb{R}$ and $x \in D$. We have*

$$
\int_{B(x,1)} \frac{1}{|y - x|^{\alpha\gamma}} M_\gamma(dy) < \infty, \quad a.s.
$$

if and only if $\alpha < \frac{d}{\gamma} + \frac{\gamma}{2}$.

Proof We only prove the if part; for the only if part, we refer to David et al. (2016). With no loss of generality, we suppose that $x = 0$. We consider $\eta \in]0, 1[$. We have

$$
\mathbb{E}\left[\left(\int_{B(0,1)} \frac{1}{|y|^{\alpha\gamma}} M_\gamma(dy)\right)^\eta\right] \le \sum_{n=1}^\infty \mathbb{E}\left[\left(\int_{\frac{1}{2^n} \le |y| \le \frac{1}{2^{n-1}}} \frac{1}{|y|^{\alpha\gamma}} M_\gamma(dy)\right)^\eta\right]
$$

$$
\le \sum_{n=1}^\infty 2^{n\alpha\eta\gamma} \mathbb{E}[(M_\gamma(\{y; \frac{1}{2^n} \le |y| \le \frac{1}{2^{n-1}}\}))^\eta]
$$

$$
\le \sum_{n=1}^\infty 2^{n\alpha\eta\gamma} \mathbb{E}[(M_\gamma(\{y; |y| \le \frac{1}{2^{n-1}}\}))^\eta]
$$

$$
\le C \sum_{n=1}^\infty 2^{n\alpha\gamma\eta} 2^{-n\zeta(\eta)},
$$

where recall that $\zeta(q) = (d + \frac{\gamma^2}{2})q - \frac{\gamma^2 q^2}{2}$. Now, since $\alpha < \frac{d}{\gamma} + \frac{\gamma}{2}$, one can choose $\eta > 0$ small such that $\alpha\gamma\eta - \zeta(\eta) < 0$; hence, we get the conclusion.

12.3   Liouville quantum gravity on the Riemann sphere

Now, in the second part of this chapter, we show how to use GMC theory to construct Liouville quantum gravity on the Riemann sphere. LQG was introduced in Polyakov's seminal 1981 paper (Polyakov 1981). In the paper, Polyakov builds a theory of summation of 2d random surfaces in the spirit of Feynman's theory of summation of random paths. On the Riemann sphere, LQG is, in fact, equivalent to Liouville quantum field theory; for a complete review on LQFT in the physics literature, we refer to Nakayama (2004). However, LQG is a general theory of random surfaces which can be defined on any 2d surface. In the case of higher genus surfaces, LQFT is a building block of LQG and they are not equivalent. For the sake of simplicity, we will restrict ourselves here to the case of the sphere where we identify LQG and LQFT: in this context, we explain the construction of LQFT following David et al. (2016).

LQFT is not only a quantum field theory but since it has extra symmetries it is also a conformal field theory (CFT). Quantum field theory and conformal field theory is a very wide topic in mathematical physics which can be approached in different ways: by algebraic methods, geometric methods, and probabilistic methods. Of course, all these approaches can be related but for reasons of simplicity (and the knowledge of the authors!) we will restrict to the probabilistic setting. Before we describe the theory, we first give a brief introduction to what is a CFT on the Riemann sphere. Then, we introduce a few notations and definitions from elementary Riemannian geometry.

12.3.1   Elementary Riemannian geometry on the sphere

We consider the standard Riemann sphere $\mathbb{S} = \mathbb{C} \cup \{\infty\}$. The Riemann sphere $\mathbb{S}$ is just the complex plane $\mathbb{C}$ with a point at infinity and is obtained as the image of the

standard $2d$ sphere by stereographic projection. We equip $\mathbb{S}$ with the standard round metric. On $\mathbb{S}$, the round metric is given in Riemannian geometry notations by $g(z)|\mathrm{d}z|^2$ where $g(z) = \frac{4}{(1+|z|^2)^2}$. This means that the length $\mathcal{L}(\sigma)$ of a curve $\sigma : [0, 1] \to \mathbb{S}$ is given by

$$\mathcal{L}(\sigma) = \int_0^1 g(\sigma(t))^{1/2}|\sigma'(t)|\mathrm{d}t.$$

One then gets the distance between two points $z_1, z_2 \in \mathbb{S}$ by taking the infimum of $\mathcal{L}(\sigma)$ over all curves σ which join z_1 to z_2. The volume form is simply given by the measure $g(z)\mathrm{d}z$ where $\mathrm{d}z$ is the Lebesgue measure on $\mathbb{R}^2$ (by using polar coordinates, it is easy to see that $\int_{\mathbb{S}} g(z)\mathrm{d}z = 4\pi$, thereby recovering the well-known fact that the surface of the sphere is 4π!). In this context, one can of course do differential calculus and C^k functions on $\mathbb{S}$ are just functions ϕ defined on $\mathbb{C}$ which are such that ϕ is C^k on $\mathbb{C}$ and $z \mapsto \phi(\frac{1}{z})$ admits a continuous extension on $\mathbb{C}$ which is C^k. The gradient ∇_g of a function ϕ is given by the simple formula

$$\nabla_g\phi(z) = \frac{1}{g(z)}\nabla_z\phi(z),$$

where ∇_z is the standard Euclidean gradient on $\mathbb{C}$. Finally, the (Ricci) curvature R_g is given by

$$R_g(z) = -\frac{1}{g(z)}\Delta_z \ln g(z),$$

where Δ_z is the standard Euclidean Laplacian. In the specific case of the round metric $(g(z) = \frac{4}{(1+|z|^2)^2})$, one finds by a simple computation a constant curvature $R_g = 2$.

12.3.2   An introduction to CFT on the Riemann sphere

The general formalism of CFT was built in the celebrated 1984 work of Belavin et al. (1984). Here we give an elementary (and incomplete) exposition of this formalism. A CFT on the Riemann sphere is usually defined by the following:

1. a real parameter c_{CFT} called the central charge
2. (primary) local fields $(\phi_\alpha)_{\alpha \in \mathcal{A}}$ defined in the complex plane $\mathbb{C}$.
3. prescribed symmetries (conformal covariance, diffeomorphism invariance, Weyl anomaly: see Gawedzki's lecture notes (Gawedzki 1999) for further details): see equality (12.9) for the conformal covariance statement.

It is not obvious to give a simple definition of the central charge but we will see in the example of LQFT how it appears. For now, let us just mention that the central charge of a CFT determines the symmetries of the theory; however, it is very important to stress that two CFTs with same central charge can be very different because the set of primary local fields plays an essential role too. In a CFT theory, what makes sense are the correlation functions $< \phi_{\alpha_1}(z_1) \cdots \phi_{\alpha_n}(z_n) >$ at noncoincident points z_i (i.e., $z_i \neq z_j$ for $i \neq j$) and for certain values of the α_i where $< . >$ should be

viewed as some underlying measure (however, this is a view as the measure does not necessarily exist). The correlation functions of primary local fields satisfy the following conformal covariance: if ψ is a Möbius transform on the sphere $\mathbb{S}$, i.e., $\psi(z) = \frac{az+b}{cz+d}$ where $a, b, c, d \in \mathbb{C}$ are such that $ad - bc = 1$, then

$$< \phi_{\alpha_1}(\psi(z_1)) \cdots \phi_{\alpha_n}(\psi(z_n)) >= \prod_{i=1}^{n} |\psi'(z_i)|^{-2\Delta_{\alpha_i}} < \phi_{\alpha_1}(z_1) \cdots \phi_{\alpha_n}(z_n) >, \quad (12.9)$$

where the real number Δ_{α_i} is called the conformal weight of the field ϕ_{α_i}. One of the successes of CFT is that it describes (conjecturally in mathematical standards) the scaling limit of correlation functions of statistical physics models at critical temperature. It is a major program in mathematical physics to make these predictions from CFT rigorous mathematical statements.

In some cases, one can also define ϕ_α in a strong sense as a random distribution in the sense of Schwartz: in that case, if $\mathcal{D}$ denotes the set of smooth functions with compact support one can consider the random distribution $\varphi \in \mathcal{D} \to \int_\mathbb{C} \phi_\alpha(x)\varphi(x)\mathrm{d}x$. In this case, the underlying measure really exists (this will be the case for some but not all primary local fields in the two examples we will consider in these notes: LQFT and the Ising model at critical temperature) and one can compute the moments of the variable $\int \phi_\alpha(z)\varphi(z)\mathrm{d}z$ (if they exist) in terms of the correlation functions by the obvious formula

$$< \left(\int_\mathbb{C} \phi_\alpha(z)\varphi(z)\mathrm{d}z \right)^n >= \int_\mathbb{C} \cdots \int_\mathbb{C} < \phi_\alpha(z_1) \cdots \phi_\alpha(z_n) > \varphi(z_1) \cdots \varphi(z_n)\,\mathrm{d}z_1 \cdots \mathrm{d}z_n.$$

Hence, in many cases, the correlation functions determine the joint laws of the collection $\left(\int_\mathbb{C} \phi_\alpha(z)\varphi(z)\mathrm{d}z \right)_{\varphi \in \mathcal{D}}$.

12.3.3   Introduction to LQFT on the Riemann sphere

LQFT is a family of CFTs parametrized by two constants $\gamma \in]0, 2]$ and $\mu > 0$; in this chapter, we will only consider the case $\gamma \in]0, 2[$. In the probabilistic setting, the goal of LQFT is to make sense of and compute as much as possible the following correlation functions which arise in theoretical physics under the heuristic form

$$< e^{\alpha_1 X(z_1)} \cdots e^{\alpha_n X(z_n)} >:= \int e^{\alpha_1 X(z_1)} \cdots e^{\alpha_n X(z_n)} e^{-S_L(X,g)} DX,$$

where DX is the 'Lebesgue' measure on functions $\mathbb{S} \to \mathbb{R}$ and S_L is the Liouville action

$$S_L(X, g) := \frac{1}{4\pi} \int_\mathbb{S} (|\nabla_g X|^2(z) + QR_g(z)X(z) + 4\pi\mu e^{\gamma X(z)}) g(z)\mathrm{d}z, \quad (12.10)$$

where recall that g is the round metric, the constant Q is defined by $Q = \frac{\gamma}{2} + \frac{2}{\gamma}$, and $\mu > 0$. LQFT is therefore an *interacting* quantum field theory where the interaction term is

$$\mu \int_{\mathbb{S}} e^{\gamma X(z)} g(z) dz. \tag{12.11}$$

The positive parameter μ, called the cosmological constant, is necessary for the exist-ence of LQFT. However, a remarkable feature of LQFT is that the parameter γ is the essential parameter of the theory as it completely determines the conformal properties of the theory (in CFT language, the parameter γ determines the central charge: we will come back to this point later in more detail). Following the standard terminology of CFT (see section 12.3.2), the $e^{\alpha_i X(z_i)}$ are local primary fields (the conformal co-variance property will be stated in section 12.3.5) in fact, in the context of LQFT, the $e^{\alpha_i X(z_i)}$ are also called vertex operators.

It is a well-known fact that the 'Lebesgue measure' DX does not exist since the space of functions $\mathbb{S} \to \mathbb{R}$ is infinite dimensional; however, it is a standard procedure in the probabilistic approach to quantum field theory (see Simon (1974) on the topic) to interpret the term $e^{-\frac{1}{4\pi} \int_{\mathbb{R}^2} |\nabla_g X|^2(z) g(z) dz} DX$ as the Gaussian free field (GFF), i.e., the Gaussian field whose covariance is given by the Green function on $\mathbb{S}$. One way to see that this is the proper definition is to perform the following integration by parts:

$$e^{-\frac{1}{4\pi} \int_{\mathbb{R}^2} |\nabla_g X|^2(z) dz} DX = e^{\frac{1}{4\pi} \int_{\mathbb{R}^2} X(z) \Delta_g X(z) g(z) dz} DX.$$

Formally, this corresponds to a Gaussian with covariance $2\pi(-\Delta_g)^{-1}$. In fact, thanks to the theory of probability, one can define the GFF rigorously with the following.

Definition 12.3.1. *The GFF with vanishing mean on the sphere X_g is the Gaussian field living in the space of distributions such that for all smooth functions f, h on $\mathbb{S}$*

$$\mathbb{E}\left[\left(\int_{\mathbb{S}} f(z) X_g(z) g(z) dz\right)\left(\int_{\mathbb{S}} h(z') X_g(z') g(z') dz'\right)\right]$$
$$= \int_{\mathbb{S}} \int_{\mathbb{S}} G_g(z, z') f(z) h(z') g(z) g(z') dz dz',$$

where G is the Green function for the Laplacian on the sphere defined for all $z \in \mathbb{S}$ by

$$-\Delta_g G(z, .) = 2\pi(\delta_z - \frac{1}{4\pi}), \quad \int_{\mathbb{S}} G_g(z, z') g(z') dz' = 0.$$

The random variable X_g lives in the space of random distributions but, in fact, it exists in a Sobolev space and $\int_{\mathbb{S}} f(z) X_g(z) g(z) dz$ makes sense for many functions f (with less regularity than C^∞). In particular, $\int_{\mathbb{S}} X_g(z) g(z) dz$ makes sense and is equal to 0 actually: this is why we call X_g the GFF with vanishing mean on the sphere. It turns out that the Green function has the following explicit form on $\mathbb{S}$,

$$G_g(z, z') = \ln \frac{(1 + |z|^2)^{1/2}(1 + |z'|^2)^{1/2}}{|z - z'|},$$

where recall that $|.|$ is the standard Euclidean distance.

Now, in the spirit of probabilistic quantum field theory, since formally we have

$$\mathrm{e}^{-S_L(X,g)}DX = \mathrm{e}^{-\frac{1}{4\pi}\int_{\mathbb{S}}(QR_g(z)X(z)+4\pi\mu\mathrm{e}^{\gamma X(z)})g(z)\mathrm{d}z} \times \mathrm{e}^{-\frac{1}{4\pi}\int_{\mathbb{R}^2}|\nabla_g X|^2(z)g(z)\mathrm{d}z}DX$$

and since we interpret $\mathrm{e}^{-\frac{1}{4\pi}\int_{\mathbb{R}^2}|\nabla_g X|^2(z)g(z)\mathrm{d}z}DX$ as the GFF measure, it is natural to interpret the formal measure $\mathrm{e}^{-S_L(X,g)}DX$ as follows, for all functions F (up to a global constant)

$$\int F(X)\mathrm{e}^{-S_L(X,g)}DX = \lim_{\epsilon\to 0}\mathbb{E}[F(X_g)\mathrm{e}^{-\frac{Q}{4\pi}\int_{\mathbb{S}}R_g(x)X_g(z)g(z)\mathrm{d}z-\mu\epsilon^{\gamma^2/2}\int_{\mathbb{S}}\mathrm{e}^{\gamma\bar{X}_{\epsilon,g}(z)}g(z)\mathrm{d}z}],$$

$$(12.12)$$

where $\bar{X}_{\epsilon,g}$ is the average of X_g in a ball of radius ϵ with respect to the metric g. However, there is something wrong with definition (12.12); though it can be used to define a standard quantum field theory in the spirit of Simon (1974), it will lack symmetry to define a CFT. The reason is that we have not taken into account the contribution of constant functions in the Gaussian measure. This omission reflects in the fact that there is something arbitrary in the choice of X_g: indeed, X_g has vanishing mean on the sphere but we could have chosen another GFF on the sphere. In particular, X_g is not conformally invariant since for all Möbius transform ψ the following equality holds in distribution:

$$X_g\circ\psi - \int_{\mathbb{S}}(X_g\circ\psi(z))g(z)\mathrm{d}z = X_g.$$

Now, the average $\int_{\mathbb{S}}X_g\circ\psi(z)g(z)\mathrm{d}z$ is a Gaussian random variable which is nonzero (unless ψ is an isometry of $\mathbb{S}$) and hence $X_g\circ\psi$ does not have the same distribution as X_g. One very natural way to get rid of this average dependence is to replace X_g by X_g+c where c is distributed according to the Lebesgue measure (and stands for the mean value of the field). This leads to the following correct definition (up to some global constant):

$$\int F(X)\mathrm{e}^{-S_L(X,g)}DX$$

$$= \lim_{\epsilon\to 0}\int_{\mathbb{R}}\mathbb{E}[F(X_g+c)\mathrm{e}^{-\frac{Q}{4\pi}\int_{\mathbb{S}}R_g(z)(X_g(z)+c)g(z)\mathrm{d}z-\mu\epsilon^{\gamma^2/2}\mathrm{e}^{\gamma c}\int_{\mathbb{S}}\mathrm{e}^{\gamma\bar{X}_{\epsilon,g}(z)}g(z)\mathrm{d}z}]\mathrm{d}c$$

$$(12.13)$$

A standard computation shows that

$$\mathbb{E}[\bar{X}_{\epsilon,g}(z)^2] = \ln\frac{1}{\epsilon} - \frac{1}{2}\ln g(z) + C + o(1),$$ $$(12.14)$$

where C is some global constant; therefore, the measure $\epsilon^{\gamma^2/2}\mathrm{e}^{\gamma\bar{X}_{\epsilon,g}(z)}g(z)\mathrm{d}z$ converges to $\mathrm{e}^{\frac{\gamma^2}{2}C}$ times the GMC measure M_γ associated with X_g and $g(z)\mathrm{d}z$ which we write

$$M_\gamma(\mathrm{d}z) = \mathrm{e}^{\gamma X_g(z)}g(z)\mathrm{d}z.$$ $$(12.15)$$

By the previous results on GMC theory, this GMC measure is well defined and nontrivial. In the sequel, we will exclusively work with this GMC measure.

12.3.4 Construction of LQFT

With the preliminary remarks of the previous subsection, we are ready to introduce the correlation functions of LQFT on the sphere and recover many known properties in the physics literature. In fact, it is standard in the physics literature to express the correlations of LQFT in the complex plane and therefore to shift the metric dependence of the theory in the field $X_g + c$: this simplifies many computations. Let us describe how to do so. If ϵ is small then a ball $B_g(z, \epsilon)$ of centre z and radius ϵ in the round metric g is to first order in ϵ the same as an Euclidean ball $B(z, \frac{\epsilon}{g(z)^{1/2}})$ of center z and radius $\frac{\epsilon}{g(z)^{1/2}}$. Hence, the average $\bar{X}_{\epsilon,g}(z)$ (with respect to balls in the round metric) is roughly the same as $X_{\frac{\epsilon}{g(z)^{1/2}},g}(z)$, where $X_{\epsilon,g}(z)$ is the average of X_g on an Euclidean ball of radius ϵ. Finally, note that we can write for all $\epsilon' > 0$

$$(\epsilon')^{\gamma^2/2} \int_{\mathbb{S}} e^{\gamma \bar{X}_{\epsilon',g}(z)} g(z) \mathrm{d}z = \int_{\mathbb{S}} \left(\frac{\epsilon'}{g(z)^{1/2}} \right)^{\gamma^2/2} e^{\gamma(\bar{X}_{\epsilon',g}(z) + \frac{Q}{2} \ln g(z))} \mathrm{d}z, \qquad (12.16)$$

where recall that $Q = \frac{\gamma}{2} + \frac{2}{\gamma}$. Since $\bar{X}_{\epsilon',g}(z) \approx X_{\frac{\epsilon'}{g(z)^{1/2}},g}(z)$, by making the change of variable $\epsilon = \frac{\epsilon'}{g(z)^{1/2}}$ in (12.16), it is not suprising that one can prove that the random measures

$$e^{\gamma(X_{\epsilon,g}(z) + \frac{Q}{2} \ln g(z))} \mathrm{d}z$$

converge in probability as ϵ goes to 0 toward $e^{\frac{\gamma^2}{2}C} M_\gamma(\mathrm{d}z)$, where C is defined by (12.14) and M_γ is defined by (12.15). Therefore, instead of working with $X_g + c$, we will work with the shifted field $\phi(z) = X_g(z) + c + \frac{Q}{2} \ln g(z)$ and the approximations $\phi_\epsilon(z) = X_{\epsilon,g} + c + \frac{Q}{2} \ln g(z)$. The field ϕ under the probability measure (12.13) is called the *Liouville field*. Finally, we set formally $V_\alpha(z) = e^{\alpha\phi(z)}$ and define the associated approximate vertex operators

$$V_{\alpha,\epsilon}(z) := \epsilon^{\alpha^2/2} e^{\alpha\phi_\epsilon(z)}. \qquad (12.17)$$

The correlation functions of LQFT are now defined by the formula

$$< \prod_{i=1}^{n} V_{\alpha_i}(z_i) >$$

$$:= Z_{\mathrm{GFF}}(g) \lim_{\epsilon \to 0} \int_{\mathbb{R}} \mathbb{E}\left[\prod_{i=1}^{n} V_{\alpha_i,\epsilon}(z_i) e^{-\frac{Q}{4\pi} \int_{\mathbb{S}} R_g(z)(X_g(z)+c)g(z)\mathrm{d}z - \mu\epsilon^{\gamma^2/2} \int_{\mathbb{S}} e^{\gamma\phi_\epsilon(z)} \mathrm{d}z} \right] \mathrm{d}c,$$

$$(12.18)$$

where one can note the presence of the partition function of the GFF $Z_{\mathrm{GFF}}(g)$ given by $\mathrm{Det}\,\Delta_g^{-1/2}$, where $\mathrm{Det}\,\Delta_g$ is the standard determinant of the Laplacian (this determinant is, in fact, nontrivial to define since the Laplacian is defined on an infinite dimensional space: see Dubédat (2011) for background). The constant $Z_{\mathrm{GFF}}(g)$ is a global constant and plays no role here so it is not important to understand exactly

how it is defined. For the readers who are unfamiliar with $\mathrm{Det}\, \Delta_g$ they can take out this term in definition (12.18) and remember that it only plays a role in the Weyl anomaly formula (see Proposition 12.3.4).

Of course, it is crucial to enquire when the limit (12.18) exists. This is the object of the following:

Proposition 12.3.2 (David et al. 2016). The correlation functions (12.18) exist and are not equal to 0 if and only if the following Seiberg bounds hold

$$\forall i, \; \alpha_i < Q \quad \text{and} \quad \sum_{i=1}^{n} \alpha_i > 2Q. \tag{12.19}$$

In particular, the number of vertex operators n must be greater or equal to 3 for the correlation functions to exist and be non trivial. If the Seiberg bounds hold then we get the following expression (up to some multiplicative constant which plays no role and depends on the C of (12.14), $\alpha_1, \cdots, \alpha_n$ and γ)

$$< \prod_{i=1}^{n} V_{\alpha_i}(z_i) >$$

$$= Z_{\mathrm{GFF}}(g) \, e^{\frac{1}{2} \sum_{i \neq j} \alpha_i \alpha_j G_g(z_i, z_j)} \prod_{i=1}^{n} g(z_i)^{\frac{\alpha_i Q}{2} - \frac{\alpha_i^2}{4}} \Gamma\left(\frac{\sum_i \alpha_i - 2Q}{\gamma}, \mu\right)$$

$$\mathbb{E}[(Z_{(z_i, \alpha_i)}(\mathbb{S}))^{-\frac{\sum_i \alpha_i - 2Q}{\gamma}}] \tag{12.20}$$

where

$$\Gamma\left(\frac{\sum_i \alpha_i - 2Q}{\gamma}, \mu\right) = \int_0^{\infty} u^{\frac{\sum_i \alpha_i - 2Q}{\gamma} - 1} e^{-\mu u} du$$

and

$$Z_{(z_i, \alpha_i)}(dz) = e^{\gamma \sum_{i=1}^{n} \alpha_i G_g(z_i, z)} M_\gamma(dz).$$

Proof Here, we give a sketch of the proof of the if part of proposition 12.3.2: therefore, we suppose that the $(\alpha_i)_i$ satisfy the Seiberg bounds (12.19). We denote $< \prod_{i=1}^{n} V_{\alpha_i, \epsilon}(z_i) >$ the right-hand side of (12.18). Since $R_g = 2$ and X_g has vanishing mean on the sphere one has

$$< \prod_{i=1}^{n} V_{\alpha_i, \epsilon}(z_i) > / Z_{\mathrm{GFF}}(g)$$

$$= \int_{\mathbb{R}} \mathbb{E}[\prod_{i=1}^{n} V_{\alpha_i, \epsilon}(z_i) e^{-\frac{Q}{4\pi} \int_{\mathbb{S}} R_g(z)(X_g(z)+c)g(z)dz - \mu \epsilon^{\gamma^2/2} \int_{\mathbb{S}} e^{\gamma \phi_\epsilon(z)} dz}] dc$$

$$= \int_{\mathbb{R}} \mathbb{E}[e^{(\sum_i \alpha_i - 2Q)c} \prod_{i=1}^{n} \epsilon^{\alpha_i^2/2} e^{\alpha_i (X_{g,\epsilon}(z_i) + \frac{Q}{2} \ln g(z_i))} e^{-\mu \epsilon^{\gamma^2/2} e^{\gamma c} \int_{\mathbb{S}} e^{\gamma (X_{g,\epsilon}(z) + \frac{Q}{2} \ln g(z))} dz}] dc$$

Now, the first step is to get rid of the vertex fields $V_{\alpha_i, \epsilon}(z_i)$ in this expression since they do not converge pointwise as ϵ goes to 0. First, we have by (12.14) that

$$\mathbb{E}[(\sum_{i=1}^{n} \alpha_i X_{g,\epsilon}(z_i))^2] = (\sum_{i=1}^{n} \alpha_i^2) \ln \frac{1}{\epsilon} - \frac{1}{2} \sum_{i=1}^{n} \alpha_i^2 \ln g(z_i) + \sum_{i \neq j} \alpha_i \alpha_j G_g(z_i, z_j)$$

$$+ (\sum_{i=1}^{n} \alpha_i^2) C + o(1) \tag{12.21}$$

where $o(1)$ converges to 0 when ϵ goes to 0. We set

$$Y_\epsilon = \sum_{i=1}^{n} \alpha_i X_{g,\epsilon}(z_i).$$

If we apply the Girsanov theorem with the variable Y_ϵ and the field $X_{g,\epsilon}(z)$, we get using (12.21) that up to $e^{O(1)}$ terms we have

$$< \prod_{i=1}^{n} V_{\alpha_i, \epsilon}(z_i) > /Z_{\mathrm{GFF}}(g)$$

$$= e^{\frac{1}{2} \sum_{i \neq j} \alpha_i \alpha_j G_g(z_i, z_j)} \prod_{i=1}^{n} g(z_i)^{\frac{\alpha_i Q}{2} - \frac{\alpha_i^2}{4}}$$

$$\times \int_{\mathbb{R}} \mathbb{E}[e^{(\sum_i \alpha_i - 2Q)c} e^{-\mu e^{\gamma^2/2} e^{\gamma c} \int_{\mathbb{S}} e^{\gamma (X_{g,\epsilon}(z) + H_{(z_i, \alpha_i), \epsilon}(z) + \frac{Q}{2} \ln g(z))} dz}] dc,$$

where $H_{(z_i, \alpha_i), \epsilon}(z) = \gamma \sum_{i=1}^{n} \alpha_i G_{g,\epsilon}(z_i, z)$ with $G_{g,\epsilon}(z, y) = \mathbb{E}[X_{g,\epsilon}(z) X_{g,\epsilon}(y)]$. We set

$$Z_{(z_i, \alpha_i), \epsilon}(dz) = e^{\gamma \sum_{i=1}^{n} \alpha_i G_{g,\epsilon}(z_i, z)} M_{\gamma, \epsilon}(dz),$$

where $M_{\gamma, \epsilon}(dz) = e^{\gamma (X_{g,\epsilon}(z) + \frac{Q}{2} \ln g(z))} dz$. Now, we make the change of variables

$$u = \epsilon^{\gamma^2/2} e^{\gamma c} Z_{(z_i, \alpha_i), \epsilon}(\mathbb{S})$$

in the above formula, which leads to

$$< \prod_{i=1}^{n} V_{\alpha_i, \epsilon}(z_i) > /Z_{\mathrm{GFF}}(g)$$

$$= \frac{1}{\gamma} e^{\frac{1}{2} \sum_{i \neq j} \alpha_i \alpha_j G_g(z_i, z_j)} \prod_{i=1}^{n} g(z_i)^{\frac{\alpha_i Q}{2} - \frac{\alpha_i^2}{4}} \Gamma(\frac{\sum_i \alpha_i - 2Q}{\gamma}, \mu) \mathbb{E}[(Z_{(z_i, \alpha_i), \epsilon}(\mathbb{S})^{-\frac{\sum_i \alpha_i - 2Q}{\gamma}}].$$

In particular, since $G_{g,\epsilon}(z, y)$ converges pointwise to $G_g(z, y)$ for $z \neq y$, it is natural to expect in view of Lemma 12.2.7 that $\mathbb{E}[(Z_{(z_i, \alpha_i), \epsilon}(\mathbb{S})^{-\frac{\sum_i \alpha_i - 2Q}{\gamma}}]$ converges to $\mathbb{E}[(Z_{(z_i, \alpha_i)}(\mathbb{S})^{-\frac{\sum_i \alpha_i - 2Q}{\gamma}}]$ as ϵ goes to 0 (we do not prove this here): if we admit this convergence, we get (12.20).

12.3.5 Properties of the theory

Now, we state that the vertex operators are indeed primary local fields (these relations are called the KPZ relations after Knizhnik et al. (1988)).

Proposition 12.3.3 (KPZ relation, David et al. 2016). If ψ is a Möbius transform, we have

$$< \prod_{i=1}^{n} V_{\alpha_i}(\psi(z_i)) > = \prod_{i=1}^{n} |\psi'(z_i)|^{-2\Delta_{\alpha_i}} < \prod_{i=1}^{n} V_{\alpha_i}(z_i) >,$$

where $\Delta_{\alpha_i} = \frac{\alpha_i}{2}(Q - \frac{\alpha_i}{2})$.

Hence, in CFT language, the vertex operators V_α are primary local fields with conformal weight $\frac{\alpha_i}{2}(Q - \frac{\alpha_i}{2})$. Therefore, in LQFT, there is an infinite number of primary local fields; hence, it is a very rich theory. The KPZ relation in Proposition 12.3.3, which is an exact conformal covariance statement, should not be confused with the geometric KPZ relations proved in Duplantier–Sheffield (2011) and Rhodes and Vargas (2011) for the GMC measures defined in Theorem 12.2.3. In particular, these geometric formulations of KPZ are very general and do not rely specifically on conformal invariance: they are valid in all dimensions and for all GMC measures defined in Theorem 12.2.3.

Finally, as is common in CFT, one would like to understand the background metric dependence of the theory and express it in terms of the central charge. More specifically, if φ is a smooth bounded function on $\mathbb{S}$, we can consider the metric $e^{\varphi(z)}g(z)|dz|^2$. Then all the formulas of Riemannian geometry of subsection 12.3.1 are valid in this new metric by replacing the function $g(z)$ by the function $e^{\varphi(z)}g(z)$. One can also define a GFF with vanishing mean $X_{e^\varphi g}$ in this new metric, etc. Therefore, one can similarly define correlations $< \prod_{i=1}^{n} V_{\alpha_i}(z_i) >_{e^\varphi g}$ by formula (12.18) where one replaces g with the metric $e^\varphi g$. The relation between the two correlation functions is given by the so-called Weyl anomaly formula.

Proposition 12.3.4 (Weyl anomaly, David et al. 2016). If φ is a smooth bounded function on $\mathbb{S}$, we have

$$< \prod_{i=1}^{n} V_{\alpha_i}(z_i) >_{e^\varphi g} = e^{\frac{c_L}{96\pi} \int_{\mathbb{S}} (|\nabla_g \varphi|^2(z) + 2R_g(z)\varphi(z)) g(z)dz} < \prod_{i=1}^{n} V_{\alpha_i}(z_i) >, \quad (12.22)$$

where $c_L = 1 + 6Q^2$. Hence, LQFT is a CFT with central charge c_L.

In CFT, this property can be seen as a definition of the central charge. There are other ways to see the central charge of the model but we will not present them here. Since the function $\gamma \mapsto 1 + 6(\frac{\gamma}{2} + \frac{2}{\gamma})^2$ is a bijection from $]0, 2[$ to $]25, \infty[$, the Weyl anomaly formula (12.22) shows that LQFT can be seen as a family of CFTs with central charge varying continuously in the range $]25, \infty[$. Hence, LQFT is an

interesting laboratory to check rigorously the general CFT formalism developped in physics following the seminal work of Belavin et al. (1984). LQFT should also arise as the scaling limit of many models in statistical physics (just like the SLE introduced by Schramm (2000), which is a family of continuous random curves corresponding to a geometrical construction of CFTs with central charge ranging continuously in $]-\infty, 1]$).

12.3.6 The Liouville measures

As mentioned in subsection 12.3.2, one can usually (but not always) define primary local fields as random distributions. In the context of LQFT, one can indeed construct the vertex operators $V_\alpha(z)$ as random distributions in the sense of Schwartz; in fact, since the approximate vertex operators (12.17) are positive random functions, one can, show that they converge in the space of random measures. Hence, $V_\alpha(z)$ can be defined as random measures. Of particular interest is the case $\alpha = \gamma$ on which we will focus in this subsection. To be more precise, let us fix n points z_i with $n \geq 3$. We want to define the random measure $V_\gamma(z)dz$ under the formal probability measure $F \mapsto <F \prod_{i=1}^{n} V_{\alpha_i}(z_i) > / < \prod_{i=1}^{n} V_{\alpha_i}(z_i) >$. In this context, we denote the underlying probability space $\mathbb{E}^{(z_i, \alpha_i)}[.]$. In view of the definition (12.18), this leads to the following definition of the Liouville measure (where one just inserts a functional of the measure in the correlation function): if F is a functional defined on measures we have

$$\mathbb{E}_\mu^{(z_i, \alpha_i)}[F(V_\gamma(z)dz)] = Z_{GFF}(g) \lim_{\epsilon \to 0} \int_\mathbb{R} \mathbb{E}[F(V_{\gamma, \epsilon}(z)dz)$$

$$\prod_{i=1}^{n} V_{\alpha_i, \epsilon}(z_i) e^{-\frac{Q}{4\pi} \int_\mathbb{S} R_g(z)(X_g(z)+c) - \mu \epsilon^{\gamma^2/2} \int_\mathbb{S} e^{\gamma \phi_\epsilon(z)} dz}]dc/ < \prod_{i=1}^{n} V_{\alpha_i}(z_i) > .$$

Like for the correlation functions, we can obtain a very explicit expression for these Liouville measures in terms of GMC measures. Along the same line as the proof of the correlations, one can show the following explicit expression for the Liouville measure (with the notations of Proposition 12.3.2)

$$\mathbb{E}_\mu^{(z_i, \alpha_i)}[F(V_\gamma(z)dz)] = \frac{\mathbb{E}\left[F(\xi \frac{Z_{(z_i, \alpha_i)}(dz)}{Z_{(z_i, \alpha_i)}(\mathbb{S})}) Z_{(z_i, \alpha_i)}(\mathbb{S})^{-\frac{\sum_i \alpha_i - 2Q}{\gamma}}\right]}{\mathbb{E}\left[Z_{(z_i, \alpha_i)}(\mathbb{S})^{-\frac{\sum_i \alpha_i - 2Q}{\gamma}}\right]}, \tag{12.23}$$

where ξ is an independent variable with density the standard Γ-law density

$$\frac{1}{Z} e^{-\mu x} x^{\frac{\sum_i \alpha_i - 2Q}{\gamma} - 1} dx$$

on $\mathbb{R}_+$ (where Z is a normalization constant to make the integral of mass 1). We can get rid of the ξ variable by conditioning the measure to have volume 1. This leads to

the unit volume Liouville measures we will denote $V_\gamma^1(z)\mathrm{d}z$:

$$\mathbb{E}^{(z_i,\alpha_i)}[F(V_\gamma^1(z)\mathrm{d}z)] = \frac{\mathbb{E}\left[F\left(\frac{Z_{(z_i,\alpha_i)}(\mathrm{d}z)}{Z_{(z_i,\alpha_i)}(\mathbb{S})}\right)Z_{(z_i,\alpha_i)}(\mathbb{S})^{-\frac{\sum_i \alpha_i - 2Q}{\gamma}}\right]}{\mathbb{E}\left[Z_{(z_i,\alpha_i)}(\mathbb{S})^{-\frac{\sum_i \alpha_i - 2Q}{\gamma}}\right]} \qquad (12.24)$$

One can note that the μ dependence has disappeared in the expression of the unit volume Liouville measure. However, the unit volume Liouville measure is not a specific GMC measure (divided by its total mass to have volume 1) as there is still the $Z_{(z_i,\alpha_i)}(\mathbb{S})^{-\frac{\sum_i \alpha_i - 2Q}{\gamma}}$ term in expression (12.24): this term really comes from the interaction term (12.11) in the Liouville action (12.10). Though the Liouville measures are defined when the $(\alpha_i)_{1\le i\le n}$ satisfy the Seiberg bounds (12.19), one can show that the unit volume measures exist under the less restrictive conditions

$$\forall i, \ \alpha_i < Q \quad \text{and} \quad Q - \frac{\sum_{i=1}^n \alpha_i}{2} < \frac{2}{\gamma} \wedge \min_{1\le i\le n}(Q - \alpha_i), \qquad (12.25)$$

where $x \wedge y$ denotes the minimum of x and y.

Among the unit volume Liouville measures, one has a very special importance in relation to planar maps: the one where $n = 3$ and for all i we have $\alpha_i = \gamma$ (one can check that for all γ in $]0,2[$, this choice of $(\alpha_i)_{1\le i\le n}$ satisfies (12.25)). By conformal invariance, we can consider the case $z_1 = 0$, $z_2 = 1$, and $z_3 = \infty$. In this case, the measure has a very special conformal invariance conjectured on the limit of planar maps called invariance by rerooting. In words, if you sample a point x according to the measure and send 0 to 0, the point x to 1, and ∞ to ∞ by a Möbius transform then the image of the measure by the map has same distribution as the initial measure. More precisely, for a point x different from 0 and ∞ let $\psi_x(z) = z/x$ be the unique Möbius transform of $\mathbb{S}$ which sends 0 to 0, the point x to 1, and ∞ to ∞. Then we have the following equality for any functional F defined on measures[3]

$$\mathbb{E}^{(0,\gamma),(1,\gamma),(\infty,\gamma)}\left[\int_{\mathbb{S}} F((V_\gamma^1(z)\mathrm{d}z) \circ \psi_x^{-1})V_\gamma^1(x)\mathrm{d}x\right] = \mathbb{E}^{(0,\gamma),(1,\gamma),(\infty,\gamma)}[F(V_\gamma^1(z)\mathrm{d}z)],$$

$$(12.26)$$

where if ν is a measure on $\mathbb{S}$ and $f : \mathbb{S} \to \mathbb{S}$ some function, the measure $\nu \circ f^{-1}$ is defined by $(\nu \circ f^{-1})(A) = \nu(f^{-1}(A))$ for all Borel sets A.

Finally, we mention that a variant to LQG was developped in a series of works by Duplantier et al. (2014) and Sheffield (2010b). The framework of these works is a bit different than the one we consider in this chapter. Duplantier–Miller–Sheffield consider a GFF version of LQG with no cosmological constant μ and in particular no correlation functions. In this approach based on a coupling between the GFF and SLE, they construct equivalence classes of random measures (called quantum cones, spheres, etc.) with two marked points and coupled to space-filling variants of SLE curves. In

---

[3] A simple and elegant proof of this property was communicated to us by Julien Dubédat.

some sense, their framework is complementary with that of David et al. (2016) which considers random measures with three or more marked points. The framework of Duplantier et al. (2014) is interesting because it establishes nontrivial links between (decorated) random planar maps and the so-called quantum cones, spheres, etc.

12.3.7 Conjectured relation with planar maps

Following Polyakov's work (Polyakov 1981), it was soon acknowledged by physicists that one should recover LQG as some kind of discretized 2d quantum gravity given by finite triangulations of size N as N goes to ∞ (see, for example, the classical physics textbook Ambjorn et al. (2005) for a review on this problem). From now on, we assume that the reader is familiar with the definition of a triangulation of the sphere equipped with a conformal structure: otherwise, he can have a look at the Appendix where we gathered the required background. More precisely, let $\mathcal{T}_N$ be the set of triangulations of $\mathbb{S}$ with N faces and $\mathcal{T}_{N,3}$ be the set of triangulations with N faces and three marked faces (see Fig. 12.2 for a simulation of a random triangulation with $N = 10^5$ and sampled according to the uniform measure on $\mathcal{T}_N$). We will choose a point in each each marked face: these points are called roots. We equip $T \in \mathcal{T}_N$ with a standard conformal structure where each triangle is given volume $1/N$ (see the Appendix). The uniformization theorem tells us that we can then conformally map the triangulation onto the sphere $\mathbb{S}$ and the conformal map is unique if we demand the map to send the three roots to prescribed points $z_1, z_2, z_3 \in \mathbb{S}$. Concretely, the uniformization provides for each face $t \in T$ a conformal map $\psi_t : t \to \mathbb{S}$ where t is an equilateral triangle of volume $\frac{1}{N}$. Then, we denote by $\nu_{T,N}$ the corresponding deterministic measure on $\mathbb{S}$ where $\nu_{T,N}(\mathrm{d}z) = |(\psi_t^{-1})'|^2 \mathrm{d}z$ on each distorted triangle $\tilde{t}$ image of a triangle t by ψ_t.

Fig. 12.2 Random triangulation with 10^5 faces (no isometric embedding into the space). Courtesy of F. David.

In particular, the volume of the total space $\mathbb{S}$ is $N \times \frac{1}{N} = 1$. Now, we consider the random measure ν_N defined by

$$\mathbb{E}^N[F(\nu_N)] = \frac{1}{Z_N} \sum_{T \in \mathcal{T}_{N,3}} F(\nu_{T,N}), \qquad (12.27)$$

for positive bounded functions F where Z_N is a normalization constant given by $\#\mathcal{T}_{N,3}$ (the cardinal of the set $\mathcal{T}_{N,3}$). We denote by $\mathbb{P}^N$ the probability law associated with $\mathbb{E}^N$.

We can now state a precise mathematical conjecture.

Conjecture 12.3.5.   *Under $\mathbb{P}^N$, the family of random measures $(\nu_N)_{N \geq 1}$ converges in law as $N \to \infty$ in the space of Radon measures equipped with the topology of weak convergence toward the law of the unit volume Liouville measure given by (12.24) with parameter $\gamma = \sqrt{\frac{8}{3}}$, where $n = 3$ and $(z_i, \alpha_i) = (z_i, \gamma)$.*

Though such a precise conjecture was first stated in David et al. (2016), it is fair to say that such a conjecture is just a clean mathematical formulation of the link between discrete gravity and LQG understood in the 1980s by physicists. As of today, conjecture 12.3.5 is still completely open (though partial progress has been made on a closely related question in Curien (2015), One should also mention that a weaker and less explicit variant of conjecture 12.3.5 appears in Sheffield (2010b). More precisely, Sheffield proposed a limiting procedure involving the GFF to define a candidate measure for the limit of $(\nu_N)_{N \geq 1}$ as $N \to \infty$ (see the introduction of section 6 and conjecture 1.(a)); however, he left open the question of convergence of this limiting procedure. Recently, Aru et al. (2015) proved that the limiting procedure does converge and that the limit is the unit volume Liouville measure given by (12.24) with parameter $\gamma = \sqrt{\frac{8}{3}}$, where $n = 3$ and $(z_i, \alpha_i) = (z_i, \gamma)$.

Let us consider the case $z_1 = 0$, $z_2 = 1$, and $z_3 = \infty$ (by conformal invariance, this is no restriction). In this case, one could also consider triangulations with a fourth marked point and send the fourth marked point to z_3 in place of the third. Of course, this should not change the limit measure and therefore the limit measure should satisfy the invariance by rerooting property (12.26).

Finally, we could also state many variants of conjecture 12.3.5 as it is expected that some form of universality should hold. More precisely, conjecture 12.3.5 should not really depend on the details to define the measure ν_N in (12.27). For instance, one expects the same conjecture to hold where $\nu_{T,N}$ could be defined by putting uniform volume $1/N$ in each triangle of the circle packed triangulation: see Fig. 12.3 for a circle packed triangulation with large N (however, in this situation, there is a subtelty in the way one fixes the circle packing in a unique way: indeed, Möbius transforms send circle packings to circle packings but the centers of the circles of the latter are not necessarily the image of the centers of the former by the Möbius transforms).

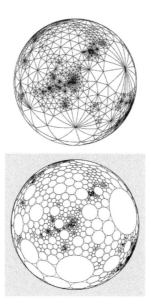

Fig. 12.3 Circle packing of a triangulation (top), and Corresponding adjacency circles (bottom). Courtesy of F. David.

12.3.8 On the Ising model at critical temperature

In this section, we give an account on the recent breakthroughs which occured in the understanding of the Ising model in the plane at critical temperature. This will provide the reader with another example of model where CFT can be made rigorous. Let us start with a few notations.

On the lattice $\mathbb{Z}^2$ and if x, y are in $\mathbb{Z}^2$ we denote $x \sim y$ the standard adjacency relation. Let N be a positive integer. We consider the box $\Lambda_N = [|-N, N|]^2$ and its frontier $\partial \Lambda_N = \{x \in {}^c\Lambda_N, \exists y \in \Lambda_N, x \sim y\}$. The state space of the model is $\{-1, 1\}^{\Lambda_N}$ and the energy of a spin configuration is given by

$$H_N^+(\sigma) = - \sum_{x \in \Lambda_N, \, x \sim y} \sigma_x \sigma_y,$$

where we will consider $+$ boundary conditions; i.e., we set the spins in $\partial \Lambda_N$ equal to 1.

The Ising model on Λ_N is then the Gibbs measure μ_N on the state space $\{-1, 1\}_N^{\Lambda}$ where the expectation of a functional F is given by

$$\mu_{N,\beta}^+(F(\sigma)) = \frac{1}{Z_{N,\beta}} \sum_{\sigma \in \{-1,1\}^{\Lambda_N}} F(\sigma) e^{-\beta H_N^+(\sigma)},$$

where $\beta > 0$ is the inverse temperature of the model and $Z_{N,\beta}$ a normalization constant ensuring that $\mu_{N,\beta}^+$ is a probability measure. The model undergoes a phase transition and the critical temperature is explicitly given by $\beta_c = \frac{1}{2} \ln(1 + \sqrt{2})$. One can show

that the measure μ_{N,β_c}^+ converges as N goes to infinity toward a measure μ_{β_c} defined in the full plane, i.e., with state space $\{-1,1\}^{\mathbb{Z}^2}$ (one can note that we have removed the superscript $+$ in the full plane measure; indeed one can show that this limit does not depend on the boundary conditions used to define the approximation measures on Λ_N).

The model was conjectured by physicists to be described by a specific CFT with central charge $c = \frac{1}{2}$ with two primary fields (to be precise there are three primary fields in the theory but the third one is just the constant 1). We will denote the two primary fields $\sigma(z)$ (the spin field) and $\epsilon(z)$ (the energy density field). We consider the spin field first and set the following definition for noncoincident points $z_1, \cdots, z_n$ and n even:

$$< \sigma(z_1) \cdots \sigma(z_n) > := \left(2^{-n/2} \sum_{\mu \in \{-1,1\}^n, \sum_i \mu_i = 0} \prod_{i<j} |z_i - z_j|^{\mu_i \mu_j / 2} \right)^{1/2}.$$

If ψ is a Mobius transform on the sphere then $|\psi(z) - \psi(y)| = |\psi'(z)|^{1/2} |\psi'(y)|^{1/2} |z - y|$ and therefore

$$< \sigma(\psi(z_1)) \cdots \sigma(\psi(z_n)) > = \prod_{i=1}^n |\psi'(z_i)|^{-1/8} < \sigma(z_1) \cdots \sigma(z_n) >; \qquad (12.28)$$

hence in CFT langage σ has conformal weight $\frac{1}{16}$.

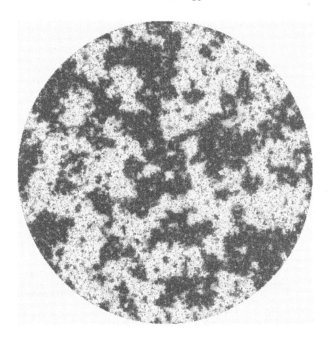

Fig. 12.4 Simulated Ising model at critical temperature with free boundary conditions. Courtesy of C. Hongler.

Let $\lfloor . \rfloor$ denote the integer part. For $\epsilon > 0$, we are now interested in the scaling limit as ϵ goes to 0 of the discrete spin field $x \mapsto \sigma_{\lfloor \frac{x}{\epsilon} \rfloor}$ defined on the rescaled lattice $\epsilon \mathbb{Z}^2$ under the measure μ_{β_c} (see Fig. 12.4 for a simulation of the spin field). In view of (12.28), it is natural to rescale the field by the factor $\epsilon^{-1/8}$.

Now the following convergence holds for the rescaled correlations,

$$\mu_{\beta_c} \Big[\prod_{i=1}^n (\epsilon^{-1/8} \sigma_{\lfloor \frac{z_i}{\epsilon} \rfloor}) \Big] \xrightarrow[\epsilon \to 0]{} C^n < \sigma(z_1) \cdots \sigma(z_n) >, \qquad (12.29)$$

where C is a lattice-specific constant. This important theorem was proved by Chelkak et al. (2015) building on the fermionic observable first studied by Smirnov (2010) and Chelkak and Smirnov (2012); in fact, the main theorem in Chelkak et al. (2015) shows the convergence of the rescaled correlations to an explicit expression in any domain (not just the full plane). The convergence result (12.29) was also proved independently by Dubédat (2011) by an exact bosonization procedure (roughly, bosonization means in this context that there exists an exact relation between the squared correlation functions of the Ising model on a lattice and the correlations of the exponential of the discrete GFF on a lattice). As is standard in rigorous CFT, one can define the limit σ as a random distribution. More precisely, Camia et al. (2015) proved that there exists a random distribution σ defined on some probability space such that $\epsilon^{-1/8} \sigma_{\lfloor \frac{x}{\epsilon} \rfloor}$ converges in law in the space of distributions toward σ.

Finally, let us mention that similar results can be proved for the energy density field ϵ. In this case, the properly rescaled (and recentered) energy $\sigma_i \sigma_j$ of a bond between two adjacent vertices $i \sim j$ converges toward the field ϵ (in the sense of the correlation functions): this is proved in Hongler (2013) and Hongler and Smirnov (2013) (in any domain and not just the full plane). It was also proved independently in the full plane by Boutillier and De Tilière (2010,2011) on general periodic isoradial graphs. There also exist explicit formulas for the correlations $< \varepsilon(z_1) \cdots \varepsilon(z_n) >$ of the field ϵ (but we will not write them here: see Hongler (2013)) and the field ϵ has conformal weight $\frac{1}{2}$, i.e.

$$< \varepsilon(\psi(z_1)) \cdots \varepsilon(\psi(z_n)) > = \prod_{i=1}^n |\psi'(z_i)|^{-1} < \varepsilon(z_1) \cdots \varepsilon(z_n) > . \qquad (12.30)$$

Let us further mention that the energy density field cannot be understood as a random distribution, hence, $< . >$ is not a real measure in (12.30).

12.3.9   Final remarks and conclusion

In this chapter, we introduced the theory of LQFT based on Kahane's GMC theory. More precisely, we introduced the correlation functions and the random measures of the theory. We stated that they satisfy the main assumptions of a CFT on the Riemann sphere. As a comparison and to illustrate the full power of CFT, we also presented in CFT language the recent developments around the Ising model in 2d at the critical point. We would like to stress as a final remark the conceptual difference in

the mathematical treatment of the two CFTs. The methods of probabilistic quantum field theory developed in 1970–1980 around path integral formulations have been up to now unsuccessful in constructing the CFT which describes the scaling limit of the Ising model at critical temperature; it is conjectured that such a construction should exist. Nonetheless, this CFT has been rigorously constructed mathematically by taking the scaling limit of the discrete Ising model, hence leaving open the other approach. On the LQFT side, recall that random planar maps (which correspond to discrete gravity) were introduced because defining LQFT by path integral formulations seemed troublesome. The idea was to construct LQFT by taking the scaling limit of large planar maps. However, as we have seen in this chapter, a direct construction of LQFT by path integral formulation is feasible, whereas proving the convergence of large planar maps is a very difficult topic. Indeed, the convergence has only been established up to now for very specific topologies (of convergence).

12.4 Appendix

12.4.1 The conformal structure on planar maps

In this appendix, we recall basic definitions and facts on triangulations equipped with a conformal structure. This part is mostly based on Gill and Rhode (2013). A finite triangulation T is a graph you can embed in the sphere such that each inner face has three adjacent edges (the edges do not cross and intersect only at vertices). The triangulation T has size N if it has N faces. We see each triangle $t \in T$ as an equilateral triangle of fixed volume a^2 say that we glue topologically according to the edges and the vertices. This defines a topological structure (and even a metric structure). Now, we put a conformal structure on T. We need an atlas, i.e., a family of compatible charts. We map the inside of each triangle t to the same triangle in the complex plane. If two triangles are adjacent in the triangulation, we map them to two adjacent equilateral triangles in the complex plane. Now, we need to define an atlas in the neighborhood of a vertex a. The vertex a is surrounded by n triangles. We first map these triangles in the complex plane in counterclockwise order such that each is equilateral. Then we use the map $z \mapsto z^{6/n}$ to 'unwind' the triangles (in fact, this unwinds the triangles only if $n > 6$) to define a homeomorphism around the vertex a. By the uniformization theorem, we can find a conformal map $\psi : T \mapsto \mathbb{C}$ where we send three points in T called roots to fixed points $x_1, x_2, x_3 \in \mathbb{S}$. For each triangle t, we can consider ψ_t, the restriction of ψ to t, as a standard conformal map from t to a distorted triangle $\tilde{t} \subset \mathbb{S}$. It is then natural to equip $\mathbb{C}$ with the standard pullback metric. More precisely, in each triangle $\tilde{t}$ the metric is given by $|(\psi_t^{-1})'(z)|^2 \mathrm{d}z$ and then one can define the metric in $\mathbb{C}$ by gluing the metric of each distorted triangle $\tilde{t}$. This metric has conical singularities at the points α of the form $\alpha = \psi(a)$ where a is a vertex of T.

Since ψ^{-1} is analytic, we have $|\psi^{-1}(z)| \approx |z - \alpha|^{n/6}$ around α (to see this compose ψ^{-1} with the chart $z \mapsto z^{6/n}$). Recall that the metric on $\mathbb{S}$ around α is of the form $|(\psi^{-1})'(z)|^2 \mathrm{d}z = e^{\lambda(z)} \mathrm{d}z$. We have $|(\psi^{-1})'(z)|^2 \approx |z - \alpha|^{2(n/6-1)}$. Therefore, there is little mass around points $n > 6$ and big mass around points $n < 6$. This metric has a cone interpretation. If $\theta > 0$ is some angle and C_θ is the corresponding cone, one can

put a conformal structure on the cone by the function $\psi : z \mapsto z^{\frac{2\pi}{\theta}}$ in which case the metric is

$$|(\psi^{-1})'(z)|^2 \mathrm{d}z = \frac{\theta}{2\pi}|z|^{2(\frac{\theta}{2\pi}-1)}\mathrm{d}z = \frac{\theta}{2\pi}|z|^{2\beta}\mathrm{d}z,$$

where $\beta = \frac{\theta}{2\pi} - 1$ is in $]-1, \infty[$. Therefore, around 0, the average Ricci curvature is then given by

$$-2\beta \int_{|z|\leq 1} \Delta_z \ln|z|\mathrm{d}z = -4\pi\beta = 2(2\pi - \theta).$$

In the case of triangulations, the angle θ is related to n by the formula $\theta = \frac{n\pi}{3}$: this means that there is negative curvature (and little mass) around α if $n > 6$ and the opposite if $n < 6$.

Acknowledgments

We would like to thank D. Chelkak for useful discussions on the Ising model and C. Hongler and F. David for the images. We also thank Y. Huang for reading carefully a prior draft of these lecture notes. R.R. acknowledges partial financial support from Grant ANR-11-JCJC CHAMU.

References

Ambjorn, J., Durhuus B., Jonsson T. (2005). *Quantum Geometry: A Statistical field theory Approach Cambridge Monographs on Mathematical Physics*. Combridge University Press, Cambridge, UK.

Aru, J., Huang, Y., Sun X. (2015). Two perspectives of the unit area quantum sphere and their equivalence, arXiv:1512.06190.

Bacry, E., Kozhemyak, A., Muzy. J.-F. (2008). Continuous cascade models for asset returns. *J. Econ. Dyn. Control* **32**(1), 156–99.

Berestycki, N. (2015). An elementary approach to Gaussian multiplicative chaos. arXiv:1506.09113.

Belavin, A.A., Polyakov. A.M., Zamolodchikov, A.B. (1984). Infinite conformal symmetry in two-dimensional quantum field theory, *Nucl. Phys. B* **241**(2), 333-80.

Bramson, M., Ding, J., Zeitouni, O. (2015). Convergence in law of the maximum of the two-dimensional discrete Gaussian free eield. *Commun. Pure Appl. Math.* **69**(1), 62-123.

Biskup, M., Louidor, O. (2016). Extreme local extrema of two-dimensional discrete Gaussian free field. *Commun. Math. Phys.* **345**(1), 271–304.

Boutillier, C., De Tilière, B. (2010). The critical Z-invariant Ising model via dimers: the periodic case. *Probab. Theory Related fields* **147**, 379-413.

Boutillier, C., De Tilière, B. (2011). The critical Z-invariant Ising model via dimers: locality property. *Commun. Math. Phys.* **301**, 473-516.

Carpentier, D., Le Doussal, P. (2001). Glass transition of a particle in a random potential, front selection in nonlinear RG and entropic phenomena in Liouville and Sinh-Gordon models. *Phys. Rev. E* **63**, 026110.

Camia, F., Garban, C., Newman, C. (2015). Planar Ising magnetization field I. Uniqueness of the critical scaling limit, *Ann. Probab.* **43** (2), 528-71.

Chelkak, D., Smirnov, S. (2012). Universality in the 2D Ising model and conformal invariance of fermionic observables, *Invent. Math.* **189**, 515-80.

Chelkak, D. , Hongler, C., Izyurov, K. (2015). Conformal invariance of spin correlations in the planar Ising model. *Ann. Math.* **181**, 1087-138.

Chen, L., Jakobson, D. (2014). Gaussian Free Fields and KPZ Relation in $\mathbb{R}^4$. *Annales I.H.P.* **15** (7), 1245-83.

Chevillard, L., Robert, R., Vargas, V. (2010). A stochastic representation of the local structure of turbulence. *Europhys. Lett.* **89**, 54002.

Curien, N. (2015) A glimpse of the conformal structure of random planar maps, *Commun. Math. Phys.* **333** (3), 1417-63.

Daley, D.J., Vere-Jones, D. (2007). *An Introduction to the Theory of Point Processes* Volume 2; *Probability and Its Applications* 2nd edn. Springer, Berlin.

David, F., Kupiainen, A., Rhodes, R., Vargas V. (2016). Liouville quantum gravity on the Riemann sphere. *Commun. Math. Phys.* **342**(3), 869–907.

Dubédat, J. (2011). Exact bosonization of the Ising model. arXiv:1112.4399.

Duplantier, B., Miller, J., Sheffield, S. (2014). Liouville quantum gravity as a mating of trees. arXiv:1409.7055.

Duplantier, B., Sheffield, S. (2011). Liouville quantum gravity and KPZ. *Invent. Math.* **185**(2), 333-93.

Fyodorov, Y., Bouchaud, J.P. (2008). Freezing and extreme-value statistics in a random energy model with logarithmically correlated potential. *J. Phys. A* **41**, 372001.

Fyodorov, Y., Le Doussal, P., Rosso, A. (2009). Statistical mechanics of logarithmic REM: duality, freezing and extreme value statistics of $1/f$ noises generated by Gaussian free fields. *J. Stat. Mech.* P10005.

Fyodorov, Y., Le Doussal, P., Rosso, A. (2010). Freezing transition in decaying Burgers turbulence and random matrix dualities. *Europhys. Lett.* **90**, 60004.

Gawedzki, K. (1999). Lectures on conformal field theory. In *Quantum Fields and Sstrings: A Course for Mathematicians*, Vols. 1, 2 (Princeton, NJ, 1996/1997), 72–805. Amer. Math. Soc., Providence, RI.

Gill, J., Rhode, S. (2013). On the Riemann surface type of random planar maps. *Revista Mat. Iberoamericana* **29** 1071-90.

Hoegh-Krohn, R. (1971). A general class of quantum fields without cut offs in two space-time dimensions. *Commun. Math. Phys.* **21** (3), 244-55.

Hongler, C. (2013). Conformal invariance of Ising model correlations. PhD dissertation. Available at http://archive-ouverte.unige.ch/unige:18163.

Hongler, C., Smirnov, S. (2013). The energy density in the planar Ising model, *Acta Math.* **211**(2), 191-225.

Hu, X., Miller, J., Peres, Y. (2010). Thick points of the Gaussian free field, *Ann. Probab.* **38**, 896-926.

Junnila, J., Saksman, E. (2015). The uniqueness of the Gaussian multiplicative chaos revisited. arXiv:1506.05099.

Kahane, J.-P. (1985). Sur le chaos multiplicatif. *Ann. Sci. Math. Québec* **9**(2), 105-50.

Knizhnik, V.G., Polyakov, A.M., Zamolodchikov, A.B. (1988). Fractal structure of 2D-quantum gravity, *Modern Phys. Lett A* **3**(8), 819-26.

Kolmogorov, A.N. (1962). A refinement of previous hypotheses concerning the local structure of turbulence, *J. Fluid. Mech.* **13**, 83-5.

Madaule, T. (2015). Maximum of a log-correlated Gaussian field. *Ann. Inst. H. Poincaré*, **51**(4), 1369-431.

Madaule, T., Rhodes, R., Vargas, V. (2016). Glassy phase and freezing of log-correlated Gaussian potentials. *Ann Appl. Probab.* **26**(2), 643–90.

Mandelbrot, B.B. (1972). A possible refinement of the lognormal hypothesis concerning the distribution of energy in intermittent turbulence. In *Statistical Models and Turbulence*, La Jolla, CA, Lecture Notes in Phys. no. 12, 333-51. Springer, Berlin.

Nakayama, Y. (2004) Liouville field theory: a decade after the revolution. *Int.J.Mod.Phys. A* **19**, 2771-930.

Polyakov, A.M. (1981). Quantum geometry of bosonic strings. *Phys. Lett. B* **103** (3), 207-10.

Rhodes, R. Vargas, V. (2011). KPZ formula for log-infinitely divisible multifractal random measures, *ESAIM Probab. Stat.* **15**, 358-71.

Rhodes R., Vargas, V. (2014). Gaussian multiplicative chaos and applications: a review. *Probab. Surv.* **11**, 315-92.

Rhodes R., Vargas, V. (2010). Multidimensional multifractal random measures. *Elec. J. Probab.* **15**, 241-58.

Robert, R., Vargas, V. (2010). Gaussian multiplicative chaos revisited. *Ann. Probab.* **38**(2), 605-31.

Shamov, A. (2014). On Gaussian multiplicative chaos. arXiv:1407.4418.

Schramm, O. (2000). Scaling limits of loop-erased random walks and uniform spanning trees. *Isr. J. math.* **118**(1), 221-88.

Sheffield, S. (2010b) Conformal weldings of random surfaces: SLE and the quantum gravity zipper. arXiv:1012.4797.

Simon, B. (1974). The $P(\phi)_2$ Euclidean Quantum Field theory, Princeton University Press, Princeton , NJ.

Smirnov, S. (2010). Conformal invariance in random cluster models. I. Holomorphic fermions in the Ising model. *Ann. Math.* **172**, 1435-67.

Webb, C. (2014). The characteristic polynomial of a random unitary matrix and Gaussian multiplicative chaos—the L^2-phase. arXiv:1410.0939.

13

Quantum spin chains and classical integrable systems

Anton ZABRODIN

Institute of Biochemical Physics RAS, 4 Kosygina st., Moscow 119334, Russia;
and ITEP, 25 B.Cheremushkinskaya, Moscow 117218, Russia; and Laboratory
of Mathematical Physics, National Research University Higher School of Economics,
20 Myasnitskaya Ulitsa, Moscow 101000, Russia

Zabrodin, A., 'Quantum Spin Chains and Classical Integrable Systems' in *Stochastic Processes and Random Matrices*. Edited by: Grégory Schehr et al, Oxford University Press (2017). © Oxford University Press 2017. DOI 10.1093/oso/9780198797319.003.0013

Chapter Contents

13.1   Introduction

The construction of quantum integrable spin chains (magnets) is based on solutions of the Yang–Baxter equation [1]. They are called (quantum) R-matrices. The R-matrices satisfy the famous Yang–Baxter $RRR = RRR$ relation. An important class of solutions are $gl(N)$-invariant R-matrices taken in finite-dimensional representations which are simple rational functions of the spectral parameter x:

$$\mathbf{R}(x) = \mathbf{1} \otimes \mathbf{1} + \frac{\eta}{x}\,\mathbf{P}. \qquad (13.1)$$

Here $\mathbf{P} \in \mathrm{End}\big(\mathbb{C}^N \otimes \mathbb{C}^N\big)$ is the permutation operator. We will focus on magnets constructed using the R-matrices from this class. They are called (generalized) spin chains of the XXX type, or simply XXX spin chains.

   For our purpose we need inhomogeneous XXX spin chains or the corresponding integrable lattice models of statistical mechanics on inhomogeneous lattices, with quasiperiodic (twisted) boundary conditions. Their quantum monodromy matrices $\mathbf{S}(x)$ are products of the type

$$\mathbf{S}(x) = \mathbf{R}^{0L}(x - x_L)\ldots\mathbf{R}^{02}(x - x_2)\mathbf{R}^{01}(x - x_1)\,(\mathbf{g} \otimes \mathbf{1}^{\otimes L}) \qquad (13.2)$$

along the chain, with x_i being inhomogeneity parameters assumed to be distinct and $\mathbf{g} \in GL(N)$ being the twist matrix assumed to be diagonal. (The label 0 corresponds to the auxiliary space, where the product is taken.) The trace $\mathbf{t}(x) = \mathrm{tr}_0\,\mathbf{S}(x)$ of the quantum monodromy matrix in the auxiliary space is what is called the quantum transfer matrix or the T-operator. The Yang–Baxter equation implies that the T-operators commute for all x, so $\mathbf{t}(x)$ is a generating function of commuting Hamiltonians $\mathbf{H}_j$:

$$\mathbf{t}(x) = \mathrm{tr}\,\mathbf{g} \cdot \mathbf{1}^{\otimes L} + \sum_{j=1}^{L} \frac{\eta\,\mathbf{H}_j}{x - x_j}. \qquad (13.3)$$

These Hamiltonians are nonlocal, i.e., involve interaction between operators on all lattice sites. However, such models still make sense as generalized spin chains with long-range interactions. Alternatively, one may prefer to keep in mind integrable lattice models of statistical mechanics rather than spin chains as such. In either case the final goal of the theory is diagonalization of the T-operators. This is usually achieved by the Bethe ansatz method in one form or another. What we are going to do is to present an alternative approach based on a hidden connection with classical many-body integrable systems explained in the following.

   As an intermediate step, we need to recall that there exists a broader family of commuting T-operators which includes $\mathbf{t}(x)$ as a subset. Using the fusion procedure in the auxiliary space, one can construct an infinite family of commuting T-operators $\mathbf{t}_\lambda(x)$ indexed by Young diagrams λ, with $\mathbf{t}_\square(x) = \mathbf{t}(x)$. Following [2, 3], we construct the master T-operator as their generating (operator-valued) function of a special form. Let $\mathbf{t} = \{t_1, t_2, t_3, \ldots\}$ be an infinite set of auxiliary 'time variables' and $s_\lambda(\mathbf{t})$ be the

Schur polynomials. The master T-operator, $\mathsf{T}(x, \mathbf{t})$, for the generalized magnets[1] is introduced in the same way as in [2] and subsequent works [5–8]:

$$\frac{\mathsf{T}(x, \mathbf{t})}{\mathsf{T}(x, \mathbf{0})} = \sum_\lambda t_\lambda(x) s_\lambda(\mathbf{t}). \tag{13.4}$$

By construction, this family of operators is commutative for all x, $\mathbf{t}$ and can be simultaneously diagonalized: $\mathsf{T}(x, \mathbf{t}) |\Psi\rangle = T(x, \mathbf{t}) |\Psi\rangle$. The main fact about the master T-operator, which makes the whole construction interesting, is that the so-defined $\mathsf{T}(x, \mathbf{t})$ satisfies the bilinear identity for the classical modified Kadomtsev–Petviashvili (mKP) hierarchy, with x being identified with the '0th time' t_0:

$$\oint_C z^{(x-x')/\eta} e^{\sum_{k \geq 1} (t_k - t'_k) z^k} \mathsf{T}\left(x, \mathbf{t} - [z^{-1}]\right) \mathsf{T}\left(x', \mathbf{t}' + [z^{-1}]\right) dz = 0 \tag{13.5}$$

for all $\mathbf{t}$, $\mathbf{t}'$, x, x' and for a properly chosen integration contour. Here $\mathbf{t} \pm [z^{-1}] :=$ $\{t_k \pm \frac{1}{k} z^{-k}\}$. This means that any eigenvalue $T(x, \mathbf{t})$ of $\mathsf{T}(x, \mathbf{t})$ is a tau-function of the mKP hierarchy. In this way, the commutative algebras of XXX spin chain Hamiltonians appear to be embedded in the infinite integrable hierarchy of nonlinear differential-difference equations, the mKP hierarchy [9–11]. This is a further development of the earlier studies [12–14] clarifying the role of classical integrable hierarchies in quantum integrable models.

The next step depends on analytical properties of the eigenvalues $T(x, \mathbf{t})$ as functions of the variable x. For finite spin chains each eigenvalue is a polynomial in x of degree L for any $\mathbf{t}$:

$$T(x, \mathbf{t}) = e^{\operatorname{tr} \xi(\mathbf{t}, \mathbf{g})} \prod_{j=1}^{L} (x - x_j(\mathbf{t})). \tag{13.6}$$

The roots depend on the times $\mathbf{t}$. At this point, a surprising link to integrable many-body systems of classical mechanics comes into play. Namely, from the fact that $T(x, \mathbf{t})$ is a tau-function of the mKP hierarchy, it follows [15, 16] that the roots x_i move in the times t_k as particles of the Ruijsenaars–Schneider (RS) L-body system [17] subject to the equations of motion corresponding to the kth Hamiltonian $\mathcal{H}_k$ of the RS model. For example, the equations of motion for the $t = t_1$ flow are

$$\ddot{x}_i = -\sum_{k \neq i} \frac{2\eta^2 \dot{x}_i \dot{x}_k}{(x_i - x_k)\left[(x_i - x_k)^2 - \eta^2\right]}, \qquad i = 1, \ldots L. \tag{13.7}$$

This link prompts to reformulate the spectral problem for the XXX spin chain Hamiltonians $\mathbf{H}_j$ in terms of the integrable model of classical mechanics. The role of the quantum-classical (QC) correspondence in supersymmetric gauge theories and branes was discussed in [18–20].

---

[1] A preliminary form of the master T-operator for these models appeared in an earlier work [4].

The RS system is often referred to as an integrable relativistic deformation of the famous Calogero–Moser system. Similarly to the latter, it admits the Lax representation; i.e., the dynamics can be translated into isospectral deformations of a matrix $Z(\{x_i(t)\}, \{\dot{x}_i(t)\})$, which is called the Lax matrix. The essence of the QC correspondence of integrable systems lies in the fact that the spectra of the quantum Hamiltonians $\mathbf{H}_j$ are encoded in the Lax matrix $Z(\{x_i(0)\}, \{\dot{x}_i(0)\}) \equiv Z_0$ for the RS system at $t = 0$ after the identifications $x_i(0) := x_i$ (the inhomogeneity parameters of the spin chain) and $\dot{x}_i(0) := -\eta H_i$ (the eigenvalues of the quantum Hamiltonians):

$$(Z_0)_{ij} = \frac{\eta H_i}{x_j - x_i + \eta}. \tag{13.8}$$

Given the x_i values, possible values of H_i are determined from the condition that the matrix Z_0 has a prescribed set of eigenvalues which is a subset of $\{g_1, g_2, \ldots, g_N\}$, where g_i are elements of the (diagonal) twist matrix $\mathbf{g}$, taken with certain multiplicities. In this way the spectral problem for the quantum Hamiltonians is reduced to a sort of inverse spectral problem for the Lax matrix.

We give two different proofs of this remarkable correspondence. One (indirect) is through the mKP hierarchy and its polynomial solutions. The other proof (based on the technique developed in [20]) is by a direct computation using the description of the spectrum in terms of the (nested) Bethe ansatz equations. Both proofs are rather technical. It is of value to find a more conceptual proof.

As a corollary, computing the spectral determinant for the RS Lax matrix, we find that the eigenvalues of the quantum Hamiltonians for all XXX spin chains on L sites are encoded in the following system of algebraic equations:

$$\sum_{1 \leq i_1 < \ldots < i_n \leq L} H_{i_1} \ldots H_{i_n} \prod_{1 \leq \alpha < \beta \leq n} \left(1 - \frac{\eta^2}{(x_{i_\alpha} - x_{i_\beta})^2} \right)^{-1} = e_n(g_1, \ldots, g_L), \quad n = 1, \ldots, L.$$
$$\tag{13.9}$$

Here e_n are elementary symmetric polynomials of L parameters g_i: $e_1 = \sum_i g_i$, $e_2 = \sum_{i<j} g_i g_j$, etc. Identifying them with elements of the twist matrix in a proper way (at $L > N$ some g_i values must be merged), one finds a part of the spectrum $(H_1, \ldots, H_L)$ for a particular spin chain among solutions to the system. Other solutions of the same system correspond to some other spin chain. In order to find the full spectrum of a given model, one should solve systems of the form (13.9), where g_i values are taken with different possible multiplicities from a given set. These are equations for the spectrum itself, not for any auxiliary parameters like in the Bethe ansatz solution. The detailed structure of solutions and their precise correspondence with spectra of particular spin chains are a subject of further study.

From the algebro-geometric point of view, Eqs (13.9) define a $2L$-dimensional algebraic variety $\mathbb{S}_L$ which can be called the *universal spectral variety* for spin chains of the XXX type. It contains comprehensive information about spectra of spin chains on L sites based on the $gl(N)$ algebras. The variety $\mathbb{S}_L$ given by Eqs (13.9) is not compact. These equations only define its affine part embedded into the $3L$-dimensional space

with coordinates $(H_1, \ldots, H_L; x_1, \ldots, x_L; g_1, \ldots, g_L)$. Presumably, a proper compact-ification of the universal spectral variety encodes information about the spectra of homogeneous spin chain Hamiltonians (when $x_i \to 0$).

13.1.1 Organization of the paper

In Section 13.2 we recall the construction of the integrable XXX spin chains start-ing with the quantum R-matrices. For our purpose we need a fully inhomogeneous model with twisted boundary conditions. We introduce nonlocal commuting Hamil-tonians, which are the main observables in the system, and such attendant objects like the higher T-operators (transfer matrices). The master T-operator is introduced in Section 13.2.4 as their generating function. In Section 13.2.5 we present the bilin-ear identity satisfied by the master T-operator which makes it possible to embed the quantum stuff into the context of classical integrable hierarchies of nonlinear PDEs. In particular, we define the classical Baker–Akhiezer function in terms of the quantum T-operators.

In Section 13.3 we establish and exploit the link to the RS L-body system. The main point here is the reformulation of the eigenvalue problem for the spin chain Hamiltonians in terms of coordinates and velocities of the RS particles. The Lax pair for the RS system is derived from the poles dynamics of the Baker–Akhiezer function in Section 13.3.2.

Section 13.4 contains some details of the QC correspondence which is based on the identification of the twist parameters with eigenvalues of the Lax matrix for the RS model (Section 13.4.1). In Section 13.4.3 the algebraic equations for eigenvalues of the spin chain Hamiltonians are obtained and the notion of the universal spectral variety is described. In Section 13.5 we give a direct proof of the QC correspondence, using the nested Bethe ansatz solution. In Section 13.6 some unsolved problems are listed.

There are also two appendices. In Appendix A we give some technical details needed for deriving the higher T-operators in a more or less explicit form as derivatives of characters. Appendix B is a reference source for the Hamiltonian approach to the RS system.

13.1.2 The notation

Throughout the paper, we use the following notation.

$\mathbf{e}_{ab}$: generators of $gl(N)$ identified with matrix units, $(\mathbf{e}_{ab})_{a'b'} = \delta_{aa'}\delta_{bb'}$

$\mathbf{v}_a$: orthonormal basis vectors in $\mathbb{C}^N$ such that $\mathbf{e}_{ab}\mathbf{v}_c = \delta_{bc}\mathbf{v}_a$

$\mathbf{g}$: a diagonal group element of $GL(N)$, $\mathbf{g} = \operatorname{diag}(g_1, \ldots, g_N)$

λ: a Young diagram with rows $\lambda_1 \geq \lambda_2 \geq \ldots \geq \lambda_\ell > 0$

λ': the transposed Young diagram (λ reflected in the main diagonal)

$\mathbf{1}$: the identity element in $\operatorname{End}(\mathbb{C}^N)$ or $\operatorname{End}(\mathbb{C}^L)$

$\mathbf{I}$: the identity operator in the tensor product spaces like $(\mathbb{C}^N)^{\otimes L}$, etc

$\mathbf{O}$: an operator in $(\mathbb{C}^N)^{\otimes L}$ or $\mathbb{C}^L \otimes (\mathbb{C}^N)^{\otimes L}$

$\mathbf{O}(x)$: an operator-valued rational function of x

$O(x)$: an operator-valued polynomial function of x

We use the notation

$$\overrightarrow{\prod_{j=1}^{L}} \mathbf{O}_j = \mathbf{O}_1 \mathbf{O}_2 \ldots \mathbf{O}_L$$

and

$$\overleftarrow{\prod_{j=1}^{L}} \mathbf{O}_j = \mathbf{O}_L \ldots \mathbf{O}_2 \mathbf{O}_1$$

for the ordered product of the operators $\{\mathbf{O}_j\}_{j=1}^{L}$.

13.2   The master T-operator for spin chains

13.2.1   Quantum R-matrices

The simplest $gl(N)$-invariant R-matrix has the form

$$\mathbf{R}(x) = 1 \otimes 1 + \frac{\eta}{x} \sum_{a,b=1}^{N} \mathbf{e}_{ab} \otimes \mathbf{e}_{ba}. \tag{13.10}$$

The variable x is the spectral parameter. The extra parameter η is not actually essential because it can be eliminated by a rescaling of x (unless one tends η to 0 as in the limit to the Gaudin model). The R-matrix (13.11) is an operator in the space $\mathbb{C}^N \otimes \mathbb{C}^N$. It can be represented as $\mathbf{R}(x) = 1 \otimes 1 + \frac{\eta}{x}\mathbf{P}$, where $\mathbf{P}$ is the permutation operator given by $\mathbf{P} = \sum_{a,b=1}^{N} \mathbf{e}_{ab} \otimes \mathbf{e}_{ba}$. It acts on homogeneous vectors as follows: $\mathbf{P}\mathbf{x} \otimes \mathbf{y} = \mathbf{y} \otimes \mathbf{x}$.

Having in mind the construction of the spin chain on L sites, one can realize the R-matrix as an operator in the space $\mathbb{C}^N \otimes (\mathbb{C}^N)^{\otimes L}$,

$$\mathbf{R}^{0j}(x) = 1 \otimes 1^{\otimes L} + \frac{\eta}{x} \sum_{a,b=1}^{N} \mathbf{e}_{ab} \otimes \mathbf{e}_{ba}^{(j)}, \tag{13.11}$$

where $\mathbf{e}_{ba}^{(j)} := 1^{\otimes(j-1)} \otimes \mathbf{e}_{ba} \otimes 1^{\otimes(L-j)}$ for $j \in \{1, 2, \ldots, L\}$. The first space $\mathbb{C}^N$ labeled by the index 0 is called the auxiliary space while the space $V = (\mathbb{C}^N)^{\otimes L}$ is the quantum space of the model. The matrix elements $(\mathbf{R}^{0j}(x))_{ab}$ of the operator $\mathbf{R}^{0j}(x)$ with respect to the auxiliary space are operators in the quantum space. They are defined by

$$\mathbf{R}^{0j}(x)(\mathbf{v}_a \otimes 1^{\otimes L}) = \sum_{b=1}^{N}(\mathbf{v}_b \otimes 1^{\otimes L})(1 \otimes (\mathbf{R}^{0j}(x))_{ba}) = \sum_{b=1}^{N} \mathbf{v}_b \otimes (\mathbf{R}^{0j}(x))_{ba},$$

where $\mathbf{v}_a$ are orthonormal basis vectors in $\mathbb{C}^N$. From (13.11) we obtain

$$(\mathbf{R}^{0j}(x))_{ab} = \delta_{ab}\mathbf{1}^{\otimes L} + \frac{\eta}{x}\mathbf{e}_{ba}^{(j)}, \qquad (13.12)$$

where we have used

$$(\mathbf{e}_{ac} \otimes \mathbf{e}_{ca}^{(j)})(\mathbf{v}_b \otimes \mathbf{1}^{\otimes L}) = \mathbf{e}_{ac}\mathbf{v}_b \otimes \mathbf{e}_{ca}^{(j)} = \delta_{cb}\mathbf{v}_a \otimes \mathbf{e}_{ca}^{(j)}.$$

For example, in the $gl(2)$-case the block matrix representation reads

$$\mathbf{R}^{0j}(x) = \begin{pmatrix} \mathbf{I} + \frac{\eta}{x}\mathbf{e}_{11}^{(j)} & \frac{\eta}{x}\mathbf{e}_{21}^{(j)} \\ \frac{\eta}{x}\mathbf{e}_{12}^{(j)} & \mathbf{I} + \frac{\eta}{x}\mathbf{e}_{22}^{(j)} \end{pmatrix}. \qquad (13.13)$$

Here $\mathbf{I} \equiv \mathbf{1}^{\otimes L}$. In the following we will keep the notation $\mathbf{1}$ for identity elements of $\mathrm{End}(\mathbb{C}^N)$ and $\mathrm{End}(\mathbb{C}^L)$ and will often write $\mathbf{I}$ for the identity operator in any other spaces involved.

One may also extend the definition of the permutation to any two tensor factors of the space $\mathcal{V}$:

$$\mathbf{P}_{ij} = \sum_{a,b=1}^{N} \mathbf{e}_{ab}^{(i)} \otimes \mathbf{e}_{ba}^{(j)}. \qquad (13.14)$$

On tensor products of the basis vectors it acts as follows (here $i < j$):

$$\mathbf{P}_{ij}(\mathbf{v}_{a_1} \otimes \cdots \otimes \mathbf{v}_{a_i} \otimes \cdots \otimes \mathbf{v}_{a_j} \otimes \cdots \otimes \mathbf{v}_{a_L})$$
$$= \mathbf{v}_{a_1} \otimes \cdots \otimes \mathbf{v}_{a_j} \otimes \cdots \otimes \mathbf{v}_{a_i} \otimes \cdots \otimes \mathbf{v}_{a_L}. \qquad (13.15)$$

The aforesaid is related to the R-matrix in the vector representation of $gl(N)$. More generally, one can consider other irreducible finite-dimensional representations. Any partition λ (identified with the Young diagram) labels a representation π_λ of the universal enveloping algebra $U(gl(N))$ if $\lambda_{N+1} = 0$. For any such representation one can construct an R-matrix acting in the tensor product of two spaces, one of which being the representation space V_λ where the representation π_λ is realized, while the other one is still $\mathbb{C}^N$. We distinguish two R-matrices of this type depending on the order of the spaces. One is the R-matrix with the auxiliary space V_λ. It has the form

$$\mathbf{R}_\lambda(x) = \mathbf{1} \otimes \mathbf{1} + \frac{\eta}{x} \sum_{a,b=1}^{N} \pi_\lambda(\mathbf{e}_{ab}) \otimes \mathbf{e}_{ba}. \qquad (13.16)$$

The auxiliary space of the other one is $\mathbb{C}^N$:

$$\mathbf{R}^\lambda(x) = \mathbf{1} \otimes \mathbf{1} + \frac{\eta}{x} \sum_{a,b=1}^{N} \mathbf{e}_{ab} \otimes \pi_\lambda(\mathbf{e}_{ba}). \qquad (13.17)$$

Clearly, $\mathbf{R}_\square(x) = \mathbf{R}^\square(x) = \mathbf{R}(x)$.

It is convenient to denote

$$\mathbf{P}_\lambda^{0j} = \sum_{a,b=1}^{N} \pi_\lambda(\mathbf{e}_{ab}) \otimes \mathbf{e}_{ba}^{(j)}, \qquad \mathbf{P}_{0j}^\lambda = \sum_{a,b=1}^{N} \mathbf{e}_{ab} \otimes \pi_\lambda(\mathbf{e}_{ba}^{(j)}), \qquad (13.18)$$

then the R-matrix acting nontrivially in the tensor product of the auxiliary space V_λ and the jth space $\mathbb{C}^N$ is $\mathbf{R}_\lambda^{0j}(x) = \mathbf{I} + \dfrac{\eta}{x} \mathbf{P}_\lambda^{0j}$ while the R-matrix acting non-trivially in the tensor product of the auxiliary space $\mathbb{C}^N$ and the jth space $V_{\Lambda^{(j)}}$ is $\mathbf{R}_{0j}^{\Lambda^{(j)}}(x) = \mathbf{I} + \dfrac{\eta}{x} \mathbf{P}_{0j}^{\Lambda^{(j)}}$. The R-matrix $\mathbf{R}^\lambda(x)$ obeys the Yang–Baxter equation

$$\mathbf{R}_{12}^\square(x_1 - x_2) \mathbf{R}_{13}^\lambda(x_1 - x_3) \mathbf{R}_{23}^\lambda(x_2 - x_3) = \mathbf{R}_{23}^\lambda(x_2 - x_3) \mathbf{R}_{13}^\lambda(x_1 - x_3) \mathbf{R}_{12}^\square(x_1 - x_2) \quad (13.19)$$

and possesses the invariance property

$$\pi_\square(\mathbf{g}) \otimes \pi_\lambda(\mathbf{g})\, \mathbf{R}^\lambda(x) = \mathbf{R}^\lambda(x)\, \pi_\square(\mathbf{g}) \otimes \pi_\lambda(\mathbf{g}) \qquad (13.20)$$

valid for any $\mathbf{g}$. The Yang–Baxter equation for the R-matrix $\mathbf{R}_\lambda(x)$ reads

$$\mathbf{R}_{\lambda,\mu}^{12}(x_1 - x_2) \mathbf{R}_\lambda^{13}(x_1 - x_3) \mathbf{R}_\mu^{23}(x_2 - x_3) = \mathbf{R}_\mu^{23}(x_2 - x_3) \mathbf{R}_\lambda^{13}(x_1 - x_3) \mathbf{R}_{\lambda,\mu}^{12}(x_1 - x_2), \quad (13.21)$$

where $\mathbf{R}_{\lambda,\mu}^{12}(x) \in \mathrm{End}(V_\lambda \otimes V_\mu)$ is a yet more general R-matrix. Its explicit form is complicated. The invariance property for $\mathbf{R}_\lambda(x)$ is similar to that for (13.20) with the opposite order of the tensor factors.

13.2.2   Inhomogeneous XXX spin chains

Here we construct, using the R-matrices from the previous subsection, the inhomogeneous integrable XXX spin chains with twisted boundary conditions.

T-operators, nonlocal Hamiltonians and integrals of motion

Let $\mathbf{g} \in GL(N)$ be a group element represented by a diagonal matrix

$$\mathbf{g} = \mathrm{diag}\,(g_1, g_2, \dots, g_N) = \sum_{a=1}^{N} g_a \mathbf{e}_{aa}.$$

We call it the twist matrix with the twist parameters g_i. It is used for the construction of an integrable spin chain with twisted boundary conditions. The T-operator (the transfer matrix) of the inhomogeneous spin chain with twisted boundary conditions is defined by

$$t(x) = \mathrm{tr}_0 \Big(\mathbf{R}^{0L}(x - x_L) \dots \mathbf{R}^{02}(x - x_2) \mathbf{R}^{01}(x - x_1)\, (\mathbf{g} \otimes \mathbf{I}) \Big), \qquad (13.22)$$

where $x_1, x_2, \dots, x_L$ are inhomogeneity parameters. We assume that they are in general position, meaning that $x_i \neq x_j$ and $x_i \neq x_j \pm \eta$ for all $i \neq j$. As is known, the

Yang–Baxter equation implies that the T-operators with fixed inhomogeneous and twist parameters commute: $[\mathbf{t}(x), \mathbf{t}(x')] = 0$ for any x, x'.

The dynamical variables of the model (which we call 'spins' in analogy with the rank 1 case) are vectors in the vector representation of $gl(N)$ realized in the spaces $\mathbb{C}^N$ attached to each site. One can define a set of nonlocal commuting Hamiltonians $\mathbf{H}_j$ as residues of $\mathbf{t}(x)$ at $x = x_j$:

$$\mathbf{t}(x) = \mathbf{I} \operatorname{tr} \mathbf{g} + \sum_{j=1}^{L} \frac{\eta \mathbf{H}_j}{x - x_j}. \tag{13.23}$$

In general, the Hamiltonians $\mathbf{H}_j$ imply a long-range interaction involving all spins in the chain (cf. [21]). Their explicit form is

$$\mathbf{H}_j = \overleftarrow{\prod_{k=1}^{j-1}} \left(\mathbf{I} + \frac{\eta \mathbf{P}_{kj}}{x_j - x_k} \right) \mathbf{g}^{(j)} \overleftarrow{\prod_{k=j+1}^{L}} \left(\mathbf{I} + \frac{\eta \mathbf{P}_{jk}}{x_j - x_k} \right) \tag{13.24}$$

$$= \sum_{I \subseteq \{1,2,\ldots,L\} \setminus \{j\}} \eta^{|I|} \left(\prod_{k \in I} \frac{1}{x_j - x_k} \right) \left(\overleftarrow{\prod_{k \in I, k<j}} \mathbf{P}_{kj} \right) \mathbf{g}^{(j)} \left(\overleftarrow{\prod_{k \in I, k>j}} \mathbf{P}_{jk} \right), \tag{13.25}$$

where $\mathbf{g}^{(j)} := \mathbf{1}^{\otimes(j-1)} \otimes \mathbf{g} \otimes \mathbf{1}^{\otimes(L-j)}$ and $\mathbf{P}_{ij}$ is the permutation operator (13.14). In the second line, the sum is taken over all subsets I of the set $\{1, 2, \ldots, L\} \setminus \{j\}$ including the empty one; $|I| \equiv \operatorname{Card} I$.

In addition to the Hamiltonians $\mathbf{H}_j$, there are other integrals of motion. It is easy to see that the operators

$$\mathbf{M}_a = \sum_{j=1}^{L} \mathbf{e}_{aa}^{(j)}, \tag{13.26}$$

referred to as *weight operators*, commute with the $\mathbf{H}_i$'s: $[\mathbf{H}_j, \mathbf{M}_a] = 0$. Therefore, the eigenstates of the Hamiltonians can be classified according to the eigenvalues $(M_1, \ldots, M_N)$ of the weight operators referred to as *weights*. For example, in the $gl(2)$ case, M_1 and M_2 are the numbers of spins with positive and negative z-projections, respectively.

Let

$$\mathcal{V} = (\mathbb{C}^N)^{\otimes L} = \bigoplus_{M_1, \ldots, M_N} \mathcal{V}(\{M_a\})$$

be the 'weight decomposition' of the quantum space into the direct sum of *weight spaces* which are eigenspaces of the weight operators with the eigenvalues $M_a \in \mathbb{Z}_{\geq 0}$, $a = 1, \ldots, N$. Then any eigenstate of the $\mathbf{H}_j$'s belongs to some weight space $\mathcal{V}(\{M_a\})$. The dimension of the weight space $\mathcal{V}(\{M_a\})$ is given by

$$\dim \mathcal{V}(\{M_a\}) = \frac{L!}{M_1! \ldots M_N!}.$$

In particular, let $1 \leq a_0 \leq N$ be some fixed index, then the space with $M_a = L\delta_{aa_0}$ is one-dimensional. It is spanned by the vector $\mathbf{v}_{a_0} \otimes \dots \otimes \mathbf{v}_{a_0}$ which is an eigenvector of the Hamiltonians (13.24). Indeed, using (13.15) in the particular case $a_1 = \dots = a_L = a_0$, one can see that

$$\mathbf{H}_j(\mathbf{v}_{a_0} \otimes \dots \otimes \mathbf{v}_{a_0}) = g_{a_0} \prod_{k=1,\neq j}^{L} \left(1 + \frac{\eta}{x_j - x_k}\right)(\mathbf{v}_{a_0} \otimes \dots \otimes \mathbf{v}_{a_0}). \qquad (13.27)$$

The weight operators are not all independent. Since $\sum_a \mathbf{e}_{aa} = \mathbf{1}$, we have $\sum_a \mathbf{M}_a = L\mathbf{I}$ and hence $\sum_a M_a = L$. Note also that

$$\sum_{j=1}^{L} \mathbf{H}_j = \sum_{j=1}^{L} \mathbf{g}^{(j)} = \sum_{a=1}^{N} g_a \mathbf{M}_a, \qquad (13.28)$$

so the model has $L+N-1$ independent commuting integrals of motion.

For completeness, we give here the definition of the T-operator for a more general inhomogeneous spin chain model with the quantum space $\otimes_{j=1}^{L} V_{\Lambda^{(j)}}$ and the auxiliary space $\mathbb{C}^N$. The spin chain is defined by the following data:

- The number of sites, L, and the inhomogeneity parameters x_i at each site;
- Tensor representations of $gl(N)$ indexed by the Young diagrams

$$\Lambda^{(j)} = (\Lambda_1^{(j)}, \dots, \Lambda_N^{(j)}) \in (\mathbb{Z}_{\geq 0})^K, \qquad \Lambda_1^{(j)} \geq \Lambda_2^{(j)} \geq \dots \geq \Lambda_N^{(j)} \geq 0 \qquad (13.29)$$

  assigned to each site $j = 1, \dots, L$; and
- Elements of the diagonal twist matrix $\mathbf{g} = \text{diag}(g_1, \dots, g_N)$ (the twist parameters).

The T-operator

$$\mathbf{t}^{\Lambda}(x) = \text{tr}_0 \left(\mathbf{R}_{0L}^{\Lambda^{(L)}}(x - x_L) \dots \mathbf{R}_{02}^{\Lambda^{(2)}}(x - x_2)\mathbf{R}_{01}^{\Lambda^{(1)}}(x - x_1)(\mathbf{g} \otimes \mathbf{I})\right) \qquad (13.30)$$

acts in the space $\otimes_{j=1}^{L} V_{\Lambda^{(j)}}$. One may also introduce a set of Hamiltonians $\mathbf{H}_j^{\Lambda}$ in the way similar to (13.23):

$$\mathbf{t}^{\Lambda}(x) = \mathbf{I} \, \text{tr} \, \mathbf{g} + \sum_{j=1}^{L} \frac{\eta \mathbf{H}_j^{\Lambda}}{x - x_j}. \qquad (13.31)$$

Our main objects of interest will be the T-operator $\mathbf{t}(x)$ and the Hamiltonians $\mathbf{H}_j$ of the model with vector representations at the sites corresponding to the choice $\Lambda^{(j)} = (1, 0, \dots, 0)$ for all $j = 1, \dots, L$.

Diagonalization of the T-operator via Bethe ansatz

The T-operators and the Hamiltonians $\mathbf{H}_j^\Lambda$ can be diagonalized by the algebraic nested Bethe ansatz [22, 23]. Although in what follows we need only the result for the choice $\Lambda^{(j)} = (1, 0, \ldots, 0)$, [2] we give here the general result for future references.

Eigenvalues of the T-operator $\mathbf{t}^\Lambda(x)$ are given by

$$T^\Lambda(x) = \sum_{b=1}^{N} g_b \prod_{k=1}^{L} \frac{x - x_k + \eta \, \Lambda_b^{(k)}}{x - x_k} \prod_{\gamma=1}^{L_{b-1}} \frac{x - \mu_\gamma^{b-1} + \eta}{x - \mu_\gamma^{b-1}} \prod_{\gamma=1}^{L_b} \frac{x - \mu_\gamma^b - \eta}{x - \mu_\gamma^b}. \tag{13.32}$$

The corresponding eigenvalues of the Hamiltonians (13.23) are

$$H_{\Lambda, i} = \eta^{-1} \mathrm{res}_{x=x_i} T^\Lambda(z) = \sum_{b=1}^{N} \Lambda_b^{(k)} g_b$$

$$\times \prod_{k \neq i}^{L} \frac{x_i - x_k + \Lambda_b^{(k)} \eta}{x_i - x_k} \prod_{\gamma=1}^{L_{b-1}} \frac{x_i - \mu_\gamma^{b-1} + \eta}{x_i - \mu_\gamma^{b-1}} \prod_{\gamma=1}^{L_b} \frac{x_i - \mu_\gamma^b - \eta}{x_i - \mu_\gamma^b}. \tag{13.33}$$

It is convenient to set $L_0 = L_N = 0$. The parameters μ_α^b with

$$\alpha = 1, \ldots, L_b, \quad b = 1, \ldots, N-1, \quad L \geq L_1 \geq L_2 \geq \ldots \geq L_{N-1} \geq 0 \tag{13.34}$$

are Bethe roots. They satisfy the system of Bethe equations which are equivalent to the conditions

$$\mathrm{res}_{x=\mu_\alpha^b} T^\Lambda(x) = 0 \qquad \text{for all } \alpha = 1, \ldots, L_b, \quad b = 1, \ldots, N-1. \tag{13.35}$$

The Bethe equations have the form

$$g_b \prod_{k=1}^{L} \frac{\mu_\beta^b - x_k + \Lambda_b^{(k)} \eta}{\mu_\beta^b - x_k + \Lambda_{b+1}^{(k)} \eta} \prod_{\gamma=1}^{L_{b-1}} \frac{\mu_\beta^b - \mu_\gamma^{b-1} + \eta}{\mu_\beta^b - \mu_\gamma^{b-1}}$$

$$= g_{b+1} \prod_{\gamma \neq \beta}^{L_b} \frac{\mu_\beta^b - \mu_\gamma^b + \eta}{\mu_\beta^b - \mu_\gamma^b - \eta} \prod_{\gamma=1}^{L_{b+1}} \frac{\mu_\beta^b - \mu_\gamma^{b+1} - \eta}{\mu_\beta^b - \mu_\gamma^{b+1}}. \tag{13.36}$$

Later we will specify these general formulae for the highest weights

$$\Lambda^{(j)} = (1, 0, \ldots, 0) \quad \text{for all} \quad j = 1, \ldots, L, \text{ i.e., } \Lambda_b^{(j)} = \delta_{b1}. \tag{13.37}$$

With this choice, the first product in the left-hand side of (13.36) disappears for $b \geq 2$.

---

[2] The Bethe ansatz for trigonometric models closely related to this case was discussed in [24].

13.2.3   The higher T-operators

The R-matrix (13.16) allows one to construct a family of T-operators with the more general auxiliary space

$$\mathbf{t}_\lambda(x) = \mathrm{tr}_{V_\lambda}\Big(\mathbf{R}_\lambda^{0L}(x - x_L)\dots \mathbf{R}_\lambda^{02}(x - x_2)\mathbf{R}_\lambda^{01}(x - x_1)\,(\pi_\lambda(\mathbf{g}) \otimes \mathbf{I})\Big). \qquad (13.38)$$

Obviously, $\mathbf{t}_\square(x)$ coincides with the T-operator $\mathbf{t}(x)$ introduced previously. The T-operators with fixed inhomogeneous and twist parameters commute with $\mathbf{t}(x)$ and among themselves,

$$[\mathbf{t}_\lambda(x),\, \mathbf{t}_\mu(x')] = 0, \qquad (13.39)$$

for any x, x', λ, μ. At $\lambda = \varnothing$ we put $\mathbf{t}_\varnothing(x)$ equal to the identity operator: $\mathbf{t}_\varnothing(x) = \mathbf{I}$.

It is clear from (13.16) and (13.38) that at $L = 0$ (the empty quantum space) as well as in the limit $x \to \infty$ for any L the T-operators become equal to the characters $\chi_\lambda(\mathbf{g}) = \mathrm{tr}_{V_\lambda}\mathbf{g}$. In what follows we also need the next-to-leading term of the expansion of $\mathbf{t}_\lambda(x)$ as $x \to \infty$. From the definition (13.38) one obtains the expansion

$$\mathbf{t}_\lambda(x) = \chi_\lambda(\mathbf{g})\,\mathbf{I} + \frac{\eta}{x}\sum_{j=1}^{L}\sum_{a,b}\frac{\partial \chi_\lambda(e^{\varepsilon \mathbf{e}_{ab}}\mathbf{g})}{\partial \varepsilon}\bigg|_{\varepsilon=0}\mathbf{e}_{ba}^{(j)} + O(1/x^2). \qquad (13.40)$$

Indeed, we have

$$\mathbf{t}_\lambda(x) = \chi_\lambda(\mathbf{g})\,\mathbf{I} + \frac{\eta}{x}\sum_{j=1}^{L}\mathrm{tr}_{V_\lambda}\Big(\mathbf{P}_\lambda^{0j}\pi_\lambda(\mathbf{g})\Big) + O(1/x^2),$$

which is converted to the form (13.40) by the chain of equalities

$$\mathrm{tr}_{V_\lambda}\Big(\mathbf{P}_\lambda^{0j}\pi_\lambda(\mathbf{g})\Big) = \sum_{a,b}\mathrm{tr}_{V_\lambda}\pi_\lambda(\mathbf{e}_{ab}\mathbf{g})\,\mathbf{e}_{ba}^{(j)} = \sum_{a,b}\frac{\partial}{\partial\varepsilon}\Big[\mathrm{tr}_{V_\lambda}\pi_\lambda(e^{\varepsilon \mathbf{e}_{ab}}\mathbf{g})\Big]\bigg|_{\varepsilon=0}\mathbf{e}_{ba}^{(j)}.$$

In fact the following explicit expression for the T-operator in terms of characters is available:

$$\mathbf{t}_\lambda(x) = \sum_{l=0}^{L}\eta^l \sum_{i_1<\dots<i_l}\sum_{\substack{a_1,\dots,a_l \\ b_1,\dots,b_l}}\overrightarrow{\prod_{\alpha=1}^{l}}\left(\frac{\mathbf{e}_{b_\alpha a_\alpha}^{(i_\alpha)}}{x - x_{i_\alpha}}\frac{\partial}{\partial\varepsilon_\alpha}\right)\chi_\lambda\big(e^{\varepsilon_l \mathbf{e}_{a_l b_l}}\dots e^{\varepsilon_1 \mathbf{e}_{a_1 b_1}}\mathbf{g}\big)\bigg|_{\varepsilon_\alpha=0}$$

$$= \overrightarrow{\prod_{l=1}^{L}}\left(\mathbf{I} + \eta\sum_{a_l,b_l}\frac{\mathbf{e}_{b_l a_l}^{(l)}}{x - x_l}\frac{\partial}{\partial\varepsilon_l}\right)\chi_\lambda\big(e^{\varepsilon_l \mathbf{e}_{a_l b_l}}\dots e^{\varepsilon_1 \mathbf{e}_{a_1 b_1}}\mathbf{g}\big)\bigg|_{\varepsilon_l=0}.$$

$$(13.41)$$

The summation over each a_α and b_α runs from 1 to N. The derivation of (13.41) is sketched in Appendix A[3]

The characters are known to satisfy the Jacobi-Trudi identities [26]:

$$\chi_\lambda(\mathbf{g}) = \det_{1 \le i,j \le \lambda_1'} \chi_{\lambda_i - i + j}(\mathbf{g}) = \det_{1 \le i,j \le \lambda_1} \chi^{\lambda_i' - i + j}(\mathbf{g}). \qquad (13.42)$$

Here $\chi_k := \chi_{(k)}$ (respectively, $\chi^k := \chi_{(1^k)}$) is the character corresponding to the one-row (respectively, one-column) diagram of length k.

There exist analogues of these identities for the T-operators, depending on the spectral parameter. These are the Cherednik–Bazhanov–Reshetikhin (CBR) determinant formulas, sometimes called the quantum Jacobi–Trudi identities:

$$\mathsf{t}_\lambda(x) = \det_{1 \le i,j \le \lambda_1'} \mathsf{t}_{\lambda_i - i + j}\big(x - (j-1)\eta\big) = \det_{1 \le i,j \le \lambda_1} \mathsf{t}^{\lambda_i' - i + j}\big(x + (j-1)\eta\big). \qquad (13.43)$$

The determinants are well defined because all the T-operators commute. Similarly to (13.42), $\mathsf{t}_k(x) := \mathsf{t}_{(k)}(x)$ and $\mathsf{t}^k(x) := \mathsf{t}_{(1^k)}(x)$ are the T-operators corresponding to the one-row and one-column diagrams, respectively. For models based on $gl(N)$-invariant R-matrices, the quantum Jacobi-Trudi identities follow from resolutions of modules for the Yangian $Y(gl(N))$ [27]. In the physical literature, they appeared in [28] for $gl(N)$, (see also [29]), in [30] for $gl(N|M)$, and in [31, 32] for some infinite dimensional representations in the context of AdS/CFT correspondence. A direct proof was given in [25].[4]

The eigenvalues of the T-operators (13.38) are rational functions of x with L poles. Another normalization, where they are polynomials in x of degree L, is also convenient and even preferable for the link to classical integrable hierarchies. The polynomial form of the T-operators is obtained as:

$$\mathsf{T}_\lambda(x) = \prod_{j=1}^{L} (x - x_j)\, \mathsf{t}_\lambda(x). \qquad (13.44)$$

In particular for $\lambda = \emptyset$, we have

$$\mathsf{T}_\emptyset(x) = \prod_{j=1}^{L} (x - x_j).$$

The CBR formulas (13.43) in the polynomial normalization acquire the form

$$\mathsf{T}_\lambda(x) = \Big(\prod_{k=1}^{\lambda_1' - 1} \mathsf{T}_\emptyset(x - k\eta) \Big)^{-1} \det_{1 \le i,j \le \lambda_1'} \mathsf{T}_{\lambda_i - i + j}\big(x - (j-1)\eta\big), \qquad (13.45)$$

---

[3] In parentheses in the second line one can recognize the co-derivative operator [25], which is a version of the matrix derivative. It proved to be a valuable technical tool for the proof of the Cherednik–Bazhanov–Reshetikhin identities and for the master T-operator construction. However, here we do not use the co-derivative explicitly.

[4] There was a minor gap in the proof given in [25] which was filled in the appendix of [2].

$$T_\lambda(x) = \left(\prod_{k=1}^{\lambda_1-1} T_\emptyset(x - k\eta) \right)^{-1} \det_{1\le i,j \le \lambda_1} T^{\lambda'_i-i+j}\big(x+(j-1)\eta\big). \tag{13.46}$$

13.2.4 The construction of the master T-operator

The master T-operator is a generating function of the T-operators $T_\lambda(x)$ of a special form [2]. (In an implicit form, the notion of the master T-operator appeared already in [4].) To present the construction, we should recall the definition of the Schur functions.

Let $\mathbf{t} = \{t_1, t_2, t_3, \ldots\}$ be an infinite set of complex parameters (we call them times) and $s_\lambda(\mathbf{t})$ be the standard Schur functions (S-functions) which can be introduced as

$$s_\lambda(\mathbf{t}) = \det_{1\le i,j \le \lambda'_1} h_{\lambda_i-i+j}(\mathbf{t}),$$

where the polynomials $h_k(\mathbf{t}) = s_{(k)}(\mathbf{t})$ (the elementary Schur functions) are defined by

$$e^{\xi(\mathbf{t},z)} = \sum_{k=0}^{\infty} h_k(\mathbf{t})z^k, \qquad \xi(\mathbf{t},z) := \sum_{n=1}^{\infty} t_k z^k. \tag{13.47}$$

It is convenient to put $h_k(\mathbf{t}) = 0$ for negative k and $s_\emptyset(\mathbf{t}) = 1$. As is obvious from the definition, the Schur functions are polynomials in the times t_i. The Schur functions are often regarded as symmetric functions of variables ξ_α such that $t_k = \frac{1}{k}\sum_\alpha \xi^k_\alpha$.

The characters can be expressed in terms of the Schur functions as follows. Set $y_k = \frac{1}{k}\operatorname{tr} \mathbf{g}^k$, where $\operatorname{tr}\mathbf{g}^k$ is the trace of $\mathbf{g}^k$ realized in the vector representation as a $N\times N$ diagonal matrix: $\operatorname{tr}\mathbf{g}^k = \sum_{a=1}^{N} g^k_a$. Then $\chi_\lambda(\mathbf{g}) = s_\lambda(\mathbf{y})$. This is equivalent to the fact that $\big(\det(1 - z\mathbf{g})\big)^{-1}$ is the generating function for the characters corresponding to one-row diagrams: $\big(\det(1 - z\mathbf{g})\big)^{-1} = \sum_{k\ge 0} h_k(\mathbf{y})z^k$. For later use we need the following identity for the characters,

$$\sum_\lambda \chi_\lambda(\mathbf{g})s_\lambda(\mathbf{t}) = \exp\left(\sum_{k\ge 1} t_k \operatorname{tr}\mathbf{g}^k\right), \tag{13.48}$$

which is simply the Cauchy–Littlewood identity for the Schur functions [26]. Here and below, the sum $\sum_\lambda$ goes over all Young diagrams λ including the empty one.

Now we are ready to introduce the master T-operator as an infinite sum over the Young diagrams:

$$T(x, \mathbf{t}) = \sum_\lambda T_\lambda(x)s_\lambda(\mathbf{t}). \tag{13.49}$$

It immediately follows from the definition that the T-operators $T_\lambda(x)$ can be restored from it by applying the differential operators in the times t_i,

$$T_\lambda(x) = s_\lambda(\tilde{\partial})T(x,\mathbf{t})\Big|_{\mathbf{t}=0}, \tag{13.50}$$

where $\tilde{\partial} := \{\partial_{t_1}, \frac{1}{2}\partial_{t_2}, \frac{1}{3}\partial_{t_3}, \ldots\}$. In particular, $\mathsf{T}_\emptyset(x) = \mathsf{T}(x,\mathbf{0}) = \prod_{j=1}^{L}(x - x_j)\,\mathbf{I}$
and $\mathsf{T}_\square(x) = \partial_{t_1}\mathsf{T}(x,\mathbf{t})\big|_{\mathbf{t}=0}$, so that the T-operator $\mathbf{t}(x) = \mathbf{t}_{(1)}(x) = \mathbf{t}_\square(x)$ (13.22) is
expressed as the logarithmic derivative of the master T-operator:

$$\mathbf{t}(x) = \partial_{t_1}\log \mathsf{T}(x,\mathbf{t})\big|_{\mathbf{t}=0}. \tag{13.51}$$

Using (13.40) and the Cauchy–Littlewood identity (13.48), one can derive the
following expansion of the master T-operator as $x \to \infty$:

$$\frac{\mathsf{T}(x,\mathbf{t})}{\mathsf{T}_\emptyset(x)} = e^{\operatorname{str} \xi(\mathbf{t},\mathbf{g})}\left(\mathbf{I} + \frac{\eta}{x}\sum_{j=1}^{L}\sum_{k\geq 1} k\,\mathbf{t}_k(\mathbf{g}^{(j)})^k + O(1/x^2)\right). \tag{13.52}$$

More generally, using (13.41) and the Cauchy–Littlewood identity one arrives, in a
similar way, to the following explicit expression for the master T-operator:

$$\frac{\mathsf{T}(x,\mathbf{t})}{\mathsf{T}_\emptyset(x)} = \sum_{l=0}^{L}\eta^l\sum_{i_1<\ldots<i_l}\sum_{\substack{a_1,\ldots,a_l \\ b_1,\ldots,b_l}}\left(\overrightarrow{\prod_{\alpha=1}^{l}}\frac{\mathbf{e}_{b_\alpha a_\alpha}^{(i_\alpha)}}{x - x_{i_\alpha}}\frac{\partial}{\partial \varepsilon_\alpha}\right)e^{\operatorname{tr}\xi\left(\mathbf{t},e^{\varepsilon_l \bullet a_l b_l}\ldots e^{\varepsilon_1 \bullet a_1 b_1}\mathbf{g}\right)}\Bigg|_{\varepsilon_\alpha=0}$$

$$= \overrightarrow{\prod_{l=1}^{L}}\left(\mathbf{I} + \eta\sum_{a_l,b_l}\frac{\mathbf{e}_{b_l a_l}^{(l)}}{x - x_l}\frac{\partial}{\partial \varepsilon_l}\right)e^{\operatorname{tr}\xi\left(\mathbf{t},e^{\varepsilon_l \bullet a_l b_l}\ldots e^{\varepsilon_1 \bullet a_1 b_1}\mathbf{g}\right)}\Bigg|_{\varepsilon_l=0}. \tag{13.53}$$

Given $z \in \mathbb{C}$, we will use the standard notation $\mathbf{t} \pm [z^{-1}]$ for the following special
shift of the time variables:

$$\mathbf{t} \pm [z^{-1}] := \left\{t_1 \pm z^{-1}, t_2 \pm \tfrac{1}{2}z^{-2}, t_3 \pm \tfrac{1}{3}z^{-3}, \ldots\right\}.$$

As we will see later, $\mathsf{T}(x, \mathbf{t} \pm [z^{-1}])$ regarded as functions of z with fixed $x, \mathbf{t}$ play an
important role. Here we only note that Eq. (13.50) implies that $\mathsf{T}(x, \mathbf{0} \pm [z^{-1}])$ are the
generating series for the T-operators corresponding to the one-row and one-column
diagrams, respectively:

$$\mathsf{T}(x, [z^{-1}]) = \sum_{s=0}^{\infty}z^{-s}\mathsf{T}_s(x), \qquad \mathsf{T}(x, -[z^{-1}]) = \sum_{a=0}^{\infty}(-z)^{-a}\mathsf{T}^a(x). \tag{13.54}$$

13.2.5   The master T-operator and the mKP hierarchy

The bilinear identity for the master T-operator

The main property of the master T-operator which provides a remarkable link to the
theory of classical nonlinear integrable equations and their hierarchies is given by the
following statement.

Theorem 13.2.1.   The master T-operator (13.49) satisfies the bilinear identity for the mKP hierarchy

$$\oint_{\mathcal{C}} z^{(x-x')/\eta} e^{\xi(t-t',z)}\, \mathsf{T}(x, \mathbf{t} - [z^{-1}])\mathsf{T}(x', \mathbf{t}' + [z^{-1}])dz = 0 \tag{13.55}$$

for all $\mathbf{t}$, $\mathbf{t}'$, x, and x'. The integration contour $\mathcal{C}$ encircles the cut $[0, \infty]$ between 0 and ∞ and does not enclose any singularities coming from the T-factors.

The proof is based on the CBR formulas (13.45) and (13.46). The definition of the master T-operator (13.49) can be interpreted as the expansion of the tau-function in Schur polynomials [9, 33, 34]. The functional relations for quantum transfer matrices [27–30] are then the Plücker-like relations for coefficients of the expansion.

The bilinear identity (13.55) is a source of various bilinear Hirota equations for the master T-operator. For example, setting $x' = x - \eta$, $\mathbf{t}' = \mathbf{t} - [z_1^{-1}] - [z_2^{-1}]$, we obtain the three-term difference Hirota equation

$$z_2 \mathsf{T}\left(x + \eta, \mathbf{t} - [z_2^{-1}]\right) \mathsf{T}\left(x, \mathbf{t} - [z_1^{-1}]\right) - z_1 \mathsf{T}\left(x + \eta, \mathbf{t} - [z_1^{-1}]\right) \mathsf{T}\left(x, \mathbf{t} - [z_2^{-1}]\right)$$

$$+ (z_1 - z_2)\mathsf{T}(x + \eta, \mathbf{t})\mathsf{T}\left(x, \mathbf{t} - [z_1^{-1}] - [z_2^{-1}]\right) = 0, \tag{13.56}$$

which is in fact, equivalent to the bilinear identity (see [35]).

The Baker-Akhiezer functions

Let $T(x, \mathbf{t})$ be any eigenvalue of the master T-operator. As it follows from (13.55), it is a tau-function of the mKP hierarchy. It is then natural to incorporate other key ingredients of the soliton theory. The most important for us are the Baker–Akhiezer (BA) function and its adjoint. In what follows we refer to both as the BA functions. They are defined as

$$\psi(x, \mathbf{t}; z) = z^{x/\eta} e^{\xi(\mathbf{t}, z)} \frac{T(x, \mathbf{t} - [z^{-1}])}{T(x, \mathbf{t})}, \tag{13.57}$$

$$\psi^*(x, \mathbf{t}; z) = z^{-x/\eta} e^{-\xi(\mathbf{t}, z)} \frac{T(x, \mathbf{t} + [z^{-1}])}{T(x, \mathbf{t})}. \tag{13.58}$$

We are going to also consider the operator-valued BA functions $\hat{\psi}(x, \mathbf{t}; z)$ defined by the same formulas with $T(x, \mathbf{t})$ substituted by $\mathsf{T}(x, \mathbf{t})$. Since these operators commute for all $x, \mathbf{t}$, the operator $\hat{\psi}(x, \mathbf{t}; z)$ is well defined.

According to the definition of the master T-operator, $\mathsf{T}(x, \mathbf{t} \mp [z^{-1}])$ is an infinite series in z^{-1}. From (13.53) one can see that this series converges to a rational function of z for any $x, \mathbf{t}$ if $|z| > \max\{|g_1|, |g_2|, \ldots, |g_N|\}$. Explicitly, we obtain

$$
\mathsf{T}(x, \mathbf{t} \mp [z^{-1}]) = \overrightarrow{\prod_{l=1}^{L}} \left((x - x_l)\mathbf{I} + \eta \sum_{a_l, b_l} \mathbf{e}_{b_l a_l}^{(l)} \frac{\partial}{\partial \varepsilon_l} \right)
$$

$$
\times \left\{ \left[\det \left(1 - z^{-1} \mathbf{g}_{\varepsilon_L, \ldots, \varepsilon_1}^{a_L b_L, \ldots, a_1 b_1} \right) \right]^{\pm 1} \mathrm{e}^{\mathrm{tr}\, \xi (\mathbf{t}, \mathbf{g}_{\varepsilon_L, \ldots, \varepsilon_1}^{a_L b_L, \ldots, a_1 b_1})} \right\} \Bigg|_{\varepsilon_l = 0} , \tag{13.59}
$$

where we have put $\mathbf{g}_{\varepsilon_n, \ldots, \varepsilon_1}^{a_n b_n, \ldots, a_1 b_1} \equiv \mathrm{e}^{\varepsilon_n \mathbf{e}_{a_n b_n}} \ldots \mathrm{e}^{\varepsilon_1 \mathbf{e}_{a_1 b_1}} \mathbf{g}$ for brevity. Therefore, the function $z^{-x/\eta} \mathrm{e}^{-\xi(\mathbf{t},z)} \psi(x, \mathbf{t}; z)$ is a rational function of z with poles at the points $z = g_a$ (the eigenvalues of the matrix $\mathbf{g}$) of at least first order. It can be also derived[5] from (13.59) that

$$
\lim_{z \to 0} z^{\pm N} \mathsf{T}(x, \mathbf{t} \mp [z^{-1}]) = (-1)^N (\det \mathbf{g})^{\pm 1} \mathsf{T}(x \pm \eta, \mathbf{t}). \tag{13.60}
$$

The left-hand side is to be understood as the analytic continuation to the point $z = 0$ of the analytic function (rational in our case) defined by the series which converges in a neighbourhood of infinity.

Since any eigenvalue $T(x, \mathbf{t})$ is a polynomial in x, the functions $z^{-x/\eta} \psi$ and $z^{x/\eta} \psi^*$, regarded as functions of x, are rational functions with L zeros and L poles which are simple in general position. From (13.49), (13.44), and (13.38), using the Cauchy–Littlewood identity, or directly from (13.59), we conclude that

$$
\lim_{x \to \infty} \frac{T(x, \mathbf{t} \mp [z^{-1}])}{T(x, \mathbf{t})} = \frac{\mathrm{e}^{\mathrm{tr}\, \xi(\mathbf{t} \mp [z^{-1}], \mathbf{g})}}{\mathrm{e}^{\mathrm{tr}\, \xi(\mathbf{t}, \mathbf{g})}} = \mathrm{e}^{\mathrm{tr}\, \xi(\mp [z^{-1}], \mathbf{g})} = \left[\det \left(1 - z^{-1} \mathbf{g} \right) \right]^{\pm 1} ,
$$

hence

$$
\lim_{x \to \infty} z^{-x/\eta} \mathrm{e}^{-\xi(\mathbf{t},z)} \psi(x, \mathbf{t}; , z) = z^{-N} \det(z\mathbf{1} - \mathbf{g}), \tag{13.61}
$$

$$
\lim_{x \to \infty} z^{x/\eta} \mathrm{e}^{\xi(\mathbf{t},z)} \psi^*(x, \mathbf{t}; z) = z^N \left(\det(z\mathbf{1} - \mathbf{g}) \right)^{-1}. \tag{13.62}
$$

---

[5] The main underlying statement is that

$$
\left(x\mathbf{I} + \eta \sum_{a,b} \mathbf{e}_{ba}^{(l)} \frac{\partial}{\partial \varepsilon} \right) \left[\det \left(\mathrm{e}^{\varepsilon \mathbf{e}_{ab}} \mathbf{g} \right) \right]^{\pm 1} \Phi(\varepsilon) \Big|_{\varepsilon = 0}
$$

$$
= [\det \mathbf{g}]^{\pm 1} \left((x \pm \eta)\mathbf{I} + \eta \sum_{a,b} \mathbf{e}_{ba}^{(l)} \frac{\partial}{\partial \varepsilon} \right) \Phi(\varepsilon) \Big|_{\varepsilon = 0}
$$

for any $\mathbf{g} \in GL(M)$ and any $\Phi \in \mathrm{End}(V_{l+1} \otimes V_{l+2} \otimes \ldots \otimes V_L)$ (here $V_i \cong \mathbb{C}^N$). It immediately follows from the Leibniz rule and the identity

$$
\frac{\partial}{\partial \varepsilon} \left[\det \left(\mathrm{e}^{\varepsilon \mathbf{e}_{ab}} \mathbf{g} \right) \right]^{\pm 1} \Big|_{\varepsilon = 0} = \mathrm{tr}(\mathbf{e}_{ab}) [\det \mathbf{g}]^{\pm 1} = \delta_{ab} [\det \mathbf{g}]^{\pm 1},
$$

which is easy to check.

The (operator-valued) functions $\hat\psi(x,z) := \hat\psi(x,\mathbf{0};z)$ and $\hat\psi^*(x,z) := \hat\psi^*(x,\mathbf{0};z)$, as well as the corresponding eigenvalues, are called *stationary* BA functions. Their explicit form directly follows from (13.59):

$$z^{-x/\eta}\hat\psi(x,z) = \sum_{l=0}^{L}\eta^l \sum_{i_1<\ldots<i_l}\sum_{\substack{a_1,\ldots,a_l \\ b_1,\ldots,b_l}}\left(\overrightarrow{\prod_{\alpha=1}^{l}} \frac{e_{b_\alpha a_\alpha}^{(i_\alpha)}}{x-x_{i_\alpha}}\frac{\partial}{\partial\varepsilon_\alpha}\right)\det\left(1-z^{-1}\mathbf{g}_{\varepsilon_l,\ldots,\varepsilon_1}^{a_l b_l,\ldots,a_1 b_1}\right)\Bigg|_{\varepsilon_\alpha=0},$$

(13.63)

$$z^{x/\eta}\hat\psi^*(x,z) = \sum_{l=0}^{L}\eta^l \sum_{i_1<\ldots<i_l}\sum_{\substack{a_1,\ldots,a_l \\ b_1,\ldots,b_l}}\left(\overrightarrow{\prod_{\alpha=1}^{l}} \frac{e_{b_\alpha a_\alpha}^{(i_\alpha)}}{x-x_{i_\alpha}}\frac{\partial}{\partial\varepsilon_\alpha}\right)\left[\det\left(1-z^{-1}\mathbf{g}_{\varepsilon_l,\ldots,\varepsilon_1}^{a_l b_l,\ldots,a_1 b_1}\right)\right]^{-1}\Bigg|_{\varepsilon_\alpha=0}.$$

(13.64)

In particular, we have the expansion of $\hat\psi(x,z)$ as $|x| \to \infty$:

$$z^{-x/\eta}\hat\psi(x,z) = \det\left(1-z^{-1}\mathbf{g}\right)\left(\mathbf{I} - \frac{\eta}{x}\sum_{j=1}^{L}\sum_{a=1}^{N}\frac{g_a e_{aa}^{(j)}}{z-g_a} + O(1/x^2)\right)$$

(13.65)

(we need it in the next section).

For calculations in the next section we also need the following general properties of the BA functions of the mKP hiertarchy.

(a) They obey the differential-difference equations of the form

$$\partial_{t_1}\psi(x,\mathbf{t};z) = \psi(x+\eta,\mathbf{t};z) + V(x,\mathbf{t})\,\psi(x,\mathbf{t};z),$$

(13.66)

$$-\partial_{t_1}\psi^*(x,\mathbf{t};z) = \psi^*(x-\eta,\mathbf{t};z) + V(x-\eta,\mathbf{t})\psi^*(x,\mathbf{t};z)$$

(13.67)

(the linear problems), where

$$V(x,\mathbf{t}) = \partial_{t_1}\log\frac{T(x+\eta,\mathbf{t})}{T(x,\mathbf{t})}$$

(see, e.g., [10, 11]).

(b) They obey the relation

$$\partial_{t_m}\log\frac{T(x+\eta,\mathbf{t})}{T(x,\mathbf{t})} = \mathrm{res}_\infty\left(\psi(x,\mathbf{t};z)\psi^*(x+\eta,\mathbf{t};z)z^m dz\right),$$

(13.68)

where the residue is normalized as $\mathrm{res}_\infty z^{-1}dz = 1$. It can be derived from (13.55) and (13.60) in the same way as in [7].

13.3 From the master T-operator to the classical RS model and back

13.3.1 Eigenvalues of the spin chain Hamiltonians as velocities of the RS particles

As we mentioned in the previous section, the eigenvalues of the master T-operator are polynomials in the spectral parameter x of degree L:

$$T(x, \mathbf{t}) = e^{\operatorname{tr} \xi(\mathbf{t}, \mathbf{g})} \prod_{k=1}^{L} (x - x_k(\mathbf{t})). \tag{13.69}$$

The roots have their own dynamics in the times. The very fact that $T(x, \mathbf{t})$ is a tau-function of the mKP hierarchy implies [15, 16] that the roots x_i move in the time t_1 as particles of the RS L-body system [17]. Moreover, their motion in the higher times t_k is the same as motion of the RS particles caused by the higher Hamiltonian flows of the RS system (see [7, 16] which extend the methods developed by Krichever [36] and Shiota [37]). We have $T(x, \mathbf{0}) = T_{\emptyset}(x) = \prod_{k=1}^{L} (x - x_k)$, where $x_k = x_k(\mathbf{0})$. This means that

(i) The inhomogeneity parameters x_k of the spin chain should be identified with *initial positions* $x_k(\mathbf{0})$ of the RS particles.

With the help of (13.51) we can write

$$\frac{T_{\square}(x)}{T_{\emptyset}(x)} = \partial_{t_1} \log T(x, \mathbf{t})|_{\mathbf{t}=0} = \operatorname{tr} \mathbf{g} - \sum_{k=1}^{L} \frac{\dot{x}_k(\mathbf{0})}{x - x_k}, \tag{13.70}$$

where $\dot{x}_k(\mathbf{0}) := \partial_{t_1} x_k(\mathbf{t})|_{\mathbf{t}=0}$. Comparing this with (13.23), we find that

$$\dot{x}_k(\mathbf{0}) = -\eta H_k, \tag{13.71}$$

where H_k is an eigenvalue of $\mathbf{H}_k$. Therefore, in addition to (i) we conclude that

(ii) The eigenvalues H_k of the XXX spin chain Hamiltonians are expressed through the *initial velocities* of the RS particles as $H_i = -\dot{x}_i(\mathbf{0})/\eta$.

In other words, any point in the phase space of the L-body RS system with coordinates $\{x_i, \dot{x}_i\}$ corresponds to an eigenstate, with the eigenvalues $H_i = -\dot{x}_i/\eta$, of the Hamiltonians of the XXX spin chain on L sites with the inhomogeneity parameters x_i.

This unexpected connection between quantum spin chains and the classical RS model was pointed out in [2] as a corollary of the Hirota bilinear equations for the master T-operator. A similar relation between quantum Hamiltonians in the Gaudin model and velocities of particles in the classical Calogero–Moser model was found in [38] using different methods (see also [39, 40] for further developments).[6] In [3] it was shown that the identifications (i) and (ii) are in fact independent of the grading: their form is the same for all spin chains of the XXX type associated with *any* (super)algebra $gl(N|M)$ including the ordinary algebras $gl(N|0) = gl(N)$.

13.3.2   Lax pair for the RS model from dynamics of poles

To make the correspondence 'quantum spin chains ↔ classical RS systems' (the QC correspondence) complete, we need the Lax matrix for the RS model. At this stage

---

[6] In [5] it was obtained from the master T-operator construction for the Gaudin model.

the setup is exactly the same as in [7]. Here we repeat the main formulas with some comments skipping the details.

Below we will derive equations of motion for the t_1-dynamics of the x_i values using Krichever's method [36], the starting point of which is the linear problem (13.66) for the BA function. Essentially, the derivation is not specific to the master T-operator case but only depends on the polynomial form of the tau-function.

One can derive equations of motion for the x_i values performing the pole expansion of the linear problem (13.66). It is convenient to denote $t_1 = t$ and put all higher times equal to 0 because they are irrelevant for this derivation. In this section we often write simply t instead of $\mathbf{t}$. According to (13.57), the general form of ψ as a function of x is

$$\psi(x, t; z) = z^{x/\eta} e^{tz} \left(c_0(z) + \sum_{j=1}^{L} \frac{c_j(z, t)}{x - x_j(t)} \right), \qquad (13.72)$$

where $c_0(z) = \det(\mathbf{1} - z^{-1}\mathbf{g})$ (see (13.61)). One should substitute it into the linear equation (13.66) with

$$V(x, t) = \partial_t \log \frac{T(x + \eta, t)}{T(x, t)} = \sum_{k=1}^{L} \left(\frac{\dot{x}_k}{x - x_k} - \frac{\dot{x}_k}{x - x_k + \eta} \right), \qquad x_k = x_k(t) \quad (13.73)$$

and cancel all the poles at $x = x_i$ and $x = x_i - \eta$ (possible poles of the second order cancel automatically). This yields an overdetermined system of linear equations for the coefficients c_i,

$$\begin{cases} (z\mathbf{1} - \mathsf{Z}) \, \vec{c} = c_0(z) \, \mathsf{X} \, \vec{1} \\[2mm] \dot{\vec{c}} = \mathsf{G} \, \vec{c}, \end{cases} \qquad (13.74)$$

where $\vec{c} = (c_1, c_2, \ldots, c_L)^{\mathsf{t}}$, $\vec{1} = (1, 1, \ldots, 1)^{\mathsf{t}}$ are L-component vectors and the $L \times L$ matrices $\mathsf{X} = \mathsf{X}(t)$, $\mathsf{Z} = \mathsf{Z}(t)$, $\mathsf{G} = \mathsf{G}(t)$ are defined by their matrix elements as follows:

$$\mathsf{X}_{ij} = x_i \delta_{ij}, \qquad \mathsf{Z}_{ij} = \frac{\dot{x}_i}{x_i - x_j - \eta} \qquad (13.75)$$

$$\mathsf{G}_{ij} = \left(\sum_{k \neq i} \frac{\dot{x}_k}{x_i - x_k} - \sum_{k \neq i} \frac{\dot{x}_k}{x_i - x_k + \eta} \right) \delta_{ij} + \left(\frac{\dot{x}_i}{x_i - x_j} - \frac{\dot{x}_i}{x_i - x_j - \eta} \right) (1 - \delta_{ij}). \quad (13.76)$$

The explicit form of the matrix G is not used in what follows. As is easy to check, the matrix $[\mathsf{X}, \mathsf{Z}] - \mathsf{Z}$ has rank 1. More precisely, these matrices satisfy the commutation relation

$$[\mathsf{X}, \mathsf{Z}] = \eta \mathsf{Z} + \dot{\mathsf{X}} \mathsf{E}, \qquad (13.77)$$

where $\mathsf{E} = \vec{1} \otimes \vec{1}^{\mathsf{t}}$ is the $L \times L$ matrix of rank 1 with all entries equal to 1. As a consequence of this commutation relation, we mention the identity $\vec{1}^{\mathsf{t}} \mathsf{Z}^k \dot{\mathsf{X}} \vec{1} = -\eta \operatorname{tr} \mathsf{Z}^{k+1}$ which holds for any $k \geq 0$ (see [7]).

The compatibility condition of the problems (13.74) is the Lax equation

$$\dot{Z} = [G, Z],\qquad(13.78)$$

which is equivalent to the equations of motion

$$\ddot{x}_i = -\sum_{k\neq i}\frac{2\eta^2 \dot{x}_i \dot{x}_k}{(x_i-x_k)\left[(x_i-x_k)^2-\eta^2\right]},\qquad i=1,\dots,L.\qquad(13.79)$$

This dynamical system called the RS model is sometimes referred to as the relativistic deformation of the Calogero–Moser model, the parameter η being the inverse 'velocity of light'. The Hamiltonian formulation is given in the Appendix. The integrability of the RS model follows from the Lax representation. The matrix Z is the Lax matrix. As it follows from (13.78), the time evolution preserves its spectrum; i.e., the coefficients $\mathcal{J}_k$ of the characteristic polynomial

$$\det(z\mathbf{1}-\mathsf{Z}(t)) = \sum_{k=0}^{L}\mathcal{J}_k z^{L-k}\qquad(13.80)$$

are integrals of motion. Equivalently, one can say that eigenvalues of the Lax matrix are integrals of motion.

For completeness, we also present here the linear problem for coefficients of the adjoint BA function

$$\psi^*(x,t;z) = z^{-x/\eta}\mathrm{e}^{-tz}\left(c_0^{-1}(z)+\sum_{j=1}^{L}\frac{c_j^*(z,t)}{x-x_j(t)}\right).\qquad(13.81)$$

As a counterpart of (13.74), we get, using the equations of motion,

$$\begin{cases}\vec{c}^{*\mathsf{t}}\dot{\mathsf{X}}^{-1}(z\mathbf{1}-\mathsf{Z}) = -c_0^{-1}(z)\vec{\mathbf{1}}^{\mathsf{t}}\\[2mm]\partial_t(\vec{c}^{*\mathsf{t}}\mathsf{X}^{-1}) = -\vec{c}^{*\mathsf{t}}\dot{\mathsf{X}}^{-1}\mathsf{G}.\end{cases}\qquad(13.82)$$

Here $\vec{c}^{*\mathsf{t}} = (c_1^*, c_2^*, \dots, c_L^*)$ and $\vec{\mathbf{1}}^{\mathsf{t}} = (1,1,\dots,1)$ (note that $\vec{\mathbf{1}}^{\mathsf{t}}\mathsf{G} = 0$). Regarding these equations as overdetermined linear problems for the (co)vector $\vec{c}^{*\mathsf{t}}\mathsf{X}^{-1}$, one comes to the same Lax equation (13.78) as their compatibility condition. The adjoint linear problems (13.82), together with general relation (13.68), are used for the extension of the time dynamics of the x_i values to the whole hierarchy, as has been done in [7, 16]; see Appendix B.

13.3.3 The BA function and the master T-operator

The solution for the vector $\vec{c}$ reads $\vec{c}(z,t) = c_0(z)(z\mathbf{1}-\mathsf{Z}(t))^{-1}\mathsf{X}\vec{\mathbf{1}}$. The BA function ψ is then given by the formula

$$\psi = c_0(z)\,z^{x/\eta}\mathrm{e}^{tz}\left(1+\vec{\mathbf{1}}^{\mathsf{t}}(x\mathbf{1}-\mathsf{X})^{-1}(z\mathbf{1}-\mathsf{Z})^{-1}\mathsf{X}\vec{\mathbf{1}}\right).\qquad(13.83)$$

Similar formulas can be obtained for the adjoint vector $\vec{c}^{*t}$ and the adjoint BA function:

$$\vec{c}^{*t}(z,t) = -c_0^{-1}(z)\,\vec{1}^{t}\,(z\mathbf{1}-Z(t))^{-1}\dot{X},$$
$$\psi^* = c_0^{-1}(z)\,z^{-x/\eta}e^{-tz}\left(1 - \vec{1}^{t}(z\mathbf{1}-Z)^{-1}(x\mathbf{1}-X)^{-1}\dot{X}\vec{1}\right). \tag{13.84}$$

It is easy to see that for nonzero values of the higher times the BA functions are given by the same formulas with the factor $e^{\pm tz}$ substituted by $e^{\pm\xi(t,z)}$. Writing $\vec{1}^{t}A\vec{1} = \mathrm{tr}(AE)$ and using the commutation relation (13.77), one can represent these expressions as ratios of determinants[7]:

$$\psi(x,t;z) = \det(1 - z^{-1}\mathbf{g})\,z^{x/\eta}\,e^{\xi(t,z)}\,\frac{\det\left[(x\mathbf{1}-X)(z\mathbf{1}-Z)-\eta Z\right]}{\det(x\mathbf{1}-X)\det(z\mathbf{1}-Z)}, \tag{13.85}$$

$$\psi^*(x,t;z) = \left[\det\left(1-z^{-1}\mathbf{g}\right)\right]^{-1}z^{-x/\eta}\,e^{-\xi(t,z)}\,\frac{\det\left[(z\mathbf{1}-Z)(x\mathbf{1}-X)+\eta Z\right]}{\det(x\mathbf{1}-X)\det(z\mathbf{1}-Z)}. \tag{13.86}$$

In particular, the stationary BA functions are given by

$$\psi(x,z) = \det(1 - z^{-1}\mathbf{g})\,z^{x/\eta}\,\frac{\det\left[(x\mathbf{1}-X_0)(z\mathbf{1}-Z_0)-\eta Z_0\right]}{\det(x\mathbf{1}-X_0)\det(z\mathbf{1}-Z_0)}, \tag{13.87}$$

$$\psi^*(x,z) = \left[\det\left(1-z^{-1}\mathbf{g}\right)\right]^{-1}z^{-x/\eta}\,\frac{\det\left[(z\mathbf{1}-Z_0)(x\mathbf{1}-X_0)+\eta Z_0\right]}{\det(x\mathbf{1}-X_0)\det(z\mathbf{1}-Z_0)}. \tag{13.88}$$

Hereafter, we use the notation $X_0 = X(0)$, $Z_0 = Z(0)$. Using (13.57), one can obtain from (13.85) an explicit determinant formula for eigenvalues of the master T-operator:

$$T(x,t) = e^{\mathrm{tr}\,\xi(t,\mathbf{g})}\det\left(x\mathbf{1} - X_0 + \eta\sum_{k\geq1}kt_k Z_0^k\right). \tag{13.89}$$

Formulas (13.85) and (13.89) are not new in the context of classical integrable hierarchies (see, e.g., [16, 37, 41]). The new observation is the close connection with quantum spin chains. It is important to stress that we can understand (13.87) and (13.89) in the operator sense, i.e. as expressions for the *quantum operators* $\hat{\psi}(x,z)$, $T(x,t)$ in terms of the matrices

$$(X_0)_{ij} = x_i\delta_{ij}\mathbf{I}, \quad (Z_0)_{ij} = \frac{\eta H_i}{x_j - x_i + \eta}.$$

The latter is the Lax matrix Z_0 with the operator-valued entries given by Eq. (13.75) with the substitution (13.71).

---

[7] The order of the factors $x\mathbf{1}-X$ and $z\mathbf{1}-Z$ under the determinant upstairs is actually not important because of the identity $\det(AB + A_1) = \det(BA + A_1)$ valid for any matrices A, B, A_1 such that $[A, A_1] = 0$.

To summarize, we have derived equations of motion of the RS model for roots of the master T-operator, together with the Lax representation. Then, embedding the initial quantum problem into the context of the classical RS model, we have obtained explicit operator expressions for the BA function and the master T-operator in terms of the quantum spin chain Hamiltonians. In the next section, we will show how one can reformulate the spectral problem for the quantum Hamiltonians in terms of an inverse spectral problem for the RS Lax matrix.

13.4   Spectrum of the spin chain Hamiltonians from the classical RS model

13.4.1   Twist parameters as eigenvalues of the Lax matrix

We have two different (but equivalent) expressions for $\hat{\psi}$: (13.63) and (13.87). Let us compare their large $|x|$ expansions. The large $|x|$ expansion of (13.63) is given by (13.65)

$$
\hat{\psi}(x, z) = c_0(z) z^{x/\eta} \left(\mathbf{I} - \frac{\eta}{x} \sum_{j=1}^{L} \sum_{a=1}^{N} \frac{g_a \mathbf{e}_{aa}^{(j)}}{z - g_a} + O(x^{-2}) \right),
$$

where g_a are the twist parameters. The expansion of (13.87) is

$$
\hat{\psi}(x, z) = c_0(z) z^{x/\eta} \left(1 - \frac{\eta}{x} \operatorname{tr} \frac{Z_0}{z\mathbf{1} - Z_0} + O(x^{-2}) \right).
$$

Equating the $O(1/x)$ terms of the two expansions leads to the relation

$$
\operatorname{tr} \frac{Z_0}{z\mathbf{1} - Z_0} = \sum_i \sum_a \frac{\mathbf{e}_{aa}^{(i)} g_a}{z - g_a}, \tag{13.90}
$$

which must be valid identically. Let us stress that its left-hand side is well defined because the entries of the matrix Z_0 are commuting operators. Using the identity $\operatorname{tr}(z\mathbf{1} - A)^{-1} = \partial_z \log \det(z\mathbf{1} - A)$ valid for any matrix A, we integrate (13.90) to obtain

$$
\det(z\mathbf{1} - Z_0) = \prod_{a=1}^{N} (z - g_a)^{M_a}, \tag{13.91}
$$

where $\mathbf{M}_a$ are the weight operators (13.26). Since the time evolution is an isospectral deformation, the same is true for the Lax matrix $Z(t)$ for any values of the times. We see that $\mathbf{M}_a$ is the 'operator multiplicity' of the eigenvalue g_a. In the weight space $\mathcal{V}(\{M_a\})$ the multiplicities become equal to the M_a's. The conclusion is:

- The Lax matrix Z has eigenvalues g_a with multiplicities $M_a \geq 0$ such that $M_1 + \ldots + M_N = L$.

Our next goal is to formulate the QC correspondence[8] between the spin chains and the RS model.

13.4.2   The QC correspondence

Consider the Lax matrix Z_0 of the L-particle RS model, where the inverse 'velocity of light', η, is identified with the parameter η introduced in the quantum R-matrix (13.11), and the initial coordinates and velocities of the particles are identified, respectively, with the inhomogeneity parameters x_i and eigenvalues of the Hamiltonians H_i through $\dot{x}_i = -\eta H_i$:

$$
Z_0 = \begin{pmatrix}
H_1 & \dfrac{\eta H_1}{x_2-x_1+\eta} & \dfrac{\eta H_1}{x_3-x_1+\eta} & \cdots & \dfrac{\eta H_1}{x_L-x_1+\eta} \\[2ex]
\dfrac{\eta H_2}{x_1-x_2+\eta} & H_2 & \dfrac{\eta H_2}{x_3-x_2+\eta} & \cdots & \dfrac{\eta H_2}{x_L-x_2+\eta} \\[2ex]
\vdots & \vdots & \vdots & \ddots & \vdots \\[2ex]
\dfrac{\eta H_L}{x_1-x_L+\eta} & \dfrac{\eta H_L}{x_2-x_L+\eta} & \dfrac{\eta H_L}{x_3-x_L+\eta} & \cdots & H_L
\end{pmatrix}.
\tag{13.92}
$$

According to the conclusion of the previous subsection, we claim that if the H_i values are eigenvalues of the Hamiltonians of the spin chain in the weight space $\mathcal{V}(\{M_a\}) \subset V$, then

$$
\mathrm{Spec}\,(Z_0) = \big(\underbrace{g_1,\ldots,g_1}_{M_1},\ \underbrace{g_2,\ldots,g_2}_{M_2},\ \cdots\ \underbrace{g_N,\ldots,g_N}_{M_N}\big).
\tag{13.93}
$$

Equivalently, let $\mathcal{H}_j = \mathrm{tr}\,(Z_0)^j$ be the higher integrals of motion of the RS model (see Appendix B), then their level set is defined by $\mathcal{H}_j = \sum_{a=1}^{N} M_a g_a^j$. In general, the matrix Z_0 with multiple eigenvalues is not diagonalizable and contains Jordan cells.

One can also say that the eigenstates of the quantum Hamiltonians correspond to the intersection points of two Lagrangian submanifolds in the phase space of the RS model. One of them is the hyperplane defined by fixing all the coordinates x_i while the other one is the Lagrangian submanifold obtained by fixing values of the L independent integrals of motion in involution $\mathcal{H}_k$, $k = 1,\ldots,L$. In general, there are many intersection points numbered by a finite set $\mathfrak{I}$, with coordinates, say $(x_1,\ldots,x_L,\,p_1^{(\alpha)},\ldots,p_L^{(\alpha)})$, $\alpha \in \mathfrak{I}$. The values of $p_j^{(\alpha)}$ give, through Eq. (B2), the spectrum of $\mathbf{H}_j$:

$$
H_j^{(\alpha)} = \mathrm{e}^{-\eta p_j^{(\alpha)}} \prod_{k=1,\neq j} \frac{x_j - x_k + \eta}{x_j - x_k}.
$$

---

[8] The QC correspondence can be traced back to [42], where joint spectra of some commuting finite-dimensional operators were linked to the classical Toda chain.

However, we cannot claim that all the intersection points correspond to the energy levels of the Hamiltonians for a given spin chain. The examples elaborated in [3] suggest that there are intersection points that do not correspond to energy levels of a particular spin chain with a fixed grading. Instead, they correspond to spectra of the Hamiltonians for supersymmetric spin chains with all possible gradings.

Summarizing, we claim that the spectral problem for the nonlocal inhomogeneous susy-XXX spin chain Hamiltonians $\mathbf{H}_j$ in the subspace $\mathcal{V}(\{M_a\})$ is closely linked to the following *inverse spectral problem* for the RS Lax matrix Z_0 of the form (13.92). Let us fix the spectrum of the matrix Z_0 to be (13.93), where $g_1, \ldots, g_N$ are eigenvalues of the (diagonal) twist matrix $\mathbf{g}$. Then we ask what is the set of possible H_j values allowed by these constraints. The eigenvalues H_j of the quantum Hamiltonians are contained in this set.

13.4.3 Algebraic equations for the spectrum

The characteristic polynomial of the matrix (13.92) can be found explicitly using the simple fact from the linear algebra that the coefficient in front of z^{L-k} in the polynomial $\det_{L \times L}(z\mathbf{1} + A)$ equals the sum of all diagonal $k \times k$ minors of the matrix A. All such minors can be found using the decomposition $Z_0 = -HQ$, where $H = \mathrm{diag}\left(H_1, H_2, \ldots, H_L\right)$ and

$$Q_{ij} = \frac{\eta}{x_i - x_j - \eta} \tag{13.94}$$

is the Cauchy matrix, and the explicit expression for the determinant of the Cauchy matrix:

$$\det_{1 \le i,j \le n} \frac{\eta}{x_i - x_j - \eta} = (-1)^n \prod_{1 \le i < j \le n} \left(1 - \frac{\eta^2}{(x_i - x_j)^2}\right)^{-1}.$$

The result is

$$\det_{L \times L}(z\mathbf{1} - Z_0) = \det_{L \times L}(z\mathbf{1} + HQ) = \sum_{n=0}^{L} \mathcal{J}_n z^{L-n}, \tag{13.95}$$

where

$$\mathcal{J}_n = (-1)^n \sum_{1 \le i_1 < \ldots < i_n \le L} H_{i_1} \ldots H_{i_n} \prod_{1 \le \alpha < \beta \le n} \left(1 - \frac{\eta^2}{(x_{i_\alpha} - x_{i_\beta})^2}\right)^{-1}. \tag{13.96}$$

In particular, the highest coefficient is given by the simple formula

$$\mathcal{J}_L = (-1)^L H_1 H_2 \ldots H_L \prod_{1 \le i < j \le L} \left(1 - \frac{\eta^2}{(x_i - x_j)^2}\right)^{-1}.$$

Let us point out that the integrals $\mathcal{H}_k$ introduced in the previous section are connected with the integrals $\mathcal{J}_k$ by the Newton formula [26]: $\sum_{k=0}^{L} \mathcal{J}_{L-k}\mathcal{H}_k = 0$ (we have set $\mathcal{H}_0 = \mathrm{tr}(\mathsf{Z})^0 = L$).

Combining (13.93) and (13.96), we see that the eigenvalues H_i of the inhomogeneous XXX Hamiltonians can be found from the system of polynomial equations

$$\sum_{1 \le i_1 < \ldots < i_n \le L} H_{i_1} \ldots H_{i_n} \prod_{1 \le \alpha < \beta \le n} \left(1 - \frac{\eta^2}{(x_{i_\alpha} - x_{i_\beta})^2}\right)^{-1} = C_n(\{M_a\}), \qquad (13.97)$$

where $n = 1, 2, \ldots, L$ and

$$C_n(\{M_a\}) = \frac{1}{2\pi i} \oint_{|z|=1} \prod_{a=1}^{N} (1 + zg_a)^{M_a} z^{-n-1} dz$$

$$= \sum_{\substack{n_1, \ldots, n_N \in \mathbb{Z}_{\ge 0}, \\ \sum_{j=1}^{N} n_j = n}} \binom{M_1}{n_1} \cdots \binom{M_N}{n_N} \prod_{a=1}^{N} g_a^{n_a}. \qquad (13.98)$$

There are L equations for L unknown quantities $H_1, \ldots, H_L$.

Let us point out some simple general properties of these equations.

1. This system does not depend on N and and is also invariant under the following transformations: (a) $\eta \to -\eta$, (b) $\{x_i\} \to \{-x_i\}$, (c) $\{H_i\} \to \{-H_i\}$ simultaneously with $\{g_a\} \to \{-g_a\}$.
2. For $M_a = L\delta_{aa_1}$ (in this case $C_n = \frac{L!}{n!(L-n)!} g_{a_1}^n$) there are two distinguished solutions

$$H_j = g_{a_1} \prod_{k=1, \ne j}^{L} \left(1 \pm \frac{\eta}{x_j - x_k}\right), \qquad (13.99)$$

which give the eigenvalues of the Hamiltonians on the vector $(\mathbf{v}_{a_1})^{\otimes L}$.
3. Assume that $L \le N$ and fix $\{a_1, \ldots, a_L\} \subseteq \{1, \ldots, n\}$ such that all the a_i values are distinct. Set $M_a = \sum_{i=1}^{L} \delta_{aa_i}$, then the right-hand sides of Eqs (13.97) are elementary symmetric polynomials[9] $e_k(g_{a_1}, \ldots, g_{a_L})$ of the twist parameters $g_{a_1}, \ldots, g_{a_L}$. Then the system of Eqs (13.97) has $L!$ solutions (counted with multiplicities). Indeed, at $\eta = 0$ the system is just

$$e_n(H_1, \ldots, H_L) = e_n(g_{a_1}, \ldots, g_{a_L}), \qquad n = 1, \ldots, L \qquad (13.100)$$

---

[9] The elementary symmetric polynomials are defined by means of the generating function as

$$\prod_{i=1}^{N} (1 + y_i z) = \sum_{k=0}^{N} e_k(y_1, \ldots, y_N) z^k.$$

All solutions of this system are given by all possible permutations of the set $(g_{a_1}, \ldots g_{a_L})$ containing L elements.

The detailed structure of solutions to (13.97) and their correspondence with spectra of particular spin chains is a subject of further study.

One can consider the system (13.97) with the right hand sides being 'in general position' meaning that there are L twist parameters g_i which are all distinct. This is the generic situation from which all other possible cases can be obtained by merging some of the twist parameters. The generic system (13.97) contains L equations of the form

$$\sum_{1 \leq i_1 < \ldots < i_n \leq L} H_{i_1} \ldots H_{i_n} \prod_{1 \leq \alpha < \beta \leq n} \left(1 - \frac{\eta^2}{(x_{i_\alpha} - x_{i_\beta})^2} \right)^{-1} = e_n(g_1, \ldots, g_L). \quad (13.101)$$

As proposed in [3], the complete information about spectra of the Hamiltonians for L-site spin chains based on all (super)algebras of the type $gl(N|M)$ is contained in the *universal spectral variety*

$$\mathbb{S}_L = \left\{ (H_1, \ldots H_L; x_1, \ldots, x_L; g_1, \ldots, g_L) \,\middle|\, \text{Eqs (13.101)} \right\}. \quad (13.102)$$

It is a $2L$-dimensional affine variety embedded into $\mathbb{C}^{3L}$. The spectra of the Hamiltonians for particular spin chains are obtained by intersecting with the hyperplanes with fixed values of x_i and g_i. The variety $\mathbb{S}_L$ is not compact. We anticipate that a proper compactification of the universal spectral variety encodes information about the spectra of Hamiltonians for spin chains when some or all x_i values coalesce.

13.5 The QC correspondence via nested Bethe ansatz

In this section we give a direct proof of the QC correspondence based on the nested Bethe ansatz solution to the XXX spin chains.

13.5.1 The nested Bethe ansatz solution

Here we specify the general nested Bethe ansatz results (13.32), (13.36) for the spin chain with vector representations at the sites. The eigenstates of the T-operator $\mathbf{t}(x)$ are obtained from the reference state $(\mathbf{v}_1)^{\otimes L}$ with $M_a = L\delta_{a1}$ by action of creation operators. In the weight space $V(M_1, \ldots, M_N)$ with $M_1 = L - L_1$, $M_2 = L_1 - L_2$, $M_3 = L_2 - L_3$, $\ldots$, $M_{N-1} = L_{N-2} - L_{N-1}$, $M_N = L_{N-1} - L_N$ such that $L_1 \geq L_2 \geq \ldots \geq L_N$ the eigenvalues of $\mathbf{t}(x)$ are given by the formula

$$T(x) = g_1 \prod_{l=1}^{L} \frac{x - x_l + \eta}{x - x_l} \prod_{\gamma=1}^{L_1} \frac{x - \mu_\gamma^1 - \eta}{x - \mu_\gamma^1}$$

$$+ \sum_{b=2}^{N} g_b \prod_{\alpha=1}^{L_{b-1}} \frac{x - \mu_\alpha^{b-1} + \eta}{x - \mu_\alpha^{b-1}} \prod_{\gamma=1}^{L_b} \frac{x - \mu_\gamma^b - \eta}{x - \mu_\gamma^b},$$

$$(13.103)$$

where the parameters μ_γ^b (the Bethe roots) obey the system of Bethe equations

$$
g_b \prod_{k=1}^{L} \frac{\mu_\beta^b - x_k + \delta_{b1}\eta}{\mu_\beta^b - x_k} \prod_{\gamma=1}^{L_{b-1}} \frac{\mu_\beta^b - \mu_\gamma^{b-1} + \eta}{\mu_\beta^b - \mu_\gamma^{b-1}}
$$

$$
= g_{b+1} \prod_{\gamma \neq \beta}^{L_b} \frac{\mu_\beta^b - \mu_\gamma^b + \eta}{\mu_\beta^b - \mu_\gamma^b - \eta} \prod_{\gamma=1}^{L_{b+1}} \frac{\mu_\beta^b - \mu_\gamma^{b+1} - \eta}{\mu_\beta^b - \mu_\gamma^{b+1}} .
$$

$$(13.104)$$

Here b runs from 1 to $N-1$ and the convention $L_0 = L_N = 0$ is implied. The Bethe equations are equivalent to the conditions that $T(x)$ given by (13.103) is regular at the points $x = \mu_\beta^b$ for all $\beta = 1, ..., L_b$, $b = 1, ..., N-1$. The corresponding eigenvalues of the Hamiltonians H_i are

$$
H_i \left(\{x_i\}_L, \{\mu_\alpha^1\}_{L_1}, g_1 \right) = g_1 \prod_{k=1}^{L} \frac{x_i - x_k + \eta}{x_i - x_k} \prod_{\gamma=1}^{L_1} \frac{x_i - \mu_\gamma^1 - \eta}{x_i - \mu_\gamma^1},
$$

$$(13.105)$$

where $\{x_i\}_L$ emphasizes the dependence on L variables x_i ($\{x_i\}_L$ means $\{x_i\}_{i=1}^{L}$, in particular, $\{x_i\}_0 = \emptyset$) and similarly for $\{\mu_\alpha^1\}_{L_1}$.

13.5.2   The QC correspondence: a direct proof

Here we follow [20].

Theorem 13.5.1.   Substitute

$$
\dot{x}_i = -\eta \, H_i \left(\{x_i\}_L, \{\mu_\alpha^1\}_{L_1}, g_1 \right), \quad i = 1, ..., L
$$

$$(13.106)$$

into the Lax matrix for the RS model (13.75); i.e., consider the matrix

$$
(Z_0)_{ij} = \frac{\eta \, H_i}{x_j - x_i + \eta}
$$

$$(13.107)$$

(see (13.92)), where H_j are eigenvalues (13.105) of the nonlocal Hamiltonians of the inhomogeneous $gl(N)$ spin chain on L-sites with $N \leq L$ and the set $\{\mu_\alpha^1\}_{L_1}$ is taken from any solution $\{\mu_\alpha^b\}_{L_b}$, $b = 1, ..., N-1$ of the Bethe equations (13.104). Then the spectrum of the Lax matrix (13.107) is of the form (13.93):

$$
\left. \mathrm{Spec}\, Z_0 \right|_{BE} = \big(\underbrace{g_1, ..., g_1}_{L-L_1}, \underbrace{g_2, ..., g_2}_{L_1-L_2}, ..., \underbrace{g_{N-1}, ..., g_{N-1}}_{L_{N-2}-L_{N-1}}, \underbrace{g_N, ..., g_N}_{L_{N-1}} \big).
$$

$$(13.108)$$

Proof   Instead of Z_0 it is more convenient to deal with the transposed Lax matrix Z_0^t given by

$$
(Z_0^t)_{ij}(\{\dot{x}_k\}_L, \{x_k\}_L, \eta) = -\frac{\dot{x}_j}{x_i - x_j + \eta} = \frac{\eta \, H_j}{x_i - x_j + \eta}, \quad i, j = 1, ..., L. \quad (13.109)
$$

Its spectrum coincides with that of Z_0.

The proof given in [20] is based on the identity

$$\det_{L \times L} \left(\mathcal{Z} \left(\{x_i\}_L, \{y_i\}_{\tilde{L}}, g \right) - \lambda \mathbf{1} \right) = (g - \lambda)^{L - \tilde{L}} \det_{\tilde{L} \times \tilde{L}} \left(\tilde{\mathcal{Z}} \left(\{y_i\}_{\tilde{L}}, \{x_i\}_L, g \right) - \lambda \mathbf{1} \right)$$

(13.110)

for the pair of $L \times L$ and $\tilde{L} \times \tilde{L}$ matrices

$$\mathcal{Z}_{ij}(\{x_k\}_L, \{y_k\}_{\tilde{L}}, g) = \frac{g\,\eta}{x_i - x_j + \eta} \prod_{k \neq j}^{L} \frac{x_j - x_k + \eta}{x_j - x_k} \prod_{\gamma = 1}^{\tilde{L}} \frac{x_j - y_\gamma}{x_j - y_\gamma + \eta}$$

(13.111)

and

$$\tilde{\mathcal{Z}}_{\alpha\beta}(\{y_i\}_{\tilde{L}}, \{x_i\}_L, g) = \frac{g\,\eta}{y_\alpha - y_\beta + \eta} \prod_{\gamma \neq \beta}^{\tilde{L}} \frac{y_\beta - y_\gamma - \eta}{y_\beta - y_\gamma} \prod_{k = 1}^{L} \frac{y_\beta - x_k}{y_\beta - x_k - \eta}$$

(13.112)

(here $L \geq \tilde{L}$). In addition, we have

$$\det_{L \times L} \left(\mathcal{Z}^0(\{x_i\}_L, g) - \lambda \mathbf{1} \right) = \det_{L \times L} \left(\tilde{\mathcal{Z}}^0(\{y_i\}_L, g) - \lambda \mathbf{1} \right) = (g - \lambda)^L,$$

(13.113)

where

$$\mathcal{Z}^0_{ij}(\{x_k\}_L, g) = \mathcal{Z}_{ij}(\{x_k\}_L, \{y_k\}_0, g) = \frac{g\,\eta}{x_i - x_j + \eta} \prod_{k \neq j}^{L} \frac{x_j - x_k + \eta}{x_j - x_k}$$

(13.114)

and

$$\tilde{\mathcal{Z}}^0_{\alpha\beta}(\{y_i\}_L, g) = \tilde{\mathcal{Z}}_{\alpha\beta}(\{y_i\}_L, \{x_i\}_0, g) = \frac{g\,\eta}{y_\alpha - y_\beta + \eta} \prod_{\gamma \neq \beta}^{L} \frac{y_\beta - y_\gamma - \eta}{y_\beta - y_\gamma}.$$

(13.115)

The idea is to calculate $\det(Z_0 - \lambda \mathbf{1})$ by sequential usage of the identity (13.110) (which allows one to pass to a smaller matrix) and the Bethe equations (BE) (13.104). Schematically,[10] the procedure of the proof is as follows:

$$\langle Z_0^{\mathsf{t}} \rangle \left(-\eta \{H_j\}_L, \{x_j\}_L, \eta \right) \stackrel{(13.105)}{=} \langle \mathcal{Z} \rangle \left(\{x_i - \eta\}_L, \{\mu_\alpha^1\}_{L_1}, g_1 \right) \stackrel{(13.110)}{\to}$$

$$\langle \tilde{\mathcal{Z}} \rangle \left(\{\mu_\alpha^1\}_{L_1}, \{x_i - \eta\}_L, g_1 \right) \stackrel{\mathrm{BE}_{b=1}}{=} \langle \mathcal{Z} \rangle \left(\{\mu_\alpha^1 - \eta\}_{L_1}, \{\mu_\alpha^2\}_{L_2}, g_2 \right) \stackrel{(13.110)}{\to}$$

$$\langle \tilde{\mathcal{Z}} \rangle \left(\{\mu_\alpha^2\}_{L_2}, \{\mu_\alpha^1 - \eta\}_{L_1}, g_2 \right) \stackrel{\mathrm{BE}_{b=2}}{=} \langle \mathcal{Z} \rangle \left(\{\mu_\alpha^2 - \eta\}_{L_2}, \{\mu_\alpha^3\}_{L_3}, g_3 \right) \stackrel{(13.110)}{\to} \dots$$

(13.116)

Each time we use (13.110) the characteristic polynomial $\det(\mathcal{Z} - \lambda \mathbf{1})$ acquires the factor $(g_b - \lambda)^{L_{b-1} - L_b}$ except for the last step when we use (13.113) to get $(g_N - \lambda)^{L_{N-1}}$.

---

[10] Here we symbolically write simply $\langle \mathcal{Z} \rangle$ for the characteristic polynomial $\det(\mathcal{Z} - \lambda \mathbf{1})$ with some overall factor.

13.6   Concluding remarks

Lastly, we would like to point out some generalizations, unsolved problems, and directions for further research.

(a) The extension to the models based on trigonometric solutions to the Yang-Baxter equation (the XXZ magnets and corresponding vertex models of statistical mechanics) is rather straightforward. It is done in the recent paper [43]. The trigonometric RS model is on the classical side of the QC correspondence.

(b) The extension to quantum integrable models with elliptic R-matrices is problematic. Conceivably, this might require new ideas. At the same time, the most natural candidate for the classical part of the QC correspondence is the elliptic RS model. The role of the spectral parameter which enters its Lax matrix is to be clarified.

(c) The QC correspondence can be formulated in terms of finite-dimensional integrable systems without mentioning hierarchies of soliton equations explicitly. Already at this level there are some open questions. What is the minimal set of properties of the model which leads to the QC correspondence? How does it depend on the choice of the Lagrangian submanifolds? Is there any information about wave functions of the quantum system encoded in intersection of the Lagrangian submanifolds?

(d) There is a well-known duality [44, 45] of the classical RS (and Calogero–Moser) type models when the action variables in a given system are treated as coordinates of particles in the dual one. Equivalently, the soliton-like tau-function whose zeros move as the RS particles becomes the spectral determinant for the dual system and vice versa. An interesting future perspective is to realize the meaning of this duality in the context of the quantun spin chains. Presumably, this duality implies some correspondence between spectra of different spin chains.

13.7   Appendix A: The higher T-operators through characters

Here we show how to derive (13.41). We have

$$t_\lambda(x) = \mathrm{tr}_{V_\lambda} \left[\left(\mathbf{I} + \frac{\eta}{x-x_L}\, \mathbf{P}_\lambda^{0L} \right) \ldots \left(\mathbf{I} + \frac{\eta}{x-x_2}\, \mathbf{P}_\lambda^{02} \right) \left(\mathbf{I} + \frac{\eta}{x-x_1}\, \mathbf{P}_\lambda^{01} \right) (\pi_\lambda(\mathbf{g}) \otimes \mathbf{I}) \right]$$

$$= \mathrm{tr}_{V_\lambda} \pi_\lambda(\mathbf{g})\, \mathbf{I} + \sum_j \frac{\eta}{x-x_j}\, \mathrm{tr}_{V_\lambda} \left(\mathbf{P}_\lambda^{0j} (\pi_\lambda(\mathbf{g}) \otimes \mathbf{I}) \right) + \sum_{i<j} \frac{\eta^2}{(x-x_i)(x-x_j)}\, \mathrm{tr}_{V_\lambda} \left(\mathbf{P}_\lambda^{0j} \mathbf{P}_\lambda^{0i} (\pi_\lambda(\mathbf{g}) \otimes \mathbf{I}) \right)$$

$$+\ \ldots\ +\ \frac{\eta^L}{(x-x_1)\ldots(x-x_L)}\, \mathrm{tr}_{V_\lambda} \left(\mathbf{P}_\lambda^{0L} \ldots \mathbf{P}_\lambda^{01} (\pi_\lambda(\mathbf{g}) \otimes \mathbf{I}) \right).$$

Plugging here the explicit form of $\mathbf{P}_\lambda^{0j}$ (13.18), we get

$$t_\lambda(x) = \mathrm{tr}_{V_\lambda} \pi_\lambda(\mathbf{g})\, \mathbf{I} + \sum_j \sum_{ab} \frac{\eta e_{ba}^{(j)}}{x-x_j}\, \mathrm{tr}_{V_\lambda} \left(\pi_\lambda(e_{ab}) \pi_\lambda(\mathbf{g}) \right)$$

$$+ \sum_{i<j} \sum_{a_1,b_1;a_2,b_2} \frac{\eta^2 e_{b_1 a_1}^{(i)} e_{b_2 a_2}^{(j)}}{(x-x_1)(x-x_2)} \mathrm{tr}_{V_\lambda} \left(\pi_\lambda(e_{a_2 b_2}) \pi_\lambda(e_{a_1 b_1}) \pi_\lambda(g) \right)$$

$$+ \ldots$$

$$+ \sum_{\{a_1,b_1;\ldots;a_L,b_L\}} \frac{\eta^L e_{b_1 a_1}^{(1)} \ldots e_{b_L a_L}^{(L)}}{(x-x_1)\ldots(x-x_L)} \mathrm{tr}_{V_\lambda} \left(\pi_\lambda(e_{a_L b_L}) \ldots \pi_\lambda(e_{a_1 b_1}) \pi_\lambda(g) \right).$$

The last step is based on the following simple lemma:

Lemma 13.7.1. Let π be a representation of $U(gl(N))$, χ its character and $h_1, h_2, \ldots, h_n$ any elements of the algebra $gl(N)$. Then for any group element $g \in GL(N)$ it holds that

$$\mathrm{tr}\left[\pi(h_1)\pi(h_2)\ldots\pi(h_n)\pi(g) \right] = \frac{\partial}{\partial \varepsilon_n} \ldots \frac{\partial}{\partial \varepsilon_1} \chi(e^{\varepsilon_1 h_1} \ldots e^{\varepsilon_n h_n} g) \Big|_{\varepsilon_i=0} . \quad (A1)$$

The proof is simple. The exponents $e^{\varepsilon_i h_i}$ are group elements for any ε_i. Therefore, we have the chain of equalities

$$\pi(e^{\varepsilon_1 h_1} \ldots e^{\varepsilon_n h_n} g) = \left[\overrightarrow{\prod_{j=1}^{n} \pi(e^{\varepsilon_j h_j})} \right] \pi(g) = \left[\overrightarrow{\prod_{j=1}^{n} e^{\varepsilon_j \pi(h_j)}} \right] \pi(g),$$

from which it follows directly that

$$\pi(h_1)\pi(h_2)\ldots\pi(h_n)\pi(g) = \frac{\partial}{\partial \varepsilon_n} \ldots \frac{\partial}{\partial \varepsilon_1} \pi(e^{\varepsilon_1 h_1} \ldots e^{\varepsilon_n h_n} g) \Big|_{\varepsilon_i=0} .$$

Taking trace of the both sides, we obtain (A1). Applying the lemma to the case $h_i = e_{a_i b_i}$, we arrive at (13.41).

13.8 Appendix B: Hamiltonian formulation of the RS model

For completeness, we give here the Hamiltonian formulation of the L-particle RS model, including the higher flows. The Hamiltonian is

$$\mathcal{H}_1 = \sum_{i=1}^{L} e^{-\eta p_i} \prod_{k=1,\neq i}^{L} \frac{x_i - x_k + \eta}{x_i - x_k}, \quad (B1)$$

with $\{p_i, x_i\}$ being the canonical variables with the standard Poisson brackets. The Hamiltonian equations of motion $\begin{pmatrix} \dot{x}_i \\ \dot{p}_i \end{pmatrix} = \begin{pmatrix} \partial_{p_i}\mathcal{H}_1 \\ -\partial_{x_i}\mathcal{H}_1 \end{pmatrix}$ give the connection between velocity and momentum

$$\dot{x}_i = -\eta e^{-\eta p_i} \prod_{k=1,\neq i}^{L} \frac{x_i - x_k + \eta}{x_i - x_k} \tag{B2}$$

and the equations of motion (13.79).

The RS model is known to be integrable, with the higher integrals of motion in involution being given by $\mathcal{H}_k = \operatorname{tr} \mathsf{Z}^k$, where Z is the Lax matrix of the model (13.75):

$$\mathsf{Z}_{ij} = \frac{\dot{x}_i}{x_i - x_j - \eta} = \frac{\eta\, e^{-\eta p_i}}{x_j - x_i + \eta} \prod_{k=1,\neq i}^{L} \left(1 + \frac{\eta}{x_i - x_k}\right). \tag{B3}$$

These integrals of motion can be regarded as Hamiltonians generating flows in the 'higher times' t_k via the Hamiltonian equations

$$\begin{pmatrix} \partial_{t_k} x_i \\ \partial_{t_k} p_i \end{pmatrix} = \begin{pmatrix} \partial_{p_i} \mathcal{H}_k \\ -\partial_i \mathcal{H}_k \end{pmatrix}, \quad k \geq 1. \tag{B4}$$

Moreover, the dynamics in the higher time t_k is precisely that induced by the mKP flow on the roots of the tau function (13.69). The fact that the integrals of motion $\mathcal{H}_k$ are in involution agrees with the commutativity of the mKP flows. The proof is based on the linear problems (13.74), (13.82), and general relation (13.68). We will not repeat it here since it is technically involved. It can be found in [7, 16].

Acknowledgements

The author thanks A. Alexandrov, M. Beketov, A. Gorsky, A. Liashyk, V. Kazakov, S. Khoroshkin, I. Krichever, S. Leurent, Z.Tsuboi, and A. Zotov for collaboration and discussions. This work was supported in part by RFBR Grant 14-02-00627 and by Grant Nsh-1500.2014.2 for support of scientific schools. The article was prepared within the framework of a subsidy granted to the HSE by the Government of the Russian Federation for the implementation of the Global Competitiveness Program.

References

1. P. Kulish and E. Sklyanin. On solutions of the Yang-Baxter equation. *Zap. Nauchn. Sem. LOMI* **95**, (1980) 129-60; Engl. transl.: *J. Soviet Math.* **19** (1982), 1956.
2. A. Alexandrov, V. Kazakov, S. Leurent, Z. Tsuboi, and A. Zabrodin, Classical tau-function for quantum spin chains. *JHEP* **09** (2013), 064 [arXiv:1112.3310 [math-ph]].
3. Z. Tsuboi, A. Zabrodin, and A. Zotov. Supersymmetric quantum spin chains and classical integrable systems. *JHEP* **05** (2015), 086. [arXiv:1412.2586]
4. V. Kazakov, S. Leurent, and Z. Tsuboi, Baxter's Q-operators and operatorial Bäcklund flow for quantum (super)-spin chains. *Commun. Math. Phys.* **311** (2012), 787-814. arXiv:1010.4022 [math-ph].

5. A. Alexandrov, S. Leurent, Z. Tsuboi, and A. Zabrodin. The master T-operator for the Gaudin model and KP hierarchy. *Nucl. Phys. B* **883** (2014), 173-223. arXiv:1306.1111 [math-ph];
   A. Zabrodin, Quantum Gaudin model and classical KP hierarchy. *J. Phys. Conf. Series* **482** (2014), 012047. arXiv:1310.6985 [math-ph].
6. A. Zabrodin, The master T-operator for vertex models with trigonometric R-matrices as classical tau-function. *Teor. Mat. Fys.* **171**(1) (2013), 59-76 (*Theor. Math. Phys.* **174** (2013), 52-67). arXiv:1205.4152.
7. A. Zabrodin, The master T-operator for inhomogeneous XXX spin chain and mKP hierarchy. *SIGMA* **10** (2014), 006. arXiv:1310.6988 [math-ph].
8. A. Zabrodin, Quantum spin chains and integrable many-body systems of classical mechanics. *Springer Proc. Phys.* **163** (2015), 29–48. arXiv:1409.4099 [math-ph].
9. M. Sato and Y. Sato, Soliton equations as dynamical systems on infinite dimensional grassmann manifold. *Lect. Notes in Num. Appl. Anal.* **5** (1982), 259-71.
10. E. Date, M. Jimbo, M. Kashiwara, and T. Miwa, Transformation groups for soliton equations. In *Nonlinear Integrable Systems—Classical and Quantum*, ed. M. Jimbo and T. Miwa, 39-120, World Scientific, Singapore, 1983.
11. M. Jimbo and T. Miwa, Solitons and infinite dimensional Lie algebras Publ. RIMS, Kyoto Univ. **19** (1983), 943-1001.
12. I. Krichever, O. Lipan, P. Wiegmann, and A. Zabrodin, Quantum integrable models and discrete classical Hirota equations. *Commun. Math. Phys.* **188** (1997), 267-304. arXiv:hep-th/9604080.
13. A. Zabrodin, Hirota equation and the Bethe ansatz, *Theor. Math. Phys.* **116** (1998), 782-819.
14. V. Kazakov, A. S. Sorin, and A. Zabrodin. Supersymmetric Bethe ansatz and Baxter equations from discrete Hirota dynamics. *Nucl. Phys. B* **790** (2008), 345-413, arXiv:hep-th/0703147;
   A. Zabrodin, Bäcklund transformations for difference Hirota equation and supersymmetric Bethe ansatz. *Teor. Mat. Fyz.* **155** (2008), 74-93 (English translation: Theor. Math. Phys. **155** (2008) 567-584), arXiv:0705.4006 [hep-th].
15. I. Krichever and A. Zabrodin. Spin generalization of the Ruijsenaars-Schneider model, non-abelian 2D Toda chain and representations of Sklyanin algebra. *Uspekhi Math. Nauk,* **50** (1995), 3-56.
16. P. Iliev, Rational Ruijsenaars-Schneider hierarchy and bispectral difference operators *Physica D* **229** (2007), no. 2, 184-90. arXiv:math-ph/0609011.
17. S.N.M. Ruijsenaars and H. Schneider. A new class of integrable systems and its relation to solitons. *Ann. Phys.* **170** (1986), 370-405.
18. N. Nekrasov, A. Rosly, and S. Shatashvili. Darboux coordinates, Yang-Yang functional, and gauge theory. *Nucl. Phys. Proc. Suppl.* **216** (2011), 69-93, arXiv:1103.3919.
19. D. Gaiotto and P. Koroteev. On three dimensional quiver gauge theories and integrability. *JHEP* **0513** (2013), 126, arXiv:1304.0779.
20. A. Gorsky, A. Zabrodin, and A. Zotov. Spectrum of quantum transfer matrices via classical many-body systems. *JHEP* **0114** (2014), 070. arXiv:1310.6958.

21. K. Hikami, P. Kulish, and M. Wadati, Construction of integrable spin systems with long-range interaction. *J. Phys. Soc. Japan* **61** (1992), 3071-6.
22. P.P. Kulish and N. Reshetikhin. Diagonalisation of GL (N) invariant transfer matrices and quantum N-wave system (Lee model). *J. Phys. A: Math. Gen.* **16** (1983), L591–6.
23. S. Belliard and E. Ragoucy. Nested Bethe ansatz for 'all' closed spin chains. *J. Phys. A* **41** (2008), 295202, arXiv:0804.2822.
24. C.L. Schultz, Eigenvectors of the multicomponent generalization of the six-vertex model. *Physica A* **122** (1983), 71-88.
25. V. Kazakov and P. Vieira. From characters to quantum (super)spin chains via fusion. *JHEP* **0810** (2008), 050. arXiv:0711.2470 [hep-th].
26. I. Macdonald, *Symmetric Functions and Hall Polynomials*, 2nd edn. Oxford University Press, Oxford, 1995.
27. I.V. Cherednik, An analogue of character formula for Hecke algebras. *Funct. Anal. and Appl.* **21** (2) (1987), 94-5 (translation: 172-4);
I. V. Cherednik, Quantum groups as hidden symmetries of classic representation theory. In *Proceedings of the XVII International Conference on Differential Geometric Methods in Theoretical Physics, Chester*, ed. A. I. Solomon 47–54. World Scientific, Singapore, 1989.
28. V. Bazhanov and N. Reshetikhin, Restricted solid-on-solid models connected with simply laced algebras and conformal field theory. *J. Phys. A: Math. Gen.* **23** (1990), 1477-92.
29. A. Kuniba, T. Nakanishi, and J. Suzuki. Functional relations in solvable lattice models I: Functional relations and representation theory. *Int. J. Mod. Phys.* **A9** (1994), 5215-66, arXiv:hep-th/9309137;
A. Kuniba, Y. Ohta, and J. Suzuki, Quantum Jacobi-Trudi and Giambelli formulae for $U_q(B_r^{(1)})$ from analytic Bethe ansatz. *J. Phys. A* **28** (1995), 6211-26. arXiv:hep-th/9506167.
30. Z. Tsuboi, Analytic Bethe ansatz and functional equations for Lie superalgebra $sl(r+1|s+1)$. *J. Phys. A* **30** (1997), 7975-91. arXiv:0911.5386 [math-ph];
Analytic Bethe Ansatz and functional equations associated with any simple root systems of the Lie superalgebra $sl(r+1|s+1)$. *Physica A* **252** (1998), 565-85. arXiv:0911.5387 [math-ph].
31. A. Hegedus, Discrete Hirota dynamics for AdS/CFT. *Nucl. Phys.* **B825** (2010), 341-65, arXiv:0906.2546 [hep-th].
32. N. Gromov, V. Kazakov, Z. Tsuboi. PSU(2, 2|4) Character of quasiclassical AdS/CFT. *JHEP* **0710** (2010), 097. arXiv:1002.3981 [hep-th].
33. A. Orlov and T. Shiota, Schur function expansion for normal matrix model and associated discrete matrix models. *Phys. Lett. A* **343** (2005), 384-96.
34. V. Enolski and J. Harnad, Schur function expansions of KP tau functions associated to algebraic curves. *Uspekhi Mat. Nauk.* **66:4** (2011), 137-78 (*Russ. Math. Surveys* **66:4** (2011), 767-807). arXiv:1012.3152.
35. Y. Shigyo, On addition formulae of KP, mKP and BKP hierarchies. *SIGMA* **9** (2013), 035. arXiv:1212.1952.

36. I.M. Krichever. Rational solutions of the Kadomtsev Petviashvili equation and integrable systems of N particles on a line. *Funct. Anal. Appl.* **12:1** (1978), 59-61; I.M. Krichever, Rational solutions of the Zakharov-Shabat equations and completely integrable systems of N particles on a line. *J. Sov. Math.* **21:3** (1983), 335-45.

37. T. Shiota. Calogero-Moser hierarchy and KP hierarchy. *J. Math. Phys.* **35** (1994), 5844-9.

38. E. Mukhin, V. Tarasov, and A. Varchenko. Gaudin Hamiltonians generate the Bethe algebra of a tensor power of vector representation of gl_N. *St. Petersburg Math. J.* **22** (2011), 463-72. arXiv:0904.2131; E. Mukhin, V. Tarasov, and A. Varchenko. KZ characteristic variety as the zero set of classical Calogero-Moser Hamiltonians. *SIGMA* **8** (2012), 072. arXiv:1201.3990.

39. E. Mukhin, V. Tarasov, and A. Varchenko. Bethe subalgebras of the group algebra of the symmetric group. arXiv:1004.4248.

40. E. Mukhin, V. Tarasov, and A. Varchenko. Spaces of quasi-exponentials and representations of the Yangian $Y(gl_N)$. arXiv:1303.1578.

41. G. Wilson, Collisions of Calogero-Moser particles and an adelic Grassmannian. *Invent. Math.* **133** (1998), 1-41.

42. A. Givental and B.-S. Kim, Quantum cohomology of flag manifolds and Toda lattices. *Commun. Math. Phys.* **168** (1995), 609-41. arXiv:hep-th/9312096.

43. M. Beketov, A. Liashik, A. Zabrodin, and A. Zotov. Trigonometric version of quantum-classical duality in integrable models. *Nucl. Phys. B*, **903** (2016), 150–63. arXiv:1510.07509.

44. S. Ruijsenaars. Action angle maps and scattering theory for some finite dimensional integrable systems. I. The pure soliton case. *Commun. Math. Phys.* **115** (1988), 127–65.

45. V. Fock, A. Gorsky, N. Nekrasov, and V. Rubtsov. Duality in integrable systems and gauge theories, *JHEP* **0700** (2000), 028.

T0177715